Mitochondria in
Pathogenesis

Mitochondria in Pathogenesis

Edited by

John J. Lemasters
University of North Carolina at Chapel Hill
Chapel Hill, North Carolina

and

Anna-Liisa Nieminen
Case Western Reserve University
Cleveland, Ohio

Kluwer Academic / Plenum Publishers
New York, Boston, Dordrecht, London, Moscow

Library of Congress Cataloging-in-Publication Data

Mitochondria in pathogenesis/edited by John J. Lemasters and Anna-Liisa Nieminen.
    p. ; cm.
    Includes bibliographical references and index.
    ISBN 0-306-46433-0
      1. Mitochondrial pathology. I. Lemasters, John J. II. Nieminen, Anna-Liisa.
    [DNLM: 1. Mitochondria—physiology. 2. Mitochondrial Myopathies. QH 603.M5
    M684127 2000]
    RB147.5 .M585 2000
    571.9′3657—dc21

                                                  00-041277

ISBN 0-306-46433-0

©2001 Kluwer Academic/Plenum Publishers, New York
233 Spring Street, New York, New York 10013

http://www.wkap.nl/

10 9 8 7 6 5 4 3 2 1

A C.I.P. record for this book is available from the Library of Congress

Printed in the United States of America

Contributors

Ravi Botla   Mayo Clinic, Rochester, Minnesota 55905

Cynthia A. Bradham   Department of Medicine, University of North Carolina at Chapel Hill, Chapel Hill, North Carolina 27599-7038

David A. Brenner   Departments of Medicine, and of Biochemistry and Biophysics, University of North Carolina at Chapel Hill, Chapel Hill, North Carolina 27599-7090

Nickolay Brustovetsky   Department of Neuroscience, University of Minnesota, Minneapolis, Minnesota 55455

Paul Burnett   Department of Pharmacology and Physiology, University of Medicine and Dentistry of New Jersey, Newark, New Jersey 07103-2714

Aaron M. Byrne   Department of Anatomy, School of Medicine, Case Western Reserve University, Cleveland, Ohio 44106-4930

Wayne E. Cascio   Department of Medicine, School of Medicine, University of North Carolina at Chapel Hill, Chapel Hill, North Carolina 27599-7038

Brad Chazotte   Department of Pharmaceutical Sciences, Campbell University, Buies Creek, North Carolina, 27506

Anna Colell   Liver Unit, Departments of Medicine and Gastroenterology, Hospital Clinic i Provincial, Instituto Investigaciones Biomedicas August Pi I Suñer, Consejo Superior Investigaciones Cientificas, Barcelona 08036, Spain

Salvatore DiMauro   Department of Neurology, Columbia Univesity, New York, 10032

Janet M. Dubinsky   Department of Physiology, University of Minnesota, Minneapolis, Minnesota, 55455

José C. Fernández-Checa   Liver Unit, Departments of Medicine and Gastroenterology, Hospital Clinic i Provincial, Instituto Investigaciones Biomedicas August Pi I Suñer, Consejo Superior Investigaciones Cientificas, Barcelona 08036, Spain

Gary Fiskum   Department of Anesthesiology, School of Medicine, University of Maryland, Baltimore, Maryland 21201

Bernard Fromenty   INSERM U 481, Hôpital Beaujon, 92118 Clichy, France

Carmen García-Ruiz   Liver Unit, Departments of Medicine and Gastroenterology, Hospital Clinic i Provincial, Instituto Investigaciones Biomedicas August Pi I Suñer, Consejo Superior Investigaciones Cientificas, Barcelona 08036, Spain

Lawrence D. Gaspers   Department of Pharmacology and Physiology, University of Medicine and Dentistry of New Jersey, Newark, New Jersey 07103-2714

Gregory J. Gores   Mayo Clinic, Rochester, Minnesota 55905

Elinor J. Griffiths   Bristol Heart Institute, University of Bristol, Bristol Royal Infirmary, Bristol BS2 8HW, United Kingdom

Andrew P. Halestrap   Department of Biochemistry, Bristol Heart Institute, University of Bristol, Bristol Royal Infirmary, Bristol BS2 8HW, United Kingdom

Teresa G. Hastings   Department of Neurology and Neuroscience, University of Pittsburgh, Pittsburgh, Pennsylvania 15261

Robert A. Haworth   Department of Surgery, University of Wisconsin Clinical Science Center, Madison, Wisconsin 53792

Kaisa M. Heiskanen   Department of Anatomy, School of Medicine, Case Western Reserve University, Cleveland, Ohio 44106-4930

Hajime Higuchi   Department of Internal Medicine, School of Medicine, Keio University, Tokyo 160, Japan

Kenneth M. Humphries   Department of Physiology and Biophysics, School of Medicine, Case Western Reserve University, Cleveland, Ohio 44106-4970

Douglas R. Hunter   Department of Surgery, University of Wisconsin Clinical Science Center, Madison, Wisconsin 53792

Hiromasa Ishii   Department of Internal Medicine, School of Medicine, Keio University, Tokyo, 160, Japan

Sabzali Javadov   Azerbaijan Medical University, Baku, Azerbaijan

Paul M. Kerr   Department of Biochemistry, Bristol Heart Institute, University of Bristol, Bristol Royal Infirmary, Bristol BS2 8HW, United Kingdom

Alicia J. Kowaltowski   Departmento de Bioquimica, Instituto de Quimica, Universidade de Sao Paolo, SP, Brazil

John J. Lemasters   Department of Cell Biology and Anatomy, University of North Carolina at Chapel Hill, Chapel Hill, North Carolina 27599-7090

Mark J. Lieser   Department of Surgery, University of Texas Southwestern Medical Center, Dallas, Texas 75235-9156

David T. Lucas   Department of Physiology and Biophysics, School of Medicine, Case Western Reserve University, Cleveland, Ohio 44106-4970

Abdellah Mansouri   INSERM U 481, Hôpital Beaujon, 92118 Clichy, France

David J. McConkey   Department of Cancer Biology, University of Texas M.D. Anderson Cancer Center, and Program in Toxicology, University of Texas–Houston Graduate School of Biomedical Sciences, Houston, Texas 77030

J. Fred Nagelkerke   Department of Toxicology, Leiden-Amsterdam Center for Drug Research, Sylvius Laboratories, 2300 RA Leiden, The Netherlands

Anna-Liisa Nieminen   Department of Anatomy, School of Medicine, Case Western Reserve University, Cleveland, Ohio 44106-4930

Hisayuki Ohata   Department of Pharmacology, School of Pharmaceutical Sciences, Showa University, Hatanodai, Shinagawa-ku, Tokyo, Japan

Joong-Won Park   Division of Gastroenterology, Department of Internal Medicine, College of Medicine, Chung-ang University, Seoul 140-757, Korea

Dominique Pessayre   INSERM U 481, Hôpital Beaujon, 99218 Clichy, France

Patrice X. Petit   Institut Cochin de Génétique Moléculaire INSERM U129—CHU Cochin Port-Royal, Paris, France

Ting Qian   Department of Cell Biology and Anatomy, University of North Carolina at Chapel Hill, Chapel Hill, North Carolina 27599-7090

Ian Reynolds   Department of Pharmacology, University of Pittsburgh, Pittsburgh, Pennsylvania 15261

Hagai Rottenberg   MCP Hahnemann University, Philadelphia, Pennsylvania 19102

Eric A. Schon   Departments of Neurology, and Genetics and Development, Columbia University, New York, New York 10032

Eric A. Shoubridge   Montreal Neurological Institute and Department of Human Genetics, McGill University, Montreal, Quebec, Canada H3A 2B4

James R. Spivey   Mayo Clinic Jacksonville, Jacksonville, Florida 32224

Konrad Streetz   Department of Gastroenterology and Hepatology, Mediziniche Hochschule Hannover, Hannover, Germany

M-Saadah Suleiman   Department of Biochemistry, Bristol Heart Institute, University of Bristol, Bristol Royal Infirmary, Bristol BS2 8HW, United Kingdom

Luke I. Szweda   Department of Physiology and Biophysics, School of Medicine, Case Western Reserve University, Cleveland, Ohio 44106-4970

Pamela A. Szweda   Department of Physiology and Biophysics, School of Medicine, Case Western Reserve University, Cleveland, Ohio 44106-4970

Andrew P. Thomas   Department of Pharmacology and Physiology, University of Medicine and Dentistry of New Jersey, Newark, New Jersey 07103-2714

Francesco Tombola   CNR Unit of Biomembranes, Department of Biomedical Sciences, University of Padova, Padova Italy

Christain Tratwein   Department of Gastroenterology and Hepatology, Mediziniche Hochschule Hannover, Hannover, Germany

Donna R. Trollinger   Department of Molecular and Cell Biology, University of California at Davis, Davis, California 95616

Lawrence C. Trost   Curriculum in Toxicology, Univesity of North Carolina at Chapel Hill, Chapel Hill, North Carolina 27599-7090

Anibal E. Vercesi   Departmento de Patologia Clínica, Faculdade de Ciências Médicas, Universidade Estadual de Campinas, Campinas, SP, Brazil

Kendall B. Wallace   Department of Biochemistry and Molecular Biology, University of Minnesota School of Medicine, Duluth, Minnesota 55812-2487

J. Paul Zoeteweij   Department of Toxicology, Leiden-Amsterdam Center for Drug Research, Sylvius Laboratories, 2300 RA Leiden, The Netherlands

Mario Zoratti   CNR Unit for Biomembranes, Department of Biomedical Sciences, University of Padova, Padova, Italy

Preface

Surprisingly, the mitochondrion has emerged as a center of attention in pathophysiology, generating both excitement and controversy. Old controversies concerning the mechanism of mitochondrial energy transduction subsided with the acceptance of Peter Mitchell's hypothesis of oxidative phosphorylation, first proposed in 1961. The chemiosmotic hypothesis elegantly described how mitochondrial respiration creates an electrochemical gradient of protons across the mitochondrial inner membrane, which in turn drives ATP synthesis through the mitochondrial ATP synthase. Tight coupling of this process requires that the mitochondrial inner membrane have an exceptionally low nonspecific permeability to protons and other charged solutions. In several pathological conditions, however, the mitochondrial permeability barrier fails, leading to cell death. Indeed, in programmed cell death (apoptosis) mitochondrial membrane failure may be a key signaling event. Moreover, the same respiratory processes that generate the proton electrochemical gradient driving oxidative phosphorylation may also lead to formation of toxic reactive oxygen species that cause cellular injury. Such oxygen radicals may contribute to the decline of bioenergetic capacity with advancing age.

Among the organelles of animal cells, mitochondria are unique in that they contain their own DNA. The complete nucleotide sequence of human mtDNA is established, and it encodes genes for several hydrophobic subunits of complexes I, III, IV, and V, as well as genes for ribosomal proteins and tRNA. Mutations of mtDNA have wide-ranging consequences for mitochondrial respiration and bioenergetics. Because scores of mtDNA copies are maternally inherited, in contrast to the single maternal and paternal copies of nuclear DNA, mitochondrial diseases have unusual features of incomplete penetrance, delayed expression, and heterogeneous tissue involvement. In addition, possible links between mtDNA mutations, mitochondrial free radical generation, and several neurodegenerative diseases, including Huntington, Parkinson, and Alzheimer diseases, are under investigation by several laboratories.

Toxic chemicals have long been known to disrupt mitochondrial metabolism and lead to cellular injury, but the mechanisms causing the injury are turning out to be more complex than expected. For instance, calcium, oxidant chemicals, ischemia/reperfusion, and a range of other agents promote onset of the mitochondrial permeability transition

(MPT) in mitochondria from liver, heart, and other tissues, as first characterized by Hunter and Haworth in the mid-1970s. The significance of the MPT to pathophysiological processes, however, is only now being recognized. Besides involvement of the transition in necrotic cell death, new findings implicate it in apoptosis as well. Indeed, it may be the exceptional form of cell death that does *not* involve transition.

This book attempts to provide an overview of recent major advances in the understanding of mitochondria's numerous roles in pathophysiology. Section headings illustrate the broad range of topics covered, and we believe this book brings together for the first time the diverse pathophysiological phenomena for which the mitochondrion is the common denominator.

We especially thank Leslie Roberts and Sherry Franklin for their diligent, patient, and expert assistance in preparing this book.

John J. Lemasters
Anna-Liisa Nieminen

Contents

Section 2. Mitochondrial Disease and Aging

Chapter 3

**Primary Disorders of Mitochondrial DNA and the Pathophysiology of
mtDNA-Related Disorders** 53

Eric A. Schon and Salvatore DiMauro

Chapter 4

Transmission and Segregation of Mammalian Mitochondrial DNA   81

Eric A. Shoubridge

Chapter 5

**Cardiac Reperfusion Injury: Aging, Lipid Peroxidation, and
Mitochondrial Function**   95

Luke I. Szweda, David T. Lucas, Kenneth M. Humphries, and Pamela A. Szweda

Section 3. Mitochondrial Ion Homeostasis and Necrotic Cell Death

Chapter 6

Ca^{2+}-Induced Transition in Mitochondria: A Cellular Catastrophe? 115

Robert A. Haworth and Douglas R. Hunter

Chapter 7

Physiology of the Permeability Transition Pore 125

Mario Zoratti and Francesco Tombola

Chapter 8

Control of Mitochondrial Metabolism by Calcium-Dependent Hormones....... 153

Paul Burnett, Lawrence D. Gaspers, and Andrew P. Thomas

Chapter 9

The Permeability Transition Pore in Myocardial Ischemia and Reperfusion ... 177

Andrew P. Halestrap, Paul M. Kerr, Sabzali Javadov, and M-Saadah Suleiman

Chapter 10

Mitochondrial Calcium Dysregulation during Hypoxic Injury to Cardiac Myocytes 201

Elinor J. Griffiths

Chapter 11

Mitochondrial Implication in Cell Death................................. 215

Patrice X. Petit

Chapter 12

Role of Mitochondria in Apoptosis Induced by Tumor Necrosis Factor-α...... 247

Cynthia A. Bradham, Ting Qian, Konrad Streetz, Christian Trautwein,
David A. Brenner, and John J. Lemasters

Chapter 13

The ATP Switch in Apoptosis 265

David J. McConkey

Section 4. Mitochondria, Free Radicals, and Disease

Chapter 14

Reactive Oxygen Generation by Mitochondria........................... 281

Alicia J. Kowaltowski and Anibal E. Vercesi

Chapter 15

Role of the Permeability Transition in Glutamate-Mediated Neuronal Injury.. 301

Ian J. Reynolds and Teresa G. Hastings

Chapter 16

Mitochondrial Dysfunction in the Pathogenesis of Acute Neural Cell Death ... 317

Gary Fiskum

Chapter 17

Varied Responses of Central Nervous System Mitochondria to Calcium 333

Nickolay Brustovetsky and Janet M. Dubinsky

Chapter 18

Mitochondrial Dysfunction in Oxidative Stress, Excitotoxicity, and Apoptosis .. 341

Anna-Liisa Nieminen, Aaron M. Byrne, and Kaisa M. Heiskanen

Section 5. Chemical Toxicity

Chapter 22

Bile Acid Toxicity.. 413

Gregory J. Gores, James R. Spivey, Ravi Botla, Joong-Won Park,
and Mark J. Lieser

Chapter 23

Chapter 24

Chapter 25

Kendall B. Wallace

Chapter 26

Dominique Pessayre, Bernard Fromenty, and Abdellah Mansouri

Evaluation of Mitochondrial Function in Intact Cells

Chapter 1

Flow Cytometric Analysis of Mitochondrial Function

Hagai Rottenberg

1. INTRODUCTION

Recent interest in mitochondria's role in cell signaling, cell injury, and cell death has created a need for new methods to study mitochondrial function *in situ*. Conventional studies of mitochondrial metabolism are based on the isolation of intact, well-coupled mitochondria. The isolation procedures select a small fraction of the mitochondrial population and discard most of the mitochondria, which, in studies of apoptosis, often amounts to discarding the baby with the bath water. Moreover, it is impossible to reconstruct the true physiological state of mitochondria *in vitro*. Recent studies implicate an increasing number of previously unknown or unsuspected proteins and metabolites as important modulators of mitochondrial function. Therefore, to elucidate the physiological functions of mitochondria it is necessary to assay mitochondrial function *in situ*.

Available methods of assaying mitochondrial function *in situ* are very limited, however. Some *in situ* mitochondrial functions can be followed by the use of radioisotopes, others by the use of nuclear magnetic resonance (NMR) (mostly ^{31}P and ^{13}C). Such methods can provide important information, but they require a large number of cells or quantity of tissue, and they provide an average of the properties of many different cell types. Often, for instance when following an unsynchronized process (e.g., apoptosis), the cell population is not homogeneous and therefore the information obtained from such measurements is very limited.

Mitochondrial function in a small number of isolated cells can be followed by fluorescence microscopy using a variety of fluorescent probes. These include membrane-

Hagai Rottenberg   MCP Hahnemann University, Philadelphia, Pennsylvania 19102.

Mitochondria in Pathogenesis, edited by Lemasters and Nieminen.
Kluwer Academic/Plenum Publishers, New York, 2001.

potential probes, calcium and pH indicators, and probes of reactive oxygen species (ROS). A serious limitation in interpretating these studies is the uncertain distribution of the probes between different cell compartments. Recently, the application of confocal microscopy has allowed for a more accurate detection of fluorescence signals directly from mitochondria *in situ* (see the chapter on confocal microscopy). The major limitation here is that only a few cells can be studied at one time, and when the cells are not homogeneous it is difficult to obtain accurate quantitative information on the different cell populations.

2. FLOW CYTOMETRY

Flow cytometry allows rapid, simultaneous measurement of light-scattering and fluorescence from a large number of individual cells (Cunningham, 1994). Current commercial instruments simultaneously measure forward light-scattering (FSC), a reliable measure of cell size; side light-scattering (SSC), a measure of cell "granularity" and useful in identifying different cell types and structural changes in cells; and fluorescence emission at up to three different wavelengths, a feature that allows the simultaneous use of up to three different fluorescent probes or, alternatively, the measurement of fluorescence emission at different wavelengths using the same probe (the fluorescence ratio method). Thousands of cells can be measured in a few seconds, providing reliable data with good kinetic resolution from very heterogeneous cell populations, including minor fractions. However, Because the excitation is provided by a laser beam, (usually an argon laser providing excitation at 488 nm), the choice of probes is more limited than in fluorescence microscopy. Another disadvantage of this method, as compared with confocal microscopy, is that only total cell fluorescence can be measured, necessitating a more careful choice of probes and a more careful interpretation of results.

3. TOOLS FOR MEASURING PLASMALEMMAL AND MITOCHONDRIAL MEMBRANE POTENTIALS

3.1. Flow Cytometry Measurements

The mitochondrial membrane potential *in situ*, $\Delta\psi_m$, is a sensitive indicator for the energetic state of mitochondria and of the cell (Brand, *et al.*, 1994) and can be used to assess the activity of mitochondrial proton pumps, electrogenic transport systems, and the activation of the mitochondrial permeability transition (Zoratti and Szabò, 1995). Many cationic fluorescence probes have been used in both fluorescence microscopy and flow cytometry to measure either $\Delta\Psi_p$ (plasma membrane potential) or $\Delta\Psi_m$. Because resting $\Delta\Psi_p$ is always negative inside (about -60 mV in most cells), the positively charged, freely permeable fluorescent probes concentrate in the cytosol until they reach the electrochemical equilibrium

$$Cin/Cout = 10^{\Delta\Psi/60} \text{ (from the Nernst equation)} \qquad (1)$$

Cell organelles with a membrane potential that is negative inside, however, will also accumulate positive probes from the cytosol, until it reaches electrochemical equilibrium with these compartments. In most cells mitochondria are the major organelles with a negative membrane potential inside ($\Delta\Psi_m = -160\,mV$ in most cells). Because $\Delta\Psi_m$ is so high, the probe concentration in mitochondria will be 2–3 orders of magnitude higher than that in the cytosol. Therefore, even though mitochondrial volume is only a fraction of the cytosol, mitochondria will contain most of the probe, and the total amount of probe in the cell will depend strongly on $\Delta\Psi_m$. In addition, because permeable probes are lipophilic, most of the probe will be membrane bound. Probe concentration is much higher in the mitochondrial matrix, and the mitochondrial membrane surface area is a large proportion of the total cell membrane surface area, so most of the bound probe will be located on the inner surface of the inner mitochondrial membrane (Rottenberg, 1989). Thus the total amount of probe accumulated in the cell, and hence the total cell fluorescence, depends on both $\Delta\Psi_p$ and $\Delta\Psi_m$. It is unfortunate that in most flow cytometry studies the fluorescence intensity of cationic probes was interpreted as indicating either the magnitude of $\Delta\Psi_p$ (most of the older studies) or the magnitude of $\Delta\Psi_m$ (most of the recent studies). With few exceptions (cf. Wilson, et al., 1985), little effort was invested in trying to support either interpretation, or in taking into account the contribution of both potentials to cell fluorescence. Extensive binding of the lipophilic probes to the inner surface of the mitochondrial inner membranes complicates measurement interpretation, because such bindings quenches fluorescence extensively (see below).

Moreover, many cationic lipophilic probes are also potent inhibitors of NADH dehydrogenase (Anderson et al., 1993) and thus strongly inhibit cell respiration (Rottenberg and Wu, 1997; Srivatava et al., 1997), which may inhibit the generation of $\Delta\Psi_m$. Also, in many cells lipophilic probes distribution between the cell and the incubation medium is not dependent only on $\Delta\Psi_p$ and $\Delta\Psi_m$ but also on the activity of multidrug resistance (MDR) transporters (e.g., P-glycoprotein) that eject the probes from the cell (Kuhnel et al., 1997; Witkowski and Miller, 1993; Petriz and Garcia-Lopez, 1997).

Finally, in a homogeneous cell population, fluorescence change in one cell is not completely independent of fluorescence change in other cells. For example, if $\Delta\Psi$ collapses in one fraction of the population, the probe released (or not taken) by that fraction will raise the concentration of free probe in the suspension and will thus be taken up by other cells that still maintain $\Delta\Psi$, resulting in an increase in their fluorescence. This is often misinterpreted as an increase in the magnitude of $\Delta\Psi$ of those cells. Selection of the appropriate probe and interpretation of the results must therefore take all of these complications into consideration.

3.2. Rhodamine 123

Rhodamine 123, a polynuclear hydrocarbon with a charged amino group, has been used extensively as a mitochondrial stain and $\Delta\Psi_m$ indicator in fluorescence microscopy and flow cytometry. Earlier studies with this probe were reviewed previously (Chen, 1988). In isolated mitochondria, Rhodamine 123 behaves much like other $\Delta\Psi$ probes: It is accumulated by energized mitochondria, which quenches the suspension fluorescence. Accumulation, and hence quenching, is proportional to the magnitude of $\Delta\Psi$, which validates its use in estimating $\Delta\Psi$ in isolated mitochondria (Emaus et al., 1986). In flow cytometry of intact cells, however, the fluorescence response of Rhodamine 123 is different. Probe is accumulated by the cells, and the rate of this process, quite slow at low concentrations (1–20 nM), depends on $\Delta\Psi_m$ and $\Delta\Psi_p$. If probe concentration is high (>100 nM), fluorescence of the mitochondrially accumulated dye will be increasingly quenched as uptake progresses, until fluorescence (but not uptake) reaches a steady state. Under these conditions, however, cell fluorescence does not indicate $\Delta\Psi_m$ magnitude. This can be verified by collapsing $\Delta\Psi_m$ with an uncoupler, which releases the accumulated dye from the mitochondria. Response varies widely depending on dye concentration and length of preincubation with the dye. If a large amount of dye is allowed to accumulate in the mitochondria, releasing it by collapsing $\Delta\Psi_m$ results, paradoxically, in increased fluorescence (Fig. 1A). Because dye transport across the plasma membrane is slow, dye released from the mitochondria, and now residing in the cytosol is no longer quenched. The fact that cells preincubated with uncouplers and mitochondrial inhibitors have much lower Rhodamine 123 fluorescence (Chen, 1988) shows that the rate of uptake depends on $\Delta\Psi_m$ (and $\Delta\Psi_p$), but does not validate use of this probe as an indicator of $\Delta\Psi_m$ magnitude. Although many investigators have pointed out that Rhodamine 123 fluorescence does not respond correctly to agents that modulate $\Delta\Psi_m$ (cf. Salvioli *et al.*, 1997), it continues to be used (and misinterpreted) extensively.

Two derivatives of Rhodamine 123, tetramethylrhodamine methyl ester (TMRM) and tetramethylrhodamine ethyl ester (TMRE), have been recommended for use instead of Rhodamine 123, because they respond more quickly to modulation of $\Delta\psi_m$ (Loew, 1993). The excitation maxima of these derivatives (~550 nm), however, is less optimal for use with an argon laser flow cytometer than is Rhodamine 123 (507 nm), thus Rhodamine 123 continues to be the most widely used indicator for $\Delta\Psi_m$ in flow cytometry despite its well-documented disadvantage.

3.3. DiOC$_6$(3)

The probe DiOC$_6$(3) has been used for many years as an indicator for $\Delta\Psi_p$ (cf. Shapiro, 1994). Carbocyanine probes also respond to $\Delta\Psi_m$, and the contribution of these two potentials to total fluorescence (as discussed above) was pointed out long ago (Wilson *et al.*, 1985) but was ignored by most investigators. More recently, DiOC$_6$(3) has been used as an indicator of $\Delta\Psi_m$ in many flow cytometry studies of apoptosis (cf. Zamzami *et al.*, 1995; reviewed in Petit *et al.*, 1996; Marchetti *et al.*, 1996; Petit *et al.*, 1995). In these and other studies, however, $\Delta\Psi_p$ contribution to the fluorescence signal has been largely ignored.

Moreover, like other lipophilic carbocyanines, DiOC$_6$(3) is a potent inhibitor of NADH dehydrogenase in isolated mitochondria (Anderson *et al.*, 1993). Cells accumulate lipophilic cations in the cytoplasm because of $\Delta\Psi_p$, so cell respiration is even more sensitive to these dyes than are isolated mitochondria. Indeed, respiration in lymphocytes is inhibited by very low concentrations of this probe (Rottenberg and Wu, 1997), well below those used routinely in assays of $\Delta\Psi_m$.

The relative contribution of $\Delta\Psi_m$ and $\Delta\Psi_p$ to the total fluorescence signal depends not only on their relative magnitude but also on cell density. In general, reducing the dye concentration, or more precisely the dye/cell ratio, increases the contribution of $\Delta\Psi_m$ to cell fluorescence (Rottenberg and Wu, 1998; Wilson *et al.*, 1985). Lowering the dye/cell

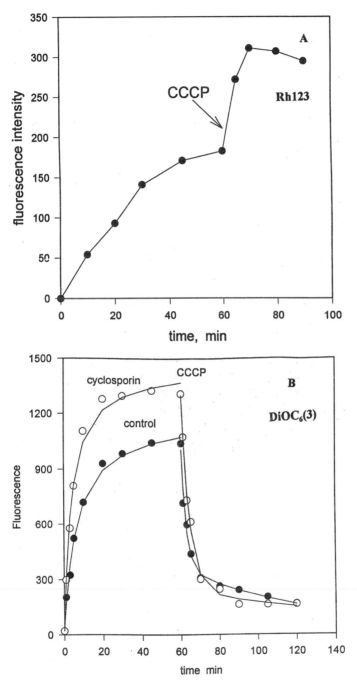

FIGURE 1. Fluorescence-change kinetics after loading lymphocytes with Rhodamine 123 and DiOC$_6$(3), and after adding uncoupler. (A) Rhodamine 123 (200 nM) was incubated with 10^6 cells/ml in MEM ($+1\%$ FCS) and average cell fluorescence was measured at the indicated time. 50 µM CCCP added where indicated (Rottenberg, unpublished). B Same as in (A), but with DiOC6 (3) (0.2 nM) and cyclosporin (1 µM) where indicated. (Reprinted from *Biochim. Biophys. Acta*: 1404, Rottenberg and Wu, "Quantitative assay by flow cytometry of the mitochondrial potential in intact cells," 393–404, Copyright 1998, with permission from Elsevier Science.)

ratio reduces the self-quenching of mitochondrially bound dye, increasing the ratio of cell-associated dye to dye that is free in the medium. Under these conditions, modulation of $\Delta\psi_p$ would mostly modulate the amount of free dye in the medium, which is a relatively small amount that would therefore have a relatively small effect on cell-associated dye quantity (Rottenberg and Wu, 1988).

To accurately correlate fluorescence intensity with $\Delta\Psi_m$ magnitude, one must measure the intensity after dye distribution has reached equilibrium. Because the rate of efflux after collapse of $\Delta\Psi_m$ is higher than the rate of uptake after dye addition, we have developed a protocol in which the dye is first incubated with the cells for 20 minutes to allow it to reach near-equilibrium distribution, then, reagents are added that modulate $\Delta\Psi_m$. After a further 20-minute incubation, which is sufficient to allow redistribution of the dye in response to $\Delta\Psi_m$ modulation, we measure cell fluorescence via flow cytometry (see Fig. 1B for kinetics of dye uptake and release at optimal dye concentration). When cells are incubated with mitochondrial inhibitors or uncouplers *before* the dye is equilibrated, as is common in most protocols (cf. Marchetti *et al.*, 1996), dye transport into the cells is inhibited and does not reach equilibrium distribution, even after prolonged incubations. Therefore fluorescence is not correlated with the $\Delta\Psi_m$ magnitude.

Figure 2A shows typical histograms of the steady-state mean fluorescence of $DiOC_6(3)$ (0.2 pmoles/10^6 cells) in mouse spleen lymphocytes obtained by this protocol. Without CCCP, most cells showed very high fluorescence intensity. A broad fraction of low and intermediate fluorescence intensity was observed as well, however. Fluorescence histograms of cells after addition of 50 μM CCCP showed a single fraction of cells with low fluorescence (about 10% of the high-potential fraction). The fluorescence histogram (without uncoupler) suggests that the cell population is heterogeneous in regard to the $\Delta\psi_m$ magnitude. Figure 3B shows a two-dimensional histogram of $DiOC_6(3)$ fluorescence and forward light-scattering (FSC, which is proportional to cell size) in spleen lympho-cytes. A small fraction of lymphocytes ($\sim$5%) exhibited both low fluorescence (i.e., low $\Delta\Psi_m$) and low FSC (i.e., small volume). Most of these cells are dead or committed to apoptosis. In addition, the major lymphocyte fraction (with normal FSC) also contained a significant fraction of low $\Delta\Psi_m$, which in most of these cells could be restored by adding cyclosporin or oligomycin (see below).

We selected 0.2 nM $DiOC_6(3)$ as an optimal concentration for measuring $\Delta\Psi_m$, based on the titration of the ratio of fluorescence intensity of the high-fluorescence fraction to fluorescence after CCCP addition (F/F_{cccp}) as a function of dye concentration. Above 1 nM, the ratio decreased as a result of increased quenching of mitochondria-accumulated dye. A test of this assay's sensitivity to modulation of $\Delta\Psi_m$ magnitude was obtained from parallel titrations (with the protonophore CCCP) of the lymphocytes' rate of respiration and fluorescence intensity. A typical example is shown in Figure 2C. Respiration was maximally stimulated at about 5 μM CCCP, which is apparently sufficient to collapse $\Delta\Psi_m$. Reducing fluorescence intensity to its lowest value required a much higher concentration of CCCP, apparently because of $\Delta\Psi_p$ collapse (see below), suggesting that the F/F_{cccp} ratio is sensitive to $\Delta\Psi_m$ magnitude over its entire range.

A more quantitative estimate of $\Delta\Psi_m$ can be obtained by analyzing the effect on fluorescence intensity of the K^+ ionophore valinomycin (Rottenberg, 1989). Figure 3A shows a titration of the $DiOC_6(3)$ fluorescence ratio (F/F_{cccp}) with valinomycin in a medium containing 5.5 mM K^+. Very low (nM) concentrations of valinomycin collapsed

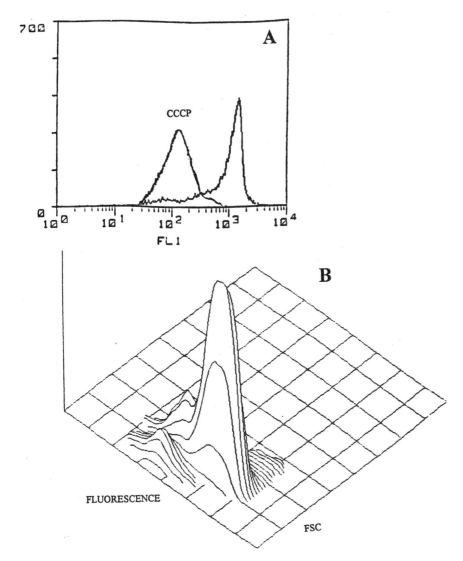

FIGURE 2. DiOC$_6$ (3) fluorescence in mouse spleen lymphocytes: Effect of CCCP. (A) Typical cell fluorescence histograms with and without CCCP. Same conditions as in Figure 1. (B) Two-dimensional histogram of DiOC$_6$(3) fluorescence and forward light-scattering (FSC) in spleen lymphocytes. (C) CCCP effect on respiration rates and DiOC$_6$ (3) fluorescence. Respiration rates of 10^7 cell/ml treated with cyclosporin and oligomycin and increasing concentrations of CCCP are shown (○). To measure fluorescence, 2×10^7 cells were incubated in 2 ml MEM medium with 2 nM DiOC$_6$(3) for 30 min. 1 μM cyclosporin and 30 ng/ml oligomycin were added, and After 10-min incubation, cells were divided into nine portions and incubated for 20 min with various CCCP concentrations, as indicated. (Reprinted from *Biochim. Biophys. Acta*:1404, Rottenberg and Wu, "Quantitative assay by flow cytometry of the mitochondrial potential in intact cells," 393–404, Copyright 1998, with permission from Elsevier Science.)

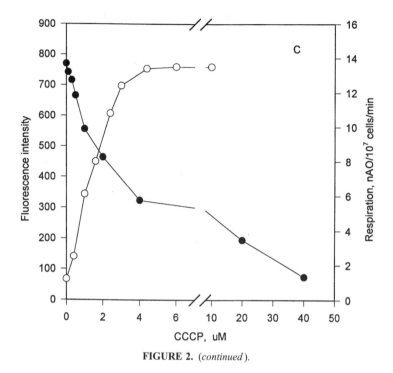

FIGURE 2. (*continued*).

$\Delta\Psi_m$ and reduced F/F_{CCCP} from a value of 10.4 in the absence of valinomycin to approximately 2.3 at 100 nM. Increasing valinomycin concentration to the μM range, however, enhanced fluorescence and increased the fluorescence ratio to 5. This complicated behavior is due to the differential effect of valinomycin on the K^+ permeability of the mitochondrial inner membrane and the plasma membrane (Felber and Brand, 1982). Low (nM) concentrations of valinomycin were sufficient to greatly increase mitochondrial K^+ permeability (Felber and Brand, 1982; Rottenberg, 1989) but not permeability of the plasma membrane. When the mitochondrial inner membrane becomes permeable to K^+ (at 0.1 μM valinomycin), $\Delta\Psi_m$ collapses, because K^+ concentration in the mitochondrial matrix is similar to K^+ concentration in the cytosol (~100 mM). When valinomycin concentration is increased, however, the plasma membrane also becomes permeable to K^+, which leaks out of the cell. At high concentrations of valinomycin (>1 μM), K^+ concentration in the cytosol is gradually reduced to the concentration in the medium (5.5 mM), partially restoring $\Delta\Psi_m$ because the K^+ gradient across the mitochondrial inner membrane has increased. The $\Delta\psi_p$ is largely determined by the K^+ gradient across the plasma membrane, thus $\Delta\psi_p = 0$ at high valinomycin concentration (Fig. 3A).

Increasing the medium K^+ concentration, at high valinomycin concentration (3 μM), increased cytosolic K^+ and therefore collapsed the $\Delta\psi_m$. Figure 3B shows that in the presence of 3 μM valinomycin, KCl reduced the F/F_{cccp} ratio. The F/F_{cccp} ratio log decreased linearly with the log $[K^+]$, a response identical to the $\Delta\Psi_m$ response obtained in isolated mitochondria treated with valinomycin and increasing concentrations of K^+ (Rottenberg, 1989). In the presence of valinomycin, $\Delta\Psi_m$ depends on the equilibrium

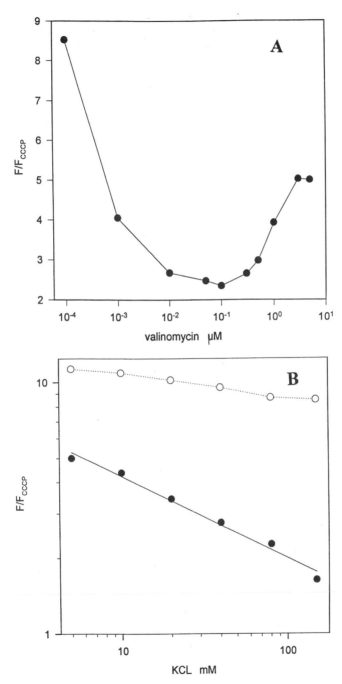

FIGURE 3. Relationships between the magnitude of $\Delta\Psi_m$ and DiOC$_6$(3) fluorescence. (A) Valinomycin effect on fluorescence ratio in MEM medium. Cells were incubated as in Fig. 2A but with different concentrations of valinomycin; CCCP was 50 μM. (B) Effect of KCl concentrations on fluorescence ratio. Protocol as in (A) except that valinomycin concentration was either ○ (empty circles) or 3 μM (full circles), and KCl as indicated. (Reprinted from *Biochim. Biophys. Acta*:1404, Rottenberg and Wu, "Quantitative assay by flow cytometry of the mitochondrial potential in intact cells," 393–404, Copyright (1998), with permission from Elsevier Science.)

distribution of K^+, and it approximately obeys the Nernst equation. Therefore, from the slope of Figure 3B we obtain

$$\Delta\Psi_m = \log (F/F_{CCCP})/0.0054 \text{ mV} \qquad (2)$$

To evaluate the effect of $\Delta\Psi_p$ on fluorescence, we examined the effect of KCl in the absence of valinomycin, because high KCl depolarizes the plasma membrane potential in lymphocytes (Felber and Brand, 1982; Wilson et al., 1985). Addition of KCl decreased the fluorescence ratio from about 10.7 to 8.7 (Fig. 3B), indicating that magnitude of plasma membrane potential contributes significantly to fluorescence ratio, even under conditions that maximize the response to $\Delta\psi_m$ (i.e., very low dye/cell ratio). A similar effect of KCl on fluorescence was observed in lymphocytes incubated with low concentrations of CCCP (5 μM), which suggests that this concentration collapses $\Delta\psi_m$ but not $\Delta\psi_p$. At high CCCP concentrations (50 μM), KCl had no effect on fluorescence, suggesting that at this concentration both $\Delta\psi_m$ and $\Delta\psi_p$ have collapsed (results not shown).

Mitochondrial membrane potential is generated by redox proton pumps under most conditions (aerobic respiration), but it can also be generated from ATP hydrolysis by ATP synthase. Under aerobic conditions, potential generated by the redox pump is used to drive ATP synthesis via proton flow through ATP synthase. Inhibition of oxidative phosphorylation by specific ATP synthase inhibitors (e.g., oligomycin) results in a small but significant increase of $\Delta\psi_m$ in isolated mitochondria. Figure 4A shows the effect of oligomycin on the fluorescence histogram. Oligomycin significantly increases fluorescence in the major lymphocyte fraction, suggesting active oxidative phosphorylation (i.e., state 3). Specific inhibitors of the proton-pumping electron transport complexes—for example, rotenone (complex I), myxothiazol and antimycin (complex III), and cyanide (complex IV)—had no effect on florescence at low concentrations unless oligomycin was also added (Figure 4A). This indicates that lymphocytes can generate sufficient ATP by glycolysis (Ardawi and Newsholme, 1985) and that reversal of mitochondrial ATP synthase can fully support $\Delta\psi_m$. In contrast, in the malarial parasite, which lacks active ATP synthase, $\Delta\psi_m$ was completely dependent on electron transport (Srivatava et al., 1997). Figure 4B shows parallel titrations of lymphocyte respiration (in the presence of oligomycin) and CCCP) and the F/F_{cccp} ratio (in presence of oligomycin) with the respiratory inhibitors rotenone, myxothiazole, and cyanide. Rotenone was the least-efficient inhibitor of the F/F_{cccp} ratio in lymphocytes (empty circles), whereas myxothiazole (full squares) and cyanide (full circles) were equally efficient. This is also true for respiration inhibition. Rotenone inhibited less than 90% of the respiration (inverted triangles), whereas myxothiazole (full triangles) and cyanide (empty triangles) inhibited respiration completely. Because $\Delta\psi_m$ does not depend strongly on respiration rate, rotenone is not an efficient inhibitor of $\Delta\psi_m$ in lymphocytes. Apparently, there is sufficient activity of alternative dehydrogenases in lymphocytes to support significant electron transport rates, at least for short incubation periods.

The maximal reduction of the F/F_{cccp} ratio obtainable by cyanide + oligomycin (80%) indicates the contribution of $\Delta\psi_p$ to the total signal; this is comparable to the reduction obtained by 0.1 μM valinomycin (Fig. 3A) and also to that obtained by a low concentration (1–5 μM) of CCCP (Fig. 2C), observations indicating that these treatments should be used, instead of 50 μM CCCP to obtain the reference F value for estimation of $\Delta\psi_m$ magnitude. The magnitudes of each of these ratios (e.g., F/F_{cccp}, (5 μm) F/F

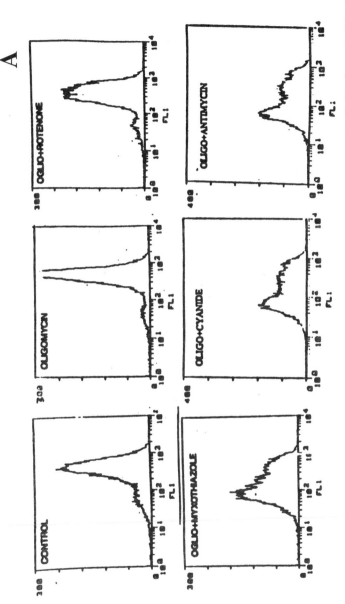

FIGURE 4. Effect of inhibitors of the mitochondrial proton pumps on $\Delta\Psi_m$. (A) Lymphocyte fluorescence histograms incubated with oligomycin (0.1 µg/ml), rotenone (10 µM), myxothiazole (10 µM), antimycin A (10 µM), and cyanide (10 mM). Assay conditions as in Figure 2A. (B) Titrations of respiration (in the presence of oligomycin [0.1 µg/ml] and CCCP [5 µM]) and the F/F_{cccp} ratio (in the presence of oligomycin) by rotenone, myxothiazole, and cyanide. Full triangles, myxothiazole (µM); empty triangles, cyanide (mM); inverted triangles, rotenone (µM). DiOC$_6$(3) fluorescence ratio assay (F/F_{cccp}) as in Figure 2A; Full squares, myxothiazole (µM); full circles, cyanide (mM); empty circles, rotenone (µM). Titration of respiration and F/F_{cccp} for each inhibitor was done with the same batch of lymphocytes. (Reprinted from *Biochim. Biophys. Acta*:1404, Rottenberg and Wu, "Quantitative assay by flow cytometry of the mitochondrial potential in intact cells," 393–404, Copyright 1998, with permission from Elsevier Science.)

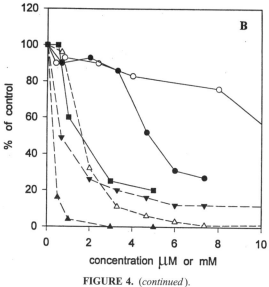

FIGURE 4. (*continued*).

[oligomycin to cyanide], and F/F [valinomycin, 0.1 μM] can be compared directly, or they can be converted to an estimate of $\Delta\psi_m$ by calibration with valinomycin and K^+ (Fig. 3B).

Activation of the mitochondrial permeability transition (MPT) abolishes $\Delta\psi_m$ (Zoratti and Szabò, 1995). Cyclosporin inhibits MPT activation and restores $\Delta\psi_m$, and it can be used to estimate the MPT activation state. The increased fluorescence of $DiOC_6(3)$ by cyclosporin in lymphocytes (Fig. 1B) is not due to inhibition of the MDR pump; increase was observed even in the presence of the MDR inhibitor reserpine. Apparently, unlike with Rhodamine 123, $DiOC_6(3)$ distribution is not strongly modulated by MDR because passive dye permeation is very fast. Therefore, the effect of cyclosporin can be attributed to inhibition of the MPT. In lymphocytes from senescent mice, membrane potential (F/F_{cccp}) and respiration were lower than that observed in young mice and both could be restored to near-normal values by the addition of cyclosporin. These observations suggest that the MPT is more activated in lymphocytes from senescent mice (Rottenberg and Wu, 1997).

Figure 5 shows a comparison of fluorescence response with various modulators of $\Delta\psi_m$ and $\Delta\Psi_p$, as sensed by 0.2 nM and 40 nM $DiOC_6(3)$, in the same batch of cells. Low concentrations of valinomycin (0.1 μM) which collapse $\Delta\psi_m$ (Fig. 3A), reduced cell fluorescence to 25% of control at low dye concentration, but actually stimulated fluorescence at high dye concentration. The combination of oligomycin and cyanide, which also collapses $\Delta\psi_m$ (Fig. 4A, B), reduced fluorescence to 31% of control at low dye concentration, but at high concentration the reduction was much smaller (79% of control). Similarly, CCCP at low concentration (1 μM, cp1), which also collapses $\Delta\psi_m$ (Fig. 2C), reduced fluorescence to 23% of control at low dye concentration, but at high dye concentration the reduction was slight, only 82% of control. In contrast, 100 mM KCl, which collapses $\Delta\psi_p$, had a small effect on fluorescence at low dye concentration but reduced it nearly 50% at the high concentration. At 50 μM, CCCP, which collapses both

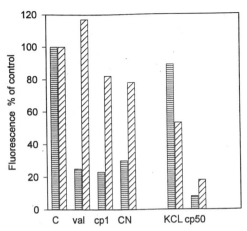

FIGURE 5. Comparison of DiOC$_6$(3) fluorescence response with modulation of $\Delta\psi_m$ and $\Delta\psi_p$ with 0.2-nM and 40-nM dye concentrations. Assay conditions as in Figure 2A, except for dye concentrations, Horizontal lines, 0.2 nM; diagonal lines, 40 nM; Control (no addition), C; valinomycin (0.1 μM), val; cccp (1 μm), cp1; cyanide (10 mM), CN; KCl (100 mM), KCl; cccp (50 μM), cp50. (Reprinted from *Biochim. Biophys. Acta*:1404, Rottenberg and Wu, "Quantitative assay by flow cytometry of the mitochindrial potential in intact cells," 393–404, Copyright 1998, with permission from Elsevier Science.)

$\Delta\psi_m$ and $\Delta\psi_p$, reduced fluorescence to 8% of control at low dye concentration and 17% of control at high dye concentration. These results indicate that at the dye concentrations routinely used for $\Delta\Psi_m$ measurements (Marchetti *et al.*, 1996; Petit *et al.*, 1996) the dye is much more sensitive to $\Delta\Psi_p$ than to $\Delta\Psi_m$.

3.4. JC-1 as a Flow Cytometry Probe for $\Delta\Psi_p$

The dye JC-1 (5,5',6,6'-tetrachloro-1,1'3,3'-tetraethylbenzimidazolyl-carbocyanine iodide) is a cationic carbocyanine dye that forms concentration-dependent aggregates associated with a florescence emission shift from the green of monomer (527 nm) to the red of the J-aggregate (590 nm). At the right concentration, the fluorescence ratio F2/F1 (e.g., aggregated dye/monomeric dye), which can be measured simultaneously for each cell, correlates with the magnitude of $\Delta\Psi_m$ (Cossarizza *et al.*, 1993). Several studies have shown that the red fluorescence of this dye (F2) is sensitive to $\Delta\Psi_m$ magnitude in several types of cells, as evidenced by its response to uncouplers and ionophores (Cossarizza *et al.*, 1993; Reers *et al.*, 1995; Salvioli *et al.*, 1997, 1998). This probe was also reported to be a good substrate of the MDR transporter, however, and expression of this protein had a strong effect on dye fluorescence in different cell lines (Kuhnel *et al.*, 1997). It is unclear to what extent this will interfere with the dye's use as a $\Delta\Psi_m$ probe under normal circumstances.

3.5. Estimation of $\Delta\Psi_m$ with the Fixable Dye CMXRos (Mitotracker Red)

Many studies in flow cytometry, in which surface receptors and other proteins are analyzed, require fixation of the cells. The cationic permeable probes discussed above are in a dynamic equilibrium with the medium and therefore must be measured in live, metabolically active cells. It was recently suggested that a fluorescence cation with a thiol-reactive moiety (CMXRos, chloromethyl-X-rosamine), a probe originally developed for labeling mitochondria with an aldehyde-fixable dye (MitoTracker), can also be used to indicate the magnitude of $\Delta\Psi_m$ *before* the cells are fixed (Macho *et al.*, 1996). Though CMXRos would not strongly label cells in which both $\Delta\Psi_m$ and $\Delta\Psi_p$ were collapsed (100 µM CCCP), there is little evidence that this is a reliable quantitative measure of $\Delta\Psi_m$ (Ferlini *et al.*, 1998). The rate of the dye's accumulation in mitochondria probably depends on $\Delta\Psi_m$ (and also $\Delta\Psi_p$), but the covalent reaction and the extent of retention after fixation may depend on a number of additional factors.

3.6. FAD Fluorescence

Intrinsic cell fluorescence can also indicate mitochondrial function. Ultraviolet cell fluorescence arises mostly from NADH and can be measured in a flow cytometer equipped with a UV laser. Both the ratio of NAD(P)H to NAD(P) and the total amount of these nucleotides are important measures of cell metabolism and can be estimated directly from the intrinsic fluorescence. Most commercial flow cytometers, however, are equipped with an argon laser (ex = 488 nm) and cannot be used for this purpose. Intrinsic fluorescence of cells illuminated with an argon laser is mostly the fluorescence of oxidized FAD—largely that of the mitochondrial α-lipoamide dehydrogenase (Kunz *et al.*, 1997). The oxidation state of mitochondrial FAD is dependent on the ratio of NAD(P)H to NAD(P), and it is possible to assess mitochondrial redox state from intrinsic FAD fluorescence. We have found that in lymphocytes and T-cells from senescent mice, intrinsic fluorescence is significantly higher than in lymphocytes from young mice, suggesting a more oxidized state in mitochondria from the senescent mice (Mather and Rottenberg; unpublished, Rottenberg and Wu, 1997).

4. MITOCHONDRIAL GENERATION OF REACTIVE OXYGEN SPECIES

Several probes that become fluorescent when oxidized by reactive oxygen species (ROS, either superoxide anion or hydrogen peroxide) are used extensively in flow cytometry studies (e.g, dihydroethidium, dihydrorhodamine, and dihydrochloro fluorescein). To the extent that most ROS are produced by the mitochondrial electron transport complexes (Chance *et al.*, 1979), and considering that electron transport inhibitors (particularly the bc_1-complex inhibitors; see below) can stimulate the generation of ROS, these can be taken as indicators of mitochondrial function. Neither the location of these indicators nor the source and nature of ROS is always clear, however. One indicator thought to be specific to superoxide anion, and which also seems localized to mitochondria when used at low concentration, is dihydroethidium (Cai and Jones, 1998). This probe has been used extensively in studies of apoptosis and appears to measure mitochondrial superoxide generation during apoptosis (cf. Castedo *et al.*, 1995). Figure 6 shows the generation of ethidium bromide from dihydroethidium bromide in lymphocytes incubated with various inhibitors of electron transport (Mather and Rottenberg, unpublished). Electron transport inhibitors, particularly inhibitors of the bc_1 complex (antimycin *a* and

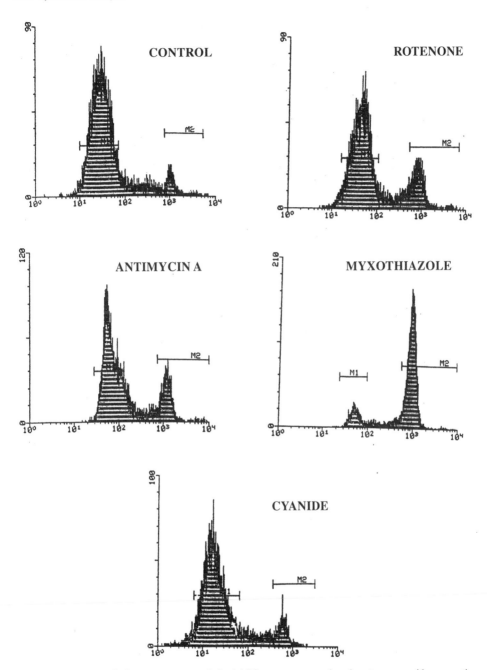

FIGURE 6. Effect of electron transport chain inhibitors on mouse lymphocyte superoxide generation. Lymphocytes incubated for 1 h with 1 um dehydroethidium and 30 μM rotenone, 15 μM antimycin A, 10 μM myxothiazole, or 10 mM NaCN, as indicated (Mather and Rottenberg, unpublished).

myxathiazole), enhance superoxide production because they induce the accumulation of semiubiquinone, believed to be the main source of superoxide generation in mitochondria (Chance *et al.*, 1979). Cyanide is an exception and appears to slightly inhibit generation of superoxide. Possibly cytochrome oxidase, perhaps through its interaction with NO, is involved in the generation of superoxide in lymphocytes.

5. MITOCHONDRIAL CALCIUM STORES

We have used Fluo-3, which indicates the free calcium concentration in the cytoplasm, to estimate mitochondrial calcium stores in different T-cell populations. Figure 7 shows dot plots FSC, which indicates cell size, against fluorescence, which is proportional to the free calcium concentration) in Fluo-3-AM-loaded T-cells. Fluorescence histograms of the same data are shown as well. In most T-cells (area R7) fluorescence is enhanced after addition of a low concentration of CCCP (5 uM), which collapses $\Delta\psi_m$ and releases calcium from the mitochondria. The CCCP did not, however cause release of

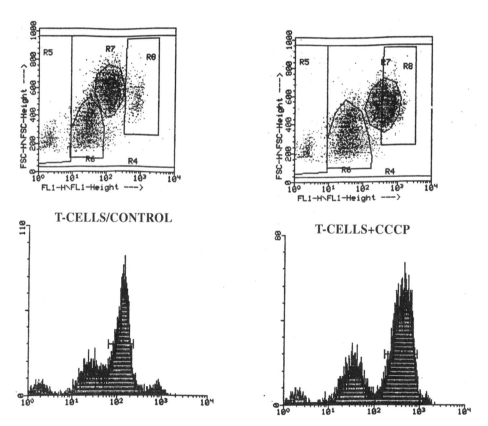

FIGURE 7. CCCP-induced release of mitochondrial calcium in mouse T-cells. T-cells were loaded with 5 μM Fluo-3-AM for 60 min. 5 μM CCCP was added and fluorescence measured again after 10 min (Mather and Rottenberg, unpublished).

calcium in shrunken cells (low FSC, region R6), which are progressing to apoptosis. Apparently the calcium was already released from mitochondria in the shrunken cells (Mather and Rottenberg, unpublished).

Several fluorescent probes have been used to measure mitochondrial matrix free calcium concentration, $[Ca^{2+}]_m$ and matrix pH (pH_m) using fluorescence microscopy (Chacon $et\ al.$, 1994, and see the chapter on mammalian DNA). There are no comparable studies using flow cytometry. The choice of probes is limited by excitation wavelength. Nevertheless, there are a number of probes that can, and probably soon will, be used to measure mitochondrial calcium and pH directly by flow cytometry.

ACKNOWLEDGEMENTS. This work was supported by PHS grant AG13779. I thank Michael Mather for critical reading of the manuscript.

REFERENCES

Anderson, W. M., Wood, J. M., and Anderson, A. C., 1993, Inhibition of mitochondrial and *Paracoccus denitrificans* NADH-ubiquinone reductase by oxacarbocyanine dyes: A structure-activity study, *Biochem. Pharmacol.* **45**:2115–2122.

Ardawi, M. S. M., and Newsholme, E. A., 1985, Metabolism in lymphocytes and its importance in the immune system, *Essays Biochem* **21**:1–44.

Brand, M. D., Chien, L. F., Ainscow, E. A., Rolfe, D. F., and Porter, R. K., 1994, The causes and functions of mitochondrial proton leak, *Biochim. Biophys. Acta* **1187**:132–139.

Cai, J., and Jones, D. P., 1998, Superoxide in apoptosis: Mitochondrial generation triggered by cytochrome *c* loss, *J. Biol. Chem.* **273**:11401–11404.

Castedo, M., Macho, A., Zamzami, N., Hirsch, T., Marchetti, P., Uriel, J., and Kroemer G., 1995, Mitochondrial perturbations define lymphocytes undergoing apoptotic depletion *in vivo, Eur. J. Immunol.* **25**:3277–3284.

Chacon, E., Reece, J. M., Nieminen, A. L., Zahrebelski, G., Herman, B., and Lemasters, J. J., 1994, Distribution of electrical potential, pH, free Ca2+, and volume inside cultured adult rabbit cardiac myocytes during chemical hypoxia: A multiparameter digitized confocal microscopy study, *Biophys. J.* **66**:942–952.

Chance, B., Sies, H., and Boveris, A., 1979, Hydroperoxide metabolism in mammalian organs, *Physiol. Rev.* **59**: 527–605.

Chen, L. B., 1988, Mitochondrial membrane potential in living cells, *Ann. Rev. Cell Biol.* **4**:155–182.

Cossarizza, A., Baccarani-Contri, M., Kalashnikova, G., Franceschi, C., 1993, A new method for the cytofluorimetric analysis of mitochondrial membrane potential using the J-aggregate forming lipophilic cation 5,5′, 6,6′-tetrachloro-1,1′,3,3′-tetraethylbenzimidazolylcarbocyanine iodide (JC-1), *Biochem.* **197**:40–45.

Cunningham, E. R., 1994, Overview of flow cytometry and fluorescent probes for cytometry, *Methods Mol. Biol.* **34**:219–224.

Emaus, R. K., Grunwald, R., and Lemasters, J. J., 1986, Rhodamine 123 as a probe of transmembrane potential in isolated rat-liver mitochondria, *Biochim. Biophys. Acta* **850**:436–48.

Felber, S. M., and Brand, M. D., 1982, Valinomycin can depolarize mitochondria in intact lymphocytes without increasing plasma membrane potassium fluxes, *FEBS Lett.* **150**:122–124.

Ferlini, C., Scambia, G., Fattorossi, A., 1998, Is chloromethyl-X-rosamine useful in measuring mitochondrial transmembrane potential? *Cytometry* **31**:74–75.

Kuhnel, J. M., Perrot, J. Y., Faussat, A. M., Marie, J. P., Schwaller, M. A., 1997, Functional assay of multidrug resistant cells using JC-1, a carbocyanine fluorescent probe, *Leukemia* **11**:1147–1155.

Kunz, D., Luley, C., Winkler K., Lin, H., and Kunz, W. S., 1997, Flow cytometric detection of mitochondrial dysfunction in subpopulation of human mononuclear cells, *Anal. Biochem.* **246**:218–224.

Loew, L. M., Tuft, R. A., Carrington, W., Fay, F. S., 1993, Imaging in five dimensions: Time-dependent membrane potentials in individual mitochondria, *Biophys. J.* **65**:2396–2407.

Macho, A., Decaudin, D., Castedo, M., Hirsch, T., Susin, S. A., Zamzami, N., Kroemer, G., 1996, Chloromethyl-X-Rosamine is an aldehyde-fixable potential-sensitive fluorochrome for the detection of early apoptosis, *Cytometry* **25**:333–340.

Marchetti, P., Hirsch, T., Zamzami, N., Castedo, M., Decaudin, D., Susin, S. A., Masse, B., and Kroemer, G., 1996, Mitochondrial permeability transition triggers lymphocyte apoptosis, *J. Immunol.* **157**:4830–4836.

Petit, P. X., Lecoeur, H., Zorn, E., Dauguet, C., Mignotte, B., and Gougeon, M. L., 1995, Alterations in mitochondrial structure and function are early events of dexamethasone-induced thymocyte apoptosis, *J. Cell Biol.* **130**:157–167.

Petit, P. X., Susin, S. A., Zamzami, N., Mignotte, B., and Kroemer, G., 1996, Mitochondria and programmed cell death: Back to the future, *FEBS Lett.* **396**:7–13.

Petriz, J., and Garcia-Lopez, J., 1997, Flow cytometric analysis of P-glycoprotein function using thodamine 123, *Leukemia* **11**:1124–1130.

Reers, M., Smiley, S. T., Mottola-Hartshorn, C., Chen, A., Lin, M., and Chen, L. B., 1995, Mitochondrial membrane potential monitored by JC-1 dye, *Methods Enzymol.* **260**:406–417.

Rottenberg, H., 1989, Proton electrochemical gradient in vesicles, organelles, and prokaryotic cells, *Methods Enzymol.* **172**:63–84.

Rottenberg, H., and Wu, S., 1997, Mitochondrial dysfunction in lymphocytes from old mice: Enhanced activation of the permeability transition, *Biochem. Biophys. Res. Commun.* **240**:68–74.

Rottenberg, H., and Wu, S., 1998, Quantitative assay by flow cytometry of the mitochondrial membrane potential in intact cells, *Biochim. Biophys. Acta* **1404**:393–404.

Salvioli, S., Ardizzoni, A., Franceschi, C., and Cossarizza, A., 1997, JC-1, but not $DiOC_6(3)$ or rhodamine 123, is a reliable fluorescent probe too assess $\Delta\Psi$ changes in intact cells: Implication for studies on mitochondrial functionality during apoptosis, *FEBS Lett.* **411**:77–82.

Salvioli S., Maseroli R., Pazienza T. L., Bobyleva V., and Cossarizza, A., 1998, Use of flow cytometry as a tool to study mitochondrial membrane potential in isolated, living hepatocytes, *Biochemistry* **63**:235–238.

Shapiro, H. H., 1994, Cell membrane potential analysis, *Methods Cell Biol.* **41**:121–133.

Srivatava, I. K., Rottenberg, H., and Vaidya, A. B., 1997, Atovquone, a broad spectrum antiparasitic drug, collapses mitochondrial membrane potential in a malarial parasite, *J. Biol. Chem.* **272**:3961–3966.

Wilson, H. A., Seligmann, B. E., and Chused, T. M., 1985, Voltage-sensitive cyanine dye fluorescence signals in lymphocytes: Plasma membrane and mitochondrial components, *J. Cell. Physiol.* **125**:61–71.

Witkowski, J. M., and Miller, R. A., 1993, Increased function of P-glycoprotein in T lymphocyte subsets of aging mice, *J. Immunol.* **150**:1296–1306.

Zamzami, N., Marchetti, P., Castedo, M., Zanin, C., Vayssiere, J. L., Petit, P. X., Kroemer, G., 1995, Reduction in mitochondrial potential constitutes an early irreversible step of programmed lymphocyte death *in vivo*, *J. Exp. Med.* **181**:1661–1672.

Zoratti, M., and Szabò, I., 1995, The mitochondrial permeability transition, *Biochim. Biophys. Acta* **1241**:139–176.

Chapter 2

Confocal Microscopy of Mitochondrial Function in Living Cells

John J. Lemasters, Ting Qian, Donna R. Trollinger, Wayne E. Cascio, Hisayuki Ohata, and Anna-Liisa Nieminen

1. INTRODUCTION

In highly aerobic cells, such as cardiac myocytes, hepatocytes, renal proximal tubule cells, and neurons, mitochondria occupy up to a quarter of the volume of the cytoplasm. Much has been learned of mitochondrial function after isolation of these organelles by tissue homogenization and differential centrifugation, but the behavior of mitochondria *in vitro* can differ in important ways from that *in vivo*. Incubation conditions *in vitro* are considerably different than in the normal intracellular milieu. *In vitro*, mitochondria are usually exposed to excess respiratory substrate, phosphate, and ADP in the absence of oxidized substrate and ATP. *In situ*, respiratory substrate availability appears to be rate-limiting, at least in part, for oxidative phosphorylation, and the respiratory intermediates NADH, ubiquinone, and the mitochondrial cytochromes are much more oxidized *in vivo* than *in vitro*. Similarly, ATP/ADP·Pi ratios are orders of magnitude greater *in vivo* than *in vitro*. *In vitro*, mitochondrial Ca^{2+} uptake is usually measured by exposing the isolated organelles to at least $50\,\mu M$ $CaCl_2$, whereas resting cytosolic free Ca^{2+} is submicromolar and never exceeds more than a few micromolar except under conditions of severe cellular

John J. Lemasters and Ting Qian Department of Cell Biology and Anatomy, University of North Carolina at Chapel Hill, Chapel Hill, North Carolina 27599-7090. **Donna R. Trollinger** Department of Molecular and Cell Biology, University of California at Davis, Davis, California 95616. **Wayne E. Cascio** Department of Medicine, School of Medicine, University of North Carolina at Chapel Hill, Chapel Hill, North Carolina 27599-7038. **Hisayuki Ohata** Department of Pharmacology, School of Pharmaceutical Sciences, Showa University, Hatanodai, Shinagawa-ku, Tokyo, Japan. **Anna-Liisa Nieminen** Department of Anatomy, School of Medicine, Case Western Reserve University, Cleveland, Ohio 44106-4930.

Mitochondria in Pathogenesis, edited by Lemasters and Nieminen.
Kluwer Academic/Plenum Publishers, New York, 2001.

stress. Additionally, isolation of mitochondria by tissue homogenization and differential centrifugation may cause loss of soluble factors in the cytosol that regulate mitochondrial function. Thus, methods for studying mitochondrial function within intact living cells and tissues are critically important. In recent years optical microscopy, especially confocal microscopy, has been used successfully to investigate mitochondrial physiology *in situ*.

2. IMAGE FORMATION IN CONFOCAL MICROSCOPY

2.1. Pinhole Principle

The lateral resolving power of conventional wide-field optical microscopy is close to 0.2 µm, but axial resolution is poorer because the effective depth of field is 2–3 µm even at highest magnification. In thicker specimens, light arising from out-of-focus planes is superimposed on the in-focus image. Out-of-focus light obscures image detail, decreases contrast, and interferes with applications requiring quantitative analysis. In conventional wide-field fluorescence microscopy, out-of-focus light creates a diffuse haze around objects of interest, making useful imaging of thick, densely stained specimens essentially impossible.

Confocal microscopy selectively removes out-of-focus light from the in-focus image plane to produce remarkably detailed images with a narrow depth of field. The *pinhole principle* of confocal microscopy, introduced by Minsky (1961), is elegantly simple. A microscope objective focuses the illuminating light, which is typically from a laser, to a small spot within the specimen (Fig. 1). The shape of the light beam is biconical. At the cross-over point, spot diameter is diffraction limited, or about 0.2 µm for a high numerical aperture (N.A.) lens. The objective lens also collects light reflected or fluoresced by the specimen. A partially reflecting prism or dichroic mirror separates returning light from illuminating light. Returning light is focused by the objective onto a small pinhole. Light originating from the in-focus cross-over point of the illuminating light beam is focused to a small spot precisely at the pinhole and thus passes through the pinhole to a light detector beyond, typically a photomultiplier. Light from above and below the focal plane is focused to spots behind and in front of the pinhole, respectively. As a consequence, out-of-focus light is spread out when it reaches the pinhole and very little passes through it (Fig. 1). Selective transmission of light originating from the in-focus plane allows the creation of thin optical sections through thick specimens.

2.2. Axial Resolution of Confocal Microscopy

Pinhole diameter determines the axial resolution of a confocal microscope: the smaller the pinhole, the smaller the thickness of the confocal slice. Once the pinhole becomes smaller than the diffraction-limited spot (airy disk) of in-focus light projected on to the pinhole, however, no improvement of axial resolution is gained by further decreasing the pinhole diameter. Theoretical axial resolution at the diffraction limit is (Inoué, 1995)

$$Z_{min} = 2\lambda\eta/(N.A.)^2 \tag{1}$$

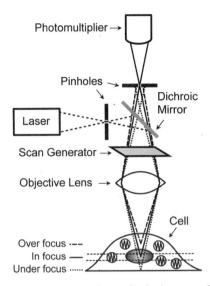

FIGURE 1. Schematic of a laser-scanning confocal microscope. See text for details.

where Z_{min} is the diffraction-limited axial resolution, λ is the wavelength of detected light, η is the refractive index of the object medium, and N.A. is the numerical aperture of the objective lens. For high-resolution imaging of living cells, λ is 0.54 µm (green light), η is 1.33 (water), and N.A. is 1.2 (water-immersion lens). The Z_{min} is then about 1 µm.

2.3. Two-Dimensional Image Formation

To create two-dimensional images, the light beam is scanned across the specimen. This is typically accomplished by placing vibrating mirrors in the light path prior to entry of light into the microscope objective (White *et al.*, 1987). This scan generator also "descans" the returning light to focus it on the stationary pinhole. To view an image, light detected by the photomultiplier is stored in computer memory and displayed on a video monitor in register with the movement of the light beam across the specimen. In theory any light source can be used in confocal microscopy, but in practice lasers are most commonly used because of their ability to produce bright point illumination. The laser, dichroic reflector, scan generator, microscope objective, pinhole, photomultiplier, and computer thus form the essential elements of a laser-scanning confocal microscope system (Fig. 1).

3. SPECIMEN PREPARATION FOR LIVE CELL IMAGING

To image living cells in culture, the cells must first be plated on glass coverslips. *Number $1\frac{1}{2}$ coverslips* (a designation that means the glass is 170-µm thick) should be used to provide a proper optical match to the microscope objective. Most cells adhere to and grow poorly on glass. Accordingly, coverslips should be treated with type I collagen or laminin to improve cell growth and adherence. After rinsing in ethanol for sterilization and drying, 1–2 drops (50–100 μl) of rat tail collagen (1 mg/ml in 0.1% acetic acid) or laminin (0.1 mg/ml in *tris*-buffered saline) are spread out across the glass surface. After air drying, the coverslips are placed in Petri dishes, culture medium is added, and the cells are plated in the usual fashion. To view the cells on the microscope, the coverslip is mounted in a special chamber, which can be built in a machine shop or purchased commercially (FCS2 chamber, Bioptechs, Butler, PA). An open chamber can be used with an inverted microscope, permitting easy access to the cells to add or replace reagents. With closed chambers, cells may be continuously perifused. Temperature regulation is achieved using heaters built into the coverslip chamber or using a stream of tempered air.

4. DEALING WITH PHOTODAMAGE AND PHOTOBLEACHING

4.1. Greater Requirement by Confocal Microscopy for Excitation Light

In confocal microscopy only light from a thin optical section is imaged, although the entire specimen is transilluminated. Consequently, in comparison with low-light, wide-field fluorescence microscopy, confocal microscopy requires brighter illumination to develop sufficiently bright fluorescence for adequate detection. The more-complicated optics of laser-scanning confocal microscopes also produce additional light losses as compared with wide-field microscopy. Finally, the quantum detection efficiency of photomultipliers is less than that of the best charge-coupled devices (CCDs) used in conventional fluorescence microscopy, though the time response of photomultipliers is much better. Consequently, specimens viewed by confocal microscopy will be exposed to more light than specimens imaged by conventional fluorescence microscopy and low-light cameras. Thus, confocal specimens are more vulnerable to photobleaching and photo-damage, especially during serial imaging of live cells.

4.2. High Numerical Aperture Objective Lenses

Several strategies help minimize photobleaching and photodamage. High-N.A. objective lenses maximize light collection, because fluoresced light collection efficiency varies with N.A. squared. For monolayers of cells, oil-immersion objective lenses with an N.A. of 1.3 to 1.4 give the brightest images. To image deeper into a living specimen than about 10 μm, a water-immersion lens of N.A. >1.0 should be used. Oil-immersion lenses should not be used when imaging deep into aqueous specimens, because the water/oil interface produces a strong spherical aberration that decreases image brightness and resolution (Keller, 1995). For ultraviolet applications especially, empirical measurement of light transmission by individual lenses is recommended, because light throughput varies with the type of glass in the lens. Ordinary glass in particular transmits ultraviolet light poorly.

4.3. Laser Attenuation

Because choices for objective lenses are usually limited, the single most important step in decreasing photodamage is attenuating the laser illumination using neutral-density filters. By increasing photomultiplier sensitivity to compensate for the weaker fluorescent signal, good images can often be collected even with attenuations of 300–3000-fold. As a rule, the laser should be attenuated just to the point where image quality starts to become unacceptable. With too much attenuation the spatial signal-to-noise ratio becomes high, resulting in a "snowy image" because too few photons are measured from each picture element (pixel).

4.4. Pinhole Size, Zoom Magnification, and Previewing

Another strategy for increasing sensitivity so that lower levels of excitation light can be used is to increase the pinhole diameter. With a larger pinhole, image brightness is greater and the spatial signal-to-noise ratio is smaller, with no increase in laser illumination. Increasing the pinhole diameter, however, diminishes axial resolution, a sometimes unacceptable trade-off. For maximum axial resolution, the pinhole should be slightly smaller than the diffraction-limited Airy disk projected onto it. A pinhole any smaller will only decrease image brightness with no corresponding improvement of axial resolution. Airy-disk diameter varies with objective-lens N.A. and magnification; thus optimal pinhole diameter is based on the magnification and N.A. of each objective lens. Many confocal microscope systems employ software to calculate the correct Airy-disk diameter and adjust the pinhole accordingly.

Laser-scanning confocal microscopes can "zoom" magnification in small increments over a 5–10 fold range. Zooming simply decreases the x and y excursion of the laser beam across the specimen. With higher zoom magnification, the same amount of excitation light becomes concentrated onto a smaller area of the specimen, which accelerates photobleaching and photodamage. Accordingly, one should use the lowest zoom magnification consistent with the particular needs of the experiment.

A common beginner's mistake is to preview the sample at high laser excitation and zoom magnification. Good images are easily obtained at first, but very soon photobleaching and photodamage make collection of any image almost impossible, even after laser attenuation. Thus, survey images should be collected judiciously and at the lowest possible illumination and zoom magnification. Only when an area of interest is found should laser intensity and zoom magnification be increased to achieve the desired image brightness and signal-to-noise characteristic.

4.5. Fluorophores

Photodamage and photobleaching also depend upon the fluorophore used. Stable fluorophores, such as many of the Rhodamine derivatives, can be repeatedly imaged with little phototoxicity. Other fluorophores, such as fluorescein and acridine orange, are unstable in the light beam; they bleach rapidly and may generate toxic photoproducts. One strategy for reducing photobleaching is to increase the concentration of fluorophore. With more fluorophore, the fluorescent signal is stronger, allowing greater attenuation of the excitation light. With ion- and membrane-potential-indicating fluorophores, however, higher fluorophore concentrations may interfere with the parameter being measured; thus an experimental determination of what is an acceptable trade-off may need to be made.

5. MULTIPARAMETER FLUORESCENCE MICROSCOPY

Lasers used for confocal microscopy emit no more than a few narrow lines of light at a time. This limited choice of laser wavelengths constrains the type of fluorophore that can be used. Commonly used lasers for confocal microscopy and their emission lines include argon, 488 and 514 nm; argon–krypton, 488, 568, and 647 nm; helium–cadmium, 442 nm; helium–neon, 543 nm; and UV–argon, 351–364 nm. The argon–krypton laser with well-separated blue, yellow, and red lines is the most versatile single laser for biological applications, but maintenance costs are high because the laser tube must be replaced every 2,000 to 3,000 hours of use. Alternatively, multiple lasers, such as relatively inexpensive and long-lived argon and helium–neon lasers, can be used simultaneously. Multiple lasers or a multiline laser can simultaneously excite spectrally distinct fluorophores. The specific fluorescence of each fluorophore can then be directed by dichroic reflectors and barrier filters to separate photomultipliers. Also, by using a transmitted-light detector, confocal fluorescence and nonconfocal brightfield images can be simultaneously collected.

6. MEMBRANE POTENTIAL IMAGING

6.1. Nernstian Uptake of Cationic Dyes

In normal living cells, the plasma membrane and mitochondrial inner membrane both maintain a negative inside electrical potential difference ($\Delta\Psi$). Lipophilic cations, such as Rhodamine 123, tetramethylrhodamine methylester, and tetramethylrhodamine ethylester, accumulate electrophoretically into the cytosol and the mitochondrial matrix in response to these membrane potentials (Ehrenberg *et al.*, 1988; Emaus *et al.*, 1986; Johnson *et al.*, 1981). At equilibrium the uptake of permeant monovalent cationic fluorophores is related to $\Delta\Psi$ by the Nernst equation

$$\Delta\Psi = -59 \log F_{in}/F_{out} \tag{2}$$

where F_{in} and F_{out} are the concentrations of fluorophore inside and outside, respectively, of the membrane in question. From confocal images of the cationic fluorophore fluorescence, maps of intracellular electrical potential can be generated using eq 2 (Chacon *et al.*, 1994).

Typically, plasmalemmal $\Delta\Psi$ is -30 to -90 mV and mitochondrial $\Delta\Psi$ is -120 to -180 mV. Residing inside cells, mitochondria are as much as 270 mV more negative than the extracellular space. From eq 2, mitochondria will accumulate cationic fluorophores to a concentration ratio of up to 30,000 to 1 relative to the cell exterior. With 256 gray levels per pixel (8 bits) of conventional computer memory, such large gradients cannot be stored using a linear scale; rather, images must be stored using either 16-bit memory (65,536 gray levels) or a nonlinear logarithmic (gamma) scale. Alternatively, 8-bit images can be collected serially at two different laser illumination intensities. An image is collected first at a low laser power setting to measure areas with bright fluorescence. Then a second image of the same field is collected at 30–100-fold higher laser power. This second image accurately records areas with weak fluorescence. Subsequent image processing combines the two images into a single 16-bit image that contains both the strong and the weak fluorescence information.

6.2. Quantifying Electrical Potential in Confocal Images

To quantify membrane potential, the cells are first loaded with a cationic, lipophilic, electrical-potential-indicating fluorophore, such as tetramethylrhodamine methyl ester (TMRM, 50–500 nM). After 15 to 30 minutes of loading, the loading buffer is replaced with experimental medium containing about a third of the loading concentration of fluorophore to maintain fluorophore equilibrium distribution between the cells and the extracellular space. Confocal images are then collected of the specific fluorescence of the fluorophore, in this case red fluorescence (>590 nm) after excitation with green light (568 nm). Subsequent image processing generates a map of intracellular electrical potential by dividing average fluorescence intensity in the extracellular space into intracellular fluorescence, on a pixel-by-pixel basis. This calculation yields F_{in}/F_{out} for each pixel. Eq 2 is then applied to calculate $\Delta\Psi$ relative to the extracellular space for each point inside the cell.

Figure 2 illustrates the distribution of electrical potential measured in a cultured hepatocyte. The cytosol and nucleus have an electrical potential between -20 and -40 mV, which represents the electrical potential difference across the plasma membrane, namely the plasmalemmal $\Delta\Psi$. Mitochondria have a much more negative potential. In Figure 2, mitochondrial electrical potential relative to extracellular space reaches a maximal magnitude of about -160 mV. This value represents the sum of both plasma-

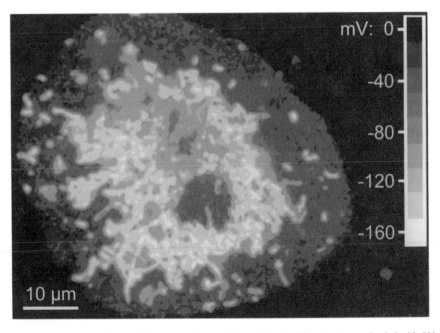

FIGURE 2. Distribution of intracellular electrical potential. A cultured rat hepatocyte was loaded with 500 nM TMRM for 15 min, and red fluorescence was imaged by laser-scanning confocal microscopy using 568 nm excitation light from an argon–krypton laser. The distribution of intracellular electrical potential relative to extracellular space was determined by application of eq 2 to each pixel in the image. (E. Chacon, G. Zahrebelski, J. M. Reece, A.-L. Nieminen, and J. J. Lemasters, unpublished.)

lemmal and mitochondrial $\Delta\Psi$. Because plasmalemmal $\Delta\Psi$ averages about $-30\,mV$, mitochondrial $\Delta\Psi$ reaches a maximum of about $-130\,mV$. Mitochondria typically show some heterogeneity in their estimated electrical potential, due, at least in part, to the fact that mitochondrial diameter is only slightly greater than confocal slice thickness. Thus, many mitochondria extend only partway into the confocal slice, so that the $\Delta\Psi$ for many mitochondria is underestimated. Therefore the maximal observed $\Delta\Psi$ for mitochondria fully sectioned by the confocal slice is the best overall estimate of mitochondrial $\Delta\Psi$ (Chacon *et al.*, 1994).

6.3. Nonideal Behavior by Potential-Indicating Fluorophores

The Nernst equation (eq 2) assumes an ideal equilibrium distribution of cationic fluorophore in response to electrical potential, but many probes do not behave ideally. The widely used green-fluorescing Rhodamine 123, for example, binds nonspecifically to the mitochondrial matrix (Emaus *et al.*, 1986). Moreover, as the probes concentrate into the mitochondrial matrix, fluorescence quenching of Rhodamine 123 and other fluorophores occurs. This concentration-dependent quenching can be decreased by using smaller concentrations of fluorophore. Cationic dyes reach millimolar concentrations in the mitochondrial matrix, where they can cause metabolic inhibition. Rhodamine 123, for example, inhibits the oligomycin-sensitive mitochondrial ATP synthase of oxidative phosphorylation at these concentrations (Emaus *et al.*, 1986); DiOC (Emaus *et al.*, 1986), another cationic fluorophore used both in flow cytometry and fluorescence microscopy, is an even more potent inhibitor of mitochondrial respiration (Rottenberg and Wu, 1997). Although TMRM and its sister molecule, tetramethylrhodamine ethylester (TMRE), seem to lack many of these undesirable characteristics (Ehrenberg *et al.*, 1988; Farkas *et al.*, 1989), low fluorophore concentrations ($< 500\,nM$) should always be used to minimize quenching and metabolic effects.

Another cationic fluorophore that accumulates in mitochondria is JC-1 (Reers *et al.*, 1995; Smiley *et al.*, 1991); JC-1 fluoresces green, but at high concentrations it forms J-aggregates, which fluoresce red. As mitochondrial $\Delta\Psi$ becomes more negative, mitochondrial accumulation of JC-1 increases, causing more J-aggregates to form. Thus, the ratio of red (J-aggregate) to green (monomeric) fluorescence is a relative indicator of $\Delta\Psi$. In confocal microscopy, individual J-aggregates can be imaged inside mitochondria, and ratioing gives the false impression that mitochondrial matrix electrical potential is inhomogeneous. Nonetheless, JC-1 is widely used to monitor relative mitochondrial $\Delta\Psi$ by nonconfocal microscopy and other fluorescence techniques, such as flow cytometry.

7. ION IMAGING

7.1. Ratio Imaging

Many of the most commonly used parameter-indicating fluorophores in confocal microscopy measure ion concentration. These membrane-impermeant fluorophores are either microinjected into cells or, more often, loaded as their membrane-permeant ester derivatives. During ester loading, intracellular esterases release and trap the ion-sensitive free acid form of the fluorophore inside the cells. Fluorescence of these fluorophores depends not only on the specific ion being measured but also on the cell shape and thickness. Confocal imaging offsets variations due to cell morphology, but not entirely, because intracellular organelles can also be responsible for heterogeneous intracellular distribution of ion-indicating fluorophores.

With some fluorophores a ratioing procedure corrects for variations in fluorophore content in the light path. This is possible when ion binding to the fluorophore causes a shift of fluorescence excitation or emission spectrum. Spectral shifts are exploited by acquiring images at two different excitation or emission wavelengths, one that increases as ion concentration increases and one that decreases or stays the same. Background images from cell-free regions must also be collected at both wavelengths. These backgrounds represent nonspecific fluorescence and are subtracted from the experimental images. After background subtraction, the image at the first wavelength is divided by the image at the second wavelength on a pixel-by-pixel basis. The resulting ratio image is proportional to ion concentration, not to the amount of fluorophore in the light path. Ratio values are compared to a standard curve. Ion concentration is often displayed using a pseudocolor or pseudo-gray-level scale to better highlight concentration gradients. Ratio imaging corrects photobleaching, dye leakage over time, and differences of regional fluorophore concentration.

7.2. pH Imaging

With monochromatic laser light sources, emission wavelengths rather than excitation wavelengths must be ratioed. A useful ratiometric fluorophore for confocal microscopy is the pH-indicating fluorophore carboxyseminaphthorhodafluor 1 (SNARF-1). With 568 nm excitation from an argon–krypton laser, SNARF-1 fluorescence emission at 640 nm increases as pH increases, whereas emission at 585 nm remains unchanged (Chacon et al., 1994). Thus, the 640-nm/585-nm ratio is proportional to pH. To calibrate the SNARF-1 signal in situ, cells are incubated at different pH values with 10 μM nigericin and 5 μM valinomycin. NaCl and KCl are replaced with their corresponding gluconate salts to minimize swelling. After ester-loading (5 μM SNARF-1 acetoxymethyl ester), SNARF-1 enters both the cytosolic and the mitochondrial compartments. Confocal ratio imaging of SNARF-1 shows heterogeneity in both hepatocytes and myocytes (Fig. 3). The pH of cytosolic and nuclear regions is 7.1–7.2, whereas mitochondrial pH is 7.8–8. Accordingly, the pH gradient across the mitochondrial inner membrane (ΔpH) is 0.6–0.8 pH units. Because protonmotive force (Δp) is $\Delta\Psi - 59\,\Delta$pH, Δp of mitochondria in intact living cells is -170 to -180 mV, in agreement with the chemiosmotic hypothesis.

7.3. Temperature Dependence of Ester Loading

Most cation-indicating fluorophores are hydrophilic multivalent carboxylic acids, which are impermeant to cellular membranes. To load them inside cells, their carboxyl groups are neutralized by forming acetate or acetoxymethyl (AM) esters. The uncharged esters cross the plasma membrane to be de-esterified and trapped in the cytoplasm. Esterases exist in several cellular compartments, including mitochondria, lysosomes, and

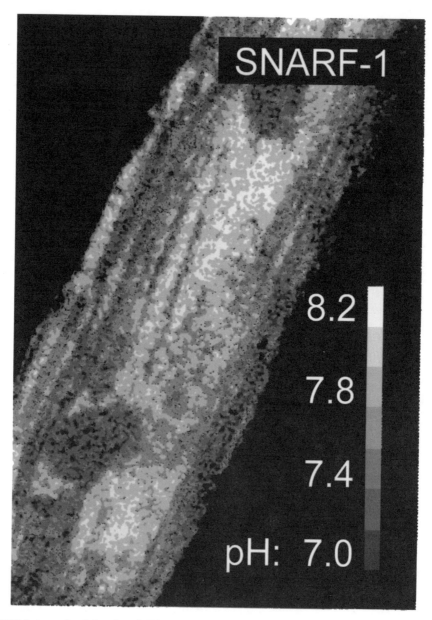

FIGURE 3. SNARF-1 ratio imaging of pH in a cardiac myocyte. An adult rabbit cardiac myocyte was loaded with 5 μM SNARF-1-AM. Confocal images of 585 and 640 nm were collected and ratioed to determine the intracellular distribution of pH. Note the heterogeneous areas of high pH that correspond to mitochondria. (Adapted from Chacon *et al.*, 1994.)

the cytosol. Thus, after ester loading, ion-indicating fluorophores can distribute into several different compartments, often in a heterogeneous fashion.

The temperature of ester loading strongly influences the intracellular distribution of many fluorophores, as illustrated for green-fluorescing calcein, a pentacarboxylic acid dye whose fluorescence is independent of phyisological changes of intracellular ions. When primary cultured hepatocytes and cardiac myocytes are incubated with calcein-AM (1–5 μM) at 37 °C, calcein enters the cytosol and nucleus but not the mitochondria (Fig. 4, upper left) (Nieminen et al., 1995). Exclusion of calcein from the mitochondria causes round voids of about 1 μm in diameter to appear in the diffuse green fluorescence of cytosolic calcein. Each void is a mitochondrion, as verified by subsequent colabeling with red-fluorescing TMRM (Fig. 4. upper right). In contrast, calcein accumulates into both

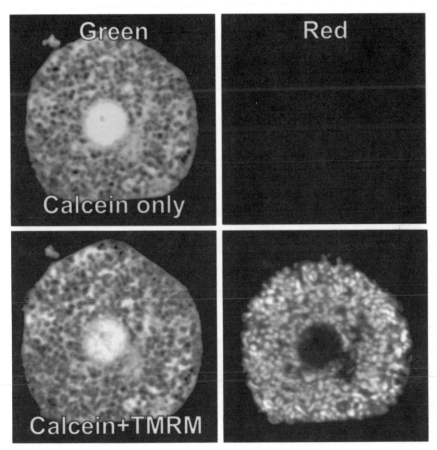

FIGURE 4. Negative-contrast confocal imaging of mitochondria in hepatocytes after warm loading of calcein. Hepatocytes were first loaded with 1 μM calcein-AM at 37 °C for 15 min. Green calcein fluorescence showed a predominantly cytosolic and nuclear distribution, leaving mitochondria as dark round voids (upper left). Red fluorescence was not detectable (upper right), but after incubation with 250 nM TMRM, individual mitochondria fluoresced red (lower right). Each TMRM-labeled mitochondria corresponded a dark void in the calcein image (lower left). (Adapted from Nieminen et al., 1995, and Qian et al., 1999.)

cytosol and mitochondria after ester loading at 4 °C (Fig. 5). Presumably during warm loading, cytosolic esterases hydrolyze the calcein-AM before it can enter mitochondria. At 4 °C, by contrast, esterase activity is slowed, which allows calcein-AM to diffuse into both the cytosol and the mitochondria before its hydrolysis to calcein free acid. In this way both mitochondrial and cytosolic esterases have the opportunity to cleave the ester and trap the free acid. After cold loading, mitochondria cannot be distinguished in the images of green calcein fluorescence, although subsequent TMRM loading proves that mitochondria are present (Fig. 5). Temperature dependence of ester loading varies with both fluorophore and cell type. For example, Fluo-3-AM, a green-fluorescing Ca^{2+} fluorophore, loads similarly to calcein, whereas Indo-1, a blue-fluorescing Ca^{2+} indicator, and SNARF-1 load into mitochondria even at 37 °C (Chacon *et al.*, 1994; Ohata, *et al.*, 1998).

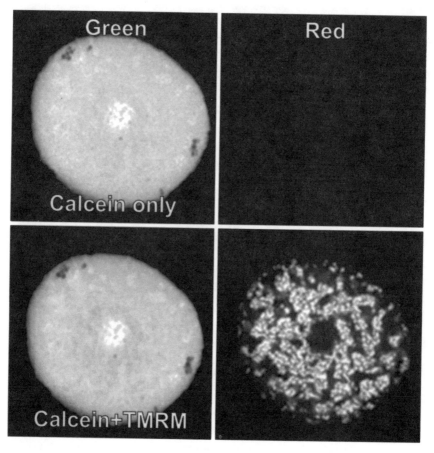

FIGURE 5. Calcein distribution into both the cytosol and the mitochondria after cold loading. Rat hepatocytes were loaded with 1 μM calcein-AM at 4 °C for 60 min. Green calcein fluorescence was evenly distributed throughout both cytosol and mitochondria (upper left). Subsequent addition of TMRM (250 nM) caused bright mitochondrial labeling in the red channel (lower right), but did not quench mitochondrial calcein fluorescence in the green channel (lower left). (Adapted from Nieminen *et al.*, 1995, and Qian *et al.*, 1999.)

7.4. The Cold-Loading/Warm-Incubation Protocol

Cytosolic fluorophores gradually leak across the plasma membrane through an organic anion carrier (Wieder et al., 1993). Fluorophores loaded into mitochondria and other organelles are released even more slowly. This selective release can be exploited to achieve selective mitochondrial localization of ester-loaded fluorophores. Cells are first ester-loaded in the cold to distribute fluorophore into both cytosol and mitochondria. Subsequently, the cells are returned to the tissue-culture incubator for several hours. During this warm incubation, cytosolic fluorophore leaks across the plasma membrane into the extracellular medium, but fluorophore in mitochondria and other organelles (chiefly lysosomes) is retained (Fig. 6) (Lemasters et al., 1999; Qian et al., 1999; Trollinger et al., 1997). Thus, by manipulating the temperature of loading and the duration of subsequent warm incubation, ester-loaded fluorophores can be directed to the cytosol, the cytosol plus organelles, or to just the organelles (Lemasters et al., 1999; Nieminen et al., 1995; Ohata et al., 1998; Qian et al., 1999; Trollinger et al., 1997).

A recent paper suggests that the dark voids observed in calcein-loaded hepatocytes are the result of fluorescent quenching by concurrent use of TMRM (Petronilli et al., 1999). When hepatocyte mitochondria are selectively loaded with calcein by the cold-loading/warm-incubation protocol, however, TMRM loading causes negligible quenching (Fig. 6) (Lemasters et al., 1999; Qian et al., 1999). Similarly, in cold-loaded hepatocytes, subsequent TMRM treatment does not cause fluorescence voids to appear (Fig. 5).

8. FREE CA^{2+} IMAGING

8.1. Nonratiometric Imaging

Green-fluorescing Fluo-3 (and Fluo-4) and red-fluorescing Rhod-2 are useful visible-wavelength fluorophores for imaging intracellular free Ca^{2+} concentration by confocal microscopy. The Ca^{2+} binds to Fluo-3 and Rhod-2 with a K_d of 300–600 nM, which causes a 50-fold or greater increase in fluorescence (Haugland, 1996; Minta et al., 1989), but it does not shift their fluorescence spectrum. Calibration therefore relies on in situ measurements of maximal and minimal fluorescence after respective additions of calcium ionophore and EGTA (Harper et al., 1993; Minta et al., 1989). Even without calibration, Fluo-3 and Rhod-2 fluorescence is a sensitive measure of relative changes of free Ca^{2+}.

8.2. Line-Scanning Confocal Microscopy of Ca^{2+} Transients

Collection of well-resolved two-dimensional images by laser-scanning confocal microscopy typically requires 2–10 seconds. To increase temporal resolution, confocal images can be collected in the line-scanning mode whereby the x-axis scan is repeated at the same y-axis position. This scanning creates x-versus-time images with a temporal resolution of 25 msec or less (Cheng et al., 1993). In Figure 7 simultaneous red and green line-scan images were collected during electrical stimulation of a cardiac myocyte coloaded with TMRM and Fluo-3. In this experiment an x-axis position was selected to cross through the cytosol, interfibrillar mitochondria, perinuclear mitochondria, and

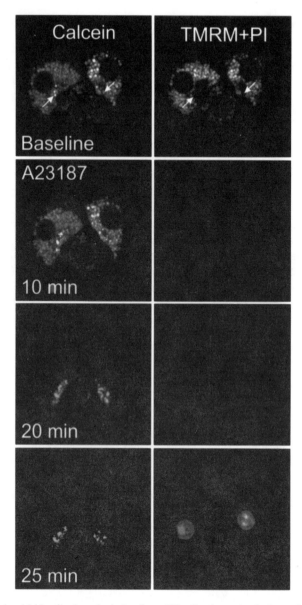

FIGURE 6. Mitochondrial localization of calcein after cold loading/warm incubation and release of mitochondrial calcein after onset of the mitochondrial permeability transition. Hepatocytes were loaded with 1 μM calcein-AM at 4 °C for 60 min, returned to a 37 °C incubator for 12 h, then loaded with 250 nM TMRM. During warm incubation, cytosolic calcein leaked from the cytosol, but noncytosolic calcein was retained. Noncytosolic calcein was predominantly in TMRM-labeled mitochondria (Baseline, upper panel). Calcein also labeled lysosomes (arrows), but lysosomes did not colabel with TMRM. After exposure to Br-A23187 (A23), a calcium ionophore, mitochondria lost TMRM fluorescence within 10 min (lower panel), indicating mitochondrial depolarization, but calcein fluorescence was initially unchanged (upper panel), indicating no change of mitochondrial permeability. Subsequently, after 20 min mitochondria released calcein into the cytosol, indicating MPT onset. Calcein release was followed quickly by loss of cell viability, shown by nuclear labeling with propidium iodide (PI) after 25 min. (Adapted from Qian *et al.*, 1999.)

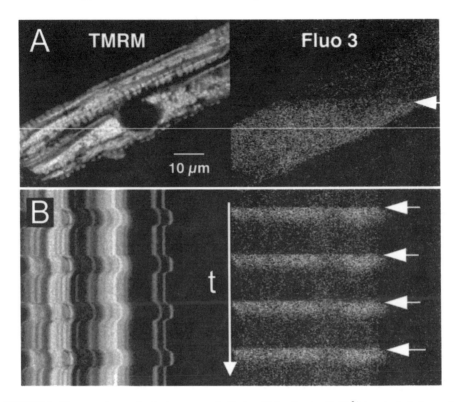

FIGURE 7. Line-scanning confocal microscopy of mitochondrial and cytosolic Ca^{2+} transients during contractions of a cardiac myocyte. A cardiac myocyte was loaded with 10 μM Fluo-3-AM for 2 hr at 4 °C and then with 600 nM TMRM at 37 °C. Simultaneous confocal images of red (TMRM, left) and green (Fluo-3) fluorescence were collected. In the x–y confocal image (A) a single field stimulation (arrow) was applied as the 1.5 sec scan was collected. In the x-vs-time line-scan confocal image (B) the myocyte was stimulated at 0.5 Hz (arrows) in the presence of 1 μM isoproterenol. The x position of the line-scan is indicated by the white line in (A). (Adapted from Ohata *et al.*, 1998.)

nucleus, as shown in Figure 7A. In the x-*versus*-time line-scan images, red TMRM fluorescence appears as wavy vertical stripes (Fig. 7B). Each stripe is a single mitochondrion, and each wave represents cell movement due to an electrically stimulated contraction. By contrast, green Fluo-3 fluorescence does not show vertical stripes, only horizontal bands that correspond to Ca^{2+} transients induced by electrical stimulation. Significantly, the banding of Fluo-3 fluorescence occurs both in pixels corresponding to TMRM-labeled mitochondria and in pixels of the TMRM-unlabeled cytosol and nucleus. These results show the simultaneous occurrence of both cytosolic and mitochondrial Ca^{2+} transients during the contractile cycle in cardiac myocytes (Chacon *et al.*, 1996; Ohata *et al.*, 1998).

To analyze the line-scan images, regions corresponding to the cytosol, the nucleus, and the interfibrillar and perinuclear mitochondria are identified by the presence or absence of TMRM fluorescence and its position within the cell. Pixels of bright red TMRM fluorescence are mitochondria, whereas pixels of low TMRM fluorescence are the cytosol and nucleus. Pixels of intermediate intensity likely represent the overlap of mitochondrial and cytosolic spaces and are excluded from analysis. Based on morphological criteria, regions of high TMRM fluorescence are further subdivided into interfibrillar and perinuclear mitochondria, and low-fluorescence areas are identified as nucleus and cytosol. The Fluo-3 fluorescence in each identified cellular compartment is then averaged for each horizontal line. Inexpensive software packages, such as NIH Image (National Institutes of Health, Bethesda, MD) and Image PC (Scion Corp., Frederick, MD), can perform this analysis. When the repeating transients of Fluo-3 fluorescence are averaged, the peak change after field stimulation is greatest in the cytosol and least in perinuclear mitochondria (Fig. 8). Isoproterenol, a β-adrenergic agonist, increases peak intensity and rate of decay of Fluo-3 fluorescence in cytosol and nucleus. The effect of isoproterenol is weaker in interfibrillar mitochondria and virtually absent in perinuclear mitochondria. Because red TMRM fluorescence and green Fluo-3 fluorescence are collected simultaneously, the calculations automatically correct for movement during myocyte contraction (Chacon *et al.*, 1996; Ohata *et al.*, 1998).

8.3. Ratio Imaging

The binding of Ca^{2+} to Fluo-3 produces increased fluorescence but no wavelength shift. Thus, ratio imaging cannot be used to estimate absolute Ca^{2+} concentrations. Indo-1 is a ratiometric Ca^{2+} indicator whose emission spectrum will shift after Ca^{2+} binding (Grynkiewicz *et al.*, 1985); Indo-1 requires excitation with ultraviolet light. When imaged by confocal microscopes equipped with an ultraviolet laser, Indo-1 fluorescence increases at 405 nm and decreases at 480 nm as Ca^{2+} increases. Like other Ca^{2+} indicators, Indo-1 is

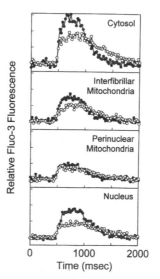

FIGURE 8. Fluo-3 fluorescence in different subcellular regions during the contractile cycle. In line-scan images (see Fig. 7B), red TMRM fluorescence was used to identify cytosol, interfibrillar and perinuclear mitochondria, and nucleus. Corresponding pixel values in images of green Fluo-3 fluorescence were averaged for each line in the line-scan in each of the four regions. Closed and open symbols are in the presence and absence, respectively, of 1 μM isoproterenol. Plotted are values averaged from four contractions. (Adapted from Ohata *et al.*, 1998.)

loaded as its acetoxymethyl ester (2–5 μM). Unlike Fluo-3 and Rhod-2, Indo-1 distributes readily into both cytosol and mitochondria of cardiac myocytes even after ester loading at 37 °C (Ohata et al., 1998). In unstimulated myocytes, ratio images of Indo-1 fluorescence show a uniform intracellular free Ca^{2+} concentration of about 200 nM in both cytosol and mitochondria (Fig. 9). This value is based on an ex situ calibration and may be an overestimation, because the binding constant of Indo-1 for Ca^{2+} apparently increases in the intracellular environment (Baker et al., 1994; Bassani et al., 1995; Roe et al., 1990). The Ca^{2+} sensitivity of Indo-1 and other BAPTA-derived probes is also affected by pH, but significant changes of the K_d for Ca^{2+} occur only when pH falls below 6.8, outside the normal range of cytosolic and mitochondrial pH (Kawanishi et al., 1991; Lattanzio and Bartschat, 1991). In any event, Indo-1 ratio imaging corrects for variable loading of the indicator into various cellular compartments and confirms that Ca^{2+} transients occur in both cytosol and mitochondria during the contractile cycle (Fig. 9).

8.4. Cold Loading of Ca^{2+} Indicators Followed by Warm Incubation

Because mitochondria of many cells can be smaller than the three-dimensional resolving power of confocal microscopy, techniques to load ion indicators into mitochondria only are desirable. The cold-loading/warm-incubation protocol allows such selective mitochondrial, noncytosolic loading. After cold loading/warm incubation of cardiac myocytes with Rhod-2-AM (5 μM), red Rhod-2 fluorescence shows a distinctly mitochondrial pattern, comparable to that observed in myocytes loaded with TMRM or Rhodamine 123 (Fig. 10, compare Fig. 13A) (Trollinger et al., 1997). Subsequently, when myocytes are electrically stimulated, mitochondrial Rhod-2 fluorescence increases with each stimulation to produce horizontal bands as 16-second confocal scans are collected (Fig. 10B). These fluorescence transients occur in the bright regions corresponding to mitochondria and are inhibited by ruthenium red, a blocker of electrogenic mitochondrial Ca^{2+} uptake (Fig. 10C) (Trollinger et al., 2000). Subsequent addition of Br-A23187, a Ca^{2+} ionophore, to saturate Rhod-2 with Ca^{2+} in all compartments verifies the mitochondrial, noncytosolic localization of Rhod-2 (Fig. 10D). The powerful protonophoric uncoupler CCCP was also added to depolarize mitochondria and prevent cycling of free Ca^{2+} that might lead to intracellular Ca^{2+} gradients after Br-A23187 treatment. Glucose and oligomycin were also added, to provide a glycolytic source of ATP and to prevent ATP depletion by the uncoupler-stimulated, oligomycin-sensitive mitochondrial ATPase (Bers et al., 1993; Nieminen et al., 1990a, 1994). After Br-A23187 treatment under these conditions, Rhod-2 fluorescence is punctate instead of diffuse, although mitochondrial morphology is distorted by the Ca^{2+}-induced contracture (Fig. 10D). When contracture is prevented by butanedione monoxime (Armstrong and Ganote, 1991), the pattern of punctate Rhod-2 fluorescence after Br-A23187 is virtually identical to that of Rhodamine 123-and TMRM-labeled mitochondria (Trollinger et al., 1997). Cytosolic areas between mitochondria remain dark, directly demonstrating the absence of significant amounts of Rhod-2 in the cytosol.

Measurement of mitochondrial Ca^{2+} via Rhod-2 localized to mitochondria by the cold-loading/warm-incubation protocol has advantages over mitochondrial Ca^{2+} measurement in cells coloaded with Fluo-3 and TMRM. With the latter measurement, only Fluo-3 pixels corresponding to the brightest and darkest TMRM pixels can be relied upon to

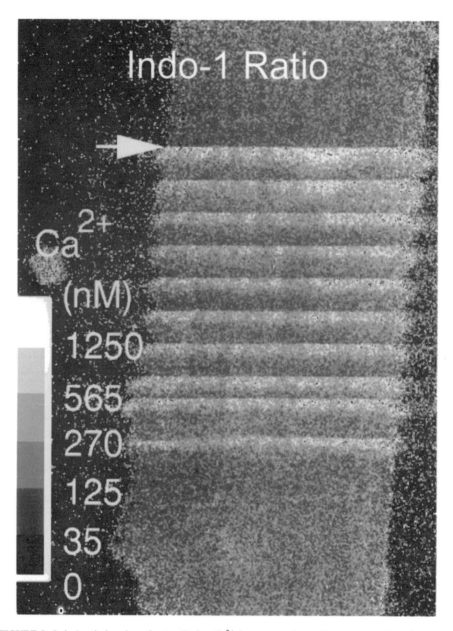

FIGURE 9. Indo-1 ratio-imaging of cytosolic free Ca^{2+} in a contracting cardiac myocyte. A cardiac myocyte was loaded with 5 μM Indo-1-AM at 37 °C. Using 351 nm excitation from a UV–argon laser, confocal images were collected at emission wavelengths of 395–415 nm and 470–490 nm as the cell was stimulated at 0.5 Hz. After background subtraction, the images were ratioed and scaled to Ca^{2+} concentration. Ca^{2+} transients occurred in cytosol and mitochondria during the contractile cycle, as identified by Rhodamine 123 labeling (see Fig. 13A). (Adapted from Ohata *et al.*, 1998.)

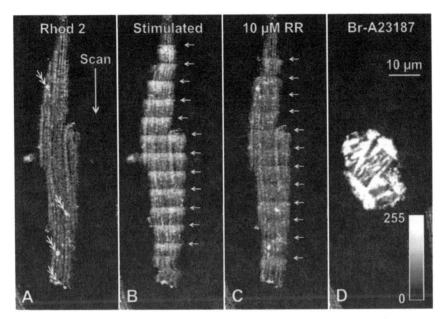

FIGURE 10. Mitochondrial Ca^{2+} transients detected after cold loading/warm incubation of Rhod-2. A cardiac myocyte was loaded with Rhod-2-AM (5 µM) at 4 °C for 30 min and incubated at 37 °C for 5 hr. Confocal images of red Rhod-2 fluorescence were then collected. (A) Without electrical stimulation Rhod 2 fluorescence showed a mitochondrial pattern of labeling. Double arrows label bright spots of lysosomal fluorescence. (B) The myocyte was stimulated at 1 Hz, and mitochondrial Rhod-2 fluorescence fluctuated with each stimulation (arrows) to produce horizontal banding in the 16 sec scans. (C) 10 µM ruthenium red (RR) was added, which suppressed the mitochondrial Ca^{2+} transients. (D) oligomycin (10 µM), CCCP (10 µM), and Br-A23187 (20 µM) were added to saturate Ca^{2+}-dependent Rhod-2 fluorescence in all compartments. The myocyte contracted, reflecting the increased cytosolic free Ca^{2+}, but Rhod-2 fluorescence was confined almost exclusively to mitochondria. (Adapted from Trollinger *et al.*, 2000.)

represent mitochondrial and cytosolic Ca^{2+}, respectively. Pixels with intermediate TMRM fluorescence must be excluded because they represent partially sectioned mitochondria. After cold loading of Rhod-2 followed by warm incubation, all Rhod-2 fluorescence is mitochondrial, and all pixels can be used for analysis, which increases signal strength and improves the signal-to-noise ratio.

8.5. Simultaneous Labeling of the Cytosol and Mitochondria with Different Ca^{2+}-Indicating Fluorophores

Independent measurement of both cytosolic and mitochondrial Ca^{2+} can be achieved by first labeling cells with Rhod-2 by the cold-loading/warm-incubation protocol. At the end of warm incubation, after cytosolic Rhod-2 has leaked from the cytosol, the cells are loaded with Fluo-3-AM (10 µM) at 37 °C. During warm loading, Fluo-3 enters the cytosol but not the mitochondria. In this dual-labeling procedure, Fluo-3 labels the cytosol, whereas Rhod-2 is selectively retained by the mitochondria (Trollinger *et al.*, 2000). After coloading with Rhod-2 and Fluo-3, red Rhod-2 fluorescence shows a mitochondrial pattern, whereas green Fluo-3 fluorescence is diffuse (Fig. 11A and 11A′), except for lysosomal uptake (see below). In cardiac myocytes loaded by this procedure, horizontal banding occurs in confocal images collected during electrical stimulation for both fluorescences (Fig. 11B and 11B′). These signals demonstrate mitochondrial and cytosolic Ca^{2+} transients, respectively, during the contractile cycle. Ruthenium red, however, inhibits only Rhod-2 fluorescence transients, not Fluo-3 transients (Fig. 11C and 11C′). In the experiment illustrated in Figure 11, low Na^+ buffer is then added to increase intracellular Ca^{2+} by reversal of the plasmalemmal Na^+/Ca^{2+} exchanger. After intracellular Ca^{2+} overloading, Rhod-2 fluorescence increases in mitochondria but not in toxic blebs appearing at the cell surface (Fig. 11D, double arrows), whose contents are an extension of the cytosol (Lemasters, *et al.*, 1981). By contrast, Fluo-3 fluorescence increases diffusely throughout the cytosol, including inside the blebs (Fig. 11D′). Thus it is possible to monitor cytosolic and mitochondrial free Ca^{2+} independently in the same cell by selective loading of different Ca^{2+} indicators into the different subcellular compartments.

8.6. Lysosomal Localization of Ca^{2+}-Indicating Fluorophores

As alluded to above, ester-loaded Ca^{2+} indicators also accumulate into small, bright fluorescence spots scattered throughout the cytoplasm (see Fig. 10A, double arrows). These bright spots represent lysosomes, because LysoTracker Red, a red-fluorescing fluorophore that labels acidic lysosomal/endosomal compartments, colocalizes to these spots (Fig. 12). Indeed, the majority of total ester-loaded Fluo-3 and Rhod-2 may localize to lysosomes.

How then can mitochondrial Ca^{2+} be measured, if most fluorescence arises from lysosomes? The answer involves both oversaturation of lysosomal pixels and imaging around the lysosomes. Typical confocal microscopes record only 256 levels of gray (0–255); intensity levels exceeding this range are recorded as a value of 255. At a sufficiently low intensity of laser illumination, images can be collected in which no pixel is saturated, and even the brightest spots of fluorescence arising from lysosomes do not exceed an intensity value of 254. In a cardiac myocyte labeled with Rhod-2 by the cold-loading/warm-incubation protocol, such an image represents fairly the intracellular distribution of total Rhod-2 fluorescence arising from the cell (Fig. 13B). Under these conditions, red Rhod-2 fluorescence has no relation to the distribution of Rhodamine 123-labeled mitochondria (compare Fig. 13A and 13B). Rather, Rhod-2 fluorescence corresponds to the pattern of lysosomal fluorescence shown in Figure 12. Moreover, no fluorescence transients are observed at this low level of laser illumination during electrical stimulation (Fig. 13B).

When the confocal scan is repeated using 100 times more laser excitation intensity, fluorescence arising from lysosomes becomes saturated and is recorded as gray-level intensities of 255. At this higher excitation, a mitochondrial pattern of fluorescence emerges in the background of the saturated lysosomal fluorescence (Fig. 13C). Moreover, this background fluorescence shows transients during each contractile cycle (Fig. 13D). Thus, signals from the more weakly labeled mitochondria can be monitored by saturating and imaging around the lysosomal fluorescence. These observations resolve the discrepancy between studies using confocal microscopy, which describe rapid mitochondrial

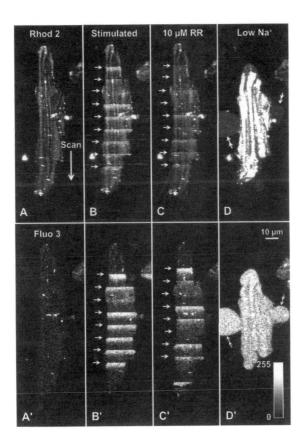

FIGURE 11. Measurement of mitochondrial and cytosolic Ca^{2+} in a myocyte simultaneously loaded with Rhod-2 in mitochondria and Fluo-3 in the cytosol. Myocytes were labeled with Rhod-2-AM by cold loading/warm incubation and then incubated with Fluo-3-AM ($10\,\mu M$) for 15 min at $37\,^{\circ}C$. After washing, confocal images of red Rhod-2 and green Fluo-3 fluorescence were collected in sequential 16 sec scans. (A) Prior to stimulation, Rhod-2 fluorescence showed a mitochondrial pattern. (A′) Fluo-3 fluorescence was diffuse, except for small bright spots of lysosomal fluorescence. (B) and (B′) The cell was electrically stimulated (arrows) as confocal images were collected, producing horizontal banding in both Rhod-2 and Fluo-3 fluorescence images, indicating mitochondrial and cytosolic Ca^{2+} transients, respectively. Electrical stimulation was near threshold, and occasional beats were missed. (C) and (C′) $10\,\mu M$ ruthenium red (RR) was added, which suppressed Rhod-2 fluorescence transients, but Fluo-3 transients remained. (D) and (D′) Buffer containing 5 mM Na^+ was added to increase intracellular Ca^{2+} by reverse Na^+/Ca^{2+} exchange. Rhod-2 fluorescence increased in mitochondria but not in toxic blebs (double arrows), whereas Fluo-3 fluorescence diffusely increased in cytosol and blebs. (Adapted from Trollinger *et al.*, 2000.)

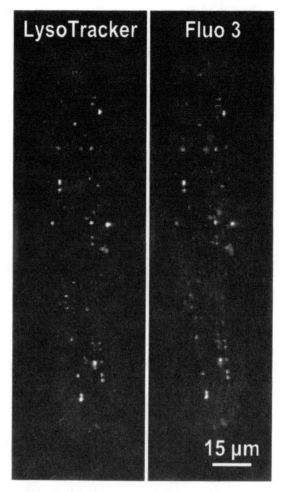

FIGURE 12. Lysosomal uptake of Ca^{2+}-indicating fluorophores after ester loading. A rabbit cardiac myocyte was loaded with 10 μM Fluo-3-AM by the cold-loading/warm-incubation protocol and then incubated with Lyso Tracker Red (50 nM) at 37 °C to label lysosomes. Red and green confocal images were collected at a laser intensity setting that prevented saturation of any pixels. LysoTracker Red fluorescence (left) and Fluo-3 fluorescence (right) colocalized in heterogeneous punctate structures. (Adapted from Trollinger *et al.*, 2000.)

Ca^{2+} transients during the contractile cycle (Chacon *et al.*, 1996; Ohata *et al.*, 1998; Trollinger *et al.*, 1997, 2000) and studies using microfluorometry, which fail to detect mitochondrial transients (Di Lisa *et al.*, 1993; Griffiths *et al.*, 1997; Miyata *et al.*, 1991). Although lysosomes are small and few, they account for the majority of noncytosolic fluorescence after ester loading. Consequently, after selective depletion or quenching of cytosolic fluorophore, spot microfluorometry shows predominantly lysosomal fluorescence. These findings underscore the need to use confocal microscopy to distinguish mitochondria from other organelles to study mitochondrial Ca^{2+} fluxes in intact living cells.

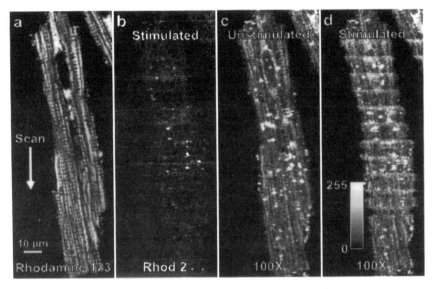

FIGURE 13. Discrimination of lysosomal and mitochondrial loading of Ca^{2+}-indicating fluorophores by confocal microscopy. A myocyte was loaded with Rhod-2-AM (5 μM) by cold loading/warm incubation and then with Rhodamine 123 (1 μM). (A) Rhodamine 123 green fluorescence shows mitochondrial distribution. (B) Rhod-2 red fluorescence was collected during electrical stimulation at 1 Hz. Laser excitation was attenuated in (B) so that no pixel was saturated. Rhod-2 fluorescence showed heterogeneous punctate fluorescence that did not correspond to Rhodamine 123-labeled mitochondria, and no fluorescence banding occurred from electrical stimulation. (C) A confocal image was collected at 100 times more excitation energy, but without electrical stimulation. Fluorescence arising from lysosomes was now saturated, and a mitochondrial pattern of fluorescence emerged in the background. (D) Another image was collected at the same instrument setting during electrical stimulation at 1 Hz. Rhod-2 fluorescence showed horizontal banding indicative of Ca^{2+} transients. (Adapted from Trollinger *et al.*, 2000.)

9. REACTIVE OXYGEN SPECIES

Generation of reactive, incompletely reduced forms of oxygen (superoxide, hydrogen peroxide, hydroxyl radical, singlet oxygen) may contribute to the pathophysiology of many diseases. Mitochondria are an important source of such reactive oxygen species (ROS) (Chance *et al.*, 1979; Forman and Boveris, 1982), and mitochondrial ROS formation occurs in living cells and can contribute to lethal cell injury (Dawson *et al.*,1993; Gores *et al.*, 1989). The ROS and their metabolic byproducts react with certain nonfluorescent compounds (such as dichlorofluorescin, dihydrorhodamine 123, and hydroethidium) to form highly fluorescent derivatives (Cathcart *et al.*, 1983; Perticarari *et al.*, 1991; Rothe *et al.*, 1988). These reactions can be adapted to image the rate and location of ROS formation in single cells. Other compounds, such as monochlorobimane and monobro- mobimane, can be used to image intracellular distribution of glutathione and protein thiols, respectively, important targets of ROS oxidation (Bellomo *et al.*, 1992; Nieminen *et al.*, 1990b).

Dichlorofluore*scin* (hydrodichlorofluorescein) reacts with hydroperoxides to form highly fluorescent dichlorofluore*scein*. To monitor hydroperoxide formation, cells are loaded with dichlorofluorescin diacetate (10 μM), which is cleaved intracellularly to trap free dichlorofluorescin in the cytoplasm. Even before imposition of cellular stress, green dichlorofluorescein fluorescence increases inside the mitochondria, directly demonstrating mitochondrial ROS generation during normal metabolism. During oxidative stress with *t*-butylhydroperoxide, however, formation of dichlorofluore*scin* from dichlorofluore*scein* increases several-fold (Byrne *et al.*, 1999; Nieminen *et al.*, 1997). Most fluorescence develops in mitochondria, indicating that mitochondria are the principal source of ROS in this model of oxidative stress. Subsequently, dichlorofluorescein leaks from the cells as viability is lost. Mitochondrial ROS formation promotes onset of the mitochondrial permeability transition, whose detection by confocal microscopy is described in the next section.

The oxidation-reduction status of mitochondria can also be conveniently monitored by confocal imaging of mitochondrial NAD(P)H autofluorescence using excitation from a UV–argon laser (Nieminen *et al.*, 1997). Because pyridine nucleotide fluorescence is quenched in the cytosol, nearly all NAD(P)H fluorescence arises from mitochondria. The NAD(P)H fluorescence responds quickly to addition of mitochondrial reductants (such as β-hydroxybutryate and cyanide) and oxidants (such as *tert*-butylhydroperoxide and uncoupler). The weak green fluorescence of oxidized flavin can also be used to monitor mitochondrial redox status.

10. THE MITOCHONDRIAL PERMEABILITY TRANSITION

Movement of calcein across membranes allows direct observation in cells of alterations in membrane permeability. At the onset of necrotic cell death, for example, cells lose trapped cytosolic calcein almost instantaneously as the permeability barrier of the plasma membrane abruptly fails (Fig. 14) (Zahrebelski *et al.*, 1995). Similarly, when calcein is placed in the extracellular space, the fluorophore penetrates cells at onset of cell death (Zahrebelski *et al.*, 1995). Other extracellular fluorophores enter also, such as propidium iodide (which binds to DNA with an enhancement of fluorescence) and high-molecular-weight dextrans.

After warm ester loading of calcein, many cellular stresses cause redistribution of calcein fluorescence from the cytosol into the mitochondria, with simultaneous loss of mitochondrial membrane potential (Nieminen *et al.*, 1995, reviewed in Lemasters *et al.*, 1998). These events indicate onset of the mitochondrial permeability transition (MPT), which is caused by the opening of a pore in the mitochondrial inner membrane, conducting solutes of molecular weight up to 1500 daltons (Zoratti and Szabò, 1995). Onset of the MPT causes mitochondrial depolarization, uncoupling of oxidative phosphorylation, and large-amplitude mitochondrial swelling. The MPT is a crucial event promoting both necrotic and apoptotic cell death (Lemasters *et al.*, 1998).

A pH below 7 inhibits conductance through the PT pore (Bernardi *et al.*, 1992; Halestrap, 1991). Thus, the naturally occurring acidosis of ischemia prevents MPT onset. After reperfusion, normal intracellular pH is restored and onset of the MPT occurs, resulting in cell death (Qian *et al.*, 1997). Cyclosporin A prevents pH-dependent cell killing after simulated ischemia/reperfusion, and confocal microscopy directly shows that MPT onset precedes pH-dependent loss of cell viability after reperfusion, as shown in Figure 14. The MPT does not occur prior to reperfusion because mitochondria continue to

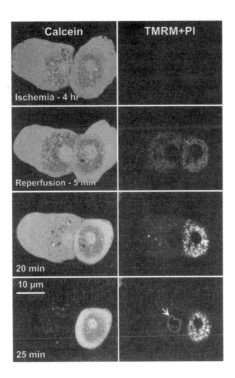

FIGURE 14. Onset of the mitochondrial permeability transition after reperfusion of ischemic hepatocytes. Green (left panels) and red (right panels) fluorescence images were collected by confocal microscopy of hepatocytes loaded with calcein and TMRM and incubated in medium containing propidium iodide. At the end of 4 h of simulated ischemia (anoxia at pH 6.2), mitochondria released most TMRM fluorescence, indicating mitochondrial depolarization, but continued to exclude cytosolic calcein, indicating that the MPT had not occurred. After reperfusion (reoxygenation at pH 7.4), mitochondria of both hepatocytes began accumulating TMRM within 5 min. After 20 min mitochondria of one hepatocyte released TMRM, indicating mitochondrial depolarization, and filled with calcein fluorescence, indicating MPT onset. After 25 min this hepatocyte subsequently lost viability, shown by cytosolic calcein fluorescence release and nuclear propidium iodide uptake (arrow). Mitochondria of the remaining hepatocyte continued accumulating TMRM and excluding cytosolic calcein. (Adapted from Qian *et al.*, 1997.)

exclude cytosolic calcein even after 4 hours of anoxia at acidotic pH (simulated ischemia) (Fig. 14). By contrast, mitochondrial depolarization does occur, and mitochondria lose virtually all their red fluorescent staining with TMRM, showing again that the voids in calcein fluorescence are not caused by TMRM quenching.

When the hepatocytes are reoxygenated at normal pH (simulated reperfusion), mitochondria initially begin to repolarize and re-accumulate TMRM, as shown in Figure 14. Subsequently, as intracellular pH rises, onset of the MPT occurs in one of the cells in the field, causing calcein to redistribute into the mitochondria. Simultaneously, TMRM is again released. In this way, increased mitochondrial membrane permeability and mitochondrial depolarization directly demonstrate onset of the MPT in a single hepatocyte after reperfusion (Fig.14). Several minutes after the MPT, the cell loses viability, as indicated by propidium iodide staining of nuclei in the red channel and release of cytosolic calcein in the green channel. In the few cells that do not undergo the MPT, viability is retained. Moreover, reperfusion in the presence of cyclosporin A or with acidotic buffer to block conductance of the PT pore prevents MPT onset and cell death (Qian *et al.*, 1997). Thus, the MPT mediates pH-dependent cell killing after ischemia/reperfusion. Cyclosporin A also protects against anoxia/reoxygenation injury to perfused hearts and prevents redistribution of hydrophilic radioactive tracers from the cytosolic space into the mitochondrial matrix space (Griffiths and Halestrap, 1995).

Many cell types have mitochondria that are much smaller than the large mitochondria of hepatocytes, myocytes, and a few other highly aerobic cells types. Small mitochondria have an average diameter of $0.3\,\mu m$ and cannot be visualized by negative contrast in confocal sections that are 3 times thicker. Indeed, negative contrast imaging of liver and heart mitochondria whose diameter is $1\,\mu m$ challenges the axial resolving power of confocal microscopy. Thus, optical alignment must be perfect to visualize the dark mitochondrial voids. An alternate approach to visualizing the MPT in .cells with small mitochondria is to use the cold-loading/warm-incubation technique to label mitochondria selectively with calcein (Fig. 6); then at onset of the MPT induced by Br-A23187, calcein will leak from the mitochondria into the cytosol. Another approach is to label both cytosol and mitochondria with calcein in the presence of $1\,mM$ $CoCl_2$, a potent quencher of calcein fluorescence. Because Co^{2+} does not readily enter mitochondria, calcein fluorescence in mitochondria is preserved until onset of the MPT (Petronilli *et al.*, 1999). The cold-loading/warm-incubation procedure may be preferable for many applications, however, because Co^{2+} causes Ca^{2+} channel blockade, dehydrogenase inhibition, genotoxicity, and other adverse effects (Beyersmann and Hartwig, 1992; Hughes and Barritt, 1989; Seghizzi *et al.*, 1994).

11. CONCLUSION

Confocal microscopy is an essential tool for studying the physiology and pathophysiology of mitochondria within single living cells. As more parameter-indicating fluorophores are discovered, the usefulness of confocal microscopy will only increase. Uniquely, confocal microscopy permits observation of $\Delta\Psi$, pH, Ca^{2+}, oxygen free radical generation, and membrane permeability in single mitochondria of living cells. In the future, confocal microscopy should provide new insights concerning the role of mitochondria in health and disease.

REFERENCES

Armstrong, S. C., and Ganote, C. E., 1991, Effects of 2,3-butanedione monoxime (BDM) on contracture and injury of isolated rat myocytes following metabolic inhibition and ischemia, *J. Mol. Cell Cardiol.* **23**:1001–1014.

Baker, A. J., Brandes, R., Schreur, J. H., Camacho, S. A., and Weiner, M. W., 1994, Protein and acidosis alter calcium-binding and fluorescence spectra of the calcium indicator indo-1, *Biophys. J.* **67**:1646–1654.

Bassani, J.W., Bassani, R.A., and Bers, D.M., 1995, Calibration of indo-1 and resting intracellular [Ca]$_i$ in intact rabbit cardiac myocytes, *Biophys. J.* **68**:1453–1460.

Bellomo, G., Vairetti, M., Stivala, L., Mirabelli, F., Richelmi, P., and Orrenius, S., 1992, Demonstration of nuclear compartmentalization of glutathione in hepatocytes, *Proc. Nat. Acad. Sci. USA* **89**:4412–4416.

Bernardi, P., Vassanelli, S., Veronese, P., Colonna, R., Szabò, I., and Zoratti, M., 1992, Modulation of the mitochondrial permeability transition pore: Effect of protons and divalent cations, *J. Biol. Chem.* **267**:2934–2939.

Bers, D. M., Bassani, J. W., and Bassani, R. A., 1993, Competition and redistribution among calcium transport systems in rabbit cardiac myocytes, *Cardiovasc. Res.* **27**:1772–1777.

Beyersmann, D., and Hartwig, A., 1992, The genetic toxicology of cobalt, *Toxicol. Appl. Pharmacol.* **115**:137–145.

Byrne, A. M., Lemasters, J. J., and Nieminen, A-L., 1999, Contribution of increased mitochondrial free Ca^{2+} to the mitochondrial permeability transition induced by *tert*-butylhydroperoxide in rat hepatocytes, *Hepatology* **29**:1523–1531.

Cathcart, R., Schwiers, E., and Ames, B. N., 1983, Detection of picomole levels of hydroperoxides using a fluorescent dichlorofluorescein assay, *Anal. Biochem.* **134**:111–116.

Chacon, E., Reece, J. M., Nieminen, A.-L., Zahrebelski, G., Herman, B., and Lemasters, J. J., 1994, Distribution of electrical potential, pH, free Ca^{2+}, and volume inside cultured adult rabbit cardiac myocytes during chemical hypoxia: A multiparameter digitized confocal microscopic study, *Biophys. J.* **66**:942–952.

Chacon, E., Ohata, H., Harper, I. S., Trollinger, D. R., Herman, B., and Lemasters, J. J., 1996, Mitochondrial free calcium transients during excitation–contraction coupling in rabbit cardiac myocytes, *FEBS Lett.* **382**:31–36.

Chance, B., Sies, H., and Boveris, A., 1979, Hydroperoxide metabolism in mammalian organs, *Physiol. Rev.* **59**:527–605.

Cheng, H., Lederer, W. J., and Cannell, M. B., 1993, Calcium sparks: Elementary events underlying excitation–contraction coupling in heart muscle, *Science* **262**:740–744.

Dawson, T. L., Gores, G. J., Nieminen, A. L., Herman, B., and Lemasters, J. J., 1993, Mitochondria as a source of reactive oxygen species during reductive stress in rat hepatocytes, *Am. J. Physiol.* **264**:C961–C967.

Di Lisa, F., Gambassi, G., Spurgeon, H., and Hansford, R. G., 1993, Intramitochondrial free calcium in cardiac myocytes in relation to dehydrogenase activation, *Cardiovasc. Res.* **27**:1840–1844.

Ehrenberg, B., Montana, V., Wei, M. D., Wuskell, J. P., and Loew, L. M., 1988, Membrane potential can be determined in individual cells from the nernstian distribution of cationic dyes, *Biophys. J.* **53**:785–794.

Emaus, R. K., Grunwald, R., and Lemasters, J. J., 1986, Rhodamine 123 as a probe of transmembrane potential in isolated rat liver mitochondria: Spectral and metabolic properties, *Biochem. Biophys. Acta* **850**:436–448.

Farkas, D. L., Wei, M.-D., Febbroriello, P., Carson, J. H., and Loew, L. M., 1989, Simultaneous imaging of cell and mitochondrial membrane potentials, *Biophys. J.* **56**:1053–1069.

Forman, H. J., and Boveris, A., 1982, Superoxide radical and hydrogen peroxide in mitochondria, *Free Radical Biol.* **5**:65–90.

Gores, G. J., Flarsheim, C. E., Dawson, T. L., Nieminen, A.-L., Herman, B., and Lemasters, J. J., 1989, Swelling, reductive stress, and cell death during chemical hypoxia in hepatocytes, *Am. J. Physiol.* **257**:C347–C354.

Griffiths, E. J., and Halestrap, A. P., 1995, Mitochondrial nonspecific pores remain closed during cardiac ischaemia, but open upon reperfusion, *Biochem. J.* **307**:93–98.

Griffiths, E. J., Stern, M. D., and Silverman, H. S., 1997, Measurement of mitochondrial calcium in single living cardiomyocytes by selective removal of cytosolic indo 1, *Am. J. Physiol.* **273**:C37–C44.

Grynkiewicz, G., Poenie, M., and Tsien, R. Y., 1985, A new generation of Ca^{2+} indicators with greatly improved fluorescence properties, *J. Biol. Chem.* **260**:3440–3450.

Halestrap, A. P., 1991, Calcium-dependent opening of a nonspecific pore in the mitochondrial inner membrane is inhibited at pH values below 7: Implications for the protective effect of low pH against chemical and hypoxic cell damage, *Biochem. J.* **278**:715–719.

Harper, I. S., Bond, J. M., Chacon, E., Reece, J. M., Herman, B., and Lemasters, J. J., 1993, Inhibition of Na^+/H^+ exchange preserves viability, restores mechanical function, and prevents the pH paradox in reperfusion injury to rat neonatal myocytes, *Bas. Res. Cardiol.* **88**:430–442.

Haugland, R. P., 1996, *Handbook of Fluorescent Probes and Research Chemicals*, 6th ed., Molecular Probes, Eugene, OR.

Hughes, B. P., and Barritt, G. J., 1989, Inhibition of the liver cell receptor-activated Ca^{2+} inflow system by metal ion inhibitors of voltage-operated Ca^{2+} channels but not by other inhibitors of Ca^{2+} inflow, *Biochim. Biophys. Acta* **1013**:197–205.

Inoué, S., 1995, Foundations of confocal scanned imaging in light microscopy, in *Handbook of Biological Confocal Microscopy*, 2nd ed. (J. B. Pawley, Ed.), Plenum, New York, pp. 1–17.

Johnson, L. V., Walsh, M. L., Bockus, B. J., and Chen, L. B., 1981, Monitoring of relative mitochondrial membrane potential in living cells by fluorescence microscopy, *J. Cell Biol.* **88**:526–535.

Kawanishi, T., Nieminen, A.-L., Herman, B., and Lemasters, J. J., 1991, Suppression of Ca^{2+} oscillations in cultured rat hepatocytes by chemical hypoxia, *J. Biol. Chem.* **266**:20062–20069.

Keller, H. E., 1995, Objective lenses for confocal microscopy, in *Handbook of Biological Confocal Microscopy*, 2nd ed. (J. B. Pawley, Ed.), Plenum, New York, pp. 111–126.

Lattanzio, F. A., and Bartschat, D. K., 1991, The effect of pH on rate constants, ion selectivity, and thermodynamic properties of fluorescent calcium and magnesium indicators, *Biochem. Biophys. Res. Commun.* **177**:184–191.

Lemasters, J. J., Ji, S., and Thurman, R. G., 1981, Centrilobular injury following low flow hypoxia in isolated, perfused rat liver, *Science* **213**:661–663.

Lemasters, J. J., Nieminen, A.-L., Qian, T., Trost, L. C., Elmore, S. P., Nishimura, Y., Crowe, R. A., Cascio, W. E., Bradham, C. A., Brenner, D. A., and Herman, B., 1998, The mitochondrial permeability transition in cell death: A common mechanism in necrosis, apoptosis, and autophagy, *Biochim. Biophys. Acta* **1366**: 177–196.

Lemasters, J. J., Trollinger, D. R., Qian, T., Cascio, W. E., and Ohata, H., 1999, Confocal imaging of Ca^{2+}, pH, electrical potential, and membrane permeability in single living cells, in *Methods in Enzymology, Green Fluorescent Protein,* Vol. 302 (P. M. Conn, Ed.), Academic, New York, pp. 341–358.

Minsky, M., 1961, *Microscopy apparatus*, U.S. Patent 3,013,467, Dec. 19, 1961 (filed Nov. 7, 1957).

Minta, A., Kao, J. P. Y., and Tsien, R. Y., 1989, Fluorescent indicators for cytosolic calcium based on rhodamine and fluorescein chromophores, *J. Biol. Chem.* **264**:8171–8178.

Miyata, H., Silverman, H. S., Sollott, S. J., Lakatta, E. G., Stern, M. D., and Hansford, R. G., 1991, Measurement of mitochondrial free Ca^{2+} contraction in living single rat cardiac myocytes, *Am. J. Physiol.* **261**:H1123–H1134.

Nieminen, A.-L., Dawson, T. L., Gores, G. J., Kawanishi, T., Herman, B., and Lemasters, J. J., 1990a, Protection by acidotic pH and fructose against lethal injury to rat hepatocytes from mitochondrial inhibition, ionophores, and oxidant chemicals, *Bio chem. Biophys. Res. Commun.* **167**:600–606.

Nieminen, A.-L., Gores, G. J., Dawson, T. L., Herman, B., and Lemasters, J. J., 1990b, Mechanisms of toxic injury by $HgCl_2$ in rat hepatocytes studied by multiparameter digitized video microscopy, in *Optical Microscopy for Biology* (B. Herman and K. Jacobson, Eds.), Alan R. Liss, New York, pp. 323–335.

Nieminen, A.-L., Saylor, A. K., Herman, B., and Lemasters, J. J., 1994, ATP depletion rather than mitochondrial depolarization mediates hepatocyte killing after metabolic inhibition, *Am. J. Physiol.* **267**:C67–C74.

Nieminen, A.-L., Saylor, A. K., Tesfai, S. A., Herman, B., and Lemasters, J. J., 1995, Contribution of the mitochondrial permeability transition to lethal injury after exposure of hepatocytes to *t*-butylhydroperoxide, *Biochem. J.* **307**:99–106.

Nieminen, A.-L., Byrne, A. M., Herman, B., and Lemasters, J. J., 1997, Mitochondrial permeability transition in hepatocytes induced by *t*-BuOOH: NAD(P)H and reactive oxygen species, *Am. J. Physiol.* **272**:C1286–C1294.

Ohata, H., Chacon, E., Tesfai, S. A., Harper, I. S., Herman, B., and Lemasters, J. J., 1998, Mitochondrial Ca^{2+} transients in cardiac myocytes during the excitation–contraction cycle: Effects of pacing and hormonal stimulation, *J. Bioenerg. Biomembr.* **30**:207–222.

Perticarari, S., Presani, G., Mangiarotti, M. A., and Banfi, E., 1991, Simultaneous flow cytometric method to measure phagocytosis and oxidative products by neutrophils, *Cytometry* **12**:687–693.

Petronilli, V., Miotto, G., Canton, M., Brini, M., Colonna, R., Bernardi, P., and Di Lisa, F., 1999, Transient and long-lasting openings of the mitochondrial permeability transition pore can be monitored directly in intact cells by changes in mitochondrial calcein fluorescence, *Biophys. J.* **76**:725–734.

Qian, T., Nieminen, A.-L., Herman, B., and Lemasters, J. J., 1997, Mitochondrial permeability transition in pH-dependent reperfusion injury to rat hepatocytes, *Am. J. Physiol.* **273**:C1783–C1792.

Qian, T., Trost, L. C., and Lemasters, J. J., 1999, Quenching or misalignment? Confocal microscopy of onset of the mitochondrial permeability transition in cultured hepatocytes, *Microsc. Microanal.* **5** (Suppl. 2): 468–469.

Reers, M., Smiley, S. T., Mottola-Hartshorn, C., Chen, A., Lin, M., and Chen, L. B., 1995, Mitochondrial membrane potential monitored by JC-1 dye, *Methods Enzymol.* **260**:406–417.

Roe, M. W., Lemasters, J. J., and Herman, B., 1990, Assessment of Fura-2 for measurements of cytosolic free calcium, *Cell Calcium* **11**:63–73.

Rothe, G., Osen, A., and Valet, G., 1988, Dihydrorhodamine 123: A new flow cytometric indicator for respiratory burst activity in neutrophil granulocytes, *Naturwissen schaften* **75**:354.

Rottenberg, H., and Wu, S., 1997, Mitochondrial dysfunction in lymphocytes from old mice: Enhanced activation of the permeability transition, *Biochem. Biophys. Res. Commun.* **240**:68–74.

Seghizzi, P., D'Adda, F., Borleri, D., Barbic, F., and Mosconi, G., 1994, Cobalt myocardiopathy: A critical review of literature, *Sci. Total Environ.* **150**:105–109.

Smiley, S. T., Reers, M., Mottola-Hartshorn, C., Lin, M., Chen, A., Smith, T. W., Steele, G. D., Jr., and Chen, L. B., 1991, Intracellular heterogeneity in mitochondrial membrane potentials revealed by a J-aggregate-forming lipophilic cation, JC-1, *Proc. Natl. Acad. Sci. USA.* **88**:3671–3675.

Trollinger, D. R., Cascio, W. E., and Lemasters, J. J., 1997, Selective loading of Rhod 2 into mitochondria shows mitochondrial Ca^{2+} transients during the contractile cycle in adult rabbit cardiac myocytes, *Biochem. Biophys. Res. Commun.* **236**:738–742.

Trollinger, D. R., Cascio, W. E., and Lemasters, J. J., 2000, Mitochondrial calcium transients in adult rabbit cardiac myocytes: Inhibition by ruthenium red and artifacts caused by lysosomal loading of Ca^{2+}-indicating fluorophores, *Biophys. J.* **79**:39–50.

White, J. G., Amos, W. B., and Fordham, M., 1987, An evaluation of confocal versus conventional imaging of biological structures by fluorescence light microscopy, *J. Cell Biol.* **105**:41–48.

Wieder, E. D., Hang, H., and Fox, M. H., 1993, Measurement of intracellular pH using flow cytometry with carboxy-SNARF-1, *Cytometry* **14**:916–921.

Zahrebelski, G., Nieminen, A.-L., Al-Ghoul, K., Qian, T., Herman, B., and Lemasters, J. J., 1995, Progression of subcellular changes during chemical hypoxia to cultured rat hepatocytes: A laser scanning confocal microscopic study, *Hepatology* **21**:1361–1372.

Zoratti, M., and Szabò, I., 1995, The mitochondrial permeability transition, *Bio chim. Biophys. Acta* **1241**: 139–176.

Section 2

Mitochondrial Disease and Aging

Chapter 3

Primary Disorders of Mitochondrial DNA and the Pathophysiology of mtDNA-Related Disorders

Eric A. Schon and Salvatore DiMauro

1. INTRODUCTION

Mitochondrial diseases are a diverse group of disorders characterized by impairment of mitochondrial function. The diversity is most easily seen in the bewildering array of clinical phenotypes that result from errors not only in maternally inherited mitochondrial DNA (mtDNA) and mendelian-inherited nuclear DNA (nDNA), but also from spontaneous mutations and environmental insults that can affect either genome.

The human mitochondrial genome is a 16,569-bp double-stranded DNA circle (Anderson *et al.*, 1981). It contains only 37 genes (Fig. 1): two encode ribosomal RNAs, 22 encode transfer RNAs, and 13 encode polypeptides. Significantly, all 13 polypeptides are components of the respiratory chain/oxidative phosphorylation system. They include seven subunits of complex I (NADH dehydrogenase–ubiquinone oxidoreductase), one subunit of complex III (ubiquinone–cytochrome *c* oxidoreductase), three subunits of complex IV (cytochrome *c* oxidase), and two subunits of complex V (ATP synthetase). The respiratory complexes also contain subunits encoded by nuclear DNA (nDNA) imported into the organelle from the cytosol and co-assembled with the mtDNA-encoded subunits into their in respective complexes in the mitochondrial inner membrane.

Eric A. Schon   Departments of Neurology, and Genetics and Development, Columbia University, New York, New York 10032.   **Salvatore DiMauro**   Department of Neurology, Columbia University, New York, New York 10032.

Mitochondria in Pathogenesis, edited by Lemasters and Nieminen.
Kluwer Academic/Plenum Publishers, New York, 2001.

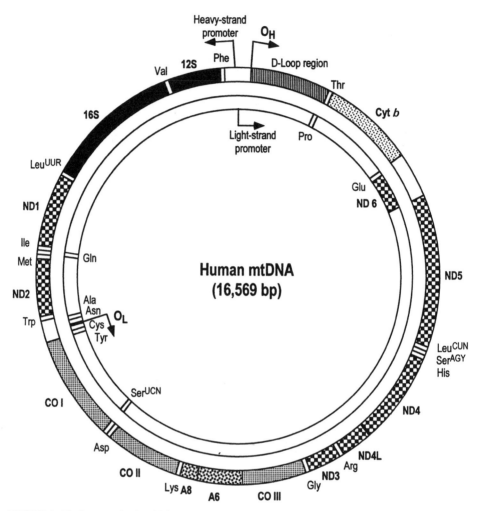

FIGURE 1. The human mitochondrial genome. Shown are the structural genes for the 12S and 16S ribosomal RNAs (rRNA), the subunits of complex I (ND1, ND2, ND3, ND4, ND4L, ND5, and ND6), complex III (Cyt *b*), complex IV (CO I, CO II, and CO III), and complex V (A6 and A8), and 22 tRNAs (amino acid nomenclature). The origins of heavy-strand (O_H) and light-strand (O_L) replication, and that of the promoters for transcription initiation from the heavy-strand (HSP) and light-strand (LSP), are shown by arrows.

Complex II (succinate dehydrogenase–ubiquinone oxidoreductase) contains only nDNA-encoded subunits.

From both a genetic and a physiological standpoint, mitochondria are semi-autonomous subcellular organelles. They have their own genome, which replicates in a unique way (Clayton, 1982), as well as equally unique transcriptional and translational apparati (Attardi *et al.*, 1990; Clayton, 1992). Replication, transcription, and translation of mtDNA are nuclear-encoded machineries, which are imported into the organelle (Schatz and Dobberstein, 1996), and translation of mtDNA-encoded polypeptides (O'Brien *et al.*, 1990) uses a modified genetic code. In addition, mitochondria import hundreds of other nDNA-encoded gene products to support a wide range of activities, almost all of which are "housekeeping functions," such as amino acid biosynthesis, lipid metabolism, and carbohydrate metabolism (e.g., the tricarboxylic acid cycle). It is estimated that mitochondria contain between 500 and 1000 imported proteins.

Mitochondria and their mtDNAs are inherited only from the mother (Giles *et al.*, 1980). Thus, mutations in mtDNA, be they neutral or pathogenic, are usually (but not always; see below) maternally inherited, meaning that though a woman can transmit the mutation to all of her children, only the daughters can pass the mutation on to their progeny. A somatic cell contains hundreds, even thousands, of mitochondria, and each mitochondrion contains an average of about five mtDNAs (Satoh and Kuroiwa, 1991). Both wild-type and mutated mtDNA may coexist within a person's tissues; this condition is known as *heteroplasmy*. In the case of a maternally transmitted pathogenic mutation in which the patient contains a heteroplasmic population of normal and mutated mtDNAs, the local concentration and distribution of mutated mtDNA may vary among tissues or may change within a tissue over time, thereby affecting the clinical phenotype arising from the mutation. This dynamism is due to the stochastic partitioning of mitochondria and mtDNAs during cell division, a phenomenon known as *mitotic segregation*. Almost every organ system can evince clinical signs and symptoms in mitochondrial disorders, but muscle and brain (which have extremely high energy requirements, and therefore relatively low thresholds for mitochondrial dysfunction) are particularly susceptible. Thus, low levels of mutation in these tissues can affect their respiratory capacity, and high levels can be devastating. This is why many mitochondrial disorders are predominantly encephalo-myopathies.

The two major biochemical features in most mitochondrial diseases are lactic acidosis and respiratory chain deficiency. From the morphological standpoint, patients with mitochondrial diseases often have ragged-red fibers (RRF) in the muscle biopsy (Bonilla *et al.*, 1992). Though RRF is an important indicator of a mitochondrial disorder, it should be emphasized that not all mitochondrial diseases display RRF.

The vast majority of mitochondrial proteins are nucleus-encoded, but almost all of the mitochondrial disorders characterized to date have been associated with errors in mtDNA. This is not to say that there are no mendelian-inherited mitochondrial diseases—in fact, as described below, a number of such pathologies have now been identified—but rather, the small size of the mitochondrial genome (coupled with the powerful molecular biological tools now available) has made the mitochondrial genome an easier field to plow than the nuclear genome.

2. DISEASES ASSOCIATED WITH mtDNA POINT MUTATIONS

More than 80 point mutations, located in all parts of the mitochondrial genome, have been described (Table I).

2.1. Ribosomal RNA Gene Mutations

Six pathogenic mutations have been found in rRNA genes. They fall into two clinical categories: aminoglycoside-induced or nonsyndromic deafness, and dilated cardiomyopathy.

Table I
Phenotypes Associated with Pathogenic MtDNA Point Mutations

Mutation[a]		Gene	"Usual" phenotype	Reference[b]
Mutations in rRNA genes (5)				
1555	A → G	12S rRNA	Aminoglycoside-induced deafness	Prezant et al., 1993
1571	T → G	12S rRNA	Dilated cardiomyopathy	Arbustini et al., 1998
1692	A → T	16S rRNA	Dilated cardiomyopathy	Arbustini et al., 1998
1703	C → T	16S rRNA	Dilated cardiomyopathy	Arbustini et al., 1998
2991	T → C	16S rRNA	Chloramphenicol resistance	Blanc et al., 1981
3228	T → G	16S rRNA	Dilated Cardiomyopathy	Arbustini et al., 1998
Mutations in tRNA genes (54)				
1606	G → A	tRNA-Val	Multisystem	Tiranti et al., 1998
1642	G → A	tRNA-Val	MELAS	Taylor et al., 1996
1644	G → T	tRNA-Val	Leigh syndrome	Chalmers et al., 1997
3243	A → G	tRNA-Leu(UUR)	MELAS/PEO/diabetes/deafness	Goto et al., 1990
3250	T → C	tRNA-Leu(UUR)	Myopathy	Goto et al., 1992
3251	A → G	tRNA-Leu(UUR)	PEO/myopathy	Sweeney et al., 1993
3252	A → G	tRNA-Leu(UUR)	MELAS	Morten et al., 1993
3254	C → G	tRNA-Leu(UUR)	Cardiomyopathy/myopathy	Kawarai et al., 1997
3256	C → G	tRNA-Leu(UUR)	Multisystem/PEO	Moraes et al., 1993b
3260	A → G	tRNA-Leu(UUR)	Cardiomyopathy/myopathy	Zeviani et al., 1991
3264	T → C	tRNA-Leu(UUR)	Diabetes	Suzuki et al., 1997
3271	T → C	tRNA-Leu(UUR)	MELAS	Goto et al., 1991
3271	Delete T	tRNA-Leu(UUR)	Encephalomyopathy	Shoffner et al., 1995
3291	T → C	tRNA-Leu(UUR)	MELAS	Goto et al., 1994
3302	A → G	tRNA-Leu(UUR)	Myopathy	Bindoff et al., 1993
3303	C → T	tRNA-Leu(UUR)	Cardiomyopathy	Silvestri et al., 1994
4269	A → G	tRNA-Ile	Encephalomyop./cardiomyop.	Taniike et al., 1992
4274	T → C	tRNA-Ile	PEO	Chinnery et al., 1997
4285	T → C	tRNA-Ile	PEO	Silvestri et al., 1996
4295	A → G	tRNA-Ile	Hypertrophic cardiomyopathy	Merante et al., 1996
4300	A → G	tRNA-Ile	Cardiomyopathy	Casali et al., 1995
4309	G → A	tRNA-Ile	PEO	Franceschina et al., 1998
4315	A → G	tRNA-Ile	Dilated cardiomyopathy	Arbustini et al., 1998
4320	C → T	tRNA-Ile	Hypertrophic cardiomyopathy	Santorelli et al., 1995
4409	T → C	tRNA-Met	Myopathy, dystrophy	Vissing et al., 1998
4450	G → A	tRNA-Met	Splenic lymphoma	Lombes et al., 1998
5521	G → A	tRNA-Trp	Myopathy	Silvestri et al., 1998
5537	Insert T	tRNA-Trp	Leigh syndrome	Santorelli et al., 1997
5549	G → A	tRNA-Trp	Chorea/encephalomyopathy	Nelson et al., 1995
5600	A → T	tRNA-Ala	Dilated cardiomyopathy	Arbustini et al., 1998
5703	G → A	tRNA-Asn	Myopathy/PEO	Moraes et al., 1993b
5814	T → C	tRNA-Cys	Encephalopathy	Manfredi et al., 1996
5877	G → A	tRNA-Tyr	CPEO	Sahashi et al., 1997
7445	A → G	tRNA-Ser(UCN)	Deafness	Reid et al., 1994
7471	Insert C	tRNA-Ser(UCN)	Deafness/myoclonus	Tiranti et al., 1995
7512	T → C	tRNA-Ser(UCN)	MERRF/MELAS	Nakamura et al., 1995
7543	A → G	tRNA-Asp	Myoclonus	El-Schahawi et al., 1995
7581	T → C	tRNA-Asp	Dilated cardiomyopathy	Arbustini et al., 1998
8296	A → G	tRNA-Lys	Diabetes	Kameoka et al., 1998
8344	A → G	tRNA-Lys	MERRF	Shoffner et al., 1990
8356	T → C	tRNA-Lys	MERRF	Silvestri et al., 1992
8363	G → A	tRNA-Lys	MERRF/deafness/cardiopathy	Santorelli et al., 1996

Table I   (*continued*)

Mutation[a]		Gene	"Usual" phenotype	Reference[b]
9997	T → C	tRNA-Gly	Cardiomyopathy	Merante *et al.*, 1994
12301	G → A	tRNA-Leu(CUN)	Sideroblastic anemia	Gattermann *et al.*, 1996
12315	G → A	tRNA-Leu(CUN)	Encephalomyopathy	Fu *et al.*, 1996
12320	A → G	tRNA-Leu(CUN)	Myopathy	Weber *et al.*, 1997
14684	C → T	tRNA-Glu	Dilated cardiomyopathy	Arbustini *et al.*, 1998
14709	T → C	tRNA-Glu	Encephalomyopathy	Hanna *et al.*, 1995
15889	T → C	tRNA-Thr	Dilated cardiomyopathy	Arbustini *et al.*, 1998
15902	A → G	tRNA-Thr	Dilated cardiomyopathy	Arbustini *et al.*, 1998
15915	G → A	tRNA-Thr	Encephalomyopathy	Nishino *et al.*, 1996
15923	A → G	tRNA-Thr	Fatal infantile resp. def.	Yoon *et al.*, 1991
15935	A → G	tRNA-Thr	Dilated cardiomyopathy	Arbustini *et al.*, 1998
15990	C → T	tRNA-Pro	Myopathy	Moraes *et al.*, 1993a
Mutations in polypeptide-coding (i.e., respiratory chain) genes (22)				
3308	T → C	ND1	MELAS	Campos *et al.*, 1997
3316	G → A	ND1	NIDDM/dilated cardiomyopathy	Huoponen *et al.*, 1991
3460	G → A	ND1	LHON	Huoponen *et al.*, 1991
4160	T → C	ND1	LHON	Howell *et al.*, 1991
5510	A → C	ND2	Dilated cardiomyopathy	Arbustini *et al.*, 1998
6721	T → C	Cox I	Sideroblastic anemia	Gattermann *et al.*, 1997
6742	T → C	Cox I	Sideroblastic anemia	Gattermann *et al.*, 1997
8851	T → C	ATPase 6	FBSN	De Meirleir *et al.*, 1995
8993	T → C	ATPase 6	NARP/MILS	Holt *et al.*, 1990
8993	T → C	ATPase 6	NARP/MILS	de Vries *et al.*, 1993
9176	T → C	ATPase 6	FBSN	Thyagarajan *et al.*, 1995
9952	G → A	Cox III	Encephalomyopathy	Hanna *et al.*, 1998
9957	T → C	Cox III	MELAS	Manfredi *et al.*, 1995a
11696	A → G	ND4	LHON/dystonia	de Vries *et al.*, 1996
11778	G → A	ND4	LHON	Wallace *et al.*, 1988
13513	G → A	ND5	MELAS	Santorelli *et al.*, 1997
14459	G → A	ND6	LHON/dystonia	Jun *et al.*, 1994
14482	C → G	ND6	LHON	Howell *et al.*, 1998
14484	T → C	ND6	LHON	Johns *et al.*, 1992
15059	G → A	Cyt *b*	Myopathy	Andreu, 1999
15615	G → A	Cyt *b*	Myopathy	Dumoulin *et al.*, 1996
15762	G → A	Cyt *b*	Myopathy	Andreu *et al.*, 1998

[a] L-strand sequence
[b] First published article

2.1.1. Sensorineural Deafness

Sensorineural deafness is a frequent symptom in mitochondrial disease and has been associated mainly with point mutations in tRNA genes (see below). One mutation, however, an A → G transition at nt-1555 of the 12S rRNA gene (all mtDNA map positions are according to Anderson *et al.*, 1981), causes maternally inherited deafness, usually but not invariably through interaction with aminoglycoside drugs such as kanamycin, streptomycin, and gentamicin. The A1555G mutation is located near a stem-loop structure in 12S rRNA required for binding of these drugs; thus, the pathogenetic mechanism probably involves interference with mitochondrial translation (Hutchin *et al.*, 1993; Inoue *et al.*, 1996; Prezant *et al.*, 1993). The mutation is often, but not always (El-Schahawi *et al.*, 1997) homoplasmic. Although aminoglycosides can induce deafness even in normal people, it appears that individuals carrying the 1555 mutation are acutely susceptible to the drug's effects (Estivill *et al.*, 1998). Thus, aminoglycoside-induced deafness is a beautiful example of one aspect of mitochondrial diseases: they can result from an *interaction* between a predisposing mtDNA genotype and an environmental insult.

It is noteworthy that *plasma cell dyscrasia* is a hematological disorder caused by using chloramphenicol to treat bacterial infections. Chloramphenicol binds specifically to the mitochondrial 16S rRNA and thereby interferes with mitochondrial translation, similar to the behavior of aminoglycosides when they bind to the 12S rRNA. The binding site is at nt-2991, and a mutation at this location causes chloramphenicol resistance (Blanc *et al.*, 1981).

2.1.2. Cardiomyopathy

Although cardiomyopathy, either hypertrophic or dilated, is but one of many symptoms seen in mitochondrial disorders, a number of mtDNA point mutations have been found selectively in patients with maternally inherited cardiomyopathies. Four such mutations are located in rRNA genes. One (at nt-1571) is in the 12S rRNA; the other three (at nt-1692, -1703, and -3228) are in the 16S gene. All four are associated with dilated cardiomyopathy (Arbustini *et al.*, 1998). Interestingly, the nt-1555 mutation in 12S rRNA has now been found in a patient with cardiomyopathy alone, sans deafness (Santorelli *et al.*, 1999).

2.2. Transfer RNA Gene Mutations

The vast majority of mtDNA point mutations are in tRNAs. Of the 54 known mutations, more than half fall into delineated clinical categories.

2.2.1. Mitochondrial Encephalomyopathy with Lactic Acidosis and Stroke-Like Episodes (MELAS)

At least five mutations are associated with *mitochondrial encephalomyopathy with lactic acidosis and stroke-like episodes* (MELAS), the most common maternally inherited mitochondrial disease. It is defined by stroke (usually before age 40); encephalopathy characterized by seizures, dementia or both; and evidence of mitochondrial dysfunction, such as RRF in muscle, lactic acidosis in blood or CSF (cerebrospinal fluid), or both (Hirano *et al.*, 1992). Other frequent clinical features include migraine-like headaches, recurrent vomiting, exercise intolerance, limb weakness, and short stature. Complex I activity is particularly reduced in this disorder (Kobayashi *et al.*, 1987). Mitochondria also accumulate in the vasculature (Hasegawa *et al.*, 1991; Ohama and Ikuta 1987) and can be visualized morphologically as strongly SDH-reactive blood vessels (SSVs); this feature, almost unique to MELAS, may cause the strokes that commonly lead to cortical blindness and hemianopia. Surprisingly, however, the brain lesions visualized by MRI do not conform to the distribution of major cerebral arteries (Matthews *et al.*, 1991). Another morphological feature unique to MELAS is a positive stain for cytochrome c oxidase (Cox) activity in RRF, which contrasts with the diminished or absent Cox activity found in most other syndromes due to tRNA mutations (Moraes *et al.*, 1993b).

Four MELAS mutations are located in the same tRNA gene, tRNA$^{Leu(UUR)}$, including the most common mutation, an A → G transition at nt-3243 (see Table 1). The reason for the prevalence of MELAS mutations in the tRNA$^{Leu(UUR)}$ gene is unknown. A fifth mutation has been found in tRNA^{Val}.

The mutations at nt-3243 and nt-3271 have been analyzed in cytoplasmic hybrids, or *cybrids*. This is a tissue culture system (King and Attardi, 1989) in which mitochondria (and mtDNA) from heteroplasmic patients can be transferred to immortalized human cells that have been depleted of their own mtDNA(ρ^0 cells), thereby allowing for side-by-side analysis of pure homoplasmic populations of wild-type and mutant mtDNA, both derived from the patient. Analysis of paired MELAS-3243 and MELAS-3271 cybrids showed protein synthesis reduction and a particularly reduced complex I activity (Bentlage and Attardi, 1996; Dunbar *et al.*, 1996; King *et al.*, 1992; Koga *et al.*, 1993, 1995). In addition, the amount of a partially processed polycistronic transcript called RNA 19 (composed of 16S rRNA, tRNA$^{Leu(UUR)}$, and ND1, and whose genes are contiguous in the DNA; see Fig. 1) was significantly increased, in both cybrids (King *et al.*, 1992; Koga *et al.*, 1993) and tissues (Kaufmann *et al.*, 1996). It is possible that RNA 19, which contains 16S rRNA, is incorporated into ribosomes, thereby interfering with translation (Schon *et al.*, 1992).

2.2.2. Progressive External Ophthalmoplegia (PEO)

Interestingly, many patients with *maternally inherited progressive external ophthalmoplegia* (PEO) also harbor the nt-3243 mutation (Moraes *et al.*, 1993c). This disease is characterized by ptosis (droopy eyelids) and by weakness of the extraocular muscles, and often of the limbs. Though most patients with PEO harbor large-scale deletions of mtDNA, which can arise sporadically or can be inherited as a mendelian trait (see below), many patients inherit PEO as a maternal trait, due either to mutations elsewhere in tRNA$^{Leu(UUR)}$, or in other tRNA genes, including tRNA^{Ile}, tRNA^{Asn}, and tRNA^{Tyr}.

Whereas muscle biopsies from MELAS-3243 patients contain mostly Cox-positive RRF, muscle in PEO-3243 patients contains numerous Cox-*negative* RRF. Analyses of individual muscle fibers from these patients have shown that abundance of the nt-3243 mutation is significantly higher in the Cox-negative RRFs in PEO-3243 than in the Cox-positive RRFs in MELAS-3243, implying that the spatial distribution of mutant mtDNAs in individual cells leads to clinically distinct phenotypes (Petruzzella *et al.*, 1994).

2.2.3. Diabetes

Yet a third mitochondrial disorder associated with the nt-3243 mutation is *maternally inherited diabetes mellitus* (MIDM) alone or together with deafness (*diabetes and deafness* [DAD]. A prominent feature of mitochondrial disease is endocrinopathy (e.g., diabetes, hypogonadism, hypoparathyroidism, adrenal insufficiency, hypothyroidism, and short stature due to growth hormone deficiency). Thus, it is perhaps not surprising to find that mtDNA mutations, such as those at nt-3264 in tRNA$^{Leu(UUR)}$ and at nt-8296 in tRNA^{Lys}, are associated with endocrine disorders. It *is* surprising, however, that the nt-3243 mutation typically associated with MELAS is the major cause of maternally inherited adult type II, non-insulin-dependent diabetes mellitus (NIDDM). Interestingly, though diabetes is one symptom in MELAS-3243 patients (Onishi *et al.*, 1993), there are nt-3243 pedigrees in which diabetes is essentially the only feature; no signs of MELAS (or PEO, for that matter) are present (Gerbitz *et al.*, 1993 Reardon *et al.*, 1992). Approximately 1–2% of all NIDDM patients harbor the nt-3243 mutation (Kadowaki *et al.*, 1993).

2.2.4. Myoclonus Epilepsy with Ragged-Red Fibers (MERRF)

A specific group of maternally inherited tRNA mutations is associated with *myoclonus epilepsy with ragged-red fibers* (MERRF). This disease is characterized by myoclonus, ataxia, generalized seizures, and myopathy with RRF (Fukuhara *et al.*, 1980). Other clinical features include dementia, hearing loss, neuropathy, optic atrophy, and short stature (Silvestri *et al.*, 1993). Pathological studies reveal neuronal loss in the inferior olivary nucleus and dentate nucleus, degeneration of the spinal cord posterior columns, and diffuse gliosis of cerebellar and brain-stem white matter (Fukuhara *et al.*, 1980), and immunohistochemical studies show a selective loss of mtDNA-encoded respiratory chain subunits in frontal cortex, medulla, and cerebellum (Sparaco *et al.*, 1994). An interesting feature of some MERRF patients is multiple lipomas; this can be an isolated manifestation (Holme *et al.*, 1993), it can be part of the overall MERRF syndrome (Berkovic *et al.*, 1991; Gámez *et al.*, 1998, Silvestri *et al.*, 1993), or it can be seen in a disorder called *Ekbom syndrome* (Calabresi *et al.*, 1994).

The "classical" MERRF mutation is an A → G transition at nt-8344 in the tRNA^{Lys} gene, but two other MERRF mutations are also located in this gene, at nt-8356 and nt-8363. Both Leigh syndrome (Howell *et al.*, 1996; Santorelli *et al.*, 1998; Silvestri *et al.*, 1993) and limb-girdle myopathy (Silvestri *et al.*, 1993) have also been associated with the nt-8344 mutation. In addition, four other mutations, in tRNA$^{Leu(UUR)}$ (Folgero *et al.*, 1995; Moraes *et al.*, 1993b), in tRNA^{Lys} (Zeviani *et al.*, 1993), and in tRNA$^{Ser(UCN)}$ (Nakamura *et al.*, 1995), have been found in patients with overlap syndromes of MELAS and MERRF. Last, a patient with myoclonus and deafness had a single-base insertion in tRNA$^{Ser(UCN)}$ (Tiranti *et al.*, 1995). All the mutations were heteroplasmic.

As with MELAS, detailed genetic analyses of the main MERRF mutations (i.e., at nt-8344 and nt-8356) have been performed in cybrids. They show that the mutated tRNA cause a severe reduction in mitochondrial protein synthesis. Moreover (and as distinct from MELAS mutations), mutated RNA also result in the synthesis of a discrete set of aberrant translation products, the functional significance of which is unknown (Chomyn, 1998; Chomyn *et al.*, 1991; Enriquez *et al.*, 1995; Hao and Moraes, 1996; Masucci *et al.*, 1995). In addition, the tRNA^{Lys} with the nt-8344 mutation is not well-aminoacylated by its cognate lysyl-tRNA synthetase (Enriquez *et al.*, 1995). Though this may cause some of the translation defects, the quantitative decrease in aminoacylation seems insufficient to explain all of the biochemical findings.

It was noted above that rRNA mutations have been associated with either deafness or cardiomyopthy, "generic" features of many mitochondrial diseases. Thus it may be no surprise that tRNA mutations can also cause these features, including the MELAS mutation at nt-3243 in tRNA$^{Leu(UUR)}$ (Kadowaki *et al.*, 1993; Remes *et al.*, 1993; van der Ouweland *et al.*, 1992). Besides being common in MELAS and MERRF, hearing loss was the major, even exclusive, symptom in two patients with maternally inherited mutations. Both mutations were in the tRNA$^{Ser(UCN)}$ gene: a transition at nt-7445 (Reid *et al.*, 1994) and a single nucleotide insertion at nt-7471 (Tiranti *et al.*, 1995).

2.2.5. Cardiomyopathy

There are also a number of cardiomyopathy-specific tRNA mutations. Remarkably, five are located in one gene, tRNA^{Ile} (at nt-4269, -4295, -4300, -4315, and -4320). In addition, three (at nt-15889, -15902, and -15935) are located in tRNA^{Thr}, and three others (at nt-3254, -3260 and -3303) in tRNA$^{Leu(UUR)}$. As with the "hot spot" of MELAS mutations in tRNA$^{Leu(UUR)}$, why these three genes are cardiomyopathy hot spots is unknown. Other cardiomyopathy mutations are at nt-5600 in tRNA^{Ala}, at nt-7581 in tRNA^{Asp}, at nt-9997 in tRNA^{Gly}, and at nt-14684 in tRNA^{Glu}.

2.2.6. Other Mutations

Other tRNA mutations have been associated with a grab bag of syndromes, including multisystem disorders not associated with any of the previously described syndromes (at nt-1606 in tRNA^{Val}), Leigh syndrome (at nt-1644 in tRNA^{Val} and nt-5537 in tRNA^{Trp}), various myopathies (at nt-3250 and -3302 in tRNA$^{Leu[UUR]}$, nt-4409 in tRNA^{Met}, and nt-5521 in tRNA^{Trp}) and encephalomyopathies (at nt-3271 in tRNA$^{Leu[UUR]}$, nt-5549 in tRNA^{Trp}, nt-5814 in tRNA^{Cys}, nt-12315 in tRNA$^{Leu[CUN]}$, nt-14709 in tRNA^{Glu}, and nt-15915 in tRNA^{Thr}), and hematological disorders, such as splenic lymphoma (at nt-4450 in tRNA^{Met}) and sideroblastic anemia (at nt-12301 in tRNA$^{Leu[CUN]}$).

2.3. Polypeptide-Coding Gene Mutations

2.3.1. Leber Hereditary Optic Neuropathy (LHON)

Leber hereditary optic neuropathy (LHON) is a form of maternally inherited blindness due to optic neuropathy (Leber, 1871). It begins in the second or third decade of life, with loss of vision (starting as a central scotoma) in one eye, followed within a short period (weeks to a few months) by loss of vision in the other eye. Numerous mutations in mtDNA have been associated with LHON, but only four (at nt-3460, -4160, -11778, and -14484) have been considered primary mutations that are authentically pathogenic (although recently, a fifth mutation, at nt-14482, has also been implicated) (Howell *et al.*, 1998). It is noteworthy that all five mutations are located in complex I genes, but unequivocal demonstration of a reduction in complex I activity has been elusive (Carelli *et al.*, 1997; Hofhaus *et al.*, 1996; Vergani *et al.*, 1995). Because LHON penetrance and severity is much greater in men than in women, nuclear background almost certainly influences the expression of LHON (Cock *et al.*, 1998; Torroni *et al.*, 1997). An autoimmune component has also been suggested (Keller-Wood *et al.*, 1994), a possibility supported by the finding that in some cases the visual loss can reverse spontaneously (Mackey and Howell, 1992;

Stone *et al.*, 1992). Two mutations, at nt-11696 in ND4 and nt-14459 in ND6, have been associated with an unusual combined phenotype of LHON plus dystonia (de Vries *et al.*, 1996; Jun *et al.*, 1994), but in fact dystonia has been found in so-called Leber plus patients harboring the classic mutations, such as those at nt-3460 and -11778 (Meire *et al.*, 1995; Nikoskelainin *et al.*, 1995; Thobois *et al.*, 1997).

2.3.2. Striatal Necrosis

One group of maternally inherited clinical syndromes, which we have dubbed the *striatal necrosis syndromes* (Schon *et al.*, 1997), are all caused by mutations in a single gene, subunit 6 of ATP synthetase (complex V). Complex V is a lollipop-shaped structure (see Fig. 2) composed of (1) the F_0 segment, which lies in the mitochondrial inner membrane and conducts protons from the intermembrane space to the matrix, and (2) the F_1 segment, which uses the proton gradient to convert ADP to ATP (a reversible reaction). Both of the mtDNA-encoded subunits of complex V, ATPase 6 and ATPase 8, are components of the F_0 proton channel.

Of the four ATPase mutations identified to date, two are associated with maternally inherited Leigh syndrome (MILS), a devastating degenerative encephalopathy; onset is usually in the first year of life. Leigh syndrome is characterized by developmental delay, seizures, hypotonia, brainstem dysfunction, myoclonus, ataxia, optic atrophy, peripheral neuropathy, psychomotor regression, and in the specific case of MILS, by retinitis pigmentosa. At autopsy there are characteristic pathological findings of symmetric necrotic foci in the basal ganglia, thalamus, and brain stem. Though lactic acidosis is usually present, RRF is not. MILS is associated with extremely high levels (>90%) of one of two mutations, both at Leu-156 in ATPase 6, a T $\rightarrow$ G transversion at nt-8993 that converts Leu to Arg, or a T $\rightarrow$ C transition that converts Leu to Pro (De Vries *et al.*, 1993; Santorelli *et al.*, 1993, 1994; Tatuch *et al.*, 1992). The other two mutations are associated with a related disorder, *familial bilateral striatal necrosis* (FBSN). The FBSN disorder is associated with a T $\rightarrow$ C transition at nt-8851 (De Meirlier *et al.*, 1995) and a T $\rightarrow$ C transition at nt-9176 (Thyagarajan *et al.*, 1995). All four mutations were heteroplasmic.

Interestingly, if either of the two MILS mutations at nt-8993 is present at levels between 70% and 90%, a completely different disorder presents itself: neuropathy, ataxia,

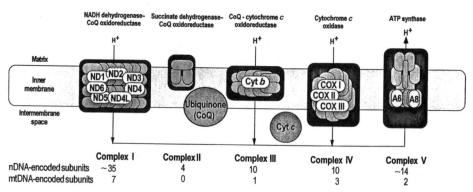

FIGURE 2. Schematic of the mitchondrial respiratory chain. White ovals: subunits encoded by mtDNA.

and retinitis pigmentosa (NARP) (Holt *et al.*, 1990). The NARP disorder, which is both milder and later in onset than MILS, is characterized by retinal pigmentary degeneration, seizures, dementia, sensory neuropathy, ataxia, proximal neurogenic muscle weakness, and development delay.

The specific role of ATPase 6 or of the selective impairment of complex V in these disorders is unknown. Analysis of cells containing high levels of the T8993G mutation show decreases in ATP synthesis of about 50% (Houstek *et al.*, 1995; Tatuch and Robinson, 1993; Vazquez-Memije *et al.*, 1996), but there are conflicting data regarding the effect of the mutation on ATP hydrolysis, the reverse reaction catalyzed by complex V (Tatuch and Robinson, 1993; Vazquez-Memije *et al.*, 1996). In cybrids harboring the T8993G mutation, state III (ADP-stimulated) respiration was reduced, again on the order of 50% (Trounce *et al.*, 1994).

2.3.3. MELAS

Point mutations in polypeptide-coding genes have been found in patients with clinically defined MELAS, at nt-9957 in COX III (Manfredi *et al.*, 1995a), at nt-3308 in ND1 (Campos *et al.*, 1997), and at nt-13513 in ND5 (Santorelli *et al.*, 1997). Interestingly, these patients had fewer RRFs than did the MELAS patients harboring tRNA mutations, although the histochemical pattern, at least in MELAS-9957 (Manfredi *et al.*, 1995a) and in MELAS-13513 (Santorelli *et al.*, 1997), was also typical of classical MELAS-3243, with Cox-positive RRF and SSV. Another mutation in COX III is located quite close to the nt-9957 mutation (at nt-9952; Hanna *et al.*, 1998) and is also associated with encephalomyopathy.

2.3.4. Other Mutations

Two mutations, both in COX I, have been found in sideroblastic anemia (Gatterman *et al.*, 1997), and another two, in ND1 and ND2, are associated with cardiopathies (Arbustini *et al.*, 1998; Huoponen *et al.*, 1991).

Finally, only three pathogenic mutations, in three different patients, have been found in the cytochrome *b* gene and not only did all three patients have a pure myopathy, but all three cases were sporadic. It appears that the mutations (at nt-15059, -15615, and -15762) (Andreu *et al.*, 1998, 1999; Dumoulin *et al.*, 1996) arose during oogenesis or embryogenesis in the patients, but it is unclear whether the occurrence of the mutations only in muscle is due to sampling bias or is a reflection of an underlying "lethality" of this mutation in nonmuscle tissue during fetal life. It is noteworthy, however, that two other sporadic tRNA mutations, in tRNA^{Pro} (Moraes *et al.*, 1993a) and in tRNA$^{Leu(CUN)}$ (Weber *et al.*, 1997), also resulted in pure myopathies, supporting the concept of segregation of a potentially embryonic lethal mutation to a relatively "viable" tissue. On the other hand, a sixth sporadic point mutation, also in tRNA$^{Leu(CUN)}$ (Fu *et al.*, 1996), caused an encephalomyopathy; because the mutation was undetectable in blood, fibroblasts, or (surprisingly) satellite cells (a result also obtained by Weber *et al.*, 1997), presumably skewed segregation was present in only some of the primordial cells destined for muscle and brain.

3. DISEASES ASSOCIATED WITH mtDNA REARRANGEMENTS

One group of mtDNA mutations occurs spontaneously with great regularity: large-scale partial deletions and duplications of mtDNA (Table II). Since they were first discovered by Holt *et al.* in 1988, almost 200 species of rearranged mtDNAs have been reported.

3.1. Sporadic Rearrangements

Kearns–Sayre syndrome (KSS) was first recognized as a distinct clinical entity more than 40 years ago (Kearns and Sayre, 1958), and it is the most prominent disorder associated with mtDNA rearrangements. Kearns–Sayre syndrome is a devastating disorder defined by onset before age 20, pigmentary retinopathy, and PEO, with at least one of these additional features: cardiac conduction block, cerebellar syndrome, or CSF protein greater than 100 mg/dl (Rowland *et al.*, 1983). Other symptoms include hearing loss, proximal limb weakness, endocrinopathies, and renal tubular dysfunction. Blood and CSF have elevated levels of lactate, and muscle biopsies show RRF.

Though a sporadic disorder, KSS is characterized by giant deletions of mtDNA (Δ-mtDNAs) (Holt *et al.*, 1989; Lestienne and Ponsot, 1988; Zeviani *et al.*, 1988). Two other disorders, both sporadic—PEO with RRF (Moraes *et al.*, 1989) and Pearson marrow/pancreas syndrome, an early-onset hematopoietic disorder (Rötig *et al.*, 1991; Pearson *et al.*, 1979)—are also associated with these deletions (Table II). In all three disorders, there is a single population of Δ-mtDNA co-existing with wild-type molecules. The specific type of deletion varies among patients, but about one-third harbor the same deletion, called the *common deletion* (Mita *et al.*, 1990; Schon *et al.*, 1989), which removes 4977 bp of mtDNA between the ATPase 8 and ND5 genes (see Fig. 1).

Table II
Phenotypes Associated with Alterations in MtDNA Structure

Disorder	Comment	Reference[a]
Rearrangements of mtDNA		
Kearns–Sayre syndrome	Sporadic single deletions and/or duplications	Zeviani *et al.*, 1988
PEO	Sporadic single deletions and/or duplications	Moraes *et al.*, 1989
Pearson syndrome	Sporadic single deletions and/or duplications	Rötig *et al.*, 1991
Tubulopathy/diabetes/ataxia	Sporadic single deletions and/or duplications	Rötig *et al.*, 1992
Diabetes and deafness	Maternally inherited single duplications	Dunbar *et al.*, 1993
AD-PEO	AD[b]-inherited multiple deletions (chr. 10)	Suomalainen *et al.*, 1995
AD-PEO	AD-inherited multiple deletions (chr. 3)	Kaukonen *et al.*, 1996
ARCO	AR-inherited multiple deletions	Bohlega *et al.*, 1996
MNGIE	AR-inherited multiple deletions	Nishino *et al.*, 1999
Wolfram syndrome	AR-inherited multiple deletions (chr. 4)	Strom *et al.*, 1998
Depletions of mtDNA		
mtDNA depletion syndrome	Autosomal recessive	Moraes *et al.*, 1991
Zidovudine myopathy	Iatrogenic	Arnaudo *et al.*, 1991

[a] First published article
[b] AD, autosomal dominant; AR, autosomal recessive

Along with Δ-mtDNAs, duplications of mtDNA (dup-mtDNAs) have also been found in patients with sporadic KSS (Brockington *et al.*, 1995; Poulton *et al.*, 1989, 1993, 1995) and Pearson syndrome (Rötig *et al.*, 1995); in other words, these patients are "triplasmic." Importantly, the species of dup-mtDNA and of Δ-mtDNA found in any one patient are related to each other topologically: the partially duplicated molecule can be considered a tandem duplication of wt-mtDNA, in which the material missing in the analogous Δ-mtDNA has been removed. In other words, the duplicated molecule is a wt-mtDNA into which the Δ-mtDNA has been inserted. For example, for the "common deletion" of 4977 bp, the Δ-mtDNA is 11,592 bp long (i.e., 16,569 − 4977) and the dup-mtDNA is 28,161 bp long (i.e., 16,569 + 16,569 − 4977, equivalent to 16,569 + 11,592). Thus, the duplications are likely recombination intermediates that can be resolved into deletions (Holt *et al.*, 1997; Poulton *et al.*, 1993).

There is only a single type of rearranged molecule in any one of these patients, implying that the population of rearranged mtDNA molecules is a clonal expansion of a single spontaneous deletion or duplication event occurring early in oogenesis or embryogenesis (Chen *et al.*, 1995). After fertilization, there is no mtDNA replication until the blastocyst stage of development, when germ layer differentiation begins. If mutated mtDNAs entered all three germ layers, KSS (a multisystem disorder) would result, whereas muscle segregation would result in PEO, and hematopoietic lineage segregation would result in Pearson syndrome (Schon *et al.*, 1991). This would also help explain why the few who survive Pearson syndrome later develop KSS (McShane *et al.*, 1991).

The pathogenicity of mtDNA deletions is almost certainly due to the fact that deletions invariably remove the tRNA genes required for translation of mtDNA-encoded mRNA (Nakase *et al.*, 1990). Additionally the tRNA hypothesis explains why even the *undeleted* genes are not translated (Davidson *et al.*, 1995; Hayashi *et al.*, 1991; Mita *et al.*, 1989) and why KSS patients with different Δ-mtDNA species display similar biochemical and morphological signs. But tRNA genes are *not* missing in dup-mtDNA, however, so other mechanisms must be invoked to indict mtDNA duplications as being pathogenic. Perhaps the excess tRNAs (derived from the duplicated segment) cause a tRNA imbalance that affects tRNA turnover or translation in general. Another possibility is that the chimeric fusion gene straddling the abnormal duplication breakpoint is translated and incorporated into a respiratory chain complex, thereby compromising its function.

3.2. Maternally Inherited Rearrangements

Though most DNA deletions arise spontaneously, there is little evidence for maternal transmission. Maternal inheritance of mtDNA duplications has been observed, however, with diabetes as the main, or even sole, feature (Ballinger *et al.*, 1994; Dunbar *et al.*, 1993; Rötig *et al.*, 1992), usually the type II, non-insulin-dependent form, although insulin-dependent diabetes has also been described (Superti-Furga *et al.*, 1993).

Maternal inheritance of a heteroplasmic, *small*-scale mtDNA rearrangement has been described in one case. Brockington *et al.*, (1993) found a sporadic-KSS patient with a typical large-scale deletion along with an unrelated tandem duplication of a 260-bp area of the D-loop region. The duplication, but not the deletion, was present in the patient's clinically normal mother, implying that the duplication was not pathogenic, a conclusion supported by other workers (Manfredi *et al.*, 1995b; Hao *et al.*, 1997).

In this regard, two unusual *small-scale microdeletions* have also been described. Keightley *et al.*, (1996) identified a patient with recurrent myoglobinuria and Cox deficiency who harbored a heteroplasmic mtDNA microdeletion of only 15 bp located within the COX III gene; the mutation thus removed five in-frame amino acids in the polypeptide. The mutation was absent from the blood of the patient's clinically normal mother. A small 5-bp microdeletion in the COX I gene was found in a patient with motor neuron disease (Comi *et al.*, 1998). This deletion resulted in a frame shift, causing premature termination of the Cox I polypeptide 5 amino acids downstream of the lesion.

4. DISEASES ASSOCIATED WITH NUCLEAR DNA MUTATIONS

As noted above, almost all mutations associated with known mitochondrial diseases have been mutations in mtDNA. Because the vast majority of proteins in mitochondria are nucleus-encoded and are imported from the cytoplasm, one would expect that numerous mendelian-inherited disorders would have been uncovered by now, but this is not the case. On the other hand, because mitochondrial function is so heterogeneous, diseases associated with defects in various aspects of organellar function would be expected to be equally heterogeneous, and for the few genes found so far, this is indeed turning out to be so. The mendelian-inherited mitochondrial diseases fall into four broad classes, described below.

4.1. Integrity Errors

The first class is an unusual group of disorders associated with mendelian-inherited defects in intergenomic communication involved in mtDNA metabolism (Table II). In *autosomal dominant-inherited PEO* (AD-PEO), affected family members, instead of harboring a single sporadically arising mtDNA deletion, harbor *multiple* species of mtDNA deletions in their muscle, apparently generated in "real time" during the patient's life span. The clinical presentation is similar to maternally inherited PEO (i.e., due to point mutations) or sporadic PEO (i.e., due to clonal rearrangements). Two chromosomal loci have been linked to AD-PEO, on chromosomes 3 (Kaukonen *et al.*, 1996) and 10 (Suomalainen *et al.*, 1995), but the specific genes or gene products are unknown.

In *autosomal recessive-inherited PEO* (AR-PEO) the patients also harbor multiple mtDNA deletions, but age of onset is earlier (childhood or adolescence, not adulthood) and the disease is far more severe. Some patients succumb to a fatal cardiomyopathy, *autosomal-recessive cardiomyopathy and ophthalmoplegia*, or ARCO syndrome (Bohlega *et al.*, 1996), whereas others have a more systemic set of neurological, muscular, and gastrointestinal symptoms in *mitochondrial neurogastrointestinal encephalomyopathy*, or MNGIE (Hirano *et al.*, 1994).

A MNGIE locus has recently been mapped to chromosome 22 (Hirano *et al.*, 1998). The culprit gene product is thymidine phosphorylase (TP), a cytosolic protein not known to be targeted to mitochondria (Nishino *et al.*, 1999). The TP protein is required for nucleotide metabolism, and a presumed imbalance in nucleotide pools (perhaps due to the inability of TP to convert thymidine to thymine) is thought to be responsible for the multiple mtDNA deletions found in this disorder.

A third autosomal recessive disorder associated with multiple deletions is *Wolfram syndrome*, characterized by juvenile diabetes mellitus, diabetes insipidus, and optic atrophy (which is why it is sometimes called DIDMOAD), plus neurological symptoms that include deafness, ataxia, and peripheral neuropathy. A gene locus responsible for Wolfram syndrome was mapped to chromosome 4 (Barrientos *et al.*, 1996), and mutations in the gene, dubbed "wolframin" (a transmembrane protein apparently not targeted to mitochondria), have now been identified (Strom *et al.*, 1998).

There is a group of disorders in which patients have the ultimate mtDNA deletion—it removes 100% of the mtDNA. These are the *mitochondrial DNA depletion syndromes*, autosomal recessive-inherited quantitative errors in mtDNA copy number. In the depletion syndromes, specific tissues are almost completely devoid of mtDNA (they are essentially ρ^0), and as a result, they have respiratory chain defects and massive mitochondrial proliferation (e.g., RRF in muscle), but there is no observable biochemical or genetic defect in unaffected tissues. The early-onset form (Moraes *et al.*, 1991) is fatal within a few months of birth, and mtDNA depletion is severe, up to 98% in affected tissues (muscle, liver, or kidney). The late-onset forms (Tritschler *et al.*, 1992) are mainly myopathies and have have lesser degrees of depletion (70–90%). The clinical spectrum associated with partial mtDNA depletion, however, may be wider than we suspect (Vu *et al.*, 1998).

Depletion of mtDNA can also be caused by drugs. *Zidovudine myopathy* is a drug-induced form of mtDNA depletion in muscle that can affect AIDS patients treated with zidovudine (also called azidothymidine, or AZT). In severely affected patients, muscle mtDNA can be reduced by as much as 80% (Arnaudo *et al.*, 1991). The cause of the depletion (depletion is reversible upon removal of the drug) is primarily inhibition by AZT (a nucleoside analog designed to inhibit HIV DNA synthesis by reverse transcriptase) of mtDNA replication by mitochondrial DNA polymerase γ.

4.2. Importation Errors

Almost all proteins imported into mitochondria are targeted to the organelle via a cleavable presequence (Schatz and Dobberstein, 1996), so it stands to reason that mutations in importation leader sequences should cause the second class of mitochondrial diseases (Fenton, 1995). In fact, mutations in leader sequences have been identified in a number of cases (Table III). For example, a nonsense mutation in methylmalonyl-CoA mutase (an enzyme of fatty acid oxidation) ultimately leads to synthesis of a truncated protein with no leader sequence, which, therefore, cannot be imported into mitochondria; this mutation causes a disease called methylmalonic acidemia (Ledley *et al.*, 1990). A leader sequence mutation in the E1α subunit of the pyruvate dehydrogenase complex caused PDHC deficiency in an infant with Leigh syndrome (Takakubo *et al.*, 1995). Other importation mutations affect the leader peptides of isovaleryl-CoA dehydrogenase (an enzyme of leucine catabolism) and ornithine aminotransferase (an enzyme of the urea cycle) (Kobayashi *et al.*, 1995). The diseases resulting from these mutations are *isovaleryl acidemia* (Vockley *et al.*, 1992) and *gyrate atrophy* (Inana *et al.*, 1989), respectively. A mutation in the leader of the gene encoding ornithine transcarbamoylase (an enzyme required in both the urea cycle and in arginine biosynthesis) did not affect import, but caused reduced mRNA levels (Grompe *et al.*, 1989).

Table III
Phenotypes Associated with Pathogenic Mutations in Nuclear DNA

Disorder	Gene	Reference[a]
Pyruvate dehydrogenase deficiency	E1α subunit	Endo et al., 1989
Ornithine transcarbamoylase deficiency	Ornithine transcarbamoylase	Grompe et al., 1989
Gyrate atrophy	Ornithine aminotransferase	Inana et al., 1989
Methymalonic acidemia	Methylmalonyl CoA mutase	Ledley et al., 1990
Isovaleryl acidemia	Isovaleryl CoA dehydrogenase	Vockley et al., 1992
Carnitine palmitoyltransferase II deficiency	Carnitine palmitoyltransferase II	Taroni et al., 1993
Menkes disease	Copper-transporting ATPase	Vulpe et al., 1993
Wilson disease	Copper-transporting ATPase	Petrukhin et al., 1993
Complex II deficiency	Iron–sulfur subunit of SDH	Bourgeron et al., 1995
Leigh syndrome with PDHC deficiency	Pyruvate dehydrogenase E1α	Takakubo et al., 1995
Friedrich's ataxia	Frataxin	Campusano et al., 1996
Systemic carnitine deficiency	Carnitine transporter	Lamhonwah and Tein, 1998
Hereditary spastic paraplegia	Paraplegin	Casari et al., 1998
Leigh syndrome with Cox deficiency	SURF1	Zhu et al., 1998

[a] First published article

In an unusual reversal of events, L-alanine: glyoxylate aminotransferase, a protein normally targeted to peroxisomes, was misdirected to mitochondria via a mutation that created a *de novo* mitochondrial targeting sequence in the N-terminal region of the protein, thereby causing primary hyperoxaluria type I (Purdue *et al.*, 1991).

4.3. Mature Polypeptide Errors

The third class of mendelian-inherited diseases involves mutations in polypeptides. Mutations in subunits of the pyruvate dehydrogenase complex cause PDHC deficiency, a common cause of congenital lactic acidosis. This is often inherited as an X-linked trait, because the gene encoding the α chain subunit of PDHC (PDH-E1α) is located at Xp22.1-p22.2. Since the first report ten years ago (Endo *et al.*, 1989), numerous mutations in this gene have been identified (e.g., Fujii *et al.*, 1996), almost all of which cause X-linked PDHC deficiency and lactic acidosis.

In addition, *Leigh syndrome (LS)* is associated with deficits of PDHC and of respiratory chain enzymes. Mutations in at least three nuclear genes have been identified in LS. Numerous mutations in the X-linked PDH-E1α gene have been identified in children with Leigh syndrome (De Vivo, 1998). In addition, a mutation in the flavoprotein subunit of succinate dehydrogenase resulted in complex II deficiency in two siblings diagnosed with Leigh syndrome (Bourgeron *et al.*, 1995). Finally, mutations in a gene called *SURF1* cause Leigh syndrome and Cox deficiency (Zhu *et al.*, 1998). Although the exact function of SURF1 is unknown, the yeast homology of this gene, called SHY1, is required for Cox assembly in yeast (Mashkevich *et al.*, 1997).

Defects in the various housekeeping genes encoding Krebs-cycle proteins cause a wide range of mitochondrial metabolic diseases, many of which display a surprising degree of complexity and tissue variability (Rustin *et al.*, 1997). Errors in lipid metabolism, such as defects in carnitine palmitoyltransferase II (Taroni *et al.*, 1993) and the carnitine–acylcarnitine carrier, or CAC (Huizing *et al.*, 1997), can cause, among other features, fatal cardiomyopathy and hepatopathy.

Two diseases that were not anticipated to be mitochondrial in nature have turned out to be due to mutations in mitochondrially targeted proteins. In *Friedreich's ataxia*, an autosomal-recessive spinocerebellar degeneration, the mutation is a massive trinucleotide repeat expansion in an intron of the culprit gene, frataxin, causing a reduction in frataxin mRNA (Campusano *et al.*, 1996). The role of frataxin, a mitochondrial protein, is unclear, but appears to involve in organellar iron homeostasis.

Even more surprising was the finding that *hereditrary spastic paraplegia* (HSP), another recessive disorder, is due to mutations in yet another mitochondrial protein, dubbed *paraplegin* (Casari *et al.*, 1998). In this case it appears that paraplegin is a member of the "AAA" class of proteases, implying that a buildup of unwanted proteins, either normally produced or mistranslated, is the primary cause of the disorder. Paradoxically, HSP is a CNS-specific disorder, even though paraplegin is ubiquitously expressed.

4.4. Cofactor and Coenzyme Errors

The fourth class of mendelian-inherited diseases involves cytochrome *c* oxidase, a heme- and copper-containing enzyme. In *Menkes disease*, Cox deficiency and respiratory-chain dysfunction arise as a consequence of a primary defect in a copper-transporting ATPase (which is *not* a mitochondrial protein), ultimately affecting mitochondrial function due to reduced Cox activity (Vulpe *et al.*, 1993). In addition, mutations in another copper-transporting ATPase (ATP7B) cause *Wilson disease* (Bull *et al.*, 1993; Petrukhin *et al.*, 1993). One group has found that the protein is targeted to mitochondria (Lutsenko and Cooper, 1998), but its subcellular localization remains controversial.

Carnitine, a required cofactor for the import of fatty acids into mitochondria, is also involved. Carnitine deficiency (which can often be reversed by administration of exogenous carnitine) can cause fatal cardiomyopathy. A locus on chromosome 5q has been linked to *primary systemic carnitine deficiency* (SCD) (Shoji *et al.*, 1998). The syntenic locus in mice contains the sodium–carnitine transporter, and a mutation in this gene causes the mouse analog of SCD, called *juvenile visceral steatosis* (JVS) (Lu *et al.*, 1998), implying that the same gene is mutated in SCD. This hypothesis has now been confirmed by the identification of mutations in the carnitine transporter in human SCD (Lamhonwah and Tein, 1998),

4.5. Other Disorders

A *fatal myopathy of infancy* associated with Cox deficiency affects infants soon after birth (DiMauro *et al.*, 1980); it is a recessive disorder. In the *reversible myopathy of infancy*, also recessive, children are born with no Cox activity in muscle (as in the fatal myopathy), but recover Cox function spontaneously over a period of about two years (DiMauro *et al.*, 1983). Finally, *Luft disease*, the first mitochondrial disorder to be described at the biochemical level (Luft *et al.*, 1962), is caused by "loose coupling" of mitochondria (i.e., poor coupling between proton pumping and ATP synthesis). In all three cases, the genes responsible are unknown.

ACKNOWLEDGEMENTS. We thank Drs. E. Bonilla and M. Hirano for their critical comments and insight. This work was supported by grants from the National Institutes of Health (NS28828, NS11766, AG12131, HD32062) and the Muscular Dystrophy Association.

REFERENCES

Anderson, S., Bankier, A. T., Barrell, B. G., de Bruijn, M. H. L., Coulson, A. R., Drouin, J., Eperon, I. C., Nierlich, D. P., Roe, B. A., Sanger, F., Schreier, P. H., Smith, A. J. H., Staden, R., and Young, I. G., 1981, Sequence and organization of the human mitochondrial genome, *Nature* **290**:457–465.

Andreu, A. L., Bruno, C., Shanske, S., Shtilbans, A., Hirano, M., Krishna, S., Hayward, L., Systrom, D. S., Brown, R. H., and DiMauro, S., 1998, Missense mutation in the mtDNA cytochrome *b* gene in a patient with myopathy, *Neurology* **51**:1444–1447.

Andreu, A. L., Bruno, C., Dunne, T. C., Tanji, K., Shanske, S., Sue, C. M., Krishna, S., Hadjigeorgiou, G. M., Shtilbans, A., Bonilla, E., and DiMauro, S., 1999, A nonsense mutation (G15059A) in the cytochrome *b* gene in a patient with exercise intolerance and myoglobinuria, *Ann. Neurol.* **45**:127–130.

Arbustini, E., Diegoli, M., Fasani, R., Grasso, M., Morbini, P., Banchieri, N., Bellini, O., Dal Bello, B., Pilotto, A., Magrini, G., Campana, C., Fortina, P., Gavazzi, A., Narula, J., and Vigano, M., 1998, Mitochondrial DNA mutations and mitochondrial abnormalities in dilated cardiomyopathy, *Am. J. Pathol.* **153**:1501–1510.

Arnaudo, E., Dalakas, M., Shanske, S., Moraes, C. T., DiMauro, S., and Schon, E. A., 1991, Depletion of mitochondrial DNA in AIDS patients with zidovudine-induced myopathy, *Lancet* **337**:508–510.

Attardi, G., Chomyn, A., King, M. P., Kruse, B., Polosa, P. L., and Murdter, N. N., 1990, Regulation of mitochondrial gene expression in mammalian cells, *Biochem. Sci. Trans.* **18**:509–513.

Ballinger, S. W., Shoffner, J. M., Gebhart, S., Koontz, D. A., and Wallace, D. C., 1994, Mitochondrial diabetes revisited, *Nature Genet.* **7**:458–459 (letter).

Barrientos, A., Volpini, V., Casademont, J., Genis, D., Manzanares, J. M., Ferrer, I., Corral, J. F. C., Urbano-Marquez, A., Estivill, X., and Nunes, V., 1996, A nuclear defect in the 4p16 region predisposes to multiple mitochondrial DNA deletions in families with Wolfram syndrome, *J. Clin. Invest.* **97**:1570–1576.

Bentlage, H. A.C. M., and Attardi, G., 1996, Relationship of genotype to phenotype in fibroblast-derived transmitochondrial cell lines carrying the 3243 mutation: Shift towards mutant genotype and role of mtDNA copy number, *Hum. Mol. Genet.* **5**:197–205.

Berkovic, S. F., Andermann, F., Shoubridge, E.A., Carpenter, S., Robitaille, Y., Andermann, E., Melmed, C., and Karpati, G., 1991, Mitochondrial dysfunction in multiple symmetrical lipomatosis, *Ann. Neurol.* **29**:566–569.

Bindoff, L. A., Howell, N., Poulton, J., McCullough, D. A., Morten, K. J., Lightowlers, R. N., Turnbull, D. M., and Weber, K., 1993, Abnormal RNA processing associated with a novel tRNA mutation in mitochondrial DNA, *J. Biol. Chem.* **268**:19559–19564.

Blanc, H., Adams, C. W., and Wallace, D. C., 1981, Different nucleotide changes in the large rRNA gene of the mitochondrial DNA confer chloramphenicol resistance on two human cell lines, *Nucl. Acids Res.* **9**:5785–5795.

Bohlega, S., Tanji, K., Santorelli, F., Hirano, M., al-Jishi, A., and DiMauro, S., 1996, Multiple mitochondrial DNA deletions associated with autosomal recessive ophthalmoplegia and severe cardiomyopathy, *Neurology* **46**:1329–1334.

Bonilla, E., Sciacco, M., Tanji, K., Sparaco, M., Petruzzella, V., and Moraes, C.T., 1992, New morphological approaches to the study of mitochondrial encephalomyopathies, *Brain. Pathol.* **2**:113–119.

Bourgeron, T., Rustin, P., Chretien, D., Birch-Machin, M., Bourgeois, M., Viegas-Pequignot, E., Munnich, A., and Rötig, A., 1995, Mutation of a nuclear succinate dehydrogenase gene results in mitochondrial respiratory chain deficiency, *Nature. Genet.* **11**:144–149.

Brockington, M., Sweeney, M. G., Hammans, S. R., Morgan-Hughes, J. A., and Harding, A. E., 1993, A tandem duplication in the D-loop of human mitochondrial DNA is associated with deletions in mitochondrial myopathies, *Nature Genet.* **4**:67–71.

Brockington, M., Alsanjari, N., Sweeney, M. G., Morgan-Hughes, J. A., Scaravilli, F., and Harding , A. E., 1995, Kearns–Sayre syndrome associated with mitochondrial DNA deletion or duplication: A molecular genetic and pathological study, *J. Neurol. Sci.* **131**:78–87.

Bull, P. C., Thomas, G. R., Rommens, J. M., Forbes, J. R., and Cox, D.W., 1993, The Wilson disease gene is a putative copper transporting P-type ATPase similar to the Menkes gene, *Nature. Genet.* **5**:327–337.

Calabresi, P. A., Silvestri, G., DiMauro, S., and Griggs, R. C., 1994, Ekbom's syndrome: Lipomas, ataxia, and neuropathy with MERRF, *Muscle Nerve* **17**:943–945.

Campos, Y., Martin, M. A., Rubio, J. C., Gutiérrez del Olmo, M. C., Cabello, A., and Arenas, J., 1997, Bilateral striatal necrosis and MELAS associated with a new T3308C mutation in the mitochondrial ND1 gene, *Biochem. Biophys. Res. Commun.* **238**:323–325.

Campusano, V., Montermini, L., Molto, M. D., Pianese, L., Cossee, M., Cavalcanti, F., Monros, E., Rodius, F., Duclos, F., Monticelli, A., Zara, F., Cañizares, J., Koutnikova, H., Didichandi, S. I., Gellera, C., Brice, A., Trouillas, P., De Michele, G., Filla, A., De Frutos, R., Palau, F., Patel, P. I., DiDonato, S., Mandel, J.-L., Cocozza, S., Koenig, M., and Pandolfo, M., 1996, Friedreich's ataxia: Autosomal recessive disease caused by an intronic GAA triple repeat expansion, *Science* **271**:1423–1427.

Carelli, V., Ghelli, A., Ratta, M., Bacchilega, E., Sangiorgio, S., Mancini, R., Leuzzi, V., Cortelli, P., Montagna, P., Lugaresi, E., and Degli Esposti, M., 1997, Leber's hereditary optic neuropathy: Biochemical effects of 11778/ND1 and 3460/ND1 mutations and correlation with the mitochondrial genotype, *Neurology* **48**:1623–1632.

Casali, C., Santorelli, F. M., D'Amati, G., Bernucci, P., DeBiase, L., and DiMauro, S., 1995, A novel mtDNA point mutation in maternally inherited cardiomyopathy, *Biochem. Biophys. Res. Commun.* **213**:588–593.

Casari, G., De Fusco, M., Ciarmatori, S., Zeviani, M., Mora, M., Fernandez, P., De Michele, G., Filla, A., Cocozza, S., Marconi, R., Durr, A., Fontaine, B., and Ballabio, A., 1998, Spastic paraplegia and OXPHOS impairment caused by mutations in paraplegin, a nuclear-encoded mitochondrial metalloprotease, *Cell* **93**:973–983.

Chalmers, R. M., Lamont, P. J., Nelson, I., Ellison, D. W., Thomas, N. H., Harding, A. E., and Hammans, S. R., 1997, A mitochondrial tRNAVal point mutation associated with adult-onset Leigh syndrome, *Neurology* **49**:589–592.

Chen, X., Prosser, R., Simonetti, S., Sadlock, J., Jagiello, G., and Schon, E. A., 1995, Rearranged mitochondrial genomes are present in human oocytes, *Am. J. Hum. Genet.* **57**:239–247.

Chinnery, P. F., Johnson, M. A., Taylor, R. W., Durward, W. F., and Turnbull, D. M., 1997, A novel mitochondrial tRNA isoleucine gene mutation causing chronic progressive external ophthalmoplegia, *Neurology* **49**:1166–1168.

Chomyn, A., 1998, The myoclonic epilepsy and ragged-red fiber mutation provides new insights into human mitochondrial function and genetics, *Am. J. Hum. Genet.* **62**:745–751.

Chomyn, A., Meola, G., Bresolin, N., Lai, S. T., Scarlato, G., and Attardi, G., 1991, In vitro genetic transfer of protein synthesis and respiration defects to mitochondrial DNA-less cells with myopathy-patient mitochondria., *Mol. Cell. Biol.* **11**:2236–2244.

Clayton, D. A., 1982, Replication of animal mitochondrial DNA, *Cell* **28**:693–705.

Clayton, D. A., 1992, Transcription and replication of animal mitochondrial DNAs, *Int. Rev. Cytol.* **141**:217–232.

Cock, H. R., Tabrizi, S. J., Cooper, J. M., and Chapira, A. H. V., 1998, The influence of nuclear background on the biochemical expression of 3460 Leber's hereditary optic neuropathy, *Ann. Neurol.* **44**:187–193.

Comi, G. P., Bordoni, A., Salani, S., Franceschina, L., Sciacco, M., Prelle, A., Fortunato, F., Zeviani, M., Napoli, L., Bresolin, N., Moggio, M., Ausenda, C. D., Taanman, J.-W., and Scarlato, G., 1998, Cytochrome *c* oxidase subunit I microdeletion in a patient with motor neuron disease, *Ann. Neurol.* **43**:110–116.

Davidson, M., Zhang, L., Koga, Y., Schon, E. A., and King, M. P., 1995, Genetic and functional complementation of deleted mitochondrial DNA, *Neurology* **45**:A831.

De Meirleir, L., Seneca, S., Lissens, W., Schoentjes, E., and Desprechins, B., 1995, Bilateral striatal necrosis with a novel point mutation in the mitochondrial ATPase 6 gene, *Pediatr. Neurol.* **13**:242–246.

De Vivo, D. C., 1998, Complexities of the pyruvate dehydrogenase complex, *Neurology* **51**:1247–1249.

de Vries, D. D., van Engelen, B. G. M., Gabreëls, F. J.M., Ruitenbeek, W., and van Oost, B. A., 1993, A second missense mutation in the mitochondrial ATPase 6 gene in Leigh's syndrome, *Ann. Neurol.* **34**:410–412.

de Vries, D. D., Went, L. N., Bruyn, G. W., Scholte, H. R., Hofstra, R. M. W., Bolhuis, P. A., and van Oost, B. A., 1996, Genetic and biochemical impairment of mitochondrial complex I activity in a family with Leber hereditary optic neuropathy and hereditary spastic dystonia, *Am. J. Hum. Genet.* **58**:703–711.

DiMauro, S., Mendell, J. R., Sahenk, Z., Bachman, D., Scarpa, A., Scofield, R. M., and Reiner, C., 1980, Fatal infantile mitochondrial myopathy and renal dysfunction due to cytochrome *c* oxidase deficiency, *Neurology* **30**:795–804.

DiMauro, S., Nicholson, J. F., Hays, A. P., Eastwood, A. B., Papadimitriou, A., Koenisberger, R., and De Vivo, D. C., 1983, Benign infantile mitochondrial myopathy due to reversible cytochrome c oxidase deficiency, *Ann. Neurol.* **14**:226–234.

Dumoulin, R., Sagnol, I., Ferlin, T., Bozon, D., Stepien, G., and Mousson, B., 1996, A novel gly 290asp mitochondrial cytochrome b mutation linked to a complex III deficiency in progressive exercise intolerance, *Mol. Cell. Probes* **10**:389–391.

Dunbar, D. R., Moonie, P. A., Swingler, R. J., Davidson, D., Roberts, R., and Holt, I. J., 1993, Maternally transmitted partial direct tandem duplication of mitochondrial DNA associated with diabetes mellitus, *Hum. Mol. Genet.* **2**:1619–1624.

Dunbar, D. R., Moonie, P. A., Zeviani, M., and Holt, I. J., 1996, Complex I deficiency with 3243G: C mitochondria in osteosarcoma cell cybrids, *Hum. Mol. Genet.* **5**:123–129.

El-Schahawi, M., Santorelli, F. M., Malkin, E., Shanske, S., and DiMauro, S., 1995, A new mitochondrial DNA mutation in the tRNA^{Asp} at np 7543 is associated with myoclonic seizures and developmental delay, *J. Mol. Med.* **73**:B37.

El-Schahawi, M., López de Munain, A., Sarrazin, A. M., Shanske, A. L., Basirico, M., Shanske, S., and DiMauro, S., 1997, Two large Spanish pedigrees with nonsyndromic sensorineural deafness and the mtDNA mutation at nt 1555 in the 12S rRNA gene: Evidence of heteroplasmy, *Neurology* **48**:453–456.

Endo, H., Hasegawa, K. K. N., Tada, K. Y. K., and Ohta, S., 1989, Defective gene in lactic acidosis: Abnormal pyruvate dehydrogenase E1 alpha-subunit caused by a frame shift, *Am. J. Hum. Genet.* **44**:358–364.

Enriquez, J. A., Chomyn, A., and Attardi, G., 1995, MtDNA mutation in MERRF syndrome causes defective aminoacylation of tRNA^{Lys} and premature translation termination, *Nature Genet.* **10**:47–55.

Estivill, X., Govea, N., Barcelo, E., Badenas, C., Romero, E., Moral, L., Scozzari, R., D'Urbano, L., Zeviani, M., and Torroni, A., 1998, Familial progressive sensorineural deafness is mainly due to the mtDNA A1555G mutation and is enhanced by treatment of aminoglycosides, *Am. J. Hum. Genet.* **62**:27–35.

Fenton, W. A., 1995, Mitochondrial protein transport: A system in search of mutations, *Am. J. Hum. Genet.* **57**:235–238.

Folgero, T., Torbergsen, T., and Oian, P., 1995, The 3243 MELAS mutation in a pedigree with MERRF, *Eur. Neurol.* **35**:168–171.

Franceschina, L., Salani, S., Bordoni, A., Sciacco, M., Napoli, L., Comi, G. P., Prelle, A., Fortunato, F., Hadjigeorgiou, G. M., Farina, E., Bresolin, N., D'Angelo, M. G., and Scarlato, G., 1998, A novel mitochondrial tRNA^{Ile} point mutation in chronic progressive external ophthalmoplegia, *J. Neurol.* **245**:755–758.

Fu, K., Hartlen, R., Johns, T., Genge, A., Karpati, G., and Shoubridge, E.A., 1996, A novel heteroplasmic tRNA$^{leu(CUN)}$ mtDNA point mutation in a sporadic patient with mitochondrial encephalomyopathy segregates rapidly in skeletal muscle and suggests an approach to therapy, *Hum. Mol. Genet.* **5**:1835–1840.

Fujii, T., Garcia Alvarez, M. B., Sheu, K. F., Kranz-Eble, P.J., and De Vivo, D. C., 1996, Pyruvate dehydrogenase deficiency: The relation of the E1 alpha mutation to the E1 beta subunit deficiency, *Pediatr. Neurol.* **14**:328–334.

Fukuhara, N., Tokigushi, S., Shirakawa, K., and Tsubaki, T., 1980, Myoclonus epilepsy associated with ragged-red fibers (mitochondrial abnormalities): Disease entity or syndrome? Light and electron microscopic studies of two cases and review of the literature, *J. Neurol. Sci.* **47**:117–133.

Gattermann, N., Retzlaff, S., Wang, Y.-L., Berneburg, M., Heinisch, J., Wlaschek, M., Aul, C., and Schneider, W., 1996, A heteroplasmic point mutation of mitohondrial tRNA$^{Leu(CUN)}$ in nonlymphoid haemopoietic cell lineages from a patient with acquired idiopathic sideroblastic anaemia, *Br. J. Haematol.* **93**:845–855.

Gattermann, N., Retzlaff, S., Wang, Y. L., Hofhaus, G., Heinisch, J., Aul, C., and Schneider, W., 1997, Heteroplasmic point mutations of mitochondrial DNA affecting subunit I of cytochrome c oxidase in two patients with acquired idiopathic sideroblastic anemia, *Blood* **90**:4961–4972.

Gámez, J., Playán, A., Andreu, A. L., Bruno, C., Navarro, C., Cervera, C., Arbós, M. A., Schwartz, S., Enriquez, J. A., and Montoya, J., 1998, Familial multiple symmetric lipomatosis associated with the A8344G mutation of mitochondrial DNA, *Neurology* **51**:258–260.

Gerbitz, K. D., Paprotta, A., Jaksch, M., Zierz, S., and Drechsel, J., 1993, Diabetes mellitus is one of the heterogeneous phenotypic features of a mitochondrial DNA point mutation within the tRNA$^{Leu(UUR)}$ gene, *FEBS Lett.* **321**:194–196.

Giles, R. E., Blanc, H., Cann, H. M., and Wallace, D. C., 1980, Maternal inheritance of human mitochondrial DNA, *Proc. Natl. Acad. Sci. USA* **77**:6715–6719.

Goto, Y.-i., Nonaka, I., and Horai, S., 1990, A mutation in the tRNA$^{Leu(UUR)}$ gene associated with the MELAS subgroup of mitochondrial encephalomyopathies, *Nature* **348**:651–653.

Goto, Y.-i., Nonaka, I., and Horai, S., 1991, A new mtDNA mutation associated with mitochondrial myopathy, encephalopathy, lactic acidosis and stroke-like episodes (MELAS), *Biochim. Biophys. Acta* **1097**:238–240.

Goto, Y.-i., Tojo, M., Tohyama, J., Horai, S., and Nonaka, I., 1992, A novel point mutation in the mitochondrial tRNA$^{Leu(UUR)}$ gene in a family with mitochondrial myopathy, *Ann. Neurol.* **31**:672–675.

Goto, Y,-i., Tsugane, K., Tanabe, Y., Nonaka, I., and Horai, S., 1994, A new point mutation at nucleotide 3291 of the mitochondrial tRNA$^{Leu(UUR)}$ gene in a patient with mitochondrial myopathy, encephalopathy, lactic acidosis, and stroke-like episodes (MELAS), *Biochem. Biophys. Res. Commun.* **202**:1624–1630.

Grompe, M., Muzny, D. M., and Caskey, C.T., 1989, Scanning detection of mutations in human ornithine transcarbamoylase by chemical mismatch cleavage, *Proc. Natl. Acad. Sci. USA* **86**:5888–5892.

Hanna, M. G., Nelson, I., Sweeney, M. G., Cooper, J. M., Watkins, P. J., Morgan-Hughes, J. A., and Harding, A. E., 1995, Congenital encephalomyopathy and adult-onset myopathy and diabetes mellitus: Different phenotypic associations of a new heteroplasmic mtDNA tRNA glutamic acid mutation, *Am. J. Hum. Genet.* **56**:1026–1033.

Hanna, M. G., Nelson, I. P., Rahman, S., Lane, R. J. M., Land, J., Heales, S., Cooper, M. J., Schapira, A. H. V., Morgan-Hughes, J. A., and Wood, N. W., 1998, Cytochrome *c* oxidase deficiency associated with the first stop-codon point mutation in human mtDNA, *Am. J. Hum. Genet.* **63**:29–36.

Hao, H., and Moraes, C. T., 1996, Functional and molecular abnormalities associated with a C → T transition at position 3256 of the human mitochondrial genome, *J. Biol. Chem.* **271**:2347–2352.

Hao, H., Manfredi, G., Clayton, D. A., and Moraes, C. T., 1997, Functional and structural features of a tandem duplication of the human mtDNA promoter region, *Am. J. Hum. Genet.* **60**:1363–1372.

Hasegawa, H., Matsuoka, T., Goto, Y.-i., and Nonaka, I., 1991, Strongly succinate dehydrogenase-reactive blood vessels in muscles from patients with mitochondrial myopathy, encephalopathy, lactic acidosis, and stroke-like episodes, *Ann. Neurol.* **29**:601–605.

Hayashi, J.-I., Ohta, S., Kikuchi, A., Takemitsu, M., Goto, Y.-i., and Nonaka, I., 1991, Introduction of disease-related mitochondrial DNA deletions into HeLa cells lacking mitochondrial DNA results in mitochondrial dysfunction, *Proc. Natl. Acad. Sci. USA* **88**:10614–10618.

Hirano, M., Ricci, E., Koenigsberger, M. R., Defendini, R., Pavlakis, S. G., De Vivo, D. C., DiMauro, S., and Rowland, L. P., 1992, MELAS: An original case and clinical criteria for diagnosis, *Neuromusc. Disord.* **2**:125–135.

Hirano, M., Silvestri, G., Blake, D. M., Lombes, A., Minetti, C., Bonilla, E., Hays, A. P., Lovelace, R. E., Butler, I., Bertorini, T. E., Threlkeld, A. B., Mitsumoto, H., Salberg, L. M., Rowland, L. P., and DiMauro, S., 1994, Mitochondrial neurogastrointestinal encephalomyopathy (MNGIE): Clinical, biochemical, and genetic features of an autosomal recessive mitochondrial disorder, *Neurology* **44**:721–727.

Hirano, M., Garcia-de-Yebenes, J., Jones, A. C., Nishino, I., DiMauro, S., Carlo, J. R., Bender, A. N., Hahn, A. F., Salberg, L. M., Weeks, D. E., and Nygaard, T. G., 1998, Mitochondrial neurogastrointestinal encephalomyopathy syndrome maps to chromosome 22q13.32-qter, *Am. J. Hum. Genet.* **63**:526–533.

Hofhaus, G., Johns, D. R., Hurko, O., Attardi, G., and Chomyn, A., 1996, Respiration and growth defects in transmitochondrial cell lines carrying the 11778 mutation associated with Leber's hereditary optic neuropathy, *J. Biol. Chem.* **271**:13155–13161.

Holme, E., Larsson, N.-G., Oldfors, A., Tulinius, M., Sahlin, P., and Stenman, G., 1993, Multiple symmetric lipomas with high levels of mtDNA with the tRNA^{Lys} A → G$^{(8344)}$ mutation as the only manifestation of disease in a carrier of myoclonus epilepsy and ragged-red fibers (MERRF) syndrome, *Am. J. Hum. Genet.* **52**:551–556.

Holt, I. J., Harding, A. E., and Morgan-Hughes, J. A., 1988, Deletions of mitochondrial DNA in patients with mitochondrial myopathies, *Nature* **331**:717–719.

Holt, I. J., Harding, A. E., and Morgan-Hughes, J. A., 1989, Deletions of muscle mitochondrial DNA in mitochondrial myopathies: Sequence analysis and possible mechanisms, *Nucl. Acids Res.* **17**:4465–4469.

Holt, I. J., Harding, A. E., Petty, R. K. H., and Morgan-Hughes, J. A., 1990, A new mitochondrial disease associated with mitochondrial DNA heteroplasmy, *Am. J. Hum. Genet.* **46**:428–433.

Holt, I. J., Dunbar, D.R., and Jacobs, H. T., 1997, Behaviour of a population of partially duplicated mitochondrial DNA molecules in cell culture: Segregation, maintenance, and recombination dependent upon nuclear background, *Hum. Mol. Genet.* **6**:1251–1260.

Houstek, J., Klement, P., Hermanska, J., Houstkova, H., Hansokova, H., Van den Bogert, C., and Zeman, J., 1995, Altered properties of mitochondrial ATP-synthase in patients with a T → G mutation in the ATPase 6 (subunit *a*) gene at position 8993 of mtDNA, *Biochim. Biophys. Acta* **1271**:349–357.

Howell, N., Kubacka, I., Xu, M., and McCullough, D. A., 1991, Leber hereditary optic neuropathy: Involvement of the mitochondrial ND1 gene and evidence for an intragenic suppressor mutation, *Am. J. Hum. Genet.* **48**:935–942.

Howell, N., Smith, R., Frerman, F., Parks, J. K., and Parker, W. D., Jr., 1996, Association of the mitochondrial 8344 MERRF mutation with maternally inherited spinocerebellar degeneration and Leigh disease, *Neurology* **46**:219–222.

Howell, N., Bogolin, C., Jamieson, R., Marenda, D. R., and Mackey, D. A., 1998, mtDNA mutations that cause optic neuropathy: How do we know? *Am. J. Hum. Genet.* **62**:196–202.

Huizing, M., Iacobazzi, V., Ijlst, L., Savelkoul, P., Ruitenbeek, W., van den Heuvel, L., Indiveri, C., Smeitink, J., Trijbels, F., Wanders, R., and Palmieri, F., 1997, Cloning of the human carnitine-acylcarnitine carrier cDNA and identification of the molecular defect in a patient, *Am. J. Hum. Genet.* **61**:1239–1245.

Huoponen, K., Vilkki, J., Aula, P., Nikoskelain,E. K., and Savontaus, M.-L., 1991, A new mtDNA mutation associated with Leber hereditary optic neuroretinopathy, *Am. J. Hum. Genet.* **48**:1147–1153.

Hutchin, T., Haworth, I., Higashi, K., Fischel-Ghodsian, N., Stoneking, M., Saha, N., Arnos, C., and Cortopassi, G., 1993, A molecular basis for human hypersensitivity to aminoglycoside antibiotics, *Nucl. Acids Res.* **21**:4174–4179.

Inana, G., Chambers, C., Hotta, Y., Inouye, L., Filpula, D., Pulford, S., and Shiono, T., 1989, Point mutation affecting processing of the ornithine aminotransferase precursor protein in gyrate atrophy, *J. Biol. Chem.* **264**:17432–17436.

Inoué, K., Takai, D., Soejima, A., Isobe, K., Yamasoba, T., Oka, Y., Goto, Y-i., and Hayashi, J., 1996, Mutant mtDNA at 1555 A to G in 12S rRNA gene and hypersusceptibility of mitochondrial translation to streptomycin can be co-transferred to rho^0 HeLa cells, *Biochem. Biophys. Res. Commun.* **223**:496–501.

Johns, D. R., Neufeld, M. J., and Park, R. D., 1992, An ND-6 mitochondrial DNA mutation associated with Leber hereditary optic neuropathy, *Biochem. Biophys. Res. Commun.* **187**:1551–1557.

Jun, A. S., Brown, M. D., and Wallace, D. C., 1994, A mitochondrial DNA mutation at nucleotide pair 14459 of the NADH dehydrogenase subunit 6 gene associated with maternally inherited Leber hereditary optic neuropathy and dystonia, *Proc. Natl. Acad. Sci. USA* **91**:6206–6210.

Kadowaki, H., Tobe, K., Mori, Y., Sakura, H., Sakuta, R., Nonaka, I., Hagura, R., Yazaki, Y., Akanuma, Y., and Kadowaki, T., 1993, Mitochondrial gene mutation and insulin-dependent type of diabetes mellitus, *Lancet* **341**:893–894.

Kameoka, K., Isotani, H., Tanaka, K., Azukari, K., Fujimura, Y., Shiota, Y., Sasaki, E., Majima, M., Furukawa, K., Haginomori, S., Kitaoka, H., and Ohsawa, N., 1998, Novel mitochondrial DNA mutation in tRNALys (8296A → G) associated with diabetes, *Biochem. Biophys. Res. Commun.* **245**:523–527.

Kaufmann, P., Koga, Y., Shanske, S., Hirano, M., DiMauro, S., King, M.P., and Schon, E.A., 1996, Mitochondrial DNA and RNA processing in MELAS, *Ann. Neurol.* **40**:172–180.

Kaukonen, J.A., Amati, P., Suomalainen, A., Rotig, A., Piscaglia, M.-G., Salvi, F., Weissenbach, J., Fratta, G., Comi, G., Peltonen, L., and Zeviani, M., 1996, An autosomal locus predisposing to multiple deletions of mtDNA on chromosome 3p, *Am. J. Hum. Genet.* **58**:763–769.

Kawarai, T., Kawakami, H., Kozuka, K., Izumi, Y., Matsuyama, Z., Watanabe, C., Kohriyama, T., and Nakamura, S., 1997, A new mitochondrial DNA mutation associated with mitochondrial myopathy: tRNA$^{Leu(UUR)}$ 3254C-to-G, *Neurology* **49**:598–600.

Kearns, T.P., and Sayre, G.P., 1958, Retinitis pigmentosa, external ophthalmoplegia, and complete heart block, *Arch. Ophthalmol.* **60**:280–289.

Keightley, J. A., Hoffbuhr, K. C., Burton, M. D., Salas, V. M., Johnston, W. S. W., Penn, A. M. W., Buist, N. R.M., and Kennaway, N. G., 1996, A microdeletion in cytochrome *c* oxidase (COX) subunit III associated with COX deficiency and recurrent myoglobinuria, *Nature Genet.* **12**:410–416.

Kellar-Wood, H., Robertson, N., Govan, G. G., Compston, D. A. S., and Harding, A. E., 1994, Leber's hereditary optic neuropathy mitochondrial DNA mutations in multiple sclerosis, *Ann. Neurol.* **36**:109–112.

King, M. P., and Attardi, G., 1989, Human cells lacking mtDNA: Repopulation with exogenous mitochondria by complementation, *Science* **246**:500–503.

King, M. P., Koga, Y., Davidson, M., and Schon, E. A., 1992, Defects in mitochondrial protein synthesis and respiratory chain activity segregate with the tRNA[Leu(UUR)] mutation associated with mitochondrial myopathy, encephalopathy, lactic acidosis, and strokelike episodes, *Mol. Cell Biol.* **12**:480–490.

Kobayashi, M., Morishita, H., Sugiyama, N., Yocochi, K., Nakano, M., Wada, Y., Hotta, Y., Terauchi, A., and Nonaka, I., 1987, Two cases of NADH-coenzyme Q reductase deficiency: Relationship to MELAS syndrome, *J. Pediatr.* **110**:223–227.

Kobayashi, T., Ogawa, H., Kasahara, M., Shiozawa, Z., and Matsuzawa, T., 1995, A single amino acid substitution within the mature sequence of ornithine aminotransferase obstructs mitochondrial entry of the precursor, *Am. J. Hum. Genet.* **57**:284–291.

Koga, Y., Davidson, M., Schon, E. A., and King, M. P., 1993, Fine mapping of mitochondrial RNAs derived from the mtDNA region containing a point mutation associated with MELAS, *Nucl. Acids. Res.* **21**:657–662.

Koga, Y., Davidson, M., Schon, E. A., and King, M. P., 1995, Analysis of cybrids harboring MELAS mutations in the mitochondrial tRNA[Leu(UUR)] gene, *Muscle Nerve* (**Suppl. 3**):S119–S123.

Lamhonwah, A. M., and Tein, I., 1998, Carnitine uptake defect: Frameshift mutations in the human plasma-lemmal carnitine transporter gene, *Biochem. Biophys. Res. Commun.* **252**:396–401.

Leber, T., 1871, Ueber hereditäre und congenital-angelegte Sehnervenleiden, *Albrecht von Graefe's Arch. f. Ophthalmol.* **17** (Abt. II):249–291.

Ledley, F. D., Jansen, R., Nham, S.U., Fenton, W. A., and Rosenberg, L. E., 1990, Mutation eliminating mitochondrial leader sequence of methylmalonyl-CoA mutase causes mut⁰ methylmalonic acidemia, *Proc. Natl. Acad. Sci. USA* **87**:8905–8909.

Lestienne, P., and Ponsot, G., 1988, Kearns–Sayre syndrome with muscle mitochondrial DNA deletion, *Lancet* **1**:885 (letter).

Lombes, A., Bories, D., Girodon, E., Frachon, P., Ngo, M. M., Breton-Gorius, J., Tulliez, M., and Goossens, M., 1998, The first pathogenic mitochondrial methionine tRNA point mutation is discovered in splenic lymphoma, *Hum. Mutat.* (**Suppl. 1**):S175–S183.

Lu, K.-m., Nishimori, H., Nakamura, Y., Shima, K., and Kuwajima, M., 1998, A missense mutation of mouse OCTN2, a sodium-dependent carnitine cotransporter, in the juvenile visceral steatosis mouse, *Biochem. Biophys. Res. Commun.* **252**:590–594.

Luft, R., Ikkos, D., Plamieri, G., Ernster, L., and Afzelius, B., 1962, A case of severe hypermetabolism of nonthyroid origin with a defect in the maintenance of mitochondrial respiratory control: A correlated clinical, biochemical, and morphological study, *J. Clin. Invest.* **41**:1776–1804.

Lutsenko, S., and Cooper, M. J., 1998, Localization of the Wilson's disease protein product to mitochondria, *Proc. Natl. Acad. Sci.USA* **95**:6004–6009.

Mackey, D., and Howell, N., 1992, A variant of Leber hereditary optic neuropathy characterized by recovery of vision and by an unusual mitochondrial genetic etiology, *Am. J. Hum. Genet.* **51**:1218–1228.

Manfredi, G., Schon, E. A., Moraes, C. T., Bonilla, E., Berry, G. T., Sladky, J. T., and DiMauro, S., 1995a, A new mutation associated with MELAS is located in a mitochondrial DNA polypeptide-coding gene, *Neuromusc. Disord.* **5**:391–398.

Manfredi, G., Servidei, S., Bonilla, E., Shanske, S., Schon, E. A., DiMauro, S., and Moraes, C. T., 1995b, High levels of mitochondrial DNA with an unstable 260-bp duplication in a patient with mitochondrial myopathy, *Neurology* **45**:762–768.

Manfredi, G., Schon, E. A., Bonilla, E., Moraes, C. T., Shanske, S., and DiMauro, S., 1996, Identification of a mutation in the mitochondrial tRNA[Cys] gene associated with mitochondrial encephalopathy, *Hum. Mutat.* **7**:158–163.

Mashkevich, G., Repetto, B., Glerum, D. M., Jin, C., and Tzagoloff, A.,1997, *SHY1* , the yeast homolog of the mammalian *SURF-1* gene, encodes a mitochondrial protein required for respiration, *J. Biol. Chem.* **272**:14356–14364.

Masucci, J., Davidson, M., Koga, Y., DiMauro, S., Schon, E. A., and King, M. P., 1995, *In vitro* analysis of mutations causing myoclonus epilepsy with ragged-red fibers in the mitochondrial tRNA[Lys] gene: Two genotypes produce similar phenotypes, *Mol. Cell. Biol.* **15**:2872–2881.

Matthews, P. M., Tampieri, D., Berkovic, S. F., Andermann, F., Silver, K., Chityat, D., and Arnold, D. L., 1991, Magnetic resonance imaging shows specific abnormalities in the MELAS syndrome, *Neurology* **41**:1043–1046.

McShane, M. A., Hammans, S. R., Sweeney, M., Holt, I. J., Beattie, T. J., Brett, E. M., and Harding, A. E., 1991, Pearson syndrome and mitochondrial encephalomyopathy in a patient with a deletion of mtDNA, *Am. J. Hum. Genet.* **48**:39–42.

Meire, F. M., Van Coster, R., Cochaux, P., Obermaier-Kusser, B., Candaele, C., and Martin, J. J., 1995, Neurological disorders in members of families with Leber's hereditary optic neuropathy (LHON) caused by different mitochondrial mutations, *Ophthalmic Genet.* **16**:119–126.

Merante, F., Tein, I., Benson, L., and Robinson, B. H., 1994, Maternally-inherited hypertrophic cardiomyopathy due to a novel T-to-C transition at nucleotide 9997 in the mitochondrial tRNA^{glycine} gene, *Am. J. Hum. Genet.* **55**:437–446.

Merante, F., Myint, T., Tein, I., Benson, L., and Robinson, B. H., 1996, An additional mitochondrial tRNA^{Ile} point mutation (A-to-G at nucleotide 4295) causing hypertrophic cardiomyopathy, *Hum. Mutat.* **8**:216–222.

Mita, S., Schmidt, B., Schon, E. A., DiMauro, S., and Bonilla, E., 1989, Detection of "deleted" mitochondrial genomes in cytochrome *c* oxidase-deficient muscle fibers of a patient with Kearns–Sayre syndrome, *Proc. Natl. Acad. Sci. USA* **86**:9509–9513.

Mita, S., Rizzuto, R., Moraes, C. T., Shanske, S., Arnaudo, E., Fabrizi, G., Koga, Y., DiMauro, S., and Schon, E. A., 1990, Recombination via flanking direct repeats is a major cause of large-scale deletions of human mitochondrial DNA, *Nucl. Acids Res.* **18**:561–567.

Moraes, C. T., DiMauro, S., Zeviani, M., Lombes, A., Shanske, S., Miranda, A. F., Nakase, H., Bonilla, E., Wernec, L. C., Servidei, S., Nonaka, I., Koga, Y., Spiro, A., Brownell, K. W., Schmidt, B., Schotland, D. L., Zupanc, M. D., De Vivo, D. C., Schon, E. A., and Rowland, L. P., 1989, Mitochondrial DNA deletions in progressive external ophthalmoplegia and Kearns–Sayre syndrome, *N. Engl. J. Med.* **320**:1293–1299.

Moraes, C. T., Shanske, S., Tritschler, H.-J., Aprille, J. R., Andreetta, F., Bonilla, E., Schon, E., and DiMauro, S., 1991, Mitochondrial DNA depletion with variable tissue expression: A novel genetic abnormality in mitochondrial diseases, *Am. J. Hum. Genet.* **48**:492–501.

Moraes, C. T., Ciacci, F., Bonilla, E., Ionasescu, V., Schon, E. A., and DiMauro, S., 1993a, A mitochondrial tRNA anticodon swap associated with a muscle disease, *Nature Genet.* **4**:284–288.

Moraes, C. T., Ciacci, F., Bonilla, E., Jansen, C., Hirano, M., Rao, N., Lovelace, R. E., Rowland, L. P., Schon, E. A., and DiMauro, S., 1993b, Two novel pathogenic mitochondrial DNA mutations affecting organelle number and protein synthesis: Is the tRNA$^{Leu(UUR)}$ gene an etiologic hotspot?, *J. Clin. Invest.* **92**:2906–2915.

Moraes, C. T., Ciacci, F., Silvestri, G., Shanske, S., Sciacco, M., Hirano, M., Schon, E. A., Bonilla, E., and DiMauro, S., 1993c, Atypical clinical presentations associated with the MELAS mutation at position 3243 of human mitochondrial DNA, *Neuromusc. Disord.* **3**:43–50.

Morten, K. J., Cooper, J. M., Brown, G. K., Lake, B. D., Pike, D., and Poulton, J., 1993, A new point mutation associated with mitochondrial encephalomyopathy, *Hum. Mol. Genet.* **2**:2081–2087.

Nakamura, M., Nakano, S., Goto, Y.-i., Ozawa, M., Nagahama, Y., Fukuyama, H., Akiguchi, I., Kaji, R., and Kimura, J., 1995, A novel point mutation in the mitochondrial tRNA$^{Ser(UCN)}$ gene detected in a family with MERRF/MELAS overlap syndrome, *Biochem. Biophys. Res. Commun.* **214**:86–93.

Nakase, H., Moraes, C. T., Rizzuto, R., Lombes, A., DiMauro, S., and Schon, E. A., 1990, Transcription and translation of deleted mitochondrial genomes in Kearns–Sayre syndrome: Implications for pathogenesis, *Am. J. Hum. Genet.* **46**:418–427.

Nelson, I., Hanna, M. G., Alsanjari, N., Scaravilli, F., Morgan-Hughes, J. A., and Harding, A. E., 1995, A new mitochondrial DNA mutation associated with progressive dementia and chorea: A clinical, pathological, and molecular genetic study, *Ann. Neurol.* **37**:400–403.

Nikoskelainen, E. K., Marttila, R. J., Huoponen, K., Juvonen, V., Lamminen, T., Sonninen, P., and Savontaus, M. L., 1995, Leber's "plus:" Neurological abnormalities in patients with Leber's hereditary optic neuropathy, *J. Neurol. Neurosurg. Psychiat.* **59**:160–164.

Nishino, I., Seki, A., Maegaki, Y., Takeshita, K., Horai, S., Nonaka, I., and Goto, I., 1996, A novel mutation in the mitochondrial tRNA^{Thr} gene associated with a mitochondrial encephalomyopathy, *Am. J. Hum. Genet.* **59**: A400.

Nishino, I., Spinazzola, A., and Hirano, M., 1999, Thymidine phosphorylase gene mutations in MNGIE, a human mitochondrial disorder, *Science* **283**:689–692.

O'Brien, T. W., Denslow, N. D., Anders, J. C., and Courtney, B. C., 1990, The translation system of mammalian mitochondria, *Biochim. Biophys. Acta* **1050**:174–178.

Ohama, E., and Ikuta, F., 1987, Involvement of choroid plexus in mitochondrial encephalomyopathy (MELAS), *Acta Neuropathol.* **75**:1–7.

Onishi, H., Inoue, K., Osaka, H., Kimura, S., Nagatomo, H., Hanihara, T., Kawamoto, S., Okuda, K., Yamada, Y., and Kosaka, K., 1993, Mitochondrial myopathy, encephalopathy, lactic acidosis, and stroke-like episodes (MELAS), and diabetes mellitus: Molecular genetic analysis and family study, *J. Neurol. Sci.* **114**: 205–208.

Pearson, H. A., Lobel, J. S., Kocoshis, S. A., Naiman, J. L., Windmiller, J., Lammi, A. T., Hoffman, R., and Marsh, J. C., 1979, A new syndrome of refractory sideroblastic anemia with vacuolization of marrow precursors and exocrine pancreatic dysfunction, *J. Pediatr.* **95**:976–984.

Petrukhin, K., Fischer, S. G., Pirastu, M., Tanzi, R. E., Chernov, I., Devoto, M., Brzustowicz, L. M., Canyanis, E., Vitale, E., Russo, J. J., Matseoane, D., Boukhgalter, B., Wasco, W., Figus, A. L., Loudianos, J., Cao, A., Sternlieb, I., Evgrafov, O., Parano, E., Pavone, L., Warburton, D., Ott, J., Penchaszadeh, G. K., Scheinberg, I. H., and Gilliam, T. C., 1993, Mapping, cloning and genetic characterization of the region containing the Wilson disease gene, *Nature Genet.* **5**:338–343.

Petruzzella, V., Moraes, C. T., Sano, M. C., Bonilla, E., DiMauro, S., and Schon, E. A., 1994, Extremely high levels of mutant mtDNAs co-localize with cytochrome *c* oxidase-negative ragged-red fibers in patients harboring a point mutation at nt-3243, *Hum. Mol. Genet.* **3**:449–454.

Poulton, J., Deadman, M. E., and Gardiner, R. M., 1989, Tandem direct duplications of mitochondrial DNA in mitochondrial myopathy: Analysis of nucleotide sequence and tissue distribution, *Nucl. Acids Res.* **17**:10223–10229.

Poulton, J., Deadman, M. E., Bindoff, L., Morten, K., Land, J., and Brown, G., 1993, Families of mtDNA re-arrangements can be detected in patients with mtDNA deletions: Duplications may be a transient intermediate form, *Hum. Mol. Genet.* **2**:23–30.

Poulton, J., Morten, K. J., Marchington, D., Weber, K., Brown, K. G., Rötig, A., and Bindoff, L., 1995, Duplications of mitochondrial DNA in Kearns–Sayre syndrome, *Muscle Nerve* (**Suppl. 3**):S154–S158.

Prezant, T. R., Agapian, J. V., Bohlman, M. C., Bu, X., Oztas, S., Qiu, W.-Q., Arnos, K. S., Cortopassi, G. A., Jaber, L., Rotter, J. I., Shohat, M., and Fischel-Ghodsian, N., 1993, Mitochondrial ribosomal RNA mutation associated with both antibiotic-induced and nonsyndromic deafness, *Nature Genet.* **4**:289–294.

Purdue, P. E., Allsop, J., Isaya, G., Rosenberg, L. E., and Danpure, C. J., 1991, Mistargeting of peroxisomal L-alanine: glyoxylate aminotransferase to mitochondria in primary hyperoxaluria patients depends upon activation of a cryptic mitochondrial targeting sequence by a point mutation, *Proc. Natl. Acad. Sci. USA* **88**:10900–10904.

Reardon, W., Ross, R. J., Sweeney, M. G., Luxon, L. M., Harding, A. E., and Trembath, R. C., 1992, Diabetes mellitus associated with a pathogenic point mutation in mitochondrial DNA, *Lancet* **340**:1376–1379.

Reid, F. M., Vernham, G. A., and Jacobs, H. T., 1994, A novel mitochondrial point mutation in a maternal pedigree with sensorineural deafness, *Hum. Mutat.* **3**:243–247.

Remes, A. M., Majamaa, K., Herva, R., and Hassinen, I. E., 1993, Adult-onset diabetes mellitus and neuroSensory hearing loss in maternal relatives of MELAS patients in a family with the tRNA$^{\mathrm{Leu(UUR)}}$ mutation, *Neurology* **43**:1015–1020.

Rowland, L. P., Hays, A. P., DiMauro, S., De Vivo, D. C., and Behrens, M., 1983, Diverse clinical disorders associated with morphological abnormalities of mitochondria, in *Mitochondrial Pathology in Muscle Diseases* (C. Cerri and G. Scarlato, Eds.), Piccin Editore, Padova, pp. 141–158.

Rötig, A., Cormier, V., Koll, F., Mize, C. E., Saudubray, J.-M., Veerman, A., Pearson, H. A., and Munnich, A., 1991, Site-specific deletions of the mitochondrial genome in the Pearson marrow-pancreas syndrome, *Genomics* **10**:502–504.

Rötig, A., Bessis, J.-L., Romero, N., Cormier, V., Saudubray, J.-M., Narcy, P., Lenoir, G., Rustin, P., and Munnich, A., 1992, Maternally inherited duplication of the mitochondrial genome in a syndrome of proximal tubulopathy, diabetes mellitus, and cerebellar ataxia, *Am. J. Hum. Genet.* **50**:364–370.

Rötig, A., Bourgeron, T., Chretien, D., Rustin, P., and Munnich, A., 1995, Spectrum of mitochondrial DNA rearrangements in the Pearson marrow-pancreas syndrome, *Hum. Mol. Genet.* **4**:1327–1330.

Rustin, P., Bourgeron, T., Parfait, B., Chretien, D., Munnich, A., and Rötig, A., 1997, Inborn errors of the Krebs cycle: A group of unusual mitochondrial diseases in human, *Biochim. Biophys. Acta* **1361**:185–197.

Sahashi, K., Ibi, T., Yoneda, M., Tanaka, M., and Ohno, K., 1997, A mitochondrial DNA mutation in the heteroplasmic tRNA-Tyr gene associated with chronic progressive external ophthalmoplegia: Clinical and molecular biological study (in Japanese), *Nippon Rinsho* **55**:3265–3269.

Santorelli, F., Shanske, S., Macaya, A., De Vivo, D. C., and DiMauro, S., 1993, The mutation at nt 8993 of mitochondrial DNA is a common cause of Leigh's syndrome, *Ann. Neurol.* **34**:827–834.

Santorelli, F. M., Shanske, S., Jain, K. D., Tick, D., Schon, E. A., and DiMauro, S., 1994, AT → C mutation at nt 8993 of mitochondrial DNA in a child with Leigh syndrome, *Neurology* **44**:972–974.

Santorelli, F. M., Mak, S. C., Vazquez-Acevedo, M., Gonzalez-Astiazaran, A., Ridaura-Sanz, C., Gonzalez-Halphen, D., and DiMauro, S., 1995, A novel mitochondrial DNA point mutation associated with mitochondrial encephalocardiomyopathy, *Biochem. Biophys. Res. Commun.* **216**:835–840.

Santorelli, F. M., Mak, S.,-C., El-Schahawi, M., Casali, C., Shanske, S., Baram, T. Z., Madrid, R. E., and DiMauro, S., 1996, Maternally-inherited cardiomyopathy and hearing loss associated with a novel mutation in the mitochondrial tRNA^{Lys} gene (G8363A), *Am. J. Hum. Genet.* **58**:933–939.

Santorelli, F. M., Tanji, K., Kulikova, R., Shanske, S., Vilarinho, L., Hays, A. P., and DiMauro, S., 1997, Identification of a novel mutation in the mtDNA ND5 gene associated with MELAS, *Biochem. Biophys. Res. Commun.* **238**:326–328.

Santorelli, F. M., Tanji, K., Shanske, S., Krishna, S., Schnmidt, R. E., Greenwood, R. S., DiMauro, S., and De Vivo, D. C., 1998, The mitochondrial DNA A8344G mutation in Leigh syndrome revealed by analysis in paraffin-embedded sections: Revisiting the past, *Ann. Neurol.* **44**:962–964.

Santorelli, F., Tanji, K., Manta, P., Casali, C., Krishna, S., Hays, A. P., Mancini, D. M., DiMauro, S., and Hirano, M., 1999, Maternally inherited cardiomyopathy: An atypical presentation of the mtDNA 12S rRNA A1555G mutation, *Am. J. Hum. Genet.* **64**:295–300.

Satoh, M., and Kuroiwa, T., 1991, Organization of multiple nucleoids and DNA molecules in mitochondria of a human cell, *Exp. Cell Res.* **196**:137–140.

Schatz, G., and Dobberstein, B., 1996, Common principles of protein translocation across membranes, *Science* **271**:1519–1526.

Schon, E. A., Rizzuto, R., Moraes, C. T., Nakase, H., Zeviani, M., and DiMauro, S., 1989, A direct repeat is a hotspot for large-scale deletions of human mitochondrial DNA, *Science* **244**:346–349.

Schon, E. A., Bonilla, E., Miranda, A. F., and DiMauro, S., 1991, Molecular biology of mitochondrial diseases, in *Molecular Genetic Approaches to Neuropsychiatric Disease* (J. Brosius and R. Fremeau, Eds.), Academic, San Diego, CA, pp. 57–80.

Schon, E. A., Koga, Y., Davidson, M., Moraes, C. T., and King, M. P., 1992, The mitochondrial tRNA$^{Leu(UUR)}$ mutation in MELAS: A model for pathogenesis, *Biochim. Biophys. Acta* **1101**:206–209.

Schon, E. A., Bonilla, E., and DiMauro, S., 1997, Mitochondrial DNA mutations and pathogenesis, *J. Bioenerg. Biomembr.* **29**:131–149.

Shoffner, J. M., Lott, M. T., Lezza, A. M. S., Seibel, P., Ballinger, S. W., and Wallace, D. C., 1990, Myoclonic epilepsy and ragged-red fiber disease (MERRF) is associated with a mitochondrial DNA tRNA^{Lys} mutation, *Cell* **61**:931–937.

Shoffner, J. M., Bialer, M. G., Pavlakis, S. G., Lott, M., Kaufman, A., Dixon, J., Teichberg, S., and Wallace, D. C., 1995, Mitochondrial encephalomyopathy associated with a single nucleotide pair deletion in the mitochondrial tRNA$^{Leu(UUR)}$ gene, *Neurology* **45**:286–292.

Shoji, Y., Koizumi, A., Kayo, T., Ohata, T., Takahashi, T., Harada, K., and Takada, G., 1998, Evidence for linkage of human primary systemic carnitine deficiency with D5S436: A novel gene locus on chromosome 5q, *Am. J. Hum. Genet.* **63**:101–108.

Silvestri, G., Moraes, C. T., Shanske, S., Oh, S. J., and DiMauro, S., 1992, A new mtDNA mutation in the tRNA^{Lys} gene associated with myoclonic epilepsy and ragged-red fibers (MERRF), *Am. J. Hum. Genet.* **51**:1213–1217.

Silvestri, G., Ciafaloni, E., Santorelli, F. M., Shanske, S., Servidei, S., Graf, W. D., Sumi, M., and DiMauro, S., 1993, Clinical features associated with the A → G transition at nucleotide 8344 of mtDNA ("MERRF" mutation), *Neurology* **43**:1200–1206.

Silvestri, G., Santorelli, F. M., Shanske, S., Whitley, C. B., Schimmenti, L. A., Smith, S. A., and DiMauro, S., 1994, A new mtDNA mutation in the tRNA$^{Leu(UUR)}$ gene associated with maternally inherited cardiomyopathy, *Hum. Mutat.* **3**:37–43.

Silvestri, G., Servidei, S., Rana, M., Ricci, E., Spinazzola, A., Paris, E., and Tonali, P., 1996, A novel mitochondrial DNA point mutation in the tRNA (Ile) gene is associated with progressive external ophthalmoplegia, *Biochem. Biophys. Res. Commun.* **220**:623–627.

Silvestri, G., Rana, M., DiMuzio, A., Uncini, A., Tonali, P., and Servidei, S., 1998, A late-onset myopathy is associated with a novel mitochondrial DNA (mtDNA) point mutation in the tRNA^{Trp} gene, *Neuromusc. Disord.* **8**:291–295.

Sparaco, M., Schon, E. A., DiMauro, S., and Bonilla, E., 1994, Myoclonus epilepsy with ragged-red fibers (MERRF): An immunohistochemical study of the brain, *Brain Pathol.* **5**:125–133.

Stone, E. M., Newman, N. J., Miller, N. R., Johns, D. R., Lott, M. T., and Wallace, D. C., 1992, Visual recovery in patients with Leber's hereditary optic neuropathy and the 11778 mutation, *J. Clin. Neuroophthalmol.* **12**:10–14.

Strom, T. M., Hortnagel, K., Hofmann, S., Gekeler, F., Scharfe, C., Rabl, W., Gerbitz, K. D., and Meitinger, T., 1998, Diabetes insipidus, diabetes mellitus, optic atrophy, and deafness (DIDMOAD) caused by mutations in a novel gene (wolframin) coding for a predicted transmembrane protein, *Hum. Mol. Genet.* **7**:2021–2028.

Suomalainen, A., Kaukonen, J., Amati, P., Timonen, R., Haltia, M., Weissenbach, J., Zeviani, M., Somer, H., and Peltonen, L., 1995, An autosomal locus predisposing to deletions of mitochondrial DNA, *Nature Genet.* **9**:146–151.

Superti-Furga, A., Schoenle, E., Tuchschmid, P., Caduff, R., Sabato, V., deMattia, D., Gitzelmann, R., and Steinmann, B., 1993, Pearson bone marrow-pancreas syndrome with insulin-dependent diabetes, progressive renal tubulopathy, organic aciduria, and elevated fetal haemoglobin caused by deletion and duplication of mitochondrial DNA, *Eur. J. Pediatr.* **152**:44–50.

Suzuki, Y., Suzuki, S., Hinokio, Y., Chiba, M., Atsumi, Y., Hosokawa, K., Shimada, A., Asahina, T., and Matsuoka, K., 1997, Diabetes associated with a novel 3264 mitochondrial tRNA (Leu) (UUR) mutation, *Diabetes Care* **20**:1138–1140.

Sweeney, M. G., Bundey, S., Brockington, M., Poulton, J. R., Weiner, J. B., and Harding, A. E., 1993, Mitochondrial myopathy associated with sudden death in young adults and a novel mutation in the mitochondrial DNA transfer RNA$^{Leu(UUR)}$ gene, *Q. J. Med.* **86**:709–713.

Takakubo, F., Cartwright, P., Hoogenraad, N., Thorburn, D. R., Collins, F., Lithgow, T., and Dahl, H.-H. M., 1995, An amino acid substitution in the pyruvate dehydrogenase E1α gene, affecting mitochondrial import of the precursor protein, *Am. J. Hum. Genet.* **57**:772–780.

Taniike, M., Fukushima, H., Yanagihara, I., Tsukamoto, H., Tanaka, J., Fujimura, H., Nagai, T., Sano, T., Yamaoka, K., Inui, K., and Okada, S., 1992, Mitochondrial tRNA^{Ile} mutation in fatal cardiomyopathy, *Biochem. Biophys. Res. Commun.* **186**:47–53.

Taroni, F., Verderio, E., Dworzak, F., Willems, P. J., Cavadini, P., and DiDonato, S., 1993, Identification of a common mutation in the carnitine palmitoyltransferase II gene in familial recurrent myoglobinuria patients, *Nature Genet.* **4**:314–320.

Tatuch, Y., and Robinson, B. H., 1993, The mitochondrial DNA mutation at 8993 associated with NARP slows the rate of ATP synthesis in isolated lymphoblast mitochondria, *Biochem. Biophys. Res. Commun.* **192**:124–128.

Tatuch, Y., Christodoulou, J., Feigenbaum, A., Clarke, J. T. R., Wherret, J., Smith, C., Rudd, N., Petrova-Benedict, R., and Robinson, B. H., 1992, Heteroplasmic mtDNA mutation (T → G) at 8993 can cause Leigh's disease when the percentage of abnormal mtDNA is high, *Am. J. Hum. Genet.* **50**:852–858.

Taylor, R. W., Chinnery, P. F., Haldane, F., Morris, A. A. M., Bindoff, L. A., Wilson, J., and Turnbull, D. M., 1996, MELAS associated with a mutation in the valine transfer RNA gene of mitochondrial DNA, *Ann. Neurol.* **40**:459–462.

Thobois, S., Vighetto, A., Grochowicki, M., Godinot, C., Broussolle, E., and Aimard, G., 1997, Leber "plus" disease: Optic neuropathy, parkinsonian syndrome, and supranuclear ophthalmoplegia (in French), *Rev. Neurol.* **153**:595–598.

Thyagarajan, D., Shanske, S., Vasquez-Memije, M., De Vivo, D., and DiMauro, S., 1995, A novel mitochondrial ATPase 6 point mutation in familial bilateral striatal necrosis, *Ann. Neurol.* **38**:468–472.

Tiranti, V., Chariot, P., Carella, F., Toscano, A., Soliveri, P., Girlanda, P., Carrara, F., Fratta, G. M., Reid, F. M., Mariotti, C., and Zeviani, M., 1995, Maternally inherited hearing loss and myoclonus associated with a novel point mutation in mitochondrial tRNA$^{Ser(UCN)}$ gene, *Hum. Mol. Genet.* **4**:1421–1427.

Torroni, A., Petrozzi, M., D'Urbano, L., Sellitto, D., Zeviani, M., Carrara, F., Carducci, C., Leuzzi, V., Carelli, V., Borboni, P., De Negri, A., and Scozzari, R., 1997, Haplotype and phylogenetic analyses suggest that one European-specific mtDNA background plays a role in the expression of Leber hereditary optic neuropathy by increasing the penetrance of the primary mutations 11778 and 14484, *Am. J. Hum. Genet.* **60**:1107–1121.

Tritschler, H.-J., Andreetta, F., Moraes, C. T., Bonilla, E., Arnaudo, E., Danon, M. J., Glass, S., Zelaya, B. M., Vamos, E., Telerman-Toppet, N., Kadenbach, B., DiMauro, S., and Schon, E. A., 1992, Mitochondrial myopathy of childhood associated with depletion of mitochondrial DNA, *Neurology* **42**:209–217.

Trounce, I., Neill, S., and Wallace, D. C., 1994, Cytoplasmic transfer of the mtDNA nt 8993 T → G (*ATP6*) point mutation associated with Leigh syndrome into mtDNA-less cells demonstrates cosegregation with a decrease in state III respiration and ADP/O ratio, *Proc. Natl. Acad. Sci. USA* **91**:8334–8338.

van den Ouweland, J. M. W., Lemkes, H. H. P. J., Ruitenbeek, W., Sandkuijl, L. A., de Vijlder, M. F., Struyvenberg, P. A. A., van de Kamp, J. J. P., and Maassen, J. A., 1992, Mutation in mitochondrial tRNA$^{\text{Leu(UUR)}}$ gene in a large pedigree with maternally transmitted type II diabetes mellitus and deafness, *Nature Genet.* **1**:368–371.

Vazquez-Memije, M. E., Shanske, S., Santorelli, F. M., Kranz-Eble, P., Davidson, E., De Vivo, D. C., and DiMauro, S., 1996, Comparative biochemical studies in fibroblasts from patients with different forms of Leigh syndrome, *J. Inher. Metab. Dis.* **19**:43–50.

Vergani, L., Martinuzzi, A., Carelli, V., Cortelli, P., Montagna, P., Schievano, G., Carrozzo, R., Angelini, C., and E., L., 1995, MtDNA mutations associated with Leber's hereditary optic neuropathy: Studies on cytoplasmic hybrid (cybrid) cells, *Biochem. Biophys. Res. Commun.* **210**:880–888.

Vissing, J., Salamon, M. B., Arlien-Søborg, P., Nørby, S., Manta, P., DiMauro, S., and Schmalbruch, H., 1998, A new mitochondrial tRNA^{Met} gene mutation in a patient with dystrophic muscle and exercise intolerance, *Neurology* **50**:1875–1878.

Vockley, J., Nagao, M., Parimoo, B., and Tanaka, K., 1992, The variant human isovaleryl-CoA dehydrogenase gene responsible for type II isovaleric acidemia determines an RNA splicing error, leading to the deletion of the entire second coding exon and the production of truncated precursor protein that interacts poorly with mitochondrial import receptors, *J. Biol. Chem.* **267**:2494–2501.

Vu, T. H., Sciacco, M., Tanji, K., Nichter, C., Bonilla, E., Chatkupt, S., Maertens, P., Shanske, S., Mendell, J., Koenigsberger, M. R., Sharer, L., Schon, E. A., DiMauro, S., and De Vivo, D. C., 1998, Clinical manifestations of mitochondrial DNA depletion, *Neurology* **50**:1783–1790.

Vulpe, C., Levinson, B., Whitney, S., Packman, S., and Gitschier, J., 1993, Isolation of a candidate gene for Menkes disease and evidence that it encodes a copper-transporting ATPase, *Nat. Genet.* **3**:7–13.

Wallace, D. C., Singh, G., Lott, M. T., Hodge, J. A., Schurr, T. G., Lezza, A. M. S., Elsas II, L. J., and Nikoskelainen, E. K., 1988, Mitochondrial DNA mutation associated with Leber's hereditary optic neuropathy, *Science* **242**:1427–1430.

Weber, K., Wilson, J. N., Taylor, L., Brierley, E., Johnson, M. A., Turnbull, D. M., and Bindoff, L. A., 1997, A new mtDNA mutation showing accumulation with time and restriction to skeletal muscle, *Am. J. Hum. Genet.* **60**:373–380.

Yoon, K. L., Aprille, J. R., and Ernst, S. G., 1991, Mitochondrial tRNA^{Thr} mutation in fatal infantile respiratory enzyme deficiency, *Biochem. Biophys. Res. Commun.* **176**:1112–1115.

Zeviani, M., Moraes, C. T., DiMauro, S., Nakase, H., Bonilla, E., Schon, E. A., and Rowland, L. P., 1988, Deletions of mitochondrial DNA in Kearns–Sayre syndrome, *Neurology* **38**:1339–1346.

Zeviani, M., Gellera, C., Antozzi, C., Rimoldi, M., Morandi, L., Villani, F., Tiranti, V., and DiDonato, S., 1991, Maternally inherited myopathy and cardiomyopathy: Association with mutation in mitochondrial DNA tRNA$^{\text{Leu(UUR)}}$, *Lancet* **338**:143–147.

Zeviani, M., Muntoni, F., Savarese, N., Serra, G., Tiranti, V., Carrara, F., Mariotti, C., and DiDonato, S., 1993, A MERRF/MELAS overlap syndrome associated with a new point mutation in the mitochondrial tRNA^{Lys} gene, *Eur. J. Hum. Genet.* **1**:80–87.

Zhu, Z., Yao, J., Johns, T., Fu, K., De Bie, I., Macmillan, C., Cuthbert, A. P., Newbold, R. F., Wang, J.-c., Chevrette, M., Brown, G. K., Brown, R. M., and Shoubridge, E. A., 1998, *SURF1*, encoding a factor involved in the biogenesis of cytochrome *c* oxidase, is mutated in Leigh syndrome, *Am. J. Hum. Genet.* **20**:337–343.

Chapter 4

Transmission and Segregation of Mammalian Mitochondrial DNA

Eric A. Shoubridge

1. INTRODUCTION

Major advances in our understanding of the molecular bases for respiratory chain deficiencies have occurred over the last decade (DiMauro *et al.*, 1998; Grossman and Shoubridge, 1996; Larsson and Clayton, 1995; Lightowlers *et al.*, 1997; Shoubridge, 1998). Mutations in mtDNA are an important cause of these disorders, which, though multisystemic, often present with prominent involvement of the heart, skeletal muscle, and the nervous system. In most cases mutant and wild-type mtDNAs are present together in the same individual (mtDNA heteroplasmy). The proportion of mutant mtDNA in an individual determines the severity, and in some cases the nature, of the clinical phenotype. The number of mutant mtDNAs transmitted by a mother to her offspring and their distribution in different tissues are thus major determinants of recurrence risk. These observations have focused renewed attention on the rules governing the transmission of mtDNA mutations between generations and on the segregation of such mutations in different tissues. New reproductive technologies such as intracytoplasmic sperm injection (ICSI) have also forced re-evaluation of the potential contribution of male mtDNA to the zygote. In this chapter, I review the current knowledge on the factors affecting the transmission and segregation of mtDNA, the possibility of leakage of paternal mtDNA, and the implications of this information for genetic counseling.

Eric A. Shoubridge   Montreal Neurological Institute and Department of Human Genetics, McGill University, Montreal, Quebec, Canada H3A 2B4.

Mitochondria in Pathogenesis, edited by Lemasters and Nieminen.
Kluwer Academic/Plenum Publishers, New York, 2001.

2. MITOCHONDRIAL DNA: STRUCTURE AND REPLICATION

Although there is a great variety in the size and content of mtDNA in plants and animals, the mtDNA of all mammals investigated has the same basic structure, a double-stranded circular DNA molecule of ~ 16.5 kb. The two stands are referred to a heavy (H) and light (L), reflecting their behaviour in CsCl density gradients. Mammalian mtDNA codes for 13 polypeptides, all of which are subunits of the respiratory chain enzyme complexes, and the 22 tRNAs and the two rRNAs necessary for their translation. The genome is exceeding compact, containing no introns and only one noncoding region of ~ 1 kb, the D-loop. This is a triple-stranded region that contains one of the replication origins (O_H) and the H- and L-strand transcription promoters. The genome copy number of mtDNA is in the range of 10^3–10^4 in most cells, organized as ~ 5 copies (range 2–10) per organelle (Nass, 1969; Satoh and Kuroiwa, 1991). Gametes are a notable exception to this generalization: mature oocytes have approximately 10^5 mtDNAs (Piko and Taylor, 1987) and sperm about 10^2 (Hecht *et al.*, 1984).

Replication of mtDNA occurs asynchronously from two origins: one for the H-strand in the D-loop and a one for the L-strand, about two-thirds of the way around the molecule (Clayton, 1982). Replication is catalyzed by a distinct polymerase, the γ-DNA polmerase, and leading-strand synthesis at O_H is primed by a short piece of RNA generated by transcription from the L-strand promoter. Thus, replication of the genome is linked in some way to transcription of mtDNA genes. Transcription initiation of mtDNA (and therefore priming of mtDNA replication) requires the presence of the only known mitochondrial transcription factor, Tfam (formerly referred to as mtTFA) (Larsson *et al.*, 1998; Shadel and Clayton, 1997). In contrast to nuclear DNA, mtDNA replication is not linked to the cell cycle (Clayton, 1982). This "relaxed" form of replication allows some templates to replicate more than once during the cell cycle, others not at all. Thus, though mtDNA copy number is tightly regulated in a cell- and tissue-specific manner (by unknown mechanisms), sequence variants can segregate during mitosis due to unequal replication from templates in the parent cell and to random sampling of mtDNAs at cytokinesis (Birky, 1994). In the absence of selection, the segregation rate of mtDNA sequence variants depends primarily on two parameters: mtDNA copy number and the number of mitotic divisions.

3. mtDNA TRANSMISSION: THE BOTTLENECK HYPOTHESIS

In mammals mtDNA is strictly maternally inherited (at least in intraspecific matings) (Giles *et al.*, 1980; Kaneda *et al.*, 1995), so new sequence variants are transmitted along maternal lineages without the benefit of recombination with male mtDNA. Most individuals have a single mtDNA sequence variant in all of their cells (mtDNA homo-plasmy). There is, however, a great deal of sequence variability among individuals. In human populations, individuals typically differ by 0.3% or about 50 nucleotides in the mtDNA sequence. The rarity of mtDNA heteroplasmy and the high degree of population polymorphism suggest that new mtDNA sequence variants are rapidly segregated in maternal lineages. This is paradoxical given the high genome copy number in mature oocytes ($\sim 10^5$ copies of mtDNA) and the relatively small number of cell divisions in oogenesis. The accumulation of new germline mutations ought to produce significant sequence heterogeneity in the mtDNA population in oocytes, and one possible outcome would be a large number different sequence variants segregating in the germline at low frequency. The fact that this is not observed suggests that the effective number of mtDNA segregating units is much less than the copy number in the mature oocyte.

These considerations led Hauswirth and Laipis (1982; Laipis *et al.*, 1988) to hypothesize the existence of a *genetic bottleneck* for mtDNA in oogenesis or early embryogenesis. This concept was based largely on observations of a heteroplasmic D-loop sequence variant in Holstein cows, which segregates rapidly in a few generations (Ashley *et al.*, 1989; Laipis *et al.*, 1988). A similar study of the same sequence variant demonstrated complete allele switching in a single generation in 40% of more than 32 independent Holstein lineages (Koehler *et al.*, 1991), suggesting that the number of segregating units might be as small as one! Rapid segregation of mtDNA sequence variants has also been observed in human pedigrees segregating pathogenic or silent mtDNA mutations (Howell *et al.*, 1991, 1992, 1994; Larsson *et al.*, 1992; Santorelli *et al.*, 1994; Uziel *et al.*, 1997; Zhu *et al.*, 1992). In general, the segregation rate is slower than that of the D-loop sequence variant in the bovine studies, and it is quite variable between pedigrees.

Two different hypotheses were proposed to explain the bottleneck for mtDNA transmission. The first suggested that a limited number of mtDNA templates were used during the 100-fold amplification of mtDNA copy number that accompanies oocyte maturation, reducing the effective number of mtDNAs that contribute to the next generation (Hauswirth and Laipis, 1982). The second suggested that mtDNAs in the inner cell-mass cells of the blastocyst were a nonrandom sample of mtDNA in the mature oocyte (Ashley *et al.*, 1989; Laipis *et al.*, 1988; Olivo *et al.*, 1983).

Much of the controversy about the nature of the mtDNA bottleneck revolves around the definition of the term "bottleneck". Some have interpreted it as a process, others as a physical restriction in the number of mitochondrial genomes. In population genetics the term refers to a period when the number of breeding individuals in a population (and hence the amount of genetic variation) is reduced relative to pre-existing or current numbers. A population that passes through such a bottleneck will obviously have a much-reduced genetic variation, and sometimes high frequencies of rare alleles, having expanded from a limited number of founder genotypes. Adhering to the conventional usage of the term, a bottleneck for mtDNA transmission would represent a stage in oogenesis or early development where mtDNA molecules, which serve as templates for all future molecules, are reduced to a small number. It is important to appreciate that the population in this case refers to the mtDNA population in an individual oocyte.

There are no complete descriptions of the changes in the number of mitochondria or in mtDNA copy number during oogenesis and early embryogenesis in a single mammalian species. Details from studies of a number of different species, however allow for a reasonably coherent picture. Ultrastructural analyses of human primordial germ cells indicate that they contain about 10 mitochondria; this increases to about 200 in oogonia and to several thousand in the oocyte in a primordial follicle (Jansen and de Boer, 1998). Measurements of mtDNA copy number in immature mouse oocytes suggest that there are approximately 10^3 copies (Piko and Taylor, 1987), increasing by 100-fold to approximately 10^5 in the mature oocyte, organized as ~1 copy per organelle. In the cow this number is ~2×10^5 mtDNAs, but the oocytes are proportionately larger (Michaels *et al.*, 1982). The mtDNA copy number is unknown in either primordial germ cells or in oogonia. It has been suggested that oogenesis is accompanied by a reduction in copy number per organelle (Laipis *et al.*, 1988), implying that there are multiple copies per organelle in the oocyte precursors, but actually, there is very little evidence for this proposition. Others have speculated that the genome copy number may be 1 per organelle at all stages of oogenesis (Jansen and de Boer, 1998). Even if one assumes a somatic-cell mtDNA copy number per organelle in the precursor population (~5 per organelle), it is clear that there is more than a 1000-fold increase in mtDNA copy number during oogenesis, from perhaps as few as 50 in the primordial germ cells to 100,000 in the pre-ovulatory oocyte. After fertilization, cell division of the early embryo proceeds without mtDNA replication (Piko and Taylor, 1987). By the time the inner cell-mass cells are set aside in the blastocyst (~100 cells), mtDNA copy number is reduced to about 10^3, within the same range as in most somatic cells. Replication of mtDNA is thought to resume shortly after implantation, but the control of this process is not well understood.

4. TESTING THE BOTTLENECK HYPOTHESIS

The morphological evidence clearly indicates a physical bottleneck in the number of mitochondria in the primordial germ cells. This requires about a 100-fold reduction in organelle number from the inner cell-mass cells of the blastocyst. How this might be mediated is unknown. The question remains as to whether this is sufficient to account for the observed rate of segregation of mtDNA sequence variants, or whether there is a further restriction in the number of founder mtDNAs at a later stage of oogenesis, as suggested by Hauswirth and Laipis (1982).

This has recently been investigated in heteroplasmic mice constructed from two different *Mus musculus domesticus* strains, NZB and BALB/c (Jenuth *et al.*, 1996). The mtDNAs of these two strains differ at about 106 sites, predicting 15 amino acid substitutions, most of which occur at nonconserved positions in evolution. The presence of a bottleneck at any point in oogenesis should result in a large increase in genetic variance at the next stage of development. Single cell polymerase chain reaction (PCR) of mature and primary oocytes and primordial germ cells was used to investigate the evolution of genetic variance for the two mtDNA genotypes during oogenesis. Little intercellular variation was observed in the degree of mtDNA heteroplasmy in the primordial germ cell (PGC) population, suggesting that they were a more or less random sample of the mtDNA population in the zygote. Analysis of primary oocytes, however, revealed that the rapid segregation of the two mtDNA sequences in the offspring of heteroplasmic females had already occurred by the time this population was differentiated in fetal life. Segregation of mtDNA between the primary and mature oocytes was quantitatively unimportant, as was segregation between the mature oocytes and neonates. These results indicate that most of the mtDNA segregation occurs during mitosis in the migratory primordial germ cells and oogonia, supporting the concept of a single bottleneck for the transmission of mtDNA. Studies of D-loop length variants in humans are consistent with the mouse studies, showing that segregation has occurred by the time oocytes are mature (Marchington *et al.*, 1997, 1998).

This model of mtDNA segregation by random genetic drift during mitosis appears to provide an adequate explanation for the rapid amplification and fixation of new mtDNA mutations in the female germline. The model predicts that germline mutations will be most important to future generations if they arise in the oocyte precursor cells. Such mutations, by definition rare in the population of primordial germ cells or oogonia, can increase in frequency in individual oocytes due to the relatively small number of founder mitochondria in oocyte precursors. Mutations arising later in oogenesis, when mtDNA copy number is high (Chen *et al.*, 1995) will only rarely be transmitted.

Using a population genetic model, the effective number of mtDNA segregating units is estimated at about 200 in the mouse (Jenuth *et al.*, 1996). This is in broad agreement with estimates obtained from human pedigrees segregating pathogenic point mutations, suggesting that the process may be similar in humans. If anything, segregation in humans would probably be slightly faster (assuming a similar number of primordial germ cells in both species), because humans generate a larger number of primary oocytes.

A notable exception concerns a point mutation at position 8993 in the gene coding for the ATP6 subunit of ATPsynthase. This mutation, associated with the NARP syndrome (neuropathy, ataxia, retinitis pigmentosa), segregates much more rapidly than other known pathogenic mutations, and the segregation favours the mutant allele (Degoul *et al.*, 1997; Santorelli *et al.*, 1994; Seller *et al.*, 1997; Tulinius *et al.*, 1995; Uziel *et al.*, 1997). Although it is difficult to completely factor out ascertainment bias in the case of pathogenic mutations, results from numerous NARP pedigrees contrast sharply with other pedigrees segregating different point mutations, in which segregation to either the mutant or wild-type allele appears equally likely. The molecular basis for the behaviour of the NARP mutant is unclear, but it could involve positive selection for the mutant allele during oogenesis or early embryogenesis.

5. MECHANISMS OF PATERNAL mtDNA EXCLUSION

The basis for the strict maternal transmission of mammalian mtDNA has been a subject of considerable speculation. The fact that mammals are anisogamous and that there is an approximately 1000-fold difference in mtDNA copy number between sperm and ovum (10^2 vs 10^5) suggests that simple dilution of male mtDNA might be sufficient to ensure that the vast majority of mtDNAs are inherited maternally. Indeed, the down-regulation of mtDNA copy number to about 100 (organized at about 1 per organelle as in the oocyte) during spermatogenesis has been suggested as a mechanism to prevent male transmission of mtDNA (Hecht *et al.*, 1984). This is paralleled by a decrease in Tfam, the transcription factor necessary for priming mtDNA replication (Larsson *et al.*, 1997); however, it remains to be determined if this is cause or effect. Ultrastructural studies of the fate of hamster sperm post-fertilization have shown so-called microvesicular bodies fusing with mitochondria and leading to their destruction (Hiraoka and Hirao, 1988). It has been suggested that sperm mitochondria themselves may not be competent to survive in the zygote. Sperm express a very large number of specific isozymes for important metabolic enzymes in glycolysis (e.g., LDH, GAPDH) and for important factors in the mitochondrial respiratory chain, such as cytochrome c (Hecht, 1995), and it is possible that metabolic differences select against survival of sperm mitochondria. There is evidence that the efficiency of repopulation of rho^0 cells with mitochondria (and therefore mtDNA) derived from sperm is dramatically less than the number of cells containing sperm mitochondria after fusion, and less than that observed when somatic cell mitochondria are used as donors (Manfredi et al., 1997). The basis for this phenomenon remains unclear, however, and it may not be relevant to exclusion mechanisms of sperm mitochondria in the zygote.

Early studies suggested that there was a barrier to transmission of mitochondria from a non-oocyte source (Ebert et al., 1989). Recent experiments (Pinkert et al., 1997) showing the survival of hepatocyte-derived mtDNAs from *Mus spretus* mice in *M.m. domesticus* embryos demonstrate, however, that non-oocyte mitochondria can survive in early embryos. In any case, the genetic experiments described below suggest that none of these factors are critical, because male transmission of mtDNA can in fact be observed under some conditions.

6. MALE TRANSMISSION OF mtDNA IN INTERSPECIFIC MATINGS

The first indication that paternal mtDNA might be leaked to the next generation came from a study of interspecific crosses in mice involving *M.m. domesticus* and *M. spretus* (Gyllensten et al., 1991). Assuming that any potential male contribution to the mtDNA pool in the offspring of the next generation would be small, female F_1 hybrids (*domesticus* × *spretus*) were backcrossed with either *domesticus* or *spretus* males for up to 26 generations in an effort to increase the proportion of paternal mtDNA. *M. spretus* mtDNA could be detected by PCR in all tissues examined in these offspring, at frequencies of 10^{-3} to 10^{-4}. It was concluded that a small amount of paternal mtDNA was added each generation, and that it accumulated to detectable levels after many generations. This explanation is clearly untenable in light of what we now know of the mtDNA bottleneck in the female germline, however. No known mechanism exists for the slow and progressive accumulation of mtDNA in the germline, and the most likely fate of a rare sequence variant would be loss due to genetic drift.

Kaneda and co-workers (1995) extended these studies by examining the fate of male mtDNA in intraspecific and interspecific crosses in early embryos. They showed that male mtDNA was undetectable in the late pronuclear stage in intraspecific crosses, coinciding with the loss of Rhodamine 123 fluorescence, a marker used to assess the mitochondrial membrane potential of sperm. In contrast, *spretus* mtDNA could be detected in the majority of two-cell embryos and in about 50% of neonates, born after implantation of the *in vitro* fertilized embryos into pseudopregnant females. This suggests that sperm from a closely related species can escape the normal surveillance mechanisms that ensure uniparental transmission of mtDNA within a species. Elimination of paternal mtDNA was also observed in embryos carrying *spretus* mtDNA on a *domesticus* nuclear background, suggesting that the molecular target(s) for elimination of paternal mtDNA in intraspecific crosses is a nuclear-encoded molecule(s) associated with the sperm midpiece. Recent experiments (Cummins et al., 1997) in which sperm were injected directly into the oocyte demonstrated that male mitochondria, tagged with a Mitotracker dye, were present in some embryos up to the 8-cell stage. To my knowledge, no ultrastructural investigations, such as those that demonstrated fusion of male mitochondria with multivesicular bodies (Hiraoka and Hirao, 1988), have been carried out on interspecific mammalian crosses to look for morphological correlates of male mitochondrial survival.

The group of Shitara (1998) studied transmission of paternal mtDNA in the somatic tissues and female germline of F_1 hybrid (*domesticus* × *spretus*) mice and in the first backcross generation (N_2) between F_1 hybrid females and *M. Spretus* males. They demonstrated that the presence of paternal mtDNA is random in different tissues of the F_1 animals, with about 40% having some evidence of *spretus* mtDNA. Paternal mtDNA was rare in the ovaries of F_1 hybrid females and undetectable in a large sample of oocytes obtained by superovulation. No paternal mtDNA was detected in any of the N_2 backcross animals, tested by examining embryos obtained by *in vitro* fertilization, indicating a possible barrier to transmission even in F_1 hybrid animals. These results are clearly at odds with observations of the Gyllensten group (1991), and it is difficult to reconcile the two studies. One possibility is that stable heteroplasmy for *spretus* mtDNA was established in the F_1 hybrid animals generated in the Gyllensten study.

The fact that no leakage of male mtDNA was observed in the N_2 backcross generation suggests that the genetic factors responsible act as either a dominant or co-dominant maternal trait. We have carried out similar experiments (in collaboration with Dr. Fred Biddle, University of Calgary) with different results. We have seen evidence of some male leakage of *spretus* mtDNA in N_2 backcross mice (unpublished observations). The pattern of male mtDNA leakage (random tissue distribution, low frequency) does not seem remarkably different from that of the F_1 animals, suggesting that male leakage prevention is inherited as a recessive trait. It is not clear why our results differ from those of Shitara and co-workers (1998), but it may have to do with the sensitivity of detection of leaked male mtDNA.

7. SEGREGATION OF mtDNA DURING EMBRYOGENESIS AND FETAL LIFE

Early studies by Piko and colleagues established that development of the preimplantation embryo was unaffected by treatment with chloramphenicol, a compound that inhibits mitochondrial translation (Piko and Chase, 1973). Transcription of mtDNA is active, however, from the two-cell stage (Taylor and Piko, 1995). Although mtDNA expression is probably required during fetal life, and there is abundant opportunity for mitotic segregation of mutant mtDNA during this period, studies of fetal tissues in individuals carrying pathogenic mtDNA mutations show little tissue-to-tissue variation in the proprtion of mutant and wild-type mtDNA (Harding *et al.*, 1992; Matthews *et al.*, 1994). This suggests that selection for respiratory chain function is likely not strong during fetal life, and the mutant mtDNA proportion at birth is probably largely determined by the proportion in the oocyte. Similar observations have been made in heteroplasmic mice that have little variation in heteroplasmy among tissues at birth, but show a strong, tissue-specific selection for alternate mtDNA genotypes with age (Jenuth *et al.*, 1997). An apparent exception to the lack of selection for mtDNA sequence variants in embryogenesis was observed in karyoplast-reconstructed mice (Meirelles and Smith, 1998); this may represent a highly artificial situation, however. Additional evidence for the proposition that selection against pathogenic mutations is weak until postnatal life comes from pedigree

analysis of patients with segregating mtDNA mutations. Although rapid shifts in proportion of mutant mtDNA can occur between generations, shifts appears just as likely to occur
in the direction of wild-type mtDNA as in that of mutant mtDNA in most pedigrees, and in
some (NARP pedigrees) the shift seems to favour the mutant.

8. SEGREGATION OF mtDNA AFTER BIRTH

In contrast to the situation during embryonic and fetal development, mitotic
segregation of pathogenic mtDNA sequence variants occurs throughout postnatal life.
The load of mtDNA mutations inherited at birth undoubtedly plays an important role in the
clinical phenotype of patients with pathogenic mtDNA mutations, but tissue-specific
segregation can modify the proportions of mutant and wild-type mtDNA significantly, and
this is often associated with a worsening clinical course. There is good evidence for
increases in the proportions of some pathogenic mtDNA mutations, including large-scale
deletions (Larsson *et al.*, 1990) and tRNA point mutations (Poulton and Morten, 1993;
Weber *et al.*, 1997) with age in the skeletal muscle of patients with mitochondrial
encephalomyopathies. Mutant mtDNAs are often undetectable in actively dividing cells
in the same individual. This had led to the suggestion that there may be feedback
mechanisms that promote replication of mutant mtDNAs in postmitotic cells, reflecting a
futile attempt to restore oxidative phophorylation function, and selection against cells with
a growth disadvantage due to the presence of the mutants in actively mitotic cells. Though
this may be the case for some mtDNA mutations, it is not a completely general
phenomenon. There is, for instance, no evidence for an age-related increase in the
proportion of the A8344G tRNA^{Lys} mutation associated with MERRF (myoclonus epilepsy
with ragged-red fibres) in skeletal muscle, nor is there evidence for selection against blood
cells containing the mutation (Boulet *et al.*, 1992; Larsson *et al.*, 1992; Mita *et al.*, 1998).
There is also one reported incidence of segregation toward wild-type mtDNA in a patient
with the A3243G mutation in tRNA^{Leu} associated with MELAS (mitochondrial encephalomyopathy lactic acidosis and stroke-like episodes) (Kawakami *et al.*, 1994). The
different segregation behaviors of pathogenic mtDNA mutations that produce similar
biochemical phenotypes remain a mystery.

The most dramatic segregation patterns are observed in sporadic cases, where mutant
mtDNAs are often detectable only in postmitotic cells (Fu *et al.*, 1996; Holt *et al.*, 1988;
Moraes *et al.*, 1993). As the initial frequency of the mutant was likely low in such cases, its
absence in dividing cells could simply be the result of random genetic drift (Fu *et al.*,
1996).

Studies of heteroplasmic mice constructed by karyoplast fusion to enucleated zygotes
(NZB and C57BL/6 strains) have shown marked differences in levels of heteroplasmy
among tissues, suggesting the operation of stringent replicative segregation (Meirelles and
Smith, 1997). Similar studies of mice constructed by cytoplast fusion of NZB and BALB
embryos have shown a different pattern of segregation of mtDNA sequence variants.
Measurable segregation of the two genotypes was observed in only four tissues: liver,
kidney, blood, and spleen. The NZB genotype was always selected in the liver and kidney,
whereas the BALB genotype was always selected in the spleen and blood of the same
animals (Jenuth *et al.*, 1997). Although the molecular basis for these observations is

unknown, the data suggest some functional interaction between tissue-specific nuclear genes and polymorphic mtDNA sequence variants.

9. GENETIC COUNSELING

The issues discussed in this chapter have important practical implications for clinical mitochondrial genetics. Genetic counseling of women who are carrying mtDNA mutations remains a problem. If, as the data suggest, segregation of mtDNA sequence variants can be treated as a binomial sampling problem, it should be possible to calculate recurrence risks if the effective number of segregating units was known. Although it is unlikely that a single sample of a few mtDNAs reflects the biological reality of this process, the problem can be treated this way for practical purpose of determining the relative transmission risks (Marchington *et al.*, 1998). Whether this will ever be truly useful remains to be seen, as confidence limits on the risk estimates may be too large to be of real value. It is obvious from the casual examination of human pedigrees segregating mtDNA point mutations that, though the risk of having an affected child increases with the level of heteroplasmy in the mother, is not trivial even at low heteroplasmy levels.

The prospects for preimplantation or prenatal diagnosis are considerably brighter. Because the level of neonate heteroplasmy is determined by the relative proportions of wild-type and mutant mtDNA in the oocyte, the mtDNA genotype of a blastomere sampled from an 8-cell embryo (and containing more than 10^4 copies of mtDNA) should be a good predictor of ultimate mtDNA genotype. The fact that little if any segregation of mtDNA sequence variants occurs during fetal life, even in those proven to be pathogenic, suggests that sampling any fetal tissue (chorionic villus, amniocytes) ought to provide a reliable indication of the overall fetal mutation load. A word of caution here, however— these assertions are based on very limited experience. Although it is comforting that polymorphic sequence variants in mice seem to behave much like pathogenic sequence variants in humans, it is unknown whether all pathogenic mtDNA mutations will obey the same rules.

10. SUMMARY AND FUTURE PERSPECTIVES

The concept of a genetic bottleneck for the transmission of mtDNA is firmly established. The bottleneck is caused by a reduction in the number of mitochondria and mtDNAs in the primordial germ cell population. How this is accomplished remains to be determined. Both relaxed replication for mtDNA and random sampling of a small number of mtDNAs at cytokinesis in oocyte precursor cells (migratory primordial germ cells, oogonia) contribute to the rapid nature of the segregation process. The effective number of segregating units of mtDNA is nearly 3 orders of magnitude less than the copy number in a mature oocyte.

There is no evidence for significant segregation at other stages of oogenesis, embryogenesis, or fetal development, suggesting that there is no stringent selection for respiratory chain function until the postnatal period. Positive selection for pathogenic mtDNAs is common, but not universal, in postmitotic cells after birth. The process occurs

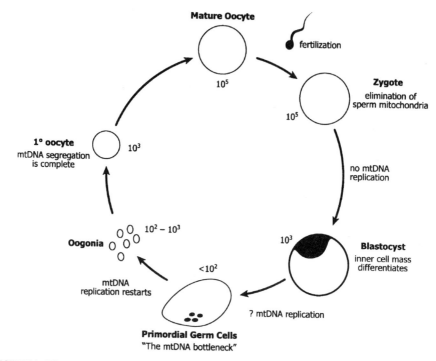

FIGURE 1. Diagrammatic representation of changes in mtDNA copy number during development of the female germline. An estimate of the number of mtDNA is indicated at each stage of development inside the circle. The genetic bottleneck for mtDNA transmission occurs in the primordial germ cells, which contain about 10 mitochondria and probably fewer than 100 copies of mtDNA. Development to the mature oocyte involves at least a 1000-fold increase in mtDNA copy number, but segregation of mtDNA sequence variants is essentially complete by the time the primary oocytes are differentiated in fetal life.

slowly over many years or decades, explaining the delayed onset and progressive course of many of these diseases. The mechanism(s) of selection appear to be mutation-specific, and is in any case unclear. There is an urgent need to develop animal models of these diseases to address these questions, but lack of a method to transform mammalian mtDNA remains a major stumbling block.

There is no strong evidence that male mtDNA can ever be transmitted to the next generation within a mammalian species. Though further work needs to be carried out, it appears from studies on mice that intracellular injection of sperm into the ovum does not alter the fate of male mitochondria, which are still actively eliminated from the embryo before the blastocyst stage. The molecular target(s) involved in this surveillance mechanism is nuclear encoded and appears to be associated with the sperm midpiece. In principle it should be possible to map the genetic elements involved in the exclusion of paternal mtDNA in these interspecific backcrosses.

Progress has been made in the clinical practice of mitochondrial genetics, and though experience is limited, there is a good indication that prenatal and preimplantation diagnosis will be useful and that it may be possible to provide some indication of recurrence risk. Until we can develop animal models, there will likely be uncertainty about the extent of mosaicism in the oocyte population and the segregation behavior of different pathogenic mutations.

ACKNOWLEDGEMENT. Work in the author's laboratory is supported by grants from the Medical Research Council of Canada, the Muscular Dystrophy Association (Canada and the USA), the March of Dimes, and the Hospital for Sick Children (Toronto) Foundation. Eric Shoubridge is an MNI Killam Scholar.

REFERENCES

Ashley, M. V., Laipis, P. J., and Hauswirth, W. W., 1989, Rapid segregation of heteroplasmic bovine mitochondria, *Nucleic Acids Res.* **17**:7325–7331.

Birky, C. W., Jr., 1994, Relaxed and stringent genomes: Why cytoplasmic genes don't obey Mendel's laws, *J. Heredity* **85**:355–365.

Boulet, L., Karpati, G., and Shoubridge, E. A., 1992, Distribution and threshold expression of the tRNA(Lys) mutation in skeletal muscle of patients with myoclonic epilepsy and ragged-red fibers (MERRF), *Am. J. Hum. Genet.* **51**:1187–1200.

Chen, X., Prosser, R., Simonetti, S., Sadlock, J., Jagiello, G., and Schon, E. A., 1995, Rearranged mitochondrial genomes are present in human oocytes [see comments], *Am. J. Hum. Genet.* **57**:239–247.

Clayton, D. A., 1982, Replication of animal mitochondrial DNA, *Cell* **28**:693–705.

Cummins, J. M., Wakayama, T., and Yanagimachi, R., 1997, Fate of microinjected sperm components in the mouse oocyte and embryo, *Zygote* **5**:301–308.

Degoul, F., Francois, D., Diry, M., Ponsot, G., Desguerre, I., Heron, B., Marsac, C., and Moutard, M. L., 1997, A near homoplasmic T8993G mtDNA mutation in a patient with atypic Leigh syndrome not present in the mother's tissues, *J. Inherit. Metab. Dis.* **20**:49–53.

DiMauro, S., Bonilla, E., Davidson, M., Hirano, M., and Schon, E. A., 1998, Mitochondria in neuromuscular disorders, *Biochim. Biophys. Acta* **1366**:199–210.

Ebert, K. M., Alcivar, A., Liem, H., Goggins, R., and Hecht, N. B., 1989, Mouse zygotes injected with mitochondria develop normally but the exogenous mitochondria are not detectable in the progeny, *Mol. Reprod. Dev.* **1**:156–163.

Fu, K., Hartlen, R., Johns, T., Genge, A., Karpati, G., and Shoubridge, E. A., 1996, A novel heteroplasmic tRNAleu (CUN) mtDNA point mutation in a sporadic patient with mitochondrial encephalomyopathy segregates rapidly in skeletal muscle and suggests an approach to therapy, *Hum. Mol. Genet.* **5**:1835–1840.

Giles, R. E., Blanc, H., Cann, H. M., and Wallace, D. C., 1980, Maternal inheritance of human mitochondrial DNA, *Proc. Natl. Acad. Sci. USA* **77**:6715–6719.

Grossman, L. I., and Shoubridge, E. A., 1996, Mitochondrial genetics and human disease, *BioEssays* **18**:983–991.

Gyllensten, U., Wharton, D., Josefsson, A., and Wilson, A. C., 1991, Paternal inheritance of mitochondrial DNA in mice [see comments], *Nature* **352**:255–257.

Harding, A. E., Holt, I. J., Sweeney, M. G., Brockington, M., and Davis, M. B., 1992, Prenatal diagnosis of mitochondrial DNA 8993 T → G disease, *Am. J. Hum. Genet.* **50**:629–633.

Hauswirth, W. W., and Laipis, P. J., 1982, Mitochondrial DNA polymorphism in a maternal lineage of Holstein cows, *Proc. Natl. Acad. Sci. USA* **79**:4686–4690.

Hecht, N. B., 1995, The making of a spermatozoon: A molecular perspective, *Dev. Genet.* **16**:95–103.

Hecht, N. B., Liem, H., Kleene, K. C., Distel, R. J., and Ho, S. M., 1984, Maternal inheritance of the mouse mitochondrial genome is not mediated by a loss or gross alteration of the paternal mitochondrial DNA or by methylation of the oocyte mitochondrial DNA, *Dev. Biol.* **102**:452–461.

Hiraoka, J., and Hirao, Y., 1988, Fate of sperm tail components after incorporation into the hamster egg, *Gamete Res.* **19**:369–380.

Holt, I. J., Harding, A. E., and Morgan-Hughes, J. A., 1988, Deletions of muscle mitochondrial DNA in patients with mitochondrial myopathies, *Nature* **331**:717–719.

Howell, N., Bindoff, L. A., McCullough, D.A., Kubacka, I., Poulton, J., Mackey, D., Taylor, L., and Turnbull, D. M., 1991, Leber hereditary optic neuropathy: Identification of the same mitochondrial ND1 mutation in six pedigrees, *Am. J. Hum. Genet.* **49**:939–950.

Howell, N., Halvorson, S., Kubacka, I., McCullough, D. A., Bindoff, L. A., and Turnbull, D. M., 1992, Mitochondrial gene segregation in mammals: Is the bottleneck always narrow? *Hum. Genet.* **90**:117–120.

Howell, N., Xu, M., Halvorson, S., Bodis-Wollner, I., and Sherman, J., 1994, A heteroplasmic LHON family: Tissue distribution and transmission of the 11778 mutation [letter], *Am. J. Hum. Genet.* **55**:203–206.

Jansen, R. P., and de Boer, K., 1998, The bottleneck: Mitochondrial imperatives in oogenesis and ovarian follicular fate, *Mol. Cell. Endocrinol.* **145**:81–88.

Jenuth, J. P., Peterson, A. C., Fu, K., and Shoubridge, E. A., 1996, Random genetic drift in the female germline explains the rapid segregation of mammalian mitochondrial DNA [see comments], *Nat. Genet.* **14**:146–151.

Jenuth, J. P., Peterson, A. C., and Shoubridge, E. A., 1997, Tissue-specific selection for different mtDNA genotypes in heteroplasmic mice, *Nat. Genet.* **16**:93–95.

Kaneda, H., Hayashi, J., Takahama, S., Taya, C., Lindahl, K. F., and Yonekawa, H., 1995, Elimination of paternal mitochondrial DNA in intraspecific crosses during early mouse embryogenesis, *Proc. Natl. Acad. Sci. USA* **92**:4542–4546.

Kawakami, Y., Sakuta, R., Hashimoto, K., Fujino, O., Fujita, T., Hida, M., Horai, S., Goto, Y., and Nonaka, I., 1994, Mitochondrial myopathy with progressive decrease in mitochondrial tRNA(Leu)(UUR) mutant genomes, *Ann. Neurol.* **35**:370–373.

Koehler, C. M., Lindberg, G. L., Brown, D. R., Beitz, D. C., Freeman, A. E., Mayfield, J. E., and Myers, A. M., 1991, Replacement of bovine mitochondrial DNA by a sequence variant within one generation, *Genetics* **129**:247–255.

Laipis, P. J., Van de Walle, M. J., and Hauswirth, W. W., 1988, Unequal partitioning of bovine mitochondrial genotypes among siblings, *Proc. Natl. Acad. Sci. USA* **85**:8107–8110.

Larsson, N. G., and Clayton, D. A., 1995, Molecular genetic aspects of human mitochondrial disorders, *Annu. Rev. Genet.* **29**:151–178.

Larsson, N. G., Holme, E., Kristiansson, B., Oldfors, A., and Tulinius, M., 1990, Progressive increase of the mutated mitochondrial DNA fraction in Kearns–Sayre syndrome, *Pediatr. Res.* **28**:131–136.

Larsson, N. G., Tulinius, M. H., Holme, E., Oldfors, A., Andersen, O., Wahlstrom, J., and Aasly, J., 1992, Segregation and manifestations of the mtDNA tRNA(Lys) A → G(8344) mutation of myoclonus epilepsy and ragged-red fibers (MERRF) syndrome, *Am. J. Hum. Genet.* **51**:1201–1212.

Larsson, N. G., Oldfors, A., Garman, J. D., Barsh, G. S., and Clayton, D. A., 1997, Down-regulation of mitochondrial transcription factor A during spermatogenesis in humans, *Hum. Mol. Genet.* **6**:185–191.

Larsson, N. G., Wang, J., Wilhelmsson, H., Oldfors, A., Rustin, P., Lewandoski, M., Barsh, G. S., and Clayton, D. A., 1998, Mitochondrial transcription factor A is necessary for mtDNA maintenance and embryogenesis in mice, *Nat. Genet.* **18**:231–236.

Lightowlers, R. N., Chinnery, P. F., Turnbull, D. M., and Howell, N., 1997, Mammalian mitochondrial genetics: Heredity, heteroplasmy, and disease, *Trends Genet.* **13**:450–455.

Manfredi, G., Thyagarajan, D., Papadopoulou, L. C., Pallotti, F., and Schon, E. A., 1997, The fate of human sperm-derived mtDNA in somatic cells, *Am. J. Hum. Genet.* **61**:953–960.

Marchington, D. R., Hartshorne, G. M., Barlow, D., and Poulton, J., 1997, Homopolymeric tract heteroplasmy in mtDNA from tissues and single oocytes: Support for a genetic bottleneck, *Am. J. Hum. Genet.* **60**:408–416.

Marchington, D. R., Macaulay, V., Hartshorne, G. M., Barlow, D., and Poulton, J., 1998, Evidence from human oocytes for a genetic bottleneck in an mtDNA disease, *Am. J. Hum. Genet.* **63**:769–775.

Matthews, P. M., Hopkin, J., Brown, R. M., Stephenson, J. B., Hilton-Jones, D., and Brown, G. K., 1994, Comparison of the relative levels of the 3243 (A → G) mtDNA mutation in heteroplasmic adult and fetal tissues, *J. Med. Genet.* **31**:41–44.

Meirelles, F. V., and Smith, L. C., 1997, Mitochondrial genotype segregation in a mouse heteroplasmic lineage produced by embryonic karyoplast transplantation, *Genetics* **145**:445–451.

Meirelles, F. V., and Smith, L. C., 1998, Mitochondrial genotype segregation during preimplantation development in mouse heteroplasmic embryos, *Genetics* **148**:877–883.

Michaels, G. S., Hauswirth, W. W., and Laipis, P. J., 1982, Mitochondrial DNA copy number in bovine oocytes and somatic cells, *Dev. Biol.* **94**:246–251.

Mita, S., Tokunaga, M., Uyama, E., Kumamoto, T., Uekawa, K., and Uchino, M., 1998, Single muscle fiber analysis of myoclonus epilepsy with ragged-red fibers, *Muscle Nerve* **21**:490–497.

Moraes, C. T., Ciacci, F., Bonilla, E., Ionasescu, V., Schon, E. A., and DiMauro, S., 1993, A mitochondrial tRNA anticodon swap associated with a muscle disease, *Nat. Genet.* **4**:284–288.

Nass, M. M., 1969, Mitochondrial DNA: I. Intramitochondrial distribution and structural relations of single- and double-length circular DNA, *J. Mol. Biol.* **42**:521–528.

Olivo, P. D., Van de Walle, M. J., Laipis, P. J., and Hauswirth, W. W., 1983, Nucleotide sequence evidence for rapid genotypic shifts in the bovine mitochondrial DNA D-loop, *Nature* **306**:400–4002.

Piko, L., and Chase, D. G., 1973, Role of the mitochondrial genome during early development in mice: Effects of ethidium bromide and chloramphenicol, *J. Cell. Biol.* **58**:357–378.

Piko, L., and Taylor, K. D., 1987, Amounts of mitochondrial DNA and abundance of some mitochondrial gene transcripts in early mouse embryos, *Dev. Biol.* **123**:364–374.

Pinkert, C. A., Irwin, M. H., Johnson, L. W., and Moffatt, R. J., 1997, Mitochondria transfer into mouse ova by microinjection, *Transgenic Res.* **6**:379–383.

Poulton, J., and Morten, K., 1993, Noninvasive diagnosis of the MELAS syndrome from blood DNA [letter], *Ann. Neurol.* **34**:116.

Santorelli, F. M., Shanske, S., Jain, K. D., Tick, D., Schon, E. A., and DiMauro, S., 1994, A T → C mutation at nt 8993 of mitochondrial DNA in a child with Leigh syndrome, *Neurology* **44**:972–974.

Satoh, M., and Kuroiwa, T., 1991, Organization of multiple nucleoids and DNA molecules in mitochondria of a human cell, *Exp. Cell Res.* **196**:137–140.

Seller, A., Kennedy, C. R., Temple, I. K., and Brown, G. K., 1997, Leigh syndrome resulting from *de novo* mutation at position 8993 of mitochondrial DNA, *J. Inherit. Metab. Dis.* **20**:102–103.

Shadel, G. S., and Clayton, D. A., 1997, Mitochondrial DNA maintenance in vertebrates, *Annu. Rev. Biochem.* **66**:409–435.

Shitara, H., Hayashi, J. I., Takahama, S., Kaneda, H., and Yonekawa, H., 1998, Maternal inheritance of mouse mtDNA in interspecific hybrids: Segregation of the leaked paternal mtDNA followed by the prevention of subsequent paternal leakage, *Genetics* **148**:851–857.

Shoubridge, E. A., 1998, Mitochondrial encephalomyopathies, *Curr. Opin. Neurol.* **11**:491–496.

Taylor, K. D., and Piko, L., 1995, Mitochondrial biogenesis in early mouse embryos: Expression of the mRNAs for subunits IV, Vb, and VIIc of cytochrome *c* oxidase and subunit 9 (P1) of H(+)-ATP synthase, *Mol. Reprod. Dev.* **40**:29–35.

Tulinius, M. H., Houshmand, M., Larsson, N. G., Holme, E., Oldfors, A., Holmberg, E., and Wahlstrom, J., 1995, *De novo* mutation in the mitochondrial ATP synthase subunit 6 gene (T8993G) with rapid segregation resulting in Leigh syndrome in the offspring, *Hum. Genet.* **96**:290–294.

Uziel, G., Moroni, I., Lamantea, E., Fratta, G. M., Ciceri, E., Carrara, F., and Zeviani, M., 1997, Mitochondrial disease associated with the T8993G mutation of the mitochondrial ATPase 6 gene: A clinical, biochemical, and molecular study in six families, *J. Neurol. Neurosurg. Psychiat.* **63**:16–22.

Weber, K., Wilson, J. N., Taylor, L., Brierley, E., Johnson, M. A., Turnbull, D. M., and Bindoff, L. A., 1997, A new mtDNA mutation showing accumulation with time and restriction to skeletal muscle, *Am. J. Hum. Genet.* **60**:373–380.

Zhu, D. P., Economou, E. P., Antonarakis, S. E., and Maumenee, I. H., 1992, Mitochondrial DNA mutation and heteroplasmy in type I Leber hereditary optic neuropathy, *Am. J. Med. Genet.* **42**:173–179.

Chapter 5

Cardiac Reperfusion Injury: Aging, Lipid Peroxidation, and Mitochondrial Function

Luke I. Szweda, David T. Lucas, Kenneth M. Humphries, and Pamela A. Szweda

1. BACKGROUND AND SIGNIFICANCE

1.1. Cardiac Reperfusion and Free Radical Formation

The steady increase in human life span over the past century has been paralleled by increases in the incidence of numerous disorders characterized by reduction, cessation, or both of blood supply to the heart. Although diagnostic advances and improved techniques to restore blood flow have resulted in declining myocardial infarction death rates, heart disease remains a leading cause of mortality. This is partly due to loss in cardiac function following re-establishment of blood flow and to an increase in the elderly population, a population particularly susceptible to complications arising from myocardial reperfusion. Potentially deleterious events that occur upon reperfusion have therefore received considerable attention. Among these are increases in the levels of free radicals such as superoxide anion and hydroxyl radical that have been observed during cardiac reperfusion (Ambrosio *et al.*, 1993; Bolli *et al.*, 1988; Das *et al.*, 1989; Ferrari, 1996; McCord, 1988; Zweier, 1988). Production of these reactive species increases rapidly during the first few minutes of reoxygenation (Ambrosio *et al.*, 1993; Das *et al.*, 1989; Ferrari, 1996; Zweier, 1988) and often remains well above basal levels long after flow has been restored (Bolli *et al.*, 1988; Ferrari, 1996; Kuzuya *et al.*, 1990). Free radicals have numerous deleterious effects on cells, organelles, and cellular components *in vitro* (Beckman and Ames, 1997;

Luke I. Szweda, David T. Lucas, Kenneth M. Humphries, and Pamela A. Szweda   Department of Physiology and Biophysics, School of Medicine, Case Western Reserve University, Cleveland, Ohio 44106-4970.

Mitochondria in Pathogenesis, edited by Lemasters and Nieminen.
Kluwer Academic/Plenum Publishers, New York, 2001.

Berlett and Stadtman, 1997; Bindoli, 1988; Henle and Linn, 1997; Rokutan *et al.*, 1987; Stadtman, 1992; Szweda and Stadtman, 1992). It has therefore been proposed that free radical damage plays a key role in myocardial reperfusion injury.

1.2 Evidence Implicating Free Radicals in Cardiac Reperfusion Injury

A major experimental strategy in finding a role for free radical damage has been to determine whether inclusion of antioxidants or inhibitors of enzymes known to produce radical species prevents free radical generation and enhances recovery of function after reperfusion. The level of protection has varied among studies, however, making it difficult to draw firm conclusions; this likely reflects the multifactorial nature of ischemia/reperfusion injury. If significant cellular necrosis occurs during the ischemic event, antioxidant interventions during reperfusion may have little effect. In addition, depending on the model studied, various sites of free radical generation may be differentially accessible to certain antioxidants or to spin and chemical traps used to detect oxygen radicals. Furthermore, antioxidants may require additional factors for optimal activity, and they may exert effects unrelated to their ability to scavenge free radicals, rendering conclusive evidence for a free radical role in cardiac reperfusion injury elusive. It is therefore critical to identify cellular and subcellular targets of free radical damage and the mechanisms and conditions under which these events contribute to reperfusion-induced myocardial dysfunction.

1.3. Mitochondria as a Source of Free Radicals during Reperfusion

Free radicals, such as hydroxyl radical, are highly reactive species. Oxidative damage would therefore be expected to occur at or near the site of free radical formation. During normal cellular metabolism, free radicals are produced at the NADH coenzyme Q reductase (complex I) and ubiquinol–cytochrome *c* reductase (complex III) components of the mitochondrial electron transport chain (Boveris, 1984; Boveris and Cadenas, 1975; Cadenas *et al.*, 1977; Giulivi *et al.*, 1995; Nohl *et al.*, 1978; Nohl and Hegner, 1978; Turrens and Boveris, 1980). Ischemia results in a reduction of iron–sulfur and ubiquinone components of the mitochondrial electron transport chain (Baker and Kalyanaraman, 1989; Grill *et al.*, 1992), inhibition of complex I (Lucas and Szweda, 1999; Veitch *et al.*, 1992), and declines in superoxide dismutase activity (Ferrari, 1996; McCord, 1988; Zweier, 1988). Each of these events would be expected to result in increases in free radical levels upon re-introduction of oxygen. Not surprisingly, mitochondria isolated from ischemia-exposed hearts exhibit elevated rates of free radical generation when allowed to respire (Otani *et al.*, 1984; Ueta *et al.*, 1990). Furthermore, when ischemic hearts are reperfused with salicylate, products formed during salicylate's interaction with oxygen radicals are found primarily within the mitochondrial fraction (Das *et al.*, 1989). Finally, inhibition of election flow during reperfusion results in diminished free radical levels in cardiac tissue (Ambrosio *et al.*, 1993). Thus, evidence indicates that mitochondria, critical to the maintenance of energy status, are a source of free radicals during reperfusion and a potential site of free radical-mediated dysfunction.

1.4. Aging and Mitochondrial Dysfunction during Ischemia and Reperfusion

Proposed mechanisms of cardiac reperfusion injury must account for observations that severity of injury increases with age. Therefore, if mitochondrial free radical production is involved, the extent of oxidative damage and mitochondrial dysfunction due to reperfusion would be more pronounced in the elderly. Age-dependent alterations in mitochondrial structure, function, and free radical production have not been fully resolved, but the general consensus is that mitochondria are altered during the normal aging process (Beckman and Ames, 1998; Hansford, 1983; Shigenaga et al., 1994). Subtle, age-related differences in mitochondrial integrity would likely become more evident under conditions of pathophysiological stress. Nevertheless, only a few studies have addressed age-dependent differences in mitochondrial structure, function, and free radical generation during myocardial ischemia (Faulk et al., 1995a, 1995b; Matsuda et al., 1997) and reperfusion (Lucas and Szweda, 1998, 1999). The few studies addressing reperfusion effects on mitochondrial function have found losses in the rate of NADH-linked ADP-dependent (state 3) respiration and in the activities of adenine nucleotide translocase and electron transport chain complexes I and III (Duan and Karmazyn, 1989; Lucas and Szweda, 1998, 1999; Veitch et al., 1992). Most of these studies used adult animals exclusively. Furthermore, reperfusion is often accompanied by poor recovery of coronary flow, making it difficult to distinguish injury resulting from reperfusion or further ischemia. Nevertheless, evidence indicates that mitochondrial functions do decline during ischemia and reperfusion (Duan and Karmazyn, 1989; Lucas and Szweda, 1998, 1999; Veitch et al., 1992), so it is important to identify the mitochondrial enzymes inactivated during reperfusion, the activity losses responsible for declines in respiration, and the effects of age on these processes. This information is required for evaluating mechanisms by which mitochondrial free radical production contributes to cardiac reperfusion injury and for identifying critical molecular determinants of mitochondrial damage.

1.5. Potential Mechanisms of Free Radical-Mediated Mitochondrial Dysfunction

To assess whether reperfusion-induced declines in mitochondrial respiration are a result of free radical events, potential mechanisms of oxidative damage must be considered. Polyunsaturated fatty acids of membrane lipids are highly susceptible to peroxidation by oxygen radicals (Esterbauer et al., 1991). Not surprisingly, reperfusion-induced increases in the level of free radicals are paralleled by elevated rates of lipid peroxidation (Ambrosio et al., 1993; Blasig et al., 1995; Cordis et al., 1993; Kramer et al., 1994). Membrane lipid peroxidation fragments polyunsaturated fatty acids and leads to production of various aldehydes, alkenals, and hydroxyalkenals (Esterbauer et al., 1991). Many of the products formed during lipid peroxidation are cytotoxic in whole animals and in cells in culture (Esterbauer et al., 1991). The aldehyde 4-hydroxy-2-nonenal (HNE), an α,β-unsaturated aldehyde and major product of lipid peroxidation, is one of the most reactive under physiological conditions (Esterbauer et al., 1991) and readily interacts with and inactivates protein (Fig. 1) (Blanc et al., 1997; Chen et al., 1995; Cohn et al., 1996; Esterbauer et al., 1991; Friguet et al., 1994; Humphries and Szweda, 1998; Keller et al., 1997; Mark et al., 1997; Sayre et al., 1997; Szweda et al., 1993; Tsai and Sokoloski, 1995;

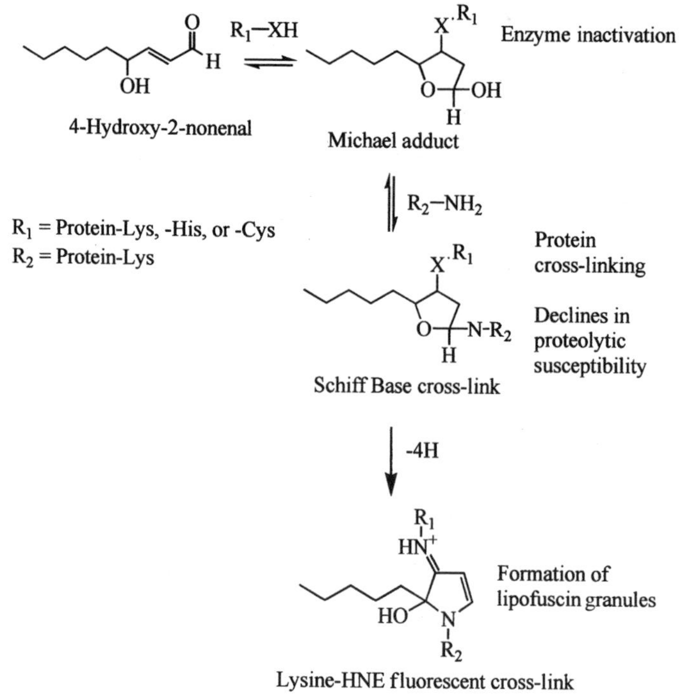

FIGURE 1. Reaction of HNE with nucleophilic amino acid residues on protein.

Tsai *et al.*, 1998; Uchida and Stadtman, 1993; Uchida *et al.*, 1993; Ullrich *et al.*, 1996). In addition HNE concentration increases during myocardial reperfusion (Blasig *et al.*, 1995), making it a key mediator of free radical damage that likely plays a critical role in reperfusion injury.

1.6. Hypothesis and Experimental Design

Based on the above considerations, it is hypothesized that mitochondrial respiration is inhibited in an age-dependent manner, during myocardial reperfusion, and that the inhibition may be mediated, in part, by HNE reaction with mitochondrial enzymes critical to electron transport and oxidative phosphorylation. Support for this hypothesis requires evidence of age-dependent mitochondrial respiration declines during reperfusion, identification of the enzymes responsible for loss of mitochondrial respiration, and proof that these enzymes are modified and inactivated by HNE. We pursued an experimental approach incorporating two complementary sets of studies, one using an *in situ* model of myocardial ischemia/reperfusion, the other employing treatment of intact cardiac mitochondrial with HNE *in vitro*. Using hearts excised from adult (8-mo) and senescent (24-mo) rats and perfused in retrograde fashion, we examined age-dependent ischemia-reperfusion effects on mitochondrial respiratory rates, and also the activities of specific enzymes involved in mitochondrial respiration and ATP synthesis (Lucas and Szweda, 1998, 1999). In parallel experiments we determined whether HNE could exert effects on overall mitochondrial respiration similar to those observed during reperfusion, identified specific enzymes highly susceptible to HNE inactivation, and defined the molecular mechanisms by which HNE inhibits these enzymes (Humphries and Szweda, 1998; Humphries *et al.*, 1998). Taken together, these results provide critical direction for ongoing experiments designed to directly determine whether reperfusion-induced loss in mitochondrial function is due to free radical-mediated inactivation of key mitochondrial enzymes.

2. RESULTS

2.1. Cardiac Ischemia/Reperfusion

2.1.1. *In Situ* Model of Myocardial Ischemia Reperfusion

To evaluate ischemia- and reperfusion-induced alterations in mitochondrial function, we perfused excised hearts from adult (8-mo) and senescent (26-mo) Fisher-344 rats in retrograde fashion (Langendorff isolated heart model) as described previously (Lucas and Szweda, 1998). Experimental protocols consisted of (1) 90 minutes of normoxic perfusion, (2) 25 minutes of perfusion followed by 25 minutes of no-flow global ischemia, or (3) 25 minutes of perfusion followed by 25 minutes of ischemia and then 40 minutes of reperfusion. We then isolated mitochondria and evaluated respiratory function (Lucas and Szweda, 1998, 1999). Hemodynamic functions (coronary flow, developed tension, contractility) and mitochondrial respiratory rates were stable throughout 90 minutes of normoxic perfusion. For ischemic periods of up to 25 minutes, reperfusion resulted in a rapid recovery of hemodynamic functions to near-maximal values for hearts from both groups of rats, ensuring that the age-related alterations in mitochondrial function were not due to inadequate reperfusion of myocardial tissue (Lucas and Szweda, 1998).

2.1.2. Age-Dependent Effects of Ischemia and Reperfusion on Mitochondrial Respiration

Mitochondria isolated from perfused hearts of both age groups exhibited similar state 3 respiration rates. Ischemia (25 minutes) resulted in an $\sim 18\%$ loss of state 3 respiration rate independent of age. In contrast, reperfusion resulted in decreases in mitochondrial respiratory rates that did depend on age, leading to further 15.0% and 28.5% declines in the state 3 respiration rate for hearts from 8- and 26-month-old rats, respectively (Table I) (Lucas and Szweda, 1999). We found no significant changes in state 4 respiration, ADP/O, or mitochondrial yield as a function of age, of ischemia, or of reperfusion (Lucas and Szweda, 1998); neither ischemia nor reperfusion resulted in global disruption of the mitochondrial membrane. These results indicate that mitochondria are important subcellular sites of ischemia and reperfusion damage and that the magnitude of reperfusion-induced mitochondrial dysfunction is greater in senescent relative to adult rat hearts.

To identify potential mitochondrial processes responsible for these effects, uncoupled respiration rates were compared to those of ADP-dependent (state 3) respiration rates. Uncoupling agents collapse the proton gradient, thereby stimulating maximum rates of electron transport and O_2 consumption independent of ADP transport and ATP synthesis.

Table I
**Effects of Ischemia and Reperfusion on Mitochondrial State 3 Respiration and
Complex Activities[a]**

Protocol (rat hearts)	State 3 nmol O/min/mg	Complex I nmol NADH/min/mg	Complex III nmol cyt c/min/mg	Complex IV nmol O/min/mg
8 month old				
Perfused (n = 7)	203.5 ± 26.2^b	136.0 ± 24.5^g	1921.5 ± 321.3	1993.0 ± 246.7
Ischemic (n = 6)	$167.0 \pm 20.3^{b,d}$	$117.8 \pm 6.3^{g,i}$	2205.4 ± 376.9	1915.5 ± 169.7
Reperfused (n = 6)	$136.5 \pm 22.6^{d,f}$	119.7 ± 8.0	1921.5 ± 322.4	2076.3 ± 412.4^k
26 month old				
Perfused (n = 6)	204.2 ± 26.2^c	131.5 ± 17.2^h	1969.3 ± 248.2	2019.4 ± 327.9
Ischemic (n = 6)	$168.3 \pm 26.6^{c,e}$	$99.5 \pm 10.2^{h,i}$	2195.0 ± 430.0	2135.0 ± 309.5^j
Reperfused (n = 6)	$110.1 \pm 11.7^{e,f}$	97.3 ± 10.7	2063.6 ± 298.4	$1709.4 \pm 52.4^{j,k}$

[a] Values are expressed as the mean ± standard deviation. P values (determined from paired t test) where like letters indicate values compared.
[b] ≤ 0.009
[c] ≤ 0.02
[d] ≤ 0.02
[e] ≤ 0.0003
[f] ≤ 0.02
[g] ≤ 0.05
[h] ≤ 0.001
[i] ≤ 0.002
[j] ≤ 0.004
[k] ≤ 0.03
(Reprinted with permission from Proc. Natl. Acad. Sci. USA: 96, Lucas and Szweda, 1999, 6689-6693, Copyright 1999, National Academy of Sciences, USA.)

Therefore, if damage to adenine nucleotide translocase or ATPase contributes to declines in ADP-dependent respiration, we would expect the state 3 respiration rate to be more affected by ischemia/reperfusion than the uncoupled respiration rate would be. We found that the rate changes of uncoupled respiration as a function of ischemia reperfusion paralleled, and were not statistically different from, those observed for ADP-dependent respiration (Lucas and Szweda, 1998). Thus, although inhibition of or damage to adenine nucleotide translocase or ATPase may occur, it does not contribute significantly to mitochondrial respiration rate declines. (Lucas and Szweda, 1998). Loss of mitochondrial respiratory function reflects damage to electron transport chain components, to a reduction in the supply of reducing equivalents (NADH), or to both.

2.1.3. Electron Transport Complexes

Complexes I, III, and IV exhibited no basal age-related differences in activity (Table I) (Lucas and Szweda, 1999). During ischemia, only complex I exhibited a decline in activity. The magnitude of inactivation was age-dependent, with losses of 13.4% and 24.3% in activity for adult and senescent rat hearts respectively. No further declines in complex I activity were observed in either adult or senescent hearts subjected to reperfusion. In contrast, reperfusion did result in inactivation of complex IV. The degree of inactivation was age-dependent, with a 19.9% loss in complex IV activity in senescent hearts and no effect on the enzyme in adult hearts. Complex III activity showed no change during ischemia or reperfusion for either age group (Table I) (Lucas and Szweda, 1999). *In vitro* studies revealed that complexes I and IV must be inhibited by approximately 50% before declines in the rate of NADH-linked respiration are observed (Lucas and Szweda, 1999). Thus, the magnitude of inactivation of these complexes during ischemia and reperfusion was not sufficient to account for observed declines in NADH-linked state 3 respiration. It was therefore important to determine whether defects in the ability to supply NADH for electron transport were responsible for losses in NADH-linked state 3 respiration.

2.1.4. NADH-Linked Dehydrogenases

We measured the activities of glutamate dehydrogenase and α-ketoglutarate dehydrogenase (KGDH) to determine whether ischemia/reperfusion-induced declines in mitochondrial respiration were due to direct effects on NADH production. These dehydrogenases were chosen because the activities of both enzymes are required for synthesis of NADH from glutamate, the respiratory substrate we used. We found no age-related differences in basal activities of glutamate dehydrogenase or KGDH following perfusion. Ischemia did not alter the activity of either glutamate dehydrogenase or KGDH. In contrast, reperfusion resulted in inactivation of KGDH with no loss in glutamate dehydrogenase activity. KGDH activity declined in an age-dependent manner, with a loss of 13.0% and 25.2% for adult and senescent hearts, respectively (Table II) (Lucas and Szweda, 1999), declines that closely paralleled those of NADH-linked state 3 respiration (Table 1). Supply of NADH by KGDH is thought to control the rate of electron transport during NADH-linked ADP-dependent mitochondrial respiration (Balaban, 1990; Brown, 1992; Cooney et al., 1981; Humphries et al., 1998; Moreno-Sanchez et al., 1990). In addition, unlike with complexes I and IV, selective in vitro inactivation of KGDH results in a decrease in the state 3 respiration rate, the magnitude of which is directly proportional to the degree of KGDH inactivation (Humphries et al., 1998). Thus reperfusion-induced loss of NADH-dependent state 3 respiratory rates appears largely due to mechanisms that lead to KGDH inactivation (Lucas and Szweda, 1999).

Table II
Effects of Ischemia and Reperfusion on Glutamate and α-Keto-glutarate Dehydrogenase Activities*

Protocol (rat hearts)	GDH nmol NADH/min/mg	KGDH nmol NADH/min/mg
8 month old		
Perfused (n = 7)	38.4 ± 4.9	98.0 ± 16.9
Ischemic (n = 6)	36.5 ± 4.0	99.6 + 3.5[b]
Reperfused (n = 6)	35.4 ± 2.7	86.7 + 8.9[b,d]
26 month old		
Perfused (n = 6)	36.0 ± 6.8	95.9 ± 2.7
Ischemic (n = 6)	35.9 ± 3.6	94.7 ± 11.0[c]
Reperfused (n = 6)	34.8 ± 5.3	70.8 ± 8.4[c,d]

[a] Values are expressed as the mean ± standard deviation. P values (determined from paired t test) where like letters indicate values compared.
[b] ≤0.005
[c] ≤0.001
[d] ≤0.007
(Reprinted with permission from Proc. Natl. Acad. Sci. USA: 96, Lucas and Szweda, 1999, 6689-6693, Copyright 1999, National Academy of Sciences, USA.)

2.2. *In Vitro* Treatment of Intact Rat Heart Mitochondria with 4-Hydroxy-2-Nonenal

2.2.1. Effects on Mitochondrial Respiration

Isolated mitochondria provide a simple *in vitro* model for highly targeted mechanistic studies of oxidative damage. Seeking support for a role of HNE in reperfusion-induced mitochondrial respiration declines, we incubated with HNE to determine whether it could exert effects similar to those observed during cardiac ischemia/reperfusion (Humphries *et al.*, 1998). Investigations of specific free radical events likely to occur help to define potential targets and damage mechanisms, thus aiding studies that use more-complex models. Incubation of cardiac mitochondria with micromolar concentrations of HNE (0–100 µM) caused rapid declines in the rate of NADH-linked ADP-dependent (state 3) mitochondrial respiration (Fig. 2). Inactivation was first order with respect to time and HNE concentration, and it was found to be irreversible under the conditions tested. No effects on ADP-independent (state 4) respiration or ADP/O were observed. In addition, at a given concentration of HNE and time of incubation, uncoupled respiration, which results in maximal oxygen consumption rates independent of adenine nucleotide translocase and ATPase activity, was inhibited to the same extent as state 3 respiration (Humphries *et al.*, 1998). Thus, HNE induces declines in mitochondrial respiration by affecting electron transport-chain components, supply of NADH, or both.

2.2.2. Sites of Inactivation

To identify specific sites of HNE inactivation responsible for loss of mitochondrial respiratory function, we explored the possibility that HNE exerts its effects by inhibiting electron transport (Humphries and Szweda, 1998). We found no HNE effect on the activity of complex I, II, III, or IV ([HNE] ≤ 50 µM) (Humphries and Szweda, 1998). Thus, HNE

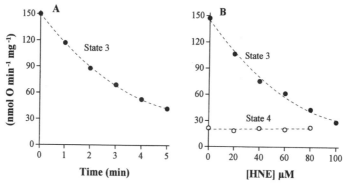

FIGURE 2. Rate of mitochondrial respiration as a function of time and 4-hydroxy-2-nonenal concentration: (A) Mitochondria (0.5 mg/ml concentration) were incubated with 15 mM glutamate for 5.0 min at 25 °C. The *abscissa* indicates the time of mitochondrial exposure to 80 µM HNE. (B) Mitochondria (0.5 mg/ml concentration) were incubated for 5.0 min at 25 °C with 15 mM glutamate (HNE concentrations indicated on the *abscissas*). State 3 respiration was initiated with addition of 0.33 mM ADP at 5.0 min. State 4 respiration was evaluated as the rate of oxygen consumption following depletion of ADP. (Reprinted with permission from *Biochemistry*: 37, Humphries *et al.*, 1998, 552–557, Copyright 1998, American Chemical Society.)

does not inhibit NADH-linked mitochondrial respiration at the level of the electron transport chain, but rather alters the supply of NADH at sites proximal to electron transport. We monitored the effects of HNE on NADH production and consumption in intact mitochondria using fluorescence spectroscopy (Estabrook, 1962), which demonstrated that mitochondria preincubated with HNE had significantly reduced NADH levels upon addition of Krebs cycle substrates (Fig. 3A) and a diminished ability to maintain steady-state NADH concentrations upon addition of HNE (Fig. 3B) (Humphries *et al.*, 1998). Stimulation of maximal rates of mitochondrial oxygen consumption by addition of ADP was accompanied by a rapid and transient drop in the level of NADH (Fig. 2), showing that the rate of NADH-linked ADP-dependent respiration is limited by production and supply of NADH rather than by electron transport (Balaban, 1990; Brown, 1992; Cooney *et al.*, 1981; Humphries *et al.*, 1998; Moreno-Sanchez *et al.*, 1990). Alterations in NADH production would therefore have profound effects on respiratory activity.

Because HNE was found to inhibit mitochondrial respiration to a similar extent with either glutamate or α-ketoglutarate as respiratory substrates, we decided to determine whether HNE limits NADH synthesis by direct inactivation of glutamate dehydrogenase, KGDH, or both (Humphries *et al.*, 1998). Under the conditions tested, the activity of glutamate dehydrogenase was unaffected by HNE. In contrast, KGDH, an enzyme critical to the supply of NADH for electron transport, was highly susceptible to HNE inactivation. The magnitude of HNE-induced KGDH inactivation was directly proportional to the loss of NADH-linked state 3 respiration (Fig. 4). Thus, HNE inhibits in NADH-linked mitochondrial respiration by inactivating KGDH, the same enzyme inactivated during cardiac reperfusion (Lucas and Szweda, 1999), making it important to identify the specific inactivation mechanisms in order to provide information critical for a rigorous assessment of the role of these processes in cardiac ischemia/reperfusion.

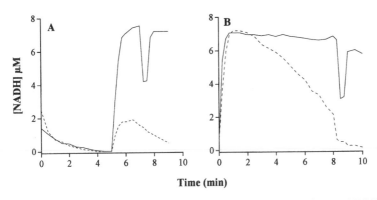

Time (min)

FIGURE 3. Effect of 4-hydroxy-2-nonenal on NADH production and on steady state NADH levels (A) Mitochondria (0.5 mg/ml concentration) were incubated for 5.0 min at 25 °C in the presence (- - - -) or absence (———) of 100 μM HNE. NADH production was then initiated by adding 15 mM α-ketoglutarate at 5.0 min. Mitochondrial ability to consume NADH was measured upon addition of 0.2 mM ADP at 7.0 min. (B) Mitochondria (0.5 mg/ml concentration) were incubated at 25 °C in the presence of 15 mM α-ketoglutarate followed by the addition of 0μM (———) and 100 μM (- - - -) HNE at 2.0 min. Mitochondrial ability to consume ADP was then measured upon addition of 0.2 mM ADP at 8.0 min. (Reprinted with permission from *Biochemistry*: 37, Humphries *et al*, 1998, 552–557, Copyright 1998, American Chemical Society.)

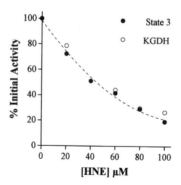

FIGURE 4. Effect of 4-hydroxy-2-nonenal on state 3 respiration and α-ketoglutarate dehydrogenase activity. Mitochondria (0.5 mg/ml) were incubated for 5.0 min at 25 °C with HNE at concentrations indicated on the *abscissas*. Following treatment with HNE, the rate of state 3 respiration and the activity of α-ketoglutarate dehydrogenase were measured as previously described. (Reprinted with permission from *Biochemistry*: 37, Humphries *et al.*, 1998, 552–557, Copyright 1998, American Chemical Society.)

2.2.3. Mechanisms of Respiration Inactivation

Using intact heart mitochondria, we evaluated the effects of HNE on a variety of NADH-linked dehydrogenases to gain information on the selectivity and potential mechanisms of HNE inactivation (Humphries and Szweda, 1998). Of the dehydrogenases tested, only KGDH and, to a lesser extent, pyruvate dehydrogenase (PDH) were inactivated by HNE (<100 μM), indicating that, under the conditions of our experiments, HNE does not unilaterally inhibit NADH-linked dehydrogenases, but rather acts specifically on KGDH and PDH (Humphries and Szweda, 1998). These enzymes are catalytically and structurally similar (Brown and Perham, 1976; Perham, 1991; Reed, 1974); both are composed of multiple copies of three enzymes: α-ketoacid decarboxylase (E1), dihydro-lipoyl transacetylase (E2), and dihydrolipoamide dehydrogenase (E3). In addition, both require the cofactors thiamine pyrophosphate bound to E1, lipoic acid covalently linked to E2, and FAD^+ for E3 activity (Fig. 5) (Brown and Perham, 1976; Perham, 1991; Reed, 1974). It is therefore likely that the susceptibilities of the KGDH and PDH to HNG inactivation are conferred by certain structural features common to both enzymes.

Inactivation of KGDH and PDH, in pure form and in intact mitochondria, was promoted under conditions in which the ratio of $NADH/NAD^+$ is high (Humphries and Szweda, 1998). The lipoic acid moieties of KGDH and PDH cycle between oxidized and reduced forms during catalysis (Brown and Perham, 1976; Perham, 1991; Reed, 1974). In the presence of NADH, the reduced form of lipoic acid containing two sulfhydryl groups is favored. Sulfhydryl groups are highly reactive toward HNE under physiological conditions (Esterbauer *et al.*, 1991), and lipoic acid is therefore a likely target of HNE modification. As judged by HNE-induced declines in the level of dithio-(bis)nitrobenzoic acid-reactive lipoic acid residues, we determined that enzyme activity loss was directly proportional to loss of lipoic acid sulfhydryl groups (Humphries and Szweda, 1998). In parallel studies, we prepared a polyclonal antibody specific to lipoic acid that exhibited no cross-reactivity toward HNE-modified lipoic acid. We then used the antibody to demonstrate that treating purified KGDH or intact cardiac mitochondria with HNE resulted in loss of lipoic acid on

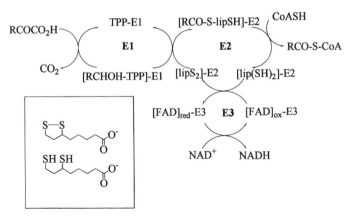

FIGURE 5. Schematic representation of the reaction mechanism of α-ketoacid dehydrogenases (adapted from Reed, 1974). E1: α-ketoacid decarboxylase; E2: dihydrolipoyl transacetylase; E3: dihydro-lipoamide dehydrogenase; TPP: thiamine pyrophosphate; lip: lipoic acid. *Inset*: Oxidized and reduced lipoic acid.

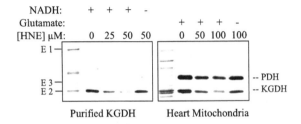

FIGURE 6. Western blot analysis of purified KGDH and mitochondria treated with HNE. Purified KGDH and mitochondria were incubated with HNE in the presence and absence of NADH (10 μM) or glutamate (15 mM) as indicated. Protein samples were separated on 10% polyacrylamide gels (1.5 μg purified KGDH containing ~40% BSA/well or 10.0 μg mitochondrial protein/well), and Western blot analysis with anti-lipoic acid as primary antibody was performed as previously described (Humphries and Szweda, 1998). The first lane in each series is a representative protein profile obtained by Coomassie staining. KGDH E1, E2, and E3 migrate at 50, 56, and 95 kDa, respectively. (Reprinted with permission from *Biochemistry: 37*, Humphries and Szweda, 1998, 15835–15841, Copyright 1998, American Chemical Society.)

the E2 subunits (Fig. 6) (Humphries and Szweda, 1998). Taken together, these findings indicate that HNE exerts its effects on KGDH and PDH by reacting with the reduced form of lipoic acid covalently bound to catalytic E2 subunits, thereby inactivating the enzymes.

3. DISCUSSION

3.1 Potential Mechanism of Reperfusion-Induced Mitochondrial Dysfunction

Cardiac reperfusion results in age-dependent declines in NADH-linked mitochondrial respiration and ATP synthesis (Lucas and Szweda, 1998, 1999). This is largely due to inactivation of KGDH, an enzyme critical for the production of NADH (Lucas and

Szweda, 1999). Treatment of intact cardiac mitochondria with HNE leads to rapid and selective inactivation of KGDH, resulting in declines in the rates of mitochondrial NADH production, electron transport, and ATP synthesis (Humphries and Szweda, 1998; Humphries *et al.*, 1998). Taken together, results of *in situ* and *in vitro* experiments strongly support a role for free radicals in cardiac reperfusion injury and offer a plausible subcellular site and mechanism of oxidative damage (Humphries and Szweda, 1998; Humphries *et al.*, 1998; Lucas and Szweda, 1998, 1999). The identification of KGDH as an enzyme both inactivated during reperfusion (Lucas and Szweda, 1999) and highly susceptible to HNE inactivation (Humphries and Szweda, 1998; Humphries *et al.*, 1998) provides the direction required to establish a direct link between reperfusion-induced free radical production and cardiac dysfunction.

3.2. Does 4-Hydroxy-2-Nonenal Mediate Free Radical Damage during Cardiac Reperfusion?

The susceptibility of KGDH to HNE inactivation is likely predicated by the chemical and structural properties of HNE and lipoic acid. This α,β-unsaturated aldehyde can react with the sulfhydryl moiety of cysteine, the imidazole nitrogens of histidine, and the ε-amino group of lysine (Cohn *et al.*, 1996; Esterbauer *et al.*, 1991; Sayre *et al.*, 1997; Szweda *et al.*, 1993; Tsai and Sokoloski, 1995; Tsai *et al.*, 1998; Uchida and Stadtman, 1993; Uchida *et al.*, 1993). The sulfhydryl moiety is the most reactive of these nucleophiles under physiological conditions (Esterbauer *et al.*, 1991). Reactivity also depends on the accessibility of nucleophilic residues to HNE (Szweda *et al.*, 1993). Lipoic acid, a strong nucleophile in its reduced form, is located on the surface of the E2 subunits of KGDH (Brown and Perham, 1976; Perham, 1991; Reed, 1974). Hydrophobic interactions are also likely between the hydrocarbon chains of lipoic acid and HNE. Thus, lipoic acid is a prime target for modification by HNE. Furthermore, HNE would be at greatest concentration in the lipid bilayer where it is formed; KGDH, which is associated with the inner mitochondrial membrane (Maas and Bisswanger, 1990; Porpaczy *et al.*, 1987), would therefore be particularly susceptible to modification and inactivation. Because many lipid peroxidation products are lipophilic electrophiles (Esterbauer *et al.*, 1991), the reaction of HNE with lipoic acid residues on KGDH is not likely to be limited to HNE and may represent a common mechanism of cytotoxicity.

3.3. Ischemic Conditions Prime Mitochondria for Free Radical Damage during Reperfusion

The ischemic environment would be expected to favor modification of lipoic acid by HNE upon subsequent reperfusion. During enzyme catalysis, the two sulfur atoms on lipoic acid cycle between reduced and oxidized states. (Brown and Perham, 1976; Perham, 1991; Reed, 1974); the reduced state is required for reaction with HNE (Humphries and Szweda, 1998). During ischemia, the ratio of NADH/NAD$^+$ increases due to the absence of O$_2$ as a terminal electron acceptor (Barlow and Chance, 1976; Barlow *et al.*, 1977), resulting in a reduction of lipoic acid. In addition, ADP levels (Reimer and Jennings,

1992) and complex I activity both decline (Lucas and Szweda, 1999; Veitch *et al.*, 1992), lowering the rate of NADH consumption which lead to an increase in the ratio of reduced to oxidized lipoic acid on the E2 subunits of KGDH, thus favoring modification by HNE. Furthermore, ischemia-induced alterations in certain mitochondrial functions are likely to predispose these organelles to free radical production during reoxygenation. Of the respiratory complexes and dehydrogenases studied, only complex I exhibited a decline in function during ischemia (Lucas and Szweda, 1999). Inhibition of complex I in isolated mitochondria results in increased free radical production (Boveris, 1984; Cadenas *et al.*, 1977). A decrease in complex I activity during ischemia would therefore likely contribute to an increase in free radical production during reperfusion. Taken together with the increase in reduction potential ($NADH/NAD^+$) (Barlow and Chance, 1976; Barlow *et al.*, 1977), this is a plausible mechanism for production of HNE and inactivation of KGDH during reperfusion.

3.4. Effects of Age on Reperfusion-Induced Mitochondrial Dysfunction

Reaching consensus on basal age-related alterations in mitochondrial structure and function has been difficult (Beckman and Ames, 1998; Hansford, 1983; Shigenaga *et al.*, 1994). Aging defects are probably subtle, and as such, apparent only under conditions of stress. Consistent with this, cardiac mitochondria isolated from adult and senescent rats exhibited no difference in function, as measured in this study (Lucas and Szweda, 1998, 1999). Following ischemia and reperfusion, however, clear age-dependent differences in certain enzyme activities, free radical damage, and mitochondrial function are evident (Lucas and Szweda, 1998, 1999). Although declines in complex I activity are not directly responsible for loss in mitochondrial respiration during ischemia (Lucas and Szweda, 1999), the process may play a role in subsequent reperfusion damage; age-dependent declines would be expected to result in increases in free radical generation upon reoxygenation (Boveris, 1984; Cadenas *et al.*, 1977). It is therefore important to determine the events responsible for this loss in activity. Use of age as an experimental variable will not only aid in identifying the mechanisms involved in ischemia-reperfusion injury, but will provide important information on changes in mitochondrial properties that occur during normal aging.

3.5. Implications for Antioxidant Interventions

The elucidation of mechanisms responsible for disrupting specific mitochondrial functions suggests strategies for intervention and provides a means for assessing the effectiveness of protection. In addition, it is likely that a number of pathophysiological processes contribute to the progression of injury. Ischemia- and reperfusion-induced alterations in Ca^{2+} homeostasis are believed to contribute to increased rates of free radical generation during cardiac reperfusion (Bagchi *et al.*, 1997). An understanding of the mechanisms by which free radicals mediate mitochondrial dysfunction would provide a basis for studies aimed at explaining the interrelationships between these factors. The results of these studies would define, at the molecular level, a likely target of free radical damage to mitochondria during cardiac ischemia/reperfusion. This information is important, both for investigating specific oxidative mechanisms at play in the progression of myocardial reperfusion injury and for gaining insight into the pathogenesis of other degenerative diseases associated with loss of mitochondrial function and increased free radical production.

ACKNOWLEDGEMENTS.   This work was supported by a Scientist Development Grant from the American Heart Association (9630025N) and a grant from the National Institutes of Health (AG16339).

REFERENCES

Ambrosio, G., Zweier, J. L., Duilio, C., Kuppusamy, P., Santoro, G., Elia, P. P., Tritto. I., Cirillo, P., Condorelli, M., Chiariello, M., and Flaherty, J. T., 1993, Evidence that mitochondrial respiration is a source of potentially toxic oxygen free radicals in intact rabbit hearts subjected to ischemia and reflow, *J. Biol. Chem.* **268**:18532–18541.

Bagchi, D., Wetscher, G. J., Bagchi, M., Hinder, P. R., Perdikis, G., Stohs, S. J., Hinder, R. A. and Das, D.K., 1997, Interrelationship between cellular calcium homeostasis and free radical generation in myocardial reperfusion injury, *Chem. Biol. Interact.* **104**:65–85.

Baker, J. E., and Kalyanaraman, B., 1989, Ischemia-induced changes in myocardial paramagnetic metabolites: Implications for intracellular oxy-radical generation, *FEBS Lett.* **244**:311–314.

Balaban, R. S., 1990, Regulation of oxidative phosphorylation in the mammalian cell, *Am. J. Physiol.* **258**:C377–389.

Barlow, C. H., and Chance, B., 1976, Ischemic areas in perfused rat hearts: Measurement by NADH fluorescence photography, *Science* **193**:909–910.

Barlow, C. H., Harken, A. H., and Chance, B., 1977, Evaluation of cardiac ischemia by NADH fluorescence photography, *Ann. Surg.* **186**:737–740.

Beckman, K. B., and Ames, B. N., 1997, Oxidative decay of DNA, *J. Biol. Chem.* **272**:19633–19636.

Beckman, K. B., and Ames, B. N. 1998, The free radical theory of aging matures, *Physiol. Rev.* **78**:547–581.

Berlett, B. S., and Stadtman, E. R., 1997, Protein oxidation in aging disease, and oxidative stress, *J. Biol. Chem.* **272**:20313–20316.

Bindoli, A., 1988, Lipid peroxidation in mitochondria, *Free Radical Biol. Med.* **5**:247–261.

Blanc, E. M., Kelly, J. F., Mark, R. J., Waeg, G., and Mattson, M. P., 1997, 4-Hydroxynonenal, an aldehydic product of lipid peroxidation, impairs signal transduction associated with muscarinic acetylcholine and metabotropic glutamate receptors: Possible action on G alpha (q/11), *J. Neurochem.* **69**:570–580.

Blasig, I. E., Grune, T., Schonheit, K., Rohde, E., Jakstadt, M., Haseloff, R. F., and Siems, W. G., 1995, 4-Hydroxynonenal, a novel indicator of lipid peroxidation for reperfusion injury of the myocardium, *Am. J. Physiol.* **269**:H14–22.

Bolli, R., Patel, B. S., Jeroudi, M. O., Lai, E. K., and McCay, P. B., 1988, Demonstration of free radical generation in 'stunned' myocardium of intact dogs with the use of the spin trap alpha-phenyl N-*tert*-butyl nitrone [published erratum appears in J. Clin. Invest. 82(5) following 1807], *J. Clin. Invest.* **82**:476–485.

Boveris, A., 1984, Determination of the production of superoxide radicals and hydrogen peroxide in mitochondria, *Methods Enzymol.* **105**:429–435.

Boveris, A., and Cadenas, E., 1975, Mitochondrial production of superoxide anions and its relationship to the antimycin insensitive respiration, *FEBS Lett.* **54**:311–314.

Brown, G. C., 1992, Control of respiration and ATP synthesis in mammalian mitochondria and cells, *Biochem. J.* **284**:1–13.

Brown, J. P., and Perham, R. N., 1976, Selective inactivation of the transacylase components of the 2-oxo acid dehydrogenase multienzyme complexes of *Escherichia coli, Biochem. J.* **155**:419–427.

Cadenas, E., Boveris, A., Ragan, C. I., and Stoppani, A. O., 1977, Production of superoxide radicals and hydrogen peroxide by NADH–ubiquinone reductase and ubiquinol–cytochrome *c* reductase from beef-heart mitochondria, *Arch. Biochem. Biophys.* **180**:248–257.

Chen, J. J., Bertrand, H., and Yu, B. P., 1995, Inhibition of adenine nucleotide translocator by lipid peroxidation products, *Free Radical Biol. Med.* **19**:583–590.

Cohn, J. A., Tsai, L., Friguet, B., and Szweda, L. I., 1996, Chemical characterization of a protein–4-hydroxy-2-nonenal cross-link: Immunochemical detection in mitochondria exposed to oxidative stress, *Arch. Biochem. Biophys.* **328**:158–164.

Cooney, G. J., Taegtmeyer, H., and Newsholme, E. A., 1981, Tricarboxylic acid cycle flux and enzyme activities in the isolated working rat heart, *Biochem. J.* **200**:701–703.

Cordis, G. A., Maulik, N., Bagchi, D., Engelman, R. M., and Das, D. K., 1993, Estimation of the extent of lipid peroxidation in the ischemic and reperfused heart by monitoring lipid metabolic products with the aid of high-performance liquid chromatography, *J. Chromatogr.* **632**:97–103.

Das, D. K., George, A., Liu, X. K., and Rao, P. S., 1989, Detection of hydroxyl radicals in the mitochondria of ischemic-reperfused myocardium by trapping with salicylate, *Biochem. Biophys. Res. Commun.* **165**:1004–1009.

Duan, J., and Karmazyn, M., 1989, Relationship between oxidative phosphorylation and adenine nucleotide translocase activity of two populations of cardiac mitochondria and mechanical recovery of ischemic hearts following reperfusion, *Can. J. Physiol. Pharmacol.* **67**:704–709.

Estabrook, R. W., 1962, Fluorometric measurement of reduced pyridine nucleotide in cellular and subcellular particles, *Anal. Biochem.* **4**:231–245.

Esterbauer, H., Schaur, R. J., and Zollner, H., 1991, Chemistry and biochemistry of 4-hydroxynonenal, malonaldehyde, and related aldehydes, *Free Radical Biol. Med.* **11**:81–128.

Faulk, E. A., McCully, J. D., Hadlow, N. C., Tsukube, T., Krukenkamp, I. B., Federman, M., and Levitsky, S., 1995a, Magnesium cardioplegia enhances mRNA levels and maximal velocities of cytochrome oxidase I in the senescent myocardium during global ischemia, *Circulation* **92**:II405–II412.

Faulk, E. A., McCully, J. D., Tsukube, T., Hadlow, N. C., Krukenkamp, I. B., and Levitsky, S., 1995b, Myocardial mitochondrial calcium accumulation modulates nuclear calcium accumulation and DNA fragmentation, *Ann. Thorac. Surg.* **60**:338–344.

Ferrari, R., 1996, The role of mitochondria in ischemic heart disease, *J. Cardiovasc. Pharmacol.* **28**:S1–10.

Friguet, B., Stadtman, E. R., and Szweda, L. I., 1994, Modification of glucose-6-phosphate dehydrogenase by 4-hydroxy-2-nonenal: Formation of cross-linked protein that inhibits the multicatalytic protease, *J. Biol. Chem.* **269**:21639–21643.

Giulivi, C., Boveris, A., and Cadenas, E., 1995, Hydroxyl radical generation during mitochondrial electron transfer and the formation of 8-hydroxydesoxyguanosine in mitochondrial DNA, *Arch. Biochem. Biophys.* **316**:909–916.

Grill, H. P., Flaherty, J. T., Zweier, J. L., Kuppusamy, P., and Weisfeldt, M. L., 1992, Direct measurement of myocardial free radical generation in an *in vivo* model: Effects of postischemic reperfusion and treatment with human recombinant superoxide dismutase, *J. Am. Coll. Cardiol.* **21**:1604–1611.

Hansford, R. G., 1983, Bioenergetics in aging, *Biochem. Biophys. Acta* **726**:41–80.

Henle, E. S., and Linn, S., 1997, Formation, prevention, and repair of DNA damage by iron/hydrogen peroxide, *J. Biol. Chem.* **272**:19095–19098.

Humphries, K. M., and Szweda, L. I., 1998, Selective inactivation of α-ketoglutarate dehydrogenase: Reaction of lipoic acid with 4-hydroxy-2-nonenal, *Biochemistry* **37**:15835–15841.

Humphries, K. M., Yoo, Y., and Szweda, L. I., 1998, Inhibition of NADH-linked mitochondrial respiration by 4-hydroxy-2-nonenal, *Biochemistry* **37**:552–557.

Keller, J. N., Mark, R. J., Bruce, A. J., Blanc, E., Rothstein, J. D., Uchida, K., Waeg, G., and Mattson, M. P., 1997, 4-Hydroxynonenal, an aldehydic product of membrane lipid peroxidation, impairs glutamate transport and mitochondrial function in synaptosomes, *Neuroscience* **80**:685–696.

Kramer, J. H., Misik, V., and Weglicki, W. B., 1994, Lipid peroxidation-derived free radical production and postischemic myocardial reperfusion injury, *Ann NY Acad Sci* **723**:180–196.

Kuzuya, T., Hoshida, S., Kim, Y., Nishida, M., Fuji, H., Kitabatake, A., Tada, M., and Kamada, T., 1990, Detection of oxygen-derived free radical generation in the canine postischemic heart during late phase of reperfusion, *Circ. Res.* **66**:1160–1165.

Lucas, D. T., and Szweda, L. I., 1998, Cardiac reperfusion injury: Aging, lipid peroxidation, and mitochondrial dysfunction, *Proc. Natl. Acad. Sci. USA* **95**:510–514.

Lucas, D. T., and Szweda, L. I., 1999, Declines in mitochondrial respiration during cardiac reperfusion: Age-dependent inactivation of alpha-ketoglutarate dehydrogenase, *Proc. Natl. Acad. Sci. USA* **96**:6689–6693.

Maas, E., and Bisswanger, H., 1990, Localization of the alpha-oxoacid dehydrogenase multienzyme complexes within the mitochondrion, *FEBS Lett.* **277**:189–190.

Mark, R. J., Lovell, M. A., Markesbery, W. R., Uchida, K., and Mattson, M. P., 1997, A role for 4-hydroxynonenal, an aldehydic product of lipid peroxidation, in disruption of ion homeostasis and neuronal death induced by amyloid beta-peptide, *J. Neurochem.* **68**:255–264.

Matsuda, H., McCully, J. D., and Levitsky, S., 1997, Developmental differences in cytosolic calcium accumulation associated with global ischemia: Evidence for differential intracellular calcium channel receptor activity, *Circulation* **96**:II233–238; discussion II238–239.

McCord, J. M., 1988, Free radicals and myocardial ischemia: Overview and outlook, *Free Radical Biol. Med.* **4**:9–14.

Moreno-Sanchez, R., Hogue, B. A., and Hansford, R. G., 1990, Influence of NAD-linked dehydrogenase activity on flux through oxidative phosphorylation, *Biochem. J.* **268**:421–428.

Nohl, H., and Hegner, D., 1978, Do mitochondria produce oxygen radicals *in vivo? Eur. J. Biochem.* **82**:563–567.

Nohl, H., Breuninger, V., and Hegner, D., 1978, Influence of mitochondrial radical formation on energy-linked respiration, *Eur. J. Biochem.* **90**:385–390.

Otani, H., Tanaka, H., Inoué, T., Umemoto, M., Omoto, K., Tanaka, K., Sato, T., Osako, T., Masuda, A., Nonoyama, A., and Kagawa, T., 1984, *In vitro* study on contribution of oxidative metabolism of isolated rabbit heart mitochondria to myocardial reperfusion injury, *Circ. Res.* **55**:168–175.

Perham, R. N., 1991, Domains, motifs, and linkers in 2-oxo acid dehydrogenase multienzyme complexes: A paradigm in the design of a multifunctional protein, *Biochemistry* **30**:8501–8512.

Porpaczy, Z., Sumegi, B., and Alkonyi, I., 1987, Interaction between NAD-dependent isocitrate dehydrogenase, alpha-ketoglutarate dehydrogenase complex, and NADH:ubiquinone oxidoreductase, *J. Biol. Chem.* **262**:9509–9514.

Reed, L. J., 1974, Multienzyme complexes, *Acc. Chem. Res.* **7**:40–46.

Reimer, K. A., and Jennings, R. B., 1992, Myocardial ischemia, hypoxia, and infarction, in *The Heart and Cardiovascular System* (H. A. Fozzard, E. Haber, R. B. Jennings, A. M. Katz, and H. E. Morgan Eds.) Raven Press, New York, pp. 1875–1973.

Rokutan, K., Kawai, K., and Asada, K., 1987, Inactivation of 2-oxoglutarate dehydrogenase in rat liver mitochondria by its substrate and *t*-butyl hydroperoxide, *J. Biochem.* (Tokyo) **101**:415–422.

Sayre, L. M., Zelasko, D. A., Harris, P. L., Perry, G., Salomon, R. G., and Smith, M. A., 1997, 4-Hydroxynonenal-derived advanced lipid peroxidation end products are increased in Alzheimer's disease, *J. Neurochem.* **68**:2092–2097.

Shigenaga, M. K., Hagen, T. M., and Ames, B. N., 1994, Oxidative damage and mitochondrial decay in aging, *Proc. Natl. Acad. Sci.* USA **91**:10771–10778.

Stadtman, E. R., 1992, Protein oxidation and aging, *Science* **257**:1220–1224.

Szweda, L. I., and Stadtman, E. R., 1992, Iron-catalyzed oxidative modification of glucose-6-phosphate dehydrogenase from *Leuconostoc mesenteroides*:Structural and functional changes, *J. Biol. Chem.* **267**:3096–3100.

Szweda, L. I., Uchida, K., Tsai, L., and Stadtman, E. R., 1993, Inactivation of glucose-6-phosphate dehydrogenase by 4-hydroxy-2-nonenal. Selective modification of an active-site lysine, *J. Biol. Chem.* **268**:3342–3347.

Tsai, L., and Sokoloski, E. A., 1995, The reaction of 4-hydroxy-2-nonenal with N alpha-acetyl-L-histidine, *Free Radical Biol. Med.* **19**:39–44.

Tsai, L., Szweda, P. A., Vinogradova, O., and Szweda, L. I., 1998, Structural characterization and immunochemical detection of a fluorophore derived from 4-hydroxy-2-nonenal and lysine, *Proc. Natl. Acad. Sci. USA* **95**:7975–7980.

Turrens, J. F., and Boveris, A., 1980, Generation of superoxide anion by the NADH dehydrogenase of bovine heart mitochondria, *Biochem. J.* **191**:421–427.

Uchida, K., and Stadtman, E. R., 1993, Covalent attachment of 4-hydroxynonenal to glyceraldehyde-3-phosphate dehydrogenase: A possible involvement of intra- and intermolecular cross-linking reaction, *J. Biol. Chem.* **268**:6388–6393.

Uchida, K., Szweda, L. I., Chae, H. Z., and Stadtman, E. R., 1993, Immunochemical detection of 4-hydroxynonenal protein adducts in oxidized hepatocytes, *Proc. Natl. Acad. Sci. USA* **90**:8742–8746.

Ueta, H., Ogura, R., Sugiyama, M., Kagiyama, A., and Shin, G., 1990, 02 partly spin trapping on cardiac submitochondrial particles isolated from ischemic and non-ischemic myocardium, *J. Mol. Cell. Cardiol.* **22**:893–899.

Ullrich, O., Siems, W. G., Lehmann, K., Huser, H., Ehrlich, W., and Grune, T., 1996, Inhibition of poly(ADP-ribose) formation by 4-hydroxynonenal in primary cultures of rabbit synovial fibroblasts, *Biochem. J.*. **315**:705–708.

Veitch, K., Hombroeckx, A., Caucheteux, D., Pouleur, H., and Hue, L., 1992, Global ischaemia induces a biphasic response of the mitochondrial respiratory chain: Anoxic pre-perfusion protects against ischaemic damage, *Biochem. J.* **281**:709–715.

Zweier, J. L., 1988, Measurement of superoxide-derived free radicals in the reperfused heart. Evidence for a free radical mechanism of reperfusion injury, *J. Biol. Chem.* **263**:1353–1357.

Mitochondrial Ion Homeostasis and Necrotic Cell Death

Chapter 6

Ca^{2+}-Induced Transition in Mitochondria: A Cellular Catastrophe?

Robert A. Haworth and Douglas R. Hunter

1. INTRODUCTION

The material presented here has in the past encountered difficulties in publication and reviewer reception. It is gratifying that evidence now exists for a role of the Ca-induced transition in some of the phenomena described below. We have not modified our original manuscript (Sections 2-6) in light of progress since 1979, though today we might choose our words differently.

Research in apoptosis has implicated the Ca-induced transition as a possible critical step in the cell death program (Zamzami *et al.*, 1996) as well as in necrosis (Lemasters *et al.*, 1998). Evidence has grown that the permeability transition pore may well incorporate interactions with outer membrane proteins (Beutner *et al.*, 1996) and that the megachannel identified as the permeability transition pore (Szabò and Zoratti 1992) is modulated by targeting protein-transport-regulating peptides (Lohret and Kinnally 1995). There is some evidence for a significant rate of spontaneous transitions under normal conditions, as measured by the ability of cyclosporin to increase retention of ^{45}Ca by isolated myocytes (Altschuld *et al.*, 1992). Such measurements are fraught with difficulty, however, because of the potential for nonhomogeneous uptake of ^{45}Ca by abnormal round cells in the preparation, which can be visualized by autoradiography (Haworth *et al.*,

Robert A. Haworth and Douglas R. Hunter   Department of Surgery, University of Wisconsin Clinical Science Center, Madison, Wisconsin 53792.

Mitochondria in Pathogenesis, edited by Lemasters and Nieminen.
Kluwer Academic/Plenum Publishers, New York, 2001.

1998). The question of the prevalence and role of spontaneous mitochondrial transitions induced by Ca under normal conditions *in vivo* thus remains open.

2. PROPERTIES OF THE Ca^{2+}-INDUCED TRANSITION

The swelling action of Ca^{2+} on mitochondria has been recognized for almost as long as the mitochondrion has been studied in isolation. Generally the phenomenon has been dismissed as pathological because of its deleterious effect on mitochondrial function and morphology. In concert with this view, a commonly advanced cause of the phenomenon is disruption of mitochondrial structural integrity by the activation of phospholipase activity (Aleksandrowicz *et al.*, 1973; Harris, 1977; Waite *et al.*, 1969). Recent work in our laboratory on the swelling action of Ca^{2+} has established, however, that the Ca^{2+}-induced permeability mechanism has properties of control and specificity that elevate it to the status of a specific controller of mitochondrial and cellular function (Haworth and Hunter, 1979; Hunter and Haworth, 1979a, 1979b; Hunter *et al.*, 1976).

The general characteristics of Ca^{2+}-induced permeability may be summarized as follows:

1. The permeability change is reversible by simply removing the Ca^{2+} via chelation with EGTA (Haworth and Hunter, 1979). This implies that the increase in permeability caused by Ca^{2+} is a direct consequence of binding to the membrane.
2. The site of action of Ca^{2+} is internal (Hunter *et al.*, 1976), is specific for Ca^{2+} over Sr^{2+} (Haworth and Hunter, 1979; Hunter and Haworth, 1979a; Hunter *et al.*, 1976), and has a high affinity for Ca^{2+} (Haworth and Hunter, 1979).
3. The Ca^{2+}-induced permeability is discontinuous in time, a feature we have described as transitional (Haworth and Hunter, 1979; Hunter *et al.*, 1976). This accounts for the sudden change in configuration of mitochondria from aggregated to orthodox, as seen by electron microscopy (Hunter *et al.*, 1976).
4. The permeability change is nonspecific. All molecules of molecular weight <1000, both charged and neutral, pass through the membrane when the transition in permeability occurs (Haworth and Hunter, 1979; Hunter *et al.*, 1976).
5. The permeability change is highly regulated by Mg^{2+}, endogenous ADP and NADH, and mitochondrial energization state (Hunter and Haworth, 1979a).
6. The permeability change is accompanied by a loss in respiratory control and an inhibition of NAD^+-linked respiration. When permeability is reversed, coupling is restored.

From this summary it is clear that we are dealing with a new dimension in membrane permeability. Whereas membrane transport is generally supposed to require carriers or channels specific for the substrate transported, the specificity in this system resides not in what moves across the membrane but rather in how the permeability change is induced and controlled. The above properties of the permeability induced by Ca^{2+} led us to infer the existence of a hydrophilic channel across the mitochondrial inner membrane that can be activated by Ca^{2+} and inhibited by the binding of ADP and NADH (Haworth and Hunter, 1979).

The discovery of such a system naturally raises a question as to what its purpose could be. The transition is not unique to heart mitochondria. Previous studies on Ca^{2+}-induced changes in configuration, permeability, and function indicate to us that the transition also occurs in mitochondria from liver (Azzi and Azzone, 1966; Binet and Volfin, 1975; Chappell and Crofts, 1965; Crofts and Chappell, 1965; Hackenbrock and Caplan, 1969), adrenal cortex (Allmann et al., 1970a, 1970b, 1970c; Pfeiffer et al., 1976), kidney (Mason and Tobes, 1977) and skeletal muscle (Wrogemann et al., 1973). In unpublished studies we have confirmed that mitochondria from heart, kidney, liver, and brain will undergo Ca^{2+}-induced transitions in a broadly similar manner. The transition may indeed be as widespread as the capability for high-affinity Ca^{2+} uptake, which would suggest that the primary function of the transition must be something useful to many cell types, although different secondary adaptations of the transition could manifest in cells with specialized functions. The question of the transition's role in pathology is now debatable. Whereas phospholipase is unambiguously catabolic in function, the reversibility of many properties of the transition raises the question of whether a transition in a mitochondrion *in vivo* needs to be catastrophic for the mitochondrion or for the cell. We showed earlier that the properties of oxidative phosphorylation, respiratory control, and membrane impermeability to solutes could be restored to post-transition mitochondria merely by removing free Ca^{2+} from the medium (Haworth and Hunter, 1979; Hunter et al., 1976). The phenomenon of energized contraction can restore mitochondria to an aggregated configuration: It was demonstrated years ago (Lehninger, 1959) that mitochondria swollen by fatty acids or phosphate, which we now know act by potentiating transitions induced by endogenous Ca^{2+} (Hunter et al., 1976), could be contracted again by the addition of ATP and Mg^{2+}. Finally, the inhibition of pyruvate–malate oxidation seen prior to the Ca^{2+}-induced transition (Haworth and Hunter, 1979) can be completely reversed by adding back cofactors (Table I). The reversibility of the transition *in vivo* thus appears to depend on the cell's energy state and ability to remove Ca^{2+} from the cytoplasm. Conditions have not yet been found for reversing ATPase activity or energized H$^+$ ejection induced by the transition, nor is it clear whether these induced properties are fundamentally irreversible or whether they could be reversed given the right conditions.

One approach to deducing the transition's role is to examine the circumstances under which controlling agents can be expected to vary. The first question is: Under normal conditions, such as state 3 or state 4 respiration with levels of Ca^{2+} realistic for mitochondria *in vivo*, how likely is the transition to occur? Estimating realistic values for mitochondrial Ca^{2+} content *in vivo* is not easy because of spurious mitochondrial Ca^{2+} uptake during homogenization, but we have found values as low as 1 nmol Ca^{2+}/mg mitochondrial protein in heart. The reason for this low level appears to be the action of the Na$^+$-induced Ca^{2+} release system (Crompton et al., 1977). We have found (Fig. 1) that at low Ca^{2+} levels, efflux induced by Na$^+$ is much more potent than even uncoupler-induced Ca^{2+} release. Only at levels of mitochondrial Ca^{2+} greater than 80 nmol/mg is uncoupler-induced efflux more effective. We have shown previously that this uncoupler-induced efflux is via the Ca^{2+}-induced transition (Hunter and Haworth, 1979b). The rate of spontaneous transition-induced Ca^{2+} efflux will be even slower than that induced by uncoupler. If conditions are further modified by adding ATP and Mg^{2+} to simulate the intracellular environment more closely, the transition can become so inhibited that even at

Table I
Reconstitution of Pyruvate–Malate Respiration in Mitochondria that have Undergone a
Ca^{2+}-Dependent Transition.

	Respiration Rate (μatom0/min/mg)			
	Without Uncoupler	With Uncoupler	Respiratory Control Index	% Reconstitution
HBHM + arsenate (before transition)[b]	0.049	0.281	5.75	100
HBHM + arsenate (after transition)[c]	0.035	0.035	1.0	12
HBHM + arsenate (after transition + co-factors)[d]	0.091	0.272	3.0[e]	97[f]

[a] HBHM (1 mg/ml) were suspended in 250 mM sucrose, 20 mM *tris* Cl pH in the presence of 0.9 mM K^+ arsenate.
[b] 5 mM K^+ pyruvate and 5 mM K^+ malate were added immediately, and oxygen consumption rate polarographically measured both before and after addition of 1 μM mCl-CCP.
[c] Mitochondria incubated 10 min in the presence of arsenate before addition of pyruvate and malate.
[d] The medium was supplemented with the following cofactors after 10 min of incubation, before addition of pyruvate and malate: coenzyme A (0.2 mM), thiamine pyrophosphate (0.4 mM), $NADP^+$ (0.3 mM), NAD^+ (0.5 mM), GDP (0.5 mM), Pi (5 mM), and $MgCl_2$ (5 mM). The two most significant cofactors were NAD^+ and Mg^{2+}.
[e] 5.8 in the absence of added phosphate.
[f] For full reconstitution, it was important to add Mg^{2+} last; see text for details

levels of 200 nmol Ca^{2+}/mg, uncoupler has difficulty in triggering Ca^{2+} efflux (Fig. 2). At the steady-state level of 1 nmol Ca^{2+}/mg mitochondria, we may therefore safely conclude that Na^+-induced Ca^{2+} efflux will throughly dominate spontaneous efflux by the Ca^{2+}-induced transition.

The ability of protective mechanisms to suppress the Ca^{2+}-induced transition so powerfully under normal conditions forces us to allow a role for the transition under only three broad circumstances. The first is under pathological conditions where cellular Ca^{2+} influx may be excessive; here the transition could serve as a death mechanism. The second is under normal conditions, where transitions could be induced by some external agent (such as a hormone) capable of specifically removing the natural inhibition of the transition. The third concerns possible roles for the transition as it occurs spontaneously at its highly inhibited rate in the normal cell.

3. TRANSITION AS A DEATH MECHANISM

A feature of the Na^+ release system illustrated in Figure 1 is its limited capacity for Ca^{2+} release. Any rate of mitochondrial Ca^{2+} accumulation in excess of 10 nmol/mg/min will result in the net accumulation of Ca^{2+}, even in the presence of Na^+. As the level of mitochondrial Ca^{2+} rises, the probability of mitochondria undergoing a transition also rises (Fig. 1). At this point, the fate of the cell is balanced on a knife edge. Conditions in the cell will tend to favor degeneration: Transition-induced mitochondrial ATPase activity will further strain the energy supplies of a struggling cell; Ca^{2+} released by the mitochondrion will be taken up by other mitochondria, increasing their chances of

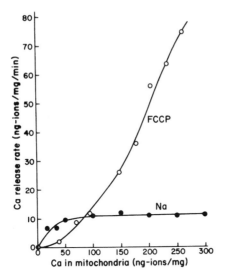

FIGURE 1. Comparison of Na$^+$-induced and uncoupler-induced Ca^{2+} release from mitochondria. Beef heart mitochondria were isolated (Hunter *et al.*, 1976) with EGTA present to obtain low values of endogenous mitochondrial Ca^{2+} (approximately 8 ng-ions Ca^{2+}/mg). Mitochondria (0.3 mg/ml) were added to a solution of 250 mM in sucrose, 20 mM in MOPS (adjusted to pH 7.1 with *tris*), 5 mM in succinate (K$^+$ salt), 2 mM in Pi (K$^+$ salt), 40 mM in KCl, and 6 µM in ADP, containing 1 mg/ml dialyzed BSA. One µM rotenone was added, followed by ^{45}Ca^{2+}. A series of experiments was done with varying amounts of ^{45}Ca^{2+}. For each experiment the amount of Ca^{2+} taken up by mitochondria within 2 min was measured. Total amount of Ca in mitochondria was determined by correcting this value for endogenous Ca content. EGTA was added and unidirectional Ca^{2+} release was measured (Hunter and Haworth, 1979b). Rate of Na$^+$-induced Ca^{2+} release was estimated from unidirectional Ca^{2+} release rate in the presence of 10 mM Na$^+$ minus the rate in the absence of Na$^+$. Rates were estimated from the first minute of release, during which time the release rates were approximately linear. Uncoupler-induced Ca^{2+} release was measured following the addition of 10 µM carbonylcyanide p-trifluoromethoxyphenylhydrazone.

undergoing a transition. The decline in cellular energy level will tend to promote the transition by decreasing the protection offered by ADP, NADH, and energization (Hunter and Haworth, 1979a). A classical positive-feedback process is thus set in motion from which the cell may never recover, as each mitochondrion becomes a utilizer of cellular ATP rather than a supplier. The process may be reversible to a point, if the cause of the initial exposure to high Ca^{2+} (such as a damaged plasma membrane) can be remedied. Otherwise, the onset of mitochondrial transitions would be expected to signal the irreversible phase of cell degeneration.

Just such a transitional change in mitochondrial configuration has been identified by Trump *et al.* (1971) as occurring at onset of the irreversible stage of cell death. "Death mechanism" would not be too strong a term for such a role of the Ca^{2+}-induced transition. The control mechanisms' elaborate nature indicates that the susceptibility of mitochondria to Ca^{2+} is no accident. There is also a certain logic to such a mechanism. By striking at the powerhouse, it assures a rapid demise for a lethally injured cell and also minimizes unnecessary oxidation for useful substrates.

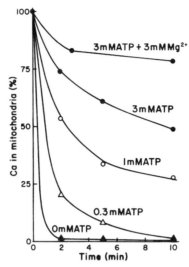

FIGURE 2. Inhibition by ATP and Mg^{2+} of uncoupler-induced Ca^{2+} release from mitochondria. Beef heart mitochondria (0.3 mg/ml) were incubated with 60 μM $^{45}Ca^{2+}$. Conditions for Ca^{2+} uptake as in Figure 1 but without rotenone. For each experiment the indicated amounts of ATP (K⁺ salt) and Mg^{+2} were added. Ten μM uncoupler was added and unidirectional Ca^{2+} release measured (Hunter and Haworth, 1979b).

4. TRANSITION INDUCED BY SPECIFIC REMOVAL OF PROTECTION

A powerful role for the transition could be as a Ca^{2+} release mechanism. If the transition is to be feasible as a mechanism of Ca^{2+} release in response to, say, hormonal stimulation (Blackmore *et al.*, 1979; Chen *et al.*, 1978), then intracellular second messengers must be postulated that could completely remove the inhibition of the transition via endogenous protective agents. Activation by removal of inhibition has many precedents in enzymology. Transitions under these conditions need not adversely affect the energy state of the cell, because the relatively small amount of Ca^{2+} would be removed by the plasma membrane pump, and mitochondria could remain coupled. The feature of transition-induced Ca^{2+} release that renders it of great potential value in control is that, like Na⁺-induced Ca^{2+} release, it is under separate regulation from high-affinity Ca^{2+} uptake. The Ca^{2+} is almost totally unable to exit through simple reversal of the ruthenium red-sensitive carrier in energized mitochondria (Hunter and Haworth, 1979b; Pozzan *et al.*, 1977; Puskin *et al.*, 1976). Nucleotides that regulate transition-mediated efflux (Haworth and Hunter, 1979; Hunter and Haworth, 1979a), and Na⁺, on the other hand, have little effect on Ca^{2+} uptake. The mitochondrion is therefore much more versatile than a simple Ca^{2+} buffer. The Ca^{2+} efflux rate from mitochondria, and hence the steady-state Ca^{2+} store in mitochondria, can in principle be regulated by the level of cytoplasmic Na⁺; the transition offers the possibility of a sudden release of that store. The importance of Ca^{2+} as a second messenger has been well established; it is implicated in a variety of cellular processes, including secretion, cell division, and muscle contraction (Berridge, 1975). The Ca^{2+}-induced transition therefore has the potential for being a link in controlling a whole range of cellular functions.

Another possible role for the transition that would require the specific binding of a cytoplasmic agent is as a pathway for the transfer of nuclear-coded proteins into the mitochondrion. On the basis of electron microscopic evidence, Butow *et al.* (1975) suggest that cytoplasmic ribosomes attach to specific points of fusion between the inner and outer membrane, and vectorially translate nascent polypeptides into the mitochondrial matrix. The size of the channel involved in the transition would suit it well for such a role.

5. ROLE OF SPONTANEOUS TRANSITIONS

We have concluded that the rate of the Ca^{2+}-induced transition is powerfully inhibited in the resting cell by virtue of the protective mechanisms and the low level of endogenous Ca^{2+}. It still appears inevitable that the transition must happen, however: as long as there is some Ca^{2+} in the mitochondria, the probability that a transition will occur cannot be reduced to zero. Of what use will this transition be to the cell, or to the mitochondrion, when it eventually does occur? One may envision several possibilities. A transition could signal the death of a mitochondrion, fitting it for lysosomal destruction. Or a transitional release of Ca^{2+} could be the event in the G_1 phase of the cell cycle that triggers the program for cell division (Berridge, 1975). Both mitochondrial turnover (Fletcher and Sanadi, 1961; Gross *et al.*, 1969) and initiation of cell division (Smith and

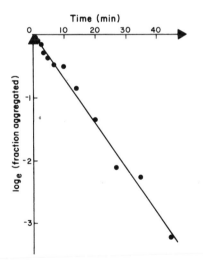

FIGURE 3. The exponential rate of the Ca^{2+}-induced transition in intact mitochondria under conditions of uniform protection. Beef heart mitochondria were suspended at 0.5 mg/ml in a buffer containing 250 mM sucrose 40 mM *tris* MOPS pH 7.1 at 30°. The Ca^{2+} ionophore A23187 (0.5 μg/mg) and uncoupler (1 μM) were added, and the suspension incubated for 4 min to deplete endogenous Ca^{2+} and burn off endogenous substrate. Then 4 nmol rotenone/mg, 2 mM β-hydroxybutyrate, and 100 μM CaCl₂ were added. Next the transition was initiated by removal of NADH protection (by adding 2 mM acetoacetate) at zero time. Twenty ml aliquots were fixed at time intervals by adding 16 ml 2% glutaraldehyde in 250 mM sucrose, 50 mM cacodylate pH 7.2. Samples were processed by conventional procedures thereafter (Wakabayashi *et al.*, 1971). Logarithmic plot linearity indicates an exponential time course of the transition in configuration, like that of radioactive decay, suggesting a uniform probability in time that any one mitochondrion will undergo a transition.

Martin, 1973) follow exponential kinetics, suggesting that they are governed by the laws of probability; both are also transitional in nature. When isolated beef heart mitochondria are incubated under conditions where the activity of transition-controlling agents is kept constant, a similar probabilistic nature can be demonstrated (Fig. 3). A third role for the transition could be as an equilibrator of cofactors essential for operation of the citric acid cycle, but for which specific membrane translocases are not known to exist. There must be a means for their entry into mitochondria, yet because of their catalytic function, very little entry activity is required.

Space prohibits discussion of possible roles for the Ca^{2+}-induced transition in the areas of control of steroid synthesis (Kadioglu and Harrison, 1971; Kahri, 1968; Pfeiffer and Tchen, 1973, 1975; Sabatini et al., 1962; Volk and Scarpelli, 1966) or of glycolysis/gluconeogenesis (Bygrave, 1966, 1967; Dorman et al., 1975). The broad scope of potential roles for the transition is at first disconcerting, but each possibility uses one or both of two features of Ca^{2+}-induced permeability: its control by protective mechanisms, and its all-or-nothing transitional nature. These two properties endow the transition with such versatility that the definition of its actual roles will doubtless occupy many hours of investigation in years to come.

ACKNOWLEDGEMENTS. We are indebted to Dr. David E. Green for his interest and support, and to John Lemasters for the invitation to present our work as a chapter in the book.

REFERENCES

Aleksandrowicz, Z., Swierczynski, J., and Wrzolkowa, T., 1973, Protective effect of nupercaine on mitochondrial structure, Biochim. Biophys. Acta 305:59–66.

Allmann, D. W., Wakabayashi, T., Korman, E. F., and Green, D. E., 1970a, Studies on the transition of the cristal membrane from the orthodox to the aggregated configuration: I. Topology of bovine adrenal cortex mitochondria in the orthodox configuration, J. Bioenerg. 1:73–86.

Allmann, D. W., Munroe, J., Wakabayashi, T., Harris, R. A., and Green, D. E., 1970b, Studies on the transition of the cristal membrane from the orthodox to the aggregated configuration: II. Determinants of the orthodox-aggregated transition in adrenal cortex mitochondria, J. Bioenerg. 1:87–107.

Allmann, D. W., Munroe, J., Wakabayashi, T., and Green, D. E., 1970c, Studies on the transition of the cristal membrane from the orthodox to the aggregated configuration: III. Loss of coupling ability of adrenal cortex mitochondria in the orthodox configuration, J. Bioenerg. 1:331–353.

Altschuld, R. A., Hohl, C. M., Castillo, L. C., Garleb, A. A., Starling, R. C., and Brierley, G. P., 1992, Cyclosporin inhibits mitochondrial calcium efflux in isolated adult rat ventricular cardiomyocytes, Am. J. Physiol. 262:H1699–H1704.

Azzi, A., and Azzone, G. F., 1966, Swelling and shrinkage phenomena in liver mitochondria: III. Irreversible swelling induced by inorganic phosphate and Ca^{2+}, Biochim. Biophys. Acta 113:438–444.

Berridge, M. J., 1975, The interaction of cyclic nucleotides and calcium in the control of cellular activity, Adv. Cyclic. Nucleotide. Res. 6:1–98.

Beutner, G., B, Beutner, RÜck, A., Riede, B., Welte, W., and Brdiczka, D., 1996, Complexes between kinases, mitochondrial porin, and adenylate translocator in rat brain resemble the permeability transition pore, FEBS Lett. 396:189–195.

Binet, A., and Volfin, P., 1975, Regulation by Mg^{2+} and Ca^{2+} of mitochondrial membrane integrity: Study of the effects of a cytosolic molecule and Ca^{2+} antagonists, Arch. Biochem. Biophys. 170:576–586.

Blackmore, P. F., Dehaye, J.P., and Exton, J. H., 1979, Studies on alpha-adrenergic activation of hepatic glucose output: The role of mitochondrial calcium release in alpha-adrenergic activation of phosphorylase in perfused rat liver, J. Biol. Chem. 254:6945–6950.

Butow, R. A., Bennett, W. F., Finkelstein, D. B., and Kellems, R. E., 1975, Nuclear–cytoplasmic interactions in the biogenesis of mitochondria in yeast, in *Membrane Biogenesis* (A. Tzagoloff, Ed.) Plenum, New York and London, pp. 155–199.

Bygrave, F. L., 1966, The effect of calcium ions on the glycolytic activity of Ehrlich ascites-tumour cells, *Biochem. J.* **101**:480–487.

Bygrave, F. L., 1967, The ionic environment and metabolic control, *Nature* **214**:667–671.

Chappell, J. B., and Crofts, A. R., 1965 Calcium ion accumulation and volume changes of isolated liver mitochondria: Calcium ion-induced swelling, *Biochem. J.* **95**:378–386.

Chen, J. L., Babcock, D. F., and Lardy, H. A., 1978, Norepinephrine, vasopressin, glucagon, and A23187 induce efflux of calcium from an exchangeable pool in isolated rat hepatocytes, *Proc. Natl. Acad. Sci. USA* **75**:2234–2238.

Crofts, A. R., and Chappell, J. B., 1965, Calcium ion accumulation and volume changes of isolated liver mitochondria: Reversal of calcium ion-induced swelling, *Biochem. J.* **95**:387–392.

Crompton, M., Kunzi, M., and Carafoli, E., 1977, The calcium-induced and sodium-induced effluxes of calcium from heart mitochondria: Evidence for a sodium–calcium carrier, *Eur. J. Biochem.* **79**:549–558.

Dorman, D. M., Barritt, G. J., and Bygrave, F. L., 1975, Stimulation of hepatic mitochondrial calcium transport by elevated plasma insulin concentrations, *Biochem. J.* **150**:389–395.

Fletcher, M. J., and Sanadi, D. R., 1961, Turnover of rat liver mitochondria, *Biochim. Biophys. Acta* **51**:356–360.

Gross, N. J., Getz, G. S., and Rabinowitz, M., 1969, Apparent turnover of mitochondrial deoxyribonucleic acid and mitochondrial phospholipids in the tissues of the rat, *J. Biol. Chem.* **244**:1552–1562.

Hackenbrock, C. R., and Caplan, A. I., 1969, Ion-induced ultrastructural transformations in isolated mitochondria: The energized uptake of calcium, *J. Cell Biol.* **42**:221–234.

Harris, E. J., 1977, The uptake and release of calcium by heart mitochondria, *Biochem. J.* **168**:447–456.

Haworth, R.A., and Hunter, D. R., 1979, The calcium-induced membrane transition in mitochondria: II. Nature of the Ca^{2+} trigger site, *J. Biol. Chem.* **195**:460–467.

Haworth, R.A., Goknur, A.B., Biggs, A.V., Redon, D., and Potter, K. T., 1998, Ca uptake by heart cells: I. Ca uptake by the sarcoplasmic reticulum of intact heart cells in suspension, *Cell Calcium* **23**:181–198.

Hunter, D. R. and Haworth, R. A., 1979a, The calcium-induced membrane transition in mitochondria: I. The protective mechanisms, *Arch. Biochem. Biophys.* **195**:453–459.

Hunter, D. R., and Haworth, R. A., 1979b, The calcium-induced membrane transition in mitochondria: III. Transitional Ca^{2+} release, *Arch. Biochem. Biophys.* **195**:468–477.

Hunter, D. R., Haworth, R. A., and Southard, J. H., 1976, The relationship between permeability, configuration, and function in calcium treated mitochondria, *J. Biol. Chem.* **251**:5069–5077.

Kadioglu, D., and Harrison, R. G., 1971 The functional relationships of mitochondria in the rat adrenal cortex, *J. Anat.* **110**:283–296.

Kahri, A. I., 1968, Effects of actinomycin D and puromycin on the ACTH-induced ultrastructural transformation of mitochondria of cortical cells of rat adrenals in tissue culture, *J. Cell Biol.* **36**:181–195.

Lehninger, A. L., 1959, Reversal of various types of mitochondrial swelling by adenosine triphosphate, *J. Biol. Chem.* **234**:2465.

Lemasters, J. J., Nieminen, A.-L., Qian, T., Trost, L. C., Elmore, S. P., Nishimura, Y., Crowe, R. A., Cascio, W. E., Bradham, C. A., Brenner, D. A., and Herman, B., 1998, The mitochondrial permeability transition in cell death: A common mechanism in necrosis, apoptosis, and autophagy, *Biochim. Biophys. Acta* **1366**:177–196.

Lohret, T. A., and Kinnally, K. W., 1995, Targeting peptides transiently block a mitochondrial channel, *J. Biol. Chem.* **270**:15950–15953.

Mason, M., and Tobes, M. C., 1977, Opposing actions of Ca^{++} and ATP plus Mg^{++} in controlling the kynurenine aminotransferase activity of isolated rat kidney mitochondria, *Biochem. Biophys. Res. Commun.* **75**:434–441.

Pfeiffer, D. R., and Tchen, T. T., 1973, The role of Ca^{2+} in control of malic enzyme activity in bovine adrenal cortex mitochondria, *Biochem. Biophys. Res. Commun.* **50**:807–813.

Pfeiffer, D. R., and Tchen, T. T., 1975, The activation of adrenal cortex mitochondrial malic enzyme by Ca^{2+} and Mg^{2+}, *Biochemistry* **14**:89–96.

Pfeiffer, D. R., Kuo, T. H., and Tchen, T. T., 1976, Some effects of Ca^{2+}, Mg^{2+}, and Mn^{2+} on the ultrastructure, light-scattering properties, and malic enzyme activity of adrenal cortex mitochondria, *Arch. Biochem. Biophys.* **176**:556–563.

Pozzan, T., Bragadin, M., and Azzone, G. F., 1977, Disequilibrium between steady-state Ca^{2+} accumulation ratio and membrane potential in mitochondria: Pathway and role of Ca^{2+} efflux, *Biochemistry* **16**:5618–5625.

Puskin, J. S., Gunter, T. E., Gunter, K. K., and Russell, P. R., 1976, Evidence for more than one Ca^{2+} transport mechanism in mitochondria, *Biochemistry* **15**:3834–3842.

Sabatini, D. D., De Robertis, E. D. P., and Bleichman, H. B., 1962, Submicroscopic study of the pituitary action on the adrenocortex of the rat, *Endocrinology* **70**:390–406.

Smith, J. A., and Martin, L., 1973, Do cells cycle? *Proc. Natl. Acad. Sci. USA* **70**:1263–1267.

Szabò, I., and Zoratti, M., 1992, The mitochondrial megachannel is the permeability transition pore, *J. Bioenerg. Biomembr.* **24**:111–117.

Trump, B. F., Croker, B. P., and Mergner, W. J., 1971, The role of energy metabolism, ion, and water shifts in the pathogenesis of cell injury, in *Cell Membranes, Biological and Pathological Aspects* (G. W. Richter and D. G. Scarpelli, Eds.), Williams and Wilkins, Baltimore, MD, pp. 84–128.

Volk, T. L., and Scarpelli, D. G., 1966 Mitochondrial gigantism in the adrenal cortex following hypophysectomy, *Lab. Invest.* **15**:707–715.

Waite, M., Scherphof, G. L., Boshouwers, F. M., and Deenen, L. L., 1969, Differentiation of phospholipase A in mitochondria and lysosomes of rat liver, *J. Lipid Res.* **10**:411–420.

Wakabayashi, T., Korman, E. F., and Green, D. E., 1971, On the structure of biological membranes: The double-tiered pattern, *J. Bioenerg.* **2**:233–247.

Wrogemann, K., Jacobson, B. E., and Blanchaer, M. C., 1973, On the mechanism of a calcium-associated defect of oxidative phosphorylation in progressive muscular dystrophy, *Arch. Biochem. Biophys.* **159**:267–278.

Zamzami, N., Susin, S. A., Marchetti, P., Hirsch, T., Gomez-Monterrey, I., Castedo, M., and Kroemer, G., 1996, Mitochondrial control of nuclear apoptosis, *J. Exp. Med.* **183**:1533–1544.

Chapter 7

Physiology of the Permeability Transition Pore

Mario Zoratti and Francesco Tombola

1. INTRODUCTION

The mitochondrial permeability transition pore (PTP) is known only indirectly. Its properties are deduced from those of the phenomenon it causes—the permeability transition (PT)—and from electrophysiological observations of a mitoplast channel, the mitochondrial megachannel (MMC), believed to coincide with the PTP. The PT has been traditionally defined as the unselective permeabilization of the inner mitochondrial membrane to solutes of MW (molecular weight) up to 1.5 kDa, caused by the operation of a proteic channel.[1] It is most often studied by following isolated mitochondria swelling caused by influx of water and solutes through the open pore(s), driven by the osmotic gradient due to the presence of indiffusible matrix proteins (i.e., a *colloidosmotic* process). Such a state of affairs poses some complications: permeabilization may conceivably result from the activation or formation of more than one kind of channel, that is, the possible existence of more than one PTP must be kept in mind. In the absence of definitive evidence on its molecular composition, the PTP must be defined operationally on the basis of its pharmacological and biophysical properties. This is no easy task, because the PT is a many-faceted, multifactor phenomenon with no absolute dependence on any one parameter. A minimum set of properties may nonetheless be subjectively chosen, such that any

---

[1]This definition actually needs updating because selective forms of the PTP are now thought to exist (see Sect. 3.1.), but it is still adopted here for clarity.

Mario Zoratti and Francesco Tombola   CNR Unit for Biomembranes, Department of Biomedical Sciences, University of Padova, Padova, Italy.

Mitochondria in Pathogenesis, edited by Lemasters and Nieminen.
Kluwer Academic/Plenum Publishers, New York, 2001.

phenomenon attributed to the PTP ought to display them. This list might include: (1) a requirement for the presence of Ca^{2+} in the matrix, (2) inhibitability by cyclosporin A (CSA) (alone or together with other inhibitors, ineffective by themselves), (3) dependence on matrix pH (at least in the presence of cyclophilin D): the PT is inhibited at acidic pH values and, (4) voltage-dependence: the PTP must open upon depolarization if the mitochondria have been suitably primed by Ca^{2+} loading or other treatments.

Most properties of the PTP have been known since Hunter and Haworth's fundamental work (Haworth and Hunter, 1979, 1980; Hunter and Haworth, 1979a, 1979b). Several reviews are available (Bernardi, 1996; Bernardi and Petronilli, 1996; Bernardi et al., 1994; Gunter and Pfeiffer, 1990; Halestrap et al., 1998; Zoratti and Szabò, 1995). In this chapter we limit ourselves to a summary presentation of established aspects of PTP physiology; our main focus is on recent (post-1994) developments. We cite some pre-1994 papers, but a more complete list of references can be found in Zoratti and Szabò (1995).

This chapter briefly discusses localization of the PTP (Sect. 2.), its intrinsic characteristics (Sect. 3.), its induction/modulation/inhibition by the factors most relevant from the point of view of pathophysiological phenomena (Sect. 4.), and various current hypotheses on its molecular composition (Sect. 5.). For obvious reasons, the distinction between these topics cannot be sharp. The discussion is limited to data obtained from subcellular preparates; the permeability transition in cells and organs is covered in other chapters of this book.

2. LOCALIZATION

Because its opening results in swelling of the mitochondria, the PTP must be considered an inner membrane (i.m.) pore. Considerable evidence suggests that it might actually be formed by a supramolecular complex, however, comprising proteins of both inner and outer (o.m.) membranes, located at contact sites. The volume of mitochondrial periplasmic space under most experimental conditions increases in isolated mitochondria, because the presence of indiffusible periplasmic proteins causes a colloidosmotic influx of water and medium osmolites through porins. Nonetheless, contact sites are present in isolated organelles. Channels resembling the PTP have been observed upon reconstitution in planar bilayers or liposomes of complexes containing proteins of both the i.m. and o.m. believed to originate from contact sites (see Sect. 5.2.). Some evidence indicates that the PTP may be identifiable as the mBzR, a complex containing AdNT (adenine nucleotide translocator), VDAC (voltage-dependent anion channel), and an 18 kDa protein of the inner membrane (Kinnally et al., 1993; Szabò and Zoratti, 1993; but see Halestrap et al., 1997a; see also Hirsch et al., 1998). Electrophysiological experiments indicate a low density of pores in mitoplast membrane (a few per mitoplast), consistent with localization at residual contact sites. Other researchers applying electrophysiological methods have concluded that the pore may be part of the i.m. protein import complex (Sect. 5.5.). Protein import takes place at contact sites. The permeability transition is strongly reduced in mitochondria overexpressing Bcl2 (Susin et al., 1998, and refs. therein), which is located in the o.m. and at contact sites (Riparbelli et al., 1995; Reed et al., 1998, and refs. therein). To reconcile i.m. permeabilization and the localization at contact sites, envisioned as zones where the two membranes are apposed rather than fused, some authors have proposed a model whereby an i.m. pore (possibly formed by the AdNT) and an o.m. pore (possibly porin, the o.m. protein import complex, or Bax) would form a "sandwich" or gap-junction-like structure (Haworth and Hunter, 1980; Kinnally *et al.*, 1992).

3. INTRINSIC CHARACTERISTICS OF THE PERMEABILITY TRANSITION PORE

3.1. Size and Selectivity

Information on the properties of the open PTP has been gleaned from classical experiments following mitochondrial swelling and the permeation of tracers, as well as from electrophysiological work. A high-conductance pore (the MMC) observed by patch-clamping swollen mitoplasts has been identified as the PTP on the basis of its pharmacological profile (Szabò and Zoratti, 1991, 1992; Szabò *et al.*, 1992), and it is considered in this review to coincide with the PTP. The PTP has classically been thought of as a large pore: solute exclusion studies suggested a diameter of 2–3 nm. As one might expect from its size, the pore is unselective, that is, it discriminates only on the basis of solute size. Electrophysiological determinations indicate a low degree of selectivity between anions and cations. Mounting evidence indicates that the PTP may come in many sizes, however with various permeability properties. Most patch-clamp recordings show channels with a maximum conductance of some 0.9–1.5 nS in 150 mM KCl, variable from experiment to experiment, with a preferred value of about 1.3 nS. This size variability suggests the possibility of underlying structural diversity as well. An evident feature is the abundance of substates, including a prominent, approximately half-size substate that suggests a mostly coordinated operation of two components, that is, a binary structure (see Fig. 1A). This "half-conductance" state appears most often as a transient state during opening or closing, and it is favored by high-salt media (Szabò and Zoratti, unpublished data). It is characterized by a fast kinetic mode, with trains of closures/openings from a nonconducting state. In some electrophysiological experiments the full-size pore appears to develop gradually from smaller pores. Furthermore, addition of CSA may induce the pore to enter low-conductance states (Zorov *et al.*, 1992; Szabò and Zoratti, unpublished data). These "precursor" or CSA-induced low-conductance pores are smaller than and distinct from the half-conductance major substate. Thus the electrophysiological evidence indicates that the MMC/PTP can adopt a variety of sizes.

Plenty of nonelectrophysiological data indicate that the PTP may exist in one or more small-pore-forming conformations. In fact, after the elegant kinetic analysis by Massari (1997), the idea is spreading that PTPs of different sizes may exist in different mitochondria and that the size of the pores may depend on the PT-inducing conditions used (e.g., Pfeiffer *et al.*, 1995; Broekemeier *et al.*, 1998). A major line of evidence consists in the observation that the time-course of mitochondria swelling does not always coincide with that of other phenomena thought to be associated with the PT. Work performed up to 1994, which points to a "first stage" of the PT involving a PTP conformation selectively permeable to protons and possibly to K^+, has been reviewed (Zoratti and Szabò, 1995, and refs. therein; Novgorodov and Gudz, 1996). In the investigations of Gogvadze and co-workers (1996)—early advocates of a small, Ca^{2+}-

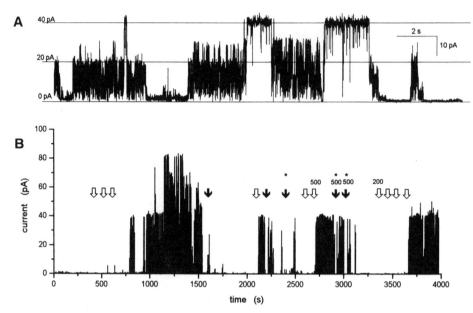

FIGURE 1. (A) MMC conductance states. An exemplificative trace from a patch-clamp experiment (the same as in B) on a rat liver mitoplast. Maximal conductance was approximately 1 nS. At least three other conductance levels are clearly identifiable, including the "half-conductance" substate (horizontal line; 20 pA). Pipette and bath medium: 150 mM KCl. Sampling frequency: 1 KHz. Filter: 500 Hz. V Pipette: +40 mV. (B) Competition between Ca^{2+} and ADP. (Same experiment and conditions as in (A). Bath volume approximately 1 ml. Current (leaks subtracted) flowing in the circuit was averaged over 5-sec consecutive intervals; means were plotted sequentially. Arrows indicate, Ca^{2+} (⇩) or ADP (⬇) added to the bath (100 nmols unless otherwise specified by numbers above arrows). The patch initially displayed only 107 pS channels. Addition of 300 nmol Ca^{2+} elicited, after a lag, activity by two MMCs, promptly inhibited upon adding 100 nmol ADP. One channel was reactivated by adding more Ca^{2+}, again inhibited by ADP, reactivated by Ca^{2+}, reinhibited by ADP, and reactivated once more by Ca^{2+} before loss of the tight seal. Asterisk indicates gaps in the record due to noise from additions (Szabò and Zoratti, unpublished data.)

selective PTP—the other phenomenon has most often been Ca^{2+} efflux. The group has long insisted, in the face of considerable skepticism, that a selective Ca^{2+} efflux from Ca^{2+}-loaded, prooxidant-treated mitochondria takes places before detectable swelling ensues and does so with no decrease in mitochondrial transmembrane potential. This efflux was sensitive to a battery of PT inhibitors, including CSA, so the authors considered it to represent the first stage of a process leading eventually to the PT. Gogvadze's group has proposed that the mechanism leading to this efflux implicates (a) NAD(P)H oxidation, (b) intramitochondrial NADP hydrolysis, and (c) ADP-ribosylation of one or more i.m. proteins. Whether steps (b) and (c) actually occur or are relevant remains to be established (see Zoratti and Szabò, 1995). There is also evidence that membrane thiol-modifying agents such as gliotoxin or phenylarsine oxide (PhAsO) induce CSA-sensitive Ca^{2+} release unaccompanied by mitochondrial swelling (Schweizer and Richter, 1994, 1996a, 1996b; Schweizer et al., 1994). Control of this phenomenon by redox events thus closely mimics that of the classical PT (Sect. 4.).

Glutathione is lost by mitochondria upon induction of the PT (Savage *et al.*, 1991). Recent reports have shown that this efflux can be partially uncoupled from PT-dependent swelling: various PT inhibitors, including CSA, polyamines, and polycations, inhibit Ca/Pi-induced swelling at concentrations that do not effectively block either glutathione loss or transmembrane-potential decrease (Rigobello *et al.*, 1993, 1995a; Savage and Reed, 1994; Reed and Savage, 1995). Higher concentrations of spermine or CSA do block glutathione release and depolarization, suggesting that the same machinery responsible for the PT may be involved.

Broekemeier and Pfeiffer (1995) have obtained evidence suggesting the gradual development of the PTP, from a K^+-permeable to Mg^{2+}-permeable to sucrose-permeable pore. The same group has reported the identification of a mitochondrial transmembrane potential- ($\Delta\psi$)-dissipating, sucrose-impermeable pore that remains active when PTP closure is induced by Ca^{2+} chelation (Broekemeier *et al.*, 1998). This conductance pathway, putatively a PTP substate, is inhibitable only in the presence of Mg^{2+} and under conditions (low oxygen tension) leading to a reduced state of the respiratory chain. This behavior suggested an association of the pore, and by inference of the PTP, with a complex of the redox chain, possibly cytochrome oxidase. As mentioned below (Sect. 4.3.), evidence has recently been presented that the full-size PTP is regulated by electron flow through complex 1.

Finally, convincing evidence has been provided by Ichas and co-workers for the formation of a Ca^{2+}- and K^+-permeable, sucrose-impermeable, Mg^{2+}-, ADP-, CSA-inhibitable, matrix pH-regulated "narrow PTP" (Ichas and Mazat, 1998, and refs. therein; Ichas *et al.*, 1994, 1997). This form of the channel, for some reason best observed in hypothonically swollen mitochondria, has been desumed to be activated not by Ca^{2+} but rather by matrix alkalinization following uptake of Ca^{2+} (or Sr^{2+}, which, being taken up more rapidly, is a better inducer of the narrow PTP even though it is an inhibitor of the full-size PTP). Its operation is prevented if the matrix pH increase is limited by a sufficiently rapid uptake (compared with Ca^{2+} uptake) of a weak acid (acetate, phosphate) or by nigericin/K^+. Under these circumstances mitochondria continue to accumulate Ca^{2+} until the full-size PTP eventually opens. If the matrix is allowed to become alkaline, instead, the narrow PTP opens, allowing Ca^{2+} and K^+ efflux (with consequent mitochondrial shrinkage) and matrix pH decrease via charge-balancing proton influx. The latter pH change would lead to reclosure of the narrow PTP, repolarization of the membrane, and resumption of Ca^{2+} uptake (see also Sect. 4.2.). An allegedly specific inhibitor of this form of the PTP, SDZ-PSC833, is available, and it should be useful in its further characterization.

In conclusion, the idea is now well supported that the same molecular machinery that forms the PTP (or a machinery closely related to it) can also form smaller-size pores capable of admitting only ions (H^+, K^+, Ca^{2+}, Mg^{2+}, and not necessarily all of these) or larger molecules still smaller than sucrose, the "classical" permeating osmolite. The exact factors leading to smaller-size versions of the PTP remain to be clarified, because they are formed upon treatment of the mitochondria with Ca^{2+}/Pi as well as with a variety of oxidizing and thiol-modifying agents and also upon matrix alkalinization. How many different small PTPs might form, and their relationship to the full-size PTP and to each other, also remains to be investigated. Unless otherwise stated, the remainder of this chapter deals with the full-size PTP.

3.2. Voltage Dependence

Voltage dependence can be summarized by stating that the pore tends to open upon depolarization, as confirmed by electrophysiological observations at the single-channel level. Halestrap's group (1998, and refs. therein) has proposed that voltage dependence reflects modulation of inhibitory ADP binding to the AdNT, which this group considers to form the PTP. The major supporting evidence is that when mitochondria are depleted of ADP by pretreatment with PPi, the PT is no longer voltage-sensitive, irrespective of whether the mitochondria were treated with SH reagents (Halestrap et al., 1997a). Voltage dependence is also observed, however, in patch-clamp experiments on excised patches and on mitoplasts that have undergone the PT and therefore presumably lost most of their soluble matrix components. We consider this a defining, intrinsic property of the PTP rather than a form of modulation. The concern that experiments supporting a role of the $\Delta\psi$ have been vitiated by uncoupler-induced production of radicals (reactive oxygen species, ROS) and consequent modification of thiol groups leading to PTP activation (Kowaltowski et al., 1996a) has been proven groundless (Scorrano et al., 1997a). Mitochondrial transmembrane potential can be modified by a variety of PT effectors (e.g., uncouplers, oligomycin), so it provides a unifying principle to explain pore behavior. Membrane surface potential is an integral part of the $\Delta\psi$ acting on the pore, thus agents modifying the surface potential may exert their action indirectly via changes of this parameter. This has been proposed as a possible or partial explanation for the effects of acyl-CoA's (Bernardi et al., 1994), of fatty acids, of long-chain acyl cations such as sphyngosine and analogous compounds, of trifluoperazine and other local anaesthetics capable of forming amphipatic cations, of external divalent or trivalent cations (Broekemeier and Pfeiffer, 1995), of spermine and polyamines in general (Broekemeier and Pfeiffer, 1995; Rigobello et al., 1995a), and of compounds shifting the conformation of the AdNT between the C and the M states (Bernardi et al., 1994; Rottenberg and Marbach, 1990), such as ATR (atractyloside) and BGK (bongkrekate). Changes in AdNT distribution between its conformers strongly influence mitrochondrial i.m. surface potential. In all cases, compounds increasing the positive (or negative) charge at the outer (or inner) surface of the mitrochondrial i.m. would act to inhibit the PT, and the opposite effect would be exerted by those decreasing it.

Bernardi's group has presented evidence that many important PTP modulators (such as Ca^{2+}, Mg^{2+}, ADP, fatty acids, thiol oxidation and cross-linking) act by modifying its sensitivity to $\Delta\psi$ decreases. The PT-inducing agents would act to increase (i.e., shift toward physiological values) the threshold potential for PTP opening, with the opposite being true for negative modulators. Whether this reflects a modification of the sensitivity of the channel voltage-sensing domain (the *gate*) due to binding by various effectors, as the authors propose, is still uncertain (see discussion in Zoratti and Szabò, 1995).

Eriksson et al. (1997, 1998) presented evidence that Arg (or possibly Lys) residues, probably located on the matrix side of the membrane and readily accessible from the matrix, play an important role in voltage-sensing by the PTP. The basis for this conclusion was provided by a chemical modification experiment employing the arginine reagents phenylglyoxal and 2,3-butanedione. The rate of AdNT-catalyzed ATP/ADP exchange was substantially unaffected by the treatment. This suggested that the critical arginine(s) was not located on the AdNT—which does contain critical Arg residues (Nelson et al.,

1993)—and that voltage sensing occurred by a mechanism unrelated to ADP binding to the AdNT, as proposed instead by Halestrap's group (1997a, see above). Other controls ruled out PT inhibitions resulting from inhibition of transport systems or of cyp-D (cyclophilin D) activity (see below).

In patch-clamp experiments, applying moderate (30–60 mV) potentials of a polarity corresponding to physiological polarity to an MCC held at low voltage is sufficient to induce long-lasting closure within several seconds. The closure is reversible in the voltage range mentioned, that is, the channel generally reopens for brief periods. The effectiveness of the voltage pulse apparently increases with its absolute value: when the voltage is applied to an open channel, the average lag time before closure becomes shorter, and the open probability in the subsequent period becomes lower as the potential increases (Szabò and Zoratti, unpublished observations). No detailed study of these aspects has been performed (and, more generally, the kinetic behavior of the PTP/MMC has not yet been well characterized). When open at potentials of physiological polarity, the pore tends to reside in its maximal conductance state, with relatively infrequent (a few per second) transitions to various substates, and occasional full closures. At potentials of the opposite polarity substate transition frequency is much higher, and the current record may assume a "noisy" appearence, but the probability of lengthy sojourns in a fully closed states(s) does not increase with voltage. The probability of the channel residing in full-conductance state thus decreases with increasing potentials of either sign, but with different modalities.

4. INDUCTION/MODULATION

In *in vitro* experiments, a variety of tools are at the researcher's disposal to cause opening or closing of the PTP. The PT in isolated mitochondria is favored by Ca^{2+} loading in conjunction with inducers such as Pi, sources of free radical species, thiol cross-linkers and acylating agents, $\Delta\psi$-reducing and surface-potential-modifying reagents (see above), fatty acids, C-conformation-stabilizing ligands of the AdNT (ATR, CATR [carboxyatractyloside], acyl-CoA's, pyridoxal-5-phosphate), and compounds that can deplete the mitochondria of protective molecules (e.g., ADP depletion by PPi). Pore opening can be inhibited, or open pores closed, by agents that counteract the action of inducers. For example, radical scavengers will protect against oxidation of thiol groups, (dithiothreitol) and DTT will reverse it; chelation of Ca^{2+} will cause the closure of open pores by sequestering matrix as well as external Ca^{2+}; acylation of some thiol groups by, for example, low concentrations of NEM (*N*-ethylmaleimide), will protect against oxidation to disulfides. In addition, the PT is inhibited by CSA, acidification of the matrix, ADP (adenine nucleotides), divalent cations, surface potential modifying agents (see above), and compounds that stabilize the M conformation of the AdNT (BGK, matrix ADP). Which parts of the electron transport chain are used may also have important consequences on PT induction (see below). Instead of describing the various PT-promoting or -inhibiting agents and their mechanisms (for which see Gunter and Pfeiffer, 1990; Zoratti and Szabò, 1995), we present here an overview of the best-understood mechanistic aspects of PT modulation presumed to have physiological relevance. This is synonymous with discussing the major factors that modify the probability of PT occurrence in a given mitochondrial population.

Unless inhibitors are administered or Ca^{2+} is sequestered, prolonged closure after prolonged opening of the full-size PTP does not readily occur *in vitro*. Endogenous protective factors such as NAD(P)H and ADP are rapidly lost though the open pores. The very presence of the pores prevents re-establishment of potential and pH gradients. Re-establishment of i.m. impermeability is favored by the rapid loss of excess matrix Ca^{2+} and by stochastic channel closures. The latter are, judging from electrophysiological work, generally brief at potentials close to zero, and they would not lead to repolarization unless all the open PTPs in a mitochondrion simultaneously closed. Phenomena such as the gradual efflux of calcein from *in situ*, calcein-loaded mitochondria (e.g., Petronilli *et al.*, 1999) might result from brief PTP openings, which would not be expected to lead to serious depletion of PT-antagonizing factors. The above reasoning does not apply to the small PTP. Because these forms of the pore display selectivity (for Ca^{2+} small cations or both), they might allow only slow or no efflux of NAD(P)H and ADP (although data on this point are lacking). Their opening might be therefore more readily reversible, particularly in the case of the pH-sensitive pore described by Ichas (1997, 1998) and co-workers; reversible, CSA-sensitive depolarizations of single mitochondria *in situ* occurring spontaneously or in response to cytosolic Ca^{2+} transients within seconds (Loew *et al.*; 1993, 1994) may be due to the operation of this type of pore.

It is generally thought that PT induction results from modification (by Ca^{2+} binding, thiol oxidation and so forth, depending on circumstances) of the structure of a pre-existing protein or complex. The possibility that it might consist instead of assembly of the PTP from components must be kept in mind, however, and indeed, current thinking is that the association of cyp-D with a membrane component is a fundamental feature of PTP formation. It is worth repeating that PT induction is influenced by many parameters, including Ca^{2+} levels, pyridine nucleotide (and some thiol) redox states, levels of inhibitors such as ADP or Mg^{2+}, and so on, which complicates analyses. The effects described in the literature are rarely all-or-nothing; their magnitude depends on conditions.

4.1. Ca^{2+}

Matrix calcium is required for PT occurrence, although high accumulation levels are not strictly necessary. The amount of Ca^{2+} needed for PT induction depends on the coadjuvating agent used and on the content of PT-antagonizing compounds such as ADP, which may vary for each mitochondrion. In fact, isolated mitochondria are heterogenous populations with respect to PT induction (see Zoratti and Szabò, 1995). In experiments on previously permeabilized, A23187-treated completely uncoupled liver mitochondria, depleted of endogenous protectants, Chernyak and Bernardi (1996) found 1 μM Ca^{2+} sufficient to induce measurable permeabilization; a maximal effect was achieved with 10–30 μM Ca^{2+} (with 1 mM Pi in the medium). The classic inducer cocktail has long been Ca^{2+}/Pi. The role of Pi remains unclear; it probably acts indirectly by buffering the matrix pH and allowing Ca^{2+} accumulation. PT induction by Ca^{2+}/Pi is antagonized by radical scavenger BHT (butylhydroxytoluene) (at high Pi concentrations; note that a nonradicalic mechanism of PTP inhibition by BHT has recently been proposed by Gudz *et al.*, [1997]) and by catalase, indicating that potentiation of free radical processes may partially account for the effect of Pi (Carbonera and Azzone, 1988; Kowaltowski *et al.*, 1996a, 1996b). An interesting mechanistic proposal explaining this potentiation has been presented by (Kowaltowski et al. 1996b).

At any rate, Pore opening by Ca^{2+} does not necessarily require addition of Pi or other coadjuvants (see Zoratti and Szabò, 1995), as shown also by electrophysiological experiments. Evidence by Haworth and Hunter (1979) indicates the presence of two Ca^{2+} binding sites, displaying cooperativity effects, on the matrix side. Ca^{2+} appears to act competitively against a variety of inhibitors, including ADP, CSA, divalent cations, protons, local anaesthetics and $\Delta\psi$ (in the sense that at higher Ca^{2+} loads, lower extents of depolarization are needed to produce the PT) (see Zoratti and Szabò, 1995). This is clearly observed in patch-clamp experiments: sequential additions of ever-increasing concentrations of Ca^{2+} and divalent cations (Sr^{2+}, Mn^{2+}, Ba^{2+}, Mg^{2+}), ADP, protons, or CSA result in the correlated activation/inhibition of the PTP (Szabò et al., 1992; Fig. 1B). Competitive inhibition by divalent cations appears to involve the same (Ca^{2+}-binding) matrix-side sites. Specific properties of each ion, such as its hydration, ionic radius, or polarizability, presumably play a crucial role in determining its effect on the PT. Inhibitory binding sites for divalent cations (including Ca^{2+}), with I_{50} in the 0.2–0.3-mM range, have been identified on the cytoplasmic side of the mitochondrial membrane (Bernardi et al., 1993), where they may or may not act via modifications of surface potential. On the other hand, Kowaltowski and Castilho (1997) reported that extramitochondrial Ca^{2+} stimulates the reaction of thiol groups with PhAsO, a thiol cross-linker, thereby facilitating the PT. They also presented evidence (Kowaltowski et al., 1997) that Ca^{2+} binding to the inner mitochondrial membrane may result in increased exposure of membrane protein thiol groups to a hydrophilic reagent (mersalyl), with a decrease of those accessible to the hydrophobic reagent NEM.

Although it is a reasonable idea that Ca^{2+} and other divalent cations (at least those that can be taken up by mitochondria) may compete for the same protein binding sites it seems unlikely that the same binding sites are involved in the case of the other agents mentioned above. An explanation for their competitive behavior must therefore be sought. For Ca^{2+}/CSA, one possibility is direct competition for another common site, possibly on cyp-D (see Sect. 4.4.). In other cases one may envision separate binding sites that reciprocally influence each other via changes in protein conformation. Also, the patch-clamp experiments mentioned above suggest that the activating Ca^{2+} binding sites are unsaturable in a reasonable concentration range (Szabò et al., 1992; Szabò and Zoratti, unpublished observations), apparently disagreeing with results showing maximal PT induction in the 10–30 μM range (see above). Since membrane phospholipids come the closest to an unsaturable Ca^{2+} binding site in mitochondria, part of the Ca^{2+} effects at high Ca^{2+} concentrations are due to interactions with the inner membrane.

4.2. Matrix pH

Matrix protons inhibit PT onset and induce PTP closure in a Ca^{2+}-competitive manner (Bernardi et al., 1992; Haworth and Hunter, 1979, 1980; Nicolli et al., 1993; Szabò et al., 1992). This inhibitory effect can be substantially eliminated by pretreatment of the mitochondria with diethylpyrocarbonate (DPC), a histidine reagent, suggesting hystidyl residue involvement (Nicolli et al., 1993). There is evidence that this effect may result from a disruption of cyp-D interaction with the inner mitochondrial membrane

(Nicolli *et al.*, 1996) (see Sect. 4.4.). Acidification reportedly reduces the amount of cyp-D bound to sub-mitochondrial particles (SMPs) a decrease prevented by DPC treatment. Halestrap's (1998) group found no such effect, ascribing the discrepancy to nonspecific cyp-D/membrane interactions under the conditions used. The mechanism by which matrix protons inhibit the PT thus remains in doubt. Nonetheless, this effect may provide an explanation for the PT-favoring effect of accumulated Pi, especially if one considers the opening of a pH-dependent narrow PTP as a first step in permeabilization (see Sect. 3.1.). Opening of this pore in Ca^{2+}-loaded mitochondria would in fact cause rapid Ca^{2+} efflux and concomitant charge-balancing proton influx (in sucrose media). The resulting matrix pH drop, together with the matrix Ca^{2+} decrease and limited loss of other matrix molecules, would block PT development. In Pi-loaded mitochondria, acidification would be prevented by buffering and by the rapid efflux of Pi on its phosphate/OH^- antiporter, and PT development may proceed more easily.

4.3. Redox Events

A distinct activation mechanism of pathological significance involves redox events. The redox state of some SH groups has long been considered an important factor in determining whether a given mitochondrion will undergo the PT or not. Thiols are thought to be the target of a vast number of radical-producing agents, such as hydroperoxides, menadione, triiodothyronine (Castilho *et al.*, 1998) or the Fe^{3+}/xanthine/xanthine oxidase system, which cause the DTT-reversible (e.g. Petronilli *et al.*, 1994a) formation of disulfide bridges via RS˙ intermediates. Induction by this pathway may not require Ca^{2+} accumulation (although it is certainly favored by it), and the pore thus formed may not have exactly the same properties as the one formed with Ca^{2+}/Pi. Furthermore, PT induction is favored by SH reagents or cross-linkers such as NEM (at concentrations >50 μM; see below), diamide, or heavy metal ions. Induction by oxidizers and SH reagents is antagonized by NEM at lower concentrations (5–20 μM) and by another SH reagent, monobromobimane (MBB) (Costantini *et al.*, 1995). These effects have been interpreted as indicating the presence in the PTP protein(s) of two vicinal cysteinyl SH groups (the *S* site; see below) whose oxidation or cross-linking, and the ensuing PT, would be prevented by their (previous) reaction with NEM or MBB (Fig. 2, paths A–C). Higher concentrations of NEM would be able to modify other thiol classes, again causing PT formation (but see below for a new model). Because MBB did not affect the characteristics of Ca^{2+} loading-dependent PTP opening, the investigators concluded that these MBB-sensitive thiol groups do not play a role in PT induction by Ca^{2+} overload (Costantini *et al.*, 1995). In 1996 Chernyak and Bernardi reported, however, that oxidation or cross-linking of MBB-reactive thiols increased apparent sensitivity of the PTP opening to Ca^{2+} levels. Results obtained by Petronilli *et al.* (1994a) are in any case consistent with thiol cross-linking or covalent modification resulting in an upward shift of the threshold potential at which the PTP opens (see above, and also Costantini *et al.*, 1996).

The ability of high concentrations of NEM or other substituted maleimides to induce the PT has been tentatively rationalized in a novel way in a recent paper (Costantini *et al.*, 1998). The model offered is essentially based on the observation that reducing agents like DTT or β-ME (β-mercaptoethanol) prevent or reverse the PT induced by high NEM,

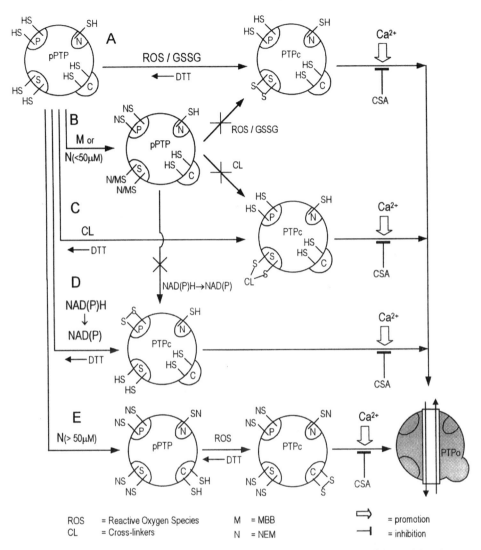

ROS = Reactive Oxygen Species   M = MBB   ⇨ = promotion
CL = Cross-linkers   N = NEM   ⊣ = inhibition

FIGURE 2. Redox modulation of the PTP. Circle: PTP precursor. Letters: putative thiol-containing sites (see text). P and N sites might coincide. The P site is considered to contain thiol groups. pPTP: PTP precursor. PTPc: PTP complex, primed by thiol oxidation or cross-linking for channel formation upon Ca^{2+} binding and/or $\Delta\psi$ decrease. PTPo: open permeability transition pore. The effects of high NEM are illustrated according to Costantini *et al's* hypothesis (1998). See text for details.

implying that high NEM somehow allows redox events to take place. Possibly NEM at high concentrations reacts with a thiol group(s) (labeled "N" in Fig. 2), inducing conformational changes that expose a criptic oxidizable site (probably another thiol pair, labeled "C" in Fig. 2) to oxidation, either "spontaneous" or induced by an added oxidizer (Fig. 2, path E). Both this oxidizable site and the site modifiable by high NEM are distinct from the S site.

Oxidizing agents need not react directly with PTP protein thiols. They may alter instead, or also, the redox state of the mitochondrial pools of glutathione and of pyridine nucleotides, with which the proteic thiols are believed to be in redox equilibrium. These two pools are connected via GSH peroxidase/reductase, an enzyme that uses NADPH to reduce GSSG to GSH. Oxidation of the NAD(P)H pool has long been considered a factor favoring the PT (e.g., Lehninger *et al.*, 1978; Prpic and Bygrave, 1980), and some evidence exists that the small PTP may be similarly modulated (Vercesi, 1984, 1987). Whether the relevant nucleotide is NADH or NADPH has been debated, with evidence presented both for the former (e.g., Haworth and Hunter, 1980; Chernyak and Bernardi, 1996) and for the latter (e.g., Prpic and Bygrave, 1980; Jung and Brierley, 1981; Roth and Dikstein, 1982; Vercesi, 1987). A $\Delta\psi$-dependent equilibrium between the redox poises of NADH and NADPH is maintained by the energy-linked transhydrogenase. Whether the GSH redox state is relevant for PT induction is uncertain, different results having been obtained by different groups.

The relationship between these redox pools and the thiols of the PTP has been investigated in recent studies (Costantini *et al.*, 1996; Chernyak and Bernardi, 1996), which have identified two dithiol sites. It appears that oxidation or cross-linking of the thiol groups in either (or both) favors mitochondrial permeabilization. The effects (PT facilitation) of modification of the two sites are additive, and could be distinguished by selective inhibition/reversal experiments. One site was identified as the NEM- and MBB-sensitive S site (see above). The redox state of this dithiol was not influenced by that of the PyNu (pyridine nucleotide) pool (Chernyak and Bernardi, 1996), being instead controlled by degree of GSH pool oxidation. The experiments excluded a direct effect of peroxides on proteic thiol groups under the conditions chosen, suggesting that glutathione obligatorily mediates S-site oxidation, a result confirmed and extended to diamide by Halestrap *et al.* (1997a).

The second (P) site sensed the redox state of the PyNu pool but not that of the GSH pool (Fig. 2, path D). In this case NEM (25 μM) could prevent the ensuing PT—the only evidence that the P site contains thiols—but MBB and other thiol reagents could not (Costantini *et al.*, 1996). Furthermore, DTT did not inhibit oxidation/swelling at the concentrations that reversed S-site cross-linking, and β-ME was also without effect. The doubt seems warranted, as stated by the authors themselves, that the NAD(P)(H) redox state might influence PTP operation by a mechanism not involving SH oxidation, possibly via direct interaction (Haworth and Hunter 1980; Hunter and Haworth, 1979a). It may be relevant in this context, and because the PTP has been proposed to be a complex containing VDAC (Szabò and Zoratti, 1993; Szabò *et al.*, 1993), that NADH and (less potently) NADPH, but not the oxidized PyNu's, have been reported to bind to VDAC, decreasing its permeability to ADP (Lee *et al.*, 1994, 1996; Zizi *et al.*, 1994). Be this as it may, these recent investigations confirm that PyNu oxidation is a PT-inducing factor, accounting at least in part for the PT-facilitating effects of compounds such as oxaloacetate, acetoacetate (Prpic and Bygrave, 1980; Jung and Brierley, 1981; Vercesi, 1984, 1987), and duroquinone (Chernyak *et al.*, 1995), which can be reduced with concomitant PyNu oxidation. The possibility that PyNu oxidation first results in the formation of a small, selective pore, which then proceeds to full-size PTP has been discussed above (Sect. 3.1.).

It appears therefore that both glutathione and pyridine nucleotides modulate the PT in an additive manner. Under normal circumstances these pools are in redox equilibrium, and their oxidation state is reflected—via reactions involving NADP(H) and the thioredoxin/thioredoxin reductase system; GSH/GSSG and glutathione reductase; and perhaps NADH, lipoate, and lipoate dehydrogenase (Rigobello *et al.*, 1995b, 1998, and refs. therein; Bindoli *et al.*, 1995)—by the oxidation state of thiol groups influencing the PT. The effect of experimental conditions and manipulations on these redox pools should thus be kept in mind when interpreting results. The extent of changes in the redox poise of both PyNu's and GSH depends markedly on experimental conditions such as the respiratory substrates used (Bindoli *et al.*, 1997; Fontaine *et al.*, 1998a, 1998b; Rigobello *et al.*, 1995b) and the presence of rotenone. A further point of relevance is that free radical processes contributing to the PT may be originated by electron leakage from the respiratory chain, particularly via oxidation of the semiquinone form of coenzyme Q (Kowaltowski *et al.*, 1995, 1996b; Turrens *et al.*, 1985). The redox state of coenzyme Q, which depends on experimental conditions, may therefore be an important factor influencing PT occurrence.

Extensive protein S-thiolation (i.e., the formation of mixed disulfides), with a decrease of reduced thiol groups and of free GSH and an increase of GSSG, occurs upon treatment of mitochondria with Ca^{2+} and oxidizing agents such as hydroperoxides and menadione (e.g., Bindoli *et al.*, 1997), but evidence for free radical involvement has been presented also for Ca/Pi-induced permeabilization (see Sect. 4.1.). Vercesi's group has maintained that the PT is the result of protein cross-linking to form high-MW (molecular weight) aggregates (e.g., Fagian *et al.*, 1990; Castilho *et al.*, 1995, 1996; Valle *et al.*, 1993). In these authors' view, the PT would proceed from a reversible process, possibly involving only intra-peptide-SS formation, to an irreversible permeabilization due to assembly of pore-forming heterogeneous protein clusters. The apparent role of the AdNT would result simply from its being the most abundant mitochondrial i.m. protein. Under even harsher conditions, permeabilization and destruction of the mitochondrial membrane arises also from lipid peroxidation. It then becomes a matter of semantics to establish what exactly constitutes the PT. We prefer to adopt a narrower definition, of a well-definable, though complicated, reproducible and reversible process that can be observed also in the absence of oxygen.

Very recently, evidence has been obtained that PTP function is modulated by electron flow through complex I of the respiratory chain (Fontaine *et al.*, 1998a, 1998b). Working with skeletal muscle mitochondria, the authors report that Ca^{2+} retention capacity (i.e., resistance to Ca^{2+}/Pi-induced PT) was much lower if the mitochondria were oxidizing site I substrates rather than succinate or ascorbate/TMPD. The effect was not a reflection of the PyNu pool, oxidation state nor was it due to H_2O_2 production. Furthermore, the ubiquinone analogs ubiquinone 0 and decylubiquinone were found to be potent PT inhibitors, irrespective of induction method or of the substrates being oxidized. These data raise the possibility that components of complex I may constitute, or be involved in a regulatory fashion with, the PTP (see Sect. 3.1. for somewhat analogous conclusions pertaining to a small PTP).

4.4. Cyclophylin D

The current interest in the PT can be traced to the discovery that CSA is a powerful inhibitor (Broekemeier *et al.*, 1989; Crompton *et al.*, 1988; Fournier *et al.*, 1987). Cyclosporin A is the most effective member of a family of cyclic endecapeptides that act as immunosuppressants. Though CSA is effective on the PT stimulated by a variety of inducing agents, the inhibitory effect is often partial, transient, or both, and its characteristics depend on the method of inducing the PT (e.g., Broekemeier and Pfeiffer, 1989, 1995). The efficacy of CSA is lower when the PT is induced by ROS or thiol cross-linking agents. Thus, for example, when the PT was induced by (besides Ca^{2+}) Pi or oxaloacetate, inhibition was and long-lasting and nearly complete, but if Hg^{2+} or t-BuOOH (t-butylhydroperoxide) were used as co-inducers, inhibition lasted just a few minutes (Brockemeier and Pfeiffer, 1995) or was low (Nemopuceno and Pereira-da-Silva, 1993). When it is scarcely effective alone, CSA displays a marked synergism with other PTP inhibitors, such as ADP, Mg^{2+}, or local anaesthetics such as trifluoperazine or butacaine. The inhibition's transitory nature under some circumstances is not understood: it may derive from loss of synergically protective matrix components, possibly via small forms of the PTP (see Sect. 3.1.). The variability of CSA effects is a strong argument for the existence of different, induction-mode-dependent, PTP forms.

Many now agree that the inhibitory effect of CSA is mediated by cyp-D, the mitochondrial member of the family of *cis-trans* peptidyl-prolyl isomerases (PPIases) known as *cyclophilins* (Halestrap *et al.*, 1998, and refs. therein; Nicolli *et al.*, 1996; Tanveer *et al.*, 1996; Zoratti and Szabò, 1995, and refs therein), which are the immediate target of cyclosporins. Cyp-D has been purified and sequenced (Connern and Halestrap, 1992; Woodfield *et al.*, 1997). Cyclosporin binds to, and inhibits, cyclophilin PPIase activity, and the potency of various derivatives as inhibitors of PPIase action parallels that as inhibitors of the PT. Scorrano *et al.*, (1997b), however, found that DPC inhibits the peptidyl prolyl-*cis/trans* isomerase activity of cyp-D, though it is a powerful PT inducer (Nicolli *et al.*, 1993). They tentatively concluded from this that cyp-D involvement in the PT might not be due to its isomerase activity. Furthermore, it does not involve an interaction of the CSA/cyclophilin complex with an as yet undiscovered mitochondrial calcineurin, because N-methyl-Val-4-CSA, a nonimmunosuppressive derivative, is as effective as CSA at PT inhibition (Nicolli *et al.*, 1996; Petronilli *et al.*, 1994b) (such an interaction is instead part of the immunosuppressive action). The favored mechanistic model is formulated instead in terms of binding of cyp-D to the PTP precursor. The view has been championed, most recently by Halestrap's group, that the PTP precursor is the AdNT. In this model, PT inducers would cause conformational changes of the AdNT, leading to increased binding of cyp-D. The cyp-D-bound AdNT would form the PTP upon Ca^{2+} triggering (Fig. 3A). CSA would inhibit by competing for cyp-D, with which it would form a complex no longer able to bind to the AdNT (e.g., Halestrap *et al.*, 1997b). Cyp-D can indeed bind to the inner mitochondrial membrane. There is evidence that treatment of mitochondria with the pore inducers t-BuOOH, diamide, and PhAsO, and decreases in the NADH/NAD ratio, increase this binding (Connern and Halestrap, 1994, 1996). The effect has been tentatively ascribed to modification of Cys56 of the AdNT (Halestrap *et al.*, 1997a). Matrix swelling has also been reported to favor the PT and to lead to more cyp-D binding (probably an ionic-strength effect) (Connern and Halestrap, 1996). According to Halestrap and co-workers (1998), ADP, BGK, CATR, matrix pH (but see Sect. 4.2.), and transmembrane potential do not influence cyp-D binding, suggesting that they modulate Ca-dependent opening by acting at independent sites. The AdNT is one of many mitochondrial membrane proteins that bind to a GST–cyp-D affinity column, and this binding is inhibited by both ATR and BGK (Halestrap *et al.*, 1998).

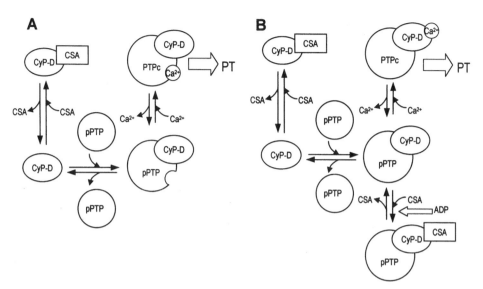

FIGURE 3. Cyclophylin D involvement in PTP formation. (A) A simple binding scheme. Cyp-D binding to the PTP precursor primes it for Ca^{2+} binding and pore formation. (B) A possible model seeking to accommodate the observed competitive effects of Ca^{2+} and CSA on the PT and MMC activity. Competition is envisioned as being for a common site (or overlapping sites) situated on cyp-D bound to PTP precursor. Though branches are not shown, CSA-bound and Ca^{2+}-bound cyp-D might also be involved in binding equilibria with the PTP precursor.

Irrespective of the molecular nature of the PTP or its "precursor," the simple binding model summarized above does not seem completely adequate to explain the observed phenomenology (as recognized by Halestrap *et al.*, 1997a, 1998). Within its framework, the fact that the PT can take place in the presence of CSA seemingly indicates that cyp-D binding is not essential for pore opening, but may rather enhance the sensitivity of the process to Ca^{2+} (Halestrap *et al.*, 1997a). Furthermore, high matrix Ca^{2+} can reverse the inhibitory effect of CSA (Connern and Halestrap, 1994, 1996; Crompton and Andreeva, 1994; Halestrap *et al.*, 1997a; Novgorodov *et al.*, 1992; Szabò *et al.*, 1992). Indeed, the effects of CSA, ADP, divalent cations, and protons can all be considered to consist of a shift toward higher Ca^{2+} of curves relating the probability of PTP opening and Ca^{2+} load. As mentioned (Sect. 4.1.), in patch-clamp experiments the sequential application of ever-higher concentrations of Ca^{2+} and CSA led to correlated opening/closing of the PTP, behavior suggestive of competition between CSA and Ca^{2+}; for (a) binding site(s), not necessarily located on the PTP complex. The results of Andreeva *et al.* (1995) raise the possibility that the competition may take place on membrane-bound cyp-D, and that the effect of CSA might be exerted not only by altering the binding between cyp-D and PTP precursor, but also by changing the properties of bound cyp-D (Fig. 3B shows one possible scheme). Photolabeling of heart mt (mitochondrial) membrane-bound PPIase with a photoactivatable, radioactive CSA derivative was in fact reportedly decreased by Ca^{2+} and increased by ADP (an effect not found when binding CSA to solubilized or purified cyp-D, implying the intervention of other components). The result might partially explain the cooperative effect of CSA and ADP in inhibiting the pore (see Zoratti and Szabò, 1995, and refs. therein). The fact that a CSA-photolabeled PPIase was found in the membrane protein fraction seems itself at odds with the model sketched above, which postulates that CSA-bound cyp-D detaches from, or cannot bind to, the PTP precursor. CSA-induced detachment of cyp-D from SMPs has been directly observed (Halestrap *et al.*, 1997a; Nicolli *et al.*, 1996), but it was not complete. Furthermore, the addition of cyp-D to SMPs under a variety of conditions failed to make them susceptible to permeabilization by PT-inducing agents (Crompton *et al.*, 1992; Halestrap *et al.*, 1998), suggesting the involvement of further soluble factors.

4.5. Adenine Nucleotides

Adenine nucleotides exert an inhibitory action on the PTP; the effectiveness decreases in the order ADP > ATP > AMP. The ADP acts from the matrix side (see discussion in Zoratti and Szabò, 1995). The protective effect of ADP is modeled as a decrease in affinity of the Ca-binding activating sites for Ca^{2+}, induced by a mechanism other than direct competition. It was classified by Haworth and Hunter (1980) as mixed-type inhibition, decreasing the V_{max} of PT propagation in a mitochondrial suspension and increasing the apparent K_m for Ca^{2+}. Analysis of their data led the authors to conclude that two cooperative binding sites per pore are present, mirroring the situation thought to exist for Ca^{2+}. Binding of either two Ca^{2+} ions or two ADP molecules would substantially inhibit binding of the other species (Haworth and Hunter, 1980). The PTP would open only if two Ca^{2+} ions, and no ADP, are bound. Halestrap *et al.* (1997a) have obtained evidence for the presence of two types of ADP binding sites. One site, with a high affinity (K_i approx. 1 µM, according to the authors), would be present only with the AdNT in the M conformation; the other, with a lower affinity (K_i approx. 25 µM), would exist independently of carrier conformation. A comparison of K_i values with K_d values for ADP binding to the AdNT suggested that both sites might reside on the AdNT, the higher-affinity site probably coinciding with the CATR binding site. The same paper reported that pretreatment of mitochondria with *t*-BuOOH, diamide, or PhAsO considerably increased the K_i for ADP at the higher-affinity site, an effect specular to and independent of that on cyp-D binding (see Sect. 4.4.). This effect has been attributed to modification of Cys159 of the AdNT (Halestrap *et al.*, 1997a). NADH behaves similarly to ADP, but it is a less powerful inhibitor, possibly acting at the same binding sites as ADP. Together the two compounds are synergistic (Haworth and Hunter, 1980).

4.6. Protease Action

Recent data suggest the possibility of protease involvement in PT induction. Gores and co-workers have shown that rat liver mitochondria contain a Ca^{2+}-activated, calpain-like cysteine protease, and that pre-incubation of mitochondria with cysteine protease inhibitors, for example, leupeptin and certain specific calpain inhibitors, strongly protects against PT induced by Ca^{2+}/Pi or *t*-BuOOH (Gores *et al.*, 1998, and refs. therein). The possibility that these inhibitors actually acted by reacting with cysteines in membrane proteins could not, however, be discounted. In fractionation studies, the calpain-like activity in question was found to reside in the intermembrane space. In assays employing isolated mitochondria suspended in a calcium-containing medium, however, its activity on a synthetic substrate was stimulated by almost one order of magnitude upon adding to the assay system a calcium ionophore, suggesting a matrix localization.

Exogenous caspase 1 can induce swelling and loss of potential by isolated mitochondria (Susin *et al.*, 1997) and caspases 2, 5, and 6–9 can cause $\Delta\psi$ dissipation and release of cytochrome *c* (Marzo *et al.*, 1998a, 1998b). Caspase 3, one of the effector caspases, is present in the intermembrane space of mitochondria and appears to be released upon induction of apoptosis (Mancini *et al.*, 1998). Conceivably, caspases may become activated and act on mitochondria during isolation, contributing, by a mechanism to be defined, to their susceptibility to PT inducers.

5. MOLECULAR IDENTITY

Which protein forms the PTP is still unknown, and identification remains the most pressing goal of research in this field. In fact, researchers might be faced with the task of identifying not one but several proteins, or one or more protein complexes, whose composition might even change depending on conditions. We briefly discuss here the current hypotheses, except the previously mentioned (Sect. 4.3) one that the PTP might arise from complex I of the respiratory chain.

5.1. Adenine Nucleotide Translocator

The idea of the AdNT as a component of the PTP was formulated about 20 years ago (Hunter and Haworth, 1979a; Le Quoc and Le Quoc, 1988; Zoratti and Szabò, 1995, and refs. therein). Because ATR favored, while BGK inhibited, PT development, the idea arose, supported by use of other AdNT ligands, that the pore would be formed by the AdNT in its C conformation (ATR-induced; BGK stabilizes the M conformation). Convincing evidence for AdNT involvement has been obtained. Recently the topic has been pursued mainly by Halestrap and co-workers (1998), who rationalize the PT-favoring effects of oxidative stress, thiol reagents, NADH oxidation, and increased matrix volume in terms of their effects on the binding of cyp-D and ADP (both considered to modulate calcium sensitivity) to different sites on the AdNT (Halestrap *et al.*, 1998; see above). Divalent cations and protons would act directly on the Ca^{2+} binding sites. The pore's voltage dependence has also been explained within this model, as an effect of voltage on carrier conformation and ADP binding (but see above). Halestrap and co-workers (1997a) explain the lack of a classical PT in yeast mitochondria (see Sect. 5.4.) by pointing out that all three AdNT isoforms in this organism lack both Cys56, the oxidation of which is proposed to favor cyp-D binding, and Pro61, proposed to be at the cyp-D binding site itself. Further relevant results are discussed in Section 5.3.

Recent electrophysiological data has shown that the isolated AdNT can form a high-conductance pore at high (1 mM) Ca^{2+} levels (Brustovetsky and Klingenberg, 1996; Tikhonova *et al.*, 1994). This strongly supports AdNT's candidacy as a PTP precursor. The properties of this channel may be compared to those of the MMC/PTP studied by patch-clamping mitoplast membranes or in suspensions of mitochondria. Specifically:

- The conductance of the pore formed by purified AdNT (from 300 to 600 pS in 100 mM KCl) is low compared with the maximal conductances of the MMC. On the other hand, it is comparable to that of the half-conductance state, so that Brustovetsky and Klingenberg might have studied a monomer whose dimer forms the MMC.
- Closures reportedly took place only at potentials much higher (>180 mV) than those effective on the MMC, and close to the physiological $\Delta\psi$ of mitochondria. This might be due to factors such as the absence of ADP and the high $[Ca^{2+}]$ levels used. Nonelectrophysiological experiments indicate that under the appropriate circumstances, PTP opening may take place even at physiological potentials.
- Gradual elimination of Ca^{2+} by perfusion with Ca-free medium resulted in a gradual decrease of the channel conductance. Comparable experiments have not been performed with the MMC, which closes completely and abruptly upon addition of excess EGTA. The MMC does display a variety of conductances, but the influence of $[Ca^{2+}]$ on their average occupancy has not been studied, nor, to our knowledge, has the dependence of the PTP size exclusion limit on $[Ca^{2+}]$. On the other hand, small-pore versions of the PTP are thought to form (see above).
- The gradual decrease of pore conductance as pH decreased has no correlation with the MMC behavior, which in our experiments was already completely closed pH 6.5, and closed abruptly upon transitions to acidic pH. At least according to Bernardi's group (Nicolli et al., 1993, 1996), the closure induced by acidification may be due to detachment of cyp-D, which was not present in the experimental system.
- The effect of ADP depended on the presence of BGK, clearly not the case with most of the relevant experiments, and again it seems to consist of conductance changes rather than full closure.

The discrepancies might be due to the presence of a single hemichannel, to the lack of interactions with other mt proteins in Brustovetsky and Klingenberg's experiments (1996), or both. The ability of the AdNT to form channels has at any rate been recently confirmed by Rueck et al. (1998). Evidence has also been presented that treatment of mitochondrial membranes with the PTP-inducers and SH cross-linkers PhAsO (Halestrap et al., 1997a) and $Cu(OP)_2$ (copper orthophenanthroline) (Costantini et al., 1998; Halestrap et al., 1997a; Majima et al., 1995) leads to covalent linking of two AdNT monomers—a suggestive finding, because the PTP likely has a dimeric structure (Szabò and Zoratti, 1993). On the other hand, formation of such a cross-linked dimer does not take place with several other PT-inducing SH reagents (Costantini et al., 1998).

5.2. The Permeability Transition Pore as a Supramolecular Complex

Though evidence in favor of AdNT involvement is substantial, it seems that the carrier might be only one component of the PTP, not the sole molecule involved. For one thing, SMPs are not permeabilized by Ca^{2+} even in the presence of cyp-D (Halestrap et al., 1998), suggesting that other nonmembranous components, proteins of the outer membrane, or both are involved in PTP formation. Further evidence for the involvement of supramolecular complexes comes from the effect of mBzR (mitochondrial benzodiazepine receptor) ligands on channel activity attributed to the PTP, suggesting that the PTP might coincide with the mBzR (see Sect. 2.).

Protein complexes have been isolated (termed "PTPC" by Kromer's group) that contain the AdNT, porin, and either hexokinase or creatine kinase, plus a variety of other proteins including Bax, Bag-1, F1 ATPase and—in the case of the HK-(hexokinase)-containing complex, but not in that of the CK-(creatine kinase)-containing complex (Beutner et al., 1998)—cyp-D (Beutner et al., 1996; Marzo et al., 1998a, 1998c). When reconstituted into liposomes, the hexokinase-containing complex induced trapped-solute release in a Ca^{2+}-, ATR-, t-BuOOH-, diamide- and caspase-inducible, ADP-, BGK-, monochlorobiman-, CSA-, N-Met-Val-4-CSA- and Bcl2-inhibitable manner, suggesting an identification with the PTP. Furthermore, when reconstituted into planar bilayers, these complexes caused the appearence of high-conductance pore activity, reminescent of the MMC (Beutner et al., 1996). The creatine kinase-containing complexes displayed permeabilizing properties only when the CK octamer was dissociated into dimers, which bind loosely to the AdNT or not at all (Beutner et al., 1998; O'Gorman et al., 1997). Permeabilization took place despite the absence of cyp-D, and N-Met-4-Val-CSA inhibited solute release from the proteoliposomes. Some indication was obtained that the inhibition might have been mediated by a CSA-induced re-association of the dissociated CK. Ligands of CK (inducing the octamer) or of HK, such as glucose and ATP, inhibited PTP-like behavior. In mitochondria, CK is thought to reside in the intermembrane space and to interact directly with the AdNT, whereas HK binds to VDAC on the cytoplasmic side of the outer membrane. Brdiczka and co-workers concluded that the AdNT was probably the pore-forming unit, presumably regulated by interactions with the kinases, cyp-D, and porin (Beutner et al., 1998; Rueck et al., 1998).

5.3. Bax

The possibility that the PTP may coincide with, or be closely associated with, Bax or other members of the Bcl2 family deserves consideration. As mentioned, Bax was present (Marzo et al., 1998a) in the isolated complexes discussed in the previous section. The PTPC (permeability transition pore complex) induced release of compounds such as calcein and ATP from liposomes upon application of PT-inducing stimuli. Immunodepletion of Bax or purification of the complex from Bax-deficient mice resulted in complexes that could not permeabilize membranes in response to ATR, but still did so when treated with t-BuOOH or diamide (Marzo et al., 1998c), again suggesting differences in the molecular processes leading to permeabilization by different agents.

When microinjected into cells, Bax induced mitochondrial $\Delta\psi$ dissipation and nuclear apoptosis. These effects were prevented by pretreating the cells with CSA, N-methyl-4-Val-CSA, and BGK (Marzo et al., 1998c). Similar effects were induced when ATR was microinjected, but ATR did not cause $\Delta\psi$ dissipation in mitochondria isolated from the livers of Bax$^-$ mice. Bax and the AdNT co-immunoprecipitated. Bax, Bak, and Bcl2 interacted with the AdNT in a yeast two-hybrid assay, and Bax was found to induce cell death in wt, but not in AdNT-null, yeast. Bax and AdNT, together, but not either protein alone, permeabilized proteoliposomes in response to ATR (in contrast to reported pore-forming properties of Bax, but this may be due to experimental conditions). Bax-derived constructs lacking the BH3 domain required for homodimerization or the pore-forming α-helices did not have the effects of wt when microinjected, nor did they induce proteoliposome permeabilization in cooperation with the AdNT, suggesting that Bax dimers are the effective species. All this indicates that Bax and the AdNT might cooperate within a supramolecular complex to form the (or *one* of the) PTP (Marzo *et al.*, 1998c).

That the PTP might, at least in some cases, be formed by Bax or other members of its "family" is furthermore suggested by their electrophysiological properties, which are reminescent, within limits, of those of the PTP itself. When purified Bax was added to planar bilayers, it incorporated to form initially low-conductance pores that progressed to (or whose activity was superimposed with) higher-conductance states ranging up to 2 nS (Antonsson *et al.*, 1997; Schlesinger *et al.*, 1997). The channel exhibited fast kinetics and a moderate anion selectivity in the hands of Schlesinger *et al.* (1997) or a slight cation selectivity in those of Antonsson *et al.* (1997). When Bax was first incorporated into liposomes and the vesicles then allowed to fuse with the planar membranes, high-conductance, mildly anion-selective activity appeared from the beginning. Bcl2 and BclxL behaved similarly, forming high-conductance multistate channels with mild cationic selectivity (Minn *et al.*, 1997; Schendel *et al.*, 1997; Schlesinger *et al.*, 1997). These pores have so far been observed only in reconstituted systems, and whether they are affected by PTP modulators or inhibitors has not been investigated. Bax-induced cytochrome *c* release from mitochondria was reportedly inhibited by CSA, even though the release took place without accompanying swelling of the organelles (Juergensmeier *et al.*, 1998).

5.4. Permeabilization of Yeast Mitochondria

The recent report (Jung *et al.*, 1997) of a "permeability transition" in yeast mitochondria, involving a pore with a size exclusion limit similar to that of the mammalian PTP but with otherwise quite different characteristics, strengthens the hypothesis that various mitochondrial permeabilization machineries may exist. The pore was induced to open by respiration (possibly via an increase of matrix pH) and external ATP (in a CATR-insensitive manner), and it was inhibited by ADP, matrix protons, and Pi. No effect was shown using Ca^{2+}, Mg, CSA, oligomycin and CATR. This pore may or may not correspond to that studied electrophysiologically by Ballarin and Sorgato (1995) or to the MCC (next section).

5.5. Electrophysiological Observations

Though many electrophysiological results have already been mentioned, additional discussion is warrented of points relevant for the molecular identity of the PTP. As mentioned, our group has established a strong correlation, based on pharmacological behavior, between the PTP and the MMC, observed by patch-clamping rat liver mitoplasts (reviewed in Zoratti and Szabò, 1994, 1995). A high-conductance channel with characteristics at least superficially similar to the MMC has also been studied by Kinnally and co-workers [who use the "MCC" acronym (Kinnally *et al.*, 1996, and refs. therein)]. Similarities between the pores observed by the two groups include high conductance, the presence of multiple substates, and inhibitability by CSA (Szabò and Zoratti, 1991;

Zorov *et al.*, 1992). On the other hand, reports on voltage dependence do not fully coincide (see Zoratti and Szabò, 1994), and the MCC has been reported to be slightly cation-selective, whereas we found no significant selectivity for the MMC. Lohret and Kinnally (1995a) have subsequently reported that a channel very similar to the mammalian MCC is present in yeast wt and porin-less mitochondria. We failed, however, to detect MMC activity in VDAC$^-$ yeast mitochondria (a different strain), which in our hands displayed a voltage-dependent, mildly cation-selective, leader-peptide-sensitive channel with a lower, voltage-dependent conductance (Bàthori *et al.*, 1996; Szabò *et al.*, 1995). We tentatively identified this channel as the PSC (peptide-sensitive channel) (Henry *et al.*, 1996 and refs. therein), recently demonstrated to correspond to the Tom40 o.m. protein import complex (Kunkele *et al.*, 1998a, 1998b). Ballarin and Sorgato (1995, 1996), using wt and VDAC$^-$ yeast mitochondria, have reported the presence of a similar channel, stabilized in the open state by ATP, which differs from the channel we observed mainly in its slight anionic selectivity. At least one additional large, cation-selective channel was identified by these colleagues via patch-clamping proteoliposomes containing mitochondrial membranes from VDAC$^-$ yeast. Lohret's group (1996) has also reported that MCC activity was seen in yeast mitochondria lacking all three AdNT isoforms (Lohret *et al.*, 1996). Yeast mitochondria do not exhibit a PT comparable to the mammalian PT (Sect. 5.4), so assuming that MMC and MCC coincide, the reports may cast doubt on the identification of the MMC as the PTP. The same researchers have gone on to report that MCC activity is specifically influenced by leader peptides (fast block) (Lohret and Kinnally, 1995b), that it is inhibited by antibodies against Tim23 (a component of the i.m. protein import apparatus), and that it is altered in yeast mitochondria carrying a mutation in Tim23 (Lohret *et al.*, 1997). There was no change in PSC activity under the latter conditions. The authors concluded that the MCC is a component of the i.m. protein import system. To reconcile available data, one must assume that either (a) MCC and MMC activities do not arise from the same protein(s) despite their similarity, (b) they do, but do not coincide with the PTP; or (c) they coincide with the PTP, but the latter is not formed by the AdNT and originates instead from one or more components of the protein import apparatus. Unfortunately, these studies performed no pharmacological characterization of the MCC, so its identification with the MMC and/or the PTP remains in question. On the other hand, the coincidence of MMC and PTP is well supported pharmacologically. One further point to keep in mind here is that other mitochondrial carriers, also members of the family of tripartite translocator proteins (to which the AdNT belongs), may substitute for the AdNT to form high-conductance channels. Indeed, the Pi carrier has been reported to form pores (Herick *et al.*, 1997), although with very different properties from those of the MMC. With regard to hypothesis (c) above, it may be relevant that the 14 a.a., positively charged bee venom peptide mastoparan (Pfeiffer *et al.*, 1995) and the signal sequence from pCoxIV (pre-cytochrome oxidase subunit IV) (Sokolove and Kinnally, 1996) can open large pores in isolated mitochondria whose relationship with the classical PTP is not clear. The permeabilization induced by low (<1 µM) mastoparan is Ca^{2+}-and Pi-dependent and CSA-sensitive and it involves a pore with a size exclusion limit similar to that of the PTP. At higher peptide concentrations the same size exclusion limit is found, but permeabilization is insensitive to Ca, Pi, and CSA. As for pCoxIV, the size exclusion limit and the pharmacological profile are suffcently different from those of the Ca/Pi-induced PT to exclude the possibility of swelling being due to PTP operation (Sokolove and Kinnally, 1996).

6. PERSPECTIVES

The PTP, long considered by many to be little more than an experimental artifact, seems on its way to being assigned an important role in cellular physiology. As far as its properties are concerned, two recent developments are important: the spreading recognition that various PTPs may exist, also as small, selective pores (at least under certain circumstances), and the discovery of modulation by the redox state of the respiratory chain itself. The major priority is still to find an exact definition of the molecular nature of the PTP, information that may be crucial for a thorough understanding of the role of mitochondria in pathological and apoptotic processes. It is entirely possible that several related, but not identical, PTPs might exist, depending on the mode of induction. A thorough investigation of this possibility is warranted, and it might be beneficial to use electrophysiological means of investigating the characteristics of pores formed under various conditions and possibly by various reconstituted molecular complexes. The same methodological approach would certainly be useful for a better understanding of conditions leading to the formation and properties of small PTP(s), as well as of the mechanistic details of activation/inhibition processes, interactions with relevant proteins (e.g., Bcl2) and the biophysical properties of the pore itself. Numerous other issues remain to be clarified. For example, could the lag time preceding mitochondrial swelling upon PT induction represent the time needed for endogenous proteases to act? And why does a subpopulation of mitochondria refuse to become permeabilized in most experiments? The current interest in such issues bodes well for continued research.

ACKNOWLEDGMENTS. We thank Dr. Ildikò Szabò for a thorough reading of the manuscript. Our work was supported by the Italian National Research Council and by Telethon—Italy (grants A.44 and A.59).

REFERENCES

Andreeva, L., Tanveer, A., and Crompton, M., 1995, Evidence for the involvement of a membrane-associated cyclosporin-A binding protein in the Ca^{2+}-activated inner membrane pore of heart mitochondria, *Eur. J. Biochem.* 230:1125–1132.

Antonsson, B., Conti, F., Ciavatta, A., Montessuit, S., Lewis, S., Martinou, I., Bernasconi, L., Bernard, A., Mermod, J.-J., Mazzei, G., Maundrell, K., Gambale, F., Sadoul, R., and Martinou, J.-C., 1997, Inhibition of Bax channel-forming activity by Bcl-2, *Science* 277:370–372.

Ballarin, C., and Sorgato, M. C., 1995, An electrophysiological study of yeast mitochondria: Evidence for two inner membrane anion channels sensitive to ATP, *J. Biol. Chem.* 270:19262–19268.

Ballarin, C., and Sorgato, M. C., 1996, Anion channels of the inner membrane of mammalian and yeast mitochondria, *J. Bioenerg. Biomembr.* 28:125–130.

Bàthori, G., Szabò, I., Wolff, D., and Zoratti, M., 1996, The high-conductance channels of yeast mitochondrial outer membranes: A planar bilayer study, *J. Bioenerg. Biomembr.* 28:191–198.

Bernardi, P., 1996, The permeability transition pore. Control points of a cyclosporin A-sensitive mitochondrial channel involved in cell death, *Biochim. Biophys. Acta* 1275:5–9.

Bernardi, P., and Petronilli, V., 1996, The permeability transition pore as a mitochondrial calcium release channel: A critical appraisal, *J. Bioenerg. Biomembr.* 28:131–138.

Bernardi, P., Vassanelli, S., Veronese, P., Colonna, R., Szabò, I., and Zoratti, M., 1992, Modulation of the mitochondrial permeability transition pore: Effect of protons and divalent cations, *J. Biol. Chem.* **267**:2934–2939.

Bernardi, P., Veronese, P., and Petronilli, V., 1993, Modulation of the mitochondrial cyclosporin A-sensitive permeability transition pore: I. Evidence for two separate Me^{2+} binding sites with opposing effects on the pore open probability, *J. Biol. Chem.* **268**:1005–1010.

Bernardi, P., Broekemeier, K. M., and Pfeiffer, D. R., 1994, Recent progress on regulation of the permeability transition pore, a cyclosporin A-sensitive pore in the mitochondrial inner membrane, *J. Bioenerg. Biomembr.* **26**:509–517.

Beutner, G., Ruck, A., Riede, B., Welte, W., and Brdiczka, D., 1996, Complexes between kinases, mitochondrial porin, and adenylate translocator in rat brain resemble the permeability transition pore, *FEBS Lett.* **396**:189–195.

Beutner, G., Ruck, A., Riede, B., and Brdiczka, D., 1998, Complexes between porin, hexokinase, mitochondrial creatine kinase, and adenylate translocator display properties of the permeability transition pore: Implication for regulation of permeability transition by the kinases, *Biochim. Biophys. Acta* **1368**:7–18.

Bindoli, A., Barzon, E., and Rigobello, M. P., 1995, Inhibitory effect of pyruvate on release of glutathione and swelling of rat heart mitochondria, *Cardiovas. Res.* **30**:821–824.

Bindoli, A., Callegaro, M. T., Barzon, E., Benetti, M., and Rigobello, M. P., 1997, Influence of the redox state of pyridine nucleotides on mitochondrial sulfhydryl groups and permeability transition, *Arch. Biochem. Biophys.* **342**:22–28.

Broekemeier, K. M., and Pfeiffer, D. R., 1989, Cyclosporin A-sensitive and insensitive mechanisms produce the permeability transition in mitochondria, *Biochem. Biophys. Res. Commun.* **163**:561–566.

Broekemeier, K. M., and Pfeiffer, D. R., 1995, Inhibition of the mitochondrial permeability transition by cyclosporin A during long time-frame experiments: Relationship between pore opening and the activity of mitochondrial phospholipases, *Biochemistry* **34**:16440–16449.

Broekemeier, K. M., Dempsey, M. E., and Pfeiffer, D. R., 1989, Cyclosporin A is a potent inhibitor of the inner membrane permeability transition in liver mitochondria, *J. Biol. Chem.* **264**:7826–7830.

Broekemeier, K. M., Klocek, C. K., and Pfeiffer, D. R., 1998, Proton selective substate of the mitochondrial permeability transition pore: Regulation by the redox state of the electron transport chain, *Biochemistry* **37**:13059–13065.

Brustovetsky, N., and Klingenberg, M., 1996, Mitochondrial ADP/ATP carrier can be reversibly converted into a large channel by Ca^{2+}, *Biochemistry* **35**:8483–8488.

Carbonera, D., and Azzone, G. F., 1988, Permeability of inner mitochondrial membrane and oxidative stress, *Biochim. Biophys. Acta* **943**:245–255.

Castilho, R. F., Kowaltowski, A. J., Meinicke, A. R., Bechara, E. J. and Vercesi, A. E., 1995, Permeabilization of the inner mitochondrial membrane by Ca^{2+} ions is stimulated by *t*-butylhydroperoxide and mediated by reactive oxygen species generated by mitochondria, *Free Radical Biol. Med.* **18**:479–486.

Castilho, R. F., Kowaltowski, A. J., and Vercesi, A. E., 1996, The irreversibility of inner mitochondrial membrane permeabilization by Ca^{2+} plus prooxidants is determined by the extent of membrane protein thiol cross-linking, *J. Bioenerg. Biomembr.* **28**:523–529.

Castilho, R. F., Kowaltowski, A. J., and Vercesi, A. E., 1998, 3,5,3′-triiodothyronine induces mitochondrial permeability transition mediated by reactive oxygen species and membrane protein thiol oxidation, *Arch. Biochem. Biophys.* **354**:151–157.

Chernyak, B. V., and Bernardi, P., 1996, The mitochondrial permeability transition pore is modulated by oxidative agents through both pyridine nucleotides and glutathione at two separate sites, *Eur. J. Biochem.* **238**:623–630.

Chernyak, B. V., Dedov, V. N., and Chernyak, V. Ya., 1995, Ca^{2+}-triggered membrane permeability transition in deenergized mitochondria from rat liver, *FEBS Lett.* **365**:75–78.

Connern, C. P., and Halestrap, A. P., 1992, Purification and N-terminal sequencing of peptidyl-prolyl *cis-trans*-isomerase from rat liver mitochondrial matrix reveals the existence of a distinct mitochondrial cyclophilin, *Biochem. J.* **284**:381–385.

Connern, C. P., and Halestrap, A. P., 1994, Recruitment of mitochondrial cyclophilin to the mitochondrial inner membrane under conditions of oxidative stress that enhance the opening of a calcium-sensitive non-specific channel, *Biochem. J.* **302**:321–324.

Connern, C. P., and Halestrap, A. P., 1996, Chaotropic agents and increased matrix volume enhance binding of mitochondrial cyclophilin to the inner mitochondrial membrane and sensitize the mitochondrial permeability transition to [Ca^{2+}], *Biochemistry* **35**:8172–8180.

Costantini, P., Chernyak, B. V., Petronilli, V., and Bernardi, P., 1995, Selective inhibition of the mitochondrial permeability transition pore at the oxidation-reduction sensitive dithiol by monobromobimane, *FEBS Lett.* **362**:239–242.

Costantini, P., Cernyak, B. V., Petronilli, V., and Bernardi, P., 1996, Modulation of the mitochondrial permeability transition pore by pyridine nucleotides and dithiol oxidation at two separate sites, *J. Biol. Chem.* **271**:6746–6751.

Costantini, P., Colonna, R., and Bernardi, P., 1998, Induction of the mitochondrial permeability transition by N-ethylmaleimide depends on secondary oxidation of critical thiol groups: Potentiation by copper-*ortho*-phenanthroline without dimerization of adenine nucleotide translocase, *Biochim. Biophys. Acta* **1365**:385–392.

Crompton, M., and Andreeva, L., 1994, On the interaction of Ca^{2+} and cyclosporin A with a mitochondrial inner membrane pore: A study using cobaltammine complex inhibitors of the Ca^{2+} uniporter, *Biochem. J.* **302**:181–185.

Crompton, M., Ellinger, H., and Costi, A., 1988, Inhibition by cyclosporin A of a Ca^{2+}-dependent pore in heart mitochondria activated by inorganic phosphate and oxidative stress, *Biochem. J.* **255**:357–360.

Crompton, M., McGuinness, O., and Nazareth, W., 1992, The involvement of cyclosporin A binding proteins in regulating and uncoupling mitochondrial energy transduction, *Biochim. Biophys. Acta* **1101**:214–217.

Eriksson, O., Fontaine, E., Petronilli, V., and Bernardi, P., 1997, Inhibition of the mitochondrial cyclosporin A-sensitive permeability transition pore by the arginine reagent phenylglyoxal, *FEBS Lett.* **409**:361–364.

Eriksson, O., Fontaine, E., and Bernardi, P., 1998, Chemical modification of arginines by 2,3-butanedione and phenylglyoxal causes closure of the mitochondrial permeability transition pore, *J. Biol. Chem.* **273**:12669–12674.

Fagian, M. M., Pereira-da-Silva, L., Martins, I. S., and Vercesi, A. E., 1990, Membrane protein thiol cross-linking associated with the permeabilization of the inner mitochondrial membrane by Ca^{2+} plus prooxidants, *J. Biol. Chem.* **265**:19955–19960.

Fontaine, E., Eriksson, O., Ichas, F., and Bernardi, P., 1998a, Regulation of the permeability transition pore in skeletal muscle mitochondria: Modulation by electron flow through the respiratory chain complex I, *J. Biol. Chem.* **273**:12662–12668.

Fontaine, E., Ichas, F., and Bernardi, P., 1998b, A ubiquinone-binding site regulates the mitochondrial permeability transition pore, *J. Biol. Chem.* **273**:25734–25740.

Fournier, N., Ducet, G., and Crevat, A., 1987, Action of cyclosporine on mitochondrial calcium fluxes, *J. Bioenerg. Biomembr.* **19**:297–303.

Gogvadze, V., Schweizer, M., and Richter, C., 1996, Control of the pyridine nucleotide-linked Ca^{2+} release from mitochondria by respiratory substrates, *Cell Calcium* **19**:521–526.

Gores, G. J., Miyoshi, H., Botla, R., Aguilar, H. I., and Bronk, S., 1998, Induction of the mitochondrial permeability transition as a mechanism of liver injury during cholestasis: A potential role for mitochondrial proteases, *Biochim. Biophys. Acta* **1366**:167–175.

Gudz, T., Eriksson, O., Kushnareva, Y., Saris, N. E., and Novgorodov, S., 1997, Effect of butylhydroxytoluene and related compounds on permeability of the inner mitochondrial membrane, *Arch. Biochem. Biophys.* **342**:143–156.

Gunter, T. E., and Pfeiffer, D. R., 1990, Mechanisms by which mitochondria transport calcium, *Am. J. Physiol.* **258**:C755–C786.

Halestrap, A. P., Woodfield, K.-Y., and Connern, C. P., 1997a, Oxidative stress, thiol reagents, and membrane potential modulate the mitochondrial permeability transition by affecting nucleotide binding to the adenine nucleotide translocase, *J. Biol. Chem.* **272**:3346–3354.

Halestrap, A. P., Connern, C. P., Griffiths, E. J., and Kerr, P. M., 1997b, Cyclosporin A binding to mitochondrial cyclophilin inhibits the permeability transition pore and protects hearts from ischemia/reperfusion injury, *Mol. Cell. Biochem.* **174**:167–172.

Halestrap, A. P., Kerr, P. M., Javadov, S., and Woodfield, K.-Y., 1998, Elucidating the molecular mechanism of the permeability transition pore and its role in reperfusion injury in the heart, *Biochim, Biophys. Acta* **1366**:79–94.

Haworth, R. A., and Hunter, D. R., 1979, The Ca^{2+}-induced membrane transition in mitochondria: II. Nature of the Ca^{2+} trigger site, *Arch. Biochem. Biophys.* **195**:460–467.

Haworth, R. A., and Hunter, D. R., 1980, Allosteric inhibition of the Ca^{2+}-activated hydrophilic channel of the mitochondrial inner membrane by nucleotides, *J. Membr. Biol.* **54**:231–236.

Henry, J.-P., Juin, P., Vallette, F., and Thieffry, M., 1996, Characterization and function of the mitochondrial outer membrane peptide-sensitive channel, *J. Bioenerg. Biomembr.* **28**:101–108.

Herick, K., Kramer, R., and Luhring, H., 1997, Patch clamp investigation into the phosphate carrier from *Saccharomyces cerevisiae* mitochondria, *Biochim. Biophys. Acta* **1321**:207–220.

Hirsch, T., Decaudin, D., Susin, S. A., Marchetti, P., Larochette, N., Resche-Rigon, M., and Kroemer, G., 1998, PK11195, a ligand of the mitochondrial benzodiazepine receptor, facilitates the induction of apoptosis and reverses Bcl-2 mediated cytoprotection, *Exp. Cell Res.* **241**:426–434.

Hunter, D. R., and Haworth, R. A., 1979a, The Ca^{2+}-induced membrane transition in mitochondria: I. The protective mechanism, *Arch. Biochem. Biophys.* **195**:453–459.

Hunter, D. R., and Haworth, R. A., 1979b, The Ca^{2+}-induced membrane transition in mitochondria: III. Transitional Ca^{2+} release, *Arch. Biochem. Biophys.* **195**:468–477.

Ichas, F., and Mazat, J.-P., 1998, From calcium signaling to cell death: Two conformations for the mitochondrial permeability transition pore: Switching from low- to high-conductance state, *Biochim. Biophys. Acta* **1366**:33–50.

Ichas, F., Jouaville, L. S., Sidash, S. S., Mazat, J.-P., and Holmuhamedov, E. L., 1994, Mitochondrial calcium spiking: A transduction mechanism based on calcium-induced permeability transition involved in cell calcium signalling, *FEBS Lett.* **348**:211–215.

Ichas, F., Jouaville, L. S., and Mazat, J.-P., 1997, Mitochondria are excitable organelles capable of generating and conveying electrical and calcium signals, *Cell* **89**:1145–1153.

Jung, D. W., and Brierley, G. P., 1981, On the relationship between the uncoupler-induced efflux of K^+ from heart mitochondria and the oxidation-reduction state of pyridine nucleotides, *J. Biol. Chem.* **256**:10490–10496.

Jung, D. W., Bradshaw, P. C., and Pfeiffer, D. R., 1997, Properties of a cyclosporin-insensitive permeability transition pore in yeast mitochondria, *J. Biol. Chem.* **272**:21104–21112.

Juergensmeier, J. M., Xie, Z., Deveraux, Q., Ellerby, L., Bredesen, D., and Reed, J. C., 1998, Bax directly induces release of cytochrome *c* from isolated mitochondria, *Proc. Natl. Acad. Sci. USA* **95**:4997–5002.

Kinnally, K. W., Zorov, D. B., Antonenko, Yu. N., and Zorov, D. B., 1992, Modulation of inner mitochondrial membrane channel activity, *J. Bioenerg. Biomembr.* **24**:99–110.

Kinnally, K. W., Zorov, D. B., Antonenko, Y. N., Snyder, S. H., McEnery, M. W., and Tedeschi, H., 1993, Mitochondrial benzodiazepine receptor linked to inner membrane ion channels by nanomolar actions of ligands, *Proc. Natl. Acad. Sci. USA* **90**:1374–1378.

Kinnally, K. W., Lohret, T. A., Campo, M. L., and Mannella, C. A., 1996, Perspectives on the mitochondrial multiple conductance channel, *J. Bioenerg. Biomembr.* **28**:115–123.

Kowaltowski, A. J., and Castilho, R. F., 1997, Ca^{2+} acting at the external side of the inner mitochondrial membrane can stimulate mitochondrial permeability transition induced by phenylarsine oxide, *Biochim. Biophys. Acta* **1322**:221–229.

Kowaltowski, A. J., Castilho, R. F., and Vercesi, A. E., 1995, Ca^{2+}-induced mitochondrial membrane permeabilization: Role of coenzyme Q redox state, *Am. J. Physiol.* **269**:C141–C147.

Kowaltowski, A. J., Castilho, R. F., and Vercesi, A. E., 1996a, Opening of the mitochondrial permeability transition pore by uncoupling or inorganic phosphate in the presence of Ca^{2+} is dependent on mitochondrial-generated reactive oxygen species, *FEBS Lett.* **378**:150–152.

Kowaltowski, A. J., Castilho, R. F., Grijalba, M. T., Bechara, E. J., and Vercesi, A. E., 1996b, Effect of inorganic phosphate concentration on the nature of inner mitochondrial membrane alterations mediated by Ca^{2+} ions: A proposed model for phosphate-stimulated lipid peroxidation, *J. Biol. Chem.* **271**:2929–2934.

Kowaltowski, A. J., Vercesi, A. E., and Castilho, R. F., 1997, Mitochondrial membrane protein thiol reactivity with N-ethylmaleimide or mersalyl is modified by Ca^{2+}: Correlation with mitochondrial permeability transition, *Biochim. Biophys. Acta* **1318**:395–402.

Kunkele, K. P., Heins, S., Dembowski, M., Nargang, F. E., Benz, R., Thieffry, M., Walz, J., Lill, R., Nussberger, S., and Neupert, W., 1998a, The preprotein translocation channel of the outer membrane of mitochondria, *Cell* **93**:1009–1019.

Kunkele, K. P., Juin, P., Pompa, C., Nargang, F. E., Henry, J. P., Neupert, W., Lill, R., and Thieffry, M., 1998b, The isolated complex of the translocase of the outer membrane of mitochondria: Characterization of the cation-selective and voltage-gated preprotein-conducting pore, *J. Biol. Chem.* **273**:31032–31039.

Lee, A.-C., Zizi, M., and Colombini, M., 1994, NADH decreases the permeability of the mitochondrial outer membrane to ADP by a factor of 6, *J. Biol. Chem.* **269**:30974–30980.

Lee, A.-C., Xu, X., and Colombini, M., 1996, The role of pyridine nucleotides in regulating the permeability of the mitochondrial outer membrane, *J. Biol. Chem.* **271**:26724–26731.

Lehninger, A. L., Vercesi, A., and Bababunmi, E. A., 1978, Regulation of Ca^{2+} release from mitochondria by the oxidation-reduction state of pyridine nucleotides, *Proc. Natl. Acad. Sci. USA* **75**:1690–1694.

Le Quoc, K., and Le Quoc, D., 1988, Involvement of the ADP/ATP carrier in calcium-induced perturbations of the mitochondrial inner membrane permeability: Importance of the orientation of the nucleotide binding site, *Arch. Biochem. Biophys.* **265**:249–257.

Loew, L. M., Tuft, R. A., Carrington, W., and Fay, F. S., 1993, Imaging in five dimensions: Time-dependent membrane potentials in individual mitochondria, *Biophys. J.* **65**:2396–2407.

Loew, L. M., Carrington, W., Tuft, R. A., and Fay, F. S., 1994, Physiological cytosolic Ca^{2+} transients evoke concurrent mitochondrial depolarizations, *Proc. Natl. Acad. Sci. USA* **91**:12579–12583.

Lohret, T. A., and Kinnally, K. W., 1995a, Multiple conductance channel activity of wild-type and voltage-dependent anion-selective channel (VDAC)-less yeast mitochondria, *Biophys. J.* **68**:2299–2309.

Lohret, T. A., and Kinnally, K. W., 1995b, Targeting peptides transiently block a mitochondrial channel, *J. Biol. Chem.* **270**:15950–15953.

Lohret, T. A., Murphy, R. C., Drgon, T., and Kinnally, K. W., 1996, Activity of the mitochondrial multiple conductance channel is independent of the adenine nucleotide translocator, *J. Biol. Chem.* **271**:4846–4849.

Lohret, T. A., Jensen, R. E., and Kinnally, K. W., 1997, Tim23, a protein import component of the mitochondrial inner membrane, is required for normal activity of the multiple conductance channel MCC *J. Cell Biol.* **137**:377–386.

Majima, E., Ikawa, K., Takeda, M., Hashimoto, M., Shinohara, Y., and Terada, H., 1995, Translocation of loops regulates transport activity of mitochondrial ADP/ATP carrier deduced from formation of a specific intermolecular disulfide bridge catalyzed by copper-*o*-phenanthroline, *J. Biol. Chem.* **270**:29548–29554.

Mancini, M., Nicholson, D. W., Roy, S., Thornberry, N. A., Peterson, E. P., Casciola-Rosen, L. A., and Rosen, A., 1998, The caspase-3 precursor has a cytosolic and mitochondrial distribution: Implications for apoptotic signaling, *J. Cell Biol.* **140**:1485–1495.

Marzo, I., Brenner, C., Zamzami, N., Susin, S. A., Beutner, G., Brdiczka, D., Remy, R., Xie, Z. H., Reed, J. C., and Kroemer, G. 1998a, The permeability transition pore complex: A target for apoptosis regulation by caspases and bcl-2-related proteins, *J. Exp. Med.* **187**:1261–1271.

Marzo, I., Susin, S. A., Petit, P. X., Ravagnan, L., Brenner, C., Larochette, N., Zamzami, N., and Kroemer, G., 1998b, Caspases disrupt mitochondrial membrane barrier function, *FEBS Lett.* **427**:198–202.

Marzo, I., Brenner, C., Zamzami, N., Juergensmeier, J. M., Susin, S. A., Vieira, H. L., Prevost, M. C., Xie, Z., Matsuyama, S., Reed, J. C., and Kroemer, G., 1998c, Bax and adenine nucleotide translocator cooperate in the mitochondrial control of apoptosis, *Science* **281**:2027–2031.

Massari, S., 1997, Kinetic analysis of the mitochondrial permeability transition, *J. Biol. Chem.* **271**:31942–31948.

Minn, A. J., VÈlez, P., Schendel, S. L., Liang, H., Muchmore, S. W., Fesik, S. W., Fill, M., and Thompson, C. B., 1997, Bcl-x_L forms an ion channel in synthetic lipid membranes, *Nature* **385**:353–357.

Nelson, D. R., Lawson, J. E., Klingenberg, M., and Douglas, M. G., 1993, Site-directed mutagenesis of the yeast mitochondrial ADP/ATP translocator: Six arginines and one lysine are essential, *J. Mol. Biol.* **230**:1159–1170.

Nemopuceno, M. F., and Pereira-da-Silva, L., 1993, Effect of cyclosporin A and trifluoperazine on rat liver mitochondria swelling and lipid peroxidation, *Braz. J. Med. Biol. Res.* **26**:1019–1023.

Nicolli, A., Petronilli, V., and Bernardi, P., 1993, Modulation of the mitochondrial cyclosporin A-sensitive permeability transition pore by matrix pH: Evidence that the pore open–closed probability is regulated by reversible histidine protonation, *Biochemistry* **32**:4461–4465.

Nicolli, A., Basso, E., Petronilli, V., Wenger, R. M., and Bernardi, P., 1996, Interactions of cyclophilin with the mitochondrial membrane and regulation of the permeability transition pore, a cyclosporin A-sensitive channel, *J. Biol. Chem.* **271**:2185–2192.

Novgorodov, S. A., and Gudz, T. I., 1996, Permeability transition pore of the inner mitochondrial membrane can operate in two open states with different selectivities, *J. Bioenerg. Biomembr.* **28**:139–146.

Novgorodov, S. A., and Gudz, T. I., Milgrom, Y. M., and Brierley, G. P., 1992, The permeability transition in heart mitochondria is regulated synergistically by ADP and cyclosporin A, *J. Biol. Chem.* **267**:16274–16282.

O'Gorman, E., Beutner, G., Dolder, M., Koretsky, A. P., Brdiczka, D., and Walliman, T., 1997, The role of creatine kinase in inhibition of mitochondrial permeability transition, *FEBS Lett.* **414**:253–257.

Petronilli, V., Costantini, P., Scorrano, L., Colonna, R., Passamonti, S., and Bernardi, P., 1994a, The voltage sensor of the mitochondrial permeability transition pore is tuned by the oxidation-reduction state of vicinal thiols: Increase of the gating potential by oxidants and its reversal by reducing agents, *J. Biol. Chem.* **269**:16638–16642.

Petronilli, V., Nicolli, A., Costantini, P., Colonna, R., and Bernardi, P., 1994b, Regulation of the permeability transition pore, a voltage-dependent mitochondrial channel inhibited by cyclosporin A, *Biochim. Biophys. Acta* **1187**:255–259.

Petronilli, V., Miotto, G., Canton, M., Brini, M., Colonna, R., Bernardi, P., and Di Lisa, F., 1999, Transient and long-lasting openings of the mitochondrial permeability transition pore can be monitored directly in intact cells by changes in mitochondrial calcein fluorescence, *Biophys. J.*, **76**:725–734.

Pfeiffer, D. R., Gudz, T. I., Novgorodov, S. A., and Erdahl, W. L., 1995, The peptide mastoparan is a potent facilitator of the mitochondrial permeability transition, *J. Biol. Chem.* **270**:4923–4932.

Prpic, V., and Bygrave, F. L., 1980, On the interrelationship between glucagon action, the oxidation-reduction state of pyridine nucleotides, and calcium retention by rat liver mitochondria, *J. Biol. Chem.* **255**:6193–6199.

Reed, D. J., and Savage, M. K., 1995, Influence of metabolic inhibitors on mitochondrial permeability transition and glutathione status, *Biochim. Biophys. Acta* **1271**:43–50.

Reed, J. C., Juergensmeier, J. M., and Matsuyama, S., 1998, Bcl-2 family proteins and mitochondria, *Biochim. Biophys. Acta* **1366**:127–137.

Rigobello, M. P., Toninello, A., Siliprandi, D., and Bindoli, A., 1993, Effect of spermine on mitochondrial glutathione release, *Biochem. Biophys. Res. Commun.* **194**:1276–1281.

Rigobello, M. P., Barzon, E., Marin, O., and Bindoli, A., 1995a, Effect of polycation peptides on mitochondrial permeability transition, *Biochem. Biophys. Res. Commun.* **217**:144–149.

Rigobello, M. P., Turcato, F., and Bindoli, A., 1995b, Inhibition of rat liver mitochondria permeability transition by respiratory substrates, *Arch. Biochem. Biophys.* **319**:225–230.

Rigobello, M. P., Callegaro, M. T., Barzon, E., Benetti, M., and Bindoli, A., 1998, Purification of mitochondrial thioredoxin reductase and its involvement in the redox regulation of membrane permeability, *Free Radical Biol. Med.* **24**:370–376.

Riparbelli, M. G., Callaini, G., Tripodi, S. A., Cintorino, M., Tosi, P., and Dallai, R., 1995, Localization of the Bcl-2 protein to the outer mitochondrial membrane by electron microscopy, *Exp. Cell Res.* **221**:363–369.

Roth, Z., and Dikstein, S., 1982, Inhibition of ruthenium red-insensitive mitochondrial Ca^{2+} release and its pyridine nucleotide specificity, *Biochim. Biophys. Res. Commun.* **105**:991–996.

Rottenberg, H., and Marback, M., 1990, Regulation of Ca^{2+} transport in rat brain mitochondria: II. The mechanism of the adenine nucleotide enhancement of Ca^{2+} uptake and retention, *Biochim. Biophys. Acta* **1016**:87–98.

Rueck, A., Dolder, M., Wallimann, T., and Brdiczka, D., 1998, Reconstituted adenine nucleotide translocase forms a channel for small molecules comparable to the mitochondrial permeability transition pore, *FEBS Lett.* **426**:97–101.

Savage, M. K., and Reed, D. J., 1994, Release of mitochondrial glutathione and calcium by a cyclosporin A-sensitive mechanism occurs without large amplitude swelling, *Arch. Biochem. Biophys.* **315**:142–152.

Savage, M. K., Jones, D. P., and Reed, D. J., 1991, Calcium- and phosphate-dependent release and loading of glutathione by liver mitochondria, *Arch. Biochem. Biophys.* **290**:51–56.

Schendel, S. L., Xie, Z., Oblatt Montal, M., Matsuyama, S., Montal, M., and Reed, J. C., 1997, Channel formation by antiapoptotic protein Bcl-2, *Proc. Natl. Acad. Sci. USA* **94**:5113–5118.

Schlesinger, P. H., Gross, A., Yin, X.-M., Yamamoto, K., Saito, M., Waksman, G., and Korsmeyer, S. J., 1997, Comparison of the ion channel characteristics of proapoptotic BAX and antiapoptotic BCL-2, *Proc. Natl. Acad. Sci. USA* **94**:11357–11362.

Schweizer, M., and Richter, C., 1994, Gliotoxin stimulates Ca^{2+} release from intact rat liver mitochondria, *Biochemistry* **33**:13401–13405.

Schweizer, M., and Richter, C., 1996a, Peroxynitrite stimulates the pyridine nucleotide-linked Ca^{2+} release from intact rat liver mitochondria, *Biochemistry* **35**:4524–4528.

Schweizer, M., and Richter, C., 1996b, Stimulation of Ca^{2+} release from rat liver mitochondria by the dithiol reagent alpha-lipoic acid, *Biochem. Pharmacol.* **52**:1815–1820.

Schweizer, M., Durrer, P., and Richter, C., 1994, Phenylarsine oxide stimulates pyridine nucleotide-linked Ca^{2+} release from rat liver mitochondria, *Biochem. Pharmacol.* **48**:967–973.

Scorrano, L., Petronilli, P., and Bernardi, P., 1997a, On the voltage dependence of the mitochondrial permeability transition pore: A critical reappraisal, *J. Biol. Chem.* **272**:12295–12299.

Scorrano, L., Nicolli, A., Basso, E., Petronilli, V., and Bernardi, P., 1997b, Two modes of activation of the permeability transition pore: The role of mitochondrial cyclophilin, *Mol. Cell. Biochem.* **174**:181–184.

Sokolove, P. M., and Kinnally, K. W., 1996, A mitochondrial signal peptide from *Neurospora crassa* increases the permeability of isolated rat liver mitochondria, *Arch. Biochem. Biophys.* **336**:69–76.

Susin, S. A., Zamzami, N., Castedo, M., Daugas, E., Wang, H.-G., Geley, S., Fassy, F., Reed, J. C., and Kroemer, G., 1997, The central executioner of apoptosis: Multiple connections between protease activation and mitochondria in Fas/APO-1/CD95- and ceramide-induced apoptosis, *J. Exp. Med.* **186**:25–37.

Susin, S. A., Zamzami, N., and Kroemer, G., 1998, Mitochondria as regulators of apoptosis: Doubt no more, *Biochim. Biophys. Acta* **1366**:151–165.

Szabò, I., and Zoratti, M., 1991, The giant channel of the inner mitochondrial membrane is inhibited by cyclosporin A, *J. Biol. Chem.* **266**:3376–3379.

Szabò, I., and Zoratti, M., 1992, The mitochondrial megachannel is the permeability transition pore, *J. Bioenerg. Biomembr.* **24**:111–117.

Szabò, I., and Zoratti, M., 1993, The mitochondrial permeability transition pore may comprise VDAC molecules: I. Binary structure and voltage dependence of the pore, *FEBS Lett.* **330**:205–210.

Szabò, I., Bernardi, P., and Zoratti, M., 1992, Modulation of the mitochondrial megachannel by divalent cations and protons, *J. Biol. Chem.* **267**:2940–2946.

Szabò, I., De Pinto, V., and Zoratti, M., 1993, The mitochondrial permeability transition pore may comprise VDAC molecules: II. The electrophysiological properties of VDAC are compatible with those of the mitochondrial megachannel, *FEBS Lett.* **330**:206–210.

Szabò, I., Bàthori, G., Wolff, D., Starc, T., Cola, C., and Zoratti, M., 1995, The high-conductance channel of porin-less yeast mitochondria, *Biochim. Biophys. Acta* **1235**:115–125.

Tanveer, A., Virji, S., Andreeva, L., Totty, N. F., Hsuan, J. J., Ward, J. M., and Crompton, M., 1996, Involvement of cyclophilin D in the activation of a mitochondrial pore by Ca^{2+} and oxidant stress, *Eur. J. Biochem.* **238**:166–172.

Tikhonova, I. M., Andreyev, A. Yu., Antonenko, Yu. N., Kaulen, A. D., Komrakov, A. Yu., and Skulachev, V. P., 1994, Ion permeability induced in artificial membranes by the ATP/ADP antiporter, *FEBS Lett.* **337**:231–234.

Turrens, J. F., Alexandre, A., and Lehninger, A. L., 1985, Ubisemiquinone is the electron donor for superoxide formation by complex III of heart mitochondria, *Arch. Biochem. Biophys.* **237**:408–414.

Valle, V. G., Fagian, M. M., Parentoni, L. S., Meinicke, A. R., and Vercesi, A. E., 1993, The participation of reactive oxygen species and protein thiols in the mechanism of mitochondrial inner membrane permeabilization by calcium plus prooxidants, *Arch. Biochem. Biophys.* **307**:1–7.

Vercesi, A. E., 1984, Dissociation of $NAD(P)^{+}$-stimulated mitochondrial Ca^{2+} efflux from swelling and membrane damage, *Arch. Biochem. Biophys.* **232**:86–91.

Vercesi, A. E., 1987, The participation of NADP, the transmembrane potential, and the energy-linked NAD(P) transhydrogenase in the process of Ca^{2+} efflux from rat liver mitochondria, *Arch. Biochem. Biophys.* **252**:171–178.

Woodfield, K. Y., Price, N. T., and Halestrap, A. P., 1997, cDNA cloning of rat mitochondrial cyclophilin, *Biochim. Biophys. Acta* **1351**:27–30.

Zizi, M., Forte, M., Blachly-Dyson, E., and Colombini, M., 1994, NADH regulates the gating of VDAC, the mitochondrial outer membrane channel, *J. Biol. Chem.* **269**:1624–1616.

Zoratti, M., and Szabò, I., 1994, Electrophysiology of the inner mitochondrial membrane, *J. Bioenerg. Biomembr.* **26**:543–553.

Zoratti, M., and Szabò, I., 1995, The mitochondrial permeability transition, *Biochim. Biophys. Acta* **1241**:139–176.

Zorov, D. B., Kinnally, K. W., and Tedeschi, H., 1992, Voltage activation of heart inner mitochondrial membrane channels, *J. Bioenerg. Biomembr.* **24**:119–124.

Chapter 8

Control of Mitochondrial Metabolism by Calcium-Dependent Hormones

Paul Burnett, Lawrence D. Gaspers, and Andrew P. Thomas

1. INTRODUCTION

A variety of hormones and other extracellular stimuli cause increases in concentration of cytosolic free Ca^{2+} ($[Ca^{2+}]_c$). In many cell types these Ca^{2+} signals are manifest as a series of transient $[Ca^{2+}]_c$ oscillations, the frequency of which is a function of agonist dose (Thomas *et al.*, 1996). Cytosolic Ca^{2+} is a second messenger of fundamental importance, regulating cellular processes as varied as muscle contraction, secretion, cell proliferation, and onset of embryogenesis (Clapham, 1995).

A primary mechanism by which hormones induce increases in $[Ca^{2+}]_c$ is through the generation of the second messenger inositol 1,4,5-trisphosphate ($InsP_3$) (Berridge, 1995; Clapham, 1995). $InsP_3$ promotes the opening of channels on the endoplasmic reticulum membrane, leading to the release of Ca^{2+} from intracellular stores (Clapham, 1995). This release is often accompanied by Ca^{2+} entry from the extracellular medium (Berridge, 1995).

Most of the processes that are activated by the increases in $[Ca^{2+}]_c$ also lead to an increase in the consumption of ATP. Consequently, the cell needs to increase the rate of respiration and oxidative metabolism (Mc Cormack *et al.*, 1990) to increase ATP production in the majority of tissues, the mitochondria are the most important sites for ATP production. The oxidation of respiratory fuels (pyruvate and fatty acyl-CoA) to produce ATP involves the formation of NADH and reduced flavoproteins by mitochondrial

Paul Burnett, Lawrence D. Gaspers, and Andrew P. Thomas   Department of Pharmacology and Physiology, University of Medicine and Dentistry of New Jersey, Newerk, New Jersey 07103-2714.

Mitochondria in Pathogenesis, edited by Lemasters and Nieminen.
Kluwer Academic/Plenum Publishers, New York, 2001.

dehydrogenases and their subsequent oxidation by molecular oxygen (Fig. 1) (Denton & McCormack, 1995).

Oxidation of reduced flavoproteins and NADH by molecular oxygen does not occur directly; rather, electrons are transferred to molecular oxygen through a series of redox reactions catalyzed by components of the respiratory chain, each with an increasing redox potential. As the electrons travel down the potential gradient of the respiratory chain, their energy is used to pump protons from the matrix side of the inner mitochondrial membrane to the cytosolic side, thus producing a chemical proton gradient (ΔpH) and a membrane potential ($\Delta\psi$). The sum of these forces, known as the proton motive force (PMF), can drive protons back into the mitochondrial matrix through the proton channel of the ATP synthase, with the production of ATP, from ADP and inorganic phosphate (P$_i$) (Fig. 1) (Brand & Murphy, 1987; Erecinska & Wilson, 1982; Nicholls & Akerman, 1984; Slater, 1987). The overall supply of reducing equivalents in the form of NADH and FADH$_2$ to the respiratory chain is largely

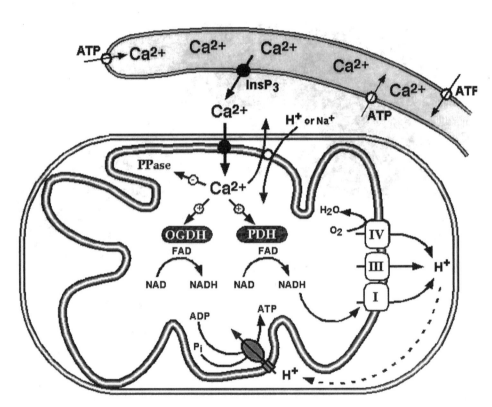

FIGURE 1. InsP$_3$-dependent regulation of mitochondrial metobolism by Ca^{2+}. Schematic diagram showing mitochondrial Ca^{2+} uptake sites juxtaposed with InsP$_3$-sensitive Ca^{2+} release channels of the ER. Intramito-chondrial Ca^{2+}-sensitive targets, pyrophosphatse (PPase), 2-oxoglutarate dehydrogenase (OGDH), and pyruvate dehydrogenase (PDH) are also illustrated, (Reprinted from *Biochim. Biophys. Acta*: 1366, Robb-Gaspers *et al.,* "Coupling between cytosolic and mitochondrial calcium oscillations: Role in the regulation of hepatic metabolism", Copyright 1988, pp. 17–32, with permission from Elsevier Science.)

determined by the activities of several key enzymes within the mitochondrial matrix. These are the pyruvate dehydrogenase complex (PDH) and two citrate cycle enzymes, 2-oxoglutarate dehydrogenase (OGDH) and NAD-isocitrate dehydrogenase (NAD-ICDH) (Mc Cormack et al., 1990; Denton and Mc Cormack, 1995).

The rate of mitochondrial ATP synthesis in vertebrate cells is regulated by alterations in key metabolite levels, namely the ATP : ADP ratio (phosphorylation potential) and the NADH : NAD$^+$ ratio (redox potential) (Erecinska and Wilson, 1982). Stimulation of cytosolic ATP-consuming processes results in a decrease in ATP : ADP in the cytosol and subsequently in the mitochondrial compartment of the cell (Denton and Halestrap, 1979; Randle, 1986; Reed and Yeaman, 1987). Within the mitochondria, a fall in ATP : ADP results in an increased flow of electrons along the respiratory chain and increased O_2 consumption (Mc Cormack et al., 1990). The increased electron flow is also expected to lower mitochondrial NADH : NAD$^+$, because this is the primary substrate for the electron transport chain (Halestrap, 1989; Mc Cormack et al., 1990). The decreased matrix redox potential and phosphorylation potential are both expected to lead to the activation of the mitochondrial dehydrogenases because their activities are enhanced by increases in [ADP] and [NAD$^+$] (Denton and Mc Cormack, 1980; Hansford, 1994; Halestrap, 1989; Mc Cormack et al., 1990). In many circumstances, however, in which cells are stimulated with agonists that increase [Ca^{2+}]$_c$, coincident with increased rates of respiration, there is an increase in the ratio of NADH : NAD$^+$ rather than the predicted decrease (Denton and Mc Cormack, 1995; Denton et al., 1996; Mc Cormack et al., 1990).

Mitochondria's role in cellular Ca^{2+} homeostasis has been a focus of considerable controversy (Crompton, 1990; Fiskum and Lehninger, 1982; Gunter and Gunter, 1994; Gunter et al., 1998). Vertebrate mitochondria contain very active systems within their inner membranes that allow the specific transfer of Ca^{2+} both into and out of the mitochondrial matrix (Fiskum and Lehninger, 1979; Gunter and Gunter, 1994). It was initially thought that mitochondria, by virtue of possessing a large capacity for Ca^{2+}-uptake (Fiskum and Lehninger, 1982), acted in the cell as mobilizable stores of Ca^{2+}, capable of regulating the extramitochondrial Ca^{2+} concentration (Exton, 1980; Hansford, 1994; Nicholls, 1978; Williamson et al., 1981). It is now clear, however, that mitochondria are not primary regulators of [Ca^{2+}]$_c$, and there is mounting evidence for a more complex role of mitochondrial Ca^{2+} transport in both cellular Ca^{2+} homeostasis and in oxidative phosphorylation regulation. Mitochondria are localized to sites of hormone-induced Ca^{2+} release from intracellular stores (Landolfi et al., 1998; Rizzuto et al., 1998; Simpson and Russel, 1996) and entry (Lawrie et al., 1996) in certain cell types. Because impairment of mitochondrial function can result in alterations in cytosolic Ca^{2+} signals (Babcock et al., 1997; Hoth et al., 1997; Simpson and Russel, 1996), mitochondria may influence the kinetics of [Ca^{2+}]$_c$ increases due to their close proximity to sites of [Ca^{2+}]$_c$ increases. Several key mitochondrial dehydrogenases are stimulated by increases of [Ca^{2+}]$_m$ in the physiological range (Mc Cormack et al., 1990), and coupling of [Ca^{2+}]$_m$ to [Ca^{2+}]$_c$ increases consequently lead to activation of NADH production (Hajnóczky et al., 1995; Robb-Gaspers et al., 1998a; Rutter et al., 1996).

This review focuses on how increases in [Ca^{2+}]$_c$ are transferred into the mitochondrial matrix in mammalian cells and how these signals are decoded to yield a final metabolic output.

2. InsP$_3$-MEDIATED Ca^{2+} SIGNALING

A common mechanism used by extracellular stimuli to exert control over metabolism relies on the mobilization of Ca^{2+} from intracellular stores, entry of Ca^{2+} from the extracellular medium, or both, resulting in [Ca^{2+}]$_c$ increase. In nonexcitable tissues such as the liver, activation of plasma membrane receptors coupled to the phosphoinositide signaling pathway results in Ca^{2+} mobilization (Berridge, 1993; Clapham, 1995; Thomas *et al.*, 1996). In resting cells [Ca^{2+}]$_c$ is maintained at low levels, approximately 100 nM. During stimulation, however, the average free [Ca^{2+}]$_c$ can rise to micromolar levels and above, depending on cell type (Jacob, 1990).

InsP$_3$ is generated as a result of phospholipase C (PLC) activation, brought about by hormone-induced activation of a receptor coupled to mobilization of [Ca^{2+}]$_c$. Which isoform of PLC is activated depends on receptor type. Receptors coupled to heterotetrameric G-proteins stimulate PLC-β, whereas receptors with intrinsic tyrosine kinase activity stimulate PLC-γ (Berridge, 1993; Clapham, 1995; Rhee and Bae, 1997). Both of the PLC isoenzymes catalyze hydrolysis of phosphatidylinositol 4,5-bisphosphate in the plasma membrane, causing corelease of the second messengers InsP$_3$ and diacylglycerol (DAG). The lipophillic DAG stimulates protein kinase C (PKC); InsP$_3$ diffuses into the cytosol, where it interacts with the InsP$_3$ receptor (InsP$_3$R) located in the endoplasmic reticulum (ER) membrane. InsP$_3$ binding promotes opening of the Ca^{2+} channel of the InsP$_3$R, causing Ca^{2+} release from the ER. An additional effect of elevating InsP$_3$ levels and the subsequent depletion of luminal Ca^{2+} from within the ER is activation of Ca^{2+} influx across the plasma membrane (Putney and Bird, 1993a, 1993b).

When examined with sufficient spatial and temporal resolution, agonist-induced [Ca^{2+}]$_c$ responses are often complex. Woods *et al.* (1986) were the first to directly demonstrate that hormones coupled to PLC activation generate periodic baseline-separated [Ca^{2+}]$_c$ spikes or oscillations in aequorin-injected hepatocytes. Workers in other laboratories using fluorescence imaging techniques later confirmed this observation (Rooney *et al.*, 1989, 1990). The [Ca^{2+}]$_c$ oscillations often originate from a discrete subcellular locus, and propagate through all or part of the cell during each oscillation in the form of an intracellular wave (Rooney and Thomas, 1993; Rooney *et al.*, 1991; Takamatsu and Wier, 1990). The amplitude of baseline-separated [Ca^{2+}]$_c$ oscillations is not affected by agonist dose; rather, increased agonist concentration results in increased oscillation frequency (Berridge, 1997; Berridge *et al.*, 1988; Rooney *et al.*, 1990; Thomas *et al.*, 1996); the overall shape and duration of individual oscillations remains constant (Mignery and Südhof, 1990; Petersen *et al.*, 1991; Rooney *et al.*, 1989, 1991; Thomas *et al.*, 1996). This phenomenon has been described as *frequency-modulated Ca^{2+} signaling* (Berridge and Galione, 1988).

Spatially resolved functional measurements that report changes in [Ca^{2+}] within the ER lumen (Alonso *et al.*, 1999; Hajnóczky *et al.*, 1994; Renard-Rooney *et al.*, 1993; Rooney Thomas, 1994) have demonstrated that InsP$_3$-sensitive Ca^{2+} stores are present throughout the path of an intracellular [Ca^{2+}]$_c$ wave, and the store correlates with ER markers. Immunohistochemical techniques have also revealed heterogeneity in the density and distribution of various components of the Ca^{2+} signaling pathway (Pozzan *et al.*, 1994; Prentki *et al.*, 1988; Tordjmann *et al.*, 1998). These differences may underlie the localized [Ca^{2+}]$_c$ responses observed in some cell types.

3. MITOCHONDRIAL Ca^{2+} TRANSPORT SYSTEMS

Mitochondria posses specific proteins that allow Ca^{2+} cycling across the inner membrane (Fig. 2) (Gunter and Gunter, 1994; Gunter *et al.*, 1998; McCormack and Denton, 1994; McCormack *et al.*, 1990). There is currently evidence for three separate Ca^{2+} transport mechanisms that function in the inner membrane of vertebrate mitochondria (Fig. 2). The structure and regulation of these transport systems is reviewed in detail by Gunter *et al.*, (1991, 1998) and elsewhere in this volume.

Briefly, mitochondrial electron transport generates a negative potential across the inner mitochondrial membrane. This negative membrane potential facilitates Ca^{2+} influx via a uniporter (Gunter and Gunter, 1994). With high levels of external Ca^{2+}, uniporter activity in isolated mitochondria can lead to the rapid uptake of large quantities of Ca^{2+}, sufficient to overwhelm the ability of the electron transport chain to maintain membrane potential (Akerman *et al.*, 1977; Nicholls and Akerman, 1984). Efflux of Ca^{2+} from mitochondria is mediated by Na$^+$-dependent and Na$^+$-independent exchange proteins (Crompton *et al.*, 1977, 1978; Fiskum and Lehninger, 1979; Puskin *et al.*, 1976). The inner membrane also contains a Ca^{2+}-activated, nonselective permeability pore, which is closed in coupled mitochondria (Gunter and Gunter, 1994; Gunter *et al.*, 1998, McCormack *et al.*, 1990). Efflux of Ca^{2+} through this pore may play a role in prolonging the hormone-induced cytosolic Ca^{2+} signal (see Section 3.2.3).

3.1. The Ca^{2+} Uniporter

The Ca^{2+} uniporter facilitates diffusion of Ca^{2+} down its electrochemical gradient into the mitochondrial matrix without directly coupling that diffusion to any other ion or molecule. The uniporter is responsible for rapid sequestration of Ca^{2+} into the mito-

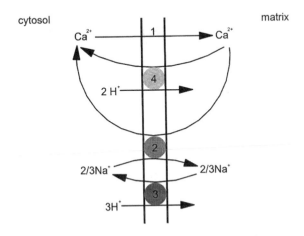

FIGURE 2. Mammalian mitochondrial Ca^{2+} transport proteins. Influx of Ca^{2+} into the mitochondrial matrix down its concentration gradient is achieved via an electroneutral uniporter (1). Efflux of Ca^{2+} is carried out by an electrophoretic Na$^+$-dependent Na$^+$/Ca^{2+} antiporter (2) and a poorly characterized H$^+$/Ca^{2+} antiporter (4). Efflux of Ca^{2+} is driven by the respiratory proton motive force via subsequent Na$^+$/H$^+$ exchange (3).

chondrial matrix (Gunter and Gunter, 1994; McCormack and Denton, 1994; McCormack *et al.*, 1990; Wingrove *et al.*, 1984). Uptake utilizes the internally negative membrane potential across the inner mitochondrial membrane ($-180\,mV$), which is generated by substrate oxidation and electron transport (Gunter and Gunter, 1994, McCormack *et al.*, 1990; Scarpa and Azzone, 1970).

Uniporter transport of Ca^{2+} consists of a series of steps: (1) binding of Ca^{2+} to the transport site on the *cis* side of the membrane, (2) translocation (conformational change), and (3) release of Ca^{2+} into the *trans* side of the membrane (Gunter and Gunter, 1994). Transport of Ca^{2+} into the mitochondrial matrix decreases mitochondrial membrane potential (Gunter and Gunter, 1994). This decrease is balanced by Ca^{2+} efflux in exchange for H^+ (see Section 3.2.). Evidence suggests that the uniporter has both an activator and a transport site (Vinogradov and Scarpa, 1973). Ca^{2+} binds to an activator site on the uniporter, enhancing subsequent transport of Ca^{2+}. Hill coefficients obtained for velocity of transport vs $[Ca^{2+}]$ are typically in the region of 2, showing some degree of positive cooperativity for Ca^{2+} (Gunter and Gunter, 1994). Binding of Ca^{2+} to the activator site either lowers the activation energy required for transport or possibly lowers the K_d for binding to the transport site (Gunter and Gunter, 1994). The uniporter can be inhibited by physiological concentrations of Mg^{2+} (Crompton and Roos, 1985; Gunter and Gunter, 1994; Nicholls and Akerman, 1984; Reed and Bygrave, 1974a, 1974b), perhaps by competition for binding at the transport site. Ruthenium red and its derivatives are among the strongest inhibitors of the uniporter, although these compounds are not thought to bind to the transport site (Matlib *et al.*, 1998; McCormack and Denton, 1994; McCormack *et al.*, 1990).

3.2. The Ca^{2+} Efflux Mechanisms

Two separate efflux mechanisms, Na^+-dependent (Crompton *et al.*, 1977) and Na^+-independent (Puskin *et al.*, 1976), have been identified (Fig. 2). Transport of Ca^{2+} out of the mitochondrial matrix against its electrochemical gradient requires energy; efflux is ultimately driven by the electron transport chain (Crompton *et al.*, 1977, 1978; Fiskum and Lehninger, 1979; Murphy and Fiskum, 1988; Puskin *et al.*, 1976).

3.2.1. The Na^+-Dependent Ca^{2+} Efflux Mechanism

Transport of Ca^{2+} by the Na^+-dependent Ca^{2+} antiporter is activated by extramitochondrial Na^+ and inhibited by extramitochondrial Ca^{2+} (Crompton *et al.*, 1980; Hayat and Crompton, 1987). This transporter is also inhibited by pharmacological agents such as verapamil and diltiazem (Chiesi *et al.*, 1987; Rizzuto *et al.*, 1987; Wolkowicz *et al.*, 1983). Until recently, there was a consensus that transport via the Na^+-dependent antiporter was electroneutral, promoting efflux of one Ca^{2+} in exchange for two Na^+ (Brand, 1985; Crompton, 1990; Li *et al.*, 1992). Studies by Jung *et al.* (1995), however, have suggested that this antiporter can promote electrophoretic rather than the electroneutral exchange. In electrophoretic exchange the efflux of one Ca^{2+} is in exchange for three Na^+, analogous to the plasma membrane Na^+/Ca^{2+} exchanger. The rate of Na^+-dependent efflux can be increased with increased respiration and inhibited with uncouplers (Baysal *et al.*, 1991;

Bragadin *et al.*, 1979; Jung *et al.*, 1995), compatible with a contribution of $\Delta\psi$ in the exchange of Na^+ for Ca^{2+}. This hypothesis is supported by kinetic evidence for three Na^+ binding sites on the Na^+-dependent antiporter (Hayat and Crompton, 1987).

3.2.2. The Na^+-Independent Ca^{2+} Efflux Mechanism

Mitochondria exhibit Ca^{2+} efflux in conditions where Na^+-dependent efflux is inhibited (Bernadi and Azzone, 1979; Gunter and Pfeiffer, 1990; Gunter *et al.*, 1991; Gunter and Gunter, 1994), suggesting a Na^+-independent efflux mechanism. This exchange pathway is separate from the Na^+-dependent pathway and has different transport kinetics (Gunter and Gunter, 1994). The Na^+-independent antiporter allows efflux of one Ca^{2+} in exchange for two H^+ (Gunter and Gunter, 1994) and is inhibited by low levels of uncoupler, such as FCCP (Bernadi and Azzone, 1979). Energy for Ca^{2+} efflux by this mechanism is also obtained from the electron transport chain (Gunter and Gunter, 1994; Gunter *et al.*, 1991; Rosier *et al.*, 1981).

3.2.3. The Mitochondrial Permeability Transition

Isolated mitochondria are able to undergo a dramatic increase in permeability to ions and small molecules (<1500 Da) when exposed to supraphysiological concentrations of Ca^{2+}, leading to an increase in mitochondrial matrix volume, a phenomenon known as the *mitochondrial permeability transition* (MPT) (Zoratti and Szabò, 1995). The ion permeability increase is brought about via formation of a nonspecific channel in the inner mitochondrial membrane. The molecular mechanism for the formation of this pore has been reviewed extensively (Halestrap *et al.*, 1998; Ichas and Mazat, 1998).

The MTP pore has been shown to exhibit two open conformations in isolated mitochondria; each conformation is believed to participate in separate cellular signaling events (see Section 4.2.) (Zoratti and Szabò, 1995). First, the pore can operate in a low-conductance state where the pore opens and closes spontaneously, allowing transient efflux of small ions from the mitochondrial matrix and leaving mitochondrial functions unimpaired. A high-conductance state exists where the pore is stabilized in the open conformation, leading to nonselective diffusion of large molecules, collapse of the PMF, and release of pro-apoptotic factors. The transition from low- to high-conductance states is dependent on saturation of Ca^{2+} binding sites on the pore (Ichas and Mazat, 1998).

4. ROLE OF MITOCHONDRIA IN $[Ca^{2+}]_c$ HOMEOSTASIS

Distribution of Ca^{2+} across the inner mitochondrial membrane depends on the relative activities of the influx and efflux pathways and on the driving forces for the transport reactions. During the subsequent discussion of the $[Ca^{2+}]_c$–$[Ca^{2+}]_m$ relationship, keep in mind that: (a) maximal activity of the Ca^{2+} uniporter is about 10 times that of the efflux systems (Garlid, 1988), (b) uniporter uptake rate is highly cooperative with respect to Ca^{2+} (Vinogradov and Scarpa, 1973), and (c) the Na^+-dependant efflux system is inhibited by micromolar external $[Ca^{2+}]$. These three observations explain mitochondria's ability to sequester large amounts of Ca^{2+} within their matrix under conditions of sufficiently elevated extramitochondrial $[Ca^{2+}]$.

The high capacity of mitochondria for Ca^{2+} accumulation *in vitro* and their ability to buffer external Ca^{2+} in the micromolar range initially led to the belief that mitochondria were the primary organelles responsible for modulating intracellular Ca^{2+} homeostasis (Nicholls, 1978). Mitochondrial buffering of external Ca^{2+} can occur when $[Ca^{2+}]_m$ is high enough to saturate mitochondrial Ca^{2+} (Denton and Mc Cormack, 1989). Additionally, it was thought that mitochondria were the store from which Ca^{2+} was mobilized in response to hormonal stimulation (Exton, 1981). These views were undermined by several observations. An early clue was derived from the Ca^{2+} sensitivity of the matrix dehydrogenases PDH, NAD-ICDH, and OGDH (2-oxyglutarate dehydrogenase) (Denton and Mc Cormack, 1980, 1989, 1995; Denton *et al.*, 1972, 1978, 1996; Hansford, 1994; Mc Cormack and Denton, 1994; Mc Cormack *et al.*, 1990) (Table I). The relatively low $K_{0.5}$ values for activation of these enzymes by Ca^{2+}, together with the fact that they are activated in a Ca^{2+}-dependent manner in cells and tissues, argued against $[Ca^{2+}]_m$ being high enough to saturate the efflux pathway. In addition, resting $[Ca^{2+}]_c$ levels are well below the affinity of the mitochondrial uniporter (Pozzan *et al.*, 1994), suggesting that $[Ca^{2+}]_c$ would be too low to load mitochondria with enough Ca^{2+} to buffer $[Ca^{2+}]_c$ (Nicholls, 1978). X-ray probe microanalysis of rapid-frozen liver showed that total mitochondrial Ca^{2+} is too low to saturate Ca^{2+} efflux and too low for mitochondria to be the principle intracellular Ca^{2+} store mobilized by Ca^{2+}-dependant hormones (Somlyo *et al.*, 1985).

With the discovery that the ER is the main intracellular Ca^{2+} store mobilized by hormones (Streb *et al.*, 1983), it is now widely accepted that the predominate role of mitochondrial Ca^{2+} transporters is regulating $[Ca^{2+}]_m$ rather than $[Ca^{2+}]_c$, although mitochondria can still play a significant role in modulating Ca^{2+} signals derived from ER or plasma membrane channels.

4.1. Quantitation of Mitochondrial Matrix $[Ca^{2+}]$

Quantitation of $[Ca^{2+}]_m$ is vital to an understanding of the physiological role of mitochondrial Ca^{2+} transport systems. A variety of approaches has yielded important information in both isolated mitochondria and intact cells.

Table I
Summary of the Properties of Ca^{2+}-Sensitive Enzymes from Rat Heart Mitochondria[a]

Enzyme system	$K_{0.5}$ for Ca^{2+} (μM)	Effect of Ca^{2+}	Effect of ↓ ATP:ADP
PDH complex	1	PDP activation	PDH-kinase activation
NAD-ICDH	10	↓ $K_{0.5}$ for isocitrate	↓ $K_{0.5}$ for isocitrate ↓ $K_{0.5}$ for Ca^{2+}
OGDH	1	↓ $K_{0.5}$ for oxoglutatate	↓ $K_{0.5}$ for oxoglutatate ↓ $K_{0.5}$ for Ca^{2+}

[a]Data from Mc Cormack *et al.*, 1990.

4.1.1. Measurement of $[Ca^{2+}]_m$ *In Vitro*

Initial efforts to determine the physiological range of $[Ca^{2+}]_m$ consisted of measuring total mitochondrial Ca^{2+}, either by atomic absorption of mitochondria fractionated under conditions where Ca^{2+} transport was inhibited (Severson *et al.*, 1976), or by X-ray probe microanalysis of snap-frozen tissue (Somlyo *et al.*, 1985). These methods, however, do not give direct information on the free Ca^{2+} levels to which mitochondrial matrix enzymes are exposed. Indirect measurements of the relationship between $[Ca^{2+}]_c$ and $[Ca^{2+}]_m$ were made by Denton *et al.* (1980) using the amount of active, dephosphorylated PDH as an indicator for $[Ca^{2+}]_m$. This study suggested that resting $[Ca^{2+}]_m$ is very low (50–100 nM), but interpretation of the effects depended on the Ca^{2+}-sensitivity of PDH in intact mitochondria being approximately equivalent to that of the purified enzyme.

Studies on isolated mitochondria loaded with Fura-2 showed that $[Ca^{2+}]_m$ is approximately 50 nM when extramitochondrial $[Ca^{2+}]$ corresponds to resting levels of $[Ca^{2+}]_c$ (100 nM) (Lukacs and Kapus, 1987; McCormack *et al.*, 1989; Moreno-Sanchez and Hansford, 1988). When extramitochondrial $[Ca^{2+}]$ was increased to approximately 1 μM, the $[Ca^{2+}]_m$ increased to levels in excess of 5 μM, implying that transfer of Ca^{2+} into the mitochondrial matrix is highly cooperative, consistent with the sigmoid kinetics of the Ca^{2+} uptake system (Gunter and Gunter, 1994; Gunter *et al.* 1998; Hansford, 1994).

The global rise of $[Ca^{2+}]_c$ observed in intact cells in response to InsP$_3$ generation is between 500 nM and 1 μM. Given that uniporter affinity for Ca^{2+} is low when measured in isolated mitochondria, seems unlikely that mitochondria would accumulate a significant amount of Ca^{2+} under these conditions without a mechanism to facilitate its uptake. A mechanism has been postulated by Gunter *et al.* (1998). They have observed that isolated mitochondria can accumulate large amounts of Ca^{2+} when exposed to rapid Ca^{2+} increases similar to those evoked by hormones in intact cells (Gunter *et al.* 1998). Uptake is faster at the beginning of each pulse and is of sufficient amount to activate the Ca^{2+}-sensitive mitochondrial dehydrogenases (7 nmoles/mg protein Ca^{2+} taken up, compared with 4 nmoles/mg protein required to activate Ca^{2+}-sensitive mitochondrial dehydrogenases). The Ca^{2+} uptake rate then rapidly decreases. These authors postulate that this fast, transient initial uptake may be carried out by the uniporter operating in a "rapid-uptake mode" (RaM), because it can be blocked by uniporter inhibitors (Gunter *et al.* 1998; Sparagna *et al.* 1995). Under RaM conditions, Ca^{2+} flux may be up to 2 orders of magnitude greater than that normally associated with the uniporter in isolated mitochondria (Gunter *et al.* 1998). There is a however, an alternative, or possibly complementary, mechanism by which Ca^{2+} uptake may be facilitated under physiological conditions in intact cells: through generation of locally high Ca^{2+} gradients in the vicinity of the mitochondria (see below).

4.1.2. Measurement of $[Ca^{2+}]_m$ in Living Cells

The relationship between $[Ca^{2+}]_c$ and $[Ca^{2+}]_m$ *in situ* has been investigated in cells loaded with Ca^{2+}-sensitive fluorescent indicators localized within the mitochondrial matrix (Hajnóczky *et al.*, 1995; Robb-Gaspers *et al.* 1998a, 1998b), or populations of cells expressing recombinant aequorin, targeted to either the mitochondrial matrix or the cytosol (Miyata *et al.*, 1991; Rutter *et al.*, 1996). These studies confirmed the nonlinear relationship between $[Ca^{2+}]_c$ and $[Ca^{2+}]_m$ initially observed in isolated mitochondria.

Rizzuto *et al.* (1992, 1993, 1994) showed that $InsP_3$-mediated mobilization of intracellular Ca^{2+} stores resulted in large, rapid $[Ca^{2+}]_m$ increases occurring in parallel with the $[Ca^{2+}]_c$ increases evoked with the same agonist in parallel cell population measurements. Mean $[Ca^{2+}]_m$ increases were several-fold higher than the underlying changes in $[Ca^{2+}]_c$. Similar relationships were observed in isolated hepatocytes loaded with Fura-2 (reporting $[Ca^{2+}]_c$) and Rhod-2 (reporting $[Ca^{2+}]_m$), which allowed hormone-induced changes in Ca^{2+} in both compartments to be monitored simultaneously in single hepatocytes (Robb-Gaspers *et al.*, 1998a, 1998b) (Fig. 3A).

Recent experimental evidence suggests that transmission of $[Ca^{2+}]_c$ oscillations into the mitochondrial matrix is supported by local calcium control between the ER $InsP_3R$ and mitochondria, resulting from the close proximity of these organelles (Csordás *et al.*, 1999; Landolfi *et al.* 1998; Rizzuto *et al.*, 1998; Simpson and Russel, 1996; Simpson *et al.*, 1997). $InsP_3R$ may be strategically positioned, allowing mitochondria to sense micro-domains of high Ca^{2+} generated in the vicinity of the $InsP_3R$. Csordás *et al.* (1999) provided evidence that microdomains of high $[Ca^{2+}]$ are generated at ER–mitochondrial

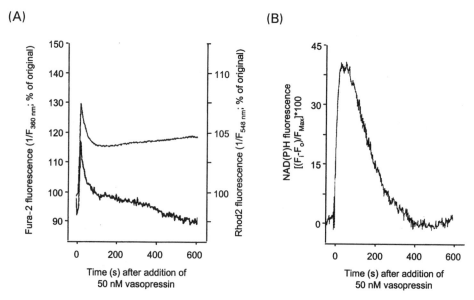

(A)   **(B)**

FIGURE 3. Determination of vasopressin-induced changes in $[Ca^{2+}]$ in populations of isolated primary hepatocytes. (A) Hepatocytes were cultured for 12 hr prior to experimental manipulation. Changes in $[Ca^{2+}]_c$ (upper trace) and $[Ca^{2+}]_m$ (lower trace) were monitored after adding 50 nM vasopressin ($t = 0$), according to methods described in Robb-Gaspers *et al.* (1998a). Upper trace shows vasopressin-mediated changes in $[Ca^{2+}]_c$ (via changes in Fura-2 fluorescence). Lower trace shows changes in $[Ca^{2+}]_m$ (via changes in Rhod-2 fluorescence). (B) NADH responses were monitored using 360 nm autofluorescence in single isolated hepato-cytes, according to methods described in Robb-Gaspers *et al.* (1998a). Trace shows the mean response from all cells in the population (Reprinted from *Biochim. Biophys. Acta*: 1366, Robb-Gaspers *et al.*, "Coupling between cytosolic and mitochondrial calcium oscillations: Role in the regulation of hepatic metabolism", Copyright 1988, pp. 17–32, with permission from Elsevier Science.)

junctions as a result of the activation of multiple $InsP_3Rs$. The $[Ca^{2+}]_c$ in these microdomains is >20-fold higher than global $[Ca^{2+}]_c$ increases, and can maximally stimulate mitochondrial Ca^{2+} uptake (Csordás et al., 1999; Rizzuto et al., 1998).

The spatial relationship between mitochondria and the ER derived from the type of functional studies described above has been confirmed in HeLa cells expressing specifically targeted green fluorescent protein (GFP) with different colors targeted to the ER and mitochondria (Rizzuto et al., 1998). Numerous close contacts between mitochondria and the ER were observed, and upon agonist-induced $InsP_3R$ opening the mitochondrial surface was exposed to a higher concentration of Ca^{2+} than was the bulk cytosol (Rizzuto et al., 1998). Maximal activation of mitochondrial Ca^{2+} uptake is evoked by these perimitochondrial $[Ca^{2+}]$ elevations, which appear to be >20-fold higher than the global increases in bulk $[Ca^{2+}]_c$ (Csordás et al., 1999).

In addition to close juxtaposition of mitochondria to regions of the ER in a variety of cells, it is now apparent in some cells types that mitochondria can accumulate Ca^{2+} during cell membrane depolarizations (Lawrie et al., 1996). In these cells mitochondria are located close to the plasma membrane, thus the mitochondrial Ca^{2+} uptake sites could be exposed to microdomains of high $[Ca^{2+}]_c$ around voltage-gated and store-operated Ca^{2+} influx channels (Lawrie et al., 1996).

4.2. Modulation of $[Ca^{2+}]_c$ Signals by Mitochondria

Given the close juxtaposition of mitochondria to Ca^{2+} release sites and their large capacity for Ca^{2+} uptake, it is feasible that they can modulate the kinetics of signals, especially transient $[Ca^{2+}]_c$ changes such as those underlying $[Ca^{2+}]_c$ waves and oscillations. In Xenopus laevis oocytes, $InsP_3$-induced $[Ca^{2+}]_c$ wave amplitude, velocity, and interwave period were enhanced by oxidizable substrates that energize mitochondria (Babcock et al., 1997; Hoth et al., 1997; Jouaville et al. 1995, 1998). These effects were abolished by ruthenium red and respiratory chain inhibitors. Similarly, mitochondria have been shown to support $[Ca^{2+}]_c$ oscillations in oligodendrocytes (Simpson et al., 1997). These authors reported colocalization of high-density patches of $InsP_3R$ with mitochondria in areas within oligodendrocyte processes. The amplitude and rate of rise of $[Ca^{2+}]_c$ responses in these areas were larger than in the surrounding cytosol. One interpretation of these data is that mitochondria can modulate the gating properties of the $InsP_3R$ to enhance the $[Ca^{2+}]_c$ response. In another example of this type of modulation, Hajnóczky et al. (1999) demonstrated that mitochondrial Ca^{2+} uptake reduces positive Ca^{2+} feedback on the $InsP_3R$, giving rise to local suppression of $InsP_3$ sensitivity in hepatocytes.

The MPT pore may also enhance $[Ca^{2+}]_c$ responses within cells. Uptake of Ca^{2+} into the mitochondrial matrix triggers transient opening of the MPT pore in isolated mitochondria and in intact cells, with subsequent efflux of Ca^{2+} into the extramitochondrial medium (Ichas et al., 1994, 1997; Selivanov et al., 1998). In intact cells, such released Ca^{2+} would amplify the increase in $[Ca^{2+}]_c$ from the ER. This phenomenon is known as mitochondrial Ca^{2+}-induced Ca^{2+} release. Uptake of Ca^{2+} via the uniporter causes a transient drop in $\Delta\psi_m$, resulting in respiratory-chain activation (Duchen, 1992; Selivanov et al., 1998) and increased matrix pH, triggering the MPT. The MPT causes a partial respiratory-chain inhibition, allowing matrix pH to fall and the MPT pore to close (Bernardi, 1996; Ichas et al., 1997; Zoratti and Szabò, 1995).

5. CONTROL OF MITOCHONDRIAL METABOLISM BY Ca^{2+}

The mechanism of Ca^{2+} regulation of mitochondrial oxidative metabolism is outlined in Figure 4A. Many extracellular stimuli increase cellular activity via an increase in [Ca^{2+}]$_c$. This increased activity requires enhanced production of ATP. The decrease in key metabolite ratios (in particular ATP:ADP) that may be expected to accompany increased cellular activity is an important effector of mitochondrial metabolism (intrinsic regulation, shown in black). An additional means of regulation exists in many mammalian cells however, namely the regulation of mitochondrial dehydrogenases by [Ca^{2+}]$_m$ (extrinsic regulation, shown in gray). The mitochondrial dehydrogenases PDH, NAD-ICDH, and OGDH, which regulate the supply of reducing equivalents to the respiratory chain are sensitive to changes in [Ca^{2+}]$_m$ in the physiological range (reviewed in McCormack *et al.*, 1990; Rutter *et al.*, 1990)

The activity of PDH is regulated by a phosphorylation–dephosphorylation cycle (Fig. 4B). Dephosphorylation, leading to activation of PDH, is catalyzed by a Ca^{2+}-sensitive phosphatase, pyruvate dehydrogenase phosphate phosphatase (PDP), and reversed by a specific kinase (PDH kinase) that is probably Ca^{2+} insensitive. The total amount of active PDH (PDH$_a$) in the cell at any given time is governed by the relative activities of these opposing enzymes (Denton *et al.*, 1996). PDP is believed to be the molecular target for the regulation of PDH activity by insulin or by hormones linked to the mobilization of [Ca^{2+}]$_c$, whereas PDH kinase is subject to regulation by intrinsic factors (Denton *et al.*, 1972, 1996; Lawson *et al.*, 1993, 1997; McCormack *et al.*, 1990; Reed and Yeaman, 1987; Rutter *et al.*, 1989; Thomas *et al.*, 1986; Yan *et al.*, 1996).

Regulation of NAD-ICDH and OGDH by intrinsic and extrinsic factors is via allosteric interactions. Both enzymes are activated by a decrease in either redox or phosphorylation potential, and these cofactors modulate Ca^{2+} sensitivity (Table I) (McCormack *et al.*, 1990). In contrast to PDH, Ca^{2+} is an allosteric activator of NAD-ICDH and OGDH (Denton *et al.*, 1978; Hansford, 1991; Lawlis and Roche, 1981; McCormack, 1985; McCormack and Denton, 1979, 1984; Nichols and Denton, 1995; Rutter, 1990) and Ca^{2+} sensitivity of NAD-ICDH is 1 order of magnitude lower than OGDH (Rutter and Denton, 1988) (Table I). The effect of Ca^{2+} is to greatly diminish the K$_{0.5}$ of the enzymes for their respective substrates. The precise molecular nature of the Ca^{2+} binding site for OGDH and NAD-ICDH remains to be fully characterized, but appears to be via a novel Ca^{2+}-binding motif (Nichols and Denton, 1995).

5.1. Regulation of the Respiratory Chain

In isolated mitochondria, control of oxidative phosphorylation is distributed along several reactions in the pathway, and the degree of control exerted by each site depends on metabolic conditions (Balaban, 1990; Moreno-Sanchez and Torres-Marquéz, 1991). Control of respiration under basal conditions resides in the phosphorylation pathways, including ATP hydrolysis, adenine nucleotide transport, and ATP synthesis (Brown *et al.*, 1990). Although direct effects of ATP on the electron transport chain have been suggested (Kadenbach, 1986), a more widely accepted mechanism for respiration regulation by the ATP:ADP ratios is via changes in the proton motive force (Nicholls and Akerman, 1984). In the absence of ADP (state 4 respiration), steady-state PMF is

(A)

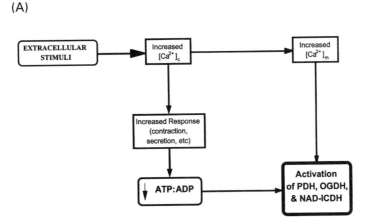

(B)

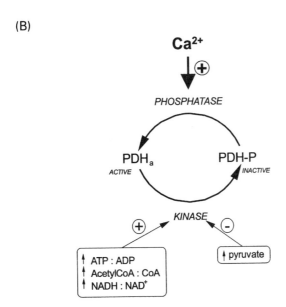

FIGURE 4. Regulation of mammalian mitochondrial metabolism. (A) Activation of pyruvate dehydrogenase (PDH), 2-oxoglutarate dehydrogenase (OGDH), and NAD-isocitrate dehydrogenase (NAD-ICDH) by intrinsic factors (black) or extrinsic factors (gray) leads to increase in supply of reducing equivalents (NADH) to the respiratory chain. This ultimately leads to enhanced ATP production. Extrinsic regulation results in enhanced ATP production independent of a decline in cellular ATP levels. (B) Regulation of PHD activity in mammalian cells. Intrinsic factors (black) and extrinsic factors (gray) incluence active pyruvate dehydrogenase (PDH$_a$) level in mammalian cells by altering the relative activities of exzymes responsible for the phosphorylation/dephosphorylation of PDH.

established due to leakage of protons into the mitochondrion, "slipping" of the proton pumps of the respiratory chain, or both. Addition of ADP (state 3 respiration) increases respiration by increasing the rate at which protons may return to the mitochondrial matrix via the ATPase. For changes in cytosolic ATP consumption to affect PMF, ADP must be imported into the mitochondrial matrix; hence it is possible for the adenine nucleotide translocase or the P_i transporter to play a role in the regulation of respiration. At high rates of respiration, the translocase may be a significant determinant of flux through the respiratory chain (Groen *et al.*, 1982).

Control of respiration and ATP production in mitochondria *in situ* is also regulated by multiple pathways. Stimulation of the respiratory chain by ADP in isolated mitochondria under saturating substrate conditions results in the decrease of NADH:NAD^+ (Hansford, 1980). In living cells, however, it is unlikely that such conditions occur. Under conditions where NADH production is insufficient to allow maximal activity of the respiratory chain, one would expect that a respiration increase could be achieved by activating the mitochondrial dehydrogenases (Fig. 1 and Fig. 4A) (Denton *et al.*, 1996; Hansford, 1994; McCormack *et al.*, 1990; Quinlan and Halestrap, 1986). In addition, Halestrap and co-workers postulated (1989, 1994) that hormonal signaling pathways may directly control flux through the respiratory chain via changes in mitochondrial volume. This is mediated by inhibition of the matrix pyrophosphatase by Ca^{2+} (Fig. 1) (Halestrap, 1989, 1994; McCormack *et al.*, 1990); the subsequent increase in pyrophosphate concentration raises the osmotic potential of the mitochondrial matrix (McCormack *et al.*, 1990; Quinlan *et al.*, 1983).

5.2. Regulation of the Mitochondrial Ca^{2+}-Sensitive Dehydrogenases by Intrinsic Factors and Ca^{2+}

Decreases in ATP:ADP occuring within the physiological range activate PDH, OGDH, and NAD-ICDH (Denton *et al.*, 1978; McCormack and Denton, 1979; Roche and Reed, 1974; Soboll and Bunger, 1981; Soboll *et al.*, 1984). The decreases activate these enzymes by increasing their sensitivity to Ca^{2+} (Table I) (McCormack *et al.*, 1990). Often, however, hormone-stimulated increases in O_2 consumption in intact tissues are accompanied by an increase in ATP:ADP; decreases in phosphorylation potential are rare (McCormack *et al.*, 1990). Thus, regulation of dehydrogenase activity by adenine nucleotide phosphorylation state probably serves as an intrinsic control to maintain basal ATP:ADP ratios, but the ratios are often not the primary means by which respiration is stimulated in response to hormones and other effectors.

Regulation of PDH, NAD-ICDH, and OGDH by Ca^{2+} *in vitro* and in isolated mitochondria, together with the kinetics of mitochondrial Ca^{2+} changes *in situ*, has led to the proposal that increases in $[Ca^{2+}]_m$ can directly stimulate oxidative metabolism via the increased supply of NADH to the respiratory chain (Fig. 1 and Fig. 4A). This extrinsic regulation by Ca^{2+} will enhance ATP production without a concomitant decline in cellular phosphorylation potential or mitochondrial redox potential (Denton *et al.*, 1996; Hansford, 1994; McCormack *et al.*, 1990).

5.3. Regulation of Mitochondrial Metabolism by Ca^{2+} in Intact Cells

5.3.1. Relationship between $[Ca^{2+}]_m$ and Mitochondrial NADH

Until recently, taking parallel measurements of agonist-induced $[Ca^{2+}]_i$ changes and their putative physiological targets within mitochondria has been problematic, but studies by Hajnóczky *et al.* (1995), Rutter *et al.* (1996), and Robb-Gaspers *et al.* (1998a, 1998b) have provided direct evidence for the ability of mitochondria to decode the often complex $[Ca^{2+}]_i$ changes into a coordinated physiological response, namely the stimulation of mitochondrial oxidative metabolism.

Hormone-induced activation of the Ca^{2+}-sensitive mitochondrial dehydrogenases (CSMDH) was demonstrated in single isolated hepatocytes (Hajnóczky *et al.*, 1995) and in populations of fibroblasts in culture (Rutter *et al.*, 1996). These findings lead to a model wherein the mitochondrial redox couple is directly regulated by $[Ca^{2+}]_m$ linked to $[Ca^{2+}]_c$. Hajnóczky *et al.* (1995) demonstrated that the periodic oscillations in $[Ca^{2+}]_c$, induced by InsP$_3$-linked hormones, are transmitted efficiently into the mitochondrial matrix as $[Ca^{2+}]_m$ oscillations in single hepatocytes. In this study the cells were loaded with Ca^{2+}-sensitive dyes localized within the cytosol (Fura-2) or mitochondrial matrix (Rhod-2). Agonist treatment resulted in oscillations in $[Ca^{2+}]_c$, the frequency of which increased with agonist dose. Oscillations in $[Ca^{2+}]_c$ evoked parallel oscillations in $[Ca^{2+}]_m$. The hormone-induced oscillations in $[Ca^{2+}]_c$ and $[Ca^{2+}]_m$ gave rise to constant amplitude oscillations in NADH fluorescence, which were the result of CSMDH activation (Hajnóczky *et al.*, 1995). These oscillations in NADH fluorescence were markedly slower to return to basal levels than were oscillations in $[Ca^{2+}]_m$. Thus as the frequency of $[Ca^{2+}]_i$ oscillations increased, the interspike period became too short for the CSMDH activities to return to basal, hence the time-averaged stimulation of the CSMDH increased with increasing agonist dose (Hajnóczky *et al.*, 1995). Non-InsP$_3$ dependent increases in $[Ca^{2+}]_c$ were less efficient at elevating $[Ca^{2+}]_m$, so the CSMDH stimulation was less for the equivalent $[Ca^{2+}]_c$ increase (Hajnóczky *et al.*, 1995). These observations are consistent with there being privileged access to the mitochondria of Ca^{2+} released by InsP$_3$-dependent mechanisms, which leads to the stimulation of mitochondrial metabolism (Rizzuto *et al.*, 1998).

5.3.2. Relationship between $[Ca^{2+}]_c$, $[Ca^{2+}]_m$, and CSMDH Activity

Studies in which CSMDH activity is monitored through global changes in NADH fluorescence do not reveal specific information regarding the activity of any single CSMDH. This information may potentially be important, given their different sensitivities to Ca^{2+} *in vitro* (Denton *et al.*, 1996; McCormack *et al.*, 1990). Furthermore, other cellular processes may be activated during hormonal stimulation, which may also alter cellular NADH levels, including alterations in respiratory chain activity (McCormack *et al.*, 1990).

Hormones, such as vasopressin, cause increases in $[Ca^{2+}]_c$ in perfused liver hepatocytes these manifest as Ca^{2+} oscillations and waves similar to those observed in isolated hepatocytes (Robb-Gaspers and Thomas, 1995). Moreover, $[Ca^{2+}]_c$ oscillations are translocated into the mitochondrial matrix and are accompanied by increases in NADH fluorescence and in the ATP : ADP ratio (Denton and McCormack, 1995; Kimura *et al.*, 1984). In this preparation the 2-oxoglutarate level is decreased, consistent with activation of OGDH. Moreover, PDH activation by hormonal stimulation in perfused liver has been demonstrated directly by increases in active PDH in rapidly isolated mitochondria extracts. Activation of PDH and OGDH is abolished by ruthenium red (McCormack, 1985).

In the heart, hormones, such as epinephrine and other positive inotrophic agents, increase contractile performance, partly by increasing the amplitude of the contractile $[Ca^{2+}]_c$ transients. Although sarcoplasmic Ca^{2+} fluctuates on a beat-to-beat basis, mean increases in $[Ca^{2+}]_c$ are within the range necessary to raise $[Ca^{2+}]_m$ (Marban *et al.*, 1980). These agents increase O_2 consumption in intact heart preparations without any concomitant change in ATP : ADP, ruling out control of respiration solely by intrinsic ATP : ADP regulation (Katz *et al.*, 1989; McCormack *et al.*, 1990; Unitt *et al.*, 1989). Moreover, adding ruthenium red to such preparations causes ATP : ADP to fall during agonist stimulation (Katz *et al.*, 1989). Significantly, increased O_2 consumption in perfused hearts is accompanied by an increased NADH level (Katz *et al.*, 1987). The role of Ca^{2+} in increasing mitochondrial respiration in perfused heart is further confirmed by the observation that ruthenium red delays the O_2 consumption increase that results from increasing the heartbeat frequency (Hak *et al.*, 1993; McCormack *et al.*, 1990). As with intact liver, direct measurements of the activation of PDH and OGDH by inotrophic agents have been made in the intact perfused heart. Increases in the active, dephosphorylated form of PDH following hormonal stimulation are prevented by adding ruthenium red to this preparation (McCormack and England, 1983).

Studies by Rutter *et al.* (1996) examined the relationship between the activity of PDH and $[Ca^{2+}]_m$ in the fibroblast cell line CHO.T. Changes in $[Ca^{2+}]_i$ following hormonal stimulation were monitored in parallel populations of cells expressing recombinant aequorin, targeted to either the cytosol or the mitochondrial matrix (Cobbold and Lee, 1991; Rizzuto *et al.*, 1994; Rutter *et al.*, 1993). Activation of purinergic receptors evoked increases in $[Ca^{2+}]_c$, which were transmitted into the mitochondrial matrix as observed in hepatocytes. The $[Ca^{2+}]_m$ response following the addition of agonist was biphasic in cells stimulated in the presence of extracellular Ca^{2+}. The $[Ca^{2+}]_m$ increased rapidly to approximately $10\,\mu M$ following stimulation with agonist; this initial large peak was followed by a smaller, but more sustained, increase in $[Ca^{2+}]_m$. The peak increase in $[Ca^{2+}]_m$ was 1 order of magnitude greater than the increase in $[Ca^{2+}]_c$, observed in parallel populations of cells transfected with cytosol-targeted aequorin.

In cells incubated in the presence of EGTA, a somewhat smaller and short-lived increase in $[Ca^{2+}]_m$ was observed. Comparison of $[Ca^{2+}]_m$ with the activity of PDH in parallel cultures of CHO. T cells confirmed that PDH was a target for hormone-induced $[Ca^{2+}]_m$ changes, in accordance with the model described above. Upon agonist stimulation, PDH activity was rapidly increased, an increase sustained for up to 5 min in the presence of extracellular Ca^{2+} (Rutter *et al.*, 1996). This is consistent with the measured time-course of $[Ca^{2+}]_m$ increases. Changes in PDH, however, like oscillations in NADH in hepatocytes, lagged behind those of $[Ca^{2+}]_m$, consistent with the relatively slow half-time for inactivation of PDH by PDH kinase (15 sec *in vitro*) (Rutter *et al.*, 1989, 1996). The authors estimated PDH sensitivity to Ca^{2+} *in situ* and found that it agreed with previous studies carried out *in vitro*, demonstrating a $k_{0.5}$ of approximately

$1 \mu M$ (McCormack et $al.$, 1990). From these data it appears likely that $[Ca^{2+}]_m$ is the primary mediator of hormone-induced PDH activation in CHO.T cells.

Regulation of PDH was found to be more complex in isolated hepatocytes (Robb-Gaspers et $al.$, 1998a, 1999b). The hepatocytes were stimulated with a maximal (50 nM) dose of vasporessin, which generated a sustained increase in $[Ca^{2+}]_c$ that was synchronized in all cells of the population (Fig. 3A, upper trace). This sustained increase also elicited a synchronous single transient increase in $[Ca^{2+}]_m$ (Fig. 3A, lower trace), accompanied by a more prolonged, but still transient, increase in NADH (Fig. 3B). Changes in $[Ca^{2+}]_m$ were compared to changes in PDH activity in parallel populations. In contrast to the CHO cell studies outlined above, PDH activity correlated with changes in NADH fluorescence and $[Ca^{2+}]_m$ only during the initial phase of these responses. Following stimulation with maximal vasopressin, PDH activity increased biphasically, and the first rapid phase accompanied the rise in $[Ca^{2+}]_m$, but this was followed by a second phase of PDH activation even as $[Ca^{2+}]_m$ and NADH levels returned to basal. This sustained secondary PDH activation was associated with a decline of cellular ATP levels, suggesting that it may reflect PDH kinase inhibition resulting from a fall in the ATP : ADP ratio (Fig. 4B) (McCormack et $al.$, 1990). Thus PDH activity in isolated hepatocytes is subject to regulation first by $[Ca^{2+}]_m$ (extrinsic regulation), with a subsequent decrease in ATP (intrinsic regulation) increasing PDH levels still further (Fig. 4). The decline in NADH levels in the face of sustained PDH activity appears to reflect Ca^{2+}-dependent activation of the respiratory chain (Halestrap, 1994; McCormack et $al.$, 1990). Direct analysis of the pH gradient and mitochondrial membrane potential components of the proton motive force in intact hepatocytes revealed a Ca^{2+}-dependent activation of the respiratory chain in response to agonist, consistent with the enhanced reoxidation of NADH (Robb-Gaspers et $al.$, 1998a, 1998b). These data show that Ca^{2+} plays an important role in signal transduction from the cytosol to the mitochondria; a single oscillation in $[Ca^{2+}]_m$ is able to evoke a complex series of changes to activate mitochondrial oxidative metabolism.

6. CONCLUSIONS

It is becoming clear that mitochondria are intimately involved in cellular Ca^{2+} signaling pathways. Mitochondria's role may be two-fold: First, the capacity of mitochondria to accumulate Ca^{2+}, coupled with the strategic localization of Ca^{2+} release channels with mitochondrial Ca^{2+} uptake sites, enables mitochondria to modulate hormone-induced $[Ca^{2+}]_c$ oscillation properties. Mitochondria can discriminate between $[Ca^{2+}]_c$ changes resulting from leaks or other effectors of cellular Ca^{2+} homeostasis and InsP$_3$-dependent Ca^{2+} signaling. Second, hormone-induced oscillations in $[Ca^{2+}]_m$ are within the physiological range for activation of Ca^{2+}-sensitive mitochondrial dehydrogenases (McCormack et $al.$, 1990; Robb-Gaspers et $al.$, 1998a, 1998b; Rutter et $al.$, 1996) and therefore provide a mechanism for coordinated regulation of mitochondrial oxidative metabolism by Ca^{2+}. The relatively slow decay of active PDH level in CHO. T cells (Rutter et $al.$, 1996) and of mitochondrial NADH and proton motive force in hepatocytes, following a $[Ca^{2+}]_m$ transient, results in a sustained increase in these parameters as the frequency of $[Ca^{2+}]_m$ oscillations increases. Hajnóczky et $al.$ (1995)

demonstrated that $[Ca^{2+}]_m$ oscillation frequencies greater than 0.5 per second elicit sustained increases in NADH. Thus, mitochondria are able to integrate oscillating Ca^{2+} signals into a graded metabolic output.

ACKNOWLEDGEMENT: Supported by NIH grant 38422.

REFERENCES

Akerman, K. E. O., Wilkstrom, M.K.E., and Saris, N.-E., 1977, Effect of inhibitors on the sigmoidicity of the calcium ion transport kinetics in rat liver mitochondria, *Biochim. Biophy. Acta* **464**: 287–294.

Alonso, M.T., Barrero, M.J., Michelena, P., Carnicero, E., Cuchillo, I., Garciá, A. G., Garciá-Sancho, J., Montero, M., and Alvarez, J., 1999, Ca^{2+}-induced Ca^{2+} release in chromaffin cells seen from the inside the ER with targeted aequorin, *J. Cell Biol.* **144**: 241–254.

Babcock, D.F., Herrington, J., Goodwin, P.C., Park, Y. B., and Hille, B., 1997, Mitochondrial participation in the intracellular Ca^{2+} network, *J. Cell Biol.* **136**: 833–844.

Balaban, R. S., 1990, Regulation of oxidative phosphorylation in the mammalian cell, *Am. J. Physiol.* **258**: C377–C389.

Baysal, K., Brierley, G. P., Novgorodov, S., and Jung, D. W., 1991, Regulation of the mitochondrial Na^+/Ca^{2+} antiport by matrix pH, *Arch. Biochem. Biophys.* **291**: 383–389.

Berridge, M. J., 1993, Cell signaling: A tale of two messengers, *Nature* **361**: 315–325.

Berridge, M. J., 1995, Capacitative calcium entry, *Biochem. J.* **312**: 1–11.

Berridge, M. J., 1997, Elementary and global aspects of calcium signaling, *J. Physiol.* **499**: 291–306.

Berridge, M.J., Galione, A., 1988, Cytosolic calcium oscillators, *FASEB J.* **2**: 3074–3082.

Berridge, M. J., Cobbold, P. H., and Cuthbertson, K. S., 1988, Spatial and temporal aspects of cell signaling, *Philos. Trans. R. Soc. Lond. Ser. A* **B320**: 325–343.

Bernardi, P., 1996, The permeability transition pore: Control points of a cyclosporin A-sensitive mitochondrial channel involved in cell death, *Biochim. Biophys. Acta* **1275**: 5–9.

Bernardi, P., and Azzone, G. F., 1979, delta pH induced calcium fluxes in rat liver mitochondria, *Eur. J. Biochem.* **102**: 555–562.

Bragadin, M., Pozzan, T., and Azzone, G. F., 1979, Kinetics of Ca^{2+} carrier in rat liver mitochondria, *Biochemistry* **18**: 5972–5978.

Brand, M. D., 1995, The stoichiometry of the exchange catalyzed by the mitochondrial calcium/sodium antiporter, *Biochem. J.* **229**: 161–166.

Brand, M. D., and Murphy, M. P., 1987, Control of electron flux through the respiratory chain in mitochondria and cells, *Biol. Rev. Cambridge Philos. Soc.* **62**: 141–193.

Brown, G. C., Lakin-Thomas, P. L., Brand, M. D., 1990, Control of respiration and oxidative phosphorylation in isolated rat liver cells, *Eur. J. Biochem.* **192**: 355–362.

Chiesi, M., Rogg, H., Eichenberger, K., Gazzotti, P., and Carafoli, E., 1987, Stereospecific action of diltiazem on the mitochondrial Na–Ca exchange system and on sarcolemmal Ca channels, *Biochem. Pharmacol.* **36**: 2735–2740.

Clapham, D. E., 1995, Calcium signaling, *Cell* **80**: 259–268.

Cobbold, P. H., and Lee, J. A. C., 1991, Aequorin measurements of cytoplasmic free calcium, in *Cellular Calcium: A Practical Approach* (J. G. McCormack and P. H. Cobbold, Eds.), Oxford University Press, pp. 55–82.

Crompton, M., 1990, in *Intracellular Calcium Regulation*. (F. Branner), Ed. R. Liss, New York, pp. 181–209.

Crompton, M., and Roos, I., 1985, On the hormonal control of heart mitochondrial Ca^{2+}, *Biochem. Soc. Trans.* **13**: 667–669.

Crompton, M., Künzi, M., and Carafoli, E., 1977, The calcium-induced and sodium-induced effluxes of calcium from heart mitochondria: Evidence for a sodium–calcium carrier, *Eur. J. Biochem.* **79**: 549–558.

Crompton, M., Moser, R., Lüdi, H., and Carafoli, E., 1978, The interrelations between the transport of sodium and calcium in mitochondria of various mammalian tissues, *Eur. J. Biochem.* **82**: 25–31.

Crompton, M., Heid, I., and Carafoli, E., 1980, The activation by potassium of the sodium–calcium carrier of cardiac mitochondria, *FEBS Lett.* **115**: 257–259.

Csordás, G., Thomas, A. P., and Hajnóczky, G., 1999, Quasi-synaptic calcium signal transmission between the endoplasmic reticulum and mitochondria, *EMBO J.* **18**: 96–108.

Denton, R. M., and Halestrap, A. P., 1979, Regulation of pyruvate metabolism in mammalian tissues, *Essays Biochem.* **15**: 37–77.

Denton, R. M., and McCormack, J. G., 1980, On the role of the calcium transport cycle in heart and other mammalian mitochondria, *FEBS Lett.* **119**: 1–8.

Denton, R. M., and McCormack, J. G., 1989, Ca^{2+} as a second messenger within mitochondria of the heart and other tissues, *Annu. Rev. Physiol.* **52**: 451–466.

Denton, R. M., and McCormack, J. G., 1995, Fuel selection at the level of mitochondria in mammalian tissues, *Proc. Nutr. Soc.* **54**: 11–22.

Denton, R. M., Randle, P. J., and Martin, B. R., 1972, Stimulation by calcium ions of pyruvate dehydrogenase phosphate phosphatase, *Biochem. J.* **128**: 161–163.

Denton, R. M., Richards, D. A., and Chin, R. J., 1978, Calcium ions and the regulation of NAD$^+$-linked isocitrate dehydrogenase from the mitochondria of rat heart and other tissues, *Biochem. J.* **176**: 899–906.

Denton, R. M., McCormack, J. G., and Edgell, N. J., 1980, Role of calcium ions in the regulation of intramitochondrial metabolism. Effects of Na$^+$, Mg^{2+} and ruthenium red on the Ca^{2+}-stimulated oxidation of oxoglutarate and on pyruvate dehydrogenase activity in intact rat heart mitochondria, *Biochim. Biophy. Acta* **190**: 107–117.

Denton, R. M., McCormack, J. G., Rutter, G. A., Burnett, P., Edgell, N. J., Moule, S. K., and Diggle, T. A., 1996, The hormonal regulation of pyruvate dehydrogenase complex, *Adv. Enzyme Regul.* **36**: 183–198.

Duchen, M. R., 1992, Ca($^{2+}$)-dependent changes in the mitochondrial energetics in single dissociated mouse sensory neurons, *Biochem. J.* **283**: 41–50.

Erecinska, M., and Wilson, D. F., 1982, Regulation of cellular energy metabolism, *J. Membr. Biol.* **70**: 1–14.

Exton, J. H., 1980, Mechanisms involved in alpha-adrenergic phenomena: Role of calcium ions in actions of catecholamines in liver and other tissues, *Am. J. Physiol.* **238**: E3–E12.

Exton, J. H., 1981, Molecular mechanisms involved in alpha-adrenergic responses, *Mol. Cell. Endocrinol.* **23**: 233–264.

Fiskum, G., and Lehninger, A. L., 1979, Regulated release of Ca^{2+} from respiring mitochondria by Ca^{2+}/H$^+$ antiport, *J. Biol. Chem.* **254**: 6236–6239.

Fiskum, G., and Lehninger, A. L., 1982, in *Calcium and Cell Function* (W. Y. Cheung, Ed.) Academic, New York, pp. 39–80.

Garlid, K. D., 1988, in *Integration of Mitochondrial Function*, Plennum, New York, pp. 257–276.

Groen, A. K., Wanders, R. J., Westerhoff, R., van der Meer, J., and Tage, J. M., 1982, Quantification of the contribution of various steps to the control of mitochondrial respiration, *J. Biol. Chem.* **257**: 2754–2762.

Gunter, K. K., and Gunter, T. E., 1994, Transport of calcium by mitochondria, *J. Bioenerg. Biomembr.* **26**: 471–485.

Gunter, T. E., and Pfeiffer, D. R., 1990, Mechanisms by which mitochondria transport calcium, *Am. J. Physiol.* **258**: C755–C786.

Gunter, K. K., Zuscik, M. J., and Gunter, T. E., 1991, The Na($^+$)-independent Ca^{2+} efflux mechanism of liver mitochondria is not a passive Ca^{2+}/H$^+$ exchanger, *J. Biol. Chem.* **266**: 21640–21648.

Gunter, T. E., Buntinas, L., Sparagna, G. C., and Gunter, K. K., 1998, The Ca^{2+} transport mechanisms of mitochondria and Ca^{2+} uptake from physiological-type Ca^{2+} transients, *Biochim. Biophys. Acta* **1366**: 5–15.

Hajnóczky, G., Lin, C., and Thomas, A. P., 1994, Luminal communication between intracellular calcium stores modulated by GTP and the cytoskeleton, *J. Biol. Chem.* **269**: 10280–10287.

Hajnóczky, G., Robb-Gaspers, L. D., Seitz, M. B., and Thomas, A.P., 1995, Decoding of cytosolic calcium oscillations in the mitochondria, *Cell* **82**: 415–424.

Hajnóczky, G., Hager, R., and Thomas, A. P., 1999, Mitochondria suppress local feedback activation of inositol 1,4,5-trisphosphate receptors by Ca^{2+}, *J. Biol. Chem.* **274**: 14157–14162.

Hak, J. B., van Beek, G. M., Eijgelshoven, M. H., and Westerhof, N., 1993, Mitochondrial dehydrogenase activity affects adaptation of cardiac oxygen consumption to demand, *Am. J. Physiol.* **264**: H448–H453.

Halestrap, A. P., 1989, The regulation of the matrix volume of mammalian mitochondria *in vivo* and *in vitro* and its role in the control of mitochondrial metabolism, *Biochim. Biophys. Acta* **973**: 355–382.

Halestrap, A. P., 1994, Regulation of mitochondrial metabolism through changes in matrix volume, *Biochem. Soc. Trans.* **22**: 522–529.

Halestrap, A. P., Kerr, P. M., Javadov, S., and Woodfield, K.-Y., 1998, Elucidating the molecular mechanism of the permeability transition pore and its role in reperfusion injury of the heart, *Biochim. Biophys. Acta* **1366**: 79–94.

Hansford, R. G., 1991, Dehydrogenase activation by Ca^{2+} in cells and tissues, *J. Bioenerg. Biomembr.* **23**: 823–830.

Hansford, R. G., 1994, Physiological role of mitochondrial Ca^{2+} transport, *J. Bioenerg. Biomembr.* **26**: 495–508.

Hayat, L. H., and Crompton, M., 1987, The effects of Mg^{2+} and adenine nucleotides on the sensitivity of the heart mitochondrial $Na^+–Ca^{2+}$ carrier to extramitochondrial Ca^{2+}: A study using arsenazo III-loaded mitochondria, *Biochem. J.* **244**: 533–538.

Hoth, M., Fanger, C. M., and Lewis, R. S., 1997, Mitochondrial regulation of store-operated calcium signaling in T lymphocytes, *J. Cell Biol.* **137**: 633–648.

Ichas, F., and Mazat, J. P., 1998, From calcium signalling to cell death: Two conformations for the mitochondrial permeability transition pore. Switching from low- to high-conductance state, *Biochim. Biophys. Acta* **1366**: 33–50.

Ichas, F., Jouaville, L. S., Sidash, S. S., Mazat, J. P., and Holmuhamedov, E. L., 1994, Mitochondrial calcium spiking: A transduction mechanism based on calcium-induced permeability transition involved in cell calcium signalling, *FEBS Lett.* **348**: 211–215.

Ichas, F., Jouaville, L. S., and Mazat, J. P., 1997, Mitochondria are excitable organelles capable of generating and conveying electrical and calcium signals, *Cell* **89**: 1145–1153.

Jacob, R., 1990, Calcium oscillations in electrically non-excitable cells, *Biochim. Biophys. Acta* **1052**: 427–448.

Jouaville, L. S., Ichas, F., Holmuhamedov, E. L., Camacho, P., and Lechleiter, J. D., 1995, Synchronization of calcium waves by mitochondrial substrates in *Xenopus laevis* oocytes, *Nature* **377**: 600–602.

Jouaville, L. S., Ichas, F., and Mazat, J. P., 1998, Modulation of cell calcium signals by mitochondria, *Mol. Cell. Biochem.* **184**: 371–376.

Jung, D. W., Baysal, K., and Brierley, G. P., 1995, The sodium–calcium antiport of heart mitochondria is not electroneutral, *J. Biol. Chem.* **270**: 672–678.

Kadenbach, B., 1986, Regulation of respiration and ATP synthesis in higher organisms: Hypothesis, *J. Bioenerg. Biomembr.* **18**: 39–54.

Katz, L. A., Koretsky, A. P., and Balaban, R. S., 1987, Respiratory control in the glucose-perfused heart: A ^{31}P NMR and NADH fluorescence study, *FEBS Lett.* **221**: 270–276.

Katz, L. A., Swain, J. A., Portman, M. A., and Balaban, R. S., 1989, Relation between phosphate metabolites and oxygen consumption of heart *in vivo, Am. J. Physiol.* **256**: H265–H274.

Kimura, S., Suzaki, T., Abe, K., Ogata, E., and Kobayashi, T., 1984, Effects of glucagon on the redox states of cytochromes in mitochondria *in situ* in perfused rat liver, *Biochem. Biophys. Res. Commun.* **119**: 212–219.

Landolfi, B., Curci, S., Debellis, L., Pozzan, T., and Hofer, A. M., 1998, Ca^{2+} homeostasis in the agonist-sensitive internal store: functional interactions between mitochondria and the ER measured *in situ* in intact cells, *J. Cell Biol.* **142**: 1235–1243.

Lawlis, V. B., and Roche, T. E., 1981, Inhibition of bovine kidney alpha-ketoglutarate dehydrogenase complex by reduced nicotinamide adenine dinucleotide in the presence or absence of calcium ion, and effect of adenosine 5′-diphosphate on reduced nicotinamide adenine dinucleotide inhibition, *Biochemistry* **20**: 2523–2527.

Lawrie, A. M., Rizzuto, R., Pozzan, T., and Simpson, A. W. M., 1996, A role for calcium influx in the regulation of mitochondrial calcium in endothelial cells, *J. Biol. Chem.* **271**: 10753–10759.

Lawson, J. E., Niu, X., Browning, K. S., Trong, H. L., Yan, J., and Reed, L. J., 1993, Molecular cloning and expression of the catalytic subunit of bovine pyruvate dehydrogenase phosphatase and sequence similarity with protein phosphatase 2C, *Biochemistry* **32**: 8987–8993.

Lawson, J. E., Park, S. H., Mattison, A. R., Yan, J., and Reed, L. J., 1997, Cloning, expression, and properties of the regulatory subunit of bovine pyruvate dehydrogenase phosphatase, *J. Biol. Chem.* **272**: 31625–31629.

Li, W., Shariat-Madar, Z., Powers, M., Sun, X., Lane, R. D., and Garlid, K. D., 1992, Reconstitution, identification, purification, and immunological characterization of the 110-kDa Na^+Ca^{2+} antiporter from beef heart mitochondria, *J. Biol. Chem.* **267**: 7983–17989.

Lukacs, G. L., and Kapus, A., 1987, Measurement of the matrix free Ca^{2+} concentration in heart mitochondria by entrapped fura-2 and quin2, *Biochem. J.* **248**: 609–613.

Marban, E., Rink, T. J., Tsien, R. W., and Tsien, R. Y., 1980, Free calcium in heart muscle at rest and during contraction measured with Ca^{2+}-sensitive microelectrodes, *Nature* **286**: 845–850.

Matlib, M. A., Zhow, Z., Knight, S., Ahmed, S., Choi, K. M., Krause-Bauer, J., Phillips, R., Altshuld, R., Katsube, Y., Sperelakis, N., and Bers, D. M., 1998, Oxygen-bridged dinuclear ruthenium amine complex specifically inhibits Ca^{2+} uptake into mitochondria *in vitro* and *in situ* in single cardiac myocytes, *J. Biol. Chem.* 273: 10223–10231.

Mc Cormack, J. G., 1985, Evidence that adrenaline activates key oxidative enzymes in rat liver by increasing intramitochondrial $[Ca^{2+}]$, *FEBS Lett.* 180: 259–264.

Mc Cormack, J. G., and Denton, R. M., 1979, The effects of calcium ions and adenine nucleotides on the activity of pig heart 2-oxoglutarate dehydrogenase complex, *Biochem. J.* 180: 533–544.

Mc Cormack, J. G., and Denton, R. M., 1984, Role of Ca^{2+} ions in the regulation of intramitochondrial metabolism in rat heart: Evidence from studies with isolated mitochondria that adrenaline activates the pyruvate dehydrogenase and 2-oxoglutarate dehydrogenase complexes by increasing the intramitochondrial concentration of Ca^{2+}, *Biochem. J.* 218: 235–247.

Mc Cormack, J. G., and Denton, R. M., 1994, Signal transduction by intramitochondrial Ca^{2+} in mammalian energy metabolism, *NIPS* 9: 71–76.

Mc Cormack, J. G., and England, P.J., 1983, Ruthenium red inhibits the activation of pyruvate dehydrogenase caused by positive inotropic agents in the perfused rat heart, *Biochem. J.* 214: 581–585.

Mc Cormack, J. G., Browne, H. M., and Dawes, H. J., 1989, Studies on mitochondrial Ca^{2+}-transport and matrix Ca^{2+} using fura-2-loaded rat heart mitochondria, *Biochim. Biophys. Acta* 973: 420–427.

Mc Cormack, J. G., Halestrap, A. P., and Denton, R. M., 1990, Role of calcium ions in regulation of mammalian intramitochondrial metabolism, *Physiol. Rev.* 70: 391–425.

Mignery, G. A., and Südhof, T. C., 1990, The ligand binding site and transduction mechanism in the inositol-1,4,5-triphosphate receptor, *EMBO J.* 9: 3893–3898.

Miyata, H., Silverman, H. S., Sollott, S. J., Lakatta, E. G., Stern, M. D., and Hansford, R. G., 1991, Measurement of mitochondrial free Ca^{2+} concentration in living single rat cardiac myocytes, *Am. J. Physiol.* 261: H1123–H1134.

Moreno-Sanchez, R., and Hansford, R. G., 1988, Dependence of cardiac mitochondrial pyruvate dehydrogenase activity on intramitochondrial free Ca^{2+} concentration, *Biochem. J.* 256: 403–412.

Moreno-Sanchez, R., and Torres-Márquez, M. E., 1991, Control of oxidative phosphorylation in mitochondria, cells, and tissues, *Int. J. Biochem.* 23: 1163–1174.

Murphy, A. N., and Fiskum, G., 1988, Abnormal Ca^{2+} transport characteristics of hepatoma mitochondria and endoplasmic reticulum, *Adv. Exp. Med. Biol.* 232: 139–150.

Nichols, B. J., and Denton, R. M., 1995, Toward the molecular basis for the regulation of mitochondrial dehydrogenases by calcium ions, *Mol. Cell. Biochem.* 149/150: 203–212.

Nicholls, D. G., 1978, The regulation of extramitochondrial free calcium ion concentration by rat liver mitochondria, *Biochem. J.* 176: 463–474.

Nicholls, D. G., and Akerman, K. E. O., 1984, Mitochondrial Ca^{2+} transport, *Biochim. Biophys. Acta.* 683: 57–88.

Petersen, C. C., Toescu, E. C., and Petersen, O. H., 1991, Different patterns of receptor-activated cytoplasmic Ca^{2+} oscillations in single pancreatic acinar cells: Dependence on receptor type, agonist concentration, and intracellular Ca^{2+} buffering, *EMBO J.* 10: 527–533.

Pozzan, T., Rizzuto, R., Volpe, P., and Meldolesi, J., 1994, Molecular and cellular physiology of intracellular calcium stores, *Physiol. Rev.* 74: 595–636.

Prentki, M., Glennon, M. C., Thomas, A. P., Moms, R. L., Matschinsky, F. M., and Corkey, B. E., 1988, Cell specific patterns of oscillating free Ca^{2+} in carbamylcholine-stimulated insulinoma cells, *J. Biol. Chem.* 263: 11044–11047.

Puskin, J. S., Gunter, T. E., Gunter, K. K., and Russel, P. R., 1976, Evidence for more than one Ca^{2+} transport mechanism in mitochondria, *Biochemistry* 15: 3834–3842.

Putney, J. W., Jr., and Bird, G. St. J., 1993a, The signal for capacitative calcium entry, *Cell* 75: 199–201.

Putney, J. W., Jr., and Bird, G. St. J., 1993b, The inositol phosphate–calcium signaling system in nonexcitable cells, *Endocrine Rev.* 14: 610–631.

Quinlan, P. T., and Halestrap, A. P., 1986, The mechanism of the hormonal activation of respiration in isolated hepatocytes and its importance in the regulation of gluconeogenesis, *Biochem. J.* 236: 789–800.

Quinlan, P. T., Thomas, A. P., Armston, A. E., and Halestrap, A. P., 1983, Measurement of the intramitochondrial volume in hepatocytes without cell disruption and its elevation by hormones and valinomycin, *Biochem. J.* 214: 395–404.

Randle, P. J., 1986, Fuel selection in animals, *Biochem. Soc. Trans.* 14: 799–806.

Reed, K. C., and Bygrave, F. L., 1974a, Accumulation of lanthanum by rat liver mitochondria, *Biochem. J.* 138: 239–252.

Reed, K. C., and Bygrave, F. L., 1974b, The inhibition of mitochondrial calcium transport by lanthanides and ruthenium red, *Biochem. J.* **140**: 143–155.

Reed, L. J., and Yeaman, S. J., 1987, in *The Enzymes*, Vol. 18, (P. D. Boyer and E. G. Krebs Eds.) Academic Press, New York, pp 77–96.

Renard-Rooney, D. C., Hajnóczky, G., Seitz, M. B., Schneider, T. G., and Thomas, A. P., 1993, Imaging of inositol 1, 4, 5-trisphosphate-induced Ca^{2+} fluxes in single permeabilized hepatocytes: Demonstration of both quantal and nonquantal patterns of Ca^{2+} release, *J. Biol. Chem.* **268**: 23601–23610.

Rhee, S. G., and Bae, Y. S., 1997, Regulation of phosphoinositide-specific phospholipase C isozymes, *J. Biol. Chem.* **272**: 15045–15048.

Rizzuto, R., Bernadi, P., Favaron, M., and Azzone, G. F., 1987, Pathways for Ca^{2+} efflux in heart and liver mitochondria, *Biochem. J.* **246**: 271–277.

Rizzuto, R., Simpson, A. W. M., Brini, M., and Pozzan, T., 1992, Rapid changes of mitochondrial Ca^{2+} revealed by specifically targeted recombinant aequorin, *Nature* **358**: 325–327.

Rizzuto, R., Brini, M., Murgia, M., and Pozzan, T., 1993, Microdomains with high Ca^{2+} close to IP3-sensitive channels that are sensed by neighboring mitochondria, *Science* **262**: 744–747.

Rizzuto, R, Bastianutto, C., Brini, M., Murgia, M., and Pozzan, T., 1994, Mitochondrial Ca^{2+} homeostasis in intact cells, *J. Cell Biol.* **126**: 1183–1194.

Rizzuto, R., Pinton, P., Carrington, W., Fay, F. S., Fogarty, K. E., Lifshitz, L. M., Tuft, R. A., and Pozzan, T., 1998, Close contacts with the endoplasmic reticulum as determinants of mitochondrial calcium responses, *Science* **280**: 1763–1766.

Robb-Gaspers, L. D., and Thomas, A. P., 1995, Coordination of Ca^{2+} signaling by intercellular propagation of Ca^{2+} waves in the intact liver, *J. Biol. Chem.* **270**: 8102–8107.

Robb-Gaspers, L. D., Burnett, P., Rutter, G. A., Denton, R. M., Rizzuto, R., and Thomas, A. P., 1998a, Integrating cytosolic calcium signals into mitochondrial metabolic responses, *EMBO J.* **17**: 4987–5000.

Robb-Gaspers, L. D., Rutter, G. A., Burnett, P., Hajnóczky, G., Denton, R. M., and Thomas, A. P., 1998b, Coupling between cytosolic and mitochondrial calcium oscillations: Role in the regulation of hepatic metabolism, *Biochim. Biophys. Acta* **1366**: 17–32.

Roche, T. E., and Reed, L. J., 1974, Monovalent cation requirement for ADP inhibition of pyruvate dehydrogenase kinase, *Biochem. Biophys. Res. Commun.* **59**: 1341–1346.

Rooney, T. A., and Thomas, A. P., 1993, Intracellular calcium waves generated by Ins(1,4,5)P3-dependent mechanisms, *Cell Calcium* **14**: 674–690.

Rooney, T. A., and Thomas, A. P., 1994, Organization of intracellular calcium signals generated by inositol lipid-dependent hormones, in *International Encyclopedia of Pharmacology and Therapeutics*, Section 139: Intracellular messengers, (C. W. Taylor, Ed.) Pergammon, Elmsford, NY, pp 407–421.

Rooney, T. A., Sass, E. J., and Thomas, A. P., 1989, Characterization of cytosolic calcium oscillations induced by phenylephrine and vasopressin in single fura-2-loaded hepatocytes, *J. Biol. Chem.* **264**: 17131–17141.

Rooney, T. A., Sass, E., and Thomas, A. P., 1990, Agonist-induced cytosolic calcium oscillations originate from a specific locus in single hepatocytes, *J. Biol. Chem.* **265**: 10792–10796.

Rooney, T. A., Renard, D. C., Sass, E., and Thomas, A. P., 1991, Oscillatory cytosolic calcium waves independent of stimulated inositol 1, 4, 5-trisphosphate formation in hepatocytes, *J. Biol. Chem.* **266**: 12272–12282.

Rosier, R. N, Tucker, D. A., Meerdink, S., Jain, I., and Gunter, T. E., 1981, Ca^{2+} transport against its electrochemical gradient in cytochrome oxidase vesicles reconstituted with mitochondrial hydrophobic proteins, *Arch. Biochem. Biophys.* **210**: 549–564.

Rutter, G. A., 1990, Ca^{2+}-binding to citrate cycle dehydrogenases, *Int. J. Biochem.* **22**: 1081–1088.

Rutter, G. A., and Denton, R. M., 1988, Regulation of $NAD^+ =$ linked isocitrate dehydrogenase and 2-oxoglurarate dehydrogenase by Ca^{2+} ions within toluene permeabilized rat heart mitochondria. Interactions with regulation by adenine nucleotides and $NADH/NAD^+$ ratios, *Biochem J.* **252**: 181–189.

Rutter, G. A., Midgley, P. J. W., and Denton, R. M., 1989, Regulation of the pyruvate dehydrogenase complex by Ca^{2+} within toluene-permeabilized heart mitochondria, *Biochim. Biophys. Acta* **1014**: 263–270.

Rutter, G. A., Theler, J.-M., Murgia, M., Wollheim, C. B., Pozzan, T., and Rizzuto, R., 1993, Stimulated Ca^{2+} influx raises mitochondrial free Ca^{2+} to supramicromolar levels in a pancreatic beta-cell line: Possible role in glucose and agonist-induced insulin secretion, *J. Biol. Chem.* **268**: 22385–22390.

Rutter, G. A., Burnett, P., Rizzuto, R., Brini, M., Murgia, M., Pozzan, T., Tavaré, J. M., and Denton, R. M., 1996, Subcellular imaging of intramitochondrial Ca^{2+} with recombinant targeted aequorin: Significance for the regulation of pyruvate dehydrogenase activity, *Proc. Natl. Acad. Sci. USA* **93**: 5489–5494.

Scarpa, A., and Azzone, G. F., 1970, The mechanism of ion translocation in mitochondria: Coupling of K$^+$ efflux with Ca^{2+} uptake, *Eur. J. Biochem.* **12**: 328–335.

Selivanov, V. A., Ichas, F., Holmuhamedov, E. L., Jouaville, L. S., Evtodienko, Y. V., and Mazat, J. P., 1998, A model of mitochondrial calcium-induced calcium release stimulating the calcium oscillations and spikes generated by mitochondria, *Biophys. Chem.* **72**: 111–121.

Severson, D. L., Denton, R. M., Bridges, B. J., and Randle, P. J., 1976, Exchangeable and total calcium pools in mitochondria of rat epididymal fat pads and isolated fat cells: Role in the regulation of pyruvate dehydrogenase activity, *Biochem. J.* **154**: 209–233.

Simpson, P. B., and Russel, J. T., 1996, Mitochondria support inositol 1,4,5-trisphosphate-mediated Ca^{2+} waves in cultured oligodendrocytes, *J. Biol. Chem.* **271**: 33493–33501.

Simpson, P. B., Mehotra, S., Lange, D. G., and Russel, J. T., 1997, High density distribution of endoplasmic reticulum proteins and mitochondria at specialized Ca^{2+} release sites in oligodendrocyte processes, *J. Biol. Chem.* **272**: 22654–22661.

Slater, E. C., 1987, The mechanism of the conservation of energy of biological oxidations, *Eur. J. Biochem.* **166**: 489–504.

Soboll, S., and Bunger, R., 1981, Compartmentaligation of adenine nucleotides in the isolated working guinea pig heart stimulated by noradrenaline, *Physiol. Chem.* **232**: 125–132.

Soboll, S., Sietz, H. J., Sies, H., Ziegler, B., and Scholtz, R., 1984, Effect of long-chain fatty acyl-CoA on mitochondrial and cytosolic ATP/ADP ratios in the intact liver cell, *Biochem. J.* **220**: 371–376.

Somlyo, A. P., Bond, M., and Somlyo, A. L., 1985, Calcium content of mitochondria and endoplasmic reticulum in liver frozen rapidly *in vivo*, *Nature* **314**: 622–625.

Sparagna, G. C., Gunter, K. K, Sheu, S., and Gunter, T. E., 1995, Mitochondrial calcium uptake from physiological-type pulses of calcium: A description of the rapid uptake mode, *J. Biol. Chem.* **270**: 27510–27515.

Streb, H., Irvine, R. F., Berridge, M. J., and Schultz, I., 1983, Release of Ca^{2+} from a nonmitochondrial intracellular store in pancreatic acinar cells by inositol-1,4,5-trisphosphate, *Nature* **306**: 67–69.

Takamatsu, T., and Wier, W. G., 1990, Calcium waves in mammalian heart: Quantification of origin, magnitude, waveform, and velocity, *FASEB J.* **4**: 1519–1525.

Thomas, A. P., Diggle, T. A., and Denton, R. M., 1986, Sensitivity of pyruvate dehydrogenase phosphate phosphatase to magnesium ions: Similar effects of spermine and insulin, *Biochem. J.* **238**: 83–91.

Thomas, A. P., Bird, G. St. J., Hajnóczky, G., Robb-Gaspers, L. D., and Putney, J. W., Jr., 1996, Spatial and temporal aspects of cellular calcium signaling, *FASEB J.* **10**: 1505–1517.

Tordjmann, T., Berthon, B., Jaquemin, E., Clair, C., Stelly, N., Guillon, G., Claret, M., and Combettes, L., 1998, Receptor-oriented intercellular calcium waves evoked by vasopressin in rat hepatocytes, *EMBO J.* **17**: 4695–4703.

Unitt, J. F., Mc Cormack, J. G., Reid, D., MacLachlan, L. K., and England, P. J., 1989, Direct evidence for a role of intramitochondrial Ca^{2+} in the regulation of oxidative phosphorylation in the stimulated rat heart: Studies using ^{31}P n.m.r. and ruthenium red, *Biochem. J.* **262**: 293–301.

Vinogradov, A., and Scarpa, A., 1973, The initial velocities of calcium uptake by rat liver mitochondria, *J. Biol. Chem.* **248**: 5527–5531.

Williamson, J. R., Cooper, R. H., and Hoek, J. B., 1981, Role of calcium in the hormonal regulation of liver metabolism, *Biochim. Biophys. Acta* **639**: 243–295.

Wingrove, D. E., Amatruda, J. M., and Gunter, T. E., 1984, Glucagon effects on the membrane potential and calcium uptake rate of rate liver mitochondria, *J. Biol. Chem.* **259**: 9390–9394.

Wolkowicz, P. E., Michael, L. H., Lewis, R. M., and Mc Millin-Wood, S., 1983, Sodium–calcium exchange in dog heart mitochondria: Effects of ischemia and verapamil, *Am. J. Physiol.* **244**: H644–H651.

Woods, N. M., Cuthbertson, K. S. R., and Cobbold, P. H., 1986, Repetitive transient rises in cytoplasmic free calcium in hormone-stimulated hepatocytes, *Nature* **319**: 600–602.

Yan, J., Lawson, J. E., and Reed, L. J., 1996, Role of the regulatory subunit of bovine pyruvate dehydrogenase phosphatase, *Proc. Natl. Acad. Sci. USA* **93**: 4953–4956.

Zoratti, M., and Szabò, I., 1995, The mitochondrial permeability transition, *Biochim. Biophys. Acta* **1241**: 139–176.

Chapter 9

The Permeability Transition Pore in Myocardial Ischemia and Reperfusion

Andrew P. Halestrap, Paul M. Kerr, Sabzali Javadov, and M-Saadah Suleiman

1. INTRODUCTION

When blood flow to the heart is stopped (global ischemia) or greatly reduced (partial ischemia) the heart ceases to beat and as the period of ischemia lengthens, the heart cells begin to die. In the case of a coronary thrombosis, unless blood flow is rapidly restored, this leads to a necrotic area of the heart, known as an *infarct*, that impairs or prevents heart function. Although restoration of blood flow is essential if permanent heart damage is to be avoided, such reperfusion can itself exacerbate the damage if the period of ischemia is too long. This phenomenon is known as reperfusion injury and has been widely studied (see Halestrap, 1994; Halestrap *et al.*, 1997a, 1998; Lemasters and Thurman, 1995; Reimer and Jennings, 1992). Initially, the heart may make a few tentative attempts to beat but then fails totally; this is accompanied by a large loss of intracellular components, including proteins. Indeed, release of proteins such as troponin and lactate dehydrogenase is often used as an indicator of the severity of damage. Associated with this cellular damage, tissue ATP levels fall to very low values, while ADP, AMP, and phosphate are all greatly elevated. Morphological studies reveal that cardiac myocytes show many of the hallmarks of necrotic cell death, including the presence of swollen amorphous mitochondria. Isolation of mitochondria from such reperfused tissue confirms that they are damaged, with impaired respiratory-chain function and oxidative phosphorylation.

Andrew P. Halestrap, Paul M. Kerr, and M-Saadah Suleiman   Department of Biochemistry, Bristol Heart Institute, University of Bristol, British Royal Infirmary, Bristol BS2 8HW, United Kingdom.   **Sabzali Javadov** Azerbaijan Medical University, Baku, Azerbaijan.

Mitochondria in Pathogenesis, edited by Lemasters and Nieminen.
Kluwer Academic/Plenum Publishers, New York, 2001.

Mitochondria are the major providers of ATP for cardiac contraction, a fact readily deduced from the rapid cessation of contraction that accompanies severe hypoxia or ischemia. Although glycolysis may be able to provide enough ATP to maintain ionic homeostasis, it cannot keep pace with the ATP demands of contraction. Thus the damage experienced by mitochondria during ischemia and reperfusion is likely to play a critical role in the impairement of heart function upon reperfusion. One means by which mitochondria can be damaged is through the opening of the mitochondrial permeability transition pore (MPTP). Under normal physiological conditions, the mitochondrial inner membrane is impermeable to all but a few selected metabolites and ions. This is essential for maintaining the membrane potential and pH gradient that drive ATP synthesis during oxidative phosphorylation. If this permeability barrier is disrupted, as is the case with MPTP opening, mitochondria become uncoupled and unable to synthesize ATP (Halestrap *et al.*, 1998; Lemasters *et al.*, 1998). The resulting decrease in ATP production impairs or prevents cardiac contraction. Indeed, if MPTP opening becomes extensive, not only will mitochondria fail to make sufficient ATP, they also hydrolyze glycolytic ATP by reversal of the proton-translocating ATPase. Unless the pore closes again, heart cells are destined to die, because ATP is required to maintain their integrity. Eventually, the permeability barrier of the plasma membrane is compromised through phospholipase A_2 action. The resulting leakage of cell contents and disruption of ion gradients then ensures cell death by necrosis (Crompton, 1990; Halestrap, 1994; Leist and Nicotera, 1997; Reimer and Jennings, 1992).

In this chapter we describe how intracellular conditions during heart reperfusion after ischemia are exactly those required for MPTP opening, and we present evidence which confirms that transition does occur during reperfusion, and that the extent to which pores remain open strongly determines heart recoversery. We also describe how we devised perfusion protocols that can significantly protect hearts from reperfusion injury (Halestrap *et al.*, 1997c, 1998; Woodfield *et al.*, 1998). These promise to be of considerable benefit in open heart surgery.

2. THE MITOCHONDRIAL PERMEABILITY TRANSITION PORE AND REPERFUSION INJURY

2.1. Intracellular Conditions during Reperfusion Favor Pore Opening

During ischemia the heart cells attempt to maintain their ATP levels through glycolysis, but this cannot keep pace with the ATP demands of contraction and the heart rapidly stops beating as ATP concentrations fall. Glycolysis remains stimulated but production of lactic acid leads to a decrease in intracellular pH (pH_i), because ischemia prevents washout of lactate and protons from the heart (Dennis *et al.*, 1991; Halestrap *et al.*, 1997b). The low pH has an inhibitory effect on glycolysis, so intracellular ATP concentrations dwindle still further, while the Na^+/H^+ antiporter is activated in an attempt to restore pH_i (Lazdunski *et al.*, 1985; Vandenberg *et al.*, 1993), which increases cellular $[Na^+]$. The Na^+ cannot be pumped out of the cell by Na^+/K^+ ATPase, however, because there is insufficient ATP to drive the process (Silverman and Stern, 1994). Thus Na^+ accumulates, which in turn prevents Ca^{2+} from being pumped out of the cell on the

Na^+/Ca^{2+} antiporter. Indeed, this process may actually reverse and allow additional Ca^{2+} to enter the cytosol from the plasma (Haigney et al., 1992; Silverman and Stern, 1994). Some of this calcium may enter the mitochondria by reversal of the Na^+/Ca^{2+} antiporter during ischemia (Griffiths et al., 1997), but upon reperfusion much more Ca^{2+} is taken up rapidly into mitochondria by means of the uniporter, overloading mitochondrial Ca^{2+} (Miyata et al., 1992; Stone et al., 1989). This alone might not be sufficient to open the MPTP, but it will do so when coupled with other factors that come into play during reperfusion. These factors include a rise in intracellular pH and oxidative stress as outlined above.

The sudden influx of oxygen into the anoxic cell causes a rapid burst of oxygen free radical production, mediated through an interaction of oxygen with ubisemiquinone formed during anoxia as a result of respiratory-chain inhibition (Boveris et al., 1976; Halestrap, 1994; Halestrap et al., 1993; Turrens et al., 1985). Xanthine oxidase may provide an additional source of oxygen free radicals. During ischemia, adenosine is catabolized to xanthine and xanthine oxidase produced by proteolytic cleavage of xanthine dehydrogenase (Nishino, 1994). Upon reperfusion, this leads to a burst of xanthine oxidation and superoxide production (Omar et al., 1991). The combination of oxidative stress and high $[Ca^{2+}]$ provides ideal conditions for the MPT, especially in the presence of elevated cellular phosphate concentrations and depleted adenine nucleotide levels. Furthermore, during reperfusion the pH_i rapidly returns to preischemic values via operation of the Na^+/H^+ antiporter, lactic acid efflux on the monocarboxylate transporter (MCT), and bicarbonate-dependent mechanisms (Vandenberg et al., 1993). When pH falls below 7.0, pore opening is strongly inhibited (Bernardi et al., 1992; Halestrap, 1991). Thus during ischemia, when pH is <6.5, the pore stays closed even in the presence of other transition inducers. When pH_i is restored upon reperfusion, however, the factors already in place to stimulate transition can now exert their full effect.

The sequence of events described above is shown diagramatically in Figure 1 and leads to conditions within the cell, summarized in Table I, that are exactly those required to induce transition. This suggests that pore opening may be a critical event in determining whether the heart recovers during reperfusion. Two experimental approaches provide data that support this. First, direct measurements of pore opening confirm that this occurs during reperfusion but not ischemia. Second, agents that protect mitochondria from MPTP opening improve the functional recovery of hearts during reperfusion.

2.2. Methods for Measuring Pore Opening in Isolated Heart Cells and in Perfused Heart

Pore opening causes the inner mitochondrial membrane to become permeable to any molecule of <1500 daltons (Crompton and Costi, 1990; Haworth and Hunter, 1979), and this provides the basis for all measurements of pore opening in situ.

2.2.1. Fluorescence Microscopy

Pore opening causes free permeation of protons, leading to mitochondrial uncoupling. Thus measurement of mitochondrial membrane potential with fluorescent dyes, such as

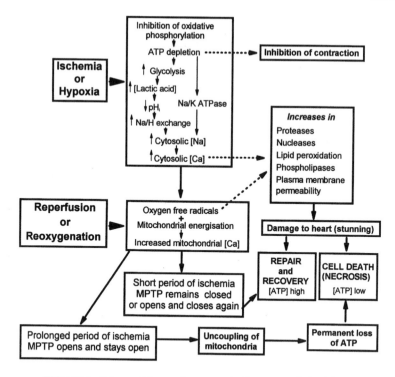

FIGURE 1. Scheme of the event sequence leading to reperfusion injury.

Rhodamine derivatives tetraethylrhodamine and tetramethylrhodamine (TMRH) or JC-1, is often used to detect the permeability transition in studies with isolated cells (Di Lisa *et al.*, 1995; Duchen *et al.*, 1993; Lemasters *et al.*, 1997, 1998). Pathways other than the MPTP can lead to uncoupling of mitochondria, however, and thus this method alone cannot give definitive evidence of pore opening. The technique can be made more rigorous when combined with a demonstration that specific MPTP inhibitors prevent mitochondrial depolarization. This approach provides a relatively easy transition assay in isolated cell preparations where dye loading and fluorescence microscopy can readily be employed.

Table I
Reperfusion Injury Factors that
Favor MPTP Opening

Mitochondrial calcium overload
Oxidative stress
Depleted mitochondrial adenine nucleotide content
High phosphate concentrations
Increased concentrations of fatty acids
Phospholipase A_2 activation
pH > 7.0 (cf. pH < 6.5 during ischemia)

Nevertheless, there are two potential pitfalls. The first is the choice of membrane-potential-sensitive fluorescent dye and concentration and conditions under which it is loaded, a matter of considerable debate (see Lemasters *et al.*, 1995; Metivier *et al.*, 1998; Salvioli *et al.*, 1997; Ubl *et al.*, 1996). The second is the choice of specific MPTP inhibitors. The immunosuppressant drug cyclosporin A (CsA) is most commonly used as a potent, specific inhibitor of pore opening. It acts by displacing mitochondrial cyclophilin D (Cyp-D) from the adenine nucleotide translocase, the probable pore-forming component of the MPTP (Connern and Halestrap, 1994, 1996; Halestrap *et al.*, 1997a, 1998; Woodfield *et al.*, 1998). The CsA may, however, have many other actions within the cell mediated by CsA/CyP-A-dependent inhibition of the calcium-sensitive protein phosphatase calcineurin (Galat and Metcalfe, 1995; Schreiber, 1991). Thus its use as a specific pore inhibitor requires demonstration that other inhibitors of calcineurin action, such as FK506, an immunosuppressive drug that does not bind to cyclophilin A (Schreiber, 1991; Galat and Metcalfe, 1995), do not affect MPTP opening. In addition, a nonimmunosuppressive CsA analogue that inhibits CyP-D and pore opening can be used, such as *N*-methyl-Ala-6-CsA or *N*-methyl-Val-4-CsA (Bernardi *et al.*, 1994; Griffiths and Halestrap, 1995; Halestrap *et al.*, 1997a). As a negative control, cyclosporin H, inactive against both calcineurin and the MPTP, is useful.

A more specific technique for measuring pore opening utilizes a fluorescent molecule normally unable to permeate the inner mitochondrial membrane, but which can permeate the pore. Lemasters and colleagues have used the green fluorescent dye calcein for this purpose. Confocal microscopy reveals that with appropriate loading conditions, the dye remains confined to the cytosol, with mitochondria showing up as dark spheres against a green fluorescent background (Lemasters *et al.*, 1997, 1998; Nieminen *et al.*, 1995). Under conditions that stimulate pore opening, however, the dye enters mitochondria and the "black holes" are lost. An advantage of calcein's green fluorescence is that it can be used in conjunction with TMRH, enabling coincident measurement of mitochondrial membrane potential. As expected, upon pore opening, calcein enters mitochondria at the same time that TMRH is lost (Nieminen *et al.*, 1995, 1996, 1997; Qian *et al.*, 1997).

Although fluorescence measurements of mitochondrial membrane potential, either alone or in conjunction with calcein, have been used to detect the pore opening in a range of cell types (Kroemer *et al.*, 1998; Lemasters *et al.*, 1998), little work has been reported using isolated heart cells subjected to simulated ischemia and reperfusion. During hypoxia, however, heart cells undergo contracture and loss of mitochondrial membrane potential. Upon re-oxygenation, some heart cells undergo hypercontracture while others recover. Both responses are associated with an initial repolarization of mitochondria, but in the hypercontracted cells a second, permanent depolarization is observed (Duchen *et al.*, 1993; Nazareth *et al.*, 1991). This process is delayed in the presence of CsA, suggesting that hypercontracture reflects pore opening (Nazareth *et al.*, 1991). It is important to note, however, isolated heart cells do not allow true reproduction of ischemia because cell concentration in the perfused organ is so much greater than what can be achieved with cells on a coverslip. Consequently, neither the intracellular accumulation of metabolites such as lactic acid and purine nucleotide breakdown products, nor the rapid intracellular pH drop characteristic of ischemia, will occur in isolated anoxic cells. Ischemic conditions can only be mimicked by use of oxygen-free media (or cyanide-containing media for chemical anoxia) supplemented with high lactate concentrations and low pH, followed by perfusion with normal medium (Griffiths *et al.*, 1998; Lemasters *et al.*, 1997; Qian *et al.*, 1997). This is more accurately referred to as *hypoxia/re-oxygenation* and is the best model accessible to fluorescence microscopy. In view of these limitations, it is important to develop a technique to measure pore opening in perfused tissues where true ischemia and reperfusion can be induced. For this purpose we have developed what is colloquially called the *hot-dog technique* (Griffiths and Halestrap, 1995) described below.

2.2.2. Measuring Pore Opening in Perfused Tissues via Mitochondrial [³H]-Deoxyglucose Entrapment

The inner mitochondrial membrane is normally impermeant to 2-deoxyglucose (DOG) and 2-deoxyglucose-6-phosphate (DOG-6P), but becomes permeant upon MPTP opening. This provides the basis of our technique for measuring pore opening in perfused tissues (Griffiths and Halestrap, 1995), the principle of which is summarized in Figure 2. Hearts are perfused in Langendorff recirculating mode with [³H]-2-DOG, which enters cardiac myocytes using the glucose carrier and is then phosphorylated to DOG-6P. No further metabolism occurs, leaving the [³H]-DOG-6P trapped within the the heart cell cytosol. Extracellular [³H]-DOG is removed from the heart by perfusion in the absence of [³H]-DOG, before hearts are subjected to various periods of ischemia and reperfusion as required. Mitochondria are then rapidly prepared in the presence of EGTA, which reseals any open mitochondria, entrapping the [³H]-DOG-6P they contain. This is determined by scintillation counting. In conjunction with measurements of citrate synthase (an indicator of mitochondrial recovery) and [³H]-DOG content in a small sample of total heart homogenate (an indicator of [³H]-DOG loading of the heart), one can estimate the percentage of mitochondria that have undergone transition.

It should be noted that this technique does not discriminate between pores that have first opened then closed, and mitochondria in which the MPTP has remained open. To determine whether some mitochondria that open early in reperfusion subsequently reseal, hearts can be loaded with [³H]-DOG after a period of reperfusion sufficient to give maximal recovery of heart function (Kerr *et al.*, 1999). If the value of mitochondrial [³H]-DOG entrapment is lower when this postloading protocol is used than when hearts are loaded with it before ischemia (preloading), this means that in some mitochondria the pore has opened and then resealed.

2.3. Pore Opening Occurs upon Reperfusion, but Not Ischemia

In control hearts not subject to ischemia/reperfusion, a small amount of [³H]-DOG is found in the mitochondrial fraction, but this apparently does not reflect basal pore opening because it does not increase with time (Griffiths and Halestrap, 1995; Halestrap *et al.*, 1997a; Kerr *et al.*, 1999). As illustrated in Figure 3, mitochondria prepared immediately after ischemia show no increase in mitochondrial DOG content, but those prepared following 2 minutes of reperfusion show a significant increase that reaches a maximum value after about 5 minutes (Griffiths and Halestrap, 1995; Halestrap *et al.*, 1997a). Thus our data confirm the prediction that extensive MPTP opening occurs only during reperfusion, not during ischemia. It may be significant that the period of MPTP opening

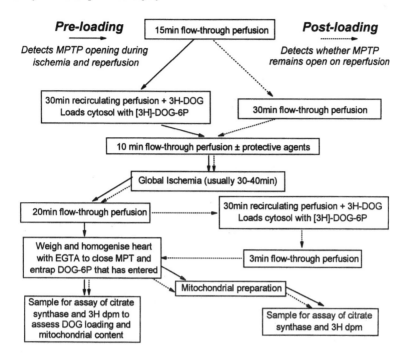

FIGURE 2. Protocol for measuring MPTP opening in perfused heart using [³H]-deoxyglucose entrapment. [3-H]-DOG-6P enters mitochondria only if the MPTP opens, thus the extent of opening is indicated by DOG content (DOG content does not allow detection of resealing after earlier opening). Detection of resealing is done via comparison of pre- and post-loeding DOG cantent

(2–5 minutes of reperfusion) is similar to that in which pH_i returns from the acid pH of ischemia (about 6.0) to preischemic values of about 7.2 (Kerr et al., 1999; Vandenberg et al., 1993). This is reflected in the pH of heart effluent during reperfusion, as shown in Figure 3. In this context, pore opening in isolated hepatocytes occurs in response to a pH jump from acid to normal pH (Lemasters et al., 1998; Qian et al., 1997).

These entrapment experiments confirm that pore opening is not, as some have argued (Piper et al., 1994), a secondary phenomenon following breakdown of the plasma membrane permeability barrier and subsequent exposure of mitochondria to extracellular [Ca^{2+}]. If the latter occured DOG would be lost from the cell before it could enter the mitochondria and thus it would not increase. Additional data are required, however, to show that pore opening is a primary cause of reperfusion injury. Such data are presented below.

2.4. Pore Closure Follows Opening in Hearts that Recover during Reperfusion

After a short period of ischemia, reperfusion leads to total recovery of left ventricular developed pressure (LVDP) and ATP/ADP ratio (Griffiths and Halestrap, 1993). Increase in mitochondrial [³H]-DOG entrapment can be seen even under these conditions, however, implying that the permeability transition *is* occurring in some mitochondria (Griffiths and

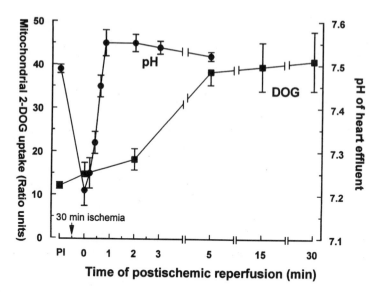

FIGURE 3. Time dependence of MPTP opening during reperfusion. Opening of the MPTP was detected using the DOG preloading technique. Hearts were subjected to 30 min of global ischemia before reperfusion for the time shown (PI: preischemic value). (Data from Halestrap *et al.*, 1997a. Parallel data for pH of heart effluent from Kerr *et al.* 1999.)

Halestrap, 1995; Halestrap *et al.*, 1997a). The full recovery of heart function suggests that this opening may be transient, with pores rapidly re-closing, allowing for normal ATP production. We confirmed this with the DOG "postloading" technique described above (Kerr *et al.*, 1999). After 40 minutes ischemia, postloading yielded about half the entrapment of preloading, as shown in Figure 4. When hearts were perfused with pyruvate in the medium, complete recovery of LVDP was observed (see below), accompanied by reduction, but not total inhibition, of the mitochondrial entrapment of preloaded [³H]-DOG. When [³H]-DOG was postloaded, however, reperfused hearts treated with pyruvate showed no DOG entrapment increase over that of control hearts not subjected to ischemia/reperfusion. Thus if the insult caused by ischemia/reperfusion is not too great, mitochondria can apparently undergo a transient MPT followed by pore closure and DOG entrapment. Indeed, degree of heart recovery during reperfusion seems best correlated with the extent to which the pores reseal after initial opening (Kerr *et al.*, 1999).

The most likely explanation for this transience is that once open, the pore allows rapid calcium loss from the mitochondria, lowering matrix [Ca²⁺] sufficiently to cause the pore to reseal. This will only occur, however, if enough "healthy" mitochondria remain in the cell to accumulate released calcium and provide sufficient ATP to maintain ionic homeostasis. The ratio of closed to open mitochondria within any cell reflects the severity of cellular insult and is critical in determining whether the cell lives or dies. Too many open mitochondria will release more calcium and hydrolyze more ATP than the closed mitochondria can accommodate. In contrast, if there are enough closed mitochondria to meet cellular ATP requirements and to accumulate released calcium without themselves undergoing the MPT the open mitochondria will close and the cell will recover. The proper

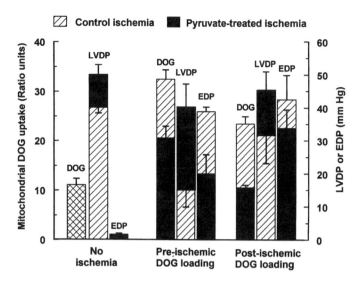

FIGURE 4. Comparison of preischemic and postischemic DOG loading shows that pore opening is followed by closure associated with recovery of heart function. Pore opening was measured using the 2-deoxyglucose entrapment procedure, with DOG loading either before ischemia (preloading) or after maximum recovery of LVDP (postloading). Heart recovery is indicated by values for LVDP and EDP at the end of reperfusion. Some hearts were perfused with 10 mM pyruvate for 10 min before ischemia and during reperfusion. Preischemic values for DOG entrapment (hatched bars) were unaffected by pyruvate, and separate data are not shown. (Data from Kerr *et al.* 1999.)

balance between open and closed mitochondria for maintaining heart function should depend on ATP demand. This may provide an explanation for why the period of normothermic global ischemia that working rat hearts can recover from on reperfusion is only about 15 minutes, compared with 30–40 minutes for Langendorff-perfused (non-working) hearts [Kerr, unpublished].

2.5. Transition Inhibitors Protect Hearts from Reperfusion Injury

2.5.1. Cyclosporin

If the permeability transition is a critical factor in developing reperfusion injury, CsA should provide some damage protection. This has been seen in isolated cardiac myocytes reoxygenated following hypoxia (Nazareth *et al.*, 1991), where there is a correlation between mitochondrial [Ca^{2+}] content and subsequent cell death (Allen *et al.*, 1992; Delcamp *et al.*, 1998; Griffiths *et al.*, 1998). We and others have shown that CsA can also protect the Langendorff-perfused heart from reperfusion injury following isothermic global ischemia (Griffiths and Halestrap, 1993, 1995; Halestrap *et al.*, 1998; Massoudy *et al.*, 1995). Our own experiments showned that hearts treated with 0.2 μM CsA before and during ischemia experienced greater recovery of LVDP, tissue ATP/ADP ratios, and functional mitochondria, and had decreased AMP levels and end diastolic pressure (EDP, which indicates contracture due to low ATP/ADP and high [Ca^{2+}]) (Griffiths and Halestrap,

1993, 1995; Halestrap *et al.*, 1997a). No protective CsA effect was seen upon loss of total adenine nucleotides (a result of purine degradation during ischemia), nor was protection from inhibition of respiratory-chain function (ADP-stimulated substrate oxidation) found (Griffiths and Halestrap, 1995; Halestrap *et al.*, 1997a). The latter may be caused by the oxygen free radicals formed during ischemia and reperfusion directly modifying respiratory-chain components (Griffiths and Halestrap, 1995, 1993; Halestrap *et al.*, 1993, 1997a) or by cytochrome *c* loss associated with swelling of mitochondria that undergo transition, (Halestrap, 1982). We have also shown that only CsA analogs that block transition in isolated mitochondria can offer protection to the reperfused heart (Griffiths and Halestrap, 1995; Halestrap *et al.*, 1997a). Others have demonstrated that CsA can protect rabbit heart from reperfusion injury following transplantation (Gatewood *et al.*, 1996). Also, CsA offers reperfusion-injury protection to other tissues, including rat liver (Kurokawa *et al.*, 1992; Shimizu *et al.*, 1994; Travis *et al.*, 1998) and brain (Folbergrova *et al.*, 1997; Li *et al.*, 1997a; Shiga *et al.*, 1992; Uchino *et al.*, 1995).

Although CsA's protective effects are consistent with its acting to prevent pore opening, direct measurement of opening in reperfused heart with the preloading technique showed no reduction in DOG entrapment by CsA (Griffiths and Halestrap, 1995), implying that CsA does not inhibit opening in the early stages of reperfusion. At this point, ATP and ADP concentrations are lowest and matrix $[Ca^{2+}]$ and oxygen free radicals are at their highest, conditions under which CsA poorly inhibitors the MPT (Halestrap *et al.*, 1997c). Thus CsA's protective effects may relate more to its ability to enhance resealing. Use of CsA as a protective agent however, has given inconsistent results; in some situations CsA actually impaired heart recovery from ischemia [Kerr, unpublished], possibly because its effects are highly concentration dependent, with optimal response at about $0.2\,\mu M$ and declining at higher concentrations (Griffiths and Halestrap, 1993). A similar concentration dependence has been observed for CsA protection of isolated cardiac myocytes subjected to reoxygenation following hypoxia (Nazareth *et al.*, 1991). It is likely that CsA has other inhibitory effects on heart function, perhaps through its well-characterized inhibition of the calcium-dependent protein phosphatase calcineurin (Galat and Metcalfe, 1995; Schreiber 1991). In view of these complications, it is unlikely that CsA is appropriate for use in open-heart surgery, so we have investigated other means of inhibiting transition during reperfusion.

2.5.2. Antioxidants and Calcium Antagonists

Oxidative stress and mitochondrial calcium overload are the two most critical factors for inducing permeability transition (Crompton, 1990; Crompton *et al.*, 1987; Halestrap *et al.*, 1993, 1998). Interventions designed to reduce them should inhibit transition during reperfusion and thus protect the heart from injury, which is in fact the case. Both free radical scavengers (Gutteridge and Halliwell, 1990; Halestrap, 1994; Halestrap *et al.*, 1993, 1998 Griffiths and Halestrap, 1993; Omar *et al.*, 1991; Reimer *et al.*, 1989; Yoshida *et al.*, 1996) and prevention of mitochondrial calcium overload with either calcium antagonists or ruthenium red (an inhibitor of mitochondrial calcium uptake) will protect hearts from reperfusion injury (Benzi and Lerch, 1992; Figueredo *et al.*, 1991; Grover *et al.*, 1990; Massoudy *et al.*, 1995; Opie, 1992; Peng *et al.*, 1980; Stone *et al.*, 1989).

These observations, however, though consistent with a critical role for the transition in reperfusion injury, might also be explained by these reagents effects on other processes within the cardiac myocyte (Piper, 1997; Piper et al., 1994).

2.5.3. Low pH$_i$

Several studies show that low pH (<7.0) can protect a variety of cells, including cardiac myocytes and hepatocytes, from oxidative stress, re-oxygenation following anoxia, or reperfusion following ischemia (Bond et al., 1993; Halestrap et al., 1993; Lemasters et al., 1998 Qian et al., 1997). Protection can be afforded either by using low extracellular pH or by adding specific inhibitors of the Na^+/H^+ antiporter, such as amiloride (Duan and Karmazyn, 1992; Dutoit and Opie, 1992; Karmazyn et al., 1993; Ladilov et al., 1995; Sack et al., 1994). Although low pH$_i$ may exert its protective effect by several means, the profound transition inhibition at pH < 7.0 (Halestrap, 1991) suggests that prevention of pore opening may be an especially important one. The observation that the pore opens during heart reperfusion over the same period that pH$_i$ is restored to pre-ischemic values supports this conclusion (Halestrap et al., 1997a; Kerr et al., 1999; Vandenberg et al., 1993). Indeed, in isolated rat hepatocytes subjected to simulated ischemia/reperfusion, confocal microscopy showed directly that MPTP opening occurs as pH$_i$ rises during the reperfusion phase (Lemasters et al., 1998; Qian et al., 1997).

2.5.4. Pyruvate

Pyruvate protects a variety of tissues from ischemia/reperfusion and anoxia/reoxygenation injury, including heart (Borle and Stanko, 1996; Bunger et al., 1989; Crestanello et al., 1998; Deboer et al., 1993; Kerr et al., 1999), intestine (Cicalese et al., 1996a, 1996b), and hepatocytes (Borle and Stanko, 1996). Its mode of action is attributed to beneficial metabolic alterations, because pyruvate is an excellent respiratory fuel which, unlike glucose and fatty acids, requires no ATP for activation prior to oxidation. In addition, as a good a respiratory substrate, pyruvate will generate a high mitochondrial NADH/NAD$^+$ ratio, preventing oxidation of protein thiol groups critical for modulation of the MPTP voltage sensor (Bernardi, 1992; Petronilli et al., 1993; Scorrano et al., 1997) and a high membrane potential, which acts as a powerful MPT inhibitor (Chernyak and Bernardi, 1996; Costantini et al., 1996; Petronilli et al., 1994). Another major factor may be pyruvate's ability to act as a free radical scavenger (Borle and Stanko, 1996; Deboer et al., 1993). By increasing the cell's defence against oxidative stress, this could provide additional protection against transition. We have proposed that pyruvate's protective effects may be further enhanced by its ability to lower pH$_i$ (Kerr et al., 1999). Pyruvate enters the cell with a proton on the monocarboxylate transporter (MCT) and under ischemic conditions is metabolized to lactate. The consequence of this is a greater accumulation of lactic acid within the cell and a lower pH$_i$ (Halestrap et al., 1997b), reflected in a drop in perfusate pH of pyruvate-treated hearts upon reperfusion that is considerably greater than that in control hearts (Kerr et al., 1999). There is also direct evidence from NMR studies that pyruvate decreases pH$_i$ in a low-flow model of ischemia (Cross et al., 1995).

With the DOG technique we have shown that the protective effect of 10 mM pyruvate (present before 40 minutes of ischemia, during ischemia, and during reperfusion) is accompanied by a reduction of mitochondrial pore opening during the initial stages of reperfusion (Kerr *et al.*, 1999). More impressive was pyruvate's ability to cause total mitochondrial resealing, determined by mitochondrial entrapment of postloaded DOG as reperfusion continued. This was associated with 100% recovery of LVDP, compared with only about 50% recovery of LVDP in the absence of pyruvate, a condition in which mitochondrial DOG entrapment implies implying that only about half the mitochondria reseal. These data, summarized in Figure 4, are the first direct evidence that mitochondrial pore opening in the initial phase of reperfusion can be reversed as hearts recover. This has implications for subsequent cell death by apoptosis, as described below.

2.5.5. Propofol

Propofol is an anesthetic frequently used during cardiac surgery and in post-operative sedation (Bryson *et al.*, 1995). It can protect hearts from injury caused by hydrogen-peroxide-induced oxidative stress (Kokita and Hara, 1996) or reperfusion injury (Ko *et al.*, 1997; Kokita *et al.*, 1998). It has been proposed that this action of propofol may be mediated by its ability to act as a free radical scavenger (Eriksson *et al.*, 1992; Green *et al.*, 1994; Murphy *et al.*, 1992, 1993) or via inhibition of plasma membrane calcium channels (Buljubasic *et al.*, 1996; Li *et al.*, 1997b). These two effects would decrease oxidative stress and cytosolic $[Ca^{2+}]$, both of which should protect mitochondria from MPTP opening. Furthermore, there are reports that propofol can inhibit transition in isolated mitochondria (Eriksson, 1991; Sztark *et al.*, 1995), although the concentrations used in these studies were considerably greater than those employed clinically. In recent experiments Javadov *et al.*, (2000) have confirmed propofol's protective effects against reperfusion injury in Langendorff-perfused hearts, but the drug used was a concentration of $2 \mu g/ml$, more typical of concentrations employed in clinical anesthesia (Bryson *et al.*, 1995; Cockshott, 1985; Servin *et al.*, 1988). This is important, because higher propofol concentrations reportedly impair oxidative phosphorylation by isolated mitochondria (Branca *et al.*, 1995; Rigoulet *et al.*, 1996; Sztark *et al.*, 1995b). Indeed, this may account for other reports suggesting that propofol has deleterious effects on reperfusion injury in dog and pig heart (Coetzee, 1996; Mayer *et al.*, 1993). We added propofol 10 minutes prior to ischemia and during reperfusion. As shown in Figure 5, recovery of propofol-treated hearts after 30 minutes of ischemia was significantly improved, with LVDP expressed as a percentage of the preischemic value (mean $\pm$ S.E.M.) increasing from $36 \pm 8\%$ ($n = 10$) in the absence of propofol to $70 \pm 11\%$ ($n = 8$; $P < 0.05$) in its presence. Both time to contracture and maximal extent of contracture decreased in the presence of propofol, as did the EDP during reperfusion (Fig. 5). These effects may be the result of propofol inhibiting calcium channels and consequently reducing calcium overload during ischemia/reperfusion.

Improvement in functional heart recovery was accompanied by a 25% decrease in mitochondrial entrapment of preloaded $[^3H]$-DOG. Furthermore, pore opening in mitochondria isolated from propofol-treated hearts was less sensitive to $[Ca^{2+}]$ than it was in control mitochondria (Fig. 5). When $2 \mu g/ml$ propofol was added directly to

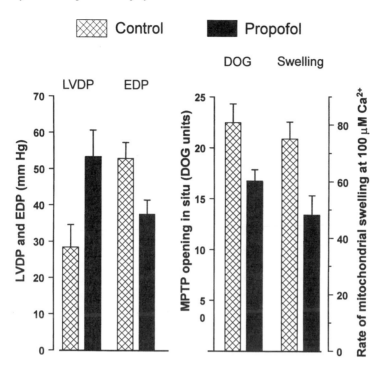

FIGURE 5. Propofol protection of heart from reperfusion injury is associated with decreased pore opening. Hearts were perfused with 2 μg/ml propofol (Intralipid as carrier) for 10 min prior to 30 min ischemia and during subsequent reperfusion. Pore opening was determined using DOG (preloading). Control heart received only Intralipid. The effects of propofol treatment on MPT calcium sensitivity was measured in subsequently isolated mitochondria Prior to mitochondrial preparation, hearts were perfused with and without propofol according to the same protocol as in DOG experiments, but without ischemia and reperfusion. Mitochondrial swelling was determined under de-energised conditions in KSCN medium at 50 μM [Ca^{2+}], as previously described. (Halestrap, 1991; Griffths and Halestrap, 1995.) Data are taken from Savador et al. (2000) and are given as means ± S.E. of five separate experiments.

isolated heart mitochondria, however, no inhibition of the permeability transition was observed. This may reflect some propofol accumulation by mitochondria *in situ* during preischemic perfusion with the drug, though this was not reflected by any change in mitochondrial oxidative phosphorylation rates. More likely the explanation of propofol's effects on mitochondria is its well-documented ability to act as a free radical scavenger (DeLaCruz et al., 1998; Eriksson et al., 1992; Green et al., 1994; Murphy et al., 1992, 1993) which lessens the oxidative stress experienced by mitochondria upon reperfusion. Indeed, in isolated mitochondria such antioxidative effects have been seen at concentrations as low as 1 μM (Eriksson et al., 1992). Oxidative stress is thought to modify thiol groups on the adenine nucleotide translocase, thus increasing the Ca^{2+}-sensitivity of the MPT (Halestrap et al., 1997c, 1998). Whatever its mechanism, propofol is another example of a reagent whose protection of the heart from reperfusion injury is accompanied by a decrease in mitochondrial pore opening *in vivo*. These data suggest that propofol may be a useful adjunct to the cardioplegic solutions used in cardiac surgery.

2.5.6. Preconditioning

Hearts can be greatly protected from reperfusion injury by subjecting them to two or three brief (3–5 minutes) ischemic periods, with intervening recovery periods before prolonged ischemia is initiated. Such preconditioning accords the heart immediate protection from reperfusion injury, which is reduced over a period of hours but then followed about 24 hours later by a second window of protection (Millar *et al.*, 1996; Schwarz *et al.*, 1997). The mechanisms responsible for preconditioning are not known in detail, but several processes are implicated. The second window of protection probably involves stress-activated protein kinase pathways (Maulik *et al.*, 1996; Mizukami and Yoshida, 1997); protein kinase *c* is implicated in short-term protection. Protein kinase *c* activation is probably a result of the release of mediators such as adenosine, bradykinin, endothelin 1, opiods, and catecholamines during the brief ischemic episodes, which then causes receptor-mediated breakdown of phosphatidylinositol-4,5-bisphosphate to produce the diacylglycerol needed to activate protein kinase *c* (Meldrum *et al.*, 1996; Millar *et al.*, 1996; Schultz *et al.*, 1997b; Yterhus *et al.*, 1994). There is also evidence for activation of K_{ATP} channel (perhaps mitochondrial) in the preconditioning mechanism because the effects are blocked by sulphonylureas, potent inhibitors of the K_{ATP} channel (Cleveland *et al.*, 1997; Liang, 1996; Schultz *et al.*, 1997a) that are mimicked by K_{ATP} channel openers (Behling Malone 1995; Garlid *et al.*, 1997; Liu *et al.*, 1998). Exactly how opening of K_{ATP}^+ channels might protect the heart is unknown. We have been unable to demonstrate any decrease in mitochondrial [^{3}H]-DOG entrapment following preconditioning, however, despite profound protection of postischemic heart function (Kerr and Halestrap, unpublished data). Thus prevention of MPTP opening by preconditioning is unlikely to provide an explanation of its effects. Another proposed preconditioning mechanism involves the mitochondrial ATPase inhibitor protein activated during brief ischemic periods (Vanderheide *et al.*, 1996; Vuorinen *et al.*, 1995), which would ensure that mitochondria in which the pore has opened would break down less of the ATP generated by glycolysis and by the remaining functional mitochondria. Although controversial (Vanderheide *et al.*, 1996; Yabe *et al.*, 1997), such a mechanism would enable hearts to stay protected from reperfusion injury even when a significant number of mitochondria remain in an open state.

3. THE PERMEABILITY TRANSITION AND APOPTOSIS IN REPERFUSION HEART INJURY

It is now recognized that in failing and reperfusion-injured hearts, some cells undergo apoptotic as opposed to necrotic cell death. This is particularly pronounced in areas surrounding a myocardial infarct, that is, areas that experience a less-pronounced ischemic insult than that which leads to necrosis (Bartling *et al.*, 1998; Bromme and Holtz, 1996; Fliss and Gattinger, 1996; Gottlieb *et al.*, 1994; Olivetti *et al.*, 1997; Umansky and Tomei, 1997). Mitochondria are required to induce apoptosis in a cell-free system, and apparently do so by releasing cytochrome *c*, which can activate the caspase cascade that initiates apoptosis (Green and Reed, 1998; Kluck *et al.*, 1997; Liu *et al.*, 1996; Reed, 1997; Yang *et al.*, 1997). Cytochrome *c* is normally located between inner and outer mitochondrial membrane and thus its release into the cytosol must involve either outer membrane rupture or a specific transport pathway. There is considerable debate as to which of these mechanisms the cell uses, but probably both are used depending on apoptotic stimulus (Green and Kroemer, 1998; Green and Reed, 1998; Reed, 1997).

Clearly, MPTP opening during reperfusion and consequent mitochondrial swelling likely rupture the outer membrane, releasing cytochrome *c*. Indeed, cytochrome *c* release occurs when mitochondria swell as a result of pore opening (Halestrap, 1982). If the pore stayed open however, mitochondria would remain uncoupled and unable to generate ATP for cellular ionic homeostasis maintenance and cellular damage repair. Under these conditions, damage continues unchecked, leading ultimately to plasma membrane rupture and cell death. This uncontrolled form of cell death—necrosis—is inflammatory, and is further exacerbated as neutrophil invasion leads to yet more damage (Halestrap, 1994; Kroemer *et al.*, 1998; Lemasters *et al.*, 1998; Reimer *et al.*, 1989). In contrast, when mitochondria open only transiently and then reseal, swelling may still rupture the outer mitochondrial membrane and release, cytochrome *c*, but, subsequent mitochondrial resealing would allow ATP production and ion gradients to be re-established. Apoptosis rather than necrosis could then be initiated. Indeed, apoptotic cells in hearts that have experienced ischemia/reperfusion show activated caspase-3 (Black *et al.*, 1998), and caspase inhibitors can protect the heart from irreversible injury (Yaoita *et al.*, 1998). Thus the decision between apoptosis and necrosis may rest on extent of pore opening and resealing, and it can account for the observation that both apoptosis and necrosis occur in the reperfused heart, with the least-damaged areas showing a preponderance of apoptosis (Bartling *et al.*, 1998; Bromme and Holtz, 1996; Fliss and Gattinger, 1996; Gottlieb *et al.*, 1994; Olivetti *et al.*, 1997; Umansky and Tomei, 1997). A diagram summarizing how the MPT may act as the decision maker between apoptosis and necrosis is given in Figure 6.

It will be of interest to establish whether hearts that recover fully on reperfusion, but in which some pore opening and re-closure has occurred, are primed for apoptosis leading

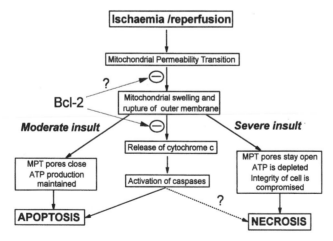

FIGURE 6. Scheme illustrating how the MPT may be involved in deciding whether a cell dies by necrosis or apoptosis.

to later impairment of function. Unfortunately, such experiments are impossible with isolated perfused heart; *in vivo* models are required. Such a model is available for investigating apoptosis in hippocampal neurons that occurs about 24 hours after ischemia/reperfusion or after brief insulin-induced hypoglycemic insult. Here a role for the MPT has been firmly established. Electron microscopy animal studies show that mitochondria in apoptotic neurons are swollen, and that both swelling and apoptosis are prevented by CsA treatment prior to insult (Friberg *et al.*, 1998). It was confirmed that CsA acts through pore inhibition, rather than through an effect on calcineurin, by demonstrating that FK506 (also active against calcineurin but not against transition) does not provide protection. The anti-apoptotic gene product Bcl2 is associated with the mitochondrial outer membrane and inhibits the MPT, cytochrome *c* release, and consequent caspase activation (Green and Kroemer, 1998; Green and Reed, 1998; Kluck *et al.*, 1997; Liu *et al.*, 1996; Reed, 1997; Yang *et al.*, 1997). Thus it is of interest that cells overexpressing Bcl2 are protected from hypoxic injury (Shimizu *et al.*, 1995; Yamabe *et al.*, 1998) and that hippocampal neurons in transgenic mouse brain overexpressing Bcl2 are protected from ischemic insults (Martinou *et al.*, 1994; Shimazaki *et al.*, 1994). No such data are available for either perfused heart or isolated heart cells, although Bcl2 is reportedly upregulated in myocytes that are salvaged after myocardial infarction (Misao *et al.*, 1996).

4. CONCLUSIONS

Opening of the permeability transition pore converts mitochondria from organelles whose supply of ATP sustains then in their normal function into organells of death. Conditions during reperfusion after ischemia are optimal for inducing this transition and thus may play a critical role in determining whether the cell recovers. From an understanding of the properties and mechanism of the MPTP, one can devise perfusion protocols that minimize pore opening and improve heart recovery following ischemia. This should lead to better cardioplegia during open-heart surgery. It remains to be established whether the short-term protection of isolated perfused heart, where damage is primarily necrotic, will also be reflected in longer-term recovery, where apoptosis may also play an important role.

ACKNOWLEDGEMENTS. This work was supported by project grants from the Medical Research Council and the British Heart Foundation.

REFERENCES

Allen, S. P., Stone, D., and McCormack, J. G., 1992, The loading of Fura-2 into mitochondria in the intact perfused rat heart and its use to estimate matrix Ca^{2+} under various conditions, *J.Mol. Cell. Cardiol.* **24**: 765–773.

Bartling, B., Holtz, J., and Darmer, D., 1998, Contribution of myocyte apoptosis to myocardial infarction? *Basic Res. Cardiol.* **93**: 71–84.

Behling, R.W., and Malone, H.J., 1995, K-ATP-channel openers protect against increased cytosolic calcium during ischaemia and reperfusion, *J.Mol. Cell. Cardiol.* **27**: 1809–1817.

Benzi, R.H., and Lerch, R., 1992, Dissociation between contractile function and oxidative metabolism in postischemic myocardium: Attenuation by ruthenium red administered during reperfusion, *Circ. Res.* **71**: 567–576.

Bernardi, P., 1992, Modulation of the mitochondrial Cyclosporin-A-sensitive permeability transition pore by the proton electrochemical gradient: Evidence that the pore can be opened by membrane depolarization, *J. Biol. Chem.* **267**: 8834–8839.

Bernardi, P., Broekemeier, K.M., and Pfeiffer, D.R., 1994, Recent progress on regulation of the mitochondrial permeability transition pore: A cyclosporin-sensitive pore in the inner mitochondrial membrane, *J. Bioenerg. Biomembr.* **26**: 509–517.

Bernardi, P., Vassanelli, S., Veronese, P., Colonna, R., Szabo, I., and Zoratti, M., 1992, Modulation of the mitochondrial permeability transition pore: Effect of protons and divalent cations, *J. Biol. Chem.* **267**: 2934–2939.

Black, S. C., Huang, J. Q., Rezaiefar, P., Radinovic, S., Eberhart, A., Nicholson, D. W., and Rodger, I. W., 1998, Co-localization of the cysteine protease caspase-3 with apoptotic myocytes after *in vivo* myocardial ischemia and reperfusion in the rat, *J. Mol. Cell. Cardiol.* **30**: 733–742.

Bond, J. M., Chacon, E., Herman, B., and Lemasters, J. J., 1993, Intracellular pH and Ca^{2+} homeostasis in the pH paradox of reperfusion injury to neonatal rat cardiac myocytes, *Am. J. Physiol.* **265**: C129–C137.

Borle, A. B., and Stanko, R. T., 1996, Pyruvate reduces anoxic injury and free radical formation in perfused rat hepatocytes, *Am. J. Physiol.* **270**: G535–G540.

Boveris, A., Cadenas, E., and Stoppani, A.O.M., 1976, Role of ubiquinone in the mitochondrial generation of hydrogen peroxide, *Biochem. J.* **156**: 435–444.

Branca, D., Vincenti, E., and Scutari, G., 1995, Influence of the anesthetic 2,6-diisopropylphenol (propofol) on isolated rat heart mitochondria, *Comp. Biochem. Physiol.* **110**: 41–45.

Bromme, H. J., and Holtz, J., 1996, Apoptosis in the heart: When and why? *Mol. Cell. Biochem.* **164**: 261–275.

Bryson, H. M., Fulton, B. R., and Faulds, D., 1995, Propofol: an update of its use in anesthesia and conscious sedation, *Drugs* **50**: 513–559.

Buljubasic, N., Marijic, J., Berczi, V., Supan, D.F., Kampine, J. P., and Bosnjak, Z. J., 1996, Differential effects of etomidate, propofol, and midazolam on calcium and potassium channel currents in canine myocardial cells, *Anesthesiology* **85**: 1092–1099.

Bunger, R., Mallet, R. T., and Hartman, D. A., 1989, Pyruvate-enhanced phosphorylation potential and inotropism in normoxic and post-ischemic isolated working heart, *Eur. J. Biochem.* **180**: 221–233.

Chernyak, B. V., and Bernardi, P., 1996, The mitochondrial permeability transition pore is modulated by oxidative agents through both pyridine nucleotides and glutathione at two separate sites, *Eur. J. Biochem.* **238**: 623–630.

Cicalese, L., Lee, K., Schraut, W., Watkins, S., Borle, A., and Stanko, R., 1996a, Pyruvate prevents ischemia reperfusion mucosal injury of rat small intestine, *Am. J. Surg.* **171**: 97–100.

Cicalese, L., Rastellini, C., Rao, A.S., and Stanko, R.T., 1996b, Pyruvate prevents mucosal reperfusion injury, oxygen free-radical production, and neutrophil infiltration after rat small bowel preservation and transplantation, *Transplant. Proc.* **28**: 2611–2611.

Cleveland, J.C., Meldrum, D.R., Cain, B.S., Banerjee, A., and Harken, A. H., 1997, Oral sulfonylurea hypoglycemic agents prevent ischemic preconditioning in human myocardium: Two paradoxes revisited, *Circulation* **96**: 29–32.

Cockshott, I.D., 1985, Propofol ("Diprivan") pharmacokinetics and metabolism: an overview, *Postgrad. Med. J.* **61**(Suppl 3): 45–50.

Coetzee, A., 1996, Comparison of the effects of propofol and halothane on acute myocardial ischaemia and myocardial reperfusion injury, *S. Afr. Med. J.* **86**(Suppl 2): C85–C90.

Connern, C. P., and Halestrap, A.P., 1994, Recruitment of mitochondrial cyclophilin to the mitochondrial inner membrane under conditions of oxidative stress that enhance the opening of a calcium-sensitive nonspecific channel, *Biochem. J.* **302**: 321–324.

Connern, C. P., and Halestrap, A. P., 1996, Chaotropic agents and increased matrix volume enhance binding of mitochondrial cyclophilin to the inner mitochondrial membrane and sensitize the mitochondrial permeability transition to [Ca^{2+}], *Biochemistry* **35**: 8172–8180.

Costantini, P., Chernyak, B. V., Petronilli, V., and Bernardi, P., 1996, Modulation of the mitochondrial permeability transition pore by pyridine nucleotides and dithiol oxidation at two separate sites, *J. Biol. Chem.* **271**: 6746–6751.

Crestanello, J. A., Lingle, D. M., Millili, J., and Whitman, G. J., 1998, Pyruvate improves myocardial tolerance to reperfusion injury by acting as an antioxidant: A chemiluminescence study, *Surgery* **124**: 92–99.

Crompton, M., 1990, The role of Ca^{2+} in the function and dysfunction of heart mitochondria, in *Calcium and the Heart* (G. A. Langer, Ed.), Raven, New York. pp. 167–198.

Crompton, M., and Costi, A., 1990, A heart mitochondrial Ca^{2+}-dependent pore of possible relevance to reperfusion-induced injury. Evidence that ADP facilitates pore interconversion between the closed and open states, *Biochem. J.* **266**: 33–39.

Crompton, M., Costi, A., and Hayat, L., 1987, Evidence for the presence of a reversible Ca^{2+}-dependent pore activated by oxidative stress in heart mitochondria., *Biochem. J.* **245**: 915–918.

Cross, H. R., Clarke, K., Opie, L. H., and Radda, G. K., 1995, Is lactate-induced myocardial ischaemic injury mediated by decreased pH or increased intracellular lactate? *J. Mol. Cell. Cardiol* **27**: 1369–1381.

Deboer, L. W. V., Bekx, P. A., Han, L. H., and Steinke, L., 1993, Pyruvate enhances recovery of rat hearts after ischemia and reperfusion by preventing free radical generation, *Am. J. Physiol.* **265**: H1571–H1576.

DeLaCruz, J. P., Villalobos, M. A., Sedeno, G., and DeLaCuesta, F. S., 1998, Effect of propofol on oxidative stress in an *in vitro* model of anoxia–reoxygenation in the rat brain, *Brain Res.* **800**: 136–144.

Delcamp, T. J., Dales, C., Ralenkotter, L., Cole, P. S., and Hadley, R. W., 1998, Intramitochondrial $[Ca^{2+}]$ and membrane potential in ventricular myocytes exposed to anoxia–reoxygenation, *Am. J. Physiol.* **275**: H484–H494.

Dennis, S. C., Gevers, W., and Opie, L. H., 1991, Protons in ischemia: Where do they come from; where do they go to? *J. Mol. Cell. Cardiol.* **23**: 1077–1086.

DiLisa, F., Blank, P. S., Colonna, R., Gambassi, G., Silverman, H. S., Stern, M. D., and Hansford, R. G., 1995, Mitochondrial membrane potential in single living adult rat cardiac myocytes exposed to anoxia or metabolic inhibition, *J. Physiol. (London)* **486**: 1–13.

Duan, J. M., and Karmazyn, M., 1992, Protective effects of amiloride on the ischemic reperfused rat heart: Relation to mitochondrial function, *Eur. J. Pharmacol.* **210**: 149–157.

Duchen, M. R., McGuinness, O., Brown, L. A., and Crompton, M., 1993, On the involvement of a Cyclosporin-A sensitive mitochondrial pore in myocardial reperfusion injury, *Cardiovasc. Res.* **27**: 1790–1794.

Dutoit, E.F., and Opie, L. H., 1992, Modulation of severity of reperfusion stunning in the isolated rat heart by agents altering calcium flux at onset of reperfusion, *Circ. Res.* **70**: 960–967.

Eriksson, O., 1991, Effects of the general anaesthetic Propofol on the Ca^{2+}-induced permeabilization of rat liver mitochondria, *FEBS Lett.* **279**: 45–48.

Eriksson, O., Pollesello, P., and Saris, N. E., 1992, Inhibition of lipid peroxidation in isolated rat liver mitochondria by the general anaesthetic propofol, *Biochem. Pharmacol.* **44**: 391–393.

Figueredo, V. M., Dresdner, K. P. Jr., Wolney, A. C., and Keller, A. M., 1991, Postischaemic reperfusion injury in the isolated rat heart: Effect of ruthenium red, *Cardiovasc. Res.* **25**: 337–342.

Fliss, H., and Gattinger, D., 1996, Apoptosis in ischemic and reperfused rat myocardium, *Circ. Res.* **79**: 949–956.

Folbergrova, J., Li, P. A., Uchino, H., Smith, M. L., and Siesjo, B.K., 1997, Changes in the bioenergetic state of rat hippocampus during 2.5 min of ischemia, and prevention of cell damage by cyclosporin A in hyperglycemic subjects, *Exp. Brain Res.* **114**: 44–50.

Friberg, H., Ferrand-Drake, M., Bengtsson, F., Halestrap, A. P., and Wieloch, T., 1998, Cyclosporin A, but not FK 506, protects mitochondria and neurons against hypoglycemic damage and implicates the mitochondrial permeability transition in cell death, *J. Neurosci.* **18**: 5151–5159.

Galat, A., and Metcalfe, S. M., 1995, Peptidylproline *cis trans* isomerases, *Prog. Biophys. Mol. Biol.* **63**: 67–118.

Garlid, K. D., Paucek, P., YarovYarovoy, V., Murray, H. N., Darbenzio, R. B., D'Alonzo, A. J., Lodge, N. J., Smith, M. A., and Grover, G. J., 1997, Cardioprotective effect of diazoxide and its interaction with mitochondrial ATP-sensitive K^+channels: Possible mechanism of cardioprotection, *Circ. Res.* **81**: 1072–1082.

Gatewood, L. B., Larson, D. F., Bowers, M. C., Bond, S., Cardy, A., Sethi, G. K., and Copeland, J. G., 1996, A novel mechanism for cyclosporine: Inhibition of myocardial ischemia and reperfusion injury in a heterotopic rabbit heart transplant model, *J. Heart Lung Transplant* **15**: 936–947.

Gottlieb, R. A., Burleson, K. O., Kloner, R. A., Babior, B. M., and Engler, R. L., 1994, Reperfusion injury induces apoptosis in rabbit cardiomyocytes, *J. Clin. Invest.* **94**: 1621–1628.

Green, D. and Kroemer, G., 1998, The Central executioners of apoptosis: Caspases or mitochondria? *Trends Cell Biol.* **8**: 267–271.

Green, D. R., and Reed, J. C., 1998, Mitochondria and apoptosis, *Science* **281**: 1309–1312.

Green, T. R., Bennett, S. R., and Nelson, V. M., 1994, Specificity and properties of propofol as an antioxidant free radical scavenger, *Toxicol. Appl. Pharmacol.* **129**: 163–169.

Griffiths, E. J., and Halestrap, A. P., 1993, Protection by Cyclosporin A of ischemia reperfusion-induced damage in isolated rat hearts, *J. Mol. Cell. Cardiol.* **25**: 1461–1469.

Griffiths, E. J., and Halestrap, A. P., 1995, Mitochondrial nonspecific pores remain closed during cardiac ischaemia, but open upon reperfusion, *Biochem. J.* **307**: 93–98.

Griffiths, E. J., Stern, M. D., and Silverman, H. S., 1997, Measurement of mitochondrial calcium in single living cardiomyocytes by selective removal of cytosolic Indo 1, *Am. J. Physiol.* **273**: C37–C44.

Griffiths, E. J., Ocampo, C. J., Savage, J. S., Rutter, G. A., Hansford, R. G., Stern, M. D., and Silverman, H. S., 1998, Mitochondrial calcium transporting pathways during hypoxia and reoxygenation in single rat cardiomyocytes, *Cardiovasc. Res.* **39**: 423–433.

Grover, G. J., Dzwonczyk, S., and Sleph, P. G., 1990, Ruthenium red improves postischemic contractile function in isolated rat hearts, *J. Cardiovasc. Pharmacol.* **16**: 783–789.

Gutteridge, J. M. C., and Halliwell, B., 1990, Reoxygenation injury and antioxidant protection: A tale of two paradoxes, *Arch. Biochem. Biophys.* **283**: 223–226.

Haigney, M. C., Miyata, H., Lakatta, E. G., Stern, M. D., and Silverman, H. S., 1992, Dependence of hypoxic cellular calcium loading on Na^+–Ca^{2+} exchange, *Circ. Res.* **71**: 547–557.

Halestrap, A. P., 1982, The nature of the stimulation of the respiratory chain of rat liver mitochondria by glucagon pretreatment of animals, *Biochem. J.* **204**: 37–47.

Halestrap, A. P., 1991, Calcium-dependent opening of a non-specific pore in the mitochondrial inner membrane is inhibited at pH values below 7: Implication for the protective effect of low pH against chemical and hypoxic cell damage, *Biochem. J.* **278**: 715–719.

Halestrap, A. P., 1994, Interactions between oxidative stress and calcium overload on mitochondrial function., in *Mitochondria: DNA, Proteins, and Disease* (V. Darley-Usmar, and A. H. V. Schapira, Eds.) Portland Press, London, pp. 113–142.

Halestrap, A. P., Griffiths, E. J., and Connern, C. P., 1993, Mitochondrial calcium handling and oxidative stress, *Biochem. Soc. Trans.* **21**: 353–358.

Halestrap, A. P., Connern, C. P., Griffiths, E. J., and Kerr, P. M., 1997a, Cyclosporin A binding to mitochondrial cyclophilin inhibits the permeability transition pore and protects hearts from ischaemia/reperfusion injury, *Mol. Cell. Biochem.* **174**: 167–172.

Halestrap, A. P., Wang, X. M., Poole, R. C., Jackson, V. N., and Price, N. T., 1997b, Lactate transport in heart in relation to myocardial ischemia, *Am. J. Cardiol.* **80**: A17–A25.

Halestrap, A. P., Woodfield, K. Y., and Connern, C. P., 1997c, Oxidative stress, thiol reagents, and membrane potential modulate the mitochondrial permeability transition by affecting nucleotide binding to the adenine nucleotide translocase, *J. Biol. Chem.* **272**: 3346–3354.

Halestrap, A. P., Kerr, P. M., Javadov, S., and Woodfield, K. Y., 1998, Elucidating the molecular mechanism of the permeability transition pore and its role in reperfusion injury of the heart, *Biochim. Biophys. Acta* **1366**: 79–94.

Haworth, R. A., and Hunter, D. S., 1979, The Ca^{2+}-induced membrane transition in mitochondria: II. Nature of the Ca^{2+} trigger site, *Arch. Biochem. Biophys.* **195**: 460–467.

Javadov, S. A., Lim, K. H. H., Kerr, P. M., Suleiman, M-S., Angelini, G. D. and Halestrap, A. P., 2000, Protection of hearts from reperfusion injury by propofol is associated with inhibition of the mitochondrial permeability transition. *Cardiovascular Research* **45**: 360–369.

Karmazyn, M., Ray, M., and Haist, J. V., 1993, Comparative effects of Na^+/H^+ exchange inhibitors against cardiac injury produced by ischemia/reperfusion, hypoxia/reoxygenation, and the calcium paradox, *J. Cardiovasc. Pharmacol.* **21**: 172–178.

Kerr, P. M., Suleiman, M.-S., and Halestrap, A. P., 1999, Reversal of the mitochondrial permeability transition during recovery of hearts from ischemia and its enhancement by pyruvate, *Am. J. Physiol.*, In Press.

Kluck, R. M., Bossy-Wetzel, E., Green, D. R., and Newmeyer, D. D., 1997, The release of cytochrome *c* from mitochondria: A primary site for Bcl-2 regulation of apoptosis, *Science* **275**: 1132–1136.

Ko, S. H., Yu, C. W., Choe, H., Chung, M. J., Kwak, Y. G., Chae, S. W., and Song, H. S., 1997, Propofol attenuates ischaemic-reperfusion injury in the isolated rat heart, *Anesth. Analg.* **85**: 719–724.

Kokita, N., and Hara, A., 1996, Propofol attenuates hydrogen-peroxide induced mechanical and metabolic derangements in the isolated rat heart, *Anesthesiol.* **84**: 117–127.

Kokita, N., Hara, A., Abiko, Y., Arakawa, J., Hashizume, H., and Namiki, A., 1998, Propofol improves functional and metabolic recovery in ischemic reperfused isolated rat hearts, *Anesth. Analg.* **86**: 252–258.

Kroemer, G., Dallaporta, B., and Resche-rigon, M., 1998, The mitochondrial death/life regulator in apoptosis and necrosis, *Annu. Rev. Physiol.* **60**: 619–642.

Kurokawa, T., Kobayashi, H., Nonami, T., Harada, A., Nakao, A., Sugiyama, S., Ozawa, T., and Takagi, H., 1992, Beneficial effects of cyclosporine on postischemic liver injury in rats, *Transplantation* **53**: 308–311.

Ladilov, Y. V., Siegmund, B., and Piper, H. M., 1995, Protection of reoxygenated cardiomyocytes against hypercontracture by inhibition of Na^+/H^+ exchange, *Am. J. Physiol.* **268**: H1531–H1539.

Lazdunski, M., Frelin, C., and Vigne, P., 1985, The sodium/hydrogen exchange system in cardiac cells: Its biochemical and pharmacological properties and its role in regulating internal concentrations of sodium and internal pH, *J. Mol. Cell. Cardiol.* **17**: 1029–1042.

Leist, M., and Nicotera, P., 1997, The shape of cell death, *Biochem. Biophys. Res. Commun.* **236**: 1–9.

Lemasters, J. J., and Thurman, R. G., 1995, The many facets of reperfusion injury, *Gastroenterology* **108**: 1317–1320.

Lemasters, J. J., Chacon, E., Ohata, H., Harper, I. S., Nieminen, A.-L., Tesfai, S. A., and Herman, B., 1995, Measurement of electrical potential, pH, and free calcium ion concentration in mitochondria of living cells by laser scanning confocal microscopy, *Methods Enzymol.* **260**: 428–444.

Lemasters, J. J., Nieminen, A.-L., Qian, T., Trost, L. C., and Herman, B., 1997, The mitochondrial permeability transition in toxic, hypoxic, and reperfusion injury, *Mol. Cell. Biochem.* **174**: 159–165.

Lemasters, J. J., Nieminen, A. L., Qian, T., Trost, L. C., Elmore, S. P., Nishimura, Y., Crowe, R. A., Cascio, W. E., Bradham, C. A., Brenner, D. A., and Herman, B., 1998, The mitochondrial permeability transition in cell death: A common mechanism in necrosis, apoptosis, and autophagy, *Biochim. Biophys. Acta* **1366**: 177–196.

Li, P. A., Uchino, H., Elmer, E., and Siesjo, B. K., 1997a, Amelioration by cyclosporin A of brain damage following 5 or 10 min of ischemia in rats subjected to preischemic hyperglycemia, *Brain Res.* **753**: 133–140.

Li, Y. C., Ridefelt, P., Wiklund, L., and Bjerneroth, G., 1997b, Propofol induces a lowering of free cytosolic calcium in myocardial cells, *Acta Anaesthesiol. Scand.* **41**: 633–638.

Liang, B. T., 1996, Direct preconditioning of cardiac ventricular myocytes via adenosine A(1) receptor and K-ATP channel, *Am J. Physiol.* **271**: H1769–H1777.

Liu, X., Kim, C. N., Yang, J., Jemmerson, R., and Wang, X., 1996, Induction of apoptotic program in cell-free extracts: Requirement for dATP and cytochrome *c*, *Cell* **86**: 147–157.

Liu, Y. G., Sato, T., O'Rourke, B., and Marban, E., 1998, Mitochondrial ATP-dependent potassium channels: Novel effectors of cardioprotection? *Circulation* **97**: 2463–2469.

Martinou, J. C., Duboisdauphin, M., Staple, J. K., Rodriguez, I., Frankowski, H., Missotten, M., Albertini, P., Talabot, D., Catsicas, S., Pietra, C., and Huarte, J., 1994, Overexpression of BCL-2 in transgenic mice protects neurons from naturally occurring cell death and experimental ischemia, *Neuron* **13**: 1017–1030.

Massoudy, P., Becker, B. F., Seligmann, C., and Gerlach, E., 1995, Preischaemic as well as postischaemic application of a calcium antagonist affords cardioprotection in the isolated guinea pig heart, *Cardiovasc. Res.* **29**: 577–582.

Maulik, N., Watanabe, M., Zu, Y. L., Huang, C. K., Cordis, G. A., Schley, J. A., and Das, D. K., 1996, Ischemic preconditioning triggers the activation of MAP kinases and MAPKAP kinase 2 in rat hearts, *FEBS Lett.* **396**: 233–237.

Mayer, N., Legat, K., Weinstabl, C., and Zimpfer, M., 1993, Effects of propofol on the function of normal, collateral-dependent, and ischemic myocardium, *Anesth. Analg.* **76**: 33–39.

Meldrum, D. R., Cleveland, J. C., Mitchell, M. B., Sheridan, B. C., Gambon-Robertson, F., Harken, A. H., and Banerjee, A., 1996, Protein kinase C mediates Ca^{2+}-induced cardioadaptation to ischemia–reperfusion injury, *Am. J. Physiol.* **271**: R718–R726.

Metivier, D., Dallaporta, B., Zamzami, N., Larochette, N., Susin, S. A., Marzo, I., and Kroemer, G., 1998, Cytofluorometric detection of mitochondrial alterations in early CD95/Fas/APO-1-triggered apoptosis of Jurkat T lymphoma cells: Comparison of seven mitochondrion-specific fluorochromes, *Immunol. Lett.* **61**: 157–163.

Millar, C. G., Baxter, G. F., and Thiemermann, C., 1996, Protection of the myocardium by ischaemic preconditioning: Mechanisms and therapeutic implications, *Pharmacol. Ther.* **69**: 143–151.

Misao, J., Hayakawa, Y., Ohno, M., Kato, S., Fujiwara, T., and Fujiwara, H., 1996, Expression of bcl-2 protein, an inhibitor of apoptosis, and Bax, an accelerator of apoptosis, in ventricular myocytes of human hearts with myocardial infarction, *Circulation* **94**: 1506–1512.

Miyata, H., Lakatta, E. G., Stern, M. D., and Silverman, H. S., 1992, Relation of mitochondrial and cytosolic free calcium to cardiac myocyte recovery after exposure to anoxia, *Circ. Res.* **71**: 605–613.

Mizukami, Y., and Yoshida, K., 1997, Mitogen-activated protein kinase translocates to the nucleus during ischaemia and is activated during reperfusion, *Biochem. J.* **323**: 785–790.

Murphy, P. G., Myers, D. S., Davies, W. J., and Webster, N. R. J. J. G., 1992, The antioxidant potential of propofol (2,6-diisopropylphenol), *Br. J. Anaesth.* **68**: 616–618.

Murphy, P. G., Bennett, J. R., Myers, D. S., Davies, M. J., and Jones, J. G., 1993, The effect of propofol anaesthesia on free radical-induced lipid peroxidation in rat liver microsomes, *Eur. J. Anaesthes.* **10**: 261–266.

Nazareth, W., Yafei, N., and Crompton, M., 1991, Inhibition of anoxia-induced injury in heart myocytes by cyclosporin-A, *J. Mol. Cell. Cardiol.* **23**: 1351–1354.

Nieminen, A.-L., Saylor, A. K., Tesfai, S. A., Herman, B., and Lemasters, J. J., 1995, Contribution of the mitochondrial permeability transition to lethal injury after exposure of hepatocytes to *t*-butylhydroperoxide, *Biochem. J.* **307**: 99–106.

Nieminen, A.-L., Petrie, T. G., Lemasters, J. J., and Selman, W. R., 1996, Cyclosporin A delays mitochondrial depolarization induced by N-methyl-D-aspartate in cortical neurons: Evidence of the mitochondrial permeability transition, *Neuroscience* **75**: 993–997.

Nieminen, A. L., Byrne, A. M., Herman, B., and Lemasters, J. J., 1997, Mitochondrial permeability transition in hepatocytes induced by *t*-BuOOH: NAD(P)H and reactive oxygen species, *Am. J. Physiol.* **271**: C1286–C1294.

Nishino, T., 1994, The conversion of xanthine dehydrogenase to xanthine oxidase and the role of the enzyme in reperfusion injury, *J. Biochem. (Tokyo)* **116**: 1–6.

Olivetti, G., Abbi, R., Quaini, F., Kajstura, J., Cheng, W., Nitahara, J. A., Quaini, E., DiLoreto, C., Beltrami, C. A., Krajewski, S., Reed, J. C., and Anversa, P., 1997, Apoptosis in the failing human heart, *N. Engl. J. Med.* **336**: 1131–1141.

Omar, B., McCord, J., and Downey, J., 1991, Ischaemia–reperfusion, in *Oxidative Stress: Oxidants and Antioxidants* (H. Sies, Ed.), Academic, San Diego, pp. 493–527.

Opie, L., 1992, Myocardial stunning: A Role for calcium antagonists during reperfusion, *Cardiovasc. Res.* **26**: 20–24.

Peng, C. F., Kane, J. J., Straus, K. D., and Murphy, M. L., 1980, Improvement of mitochondrial energy production in ischaemic myocardium by *in vivo* infusion of ruthenium red, *J. Cardiovasc. Pharmacol.* **2**: 45–54.

Petronilli, V., Cola, C., Massari, S., Colonna, R., and Bernardi, P., 1993, Physiological effectors modify voltage sensing by the Cyclosporin A-sensitive permeability transition pore of mitochondria, *J. Biol. Chem.* **268**: 21939–21945.

Petronilli, V., Costantini, P., Scorrano, L., Colonna, R., Passamonti, S., and Bernardi, P., 1994, The voltage sensor of the mitochondrial permeability transition pore is tuned by the oxidation-reduction state of vicinal thiols: Increase of the gating potential by oxidants and its reversal by reducing agents, *J. Biol. Chem.* **269**: 16638–16642.

Piper, H. M., 1997, Mechanism of myocardial injury during acute reperfusion, *News Physiol. Sci.* **12**: 53–54.

Piper, H. M., Noll, T., and Siegmund, B., 1994, Mitochondrial function in the oxygen depleted and reoxygenated myocardial cell, *Cardiovasc. Res.* **28**: 1–15.

Qian, T., Nieminen, A.-L., Herman, B., and Lemasters, J. J., 1997, Mitochondrial permeability transition in pH-dependent reperfusion injury to rat hepatocytes, *Am. J. Physiol.* **273**: C1783–C1792.

Reed, J. C., 1997, Cytochrome *c*: Can't live with it: Can't live without it, *Cell* **91**: 559–562.

Reimer, K. A., and Jennings, R. B., 1992, Myocardial ischemia, hypoxia, and infarction, in *The Heart and Cardovascular System*, 2nd ed. (H. A. Fozzard, R. B. Jennings, E. Huber, A. M. Katz, and H. E. Morgan, Eds.), Raven, New York, pp. 1875–1973.

Reimer, M. A., Murry, C. E., and Richard, V. J., 1989, The role of neutrophils and free radicals in the ischemic-reperfused heart: Why the confusion and controversy? *J. Mol. Cell. Cardiol.* **21**:1225–1239.

Rigoulet, M., Devin, A., Averet, N., Vandais, B., and Guerin, B., 1996, Mechanisms of inhibition and uncoupling of respiration in isolated rat liver mitochondria by the general anesthetic 2,6-diisopropylphenol, *Eur. J. Biochem.* **241**:280–285.

Sack, S., Mohri, M., Schwarz, E. R., Arras, M., Schaper, J., Ballagipordany, G., Scholz, W., Lang, H. J., Scholkens, B. A., and Schaper, W., 1994, Effects of a new NA^+/H^+ antiporter inhibitor on postischemic reperfusion in pig heart, *J. Cardiovasc. Pharmacol.* **23**:72–78.

Salvioli, S., Ardizzoni, A., Franceschi, C., and Cossarizza, A., 1997, JC-1, but not DiOC(6)(3) or rhodamine 123, is a reliable fluorescent probe to assess Delta Psi changes in intact cells: Implications for studies on mitochondrial functionality during apoptosis, *FEBS Lett.* **411**:77–82.

Schreiber, S., L., 1991, Chemistry and biology of the immunophilins and their immunosuppressive ligands, *Science* **251**:283–287.

Schultz, J. E. J., Yao, Z. H., Cavero, I., and Gross, G. J., 1997a, Glibenclamide-induced blockade of ischemic preconditioning is time dependent in intact rat heart, *Am. J. Physiol.* **272**: H2607–H2615.

Schultz, J. J., Hsu, A. K., and Gross, G. J., 1997b, Ischemic preconditioning is mediated by a peripheral opioid receptor mechanism in the intact rat heart, *J. Mol. Cell. Cardiol.* **29**: 1355–1362.

Schwarz, E. R., Whyte, W. S., and Kloner, R. A., 1997, Ischemic preconditioning, *Curr. Opin. Cardiol.* **12**: 475–481.

Scorrano, L., Petronilli, V., and Bernardi, P., 1997, On the voltage dependence of the mitochondrial permeability transition pore: A critical appraisal, *J. Biol. Chem.* **272**: 12295–12299.

Servin, F., Desmonts, J. M., Haberer, J. P., Cockshott, I. D., Plummer, G. F., and Farinotti, R., 1988, Pharmacokinetics and protein binding of propofol in patients with cirrhosis, *Anesthesiology* **69**: 887–891.

Shiga, Y., Onodera, H., Matsuo, Y., and Kogure, K., 1992, Cyclosporin-A protects against ischemia–reperfusion injury in the brain, *Brain Res.* **595**: 145–148.

Shimazaki, K., Ishida, A., and Kawai, N., 1994, Increase in bcl-2 oncoprotein and the tolerance to ischemia-induced neuronal death in the gerbil hippocampus, *Neurosci. Res.*.**20**: 95–99.

Shimizu, S., Kamiike, W., Hatanaka, N., Miyata, R., Inoué, T., Yoshida, Y., Tagawa, K., and Matsuda, H., 1994, Beneficial effects of cyclosporine on reoxygenation injury in hypoxic rat liver, *Transplantation* **57**: 1562–1566.

Shimizu, S., Eguchi, Y., Kosaka, H., Kamiike, W., Matsuda, H., and Tsujimoto, Y., 1995, Prevention of hypoxia-induced cell death by Bcl-2 and Bcl-xL, *Nature* **374**: 811–813.

Silverman, H. S., and Stern, M. D., 1994, Ionic basis of ischaemic cardiac injury: Insights from cellular studies, *Cardiovasc. Res.* **28**: 581–597.

Stone, D., Darley-Usmar, V., Smith, D. R., and O'Leary, V., 1989, Hypoxia–reoxygenation induced increase in cellular Ca^{2+} in myocytes and perfused hearts: The role of mitochondria, *J. Mol. Cell. Cardiol.* **21**: 963–973.

Sztark, F., Ichas, F., Ouhabi, R., Dabadie, P., and Mazat, J. P., 1995, Effects of the anaesthetic propofol on the calcium-induced permeability transition of rat heart mitochondria: Direct pore inhibition and shift of the gating potential, *FEBS Lett.* **368**: 101–104.

Travis, D. L., Fabia, R., Netto, G. G., Husberg, B. S., Goldstein, R. M., Klintmalm, G. B., and Levy, M. F., 1998, Protection by cyclosporine A against normothermic liver ischemia–reperfusion in pigs, *J. Surg. Res.* **75**: 116–126.

Turrens, J. F., Alexandre, A., and Lehninger, A. L., 1985, Ubisemiquinone is the electron donor for superoxide formation by complex III of heart mitochondria, *Arch. Biochem. Biophys.* **237**: 408–414.

Ubl, J. J., Chatton, J. Y., Chen, S. H., and Stucki, J. W., 1996, A critical evaluation of *in situ* measurement of mitochondrial electrical potentials in single hepatocytes, *Biochim. Biophys. Acta* **1276**: 124–132.

Uchino, H., Elmer, E., Uchino, K., Lindvall, O., and Siesjo, B. K., 1995, Cyclosporin A dramatically ameliorates CAI hippocampal damage following transient forebrain ischaemia in the rat, *Acta Physiol. Scand* **155**: 469–471.

Umansky, S. R., and Tomei, L. D., 1997, Apoptosis in the heart, *Adv. Pharmacol.* **41**: 383–407.

Vandenberg, J. I., Metcalfe, J. C., and Grace, A. A., 1993, Mechanisms of intracellular pH recovery following global ischaemia in the perfused heart, *Circulation Res.* **72**: 993–1003.

Vanderheide, R. S., Hill, M. L., Reimer, K. A., and Jennings, R. B., 1996, Effect of reversible ischemia on the activity of the mitochondrial ATPase: Relationship to ischemic preconditioning, *J. Mol. Cell. Cardiol.* **28**: 103–112.

Vuorinen, K., Ylitalo, K., Peuhkurinen, K., Raatikainen, P., Alarami, A., and Hassinen, I. E., 1995, Mechanisms of ischemic preconditioning in rat myocardium: Roles of adenosine, cellular energy state, and mitochondrial F_1F_0-ATPase, *Circulation* **91**: 2810–2818.

Woodfield, K.-Y., Rück, A., Brdiczka, D., and Halestrap, A. P., 1998, Direct demonstration of a specific interaction between cyclophilin-D and the adenine nucleotide translocase confirms their role in the mitochondrial permeability transition, *Biochem. J.* **336**: 287–290.

Yabe, K., Nasa, Y., Sato, M., Iijima, R., and Takeo, S., 1997, Preconditioning preserves mitochondrial function and glycolytic flux during an early period of reperfusion in perfused rat hearts, *Cardiovasc. Res.* **33**: 677–685.

Yamabe, K., Shimizu, S., Kamiike, W., Waguri, S., Eguchi, Y., Hasegawa, J., Okuno, S., Yoshioka, Y., Ito, T., Sawa, Y., Uchiyama, Y., Tsujimoto, Y., and Matsuda, H., 1998, Prevention of hypoxic liver cell necrosis by *in vivo* human bcl-2 gene transfection, *Biochem. Biophys. Res. Commun.* **243**: 217–223.

Yang, J., Liu, X. S., Bhalla, K., Kim, C. N., Ibrado, A. M., Cai, J. Y., Peng, T. I., Jones, D. P., and Wang, X. D., 1997, Prevention of apoptosis by Bcl-2: Release of cytochrome *c* from mitochondria blocked, *Science* **275**: 1129–1132.

Yaoita, H., Ogawa, K., Maehara, K., and Maruyama, Y., 1998, Attenuation of ischemia/reperfusion injury in rats by a caspase inhibitor, *Circulation* **97**: 276–281.

Yoshida, T., Watanabe, M., Engelman, D. T., Engelman, R. M., Schley, J. A., Maulik, N., Ho, Y. S., Oberley, T. D., and Das, D. K., 1996, Transgenic mice overexpressing glutathione peroxidase are resistant to myocardial ischemia reperfusion injury, *J. Mol. Cell. Cardiol.* **28**: 1759–1767.

Ytrehus, K., Liu, Y. G., and Downey, J. M., 1994, Preconditioning protects ischemic rabbit heart by protein kinase C activation, *Am. J. Physiol.* **266**: H1145–H1152.

Chapter 10

Mitochondrial Calcium Dysregulation during Hypoxic Injury to Cardiac Myocytes

Elinor J. Griffiths

1. INTRODUCTION

The heart is completely dependent on ATP from aerobic metabolism for normal contractile function. Coordinating increase in ATP demand with increase in ATP supply is essential during increase in cardiac workload, and lack of such coordination may also underlie disease states where defects in mitochondrial energy production occur, for example, diabetes, cardiomyopathies, and ischemia/hypoxia (Harding *et al.*, 1994; Marin-Garcia and Goldenthal 1994).

Mitochondrial $[Ca^{2+}]$, henceforth called ($[Ca^{2+}]_m$), is thought to be a key regulator of ATP production (Hansford, 1991, McCormack *et al.*, 1990), but the pathways involved in mitochondrial Ca^{2+} transport within intact cells have not been thoroughly characterized. Nor is the exact relationship between $[Ca^{2+}]_m$ and mitochondrial ATP production fully established. In disease states such as hypoxia and ischemia, energy metabolism is impaired and mitochondrial Ca^{2+} overload can occur. This has been associated with the transition from reversible to irreversible cell injury upon reoxygenation/reperfusion (Miyata *et al.*, 1992; Shen and Jennings, 1972). Consequently, it has been proposed that mitochondrial Ca^{2+} transport pathways might be sites for protective intervention (Cox and Matlib, 1993; Figueredo *et al.*, 1991; Miyame *et al.*, 1996; Park *et al.*, 1990). We recently found, however, that significant alterations in Ca^{2+} transport pathways can occur during hypoxia

Elinor J. Griffiths   Bristol Heart Institute, University of Bristol, Bristol Royal Infirmary, Bristol BS52 8HW, United Kingdom.

Mitochondria in Pathogenesis, edited by Lemasters and Nieminen.
Kluwer Academic/Plenum Publishers, New York, 2001.

(Griffiths *et al.*, 1998), which would clearly affect any strategies designed to confer protection by acting on Ca^{2+} transport mechanisms.

This chapter briefly reviews current knowledge about mitochondrial Ca^{2+} transport in intact myocytes under physiological conditions and methods of measuring $[Ca^{2+}]_m$, before discussing these pathways during hypoxia and reoxygenation of single cells.

2. MITOCHONDRIAL CA^{2+} TRANSPORT UNDER NORMAL CONDITIONS

The Ca^{2+} transport pathways of the sarcolemma and sarcoplasmic reticulum (SR) in cardiac cells have been well characterized, as has their role in excitation–contraction coupling. Far less is known, however, about mitochondrial Ca^{2+} transporters, despite a resurgence of interest in this area in both cardiomyocytes and other cell types.

2.1. Ca^{2+} Transport Studies in Isolated Mitochondria

The Ca^{2+} transport pathways were originally studied in isolated heart mitochondria when it was revealed that specific pathways for Ca^{2+} uptake and release were present in the inner mitochondrial membrane: Ca^{2+} influx occurred via a Ca^{2+} uniporter, driven by membrane potential and inhibited by ruthenium red, and Ca^{2+} efflux occurred via Na^+/Ca^{2+} exchange, inhibited by diltiazem, clonazepam, CGP37157, and high external Ca^{2+} (Cox *et al.*, 1993; Crompton, 1990; Gunter and Pfeiffer, 1990). These pathways are shown schematically in Figure 1, together with a third, nonspecific pathway, the mitochondrial permeability transition pore (MPTP). The MPTP is an inner membrane channel induced under nonphysiological conditions of high intramitochondrial $[Ca^{2+}]$, adenine nucleotide depletion, and oxidative stress; it is inhibited by the immunosuppressant cyclosporin A (CsA) (reviewed in Crompton, 1990; Bernardi *et al.*, 1994; Halestrap, 1994).

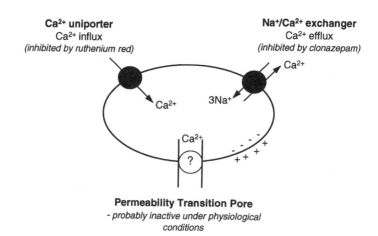

Ca²⁺ uniporter
Ca²⁺ influx
(inhibited by ruthenium red)

Na⁺/Ca²⁺ exchanger
Ca²⁺ efflux
(inhibited by clonazepam)

Permeability Transition Pore
- probably inactive under physiological conditions

FIGURE 1. Calcium transport pathways of the inner mitochondrial membrane under normal conditions.

Until recently, the Na^+/Ca^{2+} exchanger was thought to catalyze electroneutral exchange of $2Na^+:Ca^{2+}$ (Brand, 1985). Recent evidence indicates however, that under the physiological conditions of a highly negative mitochondrial membrane potential ($\Delta\psi_m$) the exchange is electrogenic, with a stoichiometry of $3Na^+:Ca^{2+}$ (Baysal et al., 1994; Jung et al., 1995). Only lately have attempts been made to purify and reconstitute Ca^{2+} transporters with, only partial success as yet (Li et al., 1992; Saris et al., 1993).

2.2. Mitochondrial Ca^{2+} Transport in Intact Cells

A major problem in studying mitochondrial Ca^{2+} transport in intact cells was the lack of suitable techniques for measuring $[Ca^{2+}]_m$; thus the relationship between $[Ca^{2+}]_m$ and $[Ca^{2+}]_c$ (cytosolic Ca^{2+}) was unknown. This is important not only for control of ATP production by mitochondria (via Ca^{2+} dehydrogenase activation) but also for determining whether $[Ca^{2+}]_m$ changes on a beat-to-beat timescale in cardiac muscle in response to changes in $[Ca^{2+}]_c$ and might therefore play a role in cellular Ca^{2+} homeostasis and excitation–contraction coupling. Experiments with regard to the latter are contradictory: Mitochondrial Ca^{2+} transients have been found in guinea pig myocytes using electron probe microanalysis (Wendt-Gallitelli and Isenberg, 1991), and in rabbit myocytes using confocal microscope imaging of cells loaded with Fluo-3 or Rhod-2 (Chacon et al., 1996; Trollinger et al., 1997). In rat cells however, no change in $[Ca^{2+}]_m$ was found during a single contraction, but a slow increase was observed upon sustained electrical stimulation (Miyata et al., 1991). This result was obtained using indo-1-AM-loaded cells where the cytosolic fluorescence signal was quenched with Mn^{2+}, a technique criticized for the possibility of Mn^{2+} interfering with mitochondrial Ca^{2+} transporters. We recently confirmed this result, however, when $[Ca^{2+}]_m$ was determined by localizing indo-1 to mitochondria following actual removal of cytosolic indo-1 by extended incubation of cells at $37\,^\circ C$ (Griffiths et al., 1997a).

This conflicting data suggests that there may be significant species differences in mitochondrial calcium handling. Two variables occur in the above studies, namely the technique employed and the species studied. A recent preliminary study compared directly myocytes from rat heart with guinea pig heart using the same technique to measure $[Ca^{2+}]_m$, namely localization of indo-1 to mitochoridria by selective removal of cytosolic indo-1 (Griffiths et al., 1997a). Upon electrical stimulation, no change in $[Ca^{2+}]_m$ occurred in rat myocytes, as previously found, but in guinea pig cells clear mitochondrial Ca^{2+} transients were seen during the contractile cycle (Griffiths, 1999). This area requires further investigation to elucidate the factors causing this dramatic difference in mitochondrial Ca^{2+} transport and the implications for cell energy-metabolism control and Ca^{2+} homeostasis.

The MPTP probably plays a minimal role in Ca^{2+} transport under normal conditions. Freshly isolated heart mitochondria can accumulate large amounts of Ca^{2+} with no deleterious effects and the MPTP occurs only if other factors, such as oxidative stress, are present (Halestrap, 1994). High rates of Ca^{2+} uptake can cause a slight drop in $\Delta\psi_m$, normally maintained at a highly negative value of about $-180\,mV$ (Crompton, 1990; Gunter and Pfeiffer, 1990). Pore opening causes membrane depolarization, which has been used as an indicator of the opening in intact cells under conditions of severe metabolic inhibition (Chacon et al., 1994). Studies measuring $\Delta\psi_m$ using fluorescent indicators such as JC-1 showed no change in $\Delta\psi_m$ in either resting cardiomyocytes or in cells electrically stimulated to contract, suggesting that the pore remained closed (Di Lisa *et al.*, 1995). One recent study, however (Duchen *et al.*, 1998), in single rat myocytes showed transient depolarizations of single mitochondria discretely localized in the cells, possibly reflecting localized release of Ca^{2+} from the sarcoplasmic reticulum, which was then taken up by adjacent mitochondria, causing a transient drop in $\Delta\psi_m$, though not pore opening *per se*.

These results highlight the importance of determining changes in both $[Ca^{2+}]_m$ and $\Delta\psi_m$ at the subcellular level. Subpopulations of mitochondria situated, for example, at sites of Ca^{2+} release (sarcoplasmic reticulum) or entry (Ca^{2+} channel) into the cell may "see" a much higher local $[Ca^{2+}]$ than the bulk phase of mitochondria and thus show different rates of Ca^{2+} uptake. If this occurs in only a small proportion of the mitochondria, studies that measure only whole-cell fluorescence might miss the changes.

3. MITOCHONDRIAL CA^{2+} TRANSPORT DURING HYPOXIA AND REOXYGENATION

3.1. Importance of $[Ca^{2+}]_m$ during Hypoxia/Reoxygenation Injury

Ultrastructural, metabolic, and ionic changes that occur upon reperfusion or reoxygenation of previously ischemic or hypoxic tissues have received extensive investigation. Elevated calcium and free radical generation have emerged as the two main damaging agents, but their exact targets and relative importance are not established (e.g., see Tani, 1990; Bagchi *et al.*, 1997: Halestrap *et al.*, 1997). They are both, however, known to cause mitochondrial dysfunction (Crompton, 1990; Halestrap, 1994), a hallmark of irreversible cell injury (Shen and Jennings, 1972).

Mitochondria can take up huge amounts of Ca^{2+} (Crompton, 1990; Gunter and Pfeiffer, 1990) and thus could potentially remove toxic levels of Ca^{2+} from the cytosol. Unfortunately, such a Ca^{2+} accumulation can eventually damage mitochondria both by competing for ATP production and, more importantly, by inducing the MPTP. Whether such accumulation is a causal factor in triggering the transition from reversible to irreversible cell damage is far from clear, however.

The following sections present, among other things, evidence that dramatic changes occur in mitochondrial Ca^{2+} transporting pathways during hypoxia, namely, that the uniporter becomes largely inactive, whereas the Na^+/Ca^{2+} exchanger reverses direction to allow Ca^{2+} entry.

3.2. Single-Cell Model of Hypoxia/Reoxygenation

To study Ca^{2+} transport pathways under hypoxic conditions, researchers in Baltimore developed a single-myocyte model of hypoxia/reoxygenation injury (Stern *et al.*, 1988). So far no work has been published using this model outside of these laboratories, highlighting the model's originality and demanding technical requirements, but it has now been described several times since the original paper (e.g. Miyata *et al.*, 1992; Silverman *et al.*, 1994, 1997) and so will be described here only briefly. Single ventricular myocytes are placed in a specially developed chamber on the stage of an inverted fluorescence microscope. This "hypoxia chamber" allows both normal superfusion of myocytes under aerobic conditions and also hypoxic superfusion. Hypoxia is achieved by a stable laminar layer of ultra-high-purity argon to prevent back-diffusion of atmospheric oxygen, yielding a $PO_2 < 0.02$ torr. The superfusion buffer is also equilibrated with argon, and octanoic acid is commonly used as a respiratory substrate to ensure that the cells are completely dependent on aerobic metabolism.

3.3. Myocyte Morphological Changes

Ventricular myocytes from normal mammalian hearts have a characteristic rod-shaped appearance and are about $120\,\mu M$ long. They can be electrically stimulated to contract under normal conditions in the hypoxia chamber, and only myocytes capable of responding in time to the stimulus with no spontaneous contractions are used for experiments. Cells in these studies therefore have intact membranes and ion transport systems, allowing coupled excitation–contraction, a clear advantage over other studies that use either cell populations (consisting of dead cells, hypercontracted cells, and rod-shaped cells with normal or abnormal ion homeostasis) or resting cells, which cannot ensure intact excitation–contraction coupling.

In the single-cell model of hypoxia/reoxygenation injury, cells are initially stimulated to contract at low rates of electrical stimulation (higher rates must be avoided because they produce oxygen at the anode). Following hypoxia induction, myocytes initially maintain their ability to contract for about 30 minutes, after which time they stop contacting due to failure of the action potential (Stern et al., 1988); the time period varies slightly for each cell and may reflect differences in glycogen content or metabolism. After approximately 2 more minutes, myocytes shorten quickly to about two-thirds their original length, a process known as *rigor-contracture*, which reflects ATP depletion beyond a certain value (Bowers et al., 1992). If myocytes are then reoxygenated within 10 minutes all cells recover, as indicated by a partial relengthening and the ability to again respond in time to the electrical stimulus. With increasing times spent in rigor, progressively fewer cells recover, instead hypercontracting into rounded, dysfunctional forms, referred to subsequently as *hypercontracted cells* because they are not actually "dead" yet, being still capable of excluding trypan blue (Cave et al., 1996). Reoxygenation is normally continued for 15 minutes, however, so we don't know whether the cells would eventually recover partially or die. Figure 2 shows the relationship between time spent in rigor and cell recovery or hypercontracture, together with a schematic representation of morphological changes that occur during hypoxia and reoxygenation.

3.4. Changes in $[Ca^{2+}]_m$ during Hypoxia and Reoxygenation

Previously we and others have shown that $[Ca^{2+}]_m$ and $[Ca^{2+}]_c$ rise after prolonged hypoxia in hearts and myocytes (Allen et al., 1993; Miyata et al., 1992; Steenbergen et al., 1990). Figure 3 shows the time course for $[Ca^{2+}]_m$ changes in isolated rat ventricular myocytes. Figure 3A shows that following hypoxia induction no increase in $[Ca^{2+}]_m$ occurred prior to rigor development, but thereafter it increased significantly. Figure 3A presents data from all myocytes studied. If the cells are divided into two groups, depending on whether they recover on reoxygenation, it is clear from Figure 3B that cells which

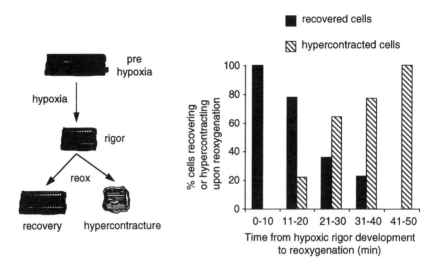

FIGURE 2. Myocyte morphology and cell recovery upon reoxygenation following hypoxic rigor development. The left panel shows a schematic of myocyte morphological changes. Approximately 30 min after induction of hypoxia, cells shortened to roughly 2/3 their original length. Upon reoxygenation (reox) cells either recovered, as indicated by maintenance of rod-shaped morphology and ability to respond to electrical stimulation, or hypercontracted into rounded, dysfunctional forms. The graph on the right shows data from myocytes subjected to hypoxia and reoxygenated at various times following rigor development. Cells were divided into groups of 10 min intervals according to time spent in rigor; $n = 99$ cells (9–22 cells in each group). Time to rigor varied slightly for each cell but there was no difference between the mean times for two groups. (Griffiths, Stern, and Silverman, unpublished data.)

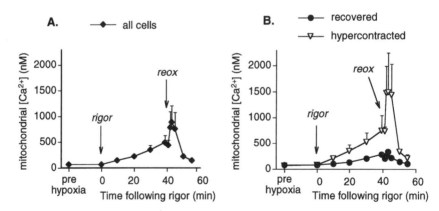

FIGURE 3. Mitochondrial [Ca^{2+}] changes during hypoxia and reoxygenation. Myocytes were subjected to hypoxia, then reoxygenated 40 min after rigor development, when approximately 50% of cells recovered. (A) Combined data from all myocytes studied. (B) Cells divided into two groups, depending on whether they recovered or hypercontracted upon reoxygenation. There is a significant difference between the two groups ($P < 0.05$, ANOVA). (Adapted from Griffiths et al., 1998).

recover maintain a much lower $[Ca^{2+}]_m$ both during hypoxia and upon reoxygenation. The same is true for $[Ca^{2+}]_c$ (Griffiths et al., 1998; Miyata et al., 1992). In fact, we found that extent of $[Ca^{2+}]_m$ increase during hypoxia determined cell recovery on reoxygenation; cells recovered if $[Ca^{2+}]_m$ remained below approximately 300 nM, whereas higher $[Ca^{2+}]$ caused hypercontracture (Griffiths et al., 1998; Miyata et al., 1992). This might at first appear to be a rather low $[Ca^{2+}]$, since it is well within the physiological range reported to occur upon inotropic stimulation of myocytes and whole hearts, and rather less than that required for half-maximal activation of mitochondrial dehydrogenases (Hansford, 1991; McCormack et al., 1990). It therefore appears that other factors compromise mitochondria during hypoxia, such as adenine nucleotide depletion and free radical generation upon reoxygenation.

A remaining question is what determines the extent of $[Ca^{2+}]$ increase for a given cell. The heterogeneity in the Ca^{2+}-handling response of individual myocytes to hypoxia has been described previously (Miyata et al., 1992) and is also seen with other ions such as Mg^{2+} (Silverman et al., 1994). The entry of Ca^{2+} into the cytosol during hypoxia is likely to occur via reversal of sarcolemmal Na^+/Ca^{2+} exchange (Haigney et al., 1992; Tani and Neely, 1989) resulting from Na^+ loading, either through Na^+ channels or through Na^+/H^+ exchange (Haigney et al., 1992; Harper et al., 1993). Differences among individual myocytes may therefore reflect alterations in one or more of these systems. Such differences are unlikely to be an artifact of the myocyte isolation procedure because this heterogenesous response is also observed in cultured myocytes (Silverman et al., 1997), where any damage to membrane proteins during the isolation procedure should have been repaired.

3.5. Route of Ca^{2+} Entry into Mitochondria during Hypoxia and Reoxygenation

It was assumed that fluxes of Ca^{2+} across the inner mitochondrial membrane followed the same pathways in hypoxic as in normoxic cells. This was based mainly on studies of isolated hearts and myocytes showing that ruthenium red (RR) was protective against ischemic/hypoxic damage at concentrations ranging from 0.1–6 µM (e.g., see Peng et al., 1980; Park et al., 1990; Figueredo et al., 1991), an effect attributed to RR's inhibition of the mitochondrial Ca^{2+} uniporter (Allen et al., 1993). Ruthenium red at such concentrations, however, also affects sarcoplasmic-reticular Ca^{2+} transport and hence inhibits cardiac contractile function (Gupta et al., 1988), which may therefore induce and "energy-sparing" effect in heart. So the protective effects of RR may be due to actions other than directly on $[Ca^{2+}]_m$. Of course, reduced $[Ca^{2+}]_m$ would then occur indirectly as a result of reduced $[Ca^{2+}]_c$. When performing experiments to determine the optimal ruthenium red concentration to use in normoxic myocytes (subjected to a Ca^{2+}-loading protocol), we found that a 20 µM concentration was required. We therefore used this concentration in hypoxic studies to determine whether ruthenium red could inhibit $[Ca^{2+}]_m$ increase.

Figure 4 shows the RR effect on $[Ca^{2+}]_m$ during hypoxia and reoxygenation of isolated rat myocytes. Control-hypercontracted data is shown in outline for comparison because all the RR-treated cells hypercontracted on reoxygenation. It is apparent that RR-treated cells actually exhibited higher $[Ca^{2+}]_m$ during rigor; the indo-1 fluorescence signal was saturated. The reasons for this are unknown, but $[Ca^{2+}]_c$ also increased in these cells,

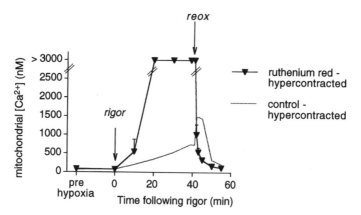

FIGURE 4. Effect of ruthenium red on mitochondrial [Ca^{2+} during hypoxia and reoxygenation. Myocytes were subjected to hypoxia as in Figure 1. When present, 20 μM ruthenium red (RR) was added 5 min prior to induction of hypoxia and remained present throughout. Data are compared with control-hypercontracted cells because all RR-treated cells hypercontracted upon reoxygenation. There was a significant RR effect both during hypoxia and upon reoxygenation ($P < 0.05$, ANOVA). (Adapted from Griffiths *et al.*, 1998).

implying a nonspecific toxic effect of ruthenium red (Griffiths *et al.*, 1998). What *is* clear is that ruthenium red is *not* inhibiting mitochondrial Ca^{2+} entry. Upon reoxygenation of the myocytes there was a very rapid fall in [Ca^{2+}]$_m$, which returned to pre-hypoxic values (Figure 4).

We then turned our attention to the mitochondrial Na$^+$/Ca^{2+} exchanger, the Ca^{2+} efflux pathway under normal conditions. We found previously that clonazepam could inhibit this pathway (mitochondrial Ca^{2+} efflux) in normoxic myocytes with very few nonspecific effects (Griffiths *et al.*, 1997b). Although the related compound CGP37157 is a more potent inhibitor in isolated mitochondria (Cox *et al.*, 1993), we found that it gave inconsistent results when used in intact myocytes (Griffiths *et al.*, 1997b). Figure 5 shows the effect of clonazepam on [Ca^{2+}]$_m$ during hypoxia and reoxygenation. Control-recovered data is shown for comparison because all clonazepam-treated cells recovered. Prior to rigor development there was a small, gradual [Ca^{2+}]$_m$ increase in the clonazepam-treated cells. This would be expected if the mitochondrial Ca^{2+} efflux pathway was inhibited, because a higher steady-state level of [Ca^{2+}]$_m$ would be achieved. Following rigor development, however, clonazepam-treated cells showed a very small further increase in [Ca^{2+}]$_m$ over the first 5 minutes, but then no further change. The hypoxic increase in [Ca^{2+}]$_c$ was unaffected by clonazepam (Griffiths *et al.*, 1998). Thus clonazepam inhibits hypoxic increase in [Ca^{2+}]$_m$, suggesting that Ca^{2+} influx during hypoxia following riaor development is via the Na$^+$/Ca^{2+} exchanger acting in reverse mode. Upon reoxygenation, however, [Ca^{2+}]$_m$ increased and remained elevated in the clonazepam group, whereas in control cells it returned to prehypoxic values. This suggests that upon reoxygenation, the Na$^+$/Ca^{2+} exchanger regains its normal directionality, because clonazepam is now inhibiting Ca^{2+} efflux. The Ca^{2+} entry upon reoxygenation is therefore again via the Ca^{2+} uniporter. In support of this, we also found that, in [Ca^{2+}]$_m$, increase in clonazepam-treated cells upon reoxygenation could be prevented by including ruthenium red in the reoxygenation buffer (Griffiths *et al.*, 1998).

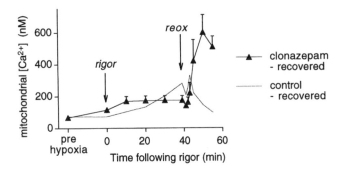

FIGURE 5. Effect of clonazepam on mitochondrial $[Ca^{2+}]$ during hypoxia and reoxygenation. Myocytes were subjected to hypoxia as in Figure 1. When present, 100 μM clonazepam was added 5 min prior to induction of hypoxia and remained present throughout. Data are compared with control-recovered cells because all clonazepam-treated cells recovered. There was a significant clonazepam effect both during hypoxia and upon reoxygenation ($P < 0.05$, ANOVA). (Adapted from Griffiths *et al.*, 1998).

3.6. Role of Ca^{2+} Transporters during Hypoxia

The above results demonstrate that Ca^{2+} entry into mitochondria during hypoxia is via the Na^+/Ca^{2+} exchanger, whereas the uniporter is inactive. These changes are summarized in Figure 6.

In retrospect our observations showing that the Ca^{2+} uniporter is largely inactive during hypoxia were perhaps not surprising. Using an identical model of myocyte hypoxia, Di Lisa *et al.* (1995) showed that $\Delta\psi_m$ (measured with fluorescent indicator JC-1) started to depolarize early in hypoxia, prior to development of rigor contracture, and reached a plateau before rising sharply again with the rigor development to a new, higher plateau that was maintained until reoxygenation. Thus the collapse of $\Delta\psi_m$ and likely collapse of the pH gradient during hypoxia provide conditions that would dramatically reduce uniporter activity. In isolated mitochondria, Ca^{2+} uptake is dependent on $\Delta\psi_m$; indeed, in one study when $\Delta\psi_m$ was dissipated, no uniporter-mediated Ca^{2+} uptake occurred even in presence

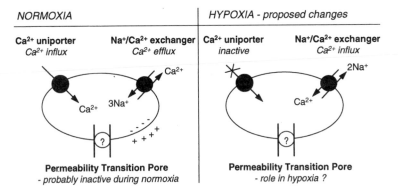

FIGURE 6. Alteraction of the mitochondrial $[Ca^{2+}]$ transport pathways during hypoxia.

of an 8-fold gradient of Ca^{2+}_{out}: Ca^{2+}_{in} (Kapus et al., 1991). In other cell types, a loss of $\Delta\psi_m$ and ΔpH elicited by uncouplers dramatically inhibited mitochondrial Ca^{2+} uptake via the Ca^{2+} uniport pathway (Rizzuto et al., 1992, 1993).

Our proposed reversal of the Na^+/Ca^{2+} exchanger during hypoxia is also entirely compatible with the properties of this channel. In isolated mitochondria, Ca^{2+} entry via the Na^+/Ca^{2+} exchanger occurs under conditions of membrane depolarization and in presence of Na^+ (Jung et al., 1995). The Ca^{2+} entry could be inhibited by diltiazem or by omission of NaCl, indicating that the Na^+/Ca^{2+} exchanger is indeed responsible. In addition, intracellular $[Na^+]$ increases during hypoxia (Haigney et al., 1992), and in de-energized mitochondria the Na^+ gradient across the membrane is approximately 2, compared with about 8 in the energized state (Jung et al., 1992). The Na^+/Ca^{2+} exchange under these conditions is likely to be electroneutral (Brand, 1985; Jung et al., 1995).

3.7. Activity of Mitochondrial Ca^{2+} Transporters upon Reoxygenation

The observation that $[Ca^{2+}]_m$ fell rapidly in RR-treated cells and rose in clonazepam-treated cells allows us to conclude that the transporters reverted to their normal directionality upon reoxygenation even in hypercontracted cells. Because mitochondria were once again capable of taking up Ca^{2+}, $\Delta\psi_m$ was probably restored. Again, this is in agreement with measurements of Di Lisa et al. (1995) showing that upon reoxygenation there was a rapid recovery of $\Delta\psi_m$ even in cells that hypercontracted, indicating that most mitochondria had intact, functioning membranes at that time. Restoration of normal directionality of both Ca^{2+} transport pathways upon reoxygenation can be accounted for by such recovery of $\Delta\psi_m$. As discussed in the preceding sections, the uniporter is highly dependent on $\Delta\psi_m$ and the Na^+/Ca^{2+} exchanger may also be driven by it, allowing electrogenic Na^+/Ca^{2+} exchange instead of electroneutral exchange in the de-energized state (Baysal et al., 1994; Jung et al., 1995).

3.8. Role of the Mitochondrial Permeability Transition Pore?

In Figure 6 the transition pore given as a Ca^{2+} transport pathway, although with many question marks about its role in both normoxia and hypoxia. As mentioned previously, the pore is unlikely to open under physiological conditions; it opens in isolated mitochondria under nonphysiological conditions of high $[Ca^{2+}]$, reduced ATP levels, and oxidative stress (Bernardi et al., 1994; Crompton, 1990; Halestrap, 1994), and also during reperfusion of ischemic rat hearts (Griffiths and Halestrap, 1995). In the latter case, however, it can reseal as the heart recovers (Halestrap et al., 1997). The pore will probably open in the isolated myocyte model, but here too it may reseal, again because $\Delta\psi_m$ is restored and mitochondrial Ca^{2+} fluxes can be accounted for by uniporter and exchanger activities. In more severe models of cell injury, however, such as prolonged chemical hypoxia, continued mitochondrial depolarization has been associated with loss of the sarcolemmal permeability barrier and cell death (Chacon et al., 1994). This may reflect permanent pore opening, which precedes cell death in hepatocytes exposed to oxidative stress induced by t-butylhydroperoxide (Nieminen et al., 1995; Qian et al., 1997). However, whether pore formation is a direct cause of cell death or merely associated with it remains to be established.

4. CONCLUSIONS

The work discussed above provides evidence that dramatic alterations occur in mitochondrial Ca^{2+} transport pathways during hypoxia; Ca^{2+} entry occurs via Na^+/Ca^{2+} exchange (the normal efflux pathway), whereas the Ca^{2+} uniporter, (the normal influx route) is largely inactive. Clonazepam, but not ruthenium red, provided protection against hypoxia/reoxygenation damage in this model. Clonazepam, however, though a useful tool for studying mitochondrial Ca^{2+} transport (at least in rat myocytes), cannot be used in whole animals because of its non-myocardial effects, mainly on the nervous system. More specific compounds are clearly needed, especially now it is apparent that Ca^{2+} transport pathways differ under normoxic and hypoxic conditions. A new ruthenium red derivative has recently been synthesized by Matlib and colleagues (1998), apparently with very few nonspecific effects in myocytes.

One drawback of the work described above is that $[Ca^{2+}]_m$ and $[Ca^{2+}]_c$ could not be measured in the same cell. Accomplishment of this, together with information regarding the subcellular behavior of individual mitochondria, will provide valuable information on the roles of Ca^{2+} transport pathways during myocardial injury. Studies using more-specific inhibitors would then determine whether these pathways are primary sites for protective intervention and would answer the question of whether abnormal mitochondrial Ca^{2+} homeostasis is a leading cause of cell injury or a secondary result.

ACKNOWLEDGEMENTS. Work described in this chapter was funded by the American Heart Association—Maryland Affiliate, and the British Heart Foundation.

REFERENCES

Allen, S. P., Darley-Usmar, V. M., McCormack, J. G., and Stone D., 1993, Changes in mitochondrial matrix free calcium in perfused rat hearts subjected to hypoxia/reoxygenation, *J. Mol. Cell. Cardiol.* **25**: 1461–1469.

Bagchi, D., Wetscher, G. J., Bagchi, M., Hinder, P. R., Perdikis, G., Stohs, S. J., Hinder, R. A., and Das, D. K., 1997, Interrelationship between cellular calcium homeostasis and free radical generation in myocardial reperfusion injury, *Chemico-Biol. Interact.* **104**: 65–85.

Baysal, K., Jung, D. W., Gunter, K. K., Gunter, T. E., and Brierley, G. P., 1994, Na^+-dependent Ca^{2+} efflux mechanism of heart mitochondria is not a passive $Ca^{2+}/2Na^+$ exchanger, *Am. J. Physiol.* **266**: C800–C808.

Bernardi, P., Broekemeier, K. M., and Pfeiffer, D. R., 1994, Recent progress on the regulation of the mitochondrial permeability transition pore; a cyclosporin sensitive pore in the inner mitochondrial membrane, *J. Bioenerg. Biomembr.* **26**: 509–517.

Bowers, K. C., Allshire, A. P., and Cobbald, P. H., 1992, Bioluminescent measurement in single cardiomyocytes of sudden cytosolic ATP depletion coincident with rigor, *J. Mol. Cell. Cardiol.* **24**: 213–218.

Brand, M. D., 1985, The stoichiometry of the exchange catalysed by the mitochondrial calcium/sodium antiporter, *Biochem. J.* **229**: 161–166.

Cave, A. C., Adrian, S., Apstein, C. S., and Silverman, H. S., 1996, A model of anoxic preconditioning in the isolated rat cardiac myocyte: Importance of adenosine and insulin, *Bas. Res. Cardiol.* **91**: 210–218.

Chacon, E., Reece, J. M., Nieminen, A.-L., Zahrebelski, G., Herman, B., and Lemasters, J. J., 1994, Distribution of electrical potential, pH, free Ca^{2+}, and volume inside cultured adult rabbit cardiac myocytes during chemical hypoxia: A multiparameter digitized confocal microscopic study, *Biophys. J.* **66**: 942–952.

Chacon, E., Ohata, H., Harper, I. S., Trollinger, D. R., Herman, B., and Lemasters, J. J., 1996, Mitochondrial free calcium transients during excitation–contraction coupling in rabbit cardiac myocytes, *FEBS Lett.* **382**: 31–36.

Cox, D. A., and Matlib, M. A., 1993, Modulation of intramitochondrial free Ca^{2+} concentration by antagonists of $Na^+–Ca^{2+}$ exchange, *Trends Pharmacol. Sci.* **14**: 408–413.

Cox, D. A., Conforti, L., Sperelakis, N., and Matlib, M. A., 1993, Selectivity of inhibition of $Na^+–Ca^{2+}$ exchange of heart mitochondria by benzothiazepine CGP-37157, *J. Cardiovasc. Pharmacol.* **21**: 595–599.

Crompton, M., 1990, The role of Ca^{2+} in the function and dysfunction of heart mitochondria, in *Calcium and the Heart* (G. A. Langer, Ed.), Raven, New York, 1990, pp. 167–198.

Di Lisa, F., Blank, P. S., Colonna, R., Gambassi, G., Silverman, H. S., Stern, M. D., and Hansford, R. G., 1995, Mitochondrial membrane potential in single living adult rat cardiomyocytes exposed to hypoxia or metabolic inhibition, *J. Physiol.* **486**: 1–13.

Duchen, M. R., Leyssens, A., and Crompton, M., 1998, Transient mitochondrial depolarizations reflect focal sarcoplasmic reticular calcium release in single rat cardiomyocytes, *J. Cell Biol.* **142**: 975–988.

Figueredo, V. M., Dresdner, K. P., Wolney, A. C., and Keller, A. M., 1991, Postischaemic reperfusion injury in the isolated rat heart: Effect of ruthenium red, *Cardiovasc. Res.* **25**: 337–342.

Griffiths, E. J., 1999, Mitochondrial calcium during the contractile cycle of isolated rat and guinea-pig cardiomyocytes, *J. Physiol, (abstract)*, in press.

Griffiths, E. J., and Halestrap, A. P., 1995, Mitochondrial non-specific pores remain closed during cardiac ischaemia but open upon reperfusion, *Biochem. J.* **307**: 93–98.

Griffiths, E. J., Stern, M. D., and Silverman, H. S., 1997a, Measurement of mitochondrial calcium in single living cardiomyocytes by selective removal of cytosolic indo-1, *Am. J. Physiol.* **273**: C37–C44.

Griffiths, E. J., Wei, S.-K., Haigney, M. C. P., Ocampo, C. J., Stern, M. D., and Silverman, H. S., 1997b, Inhibition of mitochondrial calcium efflux by clonazepam in intact single rat cardiomyocytes and effects on NADH production, *Cell Calcium* **21**: 335–343.

Griffiths, E. J., Ocampo, C. J., Savage, J. S., Rutter, G. A., Hansford, R. G., Stern, M. D., and Silverman, H. S., 1998, Hypoxia-induced alterations in mitochondrial transporting pathways in intact rat cardiomyocytes, *Cardiovasc. Res.* **39**: 423–433.

Gunter, T. E., and Pfeiffer, D. R., 1990, Mechanisms by which mitochondria transport calcium, *Am. J. Physiol.* **258**: C755–C786.

Haigney, M. C. P., Miyata, H., Lakatta, E. G., Stern, M. D., and Silverman, H. S., 1992, Sodium channel blockade reduces hypoxic sodium loading and sodium-dependent calcium loading, *Circ. Res.* **71**: 547–557.

Halestrap, A. P., 1994, Interactions between oxidative stress and calcium overload on mitochondrial function, in *Mitochondria: DNA, Proteins, and Disease* (V. Darley-Usmar and A. H. V. Schapira, Eds.), Portland Press, London, pp. 113–142.

Halestrap, A. P., Connern, C. P., Griffiths, E. J., and Kerr, P. M., 1997, Cyclosporin A binding to mitochondrial cyclophilin inhibits the permeability transition pore and protects hearts from ischaemia/reperfusion injury, *Mol. Cell. Biochem.* **174**: 167–172.

Hansford, R. G., 1991, Dehydrogenase activation by Ca^{2+} in cells and tissues, *J. Bioenerg. Biomembr.* **23**: 823–854.

Harding, S. E., Brown, L. A., Wynne, D. G., Davies, C. H., and Poole-Wilson, P. A., 1994, Mechanisms of β adrenoceptor desensitisation in the failing human heart, *Cardiovasc. Res.* **28**: 1451–1460.

Harper, I. S., Bond, J. M., Chacon, E., Reece, J. M., Herman, B., and Lemasters, J. J., 1993, Inhibition of Na^+/H^+ exchange preserves viability, restores mechanical function, and prevents the pH paradox in reperfusion injury to rat neonatal myocytes, *Bas. Res. Cardiol.* **88**: 430–442.

Jung, D. W., Apel, L. M., and Brierley, G. P., 1992, Transmembrane gradients of free Na^+ in isolated heart mitochondria estimated using a fluorescent probe, *Am. J. Physiol.* **262**: C1047–C1055.

Jung, D. W., Baysal, K., and Brierley, G. P., 1995, The sodium-calcium antiport of heart mitochondria is not electroneutral, *J. Biol. Chem.* **270**: 672–678.

Kapus, A., Szaszi, K., Kaldi, K., Ligeti, E., and Fonyo, A., 1991, Is the mitochondrial Ca^{2+} uniporter a voltage-modulated transport pathway? *FEBS Lett.* **282**: 61–64.

Li, W., Shariat-Madar, Z., Power, M., Sun, X., Lane, R. D., and Garlid, K. D., 1992, Reconstitution, identification, purification, and immunological characterization of the 110-kDa Na^+/Ca^{2+} antiporter from beef heart mitochondria, *J. Biol. Chem.* **267**: 17983–17989.

Marin-Garcia, J., and Goldenthal, M. J., 1994, Cardiomyopathy and abnormal mitochondrial function, *Cardiovasc. Res.* **28**: 456–463.

Matlib, M. A., Zhou, Z., Knight, S., Ahmed, S., Choi, K. M., Krause-Bauer, J., Philips, R., Altschuld, R., Katsube, Y., Sperelakis, N., and Bers, D. M., 1998, Oxygen-bridged dinuclear ruthenium complex specifically inhibits

Ca^{2+} uptake into mitochondria *in vitro* and *in situ* in single cardiac myocytes, *J. Biol. Chem.* **273**: 10223–10231.

McCormack, J. G., Halestrap, A. P., and Denton, R. M., 1990, The role of calcium ions in the regulation of mammalian intramitochondrial metabolism, *Physiol. Rev.* **70**: 391–425.

Miyame, M., Camacho, S. A., Weiner, M. W., and Figueredo, V. M., 1996, Attenuation of postischemic reperfusion injury is related to prevention of [Ca^{2+}]$_m$ overload in rat hearts, *Am. J. Physiol.* **40**: H2145–2153.

Miyata, H., Silverman, H. S., Sollot, S. J., Lakatta, E. G., Stern, M. D., and Hansford, R. G., 1991, Measurement of mitochondrial free Ca^{2+} concentration in living single cardiomyocytes, *Am. J. Physiol.* **261**: H1123–H1134.

Miyata, H., Lakatta, E. G., Stern, M. D., and Silverman, H. S., 1992, Relation of mitochondrial and cytosolic free calcium to cardiac myocyte recovery after exposure to anoxia, *Circ. Res.* **71**: 605–613.

Nieminen, A.-L., Saylor, A. K., Tesfai, S. A., Herman, B., and Lemasters, J. J., 1995, Contribution of the mitochondrial permeability transition to lethal injury after exposure of hepatocytes to *t*-butylhydroperoxide, *Biochem. J.* **307**: 99–106.

Park, Y., Bowles, D. K., and Kehrer, J. P., 1990, Protection against hypoxic injury in isolated-perfused rat heart by ruthenium red, *J. Pharm. Exp. Ther.* **253**: 628–635.

Peng, C. F., Kane, J. J., Straub, K. D., and Murphy, M. L., 1980, Improvement of mitochondrial energy production in ischemic myocardium by *in vivo* infusion of ruthenium red, *J. Cardiovasc. Pharmacol.* **2**: 45–54.

Qian, T., Nieminen, A.-L., Herman, B., and Lemasters, J. J., 1997, Mitochondrial permeability transition in pH-dependent reperfusion injury to rat hepatocytes. *Am. J. Physiol.* **42**: C1783–C1792.

Rizzuto, R., Simpson, A. W. M., Brini, M., and Pozzan, T., 1992, Rapid changes of mitochondrion Ca^{2+} revealed by specifically targeted recombinant aequorin, *Nature* **358**: 325–327.

Rizzuto, R., Brini, M., Murgia, M., and Pozzan, T., 1993, Microdomains with high Ca^{2+} close to IP$_3$-sensitive channels that are sensed by neighboring mitochondria, *Science* **262**: 744–747.

Saris, N.-E. L., Sirota, T. V., Virtanen, I., Niva, K., Penttila, T., Dolgachova, L. P., and Mironova, G. D., 1993, Inhibition of the mitochondrial calcium uniporter by antibodies against a 40-kD glycoprotein, *J. Bioenerg. Biomemb.* **25**: 307–312.

Shen, A. C., and Jennings, R. B., 1972, Myocardial calcium and magnesium in acute ischemic injury, *Am. J. Pathol.* **67**: 417–421.

Silverman, H. S., Di Lisa, F., Hui, R. C., Miyata, H., Sollot, S. J., Hansford, R. G., Lakatta, E. G., and Stern, M. D., 1994, Regulation of intracellular free Mg^{2+} and contraction in single adult mammalian cardiac myocytes, *Am. J. Physiol.* **266**: C222–C233.

Silverman, H. S., Wei, S.-K., Haigney, M. C. P., Ocampo, C. J., and Stern, M. D., 1997, Myocyte adaptation to chronic hypoxia and development of tolerance to subsequent acute severe hypoxia, *Circ. Res.* **80**: 699–707.

Stanley, W. C., Lopaschuk, G. D., Hall, J. L., and McCormack, J. G., 1997, Regulation of myocardial carbohydrate metabolism under normal and ischaemic conditions: Potential for pharmacological interventions, *Cardiovasc. Res.* **33**: 243–257.

Steenbergen, C., Murphy, E., Watts, J. A. and London, R. E., 1990, Correlation between cytosolic free calcium, contracture, ATP, and irreversible ischemic injury in perfused rat heart, *Circ. Res.* **66**: 135–146.

Stern, M. D., Silverman, H. S., Houser, S. G., Josephson, R. A., Capogrossi, M. C., Nichols, C. G., Lederer, W. J., and Lakatta, E. G., 1988, Hypoxic contractile faliure in rat heart myocytes is caused by failure of intracellular calcium release due to alteration of the action potential, *Proc. Natl. Acad. Sci. USA* **85**: 6954–6958.

Tani, M., 1990, Mechanisms of Ca^{2+} overload in reperfused ischemic myocardium, *Ann. Rev. Physiol.* **52**: 543–559.

Tani, M., and Neely, J. R., 1989, Role of intracellular Na^{2+} in Ca^{2+} overload and depressed recovery of ventricular function of reperfused ischemic rat hearts, *Circ. Res.* **65**: 1045–1056.

Trollinger, D. R., Cascio, W. E., and Lemasters, J. J., 1997, Selective loading of rhod-2 into mitochondria shows mitochondrial Ca^{2+} transients during the contractile cycle in adult rabbit cardiac myocytes, *Biochem. Biophys. Res. Commun.* **236**: 738–742.

Wendt-Gallitelli, M.-F., and Isenberg, G., 1991, Total and free myoplasmic calcium during a contraction cycle: X-ray microanalysis in guinea-pig ventricular myocytes, *J. Physiol.* **435**: 349–372.

Chapter 11

Mitochondrial Implication in Cell Death

Patrice X. Petit

1. INTRODUCTION

Apoptosis, ancient Greek, for "falling off" (Apo, "off" and ptôsis, "fall") is the name given to the process of physiological cell death. It was first described in 1972 (Kerr *et al.*, 1972). Apoptosis constitutes a strictly regulated ("programmed") process responsible for removal of superfluous, aged, or damaged cells. Numerous pathologies involve a deficient control of cell death. Abnormal resistance to induction of apoptosis entails malformations, autoimmune disease, or cancer due to the persistence of superfluous, self-specific, or mutated cells, respectively. In contrast, enhanced apoptotic decay of cells participates in acute diseases (infection by toxin-producing microorganisms, ischemia/reperfusion damage, infarction) as well as in chronic pathologies (neurodegenerative and neuromuscular diseases, AIDS).

It is currently assumed that the apoptotic process can be divided into at least three functionally distinct phases: induction, execution, and degradation (Kroemer *et al.*, 1995; Martin and Green 1995; Oltvai and Korsmeyer 1994; Thompson 1995). During the induction phase, cells receive apoptosis-triggering stimuli. Such death-inducers include ligation of specific receptors (Fas/APO-1/CD95, tumor necrosis factor receptor, glucocorticoid receptor, and so forth) or, in the case of obligate growth-factor receptors, the absence of receptor occupancy. In addition, apoptosis can be induced by interventions on second-messenger systems (Ca^{2+}, ceramide, kinases, and so on), suboptimal growth conditions (shortage of essential nutrients, hypoxia), mild physical damage (radiotherapy), and numerous toxins (reactive oxygen species, chemotherapy, and toxins *stricto sensu*) (Kroemer *et al.*, 1995; Thompson, 1995). Non-specific or receptor-mediated death

Patrice X. Petit   Institut Cochin de Génétique Moléculaire, INSERM U129—CHU Cochin Port-Royal, Paris, France.

Mitochondria in Pathogenesis, edited by Lemasters and Nieminen.
Kluwer Academic/Plenum Publishers, New York, 2001.

induction involves a stimulus-dependent ("private") biochemical pathway, and only after this initiation phase do the common pathways come into action. It is generally assumed that the execution phase of apoptosis defines the "decision to die" at the "point of no return" of the apoptotic cascade. At this level the private pathways coverage into one (or several) common pathway(s), and the cellular processes (redox potentials, expression levels of oncogene products including Bcl2-related proteins) have a decisive regulatory function. Once the cell is irreversibly committed to death, the various manifestations classically associated with apoptosis, such as DNA fragmentation, become detectable. This degradation phase is similar in all cell types. It is characterized by the action of catabolic enzymes, mostly specific proteases (caspases) and endonucleases, within the limits of a nearly intact plasma membrane. Thus the cell actively contributes to its own removal in a suicidal fashion and undergoes stereotyped biochemical and ultrastructural alterations. (Figure 1).

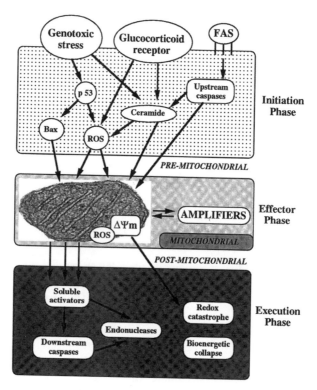

FIGURE 1. Phases of the apoptotic process. During the premitochondrial induction phase, different pro-apoptotic signal transduction "private" pathways are activated, private in the sense that the molecular events involved depend on initial apoptosis trigger. For instance, genotoxic stress involves p53 transcription and generation of ROS, which are not involved in Fas-induced apoptosis. The common mitochondrial execution phase is characterized by changes in mitochondrial function with increased membrane permeability. Amplifiers of these changes engage in a self-amplifying feedback loop. Amplifiers include caspase activation and local changes in voltage and redox potentials. The postmitochondrial degradation phase is characterized by the action of caspase and nuclease activators released from mitochondria, as well as by bioenergetic and redox changes collapsed mitochondrial functions.

This chapter examines recents knockouts and other complementary studies that shed light on mitochondrial involvement in apoptotic cell death. Through this renewed interest, mitochondria are coming once again to the forefront of research (Henkart and Grinstein, 1996) (mitochondria were pivotal in Nobel Prize-winning research defining the fundamental principles of cellular energy metabolism) this time in fields as diverse as cell death, evolutionary biology, molecular medicine—even forensic science (Kiberstis, 1999).

2. REACTIVE OXYGEN SPECIES AS MEDIATORS OF PROGRAMMED CELL DEATH

Reactive oxygen species (ROS) such as superoxide anions, hydrogen, organic peroxides, and radicals are naturally generated by all aerobic cells as by-products of many metabolite reactions (Fridovic, 1978). The ROS play an important role in physiological systems: They are responsible for the inductible expression of genes associated with inflammatory and immune responses. Evidence indicates that different stimuli use ROS as signaling messengers to activate transcription factors, such as AP-1 and nuclear factor-κB (NFκB), and to induce gene expression (Pinkus et al., 1996). Mitochondria are thought to be a major site of ROS production: Superoxide radicals are produced by a single electron transfer to molecular oxygen at the level of the respiratory chain, mainly at the ubiquinone site in complex III.

Since the initial observations of the ROS contribution to TNFα-induced cytotoxicity, evidence has grown that these products are central to cell death transduction pathways. Indeed, several recent studies suggest that ROS might mediate programmed cell death (PCD): Addition of ROS or depletion of endogenous antioxidants can promote cell death (Guénal et al., 1997; Kane et al., 1993; Lennon et al., 1991; Ratan et al., 1994; Sato et al., 1995); PCD can be delayed or inhibited by antioxidants (Greenlund et al 1995; Mayer, Noble 1994; Mehlen et al., 1996; Sandstrom and Buttke 1993; Wong et al., 1989); and an increase of intracellular ROS is sometimes associated with PCD (Martin and Cotter 1991; Quillet-Mary et al., 1997; Uckun et al., 1992). In addition, Bcl2 may act in an antioxidant pathway to block a putative ROS-mediated step in the events cascade required for PCD. The ROS contribute to the activation of execution machinery, and this contribution has been extended to PCD triggered by a wide range of influences, including UV light, ionizing radiation, anthracyclines, ceramides, glucocorticoids, and growth-factor withdrawal (Jacobson, 1996).

The contribution of ROS must be considered with caution. In most of these systems, it is difficult to know whether ROS accumulation corresponds to a causal effect and or is a postmortal manifestation of entropic processes. Moreover, in these cases ROS increase is often seen during the latter stage of the cell death program, that is, during the degradation phase when the cell is broken down, and may be associated with a necrotic-like terminal degradation of the cell. Exogenous sources of ROS such as hydrogen peroxide either can induce apoptosis or necrosis, depending upon the dose (Guénal et al., 1997; Lin et al., 1997). Because an ROS outburst in response to dramatic perturbation of the dying cell's of the physiology could convert late apoptotic steps into necrotic cell death, it appears that intracellular ROS level can determine the cell's fate at any moment: Low levels of ROS can induce PCD, and high levels promote necrosis or lead PCD-committed cells toward necrotic-like destruction.

2.1. Programmed-Cell-Death-Mediating ROS Are Produced Mainly by Mitochondria

One mode of mitochondria-mediated cell injury that has received considerable attention is the liberation of partially reduced species of molecular oxygen. Due to their high chemical reactivity, free radicals escaping from the respiratory chain or formed otherwise in the mitochondria should produce deleterious effects. Even under well-coupled conditions, as much as 2–4% of reducing equivalents escape the respiratory chain, liberating superoxide anion free radicals and subsequently, hydrogen peroxide. In addition, alternate electron acceptors (dioxygen) are reduced in the respiratory chain to form unstable intermediates that shuttle the unpaired electrons to molecular oxygen to complete a redox cycle. The ultimate products of the redox cycling are oxygen free radicals. Regardless of electron acceptor, availability of reducing equivalents in the respiratory chain may be the limiting factor for free radical generation (Boveris and Chance, 1973). As a consequence, the more reduced the respective electron transport components, the greater the oxygen free radical generation rate. The best-documented examples are situations where agents such as rotenone (complex I), antimycin A (complex III), and cyanide (complex IV) enhance redox potential via inhibition of the respiratory chain (Fig. 2). Adding these compounds to respiring mitochondria stimulates free radical generation.

The paradigm for poisoning by respiratory chain inhibitors and oxidative phosphorylation uncoupler is well established. Only recently, however, have we fully appreciated how complex the mitochondrial bioenergetics are, and realized that mitochondria contribute more to cell sustenance than just ATP synthesis. The multifaceted role of mitochondria in cell homeostasis is rooted in the protonmotive force (Δp) (Fig. 2C). Active proton extrusion from the mitochondrial matrix, which is coupled to electrons transport down a gradient of univalent redox potentials, creates both a chemical (ΔpH) and an electrical ($\Delta \Psi$) potential that enables bioenergetic activity. This provides not only the driving force for ATP synthesis but also the energy for phosphate uptake; it supports the transmembrane exchange of many intermediate metabolites, including reduced substrates, fatty acids, and amino acids for aerobic metabolism and exchange of matrix ATP with cytosolic ADP. It also supplies the energy for regulating mitochondrial calcium concentration and for importing proteins encoded by the nuclear genome. The significance of this shared bioenergetics is manifested at different levels: (1) The transmembrane movement of metabolites, ions, and small proteins produces a depolarizing current that competes with ATP synthase for the Δp (Fig. 2C). The spectrum of mitochondrial poisons is much broader than just the classic agents directly affecting components of the respiratory chain (Fig. 2A). Interference with these intricately regulated membrane transport and homeostatic processes is implicated in mitochondrial dysfunction and cell injury caused by a number of xenobiotic agents (Jones and Lash, 1993). (2) Competitive electrophoretic transport serves as the basis for linking mitochondria and important transducers in bioenergetic cell-signaling pathways. For example, mitochondria may actively participate in the mechanism of calcium signaling. Transient increases in cytosolic calcium are mirrored by increases in mitochondrial calcium and by episodic mitochondrial depolarizations (Ichas *et al.*, 1997). Associated with this are oscillating rates of oxidative

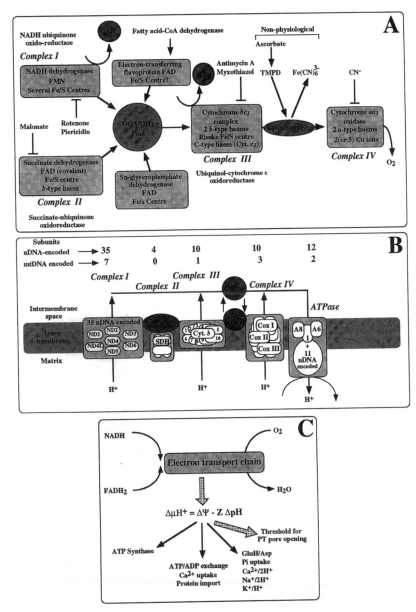

FIGURE 2. The mitochondrial electron transport chain and role of the mitochondrial protonmotive force within the cell. (A) Schematic of the electron transport chain, with emphasis on respiratory complexes I to IV, inhibitors of the respiratory chain and superoxide anion production sites. (B) The mitochondrial respiratory chain. All nuclear-DNA-encoded subunits are shown but only some mtDNA encoded subunits are shown (ND1–ND6 for complex I; Cyt = c for complex III; Cox I–III for complex IV; A8 and A9 for the ATPase). Protons are pumped from matrix to intermembrane space through complexes I, II, and IV, flowing into the matrix through the ATP synthetase, with concomitant ATP production. Coenzyme Q10 (CoQ10) and cytochrome c (Cyt-c) are fully nuclear-encoded electron transfer proteins. Small arrows show electrostatic Cyt-c interaction with inner mitochondrial membrane. (C) Mitochondrial protonmotive force supports a variety of bioenergetic and membrane transport processes and also acts as a PT-regulating threshold.

phosphorylation. Compelling kinetic evidence exists for the regulation of mitochondrial respiration by transient changes in cytosolic calcium, which manifest as altered activities of calcium-dependent, rate-limiting dehydrogenase enzymes within the matrix. A comparable paradigm for the strong influence of cytosolic pH on mitochondrial bioenergetics has also been postulated (Nieminen et al., 1990).

The nature of ROS involved in PCD is a question that brings us back to our main subject. Indeed, two opposite models have emerged concerning the source of ROS signaling in reaction to a variety of metabolic reactions and intracellular sites that can generate ROS (Jacobson, 1996). Most investigators believe that oxidants are produced by the electron transport chain (Fig. 2A). Fatty acid metabolites, such as those produced from arachidonic acid metabolites and issuing from the lipoxygenase pathway, may also mediate PCD (O'Donnell et al., 1995). On one hand it is argued that these molecules harbor a more specific reactivity than do superoxide anions and their byproducts, specificity being assumed necessary for a signaling role in PCD transduction pathways. On the other hand, exogenous fatty acid metabolites can themselves promote cell death. (The latter process is limited to systems where the death signal is mediated by surface receptors.) These observations, however, do not refute the compelling evidence of electron-chain-produced ROS involvement in cell death signaling. It certainly depends both on cell death stimulus and on cell type. Both types of ROS signaling can mediate PCD and contribute to the activation of execution machinery, as suggested by studies of TNFα-induced PCD.

Where are the electron transport chains situated that produce cell-death-signaling ROS, such as hydrogen peroxide? Here, one must consider the different intracellular compartments, such as nuclear membranes, endoplasmic reticulum, and especially the mitochondria. The most convincing arguments arise from indirect studies examining the consequences of an electron transport chain alteration upon initiation of programmed cell death. Both ROS accumulation and PCD processes require the presence of a functional respiratory chain in most ROS-dependent cell death systems (Higuchi et al., 1997; Quillet-Mary et al., 1997; Schulze-Osthoff et al., 1993). Inhibiting complex I (Quillet-Mary et al., 1997) or altering the electron transport chain by mtDNA depletion (Quillet-Mary et al., 1997; Schulze-Osthoff et al., 1993) prevent ROS production and accumulation and thus protect against PCD. The scavenger role of mitochondrial glutathione in regulating ROS-mediated PCD adds another indirect argument in favor of a mitochondrial origin of toxic ROS (Goossens et al., 1995). The ubiquinone site linked to complex III appears to be the major site of mitochondrial ROS production because it catalyzes the conversion of molecular oxygen to superoxide anion, which can lead to production of other potent ROS, such as the more diffuse hydrogen peroxide or hydrogen radical. This model (Fig. 2A) is supported by the clear potentialization of PCD that occurs when electron flow is inhibited at the ubiquinone-pool level.

2.2. Reactive Oxygen Species: Effects and Signaling

Much data converge upon the hypothesis that ROS will increase in response to impairment of the mitochondrial respiratory chain (Gudz et al., 1997; Quillet-Mary et al., 1997; Lin et al., 1997; Schulze-Osthoff et al., 1992). Related to the above considerations, observed electron transport alterations take place at complex III, which is proximal to the ubiquinone site, but the origin of electron flow disturbances is unclear. The only solid clue comes from the study of ceramide-induced PCD, in which increased H_2O_2 production (detected by 6-carbonyl-2′,7′-dichlorodihydrofluorecein diacetate, di[acetoxymethyl ester] [DCFH-DA]) was linked to mitochondrial Ca^{2+} perturbation. Inhibition of mitochondrial Ca^{2+} uptake abolished ROS accumulation and cell death (Gudz et al., 1997; Quillet-Mary et al., 1997).

In this context, the release of cytochrome c into the cytoplasm (Kluck et al., 1997a, 1997b; Liu et al., 1996; Yang et al., 1997) during the early phases of apoptosis (see below) could suggest a path to understanding. Indeed, cytochrome c release must lead to a breakdown of mitochondrial electron flow downstream of the ubiquinone site, in turn resulting in increased generation of superoxide anions and derived ROS. This hypothesis is reinforced by the correlation between cytochrome c loss and respiratory failure in Fas-induced apoptosis (Krippner et al., 1996).

Besides the question of mitochondrial ROS accumulation, there is the consideration of how ROS mediate PCD. Two models can be proposed. The first assumes that ROS are signaling molecules that activate crucial steps of the PCD machinery. The second proposes that ROS act indirectly by modifying cellular redox potential, which in turn activates key regulatory proteins involved in PCD. Most data suggest the indirect action.

Unlike fatty acid metabolites, which harbor specific reactivity and mediate particular surface-receptor signals, the mitochondrial superoxide anions and the hydroxyl radicals and hydroperoxide derivates are characterized by lack of biological specificity or even extreme, very localized reactivity—all features contrary to the requirements of specific signaling molecules (Jacobson, 1996). In that respect, a direct mitochondrially produced ROS influence would correlate with a general damaging effect on cellular structures, resulting in necrotic death or perhaps a more limited, localized action directly on mitochondria near their site of generation, which in turn could activate some mitochondria-dependent downstream cascades leading to apoptosis. Alternatively, despite compelling evidence of mitochondrial ROS's role in the PCD-signaling pathway, the prevailing idea is that they do not represent a general mediator of cell death, as suggested by the fact that PCD can occur at very low oxygen tension (Jacobson and Raff, 1995; Shimizu et al., 1995).

One approach to this problem is to consider that the major effect of increased ROS production is a subsequent decrease in availability of intracellular antioxidants, which leads to an imbalance of cell redox status, possibly the major effector of PCD. A nice illustration of this is the death caused by CD95 cross-linking or interleukin-3 (IL-3) withdrawal, which leads to GSH depletion (Bojes et al., 1997; Van den Dobbelsteen et al., 1996).

In addition, the observation that oxidation of thiols rather than glutathione can mediate apoptosis induction suggests that intracellular thiol redox status is the key factor of the cell-death-signaling pathway (Kane et al., 1993; Marchetti et al., 1997; Mirkovic et al., 1997; Sato et al., 1995). In this model the redox state of glutathione or other cellular antioxidants would be in equilibrium with that of thiols resident in crucial redox-sensitive components of the execution machinery (Kroemer et al., 1997b). Increased mitochondrial production of superoxide anions and reactive derivatives would result in a shift of SH-sensor-group(s) redox state toward increased oxidation, either directly through modification of thiols or indirectly via depletion of the intracellular antioxidant pool. The nature of the ROS and level of the intracellular antioxidant defenses would determine how regulatory components were activated to commit the cell to apoptosis or to necrosis.

3. MITOCHONDRIAL MEMBRANE POTENTIAL, PERMEABILITY TRANSITION, AND EARLY APOPTOSIS

3.1. Membrane-Potential Decrease: Universal Apoptotic Event

Multiple and major changes in mitochondrial biogenesis and function are associated with commitment to apoptosis. A drop in $\Delta\Psi_m$ occurs before DNA fragmentation in oligonucleosomal fragments (Petit et al., 1995; Vayssière et al., 1994; Zamzami et al., 1995b). This drop is responsible for a defect in maturation of mitochondrial proteins synthesized in the cytoplasm (Mignotte et al., 1990), inhibition of mitochondrial translation, and uncoupling of oxidative phosphorylations (Vayssière et al., 1994). The $\Delta\Psi_m$ drop has been extensively observed. Pioneering studies refer to apoptosis induced by dexamethasone (Petit et al., 1995); apoptosis induced by activation of peripheral T-cells, T-hybrids, and pre–B cells (Zamzami et al., 1995a, 1995b); apoptosis induced by TNFα in U937 or HeLa cells (Marchetti et al., 1996a); and apoptosis of neurons deprived of nerve growth factor (Deckwerth and Johnson 1993; Marchetti et al., 1996a). Concerning neuronal death, one might remember a pioneering work entitled *Temporal Analysis of Events Associated with Programmed Cell Death (Apoptosis) of Sympathetic Neurons Deprived of Nerve Growth Factor*, by Deckwerth and Johnson (1993), the first observation of implied mitochondrial metabolism in terms of $\Delta\Psi_m$ loss. They write, "Sympathetic neurons accumulated Rh123 in tubular cytoplamic organelles reminiscent of mitochondria as imaged by confocal microscopy. The initial rate of Rh123 uptake decreased with a time course virtually identical to the MTT reduction and preceded the onset of neuronal death." These two independent measures of mitochondrial function suggest a modest but reproducible decrease of $\Delta\Psi_m$ before loss of viability implying that generation rates of reduction equivalents (NADH) and high-energy phosphates (ATP, GTP) are diminished. It is therefore detectable, whatever the induction signal. In conclusion, it appears that the $\Delta\Psi_m$ drop, which can be slowed by the presence of cyclosporin A, is a universal characteristic that accompanies apoptosis independent of induction signal and of cell type (Kroemer et al., 1995). Nuclear fragmentation is a late event compared with $\Delta\Psi_m$ loss; however, this loss determines the point of no return of the cell death process.

3.2. Kinetic Data Implicating Mitochondria in Apoptosis

Lipophilic cationic fluorescent probes accumulate in the mitochondrial matrix, driven by the electrochemical gradient, first at the plasma membrane, next at the Δp through the mitochondrial membrane, according to the Nernst equation, where every 61.5-mV increase in membrane potential (usually between 120 and 170 mV, negative inside) corresponds to a ten-fold accumulation of dye within the mitochondria. Therefore, the concentration of such cations is 2 to 3 logs higher in the matrix than in the cytosol. Lipophylic cations such as 3,3′-dihexyloxacarbocyanine iodide [DiOC(6)3]; Rhodamine 123; 5,5′,6,6′-tetrachloro-1,1′,3,3′-tetraethylbenzimidazol carbocyanine iodide (JC-1); chloromethyl-X-rosamine

(CMXRos); or tetramethylrhodamine methylester (TMRM) can be employed for flow-cytometric determination of $\Delta\Psi_m$ (Métivier *et al.*, 1998; Petit *et al.*, 1995; Rottenberg and Wu, 1998). Cells are equilibrated with very low dye concentrations around the nM range, fully adequate to first, stain cells and second, accurately detect subtle $\Delta\Psi_m$ changes. In these conditions cell fluorescence effectively appears to correlate with $\Delta\Psi_m$ magnitude, as shown by fluorescence sensitivity to low concentrations of uncouplers, ionophores, and inhibitors of the mitochondrial proton pumps. It is evident that plasma membrane potential also affects cell fluorescence and should be taken into account (Rottenberg and Wu, 1998).

A disruption of $\Delta\Psi_m$ resulting in diminished dye uptake precedes all major changes in cell morphology and biochemistry, including DNA fragmentation and phosphatidylser-ine exposure in the outer leaflet of the plasma membrane (see Rottenberg, this book) (Fig. 3).

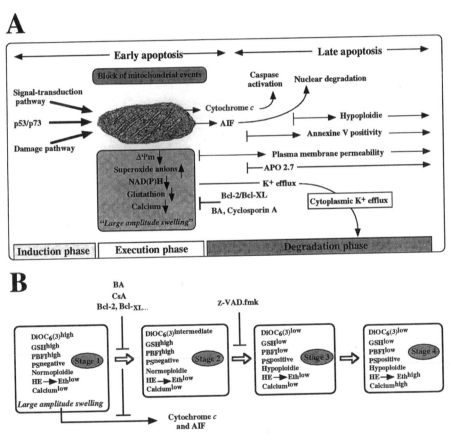

FIGURE 3. Two different representations of the sequence of events during apotosis. (A) Successive events leading to final DNA fragmentation with early apoptosis (induction phase, execution phase, and late-apoptosis events corresponding to the degradation phase, which ends "late nuclear apoptosis"). (B) The sequence of events easily measured by flow cytometric analysis. GSH, Glutathion; DiOC$_6$ (3), the representative mitochondrial membrane-potential dye; PBFI, potassium; marker; PS; phosphatidyl residues detected with annexine V-FITC; HE; hydroethydine. Calcium content measured by Fluo-3-AM.

3.3. Permeability-Transition Modulation and Apoptosis

When studying loss of $\Delta\Psi_m$ and of other mitochondrial functions during the early phase of apoptotic commitment, we considered at least two hypotheses. Alteration of $\Delta\Psi_m$ could be attributed to nonspecific damage to mitochondrial membranes (outer and inner) or to a more specific process. Such a process is the *permeability transition* (PT), which is due to the opening of a regulated proteaceous pore or "megachannel", also called the *permeability transition pore* (PTP) (Bernardi and Petronilli, 1996; Kinnaly et al., 1996; Zoratti and Szabò, 1995). The PTP is a dynamic multiprotein complex probably located at the contact sites between mitochondrial membranes (Beutner et al., 1998, 1996; Marzo et al., 1996b; O'Gorman et al., 1997). The PTP complex is believed to involve cytosolic protein (hexokinase), outer-membrane proteins (peripheral benzodiazepin receptor, PBR; porin, also called *voltage-dependent anion channel*, or VDAC), intermembrane protein (creatine kinase), at least one inner membrane protein (the adenine nucleotide translocator, ANT), and at least one matrix protein (cyclophilin D) (Bernardi and Petronilli, 1996; Beutner et al., 1996; Brustovetsky, 1996; McEnery, 1992; Zoratti and Szabò, 1995) (Fig. 4). Depending on the specific physiological effectors, the PTP complex can adopt an open or closed conformation. Substances that act specifically on mitochondrial structures to induce transition can trigger cell death. Thus protoporphyrin IX (PPIX), a ligand of the peripheral benzodiazepin receptor (PBR) and a constituent of the PTP complex, induces apoptosis in lymphoid cells (Marchetti, 1996a, 1996b) and necrosis in hepatocytes (Pastorino et al., 1994). Similarly, the protonophore carbonyl cyanide m-chlorophenylhydrazone (mCICCP), which incorporates itself into the inner mitochondrial membrane to disrupt $\Delta\Psi_m$, causes apoptosis in lymphoid cells (Susin et al., 1996. Pharmacological transition inhibitors can prevent cell death in many different models. Thus, drugs designed to prevent transition via specific interaction with the adenine nucleotide translocator (bongkrekic acid), matrix cyclophilin D (cyclosporin A and N-methyl-4-Val-cyclosporin A), or matrix thiols (chloromethyl-X-rosamine) can also inhibit cell death in different cell types. In addition to preventing death-associated $\Delta\Psi_m$ disruption, these drugs prevent all postmitochondrial manifestations of apoptotic cell death at the levels of redox balance, phosphatidyl-serine exposure, and activation of proteases and nucleases (Castedo, 1996; Marchetti, 1996a, 1996b). These pharmacological studies indicate that transition-mediated $\Delta\Psi_m$ disruption is a critical coordinating event of the death process.

4. PERMEABILITY TRANSITION AND APOPTOSIS-INDUCING FACTOR

Mitochondrial alterations can directly induce apoptosis. (Hartley et al., 1994; Wolvetang et al., 1994). The relationship between mitochondrial perturbations and subsequent nuclear alterations can be studied using an acellular system where purified nuclei and purified mitochondria are confronted (Newmeyer et al., 1994) (Table I). Such experiments demonstrate that when mitochondria are treated with substances capable of triggering pore opening, they induce nuclear apoptosis characterized by chromatin condensation and DNA fragmentation (Zamzami et al., 1996). There is a strict correlation between transition induction and nuclear apoptosis as observed using a variety of transition inductors, such as calcium, atractyloside, pro-oxidants, protono-

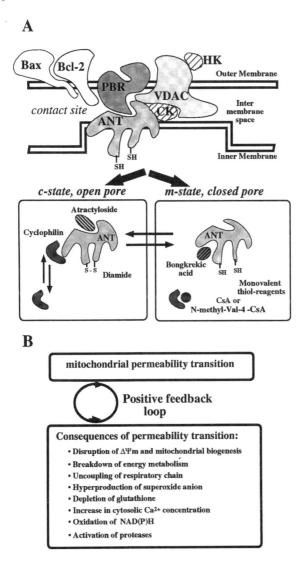

FIGURE 4. Hypothetical model of the PT pore complex and consequences of the self-amplifying mechanism. (A) The PT pore complex comprises proteins from cytosol (e.g., HK, hexokinase; which facilitates the PT), outer mitochondrial membrane (PBR, peripheral benzodiazepin receptor; VDAC, voltage-dependent anion channel) intermembrane space (CK, creatine kinase, which inhibits the PT), inner membrane (ANT, adenine nucleotide translocator) and matrix (cyclophilin D). Exact stoichiometry and interaction sites of these proteins are unknown. Additional unknown proteins (Bcl2?) may participate in complex regulation. ANT conformation, influenced by endogenous (ADP, ATP) and exogenous (atractyloside, bongkrekic acid) ligands, co-determines pore-opening probability. Cyclophilin D interaction with inner membrane proteins is also critical for pore opening; interaction is inhibited by cyclosporin A (CsA) and nonimmunosuppressive CsA derivative N-met-Val-4-CsA. It is not clear whether cyclophilin D interacts with the ANT or other inner membrane structures. Pore opening is facilitated by oxidation of vicinal thiols at the matrix site of an inner mitochondrial membrane protein. This oxidation is prevented by monovalent thiol-reactive agents such as monochlorobiman or chloromethyl-X-rosamine. It remains elusive whether these critical thiols are located within the ANT or other, unknown pore components. (B) Positive feedback loop linked to pore opening, with description of the main events related to the opening.

Table I
Cell-free Systems and Mitochondrial Apoptosis[a]

Cell-free systems	Date	Needs	References
Xenopus egg extracts	1994	Requirement of a mitochondrial-enriched fraction	(Newmeyer et al., 1994)
		Inhibition by Bcl-2	
	1995	Fas, UV, and ceramide apoptosis	(Martin et al., 1995)
	1995	Cytosolic extract of Fas-activated cells	(Enari et al., 1995)
	1995	Extract from execution-phase cells	(Lazebnik et al., 1995)
	1996	Fas-mediated apoptosis	(Enari et al., 1996)
	1997	Cytochrome c and CPP32	(Kluck et al., 1997b)
	1998	Role of the BH3 domain of the Bcl2 protein family	(Cosulich et al., 1998)
	1998	Caspase-8	(Kuwana et al., 1998)
Thymocytes	1996	Ca²⁺ thymocyte nuclei	(McConkey 1996)
HeLa cells (lacking mtDNA)	1996	Mitochondria depleted in mtDNA treated with PT pore-inducers (Ca²⁺, Atractyloside or tert-butylhydroperoxide) behave as normal mitochondria	(Marchetti et al., 1996a, 1996b).
	1996	Mitochondria treated with PT pore inducers (Ca2⁺, Atractyloside, or tert-butylhydroperoxide)	(Zamzami et al., 1996)
	1996	dATP and cytochrome c	(Liu et al., 1996)
	1996	AIF (mice)	(Susin et al., 1996)
	1996	Decreased pH or increased Ca²⁺	(Collins et al., 1996)
	1997	Bcl2 blockage of cytochrome c release	(Yang et al., 1997)
	1997	DFF	(Liu et al., 1997)
	1997	FLICE	(Muzio et al., 1997)
	1998	2-chloro-2'-deoxyadenosine-5'-trisphosphate and cytochrome c	(Leoni et al., 1998)
	1998	HL-60 cell extract (30 Gy of 137 Cs gamma rays)	(Kurihara et al., 1998)
HL-60	1997	Reactive oxygen species (ROS)	(Lin et al., 1997)
	1998	p210 Bcr-Abl protein	(Dubrez et al., 1998)
HeLa cells or 2B11 T hybridoma cells	1997	Ceramide-induced apoptosis, with AIF release	(Susin et al., 1997)

[a]For Pioneering, work on cell-free systems and apoptosis, see Jones et al., 1989, Lazebnik et al., (1993, 1994).

phores, and substances that provoke linkage of thiol residues. These substances exhibit no direct effects on nuclei in the absence of mitochondria, inducing apoptosis only when mitochondria are present. The pro-apoptotic characteristic of mitochondria treated with calcium or atractylate is abolished by transition inhibitors such as bongkrekic acid or cyclosporin A. We know that N-met-4-Val-CsA (nonimmunosuppressive analogue) can replace cyclosporine A, which shows that its inhibitory effect on the permeability transition and subsequently on nuclear apoptosis is independent of its calcineurin activity. These results suggest that pore opening is implicated in the regulation of apoptosis induced via the mitochondrial compartment.

Mitochondria originating from apoptotic cells *in vivo* can induce nuclear apoptosis in the acellular system. For example, induction of mouse hepatocyte apoptosis *in vivo* by a combination of D-galactosamine and lipopolysaccharide entails the reduction of $\Delta\Psi_m$. Mitochondria isolated in these conditions provoke apoptosis of HeLa cell nuclei *in vitro*. Similar results were obtained with mitochondria isolated from spleen cells treated with dexamethasone. In these systems mitochondria can effectively control nuclear apoptosis. (Zamzami *et al.*, 1996).

5. CYTOCHROME *c* RELEASE AND INDUCTION OF CASPASES

The dogma of cytochrome *c*'s unique simplicity of function was destroyed when Lui and colleagues (1996) discovered its involvement in apoptosis (Fig. 5). Indeed, contrary to the textbooks, which confine cytochrome *c* to the mitochondrial intermembrane space and limit its role to oxidative phosphorylation, it is now known that cytochrome *c* leaves the intermembrane space and interacts with cytosolic components, thereby stimulating the final phase of apoptotic degradation phase (Table II). This is clearly linked to a pathological situation, because cytochrome *c* is usually thought to reside in the intermembrane space, shuttling electrons between complexes III and IV on the inner membrane. Effectively, a so-far unique transport pathway is utilized by cytochrome *c* (Fig 6). After synthesis in the cytosol as a heme-free apoprotein, it binds specifically to the mitochondrial outer membrane where it is translocated. Both binding and translocation may involve participation of a specific outer-membrane component unrelated to the Tom complex (Mayer *et al.*, 1995b). As with presequence-containing preproteins (Mayer *et al.*, 1995a), the membrane passage of an apoprotein is reversible. (Mayer *et al.*, 1995b). Unidirectional transport is achieved by its stable binding to cytochrome *c* heme lyase. This enzyme is peripherally associated with the inner mitochondrial membrane, and it catalyses covalent attachment of heme to the apoprotein (Dumont *et al.*, 1991).

A cell-free system made up of HeLa cells, nuclei, and cytoplasmic extracts from apoptotic cells was at the origin of the discovery of cytochrome *c*. It was found that adding dATP to the extract resulted in prototypic apoptotic changes, including cleavage of caspase-3 precursors, yielding active caspase-3 and a subsequent DNA fragmentation associated with specific morphological changes in the cell nucleus. Fractionation of the extract revealed at least three factors necessary to the above effects. Cytochrome *c* was one. Other experiments found that treating HeLa cells with staurosporine, a potent inducer of apoptosis, causes release of cytochrome *c* from intermembrane space to

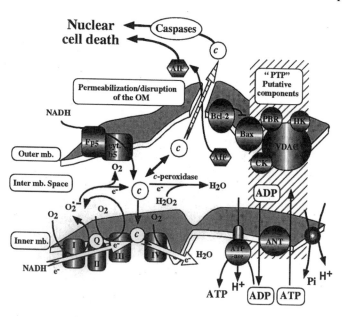

FIGURE 5. Possible role of cytochrome c in the mitochondriol antioxidant defense system of and its relation to apoptosis and the permeability transition. Presumably cytochrome c desorbed from inner mitochondrial membrane to intermembrane space oxidizes superoxides produced at complexes I and III of the respiratory chain. Cytochrome c receives electrons from an external respiratory chain composed of cytochrome b_5 (Cyt-b_5) and NADH–cytochrome b_5 reductase localized in outer membrane, by passing O_2-producing complexes I and III. Cytochrome c reduced by O_2 or b_5 is reoxidized by cytochrome c oxidase or cytochrome c peroxidase. If this fails to prevent large-scale O_2 production, Bax-mediated release of cytochrome c to the cytosol is assumed to be initiated. Release may be due to cytochrome c translocation by Bax or, more probably, to Bax-mediated outer membrane disruption. Cytosolic cytochrome c activates the apoptotic cascade, and as a result, superanions hyperproducing cells are discarded by apoptosis. AIF is also released from intermembrane (Intermb, space during the MPT. Bcl2 combines with and inactivates Bax, and may attenvate apoptosis by preventing PT opening, maybe by binding also to cytochrome c. I, III and IV: Respiratory chain complexes c and b_5, Cytochrome c and cytochrome b_5; Fp$_5$; NADH cytochrome b_5 reductase; Bcl-2 and Bax, anti- and pro-apoptotic proteins, respectively; Apaf-1, apoptosis activating factor-1. Concerning the megachannel: the ANT localized in the inner membrane, interacts with proteins inserted in the outer membrane, such as peripheral benzodiazepin receptor (PBR) and the voltage-dependent anionic channel (VDAC/porin). The structural correlate of this is the–outer membrane contact site.

cytosol. Similar results were obtained with human embryonic and monoblastic cells lines (Table I). The following year these observations (Liu *et al.*, 1996) were confirmed and extended by the groups of Wang (Bossy-Wetzel *et al.*, 1998; Kuwana *et al.*, 1998; Liu *et al.*, 1997) and Kluck *et al.*, (1997a, 1997b).

6. PERMEABILITY PERTURBATION AND CYTOCHROME c RELEASE

At least two competing theories have been advanced. The first is based on the strong correlation between the mitochondrial permeability transition and the early events of apoptosis (Kroemer *et al.*, 1995; 1997; Petit *et al.*, 1996, 1997). Transition is responsable

Table II
Cytochrome *c* Release in Apoptosis

Cell types or tissues (or acellular system)	Inducing agents	References
HL60	Staurosporine Etoposide Ara-C Paclitaxel Photodynamic therapy C2-ceramide Sphingomyelin Sphingamine	(Kim *et al.,* 1997; Yang *et al.,* 1997). (Amarante-Mendes *et al.,* 1998; Granville *et al.,* 1998)
CEM	Staurosporine Etoposide Actinomycin D Hydrogen peroxide UVB	(Kluck *et al.,* 1997a)
U937	Ionizing radiation Cisplatinum	(Kharbanda *et al.,* 1997)
Jurkat	Anti-Fas Staurosporine	(Stridh *et al.,* 1998; Van der Heiden *et al.,* 1997)
	H2O2 Tributylin	(Petit *et al.,* 1999)
FL 5.12	IL-3 withdrawal	(Van der Heiden *et al.,* 1997)
Cerebellar granule neurons (CGN)	MPP$^+$	(Du *et al.,* 1998)
Sympathetic neurons	NGF deprivation + Cyclohexamide	(Deshmuth and Johnson, 1998)
SCG neurons (rat)	NGF withdrawal	(Neame *et al.,* 1998)
Differentiated CGN	Low K$^+$ conditions	(Gleichmann *et al.,* 1998)
Rat brain	Transient focal cerebral ischemia	(Fujimura *et al.,* 1998)
HUVEC cybrids of Alzheimer disease (AD cybrids)	Defect in complex IV	(Cassarino *et al.,* 1998)
Xenopus egg extracts	Caspase-8	(Kuwana *et al.,* 1998)

for the opening of a large-conductance, for cyclosporin A-inhibitable channel located at the contact site of the mitochondrial membranes, causing dissipation of the electrochemical proton gradient and osmotic swelling of the mitochondrial matrix. Swelling ruptures the outer membrane, because the inner membrane surface area with its cristae is considerably larger than the surrounding outer membrane. Thus one model of how cytochrome *c*, normally confined to intermembrane space, escapes from mitochondria takes into account the outer membrane disruption (Petit *et al.,* 1998), that occurs as a secondary consequence of pore opening.

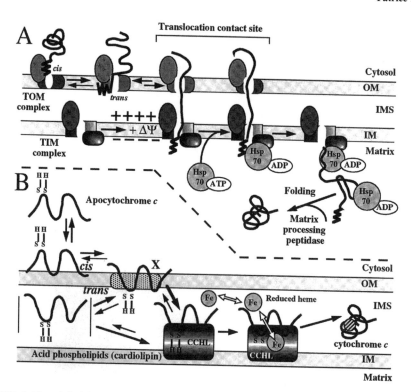

FIGURE 6. The original import pathway involved in the biogenesis of cytochrome *c*. Despite fundamental differences, common principles underlie both the apocytochrome *c* translocation and the general import routes of other mitochondrial preproteins (A). A reversible membrane passage is preceded and followed by steps introducing both specificity and unidirectionality to the reaction. Correct targeting is usually verified by the specific interaction of preproteins with *cis*-side membrane components. For apocytochrome *c*, the intermembrane-space (IMS) targeting accuracy appears guaranteed by two sequential steps, (1) binding to outer membrane (OM) and (2) to cytochrome *c* heme lyase (CCHL). Membrane translocation is reversible. This system is remarkable in that the entire preprotein can undergo a retrograde translocation, implying that apocytochrome *c* either passes through the lipid bilayer or uses a porelike structure entered and traversed from either side. Two reactions are the driving force for import into mitochondria: interaction with CCHL, and conversion to the holoform (a CCHL-catalyzed reaction). Heme attachment assures dissociation of holocytochrome *c* from CCHL, presumably triggered by flooding into the native structure. In summary, despite its unique import pathway, apocytochrome *c* obeys the general principles of other membrane translocation systems. IM, Inner membrane; SH, Thiol groups of cysteines 14 and 17.

A second model predicts a specific channel located in the outer membrane that allows release of cytochrome *c*. A good candidate was identified in Bax, a pro-apoptotic member of the Bcl2 family. The authors argue that the tridimensional structure of Bcl-x_L suggests that many Bcl2 family members are similar to the diphtheria toxin, and more precisely to toxin's pore-forming domain. The presumed function of this domain is to allow the toxin's ADP-ribosylating subunit to extrude from the lysosomes interior into the cytosol.

By analogy, some Bcl2 family proteins could form channels large enough to transport proteins, even when their helicoidal character (α-helice) is more convenient for small-ion transport. Adding recombinant Bax protein to artificial membranes results in channels that generally have a small conductance (Table III) and which may, under certain circumstances, become quite large (up to a 2-ns conductance) (Antonsson *et al.*, 1997; Schlesinger *et al.*, 1997). Moreover, when expressed in yeast, Bax induces mitochondrial cytochrome *c* release and confers a lethal phenotype (Manon *et al.*, 1997), supporting a possible direct role for Bax in controlling cytochrome *c* release.

Reports that cytochrome *c* release from the mitochondrial intermembrane space can precede dissipation of $\Delta\Psi_m$ across the inner membrane argue in favor of a specific-channel hypothesis. This suggest that cytochrome *c* can escape the intermembrane space *before* the PT opening (Kluck *et al.*, 1997a, 1997b; Yang *et al.*, 1997). It is not clear whether a specific channel is necessary or if partial swelling and simple outer membrane alteration are sufficient for these results (note that mitochondrial mitoplasts without no outer membrane can still build up a $\Delta\Psi_m$.

A different model stems from observing the direct effect of activated caspases on the mitochondrial membrane (Susin *et al.*, 1997a). This event sequence might entail cytochrome *c* release followed first by caspase activation, *then* by caspase-induced pore opening, triggered by proteolytic events (Reed, 1997). In isolated mitochondria, caspases induce first the pore opening, next the cytochrome *c*, release before destroying the membrane (Marzo *et al.*, 1998c), contrary to previous data. Moreover, isolated and purified mouse liver mitochondria undergo transition in the presence of calcium (200 nmol), tert-butylhydroperoxide (50 µM), or atractyloside (10 µM + 20 nmol Ca^{2+}), and in this study, cytochrome *c* and AIF were released only when the outer membrane was broken (Petit *et al.*, 1998).

Table III

Single-Channel Characteristics of Bcl2 Family Members Reconstituted in Planer Lipid Bilayers[a]

Protein	Ion selectivity	pH	Voltage (mV)	Salt (mM)	Conductance state (pS)	References
Bcl2	Cation	7.4	100	500 KCl	20 40 90	(Schendel *et al.*, 1997)
Bcl-x_L	Cation	7.2	20	150/50 KCl	80 134 179 276	(Minn *et al.*, 1997)
Bax	Cation	7.4	100	125 NaCl	26 80 180 250 2 (nS)	(Antonsson *et al.*, 1997)
Bax	Anion	7.2	40	450/150 KCl	329	(Schlesinger *et al.*, 1997)

[a]Channel conductance expressed in siemens (S), reciprocals of ohms. Channel conductance (g) is given by the relation: $g = I/V$, where I is current flowing through an open channel in response to applied voltage (V). Membrane potential expressed in mV. For a 20-pS conductance channel a 2-pA current would flow across the lipid bilayer at a 100-mV membrane potential. An ampere represents a flow of 6.24×10^{18} charges/sec. The result is about 10 million ions flowing through the open 20-pS channel (Clapham, 1998).

6.1. Cytochrome *c*: Execution-Phase Caspase-Activation Regulator

The major advance in understanding caspase activation has come from a recently established sequence of the main component, apoptosis protease activating-factor-1 (Apaf-1) which, in combination with cytochrome *c* (described initially as Apaf-2) and another cytosolic factor (Apaf-3), is required to induce caspase-3 activation in an apoptotic cell-free system (Zou *et al.*, 1997). The Apaf-1 sequence is similar to the sequence of CED-4 *(Caenorharditis elegans)* (Table IV). In particular, the amino-terminus of Apaf-1 probably constitutes a *ca*spase-*r*ecruitement domain (CARD), which is found in a number of cell-death proteins, such as CED-4 and CED-3, and it may bind directly to caspases (Hofmann *et al.*, 1997). It is well known that CED-4, acting as a context-dependent ATPase, promotes CED-3 autoprocessing (Chinnaiyan *et al.*, 1997a), and in mammalian cells a CED-4-like activity could act as a bridge between Bcl-x_L and caspases with large prodomains, such as caspase-1 and caspase-8 (Chinnaiyan *et al.*, 1997b). Consequently, Apaf-1 seems to be the mammalian equivalent of CED-4, but a more sophisticated one with a putative protein–protein interaction domain in the carboxy-terminus tail of the protein. This domain, absent in CED-4, could be involved in the physical interaction between Apaf-1 and cytochrome *c*. In this model binding of mitochondrially released cytochrome *c* to Apaf-1 would activate the downstream caspase or remove an intrinsic inhibitory activity (Fig. 7). Although the mechanism for mitochondrial cytochrome *c* release remains to be determined, some basic observations suggest that cytochrome *c*-mediated PCD is distinct from the above-mentioned AIF-mediated PCD (Susin *et al.*, 1996). Many arguments reinforce this. First, though AIF release occurs after transition-associated mitochondrial membrane depolarization, some authors have reported that cytochrome *c* can be extruded from mitochondria *before* the $\Delta\Psi_m$ drop occurs (Adachi *et al.*, 1997; Kluck *et al.*, 1997a, 1997b; Krippner *et al.*, 1996; Yang *et al.*, 1997). One should keep in mind, however, that a $\Delta\Psi_m$ disruption occuring in a subpopulation of mitochondria (i.e., 10–20%), or a $\Delta\Psi_m$ modulation of 10–20 mV, is undetectable by the methods used, hence cytochrome *c* release by this fraction of mitochondria could escape observation.

Table IV
Homologous Proteins in Programmed Cell Death in Nematodes and Mammals[a]

C. *elegans* proteins	Type or function	Mammalian homologues
CED-3	Initiation caspases	caspase-1 (ICE)
		caspase-4 (ICH-2)
		caspase-6 (Mch-2)
		caspase-8 (MACH/FLICE)...
	Execution caspases	caspase-2 (ICH-1)
		caspase-3 (CPP32)
		caspase-4 (ICH-2)
		caspase-7 (ICE-ALP3)...
CED-4	Apoptosome	Apaf-1 (forms complex with Apaf-2 and Apaf-3)
CED-9	Anti-apoptotic	Bcl2, Bcl-x_L, Bcl-w, BFL-1, Brag-1, Mcl-1, A1, NR13...
	Pro-apoptotic	Bax, Bak, Bcl-x, Bad, Bid, Hrk...

[a]Mammalian caspases act during either activation or initiation phase of PCD (Nicholson and Thornberry, 1997). Recently, a CED-4 homolog was found and purified in human cells (Hoffmann *et al.*, 1997; Zou *et al.*, 1997a). In mammals seom Bcl2 family members are death agonists and others are death antagonists. (For related reviews, see Kroemer, 1997b and Reed *et al.*, 1998.)

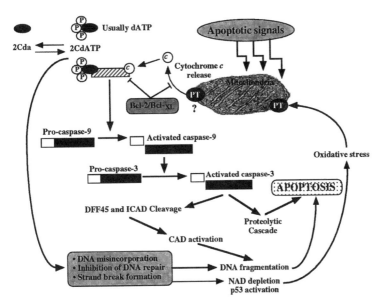

FIGURE 7. Activation of the apoptotic pathway by adenine deoxynucleotides and cytochrome c: The role of Apaf-1. In an apoptotic program artificially induced by 2-chloro-2'-deoxyadenosine-5'-triphophate and cytochrome c, an adenine desoxynucleotide such as 2CdA enters the cell and is converted to its active form, 2CdATP (usually, dATP is said to activate Apaf-1 pathway) (Zou *et al.,* 1997a), directly causing DNA strand-break formation and activating poly (ADP ribose) polymerase and p53; resulting NAD^+ and adenine nucleotides depletion is associated with a concomitant oxidative stress increase. Normally in cells exposed to elevated ROS levels mitochondria release internal dATP and cytochrome c concentrations through the PT pore (The ? marks the debate on how cytochrome c leaves mitochondria), which within the cytoplasm are *per se* insufficient to trigger caspase activation. However, binding of Apaf-1 to caspase-9 in the presence of dATP or 2CdATP and cytochrome c leads to cleavage and activation of caspase-3 (active autocatalytic protease). Active caspase-3 in turn stimulates the CAD endonuclease, which irreversibly degrades DNA, conferring upon the cell the nuclear apoptotic phenotype (Redrawn from Leoni *et al.,* 1998).

6.2. Apoptosis-Inducing Factor and Mitochondrial Cell Death

In addition to cytochrome c, mitochondrial intermembrane space contains other proteins liberated through outer membrane that participate in the degradation phase of apoptosis. Susin *et al.* (1996) first described one such protein, AIF, which can induce apoptosis of isolated nuclei in a cell-free system. The AIF flavoprotein has a relative molecular mass 57 kDa (although the translocation product of AIF, cDNA, is approximately 67 kDa *in vitro*, imported into mitochondria this gives rise to a shorter protein corresponding to the mature AIF), which shares homology with bacterial oxidoreductases. The AIF protein is normally confined to intermembrane space, as is cytochrome c but it translocates to the nucleus when apoptosis is induced and when cytochrome c translocates to the cytosol. Recombinant AIF causes chromatin condensation in isolated nuclei and large-scale fragmentation of DNA, as determined by pulse-field gel electrophoresis. Percoll-purified mitochondria are induced by AIF to release the apoptogenic proteins cytochrome c and caspase-9. Microinjection of AIF into intact cell cytoplasm induces mitochondrial membrane-potential dissipation, phosphatidylserine exposure at the outer

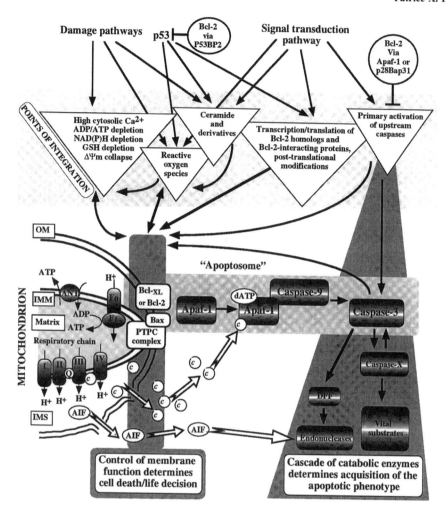

FIGURE 8. Schematic of apoptosis in mammalian cells. Activation of the different apoptotic pathways can have two consequences: (1) perturbation of mitochondrial function and integrity via activation of the mitochondrial permeability transition pore complex (PTPC) and/or of the Bax/Bcl2 complex in mitochondrial outer membrane (OM) at the contact site between membranes; or (2) primary activation of the caspase cascade. Bcl2, Bax, and their functional homologues regulate the MPT via their channel properties and/or by indirect effects on PTPC components. Perturbation of OM function and structure leads to leakage of pro-apoptotic molecules (cytochrome c) and AIF from intermembrane space (IMS) into the cytosol, where they activate caspases and endonucleases. Cytochrome c interacts with Apaf-1 released from Bcl2. The cytochrome-cApaf-1caspase-9 complex then activates caspase-3 proteolytically. These molecules and their structural homologues are the mammalian equivalent of the *Caenorhabditis elegans* apoptosome complex (CED-9, CED-4, CED-3). Subtrates upon which caspases act to perturb membrane integrity, (thereby amplifying mitochondrial structure, function, and caspase activation) are unknown. PTPC opening strongly affects energy metabolism and causes superoxide anions generation. Cytochrome c release interrupts electron transfer between respiratory chain complexes III and IV (not shown), thereby compromising respiratory function and increasing production of superoxide anions.

plasma membrane surface, and chromatin condensation. None of these effects is prevented by the wide-ranging caspase inhibitor known as *Z-VAD.fmk*. The effect of Z-VAD.fmk on mitochondrial supernatants containing AIF results from a Z-VAD.fmk-inhibitable enzymatic activity that cleaves the caspase substrate ZVAD.afc, an activity a least partly due to the presence of activated caspase-9 (Susin *et al.*, 1998). Overexpression of Bcl2 (which controls pore opening) prevents release of AIF from the mitochondrion but does not affect its apoptogenic activity; Bcl2 also inhibits AIF's translocation from mitochondria to nucleus without affecting its properties (Susin *et al.*, 1999). These results clearly show that AIF is a mitochondrial effector of apoptotic cell death, which provides a new molecular link between mitochondrial membrane permeabilization/disruption and nuclear apoptosis, and it also indicates that caspases, DEF/CAD, and AIF probably engage in complementary redundant pathways leading to nuclear apoptosis (Fig. 8).

7. THE Bcl2 FAMILY: COUNTERACTING MITOCHONDRIA PRO-APOPTOTIC SIGNALS

The Bcl2 protein and other related proteins such as Bcl-x$_L$ (CED-9) are negative regulators of cell death, able to prevent the PCD induced by various stimuli in a wide range of cells types (Korsmeyer 1992; Zhong *et al.*, 1993; reviewed in Kroemer 1997b; Reed *et al.*, 1998). Just how these proteins modulate apoptosis is not yet fully understood, however, and until now it was unclear to what extent mitochondrial Bcl2 localization accounted for its cytoprotective effects.

Nuclei and mitochondria have been extracted and purified from hybridomas of T-cells transfected by Bcl2 to study how it suppresses apoptosis *in vitro* (Zamzami *et al.*, 1996). Treatment with atractyloside (a ligand of the adenine nucleotide translocator) shows that, in contrast to mitochondria purified from control cells, mitochondria from Bcl2-transfected cells do not induce nuclear apoptosis. On the contrary, these nuclei show chromatin condensation and DNA fragmentation when confronted with control mitochondria treated with atractyloside. Furthermore, Bcl2 inhibits induction of the permeability transition by agents such as oxidants, atractyloside, and protonophores. These results show that even if Bcl2 also acts during later (postmitochondrial) cytoplasmic events (Guénal *et al.*, 1997), its major inhibitory effect is exerted at transition (Decaudin *et al.*, 1997).

The structure of one protein of the Bcl2 family, Bcl-x$_L$, was established (Muchmore *et al.*, 1996) as being similar to that of diphteria toxin, which forms a pH-sensitive transmembrane channel. Furthermore, the pro-apoptotic Bax protein can also form channels (Antonsson *et al.*, 1997), as can the anti-apoptotic proteins Bcl-x$_L$ (Minn *et al.*, 1997) and Bcl2 (Schendel *et al.*, 1997). The intrinsic properties of Bax and of Bcl-x$_L$ reveal differences, however. The channel-forming activity of Bcl-x$_L$ and Bcl2 is observed at highly acidic pH, whereas Bax forms channels at a wide range of pH levels including those close to neutral, as found in cells. It also appears that Bcl2 can block Bax's pore-forming activity. These results strengthen the hypothesis that these proteins are constituents of the mitochondrial PT pore (Marzo *et al.*, 1998a). When preparing a contact site's enriched fraction (assumed to reflect the pore's properties and containing most its components), Bax is clearly found to be part of the pore complex (Marzo *et al.*, 1998a). Bax might promote cell death by allowing flux of ions and small molecules across the mitochondrial membranes, thus triggering the permeability transition, whereas Bcl2 counteracts this effect. It has also been reported that Bax can trigger cytochrome *c* release from isolated mitochondria (Eskes *et al.*, 1998). This pathway is described as being distinct from the calcium-inductible, cyclosporin A-sensitive PT. Rather, Bax-induced cytochrome *c* release is facilitated by magnesium and cannot be blocked by transition inhibitors. This strongly suggests the existence of two mechanisms leading to cytochrome *c* release: one stimulated by calcium and cyclosporine A inhibitable, the other Bax-dependent and magnesium-sensitive but cyclosporine-insensitive. These two pathways needs further study.

Cytochrome *c* release into the cytosol induces nuclear apoptosis. We know that Bcl2 inhibits this release (Kluck *et al.*, 1997a, 1997b; Yang *et al.*, 1997). Thus one level of Bcl2's action controls the efflux of cytochrome *c*, AIF (Susin *et al.*, 1996), and the mitochondrial caspases pool (caspase-2, -3 and -9) (Mancini *et al.*, 1998; Susin *et al.*, 1998). It remains to be learned whether pore-opening consequences mediate this efflux (Petit *et al.*, 1998) or whether channel activity linked to Bax is the mediator (Eskes *et al.*, 1998).

The Bcl-x_L protein inhibits Ara C-induced pre-apoptotic accumulation of cytochrome *c* in the cytosol (Kim *et al.*, 1997). Cells that overexpress Bcl-x_L do not accumulate cytochrome *c* or undergo apoptosis in response to genotoxic stress. Co-immunoprecipitation studies show that cytochrome *c* binds to Bcl-x_L and not to pro-apoptotic Bcl-x_s. Thus, Bcl-x_s blocks binding of cytochrome *c* to Bcl-x_L (Kharbanda *et al.*, 1997). These findings support the hypothesis that both Bcl-x_L and Bcl2 protect cells from apoptosis by inhibiting release of apoptogenic intermembrane proteins.

8. TOWARD AN ENDOSYMBIOTIC THEORY OF APOPTOSIS ORIGIN AND EVOLUTION

The mitochondrion appears to fulfill the minimum requirements for being a central executioner of apoptosis, requirements based on the criteria of chronology, convergence, coordination, ubiquity, vitality, the "switch", and a final functional criteria (Kroemer, 1997a). Moreover, transition appears to be a decisive apoptosis mechanisms. This presents major implications for the phylogeny of apoptosis such as the idea that mitochondria's endosymbiotic origin may relate the appearance of symbiosis with the appearence of PCD. Mitochondria's well-known prokaryotic feature led to the hypothesis that this organelle evolved from bacteria that were endocytosed. According to the *endosymbiotic theory* (Margulis, 1996; Taylor, 1974), eukaryotic cells appear as anaerobic organisms before establishing an endosymbiotic relationship with aerobic bacteria, whose oxidative phosphorylation system they then subvert for their own use. With time the genes of these ancient organelles transferred to the nucleus of the host cell (as, for example, with the genes coding for glycolytic and Calvin-cycle glyceraldehyde-3-phosphate dehydrogenase) (Henze *et al.*, 1995). The host–invader exchange of genetic information continued after initial endosymbiosis, and in fact, continues still. Mitochondria's central role in the execution phase of eukaryotic PCD suggests that the appearance of apoptotic machinery may have been contemporary with the initial endosymbiosis.

The well-known paradigm of the *Neisseria* species *N. gonorrhoeae* and *N. meningitidis* is interesting here. These two species enter human phagocytes and epithelial cells (Kupsch *et al.*, 1993). Proteins with a crucial role in the *Neisseria* infection mechanism rely on porins (outer membrane proteins), which can translocate to the host membranes (Weel and van Putten, 1991). Surprisingly, these pathogenic bacteria porins are regulated in a manner similar to that of eukaryotic mitochondrial porins (VADC): Both interact with purine nucleoside triphosphates, down-regulating pore opening and causing a shift in voltage dependence and ion selectivity (Rudel *et al.*, 1996). Mitochondrial and *Neisseria* porins also share structural similarities in that β-sheeting is a major transmembrane motif in both, and both cases the β-sheet forms an amphipathic β-barrel structure that penetrates the lipid bilayer. These similarities suggest a common origin. Membrane-inserted porins are also tighly regulated by the host cell, suggesting the involvement of cellular factors (Rudel *et al.*, 1996). Another example in which a porinlike mechanism seems implicated in host invasion is the *Bdellovibrio species*, a group of areobic, predatory bacteria that penetrate gram-negative bacteria, causing lysis of the invaded organism. Host membrane preparations obtained within minutes of invader attack show the presence of a protein with properties similar to the *Bdellovibrio* porin OmpF (Tudor and Karp, 1994). Moreover, there is an interaction between host and invader: While *vibrio* are inside the bacterial host, signals from the soluble fraction of the host regulate activation of *Bdellovibrio's* attack phase.

About 2 billion years ago, the evolution of eukaryotes involved a merging of archaebacterial with eubacterial cells in anaerobic symbiosis. Nucleocytoplasm from the Archaea acquired a swimming motility (from the fermenting Eubacteria) becoming "mastigotes" prior to the acquisition of mitochondria or plastids. Some mastigotes further incorporated purple, Eubacteria (mitochondrial precursors) to become oxygen-respiring aerobes from which protists, fungi, and animals evolved (Margulis, 1996). The alliance was probably uncertain and unstable, and catastrophic conflicts of selection between the two genomes undoubtedly occurred. As soon as the new symbiotic organism moved into the aerobic world, life and death were likely controlled by pro-mitochondria, which provided not only antioxidants to face O_2 damages but also deadly reactive oxygen species, the by-products of oxidative phosphorylation. Conditions that favored the symbiont would have led to cell death and dissociation of pro-mitochondria from the symbiosis, assuming that at this stage pro-mitochondria were still genetically autonomous. Therefore, the symbiosis was perilous until mitochondria lost their autonomy, and before the genes essential for mitochondrial metabolism and biogenesis transferred to the nuclear genome of the host cell and ultimately resulted in obligatory symbiosis. One can hypothesize that the endosymbiotic origin of mitochondria and the evolution of aerobic metabolism in eukaryotes formed the roots of the cell death machinery, which manifests predominantly as apoptosis in metazoans (Frade and Michaelidis, 1997; Kroemer, 1997a). It is now clear that mitochondria play a central role in apoptosis, allowing an elucidation of the various molecular suspects in this process (Green, 1998). These include signal transduction pathways, the execution/degradation machinery with the intervention of multiple caspases and pro-apoptotic proteins (Liu *et al.*, 1996), and the apoptosis-inducing factor AIF (Susin *et al.*, 1996, 1999).

Possibly the porins that were translocated to host cell membrane by endosymbiotic bacteria were able to "check the health" of the host cell by sensing its ATP levels, because

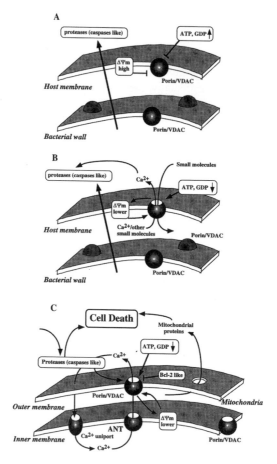

FIGURE 9. Speculative model of the evolutive origin of programmed cell death. (A) A Porinlike membrane channel (Porin/VADC) from engulfed invader bacteria that was translocated to the surrounding host membrane and was normally kept closed by both high nucleotides concentration and high $\Delta\psi_m$. The host cytosol contained inactive proteases (caspase-like) released from the invader. (B) In nonoptimal environmental conditions, nucleotide concentration decreased, porin opened, and $\Delta\psi_m$ decreased at the host-membrane level. The decreased, $\Delta\psi_m$ kept the porin open, initiating a transient destabilization in the membrane vinicity, which the host, already under potentielly fatal stress, could ill afford. A potential imbalance of Ca^{2+} concentration may have been implicated in this process, possibly via activation of inactive proteases. (C) The primitive eukaryote had to acquire proteins such as the ANT to exploit energy from the symbiont. In the newly established symbiosis, nucleotides drops could initiate an apoptosis-like process. Porins in mitochondrial outer (former host) membrane and ANT in mitochondrial inner (former bacterial) could somehow interact to form a transition pore with which other nearby proteins were associated, intervention in pore regulation was dependent on ancient mitochondrial $\Delta\psi_m$. The pore normally stayed closed under physiological conditions, but opening triggered a breakdown of $\Delta\psi_m$ that kept it open. This process was probably accompanied by calcium release into the cytosol, leading to calcium cycling with active uptake by the mitochondrial Ca^{2+} uniport, as occurs today. During this process mitochondrial components from intermembrane space and maybe from matrix were also released into the cytosol, some capable of inducing cell death, as happens now with cytochrome c and AIF. High cytosolic Ca^{2+} concentrations could also activate either caspases or endonucleases, leading to death, and caspases could also act upstream of $\Delta\psi_m$ dissipation. Molecules such as Bcl2 and anti-apoptotic Bcl2 family members in outer membrane inhibited cell death by interacting with PT pore components. During evolution the host–symbiont information exchange resulted in transfer protease-encoding genes from mitochondria to nucleus. (Redrawn from Frade and Michaelidis, 1977.)

the elevated purine nucleotide concentration kept the porin closed. High ATP levels would indicate the presence of abundant catabolites resulting from the anaerobic metabolism of the host (Fig. 9A). These catabolites could then serve as substrates for the aerobic metabolism of the invader. If ATP concentration decreased, the porins inserted into host cell membranes would open and the accompanying membrane-depolarization-associated processes would lead to the death of the host (Fig. 9B). A main consequence of porin opening could be flux of Ca^{2+} into host cytoplasm, which in turn would initiate a cascade of events including protease and endonuclease activation. Activation of the protease cascade would allow the invader to leave the dying host, take avantage of the host's liberated nutrients, or both (Fig. 9C). Thus by analogy, caspase activation during the PCD process might be a remnant of the mechanisms used by the former invaders to initiate a sequence of events leading to "digestion" of the host cell. This fits with the fundamental role of caspases during PCD, which may involve a cascade of many types of proteases (Kumar and Harvey, 1995).

One of the main assumptions of the endosymbiotic theory is that the host organism subverted the invaders' oxidative metabolism for its own benefit. Biochemical communication had to take place between host and invader. This could have been effectively established by the appearance of molecules such as adenosine nucleotide translocators in the ancient mitochondrial inner membrane (Fig. 9C), enabling the host to exploit invader-produced ATP. The special situation resulting from this could have triggered a special status favorable to the appearance of permeability transition pores (VDAC channels were already present). In such a primitive eucaryote, low ATP levels could trigger opening of the VDAC, generating permeability-transition-associated phenomena and resulting in an apoptotic-like process. Supporting this idea, modulation of ATP levels is postulated to be involved in regulating PCD (Richter *et al.*, 1996). Another important point is the appearance of Bcl2 family molecules able to regulate the permeability transition—crucial for ensuring survival of the former eukaryote. In fact, it is very likely that such molecules were already present in the host cells, probably involved in regulating Ca^{2+} release from the endoplasmic reticulum, as they do today.

Based on this hypothesis, the origin of eukaryotic PCD could be viewed in terms of molecular cross talk between bacteria and host. Eukaryotic PCD can be envisioned as the result of aerobic metabolism acquisition during early eukaryotic evolution. Nevertheless, PCD may not be limited to eukaryotes, because distinct types of PCD occur in prokaryotic cells as well. Further studies are needed on, for example, the presence or absence of apoptosis-like cell death in organisms lacking mitochondria (e.g., *Giardia intestinalis*) and on the existence of a PCD, including mitochondrial events, in fungi and plants.

9. CONCLUSIONS

First, the mitochondria's key involvement in the endosymbiotic theory provides research pathways leading to an evolutionary origin of apoptosis. According to this theory, apoptosis could have arisen simultaneously with and as a by-product of endosymbiosis (Frade and Michaelidis, 1997; Kroemer, 1997a), the genesis of mitochondria and chloroplasts.

Second, the theory provides a new method for assessing early apoptosis: determination of mitochondrial membrane-potential size. It also has advantage of measuring *in situ* events that occur long before cells are removed by heterophagy.

Third, the theory has major implications for our understanding of apoptosis regulation in both physiology and pathology. It is tempting to speculate on the links between nuclear and mitochondrial mutations. (Wong and Cortopassi, 1997) and increased apoptotic decay leading to degenerative disorders. Moreover, changes in metabolic functions induced by the Bcl2 family may explain specific features of tumors cells.

Mitochondrial involvement in apoptosis and necrosis blurs the distinction between necrosis and apoptosis. Increasing evidence suggests that mitochondria are involved in both modes of cell death. The intensity of mitochondrial disruption of structure and functions associated with changes in cellular energetics as well as with nonmitochondrial functions determines whether the cells lyse before activation of caspases and endonucleases (necrosis) or after activation of these enzymes (apoptosis), a discovery yielding important clues for the development of cytotoxic and cytoprotective drugs. Cytoprotective drugs must act at the premitochondrial or mitochondrial stage of apoptosis. Pharmacological compounds that interfere only with post-mitochondrial effects will likely be inefficient, but cytotoxic agents affecting the putative megachannel, the release of apoptogenic factors, or both—whatever the cause of their release—could facilitate the induction of cell death and overcome apoptosis resistance in cancer cells.

ACKNOWLEDGEMENTS. We would like to thank our colleagues, Drs. Castedo, Brunner, Decaudin, Marchetti, Marzo, Susin, Zamzami (CNRS, Vilejuif, France), Guénal, Mignotte, Vayssière (CNRS, CGM, Gif-sur-Yvette, France), and Mrs. Casselyn (ICGM, Paris, France), for their helpful discussions. Special thanks to Guido Kroemer for a critical reading of the manuscript. This work has been supported by ARC, ANRS, CNRS, INSERM, LNCC, the Leo Foundation, Sidaction, and the French Ministry of Science.

REFERENCES

Adachi, S., Cross, A. R., Babior, B. M., and Gottlieb, R. A., 1997, Bcl-2 and the outer mitochondrial membrane in the inactivation of cytochrome *c* during Fas-mediated apoptosis, *J. Biol. Chem.* **272**: 21878–21882.

Amarante-Mendes, G. P., Kim, C. N., Liu, L., Huang, Y., and Perkins, C. L., 1998, Bcr-Abl exerts its anti-apoptotic effect against diverse apoptotic stimuli through blockage of mitochondrial release of cytochrome *c* and activation of caspase-3, *Blood* **5**: 1700–1705.

Antonsson, B., Conti, F., Ciavatta, A., Montessuit, S., and Lewis, S., 1997, Inhibition of Bax channel-forming activity by Bcl-2, *Science* **277**: 370–372.

Bernardi, P., and Petronilli, V., 1996, The permeability transition pore as a mitochondrial calcium release channel: A critical appraisal, *J. Bioenerg. Biomembr.* **28**: 129–136.

Beutner, G., Rück, A., Riede, B., Welte, W., and Brdiczka, D., 1996, Complexes between kinases, mitochondrial porin, and adenylate translocator in rat brain resemble the permeability transition pore, *FEBS Lett.* **396**: 189–195.

Beutner, G., Rück, A., Riede, B., and Brdiczka, D., 1998, Complexes between porin, hexokinase, mitochondrial creatine kinase, and adenylate translocator display properties of the permeability transition pore: Implication for regulation of the permeability transition by the kinases, *Biochim. Biophys. Acta* **1368**: 7–18.

Bojes, H. K., Datta, K., Xu, J., Chin, A. S. P., Nunez, G., and Kehrer, J. P., 1997, Bcl-XL overexpression attenuates glutathione depletion in FL5.12 cells following interleukin-3 withdrawal, *Biochem. J.* **325**: 315–319.

Bossy-Wetzel, E., Newmeyer, D. D., and Green, D. D., 1998, Mitochondrial cytochrome *c* release in apoptosis occurs upstream of DEVD-specific caspase activation and independently of mitochondrial depolarization, *EMBO J.* **17**: 37–49.

Boveris A., and Chance, B., 1973, The mitochondial generation of hydrogen peroxide, *Biochem. J.* **134**: 707–716.

Brustovetsky, N., and Klingenberg, M., 1996, Mitochondrial ADP/ATP carrier can be reversibly converted into a large channel by Ca^{2+}. *Biochemistry* **35**: 8483–8488.

Cassarino, D. S., Swerdlow, R. H., Parks, J. K., Parker, W. D., and Bennett, J. P., 1998, Cyclosporine, a increases resting mitochondrial membrane potential in SY5Y cells and reverses the depressed mitochondrial membrane potential of Alzheimer's disease cybrids, *Biochem. Biophys. Res. Commun.* **248**: 168–173.

Castedo, M., Hirsch, T. Susin, S. A. Zamzami, N. Marchetti, P. Macho, A., and Kroemer, G., 1996. Sequential acquisition of mitochondrial and plasma membrane alterations during early lymphocyte apoptosis. *J. Immunol.* **157**: 512–521.

Chinnaiyan, A. M., Chaudhary, D., O'Rourke, K., Koonin, K., and Dixit, V. M., 1997a, Role of CED-4 in the activation of CED-3, *Nature* **388**: 728–729.

Chinnaiyan, A. M., O'Rourke, K., Lane, B. R., and Dixit, V. M., 1997b, Interaction of CED-4 with CED-3 and CED-9: A molecular framework for cell death, *Science* **275**: 1122–1126.

Clapham, D. E., 1998, At last, the structure of an ion-selective channel, *Nature Struct. Biol.* **5**: 342–344.

Collins, M. K. L., Forlong, I. J., Malde, P., Ascaso, R., Oliver, J., and Rivas, A. L., 1996, An apoptotic endonuclease activated either by decreasing pH or by increasing calcium, *J. Cell Sci.* **109**: 2393–2399.

Cosulich, S. C., Worrall, V., Hedge, P. J., Green, S., and Clarke, P. R., 1998, Regulation of apoptosis by BH3 domains in a cell-free system. *Curr. Biol.* **7**: 913–920.

Decaudin, D., Geley, S., Hirsdch, T., Castedo, M., and Marchetti, P., 1997, Bcl-2 and Bcl-XL antagonize the mitochondrial dysfunction preceding nuclear apoptosis induced by chemotherapeutic agents, *Cancer Res.* **57**: 62–69.

Deckwerth, T. L., and Johnson, E. M., 1993, Temporal analysis of events associated with programmed cell death (apoptosis) of sympathetic neurons deprived of nerve growth factor, *J. Cell Biol.* **123**: 1207–1222.

Deshmuth, M., and Johnson, E. M., 1998, Evidence of a novel event during neuronal death: Development of competence-to-die in response to cytoplasmic cytochrome *c, Neuron* **21**: 695–705.

Du, Y., Dodel, R. C., Bales, K. R., Jemmerson, E., Hamilton-Byrd, S. M., and Paul, J., 1998, Alpha2-macroglobulin attenuates beta-amyloid peptide 1-40 fibril formation and associated neurotoxicity of cultured fetal rat cortical neurons, *Neurochemistry* **69**: 1382–1388.

Dubrez, L., Eymin, B., Sordet, O., Droin, N., Turhan, A. G., and Solary, E., 1998, BCR-Abl delays apoptosis upstream of caspase-3 activation, *Blood* **91**: 2415–2422.

Dumont, M. E., Cardillo, T. S., Hayes, M. K., and Sherman, F., 1991, Role of cytochrome *c* heme lyase in mitochondrial import and accumulation of cytochrome *c* in *Saccharomyces cerevisiae, Mol. Cell. Biol.* **11**: 5487–5496.

Enari, M., Hase, A., and Nagata, S., 1995, Apoptosis by a cytosolic extract from Fas-activated cells, *EMBO J.* **14**: 5201–5208.

Enari, M., Talanian, R. V., Wong, W. W., and Nagata, S., 1996, Sequential activation of ICE-like and CPP32-like proteases during Fas-mediated apoptosis, *Nature* **380**: 723–726.

Eskes, R., Antosson, B., Osen-Sand, A., Montesuit, S., and Richter, C., 1998, Bax-induced cytochrome *c* release from mitochondria is independent of the permeability transition pore but highly dependent on Mg^{2+} ions, *J. Cell Biol.* **143**: 217–224.

Frade, J. M., and Michaelidis, T. M., 1997, Origin of eucaryotic programmed cell death: A consequence of the aerobic metabolism, *BioEssays* **19**: 827–832.

Fridovic, I., 1978, The biology of oxygen radicals, *Science* **201**: 875–880.

Fujimura, M., Morita-Fugimura, Y., Murakani, K., Kawase, M., and Chan, P. H., 1998, Cytosolic redistribution of cytochrome *c* after transient focal ischemia in rats, *J. Cereb. Blood Flow Metab.* **18**: 1239–1247.

Gleichmann, M., Beinroth, S., Reed, J. C., Krajewski, S., and Schultz, J. B., 1998, Potassium deprivation-induced apoptosis of cerebellar granule neurons: Cytochrome *c* release in the absence of altered expression of Bcl-2 family proteins, *Cell Physiol. Biochem.* **8**: 194–201.

Goossens, V., Grooten, J., De Vos, K., and Fiers, W., 1995, Direct evidence for tumor necrosis factor-induced mitochondrial reactive oxygen intermediates and their involvement in cytotoxicity, *Proc. Natl. Acad. Sci. USA* **92**: 8115–8119.

Granville, D. J., Carthy, C. M., Jiang, H., Shore, G. C., McManus, B. M., and Hunt, D. W., 1998, Rapid cytochrome *c* release, activation of caspases 3, 6, 7, and 8 followed by Bap31 cleavage in HeLa cells treated with photodynamic therapy, *FEBS Lett.* **437**: 5–10.

Green, D. R., 1998, Apoptotic pathways: The road to ruin, *Cell* **94**: 695–698.

Greenlund, L. J. S., Deckwert, T. L., and Johnson, E. M., 1995, Superoxide dismutase delays neuronal apoptosis: A role for reactive oxygen species in programmed cell death, *Neuron* **14**: 303–315.

Gudz, T. I., Tserng, K. Y., and Hoppel, C. L., 1997, Direct inhibition of mitochondrial respiratory chain complex III by cell-permeable ceramide, *J. Biol. Chem.* **272**: 24154–24158.

Guénal, I., Sidoti-de Fraisse, C., Gaumer, S., and Mignotte, B., 1997, Bcl-2 and HSP27 act at different levels to suppress programmed cell death, *Oncogene* **15**: 347–360.

Hartley, A., Stone, J. M., Heron, C., Cooper, J. M., and Schapira, A. H. V., 1994, Complex I inhibitors induce dose-dependent apoptosis in PC12 cells: Relevance to Parkinson disease, *J. Neurochem.* **63**: 1987–1990.

Henkart, P. A., and Grinstein, S., 1996, Apoptosis: Mitochondria resurrected? *J. Exp. Med.* **183**: 1293–1295.

Henze, K., Nadr, A., Wettern, M., Cerff, R., and Martin, W., 1995, A nuclear gene of eubacterial origin in *Euglena gracilis* reflects cryptic endosymbioses during protist evolution, *Proc. Natl. Acad. Sci. USA* **92**: 9122–9126.

Higuchi, M., Aggarwal, B. B., and Yeh, E. T. H., 1997, Activation of CCP32-like protease in tumor necrosis factor-induced apoptosis is dependent on mitochondrial function, *J. Clin. Invest.* **99**: 1751–1758.

Hoffmann, K., Bucher, P., and Tschopp, J., 1997, The CARD domain: A new apoptotic signaling motif, *Trends Biochem. Sci.* **22**: 155–156.

Ichas, F., Jouaville, L., and Mazat, J. P., 1997, Mitochondria are excitable organelles capable of generating and conveying electrical and calcium signals, *Cell* **89**: 1145–1153.

Jacobson, M. D., 1996, Reactive oxygen species and programmed cell death, *Trends Biochem. Sci.* **21**: 83–86.

Jacobson, M. D., and Raff, M. C., 1995, Programmed cell death and Bcl-2 protection in very low oxygen, *Nature* **374**: 814–816.

Jones, D. P., McConkey, D. J., Nicotera, P., and Orrenius, S., 1989, Calcium-activated DNA fragmentation in rat liver nuclei, *J. Biol. Chem.* **264**: 6398–6403.

Jones, D. P., and Lash, L. H., 1993, *Criteria for Assessing Normal and Abnormal Mitochondrial Functions*, Vol. 2, Academic, San Diego, pp. 1–7.

Kane, D. J., Sarafian, T. A., Anton, R., Hahn, H., and Gralla, E. B., 1993, Bcl-2 inhibition of neural death: Decreased generation of reactive oxygen species, *Science* **262**: 1274–1277.

Kerr, J. F. R., Wyllie, A. H., and Currie, A. R., 1972, Apoptosis: A basic biological phenomenon with wide-ranging implications in tissue kinetics, *Br. J. Cancer* **26**: 239–257.

Kharbanda, S., Pandey, P., Schofield, L., Israels, S., and Roncinske, R., 1997, Role of Bcl-XL as an inhibitor of cytosolic cytochrome *c* accumulation in DNA damage-induced apoptosis, *Proc. Natl. Acad. Sci. USA* **94**: 6939–6942.

Kiberstis, P. A., 1999, Mitochondria make their comeback, *Science* **283**: 1475.

Kim, C. N., Wang, X., Huang, Y., Liu, L., Fang, G., and Bhalla, K., 1997, Overexpression of Bcl-XL inhibits Ara-C-induced mitochondrial loss of cytochrome *c* and other perturbations that activate the molecular cascade of apoptosis, *Cancer Res.* **57**: 3115–3120.

Kinnaly, K. W., Lohret, T. A., Campo, M. L., and Mannella, C. A., 1996, Perspectives on the mitochondrial multiple conductance channel, *J. Bioenerg. Biomenbr.* **28**: 115–123.

Kluck, R. M., Bossy-Wetzel, E., Green, D. R., and Newmeyer, D. D., 1997a, The release of cytochrome *c* from mitochondria: A primary site for Bcl-2 regulation of apoptosis, *Science* **275**: 1132–1136.

Kluck, R. M., Martin, S. J., Hoffman, B. M., Zhou, J. S., Green, D. R., and Newmeyer, D. D., 1997b, Cytochrome *c* activation of CPP32-like proteolysis plays a critical role in *Xenopus* cell-free apoptosis system, *EMBO J.* **16**: 4639–4649.

Korsmeyer, S. J., 1992, Bcl-2: A repressor of lymphocyte death, *Immunol. Today* **13**: 285–288.

Krippner, A., Matsuno-Yagi, A., Gottlieb, R. A., and Babior, B. M., 1996, Loss of function of cytochrome *c* in Jurkat cells undergoing Fas-mediated apoptosis, *J. Biol. Chem.* **271**: 21629–21636.

Kroemer, G., 1997a, Mitochondrial implication in apoptosis: Toward an endosymbiont hypothesis of apoptosis evolution, *Cell Death Diff.* **4**: 443–456.

Kroemer, G., 1997b, The proto-oncogene Bcl-2 and its role in regulating apoptosis: The mechanism of action of Bcl-2 clues for therapeutic interventions, *Nature Medicine* **3**: 614–620.

Kroemer, G., Petit, P. X., Zamzami, N., Vayssière, J.-L., and Mignotte, B., 1995, The biochemistry of apoptosis, *FASEB J.* **9**: 1277–1287.

Kroemer, G., Zamzami, N., and Susin, S. A., 1997, Mitochondrial control of apoptosis, *Immunol. Today* **18**: 44–51.

Kumar, S., and Harvey, N. L., 1995, Role of multiple cellular proteases in the execution of programmed cell death, *FEBS Lett.* **375**: 169–173.

Kupsch, E. M., Knepper, B., Kuroki, T., Heuer, I., and Meyer, T. F., 1993, Variable opacity (Opa) outer membrane proteins account for the cell trophisms displayed by *Neisseria gonorrhoeae* for human leukocytes and epithelial cells, *EMBO J.* **12**: 641–650.

Kurihara, H., Torigoe, S., Omura, M., Saito, K., Kurihara, M., and Matsubara, S., 1998, DNA fragmentation induced by a cytoplasmic extract from irradiated cells, *Radiat. Res.* **150**: 269–274.

Kuwana, T., Smith, J. J., Muzio, M., Dixit, V., Newmeyer, D. D., and Kornbluth, S., 1998, Apoptosis induction by caspase-8 is amplified through the mitochondrial release of cytochrome *c, J. Biol. Chem.* **273**: 16589–16594.

Lazebnik, Y. A., Cole, S., Cooke, C. A., Nelson, W. G., and Earnshaw, W. C., 1993, Nuclear events of apoptosis *in vitro* in cell-free mitotic extracts: A model system for analysis of the active phase of apoptosis, *J. Cell. Biol.* **123**: 7–22.

Lazebnik, Y. A., Kaufmann, S. H., Desnoyers, S., Poirier, G. G., and Earnshaw, W. C., 1994, Cleavage of poly(ADP-ribose) polymerase by a proteinase with properties like ICE, *Nature* **371**: 346–347.

Lazebnik, Y. A., Takahashi, A., Poirier, G. G., Kaufman, S. H., and Earnshaw, W. C., 1995, Characterization of the execution phase of apoptosis *in vitro* using extracts from condemned-phase cells, *J. Cell Sci.* **S19**: 41–49.

Lennon, S. V., Martin, S. J., and Cotter, T. G., 1991, Dose-dependent induction of apoptosis in human tumor cell lines by widely diverging stimuli, *Cell Prolif.* **24**: 203–214.

Leoni, L. M., Chao, Q., Cottam, H. B., Rosenbach, M., and Carrera, C. J., 1998, Induction of an apoptotic program in cell-free extract by 2-chloro-2′-deoxyadenosine 5′-triphosphate and cytochrome *c Proc. Natl. Acad. Sci. USA* **95**: 9567–9571.

Lin, K. T., Xue, J. Y., and Wong, P. Y., 1997, Reactive oxygen species participate in peroxinitrite-induced apoptosis in HL-60 cells, *Biochem. Biophys. Res. Commun.* **230**: 115–119.

Liu, X., Kim, C. N., Yang, J., Jemmerson, R., and Wang, X., 1996, Induction of apoptic program in cell-free extracts: Requirement for dATP and cytochrome *c Cell* **86**: 147–157.

Liu, X., Zou, H., Slaughter, C., and Wang, X., 1997, DFF, a heterodimeric protein that functions downstream of caspase-3 to trigger DNA fragmentation during apoptosis, *Cell* **89**: 175–184.

Mancini, M., Nicholson, D. W., Roy, S., Thornberry, N. A., and Peterson, E. P., 1998, The caspase-3 precursor has a cytosolic and mitochondrial distribution: Implications for apoptotic signaling, *J. Cell Biol.* **140**: 1485–1495.

Manon, S., B. C., and Guérin, M., 1997, Release of cytochrome *c* and decrease of cytochrome *c* oxidase in Bax-expressing yeast cells, and prevention of these effects by coexpression of Bcl-XL, *FEBS Lett.* **415**: 29–32.

Marchetti, P., Susin, S. A., Decaudin, D., Gamen, S., and Castedo, M., 1996a, Apoptosis-associated derangement of mitochondrial function in cells lacking mitochondrial DNA, *Cancer Res.* **56**: 2033–2038.

Marchetti, P., Zamzami, N., Susin, S. A., Petit, P. X., and Kroemer, G., 1996b, Apoptosis of cells lacking mitochondrial DNA, *Apoptosis* **1**: 119–125.

Marchetti, P., Decaudin, D., Macho, A., Zamzami, N., and Hirsch, T., 1997, Redox regulation of apoptosis: Impact of thiol oxidation status on mitochondrial function, *Eur. J. Immunol.* **27**: 289–296.

Margulis, L., 1996, Archaeal–eubacterial mergers in the origin of Eukarya: Phylogenetic classification of life, *Proc. Natl. Acad. Sci. USA* **93**: 1071–1076.

Martin, S. J., and Cotter, T. G., 1991, Ultraviolet B irradiation of human leukemia HL-60 cells *in vitro* induces apoptosis, *Int. J. Radiat. Biol.* **59**: 1001.

Martin, S. J., and Green, D. R., 1995, Protease activation during apoptosis: Death by a thousand cuts? *Cell* **82**: 349–352.

Martin, S. J., Newmeyer, D. D., Mathisa, S., Farschon, D. M., and Wang, H. G., 1995, Cell-free reconstitution of Fas-, UV radiation- and ceramide-induced apoptosis, *EMBO J.* **14**: 5191–5200.

Marzo, I., Brenner, C., Zamzami, N., Jurgensmeier, J. M., and Susin, S. A., 1998a, Bax and adenine nucleotide translocator cooperate in the mitochondrial control of apoptosis, *Science* **281**: 2027–2031.

Marzo, I., Brenner, C., Zamzami, N., Susin, S. A., and Beutner, G., 1998b, The permeability transition pore complex: A target for apoptosis regulation by caspases and Bcl-2, *J. Exp. Med.* **187**: 1261–1271.

Marzo, I., Susin, S. A., Petit, P. X., Ravagnan, L., and Brenner, C., 1998c, Caspases disrupt mitochondrial membrane barrier function, *FEBS Lett.* **427**: 198–202.

Mayer, M., and Noble, M., 1994, *N*-acetyl-L-cystein is a pluripotent protector against cell death and enhancer of trophic factor-mediated cell survival *in vitro*, *Proc. Natl. Acad. Sci. USA* **91**: 7496–7500.

Mayer, A., Neupert, W., and Lill, R., 1995a, Mitochondrial protein import: Reversible binding of the presequence at the *trans* side of the outer membrane drives partial translocation and unfolding, *Cell* **80**: 127–137.

Mayer, A., Neupert, W., and Lill, R., 1995b, Translocation of apopcytochrome *c* across the outer membrane of mitochondria, *J. Biol. Chem.* **270**: 12390–12397.

McConkey, D. J., 1996, Calcium-dependent, interleukin 1 beta-converting enzyme inhibitor-insensitive degradation of lamin B-1 and DNA fragmentation in isolated thymocyte nuclei, *J. Biol. Chem.* **271**: 22398–22406.

McEnery, M. W., Snowman, A. M., Trifiletti, R. R. and Snyder, S. H., 1992, Isolation of the mitochondrial benzodiazepine receptor. Association with the voltage-dependent anion channel and the adenine nucleotide carrier. *Proc. Natl. Acad. Sci. USA* **89**: 3170–3174.

Mehlen, P., Schulze-Osthoff, K., and Arrigo, A.-P., 1996, Small stress proteins as novel regulators of apoptosis: Heat shock protein 27 blocks Fas/Apo-1 and staurosporine-induced cell death, *J. Biol. Chem.* **271**: 16510–16514.

Métivier, D., Dallaporta, B., Zamzami, N., Larochette, N., and Susin, S. A., 1998, Cytofluorometric detection of mitochondrial alterations in early CD95/Fas/APO-1-triggered apoptosis of Jurkat lymphoma cells: Comparison of seven mitochondrion-specific fluorochromes, *Immunol. Lett.*, in press.

Mignotte, B., Larcher, J. C., Zheng, D. Q., Esnault, C., Coulaud, D., and Feuteun, J., 1990, SV40 induced cellular immortalization: Phenotypic changes associated with the loss of proliferative capacity in a conditionally immortalized cell line, *Oncogene* **5**: 1529–1533.

Minn, A. J., Vélez, P., Schnedel, S. L., Liang, H., and Muchmore, S. W., 1997, Bcl-XL forms an ion channel in synthtic lipid membranes, *Nature* **385**: 353–357.

Mirkovic, N., Voehringer, D. W., Story, M. D., McConkey, D. J., McDonnell, T. J., and Meyn, R. E., 1997, Resistance to radiation-induced apoptosis in Bcl-2-expressing cells is reversed by depleting cellular thiols, *Oncogene* **15**: 1461–1470.

Muchmore, S. W., Sattler, M., Liang, H., Meadows, R. P., and Harlan, J. E., 1996, X-ray and NMR structure of human Bcl-xL, an inhibitor of programmed cell death, *Nature* **381**: 335–341.

Muzio, M., Salvesen, G. S., and Dixit, V. M., 1997, FLICE induced apoptosis in a cell-free system: Cleavage of caspase zymogens, *J. Biol. Chem.* **272**: 2952–2956.

Neame, S. J., Rubin, L. L., and Philpott, K. L., 1998, Blocking cytochrome *c* activity within intact neurons inhibits apoptosis, *J. Cell Biol.* **142**: 1583–1593.

Newmeyer, D. D., Farschon, D. M., and Reed, J. C., 1994, Cell-free apoptosis in *Xenopus* egg extracts: Inhibition by Bcl-2 and requirement for an organelle fraction enriched in mitochondria, *Cell* **79**: 353–364.

Nicholson, D. W., and Thornberry, N. A., 1997, Caspases: Killer proteases, *Trends Biochem. Sci.* **22**: 299–306.

Nieminen, A.-L., Dawson, T. L., Gores, G. J., Kawanishi, T., Herman, B., and Lemasters, J. J., 1990, Protection by acidotic pH and fructose against lethal injury to rat hepatocytes from mitochondrial inhibitors, ionophores, and oxidant chemicals, *Biochem. Biophys. Res. Commun.* **167**: 600–606.

O'Donnell, V. B., Spycher, S., and Azzi, A., 1995, Involvement of oxidants and oxidant-generating enzymes in tumor necrosis factor α-mediated apoptosis: Role for lipoxygenase pathway but not mitochondrial respiratory chain, *Biochem. J.* **310**: 133–141.

O'Gorman, E., Beutner, G., Dolder, M., Koretsky, A. P., Brdiczka, D., and Walliman, T., 1997, The role of creatine kinase in inhibition of mitochondrial permeability transition, *FEBS Lett.* **414**: 253–257.

Oltvai, Z. N., and Korsmeyer, S. J., 1994, Checkpoints of dueling dimers foil death wishes, *Cell* **79**: 189–192.

Pastorino, J. G., Simbula, G., Gilfor, E., Hoek, J. B., and Farber, J. L., 1994. Protoporphyrin IX, an endogenous ligand of the peripheral benzodiazepine receptor, potentiates induction of the mitochondrial permeability transition and the killing of cultured hepatocytes by rotenone. *J Biol Chem* **269**: 31041–31046.

Petit, P. X., Lecoeur, H., Zorn, E., Dauguet, C., Mignotte, B., and Gougeon, M. L., 1995, Alterations of mitochondrial structure and function are early events of dexamethasone-induced thymocyte apoptosis, *J. Cell. Biol.* **130**: 157–167.

Petit, P. X., Susin, S. A., Zamzami, N., Mignotte, B., and Kroemer, G., 1996, Mitochondria and programmed cell death: Back to the future, *FEBS Lett.* **396**: 7–14.

Petit, P. X., Zamzami, N., Sidoti-de Fraisse, C., Vayssiére, J. L., and Mignotte, B., 1997, Implication of mitochondria in apoptosis, *Mol. Cell. Biochem.* **174**: 185–188.

Petit, P. X., Goubern, M., Diolez, P., Susin, S. A., and Zamzami, N. G. K., 1998, Disruption of the outer mitochondrial membrane as a result of large amplitude swelling: The impact of irreversible permeability transition, *FEBS Lett.* **426**: 11–116.

Pinkus, R., Weiner, L. M., Daniel, V., 1996, Role of oxidants and antioxidants in the induction of AP-1, NFκB, and glutathion S-transferase gene expression, *J. Biol. Chem.* **271**: 13422–13429.

Quillet-Mary, A., Jaffrezou, J. P., Mansat, V., Bordier, C., Naval, J., and Laurent, G., 1997, Implication of mitochondrial hydrogen peroxide generation in ceramide-induced apoptosis, *J. Biol. Chem.* **272**: 21388–21395.

Ratan, R. R., Murphy, T. H., and Baraban, J. M., 1994, Oxidative stress induces apoptosis in embryonic cortical neurons, *J. Neurochem.* **62**: 376–379.

Reed, J. C., 1997, Cytochrome *c*: Can't live without it, *Cell* **91**: 559–562.

Reed, J. C., Jurgensmeier, J. M., and Matsuyama, S., 1998, Bcl-2 family proteins and mitochondria, *Biochim. Biophys. Acta* **1366**: 127–137.

Richter, C., Schweizer, M., Cossariza, A., and Franceschi, C., 1996, Control of apoptosis by ATO levels, *FEBS Lett.* **378**: 107–110.

Rottenberg, H., and Wu, S., 1998, Quantitative assay by flow cytometry of the mitochondrial membrane potential in intact cells, *Biochim. Biophys. Acta* **1404**: 393–404.

Rudel, T., Schmid, A., Benz, R., Kolb, H. A., Lang, F., and Meyer, T. F., 1996, Modulation of Neisseria porin (poR) by cytosolic ATP/GTP of target cells: Parallels between pathogen accomodation and mitochondria endosymbiosis, *Cell* **85**: 391–402.

Sandstrom, P. A., and Buttke, T. M., 1993, Autocrine production of extracellular catalase prevents apoptosis of the human CEM T-cell line in serum-free medium, *Proc. Natl. Acad. Sci. USA* **90**: 4708–4712.

Sato, N., Iwata, S., Nakamura, K., Hori, T., Mori, K., and Yodoi, J., 1995, Thiol-mediated redox regulation of apoptosis: Possible roles of cellular thiols other than glutathione in T cell apoptosis, *J. Immunol.* **154**: 3194–3203.

Schendel, S. L., Xie, Z., Montal, M. O., Matsuyama, S., Montal, M., and Reed, J. C., 1997, Channel formation by antiapoptotic protein Bcl-2, *Proc. Natl. Acad. Sci. USA* **94**: 5113–5118.

Schlesinger, P. H., Gross, A., Yin, X. M., Yamamoto, K., and Saito, M., 1997, Comparison of the ion channel characteristics of proapoptotic BAX and antiapoptotic BCL-2, *Proc. Natl. Acad. Sci. USA* **94**: 11357–11362.

Schulze-Osthoff, K., Bakker, A. C., Vanhaesebroeck, B., Beyaert, R., Jacob, W. A., and Fiers, W., 1992, Cytotoxic activity of tumor necrosis factor is mediated by early damage of mitochondrial functions: Evidence for the involvement of mitochondrial radical generation, *J. Biol. Chem.* **267**: 5317–5323.

Schulze-Osthoff, K., Beyaert, R., Vandevoorde, V., Haegeman, G., and Fiers, W., 1993, Depletion of the mitochondrial electron transport abrogates the cytotoxic and gene-inductive effects of TNF, *EMBO J.* **12**: 3095–3104.

Shimizu, S., Eguchi, Y., Kosaka, H., Kamiike, W., Matsuda, H., and Tsujimoto, Y., 1995, Prevention of hypoxia-induced cell death by Bcl-2 and Bcl-XL, *Nature* **374**: 811–813.

Stridh, H., Kimland, M., Jones, D. P., Orrenius, S., and Hampton, M. B., 1998, Cytochrome *c* release and caspase activation in hydrogen peroxide-and tributyltin-induced apoptosis, *FEBS Lett.* **429**: 351–355.

Susin, S. A., Zamzami, N., Castedo, M., Hirsch, T., and Marchetti, P., 1996, Bcl-2 inhibits the mitochondrial release of an apoptogenic protease, *J. Exp. Med.* **184**: 1331–1342

Susin, S. A., Zamzami, N., Castedo, M., Daugas, E., and Wang, H.-G., 1997a, The central executioner of apoptosis: Multiple connections between proteases activation and mitochondria in Fas/APO-1/CD95- and ceramide-induced apoptosis, *J. Exp. Med.* **186**: 25–37.

Susin, S. A., Zamzami, N., Larochette, N., Dallaporta, B., and Marzo, I., 1997b, A cytofluorometric assay of nuclear apoptosis in a cell-free system: Application to ceramide-induced apoptosis, *Exp. Cell Res.* **236**: 397–403.

Susin, S. A., Lorenzo, H. K., Zamzami, N., Marzo, I., and Brenner, C., 1998, Mitochondrial release of caspase-2 and caspase-9 during the apoptotic process, *J. Exp. Med.* **189**: 381–394.

Susin, S. A., Lorenzo, H. K., Zamzami, N., Marzo, I., and Snow, E. B., 1999, Molecular characterization of mitochondrial apoptosis-inducing factor, *Nature* **397**: 441–445.

Taylor, F. R. J., 1974, Implications and extensions of the serial endosymbiosis theory of the origin of eukaryotes, *Taxon* **23**: 229–258.

Thompson, C. B., 1995, Apoptosis in the pathogenesis and treatment of disease, *Science* **267**: 1456–1462.

Tudor, J. J., and Karp, M. A., 1994, Translocation of an outer membrane protein into prey cytoplasmic membranes of *Bdellovibrio bacteriovirus*, *J. Bacteriol.* **172**: 4002–4007.

Uckun, F. M., Tuel-Ahlgren, L., Song, C. W., Waddick, K., and Myers, D. E., 1992, Ionizing radiation stimulates unidentified tyrosine-specific protein kinases in human B-lymphocyte precursors, triggering apoptosis and clonogenic cell death, *Proc. Natl. Acad. Sci. USA* **89**: 9005–9009.

Van den Dobbelsteen, D. J., Nobel, C. S. I., Schlegel, J., Cotgreave, I. A., Orrenius, S., and Slater, A. F. G., 1996, Rapid and specific efflux of reduced glutathione during apoptosis induced by anti-Fas/APO-1 antibody, *J. Biol. Chem.* **271**: 15240–15427.

Van der Heiden, M. G., Chandel, N. S., Williamson, E. K., Schumacher, P. J., and Thompson, C. B., 1997, Bcl-XL regulates the membrane potential and volume homeostasis of mitochondria, *Cell* **91**: 627–637.

Vayssière, J.-L., Petit, P. X., Risler, Y., and Mignotte, B., 1994, Commitment to apoptosis is associated with changes in mitochondrial biogenesis and activity in cell lines conditionally immortalized with simian virus 40, *Proc. Natl. Acad. Sci. USA* **91**: 11752–11756.

Weel, J. F. L., and van Putten, J. P. M., 1991, Fate of the major outer membrane protein P.IA in early and late events of gonococcal infection of epithelial cells, *Res. Microbiol.* **142**: 985–993.

Wolvetang, E. J., Johnson, K. L., Krauer, K., Ralph, S. J., and Linnane, A. W., 1994, Mitochondrial respiratory chain inhibitors induce apoptosis, *FEBS Lett.* **339**: 40–44.

Wong, A., and Cortopassi, G., 1997, MtDNA mutations confer cellular sensitivity to oxidant stress that is partially rescued by calcium depletion and cyclosporine A, *Biochem. Biophys. Res. Commun.* **239**: 139–145.

Wong, G. H. W., Elwell, J. H., Oberley, L. W., and Goeddel, D. V., 1989, Manganous superoxide dismutase is essential for cellular resistance to cytotoxicity of tumor necrosis factor, *Cell* **58**: 923–931.

Yang, J., Liu, X., Bhalla, K., Kim, C. N., and Ibrado, A. M., 1997, Prevention of apoptosis by Bcl-2: Release of cytochrome *c* from mitochondria blocked, *Science* **275**: 1129–1132.

Zamzami, N., Marchetti, P., Castedo, M., Decaudin, D., and Macho, A., 1995a, Sequential reduction of mitochondrial transmembrane potential and generation of reactive oxygen species in early programmed cell death, *J. Exp. Med.* **182**: 367–377.

Zamzami, N., Marchetti, P., Castedo, M., Zanin, C., and Vayssière, J.-L., 1995b, Reduction in mitochondrial potential constitutes an early irreversible step of programmed lymphocyte death *in vivo, J. Exp. Med.* **181**: 1661–1672.

Zamzami, N., Susin, S. A., Marchetti, P., Hirsch, T., and Gómez-Monterrey, I., 1996, Mitochondrial control of nuclear apoptosis, *J. Exp. Med.* **183**: 1533–1544.

Zhong, L.-T., Sarafinn, T., Kane, D. J., Charles, A. C., and Mah, S. P., 1993, Bcl-2 inhibits death of central neural cells induced by multiple agents, *Proc. Natl. Acad. Sci. USA* **90**: 4533–4537.

Zoratti, M., and Szabò, I., 1995, The mitochondrial permeability transition, *Biochim. Biophys. Acta—Rev. Biomembr.* **1241**: 139–176.

Zou, H., Henzel, W. J., Liu, X., Lutschg, A., and Wang, X., 1997, Apaf-1, a human protein homologous to *C. elegans* CED-4, participates in cytochrome *c*-dependent activation of caspase-3, *Cell* **90**: 405–413.

Chapter 12

Role of Mitochondria in Apoptosis Induced by Tumor Necrosis Factor-α

Cynthia A. Bradham, Ting Qian, Konrad Streetz,
Christian Trautwein, David A. Brenner, and John J. Lemasters

1. APOPTOSIS SIGNALING AND TNFα

The cytokine TNFα is secreted mainly by activated macrophages. Activation of the TNFα receptor (TNFR) causes changes in gene expression that result in varying cellular effects that range from induction of apoptosis to proliferation to differentiation, depending on cell type and signal context (Wallach *et al.*, 1997). How these various responses are distinguished is an ongoing area of research. Understanding the molecular basis for the choice of apoptotic vs survival responses could have profound clinical implications, particularly in cancer therapy.

In vivo, exogenous TNFα causes generalized inflammation at low doses, and septic shock and liver injury at higher doses. Other well-established physiological effects are cytotoxicity and cachexia (Kettelhut *et al.*, 1987). *In vitro,* however, most cells, including hepatocytes, resist TNFα-mediated apoptosis unless metabolic inhibitors such as actinomycin D or cycloheximide are also present, indicating that TNFα induces expression of a protective factor (Leist *et al.*, 1994). Indeed, blockade of the transcription factor NF-κB, which is strongly induced by TNFα, sensitizes resistant cells to TNFα-mediated apoptosis *in vitro* and *in vivo* (Iimura *et al.*, 1998; Wang *et al.*, 1996).

Cynthia A. Bradham   Department of Medicine, University of North Carolina at Chapel Hill, Chapel Hill, North Carolina 27599-7038.   **Ting Qian and John J. Lemasters**   Department of Cell Biology and Anatomy, University of North Carolina at Chapel Hill, Chapel Hill, North Carolina 27599-7090.   **Konrad Streetz and Christian Trautwein**   Department of Gastroenterology and Hepatology, Mediziniche Hochschule Hannover, Hannover, Germany.   **David A. Brenner**   Departments of Medicine, and of Biochemistry and Biophysics, University of North Carolina at Chapel Hill, Chapel Hill, North Carolina 27599-7090.

Mitochondria in Pathogenesis, edited by Lemasters and Nieminen.
Kluwer Academic/Plenum Publishers, New York, 2001.

1.1. The TNFR Superfamily and Death Domains

The TNFα cytokine belongs to a family of ligands that activate a structurally similar family of receptors, including lymphotoxin-α and -β, Fas ligand (FasL), OX40L, CD40L, CD27L, CD30L, 4-1BBL, and TRAIL. These ligands are type II membrane proteins (i.e., the C-terminus is extracellular and the N-terminus is intracellular). Some of these also exist in soluble forms as a result of proteolytic release. In structural studies of soluble TNFα, the conserved receptor-binding domain folds into a β-pleated-sheet sandwich or "jelly roll" and trimerizes (Jones *et al.*, 1989). The TNFR family members are type I membrane proteins that share a conserved cysteine-rich extracellular ligand-binding domain. This family includes TNFR I and II, Fas, NGFR, OX40, CD40, CD27, CD30, 4-1BB, DR3, DR4, and DR5. The receptors apparently exist as head-to-head dimers that oligomerize upon ligand binding (Banner *et al.*, 1993).

A subset of these receptors have a conserved cytoplasmic region known as the *death domain*. This domain is restricted to apoptosis-inducing receptors, in particular Fas and TNFR I (in which it was first identified) (Itoh and Nagata, 1993; Tartaglia *et al.*, 1993). The death domain in TNFR comprises approximately 80 amino acids and is required for TNFα-induced apoptosis induction. The corresponding region in Fas is 65 amino acids long and 28% identical to the TNFR domain. The death domain mediates protein–protein interactions between TNFR and the scaffolding protein TRADD (*T*NF *r*eceptor-*a*ssociated protein with *d*eath-*d*omain) or in the case of Fas, between Fas and FADD (Chinnaiyan an *et al.*, 1995; Hsu *et al.*, 1995). The TRADD protein serves as a branching point for TNFα signaling by binding to Traf2 (*T*NF *r*eceptor-*a*ssociated *f*actor 2), RIP (*r*eceptor-*i*nteracting *p*rotein, another death-domain-containing protein) and FADD (*F*as-*a*ssociated protein with *d*eath *d*omain) (Hsu *et al.*, 1996a, 1996b).

1.2. Receptor Protein Complex

The activated TNFR forms a signal-inducing complex containing TRADD, Traf2, FADD, and RIP (Fig. 1). The Traf2 and RIP proteins activate stress- and inflammation-associated signal-transduction cascades, including the JNK/c-Jun pathway (Traf2) and the NF-κB pathway (RIP) (Natoli *et al.*, 1997; Ting *et al.*, 1996; Yeh *et al.*, 1997). NF-κB activation is particularly important in inflammatory processes and imparts a protective effect on the stimulated cell, because inhibition of NF-κB activity sensitizes the cell to TNFα-mediated apoptosis (Wang *et al.*, 1996). Normally, NF-κB is localized to the cytosol through interaction with IκB, which binds to the nuclear localization signal on NF-κB and inhibits its translocation. Activation of TNFR results in sequential activation of NIK (NF-κB-*i*nducing *k*inase) and IKK (*I*κB *k*inase) (DiDonato *et al.*, 1997; Malinin *et al.*, 1997). The IKK phosphorylates IκB, which signals for IκB ubiquitination and degradation, releasing NF-κB to translocate to the nucleus and activate κB-dependent genes (Fig. 1).

The FADD protein conveys the apoptotic signal from both Fas (through direct interaction) and TNFR (through TRADD) via its *death-effector domain*, another protein–protein interaction motif that complexes FADD with procaspase 8. The FADD-procaspase-8 complex is known as the DISC (*d*eath-*i*nducing *s*ignaling *c*omplex) and is recruited to the receptor upon ligand binding (Muzio *et al.*, 1996). Oligomerization of procaspase 8

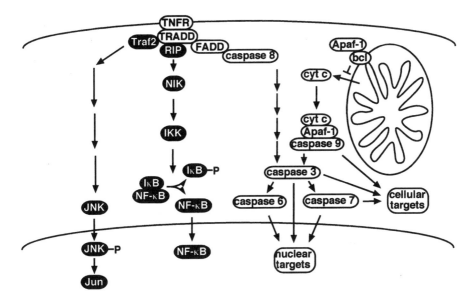

FIGURE 1. TNFα signal transduction. TNFR I binds TRADD, which forms a complex with Traf2, RIP, and FADD/caspase 8. Traf 2 is required for JNK activation; RIP is required for NF-κB activation. The sequential activation of kinases NIK and IKK leads to phosphorylation and subsequent degradation of IκB, releasing and thereby activating NF-κB. Oligomerization of procaspase-8 permits its auto-activation, initiating the caspase cascade. In addition, cytochrome *c* release from mitochondria, which is antagonized by Bcl-2, activates caspase 9, which in turn activates caspase 3 and other executionary caspases.

results in autoprocessing and activation, initiating the caspase cascade associated with apoptotic cell death (Martin *et al.*, 1998) (Fig. 1). Caspase 8 is required for induction of apoptosis by either TNFα or Fas (Juo *et al.*, 1998; Varfolomeev *et al.*, 1998).

1.3. Caspase Cascade

Caspases are a growing family of proteases that share sequence homology with the prototype CED-3, a *Caenorhabditis elegans* protease required for programmed cell death (Alnemri, 1997; Salveson and Dixit, 1997). Caspases are *c*ysteine proteases that cleave at *asp*artate residues, hence their name. The numbering system is based on their order of publication as opposed to their sequence of activation during programmed cell death. All caspases are synthesized as *pro* forms, requiring caspase-mediated cleavage for activation. Caspase 8 and 10 both contain death-effector domains, implying death-receptor-complex association and autoprocessing. Caspase 8 and caspase 10 are thought to initiate the cascade, and the final executionary caspases (e.g., caspase 3, 6, and 7) cleave key target proteins, such as nuclear lamins and cytoskeletal components (Rosen and Casciola-Rosen, 1997), causing cellular and nuclear collapse as well as inducing DNA ladder formation through cleavage of DFF/ICAD (Enari *et al.*, 1998; Liu *et al.*, 1997). Although both the caspase activation sequence and the minimum caspase number required to induce cell death remain undefined, there are clear distinctions in the actions of various caspases. For example, caspase 1-like proteases are activated prior to caspase 3-like caspases during Fas-mediated apoptosis (Enari *et al.*, 1996). In addition, specific caspases, in particular caspase 3, 8, and 9, are required for embryonic development (Kuida *et al.*, 1998, 1996; Varfolomeev *et al.*, 1998).

1.4. Cytochrome *c* Release

Recent studies have demonstrated that cytochrome *c* release from the mitochondrial intermembrane space is required to activate late-acting caspase 3 (Li *et al.*, 1997b; Liu *et al.*, 1996). It was further shown that cytochrome *c* will complex in the cytosol with Apaf-1 (*a*poptotic *p*rotease-*a*ctivating *f*actor 1), dATP, and procaspase 9, resulting in caspase 9 activation, presumably due to a conformational change (Fig. 1). Caspase 9 in turn cleaves and activates caspase 3 (Li *et al.*, 1997b; Zou *et al.*, 1997). Bcl-2, an evolutionarily conserved transmembrane protein with pore-forming and scaffolding capabilities, functionally inhibits apoptosis by blocking cytochrome *c* release from mitochondria, and by binding Apaf-1 and inhibiting caspase 9 activation (Hu *et al.*, 1998; Kluck *et al.*, 1997; Yang *et al.*, 1997).

The work of Kroemer and colleagues showed that onset of the mitochondrial permeability transition (MPT) in isolated mitochondria is sufficient to induce apoptotic changes in isolated nuclei. They further demonstrated that onset of the MPT occurs during apoptosis and results in release of a 50 kDa protease, AIF (*a*poptosis-*i*nducing *f*actor) (Marchetti *et al.*, 1996; Susin *et al.*, 1996, 1997). Other groups, however, reported that cytochrome *c* release occurred in the absence of mitochondrial depolarization, indicating that the permeability transition is not required for apoptosis (Bossy-Wetzel *et al.*, 1998; Kluck *et al.*, 1997; Yang *et al.*, 1997). These controversial results stimulated an investigation into the role of mitochondria in TNFα-mediated apoptosis in hepatocytes.

2. THE TNFα-MEDIATED APOPTOSIS MODEL IN PRIMARY CULTURE

In vivo, TNFα signaling is associated with a variety of pathological states in the liver, including alcoholic and viral hepatitis, ischemia/reperfusion injury, and injury from hepatotoxins (Bradham *et al.*, 1998a). Hepatic mitochondria lend themselves to confocal imaging, so a model for TNFα-mediated apoptosis was established in primary hepatocyte cultures by inhibiting NF-κB activation using an adenovirus expressing an IκB super-repressor (SR) mutant (Iimura *et al.*, 1998). This mutant protein, containing serine-to-alanine substitutions (residues 32 and 36), resists IKK-mediated phosphorylation and degradation, thereby inhibiting NF-κB activation. Hepatocytes expressing the IκB super-repressor (IκB SR) undergo pronounced apoptosis in response to TNFα, indicated by the hallmark signs of membrane blebbing and cellular condensation, nuclear condensation and fragmentation, caspase 3 activation, and DNA ladder formation (Fig. 2). In contrast, normal hepatocytes resist TNFα-mediated apoptosis, undergoing none of these changes (Bradham *et al.*, 1998b).

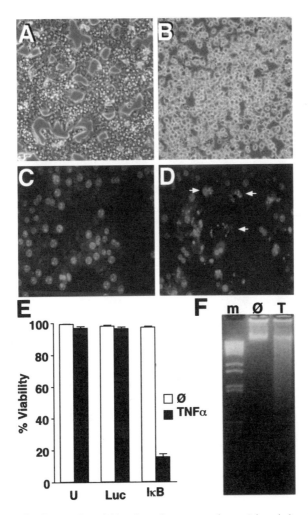

FIGURE 2. TNFα-mediated apoptosis model in primary hepatocyte cultures. Adenoviral-mediated expression of IκB S32, 36A (IκB SR) sensitizes hepatocytes to TNFα-mediated apoptosis. Hepatocytes expressing the IκB SR have normal cellular (A) and nuclear (C) morphology, but when treated with TNFα for 24 hr, they undergo cellular condensation and blebbing (B), nuclear condensation (D), and DNA fragmentation (F). This is not an adenoviral effect, because control adenovirus expressing luciferase (luc) behaved no differently from uninfected (U), normal hepatocytes (E). (Reproduced with permission from Bradham *et al.*, *Mol. Cell. Biol.* Copyright 1998, American Society for Microbiology.)

2.1. Mitochondrial Permeability Transition during TNFα-Mediated Apoptosis

Mitochondrial depolarization and the MPT can be directly visualized in living cells using fluorescent labeling and confocal imaging. *T*etra*m*ethyl*r*hodamine *m*ethylester (TMRM) is a cationic red fluorophore that accumulates in polarized mitochondria. Loss of TMRM staining indicates mitochondrial depolarization. Calcein is a relatively large, green fluorophore that is excluded from intact mitochondria. Calcein-stained cells contain dark voids representing mitochondrial outlines. Redistribution of calcein into mitochondria demonstrates increased mitochondrial permeability and reflects onset of the MPT. TNFα stimulates MPT onset and the accompanying mitochondrial depolarization in hepatocytes expressing the IκB SR. At 7 hours of TNFα treatment, mitochondria are polarized and impermeant to calcein (Fig. 3A). Onset of the MPT occurs after approximately 8 hours of

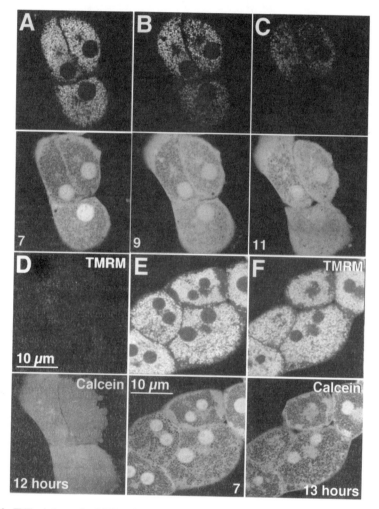

FIGURE 3. TNFα induces the MPT and mitochondrial depolarization in IκB SR-expressing hepatocytes. Primary hepatocytes expressing the IκB SR were stimulated with TNFα and mitochondria were then visualized using confocal microscopy. The cationic red fluorophore TMRM (upper panels) accumulates in polarized mitochondria, while the green fluorophore calcein (lower panels) is excluded from mitochondria. TNFα treatment of IκB SR-expressing hepatocytes induces loss of TMRM staining in a gradual manner between 8 and 12 hr of treatment, indicating mitochondrial depolarization (A–D, upper panels). Simultaneously, calcein redistributes into the mitochondria, indicating MPT onset (A–D, lower panels). In contrast, mitochondria within uninfected hepatocytes do not depolarize or undergo MPT in response to TNFα (E, F). (Reproduced with permission from Bradham *et al.*, *Mol. Cell. Biol.,* Copyright 1998, American Society for Microbiology.)

exposure to TNFα in an asynchronous manner within a given cell, so that initially only a subset of mitochondria are affected (Fig. 3B). Between 9 and 11 hours the extent of involvement increases (Fig. 3C), and by 12 hours, the MPT and depolarization are complete in the majority of mitochondria (Fig. 3D). Interestingly, the onset of the MPT is completely inhibited in the presence of NF-κB activation (Fig. 3E–F), indicating that this transcription factor protects against apoptosis at or upstream of the mitochondria.

2.2. The Mitochondrial Permeability Transition Precedes and is Required for Cytochrome *c* Release and Caspase 3 Activation

Onset of the MPT at approximately 8 hours after TNFα treatment occurs prior to several other apoptotic events. Nuclear condensation begins at 10 hours after TNFα treatment, and cytochrome *c* release and caspase 3 activation occur at 9 and 11 hours after treatment, respectively (Fig. 4). Membrane blebbing is first observed 12 hours after TNFα treatment (Fig. 3D). The results support a model in which the MPT is required for these subsequent changes.

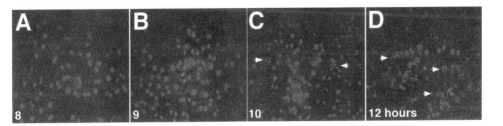

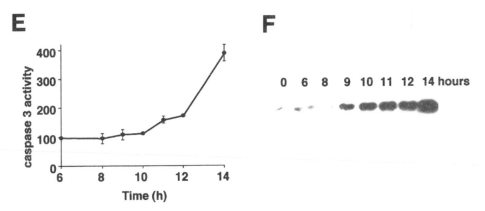

FIGURE 4. Onset of the MPT precedes nuclear condensation, caspase 3 activation, and cytochrome *c* release. A time-course of nuclear morphology (A–D) shows the first nuclear condensation detection at 10 hr after TNFα treatment. Caspase 3 activity was assessed by incubating cellular extracts with DEVD–AFC, the peptide substrate for caspase 3 (DEVD) conjugated with the fluorescent tag AFC, then measuring AFC release. A time course shows caspase 3 activation at 11 hr after TNFα treatment (E). Cytochrome *c* release was determined by preparation of S100 fractions, followed by cytochrome *c* assessment using Western blotting. A time course shows cytochrome *c* release at 9 hr after TNFα treatment. (Reproduced with permission from Bradham *et al.*, *Mol. Cell. Biol.*, Copyright 1998, American Society for Microbiology.)

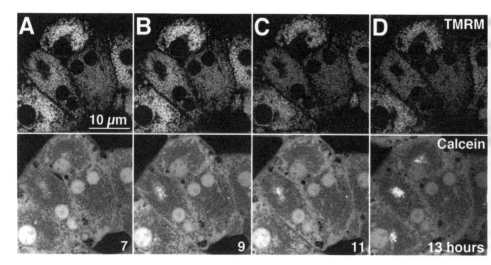

FIGURE 5. CsA blocks onset of the MPT. Primary hepatocytes expressing the IκB SR were treated with TNFα and CsA (2 μM), and mitochondria were then visualized with confocal microscopy using TMRM and calcein, as in Figure 3. CsA completely inhibits MPT onset and accompanying depolarization. (Reproduced with permission from Bradham *et al.*, *Mol. Cell. Biol.*, Copyright 1998, American Society for Microbiology.)

To test this model further, the MPT was specifically inhibited using cyclosporin A (CsA), an immunosuppressive endecapeptide that inhibits MPT onset via a cyclophilin in the mitochondrial matrix (Halestrap *et al.*, 1997) (Fig. 5). The MPT-inhibitory effect is distinct from the immunosuppressive effect of CsA, because nonimmunosuppressive variants of CsA still inhibit transition, while the immunosuppressive agent FK506 has no MPT effect (Henke and Jung, 1993; Schweizer *et al.*, 1993). Treatment of IκB SR-expressing hepatocytes with TNFα and CsA inhibits apoptotic cell death (Fig. 6A). In an inhibitor-chase experiment, hepatocytes were first treated with TNFα, then CsA was added at time points afterward and viability was assessed after 24 hours. CsA protected cells when added up to 8 hours after TNFα, but lost this effect at 10 or more hours (Fig. 6B), indicating that CsA protects via inhibition of the MPT, and confirming onset between 8 and 10 hours after TNFα treatment.

These results show that in hepatocytes, MPT onset is required for apoptosis, in contrast to reports from other laboratories, which suggested that the MPT and its accompanying mitochondrial depolarization occur after apoptosis has been executed, once the cell is officially "dead" (Bossy-Wetzel *et al.*, 1998; Kluck *et al.*, 1997; Yang *et al.*, 1997). The mitochondrial role in apoptotic signaling may differ in various cell types (Scaffidi *et al.*, 1998). The majority of apoptotic studies use transformed cell lines, which may have very different basal signaling compared with nontransformed primary cultures. In addition, many studies assess lymphocyte-derived cell lines. It is well established that many unique signaling events occur in lymphocytes, which may not reflect a "generalized" cell.

Treatment of IκB SR-expressing primary hepatocytes with TNFα and CsA blocks cytochrome *c* release and inhibits caspase 3 activation, demonstrating that transition onset is necessary for cytochrome *c* release and subsequent caspase 3 activation (Fig. 7),

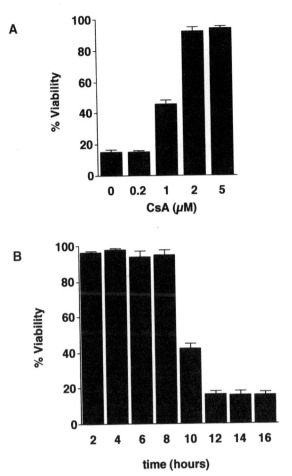

FIGURE 6. CsA inhibits TNFα-mediated apoptosis. (A) A dose–response experiment shows that 2 μM CsA completely inhibits TNFα-mediated apoptosis in hepatocytes expressing the IκB SR. (B) In an inhibitor-chase experiment, IκB SR-expressing hepatocytes were treated with TNFα, then CsA was added at time points afterward. Cell viability was assessed after 24 hr of TNFα treatment. The CsA showed a protective effect when added up to 8 hr after TNFα, but the effect is lost at 10 hr or more. (Reproduced with permission from Bradham *et al.*, *Mol. Cell. Biol.*, Copyright 1998, American Society for Microbiology.)

supporting a model for cytochrome *c* release in which the onset of the MPT causes matrix swelling that leads to outer mitochondrial membrane rupture. This rupture then releases cytochrome *c* to activate caspase 9 and then caspase 3. Several reports have shown that cytochrome *c* release occurs in cells that contain polarized mitochondria, in apparent disagreement with this model (Kluck *et al.*, 1997; Li *et al.*, 1998; Yang *et al.*, 1997), but these observations are not necessarily inconsistent and may instead reflect the gradual nature of mitochondrial changes during apoptosis, in contrast to the rapid mitochondrial changes in necrotic cells. In primary hepatocyte cultures, at 9 hours after TNFα treatment a subset of mitochondria have undergone the MPT and depolarized, and cytochrome *c*

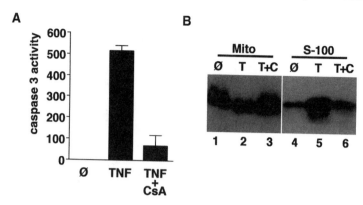

FIGURE 7. CsA inhibits caspase 3 activation and cytochrome *c* release. IκB SR-expressing hepatocytes were either untreated (Ø), or stimulated with TNFα alone or with CsA. (A) Caspase 3 activation (measured as in Fig. 4) is strongly inhibited by CsA. (B) Cytochrome *c* levels were assessed in both mitochondria and S100 fractions using Western blotting. TNFα-mediated cytochrome *c* release is blocked by CsA (lanes 3 and 6). (Reproduced with permission from Bradham, *et al.*, *Mol. Cell. Biol.*, Copyright 1998, American Society for Microbiology.)

release has also been initiated, but the larger mitochondrial fraction remains intact. Thus, though mitochondrial observations by FACS analysis suggest that the mitochondria have not undergone the MPT at the time of cytochrome *c* release, the fact that CsA treatment inhibits cytochrome *c* release indicates that the MPT is indeed required for this redistribution, which in turn is required for caspase 3 activation.

2.3. The Mitochondrial Permeability Transition as a Component of the Signaling Cascade

When a truncated version of the FADD protein (ΔFADD), which binds to TRADD but cannot interact with caspase 8 (Chinnaiyan *et al.*, 1996), is expressed along with the IκB SR, hepatocytes are protected from TNFα-mediated apoptosis and do not activate caspase 3 (Fig. 8). In these cells TNFα does not induce the MPT (Fig. 9A, 9B), demonstrating that the MPT onset lies downstream from FADD and along the apoptotic pathway.

In a parallel series of experiments, the serpin inhibitor crmA was expressed along with the IκB SR. The viral protein crmA binds and inhibits caspases (Ray *et al.*, 1992). Caspase 1 and Caspase 8, both considered "upstream" caspases, are most potently inhibited by crmA (Zhou *et al.*, 1997). Expression of crmA prevents cell death and caspase 3 activation (Fig. 8), and similar to ΔFADD, blocks the onset of the MPT (Figure 9C, 9D), showing that onset requires caspase activity, and that the transition lies downstream of one or more caspases.

This is consistent with earlier studies indicating that crmA cannot block cell death induced by the pro-apoptotic mitochondrial proteins Bax, Bik, and Bak, indicating that a crmA-sensitive caspase acts upstream of the mitochondria (Cheng *et al.*, 1997; Orth and Dixit, 1997). The most probable candidate is caspase 8. Recent studies (Li *et al.*, 1998; Luo *et al.*, 1998) show that caspase 8 cleaves the Bcl-2-interacting protein Bid. The C-terminal portion of Bid translocates to mitochondria, where it binds and inhibits Bcl-2,

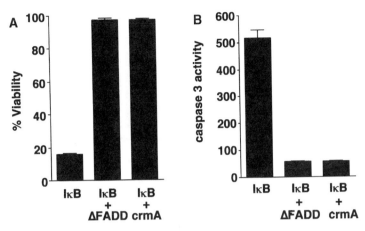

FIGURE 8. ΔFADD and crmA block TNFα-mediated apoptosis and caspase 3 activation. Adenoviral-mediated co-expression of the IκB SR and either the dominant negative ΔFADD or the caspase 8 inhibitor crmA blocks both TNFα-mediated cell death (A) and caspase 3 activation (B). (Reproduced with permission from Bradham, *et al.*, *Mol. Cell. Biol.*, Copyright 1998, American Society for Microbiology.)

thereby mediating cytochrome *c* release. Thus an unbroken pathway from the TNRF I to mitochondria has now been defined (Fig. 12).

The preferred peptide substrate for caspase 3 is DEVD (Asp-Glu-Val-Asp) (Nicholson *et al.*, 1995). When the non-cleavable aldehyde derivative DEVD-cho is incubated with cells expressing the IκB SR, cell death is prevented (Fig. 10). In contrast to ΔFADD

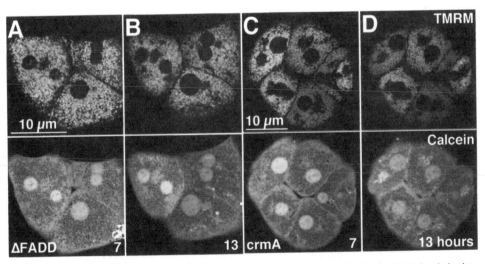

FIGURE 9. ΔFADD and crmA block TNFα-induced onset of the MPT and mitochondrial depolarization. Adenoviral-mediated co-expression of the IκB SR and either the dominant negative ΔFADD (A, B) or the caspase 8 inhibitor crmA (C, D) blocks TNFα-induced transition and depolarization (measured as in Fig. 3). (Reproduced with permission from Bradham *et al.*, *Mol. Cell. Biol.*, Copyright 1998, American Society for Microbiology.)

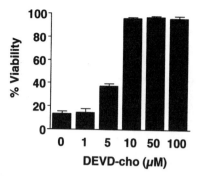

FIGURE 10. DEVD-cho blocks TNFα-mediated apoptosis. IκB SR-expressing primary hepatocytes were treated with TNFα and a dose–response of DEVD-cho. 10 μM DEVD-cho was the minimal protective dose. (Reproduced with permission from Bradham *et al.*, *Mol. Cell. Biol.*, Copyright 1998, American Society for Microbiology.)

and crmA, however, a minimal dose of DEVD-cho does not inhibit MPT onset or mitochondrial depolarization (Fig. 11), indicating that the transition lies upstream of caspase 3, and presumably other executionary caspases. These results in primary hepatocytes rule out the possibility that mitochondrial depolarization and permeability transition result from the apoptotic process, showing instead that MPT onset is an integral part of the caspase cascade in primary hepatocytes, lying between caspase 8 and caspase 3.

2.4. Role of Bcl2 Proteins

The Bcl-2 and Bcl-x$_L$ proteins inhibit both MPT onset and cytochrome *c* release (Kluck *et al.*, 1997; Susin *et al.*, 1996; Yang *et al.*, 1997; Zamzami *et al.*, 1996), implying

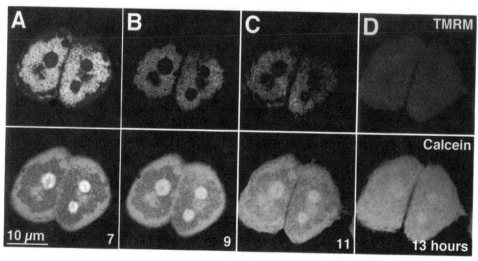

FIGURE 11. DEVD-cho does not inhibit onset of the MPT or depolarization. IκB SR-expressing primary hepatocytes were treated with TNFα and 10 μM DEVD-cho, then mitochondria were visualized as in Figure 3. TNFα-induced mitochondrial changes were not affected by inhibition of caspase-3. (Reproduced with permission from Bradham *et al.*, *Mol. Cell. Biol.*, Copyright 1998, American Society for Microbiology.)

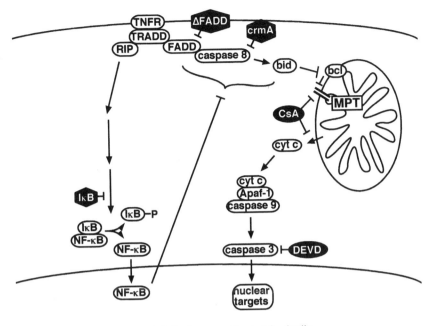

FIGURE 12. Summary. See text for details.

that the primary function for Bcl is proteins MPT inhibition, as previously suggested (Susin *et al.*, 1997). Both Bcl-2 and Bcl-x$_L$ are localized to mitochrondrial membrane contact sites (de Jong *et al.*, 1994) and can form ion channels in model membranes (Schlesinger *et al.*, 1997). This is supported by studies showing that Bcl-x$_L$ regulates mitochondrial volume homoeostasis, preventing MPT swelling (Vander Heiden *et al.*, 1997), and that Bcl-2 maintains mitochondrial polarization by enhancing proton efflux in the presence of uncouplers (Shimizi *et al.*, 1998). The pro-apoptotic Bcl-2 family member Bax induces cytochrome *c* release in a CsA-inhibitable manner (Pastorino *et al.*, 1998). Further, cells that are insensitive to injected cytochrome *c* are still protected from TNFα-mediated apoptosis by Bcl-x$_L$ (Li *et al.*, 1997a; Rosse *et al.*, 1998), indicating a role for Bcl proteins aside from regulating cytochrome *c* release.

3. CONCLUSIONS

The mitochrondrial permeability transition plays a key role in TNFα-mediated apoptosis in primary hepatocytes (Fig. 12). In this model, TNFR I activation recruits FADD and procaspase 8 to the receptor via interaction with TRADD. Auto-activated caspase 8 cleaves Bid, which then translocates to and inhibits mitochondrial Bcl-2, permitting MPT onset. The MPT causes matrix swelling and outer membrane rupture, thus releasing cytochrome *c*. Cytochrome *c* complexes with Apaf 1, procaspase 9, and dATP, which induces caspase 9 activation. Caspase 9 cleaves and activates caspase 3, which results in apoptosis execution.

Activated NF-κB protects cells from TNFα-mediated apoptosis by blocking MPT onset, as indicated by comparison of normal cells with cells expressing the IκB SR, delivered by adenoviral-mediated gene transfer (see Fig. 3). This transition-inhibitory effect is recapitulated by adenoviral-mediated co-expression of the IκB SR with either ΔFADD or crmA (see Fig. 9). Direct inhibition of the transition with CsA (Fig. 5) blocks cytochrome *c* release and caspase 3 activation (Fig. 7) as well as inhibiting apoptotic cell death (Fig. 6). Inhibition of caspase 3 with DEVD-cho did not inhibit transition onset (Fig. 11), although it completely inhibited cell death (Fig. 10), confirming that MPT onset is upstream of caspase 3 activation.

The mechanism underlying the protective role of NF-κB remains unexplained. The NF-κB induces expression of the caspase inhibitors IAP-1 and -2 (*i*nhibitors of *a*poptosis *P*rotein), which prevent apoptotic cell death (Chu *et al.*, 1997; You *et al.*, 1997). It is unlikely, however, that IAPs could block the MPT onset, because these proteins bind and inhibit the late-acting caspase 3, 6, 7, and 9, but not caspase 8 (Deveraux *et al.*, 1998; Roy *et al.*, 1997). Further, IAPs inhibit apoptosis induced by Bik and Bak, indicating that they function downstream of mitochondria (Orth and Dixit, 1997). Identification of NF-κB-inducible gene products that can interfere with the MPT onset will contribute to understanding the mechanism by which TNFα induces either proliferation or apoptosis.

REFERENCES

Alnemri, E. S., 1997, Mammalian cell death proteases: A family of highly conserved aspartate specific cysteine proteases, *J. Cell. Biochem.* **64**:33–42.

Banner, D., D'Arcy, A., Janes, W., Gentz, R., Schoenfeld, H. J., Broger, C., Loetscher, H., and Lesslauer, W., 1993, Crystal structure of the soluble human 55 kd TNF receptor–human TNF beta complex: Implications for TNF receptor activation, *Cell* **73**:431–445.

Bossy-Wetzel, E., Newmeyer, D. D., and Green, D. R., 1998, Mitochondrial cytochrome *c* release in apoptosis occurs upstream of DEVD-specific caspase activation and independently of mitochondrial transmembrane depolarization, *EMBO J.* **17**:37–49.

Bradham, C. A., Plumpe, J., Manns, M. P., Brenner, D. A., and Trautwein, C., 1998a, Mechanisms of hepatic toxicity: I TNF-induced liver injury, *Am. J. Physiol.* **275**:G387–G392.

Bradham, C. A., Qian, T., Streetz, K., Brenner, D. A., and Lemasters, J. J., 1998b, The mitochondrial permeability transition is required for TNFα-mediated apoptosis and cytochrome *c* release, *Mol. Cell. Biol.* **18**:6353–6364.

Cheng, E. H.-Y., Kirsh, D. G., Clem, R. J., Ravi, R., Kastan, M. B., Bedi, A., Ueno, K., and Hardwick, J. M., 1997, Conversion of bcl-2 to a bax-like death effector by caspases, *Science* **278**:1966–1968.

Chinnaiyan, A. M., O'Rourke, K., Tewari, M., and Dixit, V. M., 1995, FADD, a novel death domain-containing protein, interacts with the death domain of Fas and initiates apoptosis, *Cell* **81**:505–512.

Chinnaiyan, A. M., Tepper, C. G., Seldin, M. F., O'Rourke, K., Kischkel, F. C., Hellbardt, S., Krammer, P. H., Peter, M. E., and Dixit, V. M., 1996, FADD/MORT1 is a common mediator of CD95 (Fas/APO-1) and tumor necrosis factor receptor-induced apoptosis, *J. Biol. Chem.* **271**:4961–4965.

Chu, Z. L., McKinsey, T. A., Liu, L., Gentry, J. J., Malim, M. H., and Ballard, D. W., 1997, Suppression of tumor necrosis factor-induced cell death by inhibitor of apoptosis c-IAP2 is under NF-κB control, *Proc. Natl. Acad. Sci. USA* **94**:10057–10062.

de Jong, D., Prins, F. A., Mason, D. Y., Reed, J. C., van Ommen, G. B., and Kluin, P. M., 1994, Subcellular localization of the bcl-2 protein in malignant and normal lymphoid cells, *Cancer Res.* **54**:256–260.

Deveraux, Q. L., Roy, N., Stennicke, H. R., Van Arsdale, T., Zhou, Q., Srinivasula, S. M., Alnemri, E. S., Salveson, G. S., and Reed, J. C., 1998, IAPs block apoptotic events induced by caspase 8 and cytochrome *c* by direct inhibition of distinct caspases, *EMBO J.* **17**:2215–2223.

DiDonato, J. A., Hayakawa, M., Rothwarf, D. M., Zandi, E., and Karin, M., 1997, A cytokine-responsive IκB kinase that activates the transcription factor NF-κB, *Nature* **388**:548–554.

Enari, M., Talanian, R. V., Wong, W. W., and Nagata, S., 1996, Sequential activation of ICE-like and CPP32-like proteases during Fas-mediated apoptosis, *Nature* **380**:723–726.

Enari, M., Shakahira, H., Yokoyama, H., Okawa, K., Iwamatsu, A., and Nagata, S., 1998, A caspase-activated DNase that degrades DNA during apoptosis, and its inhibitor ICAD, *Nature* **391**:43–50.

Halestrap, A. P., Connern, C. P., Griffiths, E. J., and Kerr, P. M., 1997, Cyclosporin A binding to mitochondrial cyclophilin inhibits the permeability transition pore and protects hearts from ischemia/reperfusion injury, *Mol. Cell. Biochem.* **174**:167–172.

Henke, W., and Jung, K., 1993, Comparison of the effects of the immunosuppressive agents FK 506 and cyclosporin A on rat kidney mitochondria, *Biochem. Pharm.* **45**:829–832.

Hsu, H., Huang, J., Shu, H. B., Baichwal, V., and Goeddel, D. V., 1996a, TNF-dependent recruitment of the protein kinase RIP to the TNF receptor-1 signaling complex, *Immunity* **4**:387–396.

Hsu, H., Shu, H.-B., Pan, M.-G., and Goeddel, D. V., 1996b, TRADD–TRAF2 and TRADD–FADD interactions define two distinct TNF receptor 1 signal transduction pathways, *Cell* **84**:299–308.

Hsu, H., Xiong, J., and Goeddel, D. V., 1995, The TNF receptor 1-associated protein TRADD signals cell death and NF-κB activation, *Cell* **81**:495–504.

Hu, Y., Benedict, M. A., Wu, D., Inohara, N., and Nunez, G., 1998, Bcl-xL interacts with Apaf-1 and inhibits Apaf-1-dependent caspase 9 activation, *Proc. Natl. Acad. Sci. USA* **95**:4386–4391.

Iimura, Y., Nishiura, T., Hellerbrand, C., Behrns, K. E., Schoonhoven, R., Grisham, J. W., and Brenner, D. A., 1998, NF-κB prevents apoptosis and liver dysfunction during liver regeneration, *J. Clin. Invest.* **101**:802–811.

Itoh, N., and Nagata, S., 1993, A novel protein domain required for apoptosis, *J. Biol. Chem.* **268**:10932–10937.

Jones, E. Y., Stuart, D. I., and Walker, N. P., 1989, Structure of tumour necrosis factor, *Nature* **338**:225–228.

Juo, P., Kuo, C. J., Yuan, J., and Blenis, J., 1998, Essential requirement for caspase 8/FLICE in the initiation of the Fas-induced apoptotic cascade, *Curr. Biol.* **8**:1001–1008.

Kettelhut, I. C., Fiers, W., and Goldberg, A. L., 1987, The toxic effects of tumor necrosis factor *in vivo* and their prevention by cyclooxygenase inhibitors, *Proc. Natl. Acad. Sci. USA* **84**:4273–4277.

Kluck, R. M., Bossy-Wetzel, E., Green, D. R., and Newmeyer, D. D., 1997, The release of cytochrome *c* from mitochondria: A primary site for Bcl-2 regulation of apoptosis, *Science* **275**:1132–1136.

Kuida, K., Zheng, T. S., Na, S., Kuan, C., Yang, D., Karasuyama, H., Rakic, P., and Flavell, R., 1996, Decreased apoptosis in the brain and premature lethality in CPP32-deficient mice, *Nature* **384**:368–372.

Kuida, K., Haydar, T. F., Kuan, C. Y., Gu, Y., Taya, C., Karasuyama, H., Su, M. S., Rakic, P., and Flavell, R. A., 1998, Reduced apoptosis and cytochrome c-mediated caspase activation in mice lacking caspase 9, *Cell* **94**:325–337.

Leist, M., Gantner, F., Bohlinger, I., Germann, P. G., Tiegs, G., and Wendel, A., 1994, Murine hepatocyte apoptosis induced *in vitro* and *in vivo* by TNF-alpha requires transcriptional arrest, *J. Immunol.* **153**:1778–1787.

Li, F., Srinivasan, A., Wang, Y., Armstrong, R. C., Tomaselli, K. J., and Fritz, L. C., 1997a, Cell-specific induction of apoptosis by microinjection of cytochrome *c*: Bcl-x$_L$ has activity independent of cytochrome *c* release, *J. Biol. Chem.* **272**:30299–30305.

Li, P. D. N., Budihardjo, I., Srinivasula, S. M., Ahmed, M., Alnemri, E. S., and Wang, X., 1997b, Cytochrome *c* and dATP-dependent formation of Apaf-1/caspase 9 complex initiates an apoptotic protease cascade, *Cell* **91**:479–489.

Li, H., Zhu, H., Xu, C.-J., and Yuan, J., 1998, Cleavage of BID by caspase 8 mediates the mitochondrial damage in the Fas pathway of apoptosis, *Cell* **94**:491–501.

Liu, X., Kim, C. N., Yang, J., Jemmerson, R., and Wang, X., 1996, Induction of apoptosis in cell-free extracts: Requirement for dATP and cytochrome *c*, *Cell* **86**:147–157.

Liu, X., Zou, H., Slaughter, C., and Wang, X., 1997, DFF, a heteromeric protein that functions downstream of caspase 3 to trigger DNA fragmentation during apoptosis, *Cell* **89**:175–184.

Luo, X., Budihardjo, I., Zou, H., Slaughter, C., and Wang, X., 1998, Bid, a Bcl2 interacting protein, mediates cytochrome *c* release from mitochondria in response to activation of cell surface death receptors, *Cell* **94**:481–490.

Malinin, N. L., Boldin, M. P., Kovalenko, A. V., and Wallach, D., 1997, MAP3K-related kinase involved in NF-κB induction by TNF, CD-95, and IL-1, *Nature* **385**:540–544.

Marchetti, P., Castedo, M., Susin, S. A., Zamzami, N., Hirsch, T., Macho, A., Haeffner, A., Hirsch, F., Geuskens, M., and Kroemer, G., 1996, Mitochondrial permeability transition is a central coordinating event of apoptosis, *J. Exp. Med.* **184**:1155–1160.

Martin, D. A., Siegel, R. M., Zheng, L., and Lenardo, M. J., 1998, Membrane oligomerization and cleavage activates the caspase 8 (FLICE/MACHα1) death signal, *J. Biol. Chem.* **273**:4345–4349.

Muzio, M., Chinnaiyan, A. M., Kischkel, F. C., O'Rourke, K., Shevchenko, A., Ni, J., Scaffidi, C., Bretz, J. D., Zhang, M., Gentz, R., Mann, M., Krammer, P. H., Peter, M. E., and Dixit, V. M., 1996, FLICE, a novel FADD-homologous ICE/CED-3-like protease, is recruited to the CD95 (FAS/APO-1) death-inducing signaling complex, *Cell* **85**:817–827.

Natoli, G., Costanzo, A., Ianni, A., Templeton, D. J., Woodjet, J. R., Balsano, C., and Levrero, M., 1997, Activation of SAPK/JNK by TNF receptor 1 through a nontoxic TRAF2-dependent pathway, *Science* **275**:200–203.

Nicholson, D. W., Ali, A., Thornberry, N. A., Vaillancourt, J. P., Ding, C. K., Gallant, M., Gareau, Y., Griffen, P. R., Labelle, M., Lazebnik, Y. A., Munday, N. A., Raju, S. M., Smulson, M. E., Yamin, T.-T., Yu, V. L., and Miller, D. K., 1995, Identification and inhibition of the ICE/CED-3 protease necessary for mammalian apoptosis, *Nature* **376**:37–43.

Orth, K., and Dixit, V. M., 1997, Bik and Bak induce apoptosis downstream of crmA but upstream of inhibitor of apoptosis, *J. Biol. Chem.* **272**:8841–8844.

Pastorino, J. G., Chen, S. T., Tafani, M., and Farber, J. L., 1998, The overexpression of Bax produces cell death upon induction of the mitochondrial permeability transition, *J. Biol. Chem.* **273**:7770–7775.

Ray, C. A., Black, R. A., Kronheim, S. R., Greenstreet, T. A., Sleath, P. R., Galvesen, G. S., and Pickup, D. J., 1992, Viral inhibition of inflammation: Cowpox virus encodes an inhibitor of the interleukin-1β converting enzyme, *Cell* **69**:597–604.

Rosen, A., and Casciola-Rosen, L., 1997, Macromolecular substrates for the ICE-like proteases during apoptosis, *J. Cell. Biochem.* **64**:50–54.

Rosse, T., Olivier, R., Monney, L., Rager, M., Conus, S., Fellay, I., Jansen, B., and Borner, C., 1998, Bcl-2 prolongs cell survival after bax-induced release of cytochrome *c*, *Nature* **391**:496–499.

Roy, N., Deveraux, Q. L., Takahashi, R., Salveson, G. S., and Reed, J. C., 1997, The c-IAP-1 and c-IAP-2 proteins are direct inhibitors of specific caspases, *EMBO J.* **16**:6914–6925.

Salveson, G. S., and Dixit, V. M., 1997, Caspases: Intracellular signaling by proteolysis, *Cell* **91**:443–446.

Scaffidi, C., Fulda, S., Srinivasan, A., Fiesen, C., Li, F., Tomaselli, K. J., and Peter, M. E., 1998, Two CD95 (APO-1/Fas) signaling pathways, *EMBO J.* **17**:1675–1687.

Schlesinger, P. H., Gross, A., Yin, X.-M., Yamamoto, K., Saito, M., Waksman, G., and Korsmeyer, S. J., 1997, Comparison of the ion channel characteristics of proapoptotic BAX and antiapoptotic BCL-2, *Proc. Natl. Acad. Sci. USA* **94**:11357–11362.

Schweizer, M., Schengel, J., Baumgartner, D., and Richer, C., 1993, Sensitivity of mitochondrial peptidyl-prolyl *cis-trans* isomerase, pyridine nucleotide hydrolysis, and Ca2+ release to cyclosporin A and related compounds, *Biochem. Pharm.* **45**:641–646.

Shimizi, S., Eguchi, Y., Kamiike, W., Funahashi, Y., Mignon, A., Lacronique, V., Matsuda, H., and Tsujimoto, Y., 1998, Bcl-2 prevents apoptotic mitochondrial dysfunction by regulating proton flux, *Proc. Natl. Acad. Sci. USA* **95**:1455–1459.

Susin, S. A., Zamzami, N., Castedo, M., Hirsch, T., Marchetti, P., Macho, A., Daugas, E., Geuskens, M., and Kroemer, G., 1996, Bcl-2 inhibits the mitochondrial release of an apoptogenic protease, *J. Exp. Med.* **184**:1331–1341.

Susin, S. A., Zamzami, N., Castedo, M., Daugas, E., Wang, H.-G., Geley, S., Fassy, F., Reed, J. C., and Kroemer, G., 1997, The central executioner of apoptosis: Multiple connections between protease activation and mitochondria in Fas/APO-1/CD95- and ceramide-induced apoptosis, *J. Exp. Med.* **186**:25–37.

Tartaglia, L. A., Ayres, T. M., Wong, G. H. W., and Goeddel, D. V., 1993, A novel domain within the 55 kd TNF receptor signals cell death, *Cell* **74**:845–853.

Ting, A. T., Pimentel-Muinos, F. X., and Seed, B., 1996, RIP mediates tumor necrosis factor receptor 1 activation of NF-kappaB but not Fas/APO-1-initiated apoptosis, *EMBO J.* **15**:6189–6196. ·

Vander Heiden, M. G., Chandel, N. S., Williamson, E. K., Schumacker, P. T., and Thompson, C. B., 1997, Bcl-x$_L$ regulates the membrane potential and volume homeostasis of mitochondria, *Cell* **91**:627–637.

Varfolomeev, E. E., Schuchmann, M., Luria, V., Chiannilkulchai, N., Beckmann, J. S., Mett, I. L., Rebrikov, D., Brodianski, V. M., Kemper, O. C., Kollet, O., Lapidot, T., Soffer, D., Sobe, T., Avraham, K. B., Goncharov, T., Holtmann, H., Lonai, P., and Wallach, D., 1998, Targeted disruption of the mouse caspase 8 gene ablates cell death induction by TNF receptors, Fas/Apo 1, and DR3 and is lethal prenatally, *Immunity* **9**:267–276.

Wallach, D., Boldin, M., Varfolomeev, E., Beyaert, R., Vandenabeele, P., and Fiers, W., 1997, Cell death induction by receptors of the TNF family: Toward a molecular understanding, *FEBS Lett.* **410**:96–106.

Wang, C.-Y., Mayo, M. W., and Baldwin, A. S. J., 1996, TNF- and cancer therapy-induced apoptosis: Potentiation by inhibiting NF-κB, *Science* **274**:784–787.

Yang, J., Liu, X., Bhalla, K., Kim, C. N., Ibrado, A. M., Cai, J., Peng, T.-I., Jones, D. P., and Wang, X., 1997, Prevention of apoptosis by bcl-2: Release of cytochrome *c* from mitochondria blocked, *Science* **275**:1129–1132.

Yeh, W. C., Shaninian, A., Speiser, D., Kraunus, J., Billia, F., Wakeham, A., de la Pompa, J. L., Ferrick, D., Hum, B., Iscove, N., Ohashi, P., Rothe, M., Goeddel, D. V., and Mak, T. M., 1997, Early lethality, functional NF-κB activation, and increased sensitivity to TNF-induced cell death in TRAF2-deficient mice, *Immunity* **7**:715–725.

You, M., Ku, P. T., Hrdlickova, R., and Bose, H. R. J., 1997, ch-IAP, a member of the inhibitor of apoptosis protein family, is a mediator of the antiapoptotic activities of the v-Rel oncoprotein, *Mol. Cell. Biol.* **17**:7328–7341.

Zamzami, N., Susin, S. A., Marchetti, P., Hirsch, T., Gomez-Monterrey, I., Castedo, M., and Kroemer, G., 1996, Mitochondrial control of nuclear apoptosis, *J. Exp. Med.* **183**:1533–1544.

Zhou, Q., Snipas, S., Orth, K., Muzio, M., Dixit, V. M., and Salveson, G. S., 1997, Target protease specificity of the viral serpin crmA: Analysis of five caspases, *J. Biol. Chem.* **272**:7797–7800.

Zou, H., Henzel, W. J., Liu, X., Lutschg, A., and Wang, X., 1997, Apaf-1, a human protein homologous to *C. elegans* CED-4, participates in cytochrome *c*-dependent activation of caspase 3, *Cell* **90**:405–413.

Chapter 13

The ATP Switch in Apoptosis

David J. McConkey

1. INTRODUCTION

The term *apoptosis* was originally coined by Kerr, Wyllie, and Currie to describe a pattern of morphological alterations associated with normal programmed cell death and certain pathological processes *in vivo* (Kerr *et al.*, 1972). These include cell shrinkage and loss of contact with neighboring cells, formation of cytoplasmic vacuoles, plasma and nuclear membrane blebbing, and chromation condensation (Wyllie *et al.*, 1980b). In addition, apoptosis typically affects scattered cells in tissues that are removed by phagocytic cells (i.e., macrophages) before plasma membrane integrity is lost (Savill *et al.*, 1993). This clearance is triggered by specific surface changes on the apoptotic cell, the most familiar of which is the "flipping" of phosphatidylserine from the inner to outer leaflet of the plasma membrane (Fadok *et al.*, 1992). In contrast, necrosis usually occurs within tracts of contiguous cells that lose plasma membrane integrity at an early phase of cell death. This results in release of intracellular debris in the tissue, triggering an inflammatory response and collateral damage. Thus, morphologically and functionally, a strict dichotomy between apoptosis and necrosis appears to exist (Table I).

Subsequent investigation into the biochemical requirements for apoptosis helped to distinguish it from necrosis. For example, early work in many different model systems demonstrated that programmed cell death was abrogated by protein and mRNA synthesis inhibitors (i.e., cycloheximide and actinomycin D) (Cohen and Duke, 1984; Lockshin, 1969; Tata, 1966; Wyllie *et al.*, 1984), indicating that apoptosis required active gene expression. Furthermore, apoptosis was shown to be sensitive to control by levels of

David J. McConkey   Department of Cancer Biology, University of Texas M.D. Anderson Cancer Center, and Program in Toxicology, University of Texas-Houston Graduate School of Biomedical Sciences, Houston, Texas 77030.

Mitochondria in Pathogenesis, edited by Lemasters and Nieminen.
Kluwer Academic/Plenum Publishers, New York, 2001.

Table I
Morphological and Functional Distinctions Between Apoptosis and Necrosis

Apoptosis	Necrosis
Active	Passive
Shrinkage	Swelling
Intact membrane	Ruptured membrane
No inflammation	Inflammation
Multiple triggers	Toxicant-specific mechanisms
Caspase-dependent	Caspase-independent

steroid hormones (Schwartz and Truman, 1982; Wyllie *et al.*, 1980a) and cytokines acting via their specific receptors (Laster *et al.*, 1988; Williams *et al.*, 1990), and a variety of signal transduction intermediates were implicated in either stimulation or inhibition of apoptosis (McConkey and Orrenius, 1996b). Finally, Wyllie (1980a) demonstrated that apoptosis in glucocorticoid-treated thymocytes was associated with endogenous endonuclease activation, usually leading to the formation of the characteristic oligonucleosomal DNA fragments (DNA ladders) used to diagnose the process to this day. Importantly, where interrogated, these features of apoptosis were not observed during necrosis, and the idea that apoptosis is an active "suicidal" cell-death response was established.

2. MOLECULAR REGULATION OF APOPTOSIS

For years scientists remained somewhat skeptical about the relevance of apoptosis, largely because no molecular intermediates within the pathway had been identified. Although this skepticism was somewhat ameliorated when endonuclease activation emerged as a biochemical indicator of apoptosis, investigation into the genetic regulation of programmed cell death in the nematode *Caenorhabditis elegans* proved instrumental in defining the molecular control of apoptosis. During development, exactly 1090 cells are produced in the organism, and of these, 131 die with precise timing (Hengartner and Horvitz, 1994b). Mutagenesis-based molecular genetic analysis by Horvitz and colleagues identified two genes, termed CED-3 and CED-4, that are absolutely required for all but one of these cell deaths (Ellis and Horvitz, 1986), and a third gene, CED-9, which functions to block cell death in many (if not all) of the remaining cells (Hengartner *et al.*, 1992).

Molecular cloning of CED-3, CED-4, and CED-9 has subsequently demonstrated that each gene possesses at least one mammalian homologue, and functional studies have demonstrated that the mammalian protein products play central roles in regulating apoptosis. The CED-3 gene was found to be structurally similar to human interleukin-1β-converting enzyme (ICE) (Yuan *et al.*, 1993), a protease required for conversion of the precursor form of IL-1β into its active product (Thornberry *et al.*, 1992). Subsequent work showed that ICE is a member of a larger family of aspartate-specific cysteine proteases, now collectively known as *caspases* (Cohen, 1997). Cloning of CED-9 revealed that it is homologous to the human Bcl2 gene (Hengartner and Horvitz, 1994a), implicated in the suppression of multiple pathways of apoptosis in mammalian cells (Reed, 1997). Finally, work by Zou and colleagues (1997) showed that CED-4 is homologous to a protein termed *apoptosis protease activating factor-1* (Apaf-1), a polypeptide purified by virtue of its ability to indirectly promote activation of caspase-3. Peptide-based or viral caspase inhibitors block nearly all biochemical events known to be associated with apoptosis, demonstrating that caspase activation is required for the response (Cohen, 1997). Furthermore, the human Bcl2 gene can complement loss of CED-9 function in the nematode (Vaux *et al.*, 1992), establishing the functional correspondence of these central genetic regulators. Subsequent work demonstrated that caspases are also required for apoptosis in *Drosophila* (McCall and Steller, 1997). Together these seminal studies established central components of the molecular pathway leading to apoptosis and demonstrated that this pathway is evolutionarily conserved.

3. TRIGGERS FOR APOPTOSIS

Numerous extracellular stimuli can initiate apoptosis (Table II). Yet they can apparently be grouped into two general categories based upon overall mechanism of action. In the first group are polypeptides homologous to the receptor for tumor necrosis factor (TNF) and CD95/Fas, which are directly coupled to the cell death machinery. Members of this family interact physically with caspases via regions in their cytoplasmic domains, known as *death domains*, and receptor oligomerization by ligand directly initiates the caspase proteolytic cascade (Cohen, 1997). In the second group are stimuli that activate the caspases indirectly, either by upregulating cell-death receptor pathway(s) (Friesen *et al.*, 1996) or by more-direct effects on critical biochemical targets within the cell. Included in this second list are treatments that induce DNA damage (irradiation, cancer chemotherapeutic agents), a variety of different toxicants, and certain pharmacological agents (staurosporine, Ca^{2+} ionophores, thapsigargin) that target signal transduction (Thompson, 1995). (Staurosporine was popularized as a global trigger for apoptosis by Weil *et al.*, (1996) who conducted an extensive survey of mammalian cell types and found that only cells at or before the blastomere stage of development are insensitive to its

Table II
Triggers for Apoptosis[a]

Stimulus	Mechanism
Fas, TNF	Direct caspase activation
Staurosporine	Protein kinases
Ca^{2+} ionophores, thapsigargin	Cystolic Ca^{2+} increase
Proteasome inhibitors	Unknown
Ionizing-radiation	DNA damage
Cancer chemotherapeutics	DNA damage, DNA/RNA synthesis
Chemical toxicants	Various
Growth-factor withdrawal	Bad activation (?)[c]
Anoikis (cell detachment)[b]	Unknown

[a]See chapter by Niemin *et al.* for more details.
[b]In these two examples, it is the absence of survival signals that leads to cell death.
[c]Though identification of signals is just beginning, Bad (a Bcl2 family member) seems to be a common target; it is activated after withdrawal of survival factors.

effects. Although staurosporine is a protein kinase antagonist, one group (Meikrantz *et al.*, 1994) has linked its pro-apoptotic effects to activation of cyclin-dependent kinases.)

4. MECHANISMS OF CASPASE ACTIVATION

Other genetic studies in *C. elegans* revealed that CED-9 and CED-4 function upstream of CED-3 in the apoptotic pathway, suggesting that caspase activation might be rate-limiting for apoptosis (Hengartner and Horvitz, 1994b). Precisely how caspase activation is triggered in whole cells remained unclear, however. Early clues came from Hockenbery and co-workers (1990), who demonstrated that a substantial fraction of cellular Bcl2 protein localizes to mitochondria. Subsequent work (Kroemer *et al.*, 1997) showed that loss of mitochondrial transmembrane potential ($\Delta\Psi$) is an early and nearly invariant feature of apoptosis. Substantive evidence for a link between mitochondria and caspase activation has more recently come from Liu *et al.*, (1996), who demonstrated that the electron transport chain intermediate cytochrome *c* is released into the cytosol at an early stage in apoptosis and can directly promote caspase activation in cell extracts. Kroemer's group (1997) also identified another mitochondrial factor (*apoptosis-inducing factor*, or AIF) that can promote caspase activation and DNA fragmentation.

Isolation of Apaf-1 reveals a concrete mechanism for how mitochondrial changes can promote caspase activation (Li *et al.*, 1997). The Apaf-1 (and by analogy, CED-4) appears to function as a scaffold protein (Table II). Once cytochrome *c* is released from mitochondria, it directly binds to Apaf-1 in a dATP-dependent fashion and promotes a conformational change that allows caspase-9 to join the complex, termed the *apoptosome* (Li *et al.*, 1997) (Table II). The structural alterations that follow caspase-9 binding result in its enzymatic activation via a mechanism that requires ATP hydrolysis (Li *et al.*, 1997), and once activated, caspase-9 can proteolytically activate other caspases (including-3, -6, and -7) (Cohen, 1997). Moreover, other recent work indicates that Bcl2 functions in part to prevent loss of mitochondrial membrane potential and release of cytochrome *c* (Kluck *et al.*, 1997; Yang *et al.*, 1997), thereby preventing assembly of the apoptosome. In addition, Bcl2 family members may directly bind Apaf-1/CED-4 (Chinnaiyan *et al.*, 1997; Wu *et al.*, 1997) (Fig. 1), resulting in a more direct inhibition of caspase-9 activation.

Precisely how cytochrome *c* is released from mitochondria remains unclear. Although controversial, the suggestion has been put forward that cytochrome, *c* and AIF escape as a consequence of the *mitochondrial permeability transition* (MPT) (Kroemer *et al.*, 1997), which involves the opening of large transmembrane pores in response to mitochondrial damage allowing ions and polypeptides to emerge from the organelle. What makes this model attractive is that the MPT is triggered by a drop in $\Delta\Psi$, which is tightly associated with apoptosis (Kroemer *et al.*, 1997). Furthermore, known antagonists of the transition channel (i.e, cyclosporin A and bongkrekic acid) block apoptosis (Kroemer *et al.*, 1997; Pastorino *et al.*, 1996). Other work suggests, however, that cytochrome *c* escapes prior to $\Delta\Psi$ drop (Kluck *et al.*, 1997) and that mitochondrial swelling (a known MPT consequence *in vitro*) is not necessarily prominent in apoptotic cells *in vivo* (Wyllie *et al.*, 1980b).

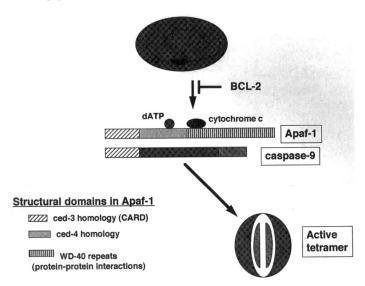

FIGURE 1. Mitochondrial control of caspase activation. Apoptotic stimuli target mitochondria via mechanisms that are still unclear, leading to a drop in $\Delta\Psi$ and release of cytochrome c. In the presence of ATP or dATP, cytochrome c binds to Apaf-1 and promotes interaction with caspase-9, forming a molecular complex sometimes called the *apoptosome*. Hydrolysis of ATP or dATP promotes activation of caspase-9, which can then promote activation of downstream caspases (such as caspase-3 and caspase-7). Anti-apoptotic members of the Bcl2 family can block caspase-9 activation by preventing release of cytochrome c, by preventing assembly or function of the apoptosome, or both.

5. MOLECULAR OVERLAP OF APOPTOSIS AND NECROSIS

Even though the morphological alterations associated with apoptosis and necrosis are (by definition) distinct, accumulating evidence indicates that certain key biochemical mechanisms are shared by the two responses. Studies by Lennon *et al.*, (1991) using a panel of toxicants demonstrated that low to moderate concentrations trigger apoptosis and higher levels trigger necrosis. Bonfoco *et al.* more recent (1995) work established a similar dose-dependent relationship between apoptosis and necrosis in cortical neuronal cells exposed to *N*-methyl-D-aspartate or nitric oxide. Other studies showed that endonuclease activation leading to formation of characteristic "DNA ladders" can occur in certain examples of necrosis (Dong *et al.*, 1997). Indeed, DNA fragmentation is a characteristic feature of necrotic, ischemic injury to the brain *in vivo*, suggesting that this type of necrosis could be physiologically important.

Some of the best evidence for mechanistic overlap of apoptosis and necrosis comes from *in vitro* models of ischemia/reperfusion injury (Lemasters *et al.*, 1987). Hypoxia results in a rapid inhibition of mitochondrial respiration, leading to a steep, quick decline in $\Delta\Psi$ and ATP. These events can be mimicked *in vitro* by incubating cells in a glucose-free medium containing a mitochondrial respiration inhibitor such as cyanide, rotenone, antimycin-A, or oligomycin. The chemical hypoxia ultimately results in necrotic cell death characterized by ATP depletion, $\Delta\Psi$ drop, mitochondrial swelling, and loss of plasma membrane integrity (Lemasters *et al.*, 1987). Strikingly, overexpression of Bcl2 can dramatically inhibit necrosis from chemical hypoxia (Shimizu *et al.*, 1996a, 1996b), challenging the notion that apoptosis and necrosis are regulated by completely independent molecular mechanisms. How then are they ultimately distinguished at the biochemical level? Recent work (Leist *et al.*, 1997; Eguchi *et al.*, 1997) indicates that ATP level involvement, and possibly also involvement of several other important intracellular mediators as well.

6. THE ATP SWITCH IN APOPTOSIS

The implication of phosphoryation and gene expression in apoptosis suggested that the process was energy dependent, an assumption confirmed in several recent studies (Leist *et al.*, 1997; Eguchi *et al.*, 1997). Although early work indicated that energy depletion might be an important pathway of cellular euthanasia (Gaal *et al.*, 1987), apoptosis is not preceded by a rapid decline in ATP level (Ankarcrona *et al.*, 1995; Shimizu *et al.*, 1996a), even though mitochondrial alterations (release of cytochrome *c*) are crucial for caspase activation. Early evidence for a link between ATP levels and apoptosis came from work by Ankarcrona *et al.*, (1995) who showed that glutamate-induced neuronal cell death involved early necrosis, triggered by a rapid $\Delta\Psi$ drop and ATP depletion, then mitochondrial recovery, restoration of ATP levels, and apoptosis.

More recently, independent studies from three laboratories documented that ATP levels can serve as a "switch" between apoptosis and necrosis (Eguchi *et al.*, 1997; Leist *et al.*, 1997; Lelli *et al.*, 1998) These groups used chemical hypoxia to deplete cellular ATP levels and anti-Fas antibodies to trigger apoptosis; Leist and co-workers also employed staurosporine, which, as noted, has been proposed as a universal trigger for apoptosis in mammalian cells. Cells in normal growth medium incubated with either anti-Fas antibody or staurosporine underwent rapid apoptosis not preceded by an early drop in ATP level (Eguchi *et al.*, 1997; Leist *et al.*, 1997). In contrast, cells treated with either stimulus under conditions of chemical hypoxia exhibited marked ATP depletion and necrosis. To determine how much ATP was required for apoptosis, Leist *et al.*, (1997) performed ATP clamping experiments by incubating oligomycin-treated cells in exogenous glucose, which allowed for production of ATP by glycolysis. The results indicated that a 50% cellular ATP depletion was sufficient to switch the mechanism of cell death from apoptosis to necrosis in staurosporine-treated cells, whereas a depletion of approximately 70% was required to switch the mode of cell death in Fas-triggered cells. Similar results were recently obtained by Lieberthal and colleagues (1998) in cultured mouse proximal tubular cells, though in this system the ATP concentration threshold required for apoptosis appears to be 15–25% of control levels. Finally, analysis of the biochemical events involved in cell death demonstrated that caspase activation and DNA fragmentation were suppressed under conditions of ATP depletion (Eguchi *et al.*, 1997; Leist *et al.*, 1997). Thus ATP is required for both protease and endonuclease activation in apoptotic cells.

One interesting observation that emerged indirectly from this concerns the relative kinetics of apoptotic vs necrotic cell death. Although it has been generally assumed that necrosis occurs more rapidly, in the Jurkat cells depletion of ATP led to a 2–3 hour lag in the kinetics of cell death, as measured by staining with fluorescent vital dyes (Leist *et al.*, 1997). Thus the ordered degradation of the cell during apoptosis may accelerate cell death.

7. OTHER BIOCHEMICAL SWITCHES

Although ATP depletion is certainly a critical mechanism that can switch the mode of cell death from apoptosis to necrosis, good evidence is emerging that other biochemical events can dictate the cell death pathway as well, including intracellular Ca^{2+} levels, oxidative stress, and thiol depletion (Table III). A common denominator is that they are often triggered by changes in mitochondrial function, and in all cases direct effects of Bcl2 family members on them have been well documented. In addition, there is significant "cross talk" among these biochemical intermediates, and they can all trigger (or be triggered by) ATP depletion. Evidence for the involvement of these "alternative switches" in apoptosis and necrosis is outlined below.

7.1. Role of Calcium

Alterations in intracellular Ca^{2+} homeostasis have been implicated in both apoptotic and necrotic cell death (Orrenius et al., 1989). Indeed, Ca^{2+} ionophores can induce either response, depending on its concentration (McConkey et al., 1989); the Ca^{2+}-dependent protease calpain has also been implicated in both forms of cell death (Squier et al., 1994; Waters et al., 1997). It appears that the level, nature (sustained vs transient), or both of the cytosolic Ca^{2+} increase is a critical determinant of outcome. Transient increases are often associated with cellular activation, proliferation, or both, whereas low-to-moderate (200–400 nM) sustained increases (McConkey and Orrenius, 1996a) and/or sustained emptying of the endoplasmic reticular (ER) Ca^{2+} pool (Baffy et al., 1993; Lam et al., 1993) are characteristic of programmed cell death. In contrast, uncontrolled, massive Ca^{2+} influx (to levels above 1 μM) is invariably associated with necrosis. The latter can trigger mitochondrial Ca^{2+} overload and subsequent $\Delta\Psi$ collapse and ATP depletion (Richter, 1993), which interferes with caspase activation as described above. In addition, although certain Ca^{2+} levels promote nuclear endonuclease activation in isolated nuclei, higher levels actually block DNA fragmentation (Jones et al., 1989; Vanderbilt et al., 1982), suggesting that the endonuclease itself may be inhibited by high Ca^{2+} levels.

Overexpression of Bcl2 can block Ca^{2+}-mediated cell death. Recent work from Lam et al., (1994) has shown that Bcl2 regulates cytosolic Ca^{2+} fluxes by preventing depletion of the ER Ca^{2+} pool, and there is evidence that the protein also acts at the level of mitochondria (Murphy et al., 1996) and nucleus (Marin et al., 1996) to prevent changes in intracellular Ca^{2+} compartmentalization. Indeed, part of Bcl2's protective effect on

Table III
Biochemical Determinants of Apoptosis and Necrosis

Apoptosis	Necrosis
ATP depletion: 25–70%	ATP depletion: 70–100%
Ca^{2+} increase: 200–400 nM	Ca^{2+} increase: >1 μM
ROI* levels: moderate	ROI levels: high
Progressive GHS depletion	Rapid, extensive thiol oxidation
Inhibition by Bcl2	Inhibition by Bcl2

*Reactive oxygen intermediates.

mitochondria appears linked to its ability to promote mitochondrial Ca^{2+} sequestration (Murphy *et al.*, 1996), which could inhibit $\Delta\Psi$ loss and subsequent oxidative stress during toxic cell killing (Richter, 1993).

7.2. Role of Reactive Oxygen Species

Several studies have shown that classical pathways of apoptosis involve moderate reactive oxygen species (ROS) accumulation (Fernandez *et al.*, 1995; Hockenbery *et al.*, 1993; Slater *et al.*, 1995). Particular attention has lately been paid to the p53-dependent pathway of apoptosis, because work by Polyak *et al.*, (1997) showed that p53 regulates the expression of many genes implicated in the response to oxidative stress. In addition, apoptosis induced in T lymphocytes by HIV infection appears to involve oxidative stress (Israel and Gougerot-Pocidalo, 1997) (and low intracellular glutathione levels independently predict poor long-term survival in HIV-infected patients [Herzenberg *et al.*, 1997]). Yet despite all of this information, precisely how ROS trigger the response remains unclear.

Like Ca^{2+}, oxidative stress promotes either apoptosis or necrosis, depending on level of insult. In early work with menadione (a redox-cycling quinone) in isolated hepatocytes, we showed that low to moderate concentrations stimulated DNA fragmentation, whereas higher levels induced necrosis without DNA fragmentation (McConkey *et al.*, 1988), results recently reproduced in rat pancreatic acinar cells (Sato *et al.*, 1997). Similar dose dependence has been documented for other oxidants, including nitric oxide (Messmer and Brune, 1996), hydrogen peroxide (Hampton and Orrenius, 1997), and 2,3-dimethoxy-1, 4-naphthoquinone (Dybukt *et al.*, 1994).

The simplest interpretation of these results is that high ROS levels directly inhibit apoptotic pathway components (Kazzaz *et al.*, 1996). Caspase activity is preserved by reducing agents (i.e., dithiothreitol) *in vitro*, most likely because the family possesses a potentially redox-sensitive cysteine residue within the active site. Indeed, hydrogen peroxide and nitric oxide can directly inhibit caspase activity *in vivo* (Hampton and Orrenius, 1997; Kim *et al.*, 1997), and other work shows that superoxide can block the Fas pathway (Clement and Stamenkovic, 1996), presumably by interfering with caspase-8/FLICE and possibly other caspases. In addition, given the tight interrelationship between oxidative stress and mitochondrial function (Richter, 1993), it is possible that excessive ROS production leads to excessive depletion of ATP.

Initially, it was thought that Bcl2 might possess direct antioxidant activity, because its overexpression can attenuate lipid peroxidation in cells exposed to oxidants or other inducers of apoptosis (Hockenbery *et al.*, 1993). Indeed, evidence documenting a Bcl2 effect on lipid peroxidation continues to accumulate (Tyurina *et al.*, 1997).

Other work, however, suggests that these effects are secondary to the regulation of intracellular glutathione levels (Kane *et al.*, 1993; Mirkovic *et al.*, 1997; Voehringer *et al.*, 1998) or Ca^{2+} (Lam *et al.*, 1994; Marin *et al.*, 1996) and to maintenance of mitochondrial transmembrane potential (Zamzami *et al.*, 1995). Importantly, the antiapoptotic Bcl2 homolog Bcl-x_L can also modulate intracellular thiols, strongly suggesting that these activities are important for suppression of cell death (Bojes *et al.*, 1997).

7.3. Role of Intracellular Thiols

Reduced glutathione (GSH) and thioredoxin are the major soluble intracellular antioxidants implicated in apoptosis and necrosis regulation. Early work on the mechanisms of toxic cell killing by menadione and other oxidants demonstrated that depletion of intracellular GSH is a common, requisite feature of necrosis (Thor et al., 1982). Similarly, later work showed that GSH depletion also occurs during apoptosis (Backway et al., 1997; Bojes et al., 1997; Fernandez et al., 1995; Macho et al., 1997; Slater et al., 1995). Antioxidants such as N-acetylcysteine (Fernandez et al., 1995; Mayer and Noble, 1994), α-tocopherol, ascorbate (Mayer and Noble, 1994), thioredoxin (Baker et al., 1997) and radical spin traps (Mayer and Noble, 1994; Slater et al., 1995) can block apoptosis. Furthermore, as noted above, the anti-apoptotic effects of Bcl2 have been linked to antioxidant properties (Hockenbery et al., 1993) that include elevation of total cellular (Kane et al., 1993; Mirkovic et al., 1997) and possibly nuclear (Voehringer et al., 1998) GSH levels.

On the other hand, excessive depletion of GSH and thioredoxin tends to predispose cells to necrosis. Lowering intracellular GSH levels with the GSH-synthesis inhibitor buthionine sulfoximine switches the mechanism of cell death from apoptosis to necrosis in human leukemic cells exposed to alkylating cancer chemotherapeutic agents (Fernandes and Cotter, 1994). Similarly, in Jurkat cells, low to moderate concentrations of the thiol-reactive agent diamide induce apoptosis, whereas higher concentrations induce necrosis, effects that appear linked to diamide's ability to deplete intracellular thioredoxin levels (Sato et al., 1995). Excessive thiol depletion probably blocks caspase activity, perhaps via inhibition of the active site cysteine. In light of these effects, it will be interesting to determine how levels of oxidized glutathione (GSSG) and thioredoxin affect caspase activation and whether these mechanisms represent reversible systems for blocking the caspases under physiological conditions.

8. CONCLUSIONS

As a result of the work outlined above and other studies, a clear-cut distinction between apoptosis and necrosis no longer exists at the biochemical level. The strongest evidence for overlap comes from studies in models of hypoxia showing that over-expression of Bcl2 or its homolog Bcl-x_L can block necrosis. The effects of Bcl2 appear largely due to direct effects on mitochondria, including stabilization of membrane potential, preservation of ATP production, prevention of oxidative stress, and enhanced Ca^{2+} uptake. In addition, Bcl2 may exert similar effects on the ER and the nucleus by regulating Ca^{2+} and GSH fluxes. If Bcl2 is not linked solely to suppression of apoptosis, what molecular distinctions between apoptosis and necrosis are we left with? A particularly attractive conclusion is that apoptosis requires caspases whereas necrosis does not (Hirsch et al., 1997). This would explain the ATP requirement for apoptosis, because Apaf-1-mediated activation of caspase-9-requires ATP hydrolysis (Li et al., 1997), and oxidation of the caspase active site cysteine would explain why excessive oxidative stress or thiol depletion inhibit apoptosis and lead to necrosis. It should be noted, however, that viral caspase inhibitors (such as the cowpox virus crmA protein) can partially attenuate necrosis that is due to chemical hypoxia (Shimizu *et al.*, 1996a, 1996b). Further work is needed to directly examine the activation status of particular caspases in additional models of necrosis.

REFERENCES

Ankarcrona, M., Dypbukt, J. M., Bonfoco, E., Zhivotovsky, B., Orrenius, S., Lipton, S. A., and Nicotera, P., 1995, Glutamate-induced neuronal death: A succession of necrosis or apoptosis depending on mitochondrial function, *Neuron* **15**:961–973.

Backway, K. L., McCulloch, E. A., Chow, S., and Hedley, D. W., 1997, Relationships between the mitochondrial permeability transition and oxidative stress during ara-C toxicity, *Cancer Res.* **57**:2446–2451.

Baffy, G., Miyashita, T., Williamson, J. R., and Reed, J. C., 1993, Apoptosis induced by withdrawal of interleukin-3 (IL-3) from an IL-3-dependent hematopoietic cell line is associated with repartitioning of intracellular calcium and is blocked by enforced BCL-2 oncoprotein production, *J. Biol. Chem.* **268**:6511–6519.

Baker, A., Payne, C. M., Briehl, M. M., and Powis, G., 1997, Thioredoxin, a gene found overexpressed in human cancer, inhibits apoptosis *in vitro* and *in vivo*, *Cancer Res.* **57**:5162–5167.

Bojes, H. K., Datta, K., Xu, J., Chin, A., Simonian, P., Nunez, G., and Kehrer, J. P., 1997, BCL-XL overexpression attenuates glutathione depletion in FL5.12 cells following interleukin-3 withdrawal, *Biochem. J.* **325**:315–319.

Bonfoco, E., Kraine, D., Ankarcrona, M., Nicotera, P., and Lipton, S. A., 1995, Apoptosis and necrosis: Two distinct events induced, respectively, by mild and intense insults with N-methyl-D-aspartate or nitric oxide/superoxide in cortical cell cultures, *Proc. Natl. Acad. Sci. USA* **92**:7162–7166.

Chinnaiyan, A. M., O'Rourke, K., Lane, B. R., and Dixit, V. M., 1997, Interaction of ced-4 with ced-3 and ced-9: A molecular framework for cell death, *Science* **275**:1122–1126.

Clement, M. V., and Stamenkovic, I., 1996, Superoxide anion is a natural inhibitor of Fas-mediated cell death, *EMBO J.* **15**:216–225.

Cohen, G. M., 1997, Caspases: The executioners of apoptosis, *Biochem. J.* **326**:1–16.

Cohen, J. J., and Duke, R. C., 1984, Glucocorticoid activation of a calcium-dependent endonuclease in thymocyte nuclei leads to cell death, *J. Immunol.* **132**:38–42.

Dong, Z., Saikumar, P., Weinberg, J. M., and Venkatachalam, M. A., 1997, Internucleosomal DNA cleavage triggered by plasma membrane damage during necrotic cell death, *Am. J. Pathol.* **151**:1205–1213.

Dybukt, J. M., Ankarcrona, M., Burkitt, M., Sjoholm, A., Strom, K., Orrenius, S., and Nicotera, P., 1994, Different prooxidant levels stimulate growth, trigger apoptosis, or produce necrosis of insulin-secreting RINm5F cells. The role of intracellular polyamines, *J. Biol. Chem.* **269**:30553–30560.

Eguchi, Y., Shimizu, S., and Tsujimoto, Y., 1997, Intracellular ATP levels determine cell death fate by apoptosis or necrosis, *Cancer Res.* **57**:1835–1840.

Ellis, H. M., and Horvitz, H. R., 1986, Genetic control of programmed cell death in the nematode *C. elegans, Cell,* **44**:817–829.

Fadok, V. A., Voelker, D. R., Campbell, P. A., Cohen, J. J., Bratton, D. L., and Henson, P. M., 1992, Exposure of phosphatidylserine on the surface of apoptotic lymphocytes triggers specific recognition and removal by macrophages, *J. Immunol.* **148**:2207–2216.

Fernandes, R. S., and Cotter, T. G., 1994, Apoptosis or necrosis: Intracellular levels of glutathione influence the mode of cell death, *Biochem. Pharmacol.* **48**:675–681.

Fernandez, A., Kiefer, J., Fosdick, L., and McConkey, D. J., 1995, Oxygen radical production and thiol depletion are required for Ca^{2+}-mediated endogenous endonuclease activation in apoptotic thymocytes, *J. Immunol.* **155**:5133–5139.

Friesen, C., Herr, I., Krammer, P. H., and Debatin, K. M., 1996, Involvement of the CD95 (APO-1/Fas) receptor/ligand system in drug-induced apoptosis in leukemia cells, *Nature Medicine* **2**:574–577.

Gaal, J. C., Smith, K. R., and Pearson, C. K., 1987, Cellular euthanasia mediated by a nuclear enzyme: A central role for nuclear ADP-ribosylation in cellular metabolism, *Trends Biochem. Sci.* **12**:129–130.

Hampton, M. B., and Orrenius, S., 1997, Dual regulation of caspase activity by hydrogen peroxide: Implications for apoptosis, *FEBS Lett.* **414**:552–556.

Hengartner, M. O., Ellis, R. E., and Horvitz, H. R., 1992, *Caenorhabditis elegans* gene ced-9 protects cells from programmed cell death, *Nature* **356**:494–499.

Hengartner, M. O., and Horvitz, H. R., 1994a, *C. elegans* cell survival gene ced-9 encodes a functional homolog of mammalian proto-oncogene bcl-2, *Cell* **76**:665–676.

Hengartner, M. O., and Horvitz, H. R., 1994b, Programmed cell death in *Caenorhabditis elegans, Curr. Opin. Genet. Dev.* **4**:581–586.

Herzenberg, L. A., Rosa, S. C. D., Dubs, J. G., Roederer, M., Anderson, M. T., Ela, S. W., Deresinski, S. C., and Herzenberg, L. A., 1997, Glutathione deficiency is associated with impaired survival in HIV disease, *Proc. Natl. Acad. Sci. USA* **94**:1967–1972.

Hirsch, T., Marchetti, P., Susin, S. A., Dallaporta, B., Zamzami, N., Marzo, I., Beuskens, M., and Kroemer, G., 1997, The apoptosis–necrosis paradox: Apoptogenic proteases activated after the mitochondrial permeability transition determine the mode of cell death, *Oncogene* **15**:1573–1581.

Hockenbery, D. M., Nunez, G., Milliman, C., Schreiber, R. D., and Korsmeyer, S. J., 1990, BCL-2 is an inner mitochondrial membrane protein that blocks programmed cell death, *Nature* **348**:334–336.

Hockenbery, D. M., Oltvai, Z. N., Yin, X. M., Milliman, C. L., and Korsmeyer, S. J., 1993, BCL-2 functions in an antioxidant pathway to prevent apoptosis, *Cell* **75**:241–251.

Israel, N., and Gougerot-Pocidalo, M. A., 1997, Oxidative stress in human immunodeficiency virus infection, *Cell Mol. Life Sci.* **53**:864–870.

Jones, D. P., McConkey, D. J., Nicotera, P., and Orrenius, S., 1989, Calcium-activated DNA fragmentation in rat liver nuclei, *J. Biol. Chem.* **264**:6398–6403.

Kane, D. J., Sarafian, T. A., Anton, R., Hahn, H., Gralla, E. B., Valentine, J. S., Ord, T., and Bredesen, D. E., 1993, BCL-2 inhibition of neural death: Decreased generation of reactive oxygen species, *Science* **262**:1274–1277.

Kazzaz, J. A., Xu, J., Palaia, T. A., Mantell, L., Fein, A. M., and Horowitz, S., 1996, Cellular oxygen toxicity: Oxidant injury without apoptosis, *J. Biol. Chem* **271**:15182–15186.

Kerr, J. F. R., Wyllie, A. H., and Currie, A. R., 1972, Apoptosis: A basic biological phenomenon with wide-ranging implications in tissue kinetics, *Br. J. Cancer* **26**:239–257.

Kim, Y. M., Talanian, R. V., and Billiar, T. R., 1997, Nitric oxide inhibits apoptosis by preventing increases in caspase-3-like activity via two distinct mechanisms, *J. Biol. Chem.* **272**:31138–31148.

Kluck, R. M., Bossy-Wetzel, E., Green, D. R., and Newmeyer, D. D., 1997, The release of cytochrome *c* from mitochondria: A primary site for bcl-2 regulation of apoptosis, *Science* **275**:1132–1136.

Kroemer, G., Zamzami, N., and Susin, S. A., 1997, Mitochondrial control of apoptosis, *Immunol. Today* **18**:44–52.

Lam, M., Dubyak, G., and Distelhorst, C. W., 1993, Effect of glucocorticoid treatment on intracellular calcium homeostasis in mouse lymphoma cells, *Mol. Endocrinol.* **7**:686–693.

Lam, M., Dubyak, G., Chen, L., Nunez, G., Miesfeld, R. L., and Distelhorst, C. W., 1994, Evidence that bcl-2 represses apoptosis by regulating endoplasmic reticulum-associated Ca^{2+} fluxes, *Proc. Natl. Acad. Sci. USA* **91**:6569–6573.

Laster, S. M., Wood, J. G., and Gooding, L. R., 1988, Tumor necrosis factor can induce both apoptotic and necrotic forms of cell lysis, *J. Immunol.* **141**:2629–2634.

Leist, M., Single, B., Castoldi, A. F., Kuhnle, S., and Nicotera, P., 1997, Intracellular adenosine triphosphate (ATP) concentration: A switch in the decision between apoptosis and necrosis, *J. Exp. Med.* **185**:1484–1486.

Lelli, J. L., Becks, L. L., Dabrowska, M. I., and Hinshaw, D. B., 1998, ATP converts necrosis to apoptosis in oxidant-injured endothelial cells, *Free Radical Biol. Med.* **25**:694–702.

Lemasters, J. J., DiGuiseppi, J., Nieminen, A.-L., and Herman, B., 1987, Blebbing, free Ca^{2+} and mitochondrial membrane potential preceding cell death in hepatocytes, *Nature* **325**:78–81.

Lennon, S. V., Martin, S. J., and Cotter, T. G., 1991, Dose-dependent induction of apoptosis in human tumour cell lines by widely divergent stimuli, *Cell Prolif.* **24**:203–204.

Li, P., Nijhawan, D., Budihardjo, I., Srinivasula, S. M., Ahmad, M., Alnemri, E. S., and Wang, X., 1997, Cytochrome *c* and dATP-dependent formation of Apaf-1/caspase-9 complex initiates an apoptotic protease cascade, *Cell* **91**:479–489.

Lieberthal, W., Menza, S. A., and Levine, J. S., 1998, Graded ATP depletion can cause necrosis or apoptosis of cultured mouse proximal tubular cells, *Am. J. Physiol.* **274**:F315–F327.

Liu, X., Kim, C. N., Yang, J., Jemmerson, R., and Wang, X., 1996, Induction of the apoptotic program in cell-free extracts: Requirement for dATP and cytochrome *c., Cell* **86**:147–157.

Lockshin, R. A., 1969, Programmed cell death: Activation of lysis by a mechanism involving the synthesis of protein, *J. Insect Physiol.* **15**:1505–1516.

Macho, A., Hirsch, T., Marzo, I., Marchetti, P., Dallaporta, B., Susin, S. A., Zamzami, N., and Kroemer, G., 1997, Glutathione depletion is an early, and calcium elevation is a late, event of thymocyte apoptosis, *J. Immunol.* **158**:4612–4619.

Marin, M. C., Fernandez, A., Bick, R. J., Brisbay, S., Buja, M., Snuggs, M., McConkey, D. J., Eschenbach, A. C. V., Keating, M. J., and McDonnell, T. J., 1996, Apoptosis suppression by Bcl-2 is correlated with the regulation of nuclear and cytosolic Ca^{2+}, *Oncogene* **12**:2259–2266.

Mayer, M., and Noble, M., 1994, N-acetyl-L-cysteine is a pluripotent protector against cell death and enhancer of trophic factor-mediated cell survival *in vitro, Proc. Natl. Acad. Sci. USA* **91**:7496–7500.

McCall, K., and Steller, H., 1997, Facing death in the fly: Genetic analysis of apoptosis in *Drosophila, Trends Genet.* **13**:222–226.

McConkey, D. J., and Orrenius, S., 1996a, The role of calcium in the regulation of apoptosis, *J. Leukocyte Biol.* **59**:775–783.

McConkey, D. J., and Orrenius, S., 1996b, Signal transduction pathways in apoptosis, *Stem Cell* **14**:619–631.

McConkey, D. J., Hartzell, P., Nicotera, P., Wyllie, A. H., and Orrenius, S., 1988, Stimulation of endogenous endonuclease activity in hepatocytes exposed to oxidative stress, *Toxicol. Lett.* **42**:123–130.

McConkey, D. J., Hartzell, P., Nicotera, P., and Orrenius, S., 1989, Calcium-activated DNA fragmentation kills immature thymocytes, *FASEB J.* **3**:1843–1849.

Meikrantz, W., Gisselbrecht, S., Tam, S. W., and Schlegel, R., 1994, Activation of cyclin A-dependent protein kinases during apoptosis, *Proc. Natl. Acad. Sci. USA* **91**:3754–3758.

Messmer, U. K., and Brune, B., 1996, Nitric oxide (NO) in apoptotic versus necrotic RAW 264.7 macrophage cell death: The role of NO-donor exposure, NAD^+ content, and p53 accumulation, *Arch. Biochem. Biophys.* **327**:1–10.

Mirkovic, N., Voehringer, D. W., Story, M. D., McConkey, D. J., McDonnell, T. J., and Meyn, R. E., 1997, Resistance to radiation-induced apoptosis in BCL-2-expressing cells is reversed by depleting cellular thiols, *Oncogene* **15**:1461–1470.

Murphy, A. N., Bredesen, D. E., Cortopassi, G., Wang, E., and Fiskum, G., 1996, BCL-2 potentiates the maximal calcium uptake capacity of neural cell mitochondria, *Proc. Natl. Acad. Sci. USA* **93**:9893–9898.

Orrenius, S., McConkey, D. J., Bellomo, G., and Nicotera, P., 1989, Role of Ca^{2+} in toxic cell killing, *Trends Pharmacol. Sci.* **10**:281–285.

Pastorino, J. G., Simbula, G., Yamamoto, K., Glascott, P. A., Rothman, R. J., and Farber, J. L., 1996, The cytotoxicity of tumor necrosis factor depends on induction of the mitochondrial permeability transition, *J. Biol. Chem.* **271**:29792–29798.

Polyak, K., Xia, Y., Zweier, J. L., Kinzler, K. W., and Vogelstein, B. A., 1997, A model for p53-induced apoptosis, *Nature* **389**:300–305.

Reed, J. C., 1997, Double identity for proteins of the BCL-2 family, *Nature* **387**:773–776.

Richter, C., 1993, Pro-oxidants and mitochondrial Ca^{2+}: Their relationship to apoptosis and oncogenesis, *FEBS Lett.* **325**:104–107.

Sata, N., Klonowski-Stumpe, H., Han, B., Haussinger, D., and Niederau, C., 1997, Menadione induces both necrosis and apoptosis in rat pancreatic acinar AR4-2J cells, *Free Radical Biol. Med.* **23**:844–850.

Sato, N., Iwata, S., Nakamura, K., Hori, T., Mori, K., and Yodoi, J., 1995, Thiol-mediated redox regulation of apoptosis: Possible roles of cellular thiols other than glutathione in T cell apoptosis, *J. Immunol.* **154**:3194–3203.

Savill, J., Fadok, V., Henson, P., and Haslett, C., 1993, Phagocytic recognition of cells undergoing apoptosis, *Immunol. Today* **14**:131–136.

Schwartz, L. M., and Truman, J. W., 1982, Peptide and steroid regulation of muscle degeneration in an insect, *Science* **215**:1420–1424.

Shimizu, S., Eguchi, Y., Kamiike, W., Waguri, S., Uchiyama, Y., Matsuda, H., and Tsujimoto, Y., 1996a, BCL-2 blocks loss of mitochondrial membrane potential while ICE inhibitors act at a different step during inhibition of death induced by respiratory chain inhibitors, *Oncogene* **13**:21–29.

Shimizu, S., Eguchi, Y., Kamiike, W., Waguri, S., Uchiyama, Y., Matsuda, H., and Tsujimoto, Y., 1996b, Retardation of chemical hypoxia-induced necrotic cell death by BCL-2 and ICE inhibitors: Possible involvement of common mediators in apoptotic and necrotic signal transductions, *Oncogene* **12**:2045–2050.

Slater, A. F., Nobel, C. S., Maellaro, E., Bustamante, J., Kimland, M., and Orrenius, S., 1995, Nitrone spin traps and a nitroxide antioxidant inhibit a common pathway of thymocyte apoptosis, *Biochem. J.* **306**:771–778.

Squier, M. K. T., Miller, A. C. K., Malkinson, A. M., and Cohen, J. J., 1994, Calpain activation in apoptosis, *J. Cell Physiol.* **159**:229–237.

Tata, J. R., 1966, Requirement for RNA and protein synthesis for induced regression of the tadpole tail in organ culture, *Dev. Biol.* **13**:77–94.

Thompson, C. B., 1995, Apoptosis in the pathogenesis and treatment of disease, *Science* **267**:1456–1462.

Thor, H., Smith, M. T., Hartzell, P., Bellomo, G., Jewell, S. A., and Orrenius, S., 1982, The metabolism of menadione (2-methyl-1, 4-naphthoquinone) by isolated hepatocytes, *J. Biol. Chem.* **257**:12419–12425.

Thornberry, N. A., Bull, H. G., Calaycay, J. R., Chapman, K. T., Howard, A. D., Kostura, M. J., Miller, D. K., Molineaux, S. M., Weidner, J. R., Aunins, J., Elliston, K. O., Ayala, J. M., Casano, F. J., Chin, J., Ding, G. J., Egger, L. A., Gaffney, E. P., Limjuco, G., Pahlha, O. C., Raju, S. M., Rolando, A. M., Salley, J. P., Yamin, T. T., Lee, T. D., Shively, J. E., MacCross, M., Mumford, R. A., Schmidt, J. A., and Tocci, M. J., 1992, A novel heterodimeric cysteine protease is required for interleukin-1b processing in monocytes, *Nature* **356**:768–774.

Tyurina, Y. Y., Tyurina, V. A., Carta, G., Quinn, P. J., Schor, N. F., and Kagan, V. E., 1997, Direct evidence for antioxidant effect of BCL-2 in PC12 rat pheochromocytoma cells, *Arch. Biochem. Biophys.* **344**:413–423.

Vanderbilt, J. N., Bloom, K. S., and Anderson, J. N., 1982, Endogenous nuclease: Properties and effects on transcribed genes in chromatin, *J. Biol. Chem.* **257**:13009–13017.

Vaux, D. L., Weissman, I. L., and Kim, S. K., 1992, Prevention of programmed cell death in *Caenorhabditis elegans* by human bcl-2, *Science* **258**:1955–1957.

Voehringer, D., McConkey, D. J., McDonnell, T., Brisbay, S., and Meyn, R. E., 1998, BCL-2 causes redistribution of glutathione to the nucleus, *Proc. Natl. Acad. Sci. USA* **95**:2956–2960.

Waters, S. L., Sarang, S. S., Wang, K. K. W., and Schnellmann, R. G., 1997, Calpains mediate calcium and chloride influx during the late phase of cell injury, *J. Pharmacol. Exp. Ther.* **283**:1177–1184.

Weil, M., Jacobson, M. D., Coles, H. S. R., Davies, T. J., Gardner, R. L., Raff, K. D., and Raff, M. C., 1996, Constitutive expression of the machinery for programmed cell death, *J. Cell Biol.* **133**:1053–1059.

Williams, G. T., Smith, C. A., Spooncer, E., Dexter, T. M., and Taylor, D. R., 1990, Haemopoietic colony stimulating factors promote cell survival by suppressing apoptosis, *Nature* **343**:76–79.

Wu, D., Wallen, H. D., and Nunez, G., 1997, Interaction and regulation of subcellular localization of ced-4 by ced-9, *Science* **275**:1126–1129.

Wyllie, A. H., 1980a, Glucocorticoid-induced thymocyte apoptosis is associated with endogenous endonuclease activation, *Nature* **284**:555–556.

Wyllie, A. H., Kerr, J. F. R., and Currie, A. R., 1980b, Cell death: The significance of apoptosis, *Int. Rev. Cytol.* **68**:251–305.

Wyllie, A. H., Morris, R. G., Smith, A. L., and Dunlop, D., 1984, Chromatin cleavage in apoptosis: Association with condensed chromatin morphology and dependence on macromolecular synthesis, *J. Pathol.* **142**:67–77.

Yang, J., Liu, X., Bhalla, K., Kim, C. N., Ibrado, A. M., Cai, J., Peng, T. I., Jones, D. P., and Wang, X., 1997, Prevention of apoptosis by bcl-2: Release of cytochrome *c* from mitochondria blocked, *Science* **275**:1129–1132.

Yuan, J., Shaham, S., Ledoux, S., Ellis, H. M., and Horvitz, H. R., 1993, The *C. elegans* cell death gene ced-3 encodes a protein similar to mammalian interleukin-1b-converting enzyme, *Cell* **75**:641–652.

Zamzami, N., Marchetti, P., Castedo, M., Decaudin, D., Macho, A., Hirsch, T., Susin, S. A., Petit, P. X., Mignotte, B., and Kroemer, G., 1995, Sequential reduction of mitochondrial transmembrane potential and generation of reactive oxygen species in early programmed cell death, *J. Exp. Med.* **182**:367–377.

Zou, H., Henzel, W. J., Liu, X., Lutschg, A., and Wang, X., 1997, Apaf-1, a human protein homologous to *C. elegans* ced-4, participates in cytochrome *c*-dependent activation of caspase-3, *Cell* **90**:405–413.

Mitochondria, Free Radicals, and Disease

Chapter 14

Reactive Oxygen Generation by Mitochondria

Alicia J. Kowaltowski and Anibal E. Vercesi

1. INTRODUCTION

The partnership between the ancestors of today's mitochondria and eukaryotic cells has undoubtedly been one of great success, and has brought to eukaryotic cells the advantage of ATP production via oxidative phosphorylation. The presence of an electron transport chain in mitochondria, however, which continuously reduces oxygen to build the transmembrane electrochemical H^+ potential necessary for ATP synthesis, has a potentially serious side effect for these cells: the constant generation of reactive oxygen species (ROS).

When taking into account the highly complex process of electron transfer through the mitochondrial respiratory chain, it is surprising that only 1–5% of the electrons entering the chain fail to generate water from oxygen. Most of these "lost" electrons combine with oxygen at intermediate steps of the respiratory chain, promoting the monoelectronic reduction of oxygen, which generates the superoxide radical ($O_2 \cdot ^-$) (Boveris and Chance, 1973; Liu, 1997; Turrens, 1997). Indeed, due to the continuous function of the electron transport chain, the 1–5% of electron leakage is sufficient to make mitochondrial $O_2 \cdot ^-$ generation the major cellular source of ROS in most tissues (Liu, 1997; Turrens, 1997). These mitochondrially generated ROS have been implicated in many degenerative life-cycle events, including aging and necrotic and apoptotic cell death (reviewed in Green and Reed, 1998; Kowaltowski and Vercesi, 1999; Lemasters *et al.*, 1998; Zamzami *et al.*, 1996). ROS (possibly mitochondrially generated) have also been implicated in the pathogenesis of specific diseases such as arteriosclerosis, Alzheimer, Parkinson, stroke,

Alicia J. Kowaltowski   Departmento de Bioquimica, Instituto de Quimica, Universidade de Sao Paulo, SP, Brazil.   **Anibal E. Vercesi**   Departamento de Patologia Clínica, Faculdade de Ciências Médicas, Universidade Estadual de Campinas, Campinas, SP, Brazil.

Mitochondria in Pathogenesis, edited by Lemasters and Nieminen.
Kluwer Academic/Plenum Publishers, New York, 2001.

and myocardial infarct (Bolli, 1998; Cadet and Brannock, 1998; Juurlink and Sweeney, 1997; Kristian and Siesjo, 1998; Mutthaup *et al.*, 1997; Visioli, 1998).

2. THE MITOCHONDRIAL ELECTRON TRANSPORT CHAIN AND SUPEROXIDE GENERATION SITES

The electron transport chain controls the redox energy necessary to generate mitochondrial membrane potential, and therefore, to promote oxidative phosphorylation. Normally, electrons from different substrates are collected in the form of NADH and transferred to the iron center of NADH dehydrogenase (complex I, see Fig. 1). Complex I transfers the electrons to the oxidized form of coenzyme Q (UQ), generating the reduced form of coenzyme Q (UQH_2). Electrons originating from succinate are fed to UQ through complex II, resulting also in coenzyme Q reduction. In some tissues coenzyme Q may also be reduced by glycerol-3-phosphate dehydrogenase (in the presence of cytosolic glycerol-3-phosphate) or by ubiquinone oxidoreductase (as a result of fatty acid β-oxidation). The UQH_2 then deprotonates, resulting in formation of the semiquinone anion species ($UQH\cdot$), the form that donates electrons to cytochrome *c*. Two separate pools of $UQH\cdot$ (one from the cytoplasmic face and the other from the matrix face of inner mitochondrial membrane) exist, and both $UQH\cdot$ forms are oxidized together when regenerating UQ and donating electrons to cytochrome *c*. Cytochrome *c* transfers electrons to cytochrome oxidase (complex IV). This complex is responsible for the transfer of electrons to oxygen, resulting in water generation, in a process involving four consecutive one-electron transfers (for a more detailed description of the mitochondrial respiratory chain, see Nicholls and Ferguson, 1982).

Different components of the respiratory chain may convert a small portion of oxygen (1–2%) to $O_2\cdot^-$ through the monoelectronic reduction (Boveris and Chance, 1973). Indeed, $O_2\cdot^-$ is generated at the level of NADH dehydrogenase (Boveris and Chance, 1973; Turrens and Boveris, 1980) and coenzyme Q (Boveris and Chance, 1973; Cadenas

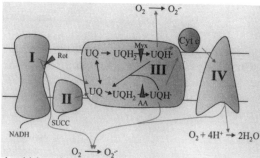

FIGURE 1. The mitochondrial respiratory chain and possible sites of electron leakage and $O_2\cdot^-$ formation. Roman numerals indicate complexes I–IV. Narrow arrows represent electron transfers. Broad arrows represent possible electron leakage sites. SUCC, succinate; Rot, rotenone; UQ, oxidized form of coenzyme Q; UQH_2, reduced form of coenzyme Q; $UQH\cdot$, semiquinone anion; Myx, myxothiazol; AA, antimycin A; Cyt-*c*, cytochrome *c*.

et al., 1977; Turrens *et al.*, 1985) (see Fig. 1). Although cytochrome oxidase is the site in which oxygen normally undergoes reduction, and this process is conducted in four one-electron steps, mitochondrial $O_2 \cdot^-$ generation at the level of cytochrome oxidase is surprisingly low (Turrens, 1997). It is low because cytochrome oxidase can bind firmly to the partially reduced oxygen intermediates, preventing their release before full reduction is achieved. Superoxide generation at the level of the Fe–S center of NADH dehydrogenase is more substantial and is promoted by NAD-based substrates such as malate, glutamate, and pyruvate and is stimulated by rotenone, an inhibitor of electron transfer from complex I to coenzyme Q (Turrens and Boveris, 1980, 1997). Coenzyme Q is also an important site for electron leakage at the respiratory chain. This probably occurs during the donation of electrons from semiquinone anions (themselves free radicals) to oxygen. Electron leakage at the level of coenzyme Q is stimulated by succinate, cyanide, and antimycin A (Boveris *et al.*, 1976; Cadenas *et al.*, 1977; Kowaltowski *et al.*, 1998a; Turrens, 1997; Turrens *et al.*, 1985). Antimycin A has a notably large stimulatory role, because blocks UQH· formation at the matrix face of the mitochondrial inner membrane, promoting an accumulation of semiquinone anions formed previously at the cytosolic face of the inner mitochondrial membrane (Fig. 1). Myxothiazol, an inhibitor of UQH· formation at the cytosolic inner-membrane face, prevents $O_2 \cdot^-$ generation at the level of coenzyme Q (Dawson, 1993; Hansford *et al.*, 1997; Kowaltowski *et al.*, 1998a; Turrens, 1997; Turrens *et al.*, 1985).

3. DIFFERENT REACTIVE OXYGEN SPECIES DETECTABLE IN MITOCHONDRIA

Many different ROS are detectable in mitochondria (see Fig 2). As mentioned, $O_2 \cdot^-$ is generated continuously by the electron transport chain, even under physiological conditions. Because $O_2 \cdot^-$ is relatively reactive and not highly diffusible, detecting this radical in mitochondrial suspensions can be problematic (Turrens, 1997). It is nevertheless

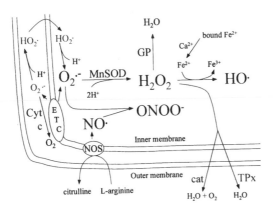

FIGURE 2. ROS present in mitochondria and antioxidant defenses. ETC, electron transport chain; MnSOD, Mn-superoxide dismutase; GP, glutathione peroxidase; NOS, nitric oxide synthase; Cyt-*c*, cytochrome *c*; cat, catalase; TPx, thioredoxin peroxidase.

possible to detect $O_2 \cdot^-$ in suspensions of submitochondrial particles (Castilho *et al.*, 1998a; Turrens and Boveris, 1980; Turrens *et al.*, 1982) and with membrane-permeable probes such as hydroethidium (Budd *et al.*, 1997; Sengpiel *et al.*, 1998). Diffusion of $O_2 \cdot^-$ through inner membrane is thought to occur mainly via the combination of $O_2 \cdot^-$ and protons, generating the perhydroxyl radical ($HO_2 \cdot^-$) (Liu, 1997).

Mitochondrially generated $O_2 \cdot^-$ is readily dismutated to hydrogen peroxide (H_2O_2) by intramitochondrial Mn-superoxide dismutase (see Fig. 2 and description below). The H_2O_2 is not a free radical because it does not present an unpaired electron but is classified as "ROS" due to its oxidative power. Because it is more stable and diffusible than $O_2 \cdot^-$, H_2O_2 can be detected more readily in mitochondrial suspensions (Boveris *et al.*, 1976; Korshunov *et al.*, 1997; Kowaltowski *et al.*, 1995, 1996a; 1996b, 1998a, 1998b, 1998c; Loschen and Azzi, 1975; Loschen *et al.*, 1973). Indeed, detection of H_2O_2 is often used as an indicator of mitochondrial ROS generation.

In the presence of Fe^{2+}, H_2O_2 generates the hydroxyl radical ($HO \cdot$) (Sutton and Winterborn, 1989). Under normal conditions, most of mitochondrial Fe^{2+} remains tightly bound to complexing molecules, but the presence of Ca^{2+} ions may increase free Fe^{2+} availability (Castilho *et al.*, 1995b; Merryfield and Lardy, 1982). Thus Ca^{2+} ions can both increase $O_2 \cdot^-$ generation at the electron transport chain and stimulate $HO \cdot$ generation by mobilizing intramitochondrial Fe^{2+} from yet unidentified Fe^{2+} pools (reviewed in Vercesi *et al.*, 1997). The $HO \cdot$ is a very reactive free radical with a short half-life (reviewed in Lubec, 1996), rendering its detection in biological systems very complex. Nevertheless, $HO \cdot$ has already been detected in submitochondrial particles (Giulivi *et al.*, 1995; Grijalba, Verceri, and Schreier, 1999).

Recently, mitochondria were found to posses a nitric oxide synthase, which permits the organelle to generate nitric oxide ($NO \cdot$) during L-arginine oxidation (Giulivi *et al.*, 1998; Tatoyan and Giulivi 1998). The importance for $NO \cdot$ generation in mitochondria is not yet known, although this radical modulates respiration and ATP synthesis by inhibiting cytochrome oxidase. Mitochondrially generated $NO \cdot$ may combine with $O_2 \cdot^-$ to produce yet another ROS—peroxynitrite ($ONOO^-$) (Pryor and Squadrito, 1995).

Finally, mitochondria may still generate singlet oxygen (Costa *et al.*, 1993; Kerver *et al.*, 1997). This oxygen form is far more reactive than molecular oxygen, which is in a triplet state, because the spin of the electrons occupying the highest energy level is in opposite directions, permitting its reaction with organic molecules in their fundamental (singlet) state (Kasha, 1991).

4. CONDITIONS THAT INCREASE OR DECREASE MITOCHONDRIAL REACTIVE OXYGEN SPECIES GENERATION

As mentioned above, using certain respiratory substrates and inhibitors may signifi-cantly alter mitochondrial $O_2 \cdot^-$ generation at different levels of the respiratory chain. These conditions are not very physiologically or pathologically relevant, however. It is nevertheless important to stress that the commonly stated affirmative that mitochondria must respire to generate ROS is a misconception. Indeed, under certain conditions, such as antimycin A poisoning, mitochondria do not respire but generate larger amounts of ROS than in the resting state (Cadenas *et al.*, 1977; Kowaltowski *et al.*, 1998a). This ROS generation is thought to cause the small "antimycin-insensitive" mitochondrial oxygen consumption.

There are many situations that may occur *in vivo* to alter the rate of mitochondrial ROS generation. One classical example is uncoupling (reviewed in Skulachev, 1997). A "mild" uncoupling of mitochondrial respiration in relation to membrane potential can significantly decrease ROS generation and is proposed as a mitochondrial defense against oxidative stress (Korshunov *et al.*, 1997; Skulachev, 1997). Mild uncoupling may occur in the presence of thyroid hormones (Boblyleva *et al.*, 1998; Skulachev, 1997), fatty acids translocated by the ATP/ADP translocator (Korshunov *et al.*, 1998), or secondarily, to the activity of mitochondrial uncoupling proteins, also promoted by fatty acids (Kowaltowski *et al.*, 1998a; Nègre-Salvayre, 1997). It is hypothesized that the decrease in ROS generation that occurs in uncoupled mitochondria is related to the increase of respiratory rates, which may shorten the life of semiquinone anions, thus preventing their donation of electrons to oxygen (Skulachev, 1997). Another possible explanation is that increased respiration may decrease oxygen tension in the mitochondrial microenvironment, thus decreasing superoxide formation (Skulachev, 1997). These hypothesis are supported by the observation that the increased respiratory rate in state III respiration (which is not related to uncoupling) is also accompanied by a reduced rate of ROS generation (Boveris and Chance, 1973). Yet whereas uncoupling of mitochondrial respiration from ADP phosphorylation usually decreases ROS generation, it may increase generation under certain specific conditions. For example, protonophores increase ROS generation when the respiratory chain is inhibited by antimycin A (Boveris and Chance, 1973; Cadenas and Boveris, 1980) or when mitochondria are loaded with Ca^{2+} and treated with an inhibitor of the Ca^{2+} uniporter to prevent Ca^{2+} efflux (Kowaltowski *et al.*, 1996a).

Another pathologically relevant situation that alters mitochondrial ROS generation is the change in oxygen tension of the mitochondrial environment. It is clear that if mitochondria are incubated in the *total* absence of molecular oxygen (anoxia), they cannot generate ROS. Mitochondria incubated under hypoxic conditions can, however, generate significant quantities of ROS, though it is unclear whether mitochondrial ROS generation increases under hypoxic conditions (Delgi Esposti and McLennan, 1998; Duranteau *et al.*, 1998; Vanden Hoek *et al.*, 1998) or decreases (Costa *et al.*, 1993; Yang and Block, 1995) in relation to normoxic conditions. Mitochondria incubated in higher-than-normal oxygen concentrations (hyperoxia) generate more ROS than in the resting state (Boveris and Chance, 1973; Jamieson *et al.*, 1986; Liu *et al.*, 1998). Interestingly, mitochondria incubated under anoxic conditions and then supplemented with oxygen (an *in vitro* experiment that mimics ischemia/reperfusion) generate large quantities of ROS soon after reoxygenation. Similar results may be seen *in vivo* during myocardial ischemia followed by reperfusion (Bolli *et al.*, 1988, 1989). This burst in ROS generation seems to be related to mitochondrial Ca^{2+} uptake and fast manipulation of the coenzyme Q redox state, because it can be mimicked by incubating the mitochondrial suspension in the presence of rotenone, then supplementing succinate, and it is prevented by Ca^{2+} chelators (Kowaltowski *et al.*, 1995). Mitochondrially generated ROS may play a large role in the pathogenesis of myocardial and cerebral damage after ischemia/reperfusion (Bolli, 1998; Fiskum *et al.*, 1998; Kristian and Siesjo, 1998).

Respiratory-chain-generated ROS increase dramatically when mitochondria are loaded with Ca^{2+} (Cadenas and Boveris, 1980; Castilho *et al.*, 1995b; Kowaltowski *et al.*, 1995, 1996a, 1996b, 1998a, 1998b, 1998c). Mitochondria are capable of accumulating Ca^{2+} ions in their matrix electrophoretically due to the mitochondrial membrane potential when extramitochondrial Ca^{2+} concentration rises to about 0.5 μM (Nicholls and Fergusson, 1982). The presence of Ca^{2+} in the mitochondrial matrix induces mitochondria to generate large amounts of ROS, in a manner stimulated by inorganic phosphate (Kowaltowski *et al.*, 1996a, 1996b, 1998b). Uptake of Ca^{2+} is necessary for this stimulation because the inhibitor of the Ca^{2+} uniporter ruthenium red prevents this Ca^{2+} effect. Also, uncoupled mitochondria only suffer a Ca^{2+}-induced increase in ROS generation when treated with high concentrations of Ca^{2+}, when supplemented with Ca^{2+} ionophores (Cadenas and Boveris, 1980; Kowaltowski *et al.*, 1998a), or both, to force cation uptake via a concentration gradient. The Ca^{2+}-stimulated ROS generation seems to occur mainly at the level of coenzyme Q (Kowaltowski *et al.*, 1995).

The cause of the Ca^{2+}-induced increase in mitochondrial ROS generation is unknown, but Ca^{2+} binding to mitochondrial membrane is a necessary step because dibucaine, a drug capable of displacing Ca^{2+} ions from their membrane binding sites (Low *et al.*, 1979), inhibits this effect (Kowaltowski *et al.*, 1998c). Indeed, there is evidence (Vercesi *et al.*, 1997) that Ca^{2+} binds preferentially to cardiolipins of the inner mitochondrial membrane, the only negatively charged phospholipids present. Electron paramagnetic resonance (EPR) spectra of spin probes incorporated into submitochondrial particles respiring on succinate indicate that Ca^{2+}–cardiolipin complexation induces lipid lateral-phase separation and increased lipid packing associated with increased rates of carbon-centered radical production (Grijalba, Vercesi, and Schreier, 1999). Under these conditions, HO· was also detected through the formation of the DMPO–OH adduct. These findings are also supported by the observation, using small-angle X-ray diffraction, that Ca^{2+} can affect the organization of cardiolipin-containing membranes in a dibucaine-sensitive manner (Kowaltowski *et al.*, unpublished).

Mitochondrial ROS generation stimulated by Ca^{2+} may play a pivotal role in the onset of cell death in conditions involving cell Ca^{2+} overload, such as glutamate neurotoxicity (Gunasekar *et al.*, 1995; Reynolds and Hastings, 1995). In fact, if mitochondrial Ca^{2+} uptake is blocked when neuronal cells are treated with excitotoxic concentrations of glutamate, cell death may be prevented (Budd and Nicholls, 1996; Castilho *et al.*, 1998b). Cell death is also prevented by cell-permeant antioxidants (Castilho *et al.*, 1999).

5. ANTIOXIDANT DEFENSES

Because mitochondrial ROS generation is a continuous process even under physiological conditions, these organelles have developed a complex antioxidant defense system (Fig. 2). As described previously, mitochondria contain an Mn-superoxide dismutase (MnSOD) similar to bacterial SOD (Doonan *et al.*, 1984) capable of dismutating $O_2\cdot^-$ into H_2O_2. Mitochondria also contain glutathione peroxidase (Sies and Moss, 1978; Zakowski and Tappel, 1978), a thiol peroxidase capable of removing H_2O_2 using reduced glutathione as substrate. The resulting oxidized form of glutathione can be recovered by NADPH through glutathione reductase. In this respect the energy-linked $NAD(P)^+$ transhydrogenase in the inner mitochondrial membrane plays a special role in the oxidative stress defense mechanism (Hoek and Rydstron, 1988). It catalyses the reversible transfer of hydrogen between NAD^+, according to the reaction

$$NADH + NADP^+ \Rightarrow NAD + NADPH$$

This enzyme can function as a proton pump, using a respiration-generated electrochemical H^+ gradient to displace the equilibrium constant of the above reaction in the direction of NADPH production to guarantee a source of reducing equivalents for regenerating reduced glutathione. This characteristic links mitochondrial coupling and membrane potential to the redox potential. Therefore, if mitochondria are not fully coupled or if membrane potential decreases, the energy-linked transhydrogenase may be unable to respond readily to an excessive NADPH depletion, and a condition of oxidative stress may arise. In fact, pyridine nucleotide oxidant treatment of mitochondria loaded with Ca^{2+} results in severe oxidative damage to the organelle (Vercesi et al., 1997). This was evidenced by early studies showing that mitochondrial Ca^{2+} accumulation and retention was favored by the reduced state of pyridine nucleotides, whereas oxidation of these nucleotides was followed by release of accumulated Ca^{2+} (Lehninger et al., 1978). This Ca^{2+} release was later associated with the mitochondrial permeability transition, a nonspecific form of inner-membrane permeabilization caused by oxidative stress (see discussion below, and Vercesi et al., 1997).

Mitochondria also contain lipid-soluble antioxidants such as α-tocopherol (vitamin E) and UQH_2, itself, both potent inhibitors of mitochondrial lipid peroxidation (Ernster et al., 1992).

Cytochrome c is another mitochondrial component that can act as an antioxidant (reviewed in Skulachev, 1998). Loss of mitochondrial cytochrome c, a trigger for apoptosis in many systems (see discussion below), stimulates mitochondrial $O_2{\cdot}^-$ generation (Cai and Jones, 1998). The oxidized form of cytochrome c can receive an electron from $O_2{\cdot}^-$, converting it back to oxygen. It is interesting that in this process the "lost" electron can be fed back into the electron transport chain.

Cytochrome oxidase (complex IV), another component of the respiratory chain, promotes catabolism of $NO\cdot$ to NO_2^- and NO_3^-, which not only removes $NO\cdot$, but also prevents $ONOO^-$ formation (Guilivi, 1998).

Mitochondrially generated H_2O_2, highly diffusible through biological membranes, can be removed by extramitochondrial antioxidants such as catalase (Valle et al., 1993; Kowaltowski et al., 1996a, 1996b, 1998b) and thioredoxin peroxidase (Kowaltowski et al., 1998b). Catalase has also been located in rat heart mitochondria (Radi et al., 1991). Indeed, H_2O_2 measurements in rat heart mitochondria require pretreatment with the catalase inhibitor aminotriazole (Korshunov et al., 1997). Also, a few isoforms of thioredoxin peroxidases (capable of removing H_2O_2 using reduced thiols as substrates) and thioredoxin reductases (capable of reducing dithiols) are located in mitochondria (Rigobello et al., 1998; Watabe et al., 1997).

Finally, the last line of mitochondrial antioxidant defense may be the methionine residues of mitochondrial proteins (Berlett and Stadtman, 1997). These residues generate methionine sulfoxide when oxidized, which does not strongly compromise protein functionality, and they may be a last mitochondrial resource to protect against protein cysteine residue oxidation, which results in dithiol cross-linkage and large alterations in protein structure and function.

Overall, mitochondrial antioxidant components protect the organelle from most oxidative damage during physiological conditions. If mitochondrial ROS generation is increased, however, or the antioxidant defenses are exhausted, mitochondria will accumulate ROS and suffer the consequences of oxidative stress.

6. EFFECTS OF MITOCHONDRIAL REACTIVE OXYGEN SPECIES UNDER OXIDATIVE STRESS CONDITIONS

Under conditions of oxidative stress, mitochondrially generated ROS accumulate, and due to their high reactivity, promote oxidative injury to mitochondrial components (Fig. 3). Most of the ROS generation occurs at the electron transport chain, so lipid and protein components of the inner mitochondrial membrane are expected to be preferred targets of these ROS. Also, mitochondrial DNA (mtDNA) is located in the mitochondrial matrix and is closely attached to the inner membrane, rendering it susceptible to mitochondrial oxidative stress.

6.1. Oxidative Damage to Mitochondrial Membrane Proteins

Possible ROS effects on proteins include cleavage of peptide bonds (resulting in protein fragmentation), oxidation of amino acid residues (mainly the sulfur-containing residues cysteine and methionine), and generation of carbonyl derivatives (Berlett and Stadtman, 1997). These alterations may explain mitochondrial impairment, such as respiratory inhibition, observed after conditions of oxidative stress (Myers *et al.*, 1995; Salganik *et al.*, 1994; Xie *et al.*, 1998; Zwicker *et al.*, 1998).

6.2. Mitochondrial Permeability Transition

A specific form of inner-membrane protein thiol oxidation results in nonspecific membrane permeabilization, known as the *mitochondrial permeability transition* (MPT) (reviewed in Zoratti and Szabò, 1995; Bernardi, 1996; Vercesi *et al.*, 1997). Mitochondrial membrane permeabilization secondary to the MPT is Ca^{2+} dependent and is inhibited by the immune suppressor cyclosporin A (Zoratti and Szabò, 1995). After permeabilization Ca^{2+} chelation may result in recovery of inner membrane permeability, hence the name "permeability transition" (Hunter and Haworth, 1979).

Almost universally, the transition depends on the presence of Ca^{2+}, but this form of mitochondrial permeabilization can be largely stimulated by the presence of many compounds called *MPT inducers* (for a list of inducers, see Zoratti and Szabò, 1995), which include inorganic phosphate (Rossi and Lehninger, 1964), oxidants of pyridine nucleotides (Bernardes *et al.*, 1986; Lehninger *et al.*, 1978; Vercesi, 1984b), dithiol reagents (Bernardes *et al.*, 1994; Lenartowics *et al.*, 1991), uncouplers (Bernardi, 1992; Harris *et al.*, 1979), and thyroid hormones (Castilho *et al.*, 1998a; Harris *et al.*, 1979), among others (for a list, see Zoratti and Szabò, 1995). The large variety of chemical characteristics of MPT inducers is one of the puzzling features of this phenomenon.

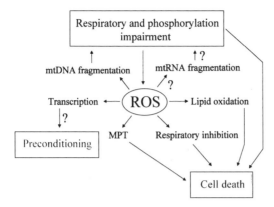

FIGURE 3. Consequences of mitochondrially generated reactive oxygen species (ROS).

Though some authors have questioned the concept (Kuzminova *et al.*, 1998; Scorrano *et al.*, 1997), there is extensive evidence that the MPT is caused by mitochondrially generated ROS (reviewed in Vercesi *et al.*, 1997). The evidence for this is summarized as follows:

1. The MPT promoted by Ca^{2+} is largely stimulated by prooxidants such as *t*-butyl hydroperoxide and acetoacetate, which deplete mitochondria of NAD(P)H, and in consequence, decrease their reducing power (see discussion above; Bernardes *et al.*, 1986; Castilho *et al.*, 1995b, 1996; Lehninger *et al.*, 1978; Vercesi, 1984b).
2. The MPT can be promoted by ROS or ROS-generating systems such as aloxan (Frei *et al.*, 1985), xanthine/xanthine oxidase (Takayama *et al.*, 1993), 5-aminolevulinic acid (Hermes-Lima, 1995; Hermes-Lima *et al.*, 1991; Vercesi *et al.*, 1994), nitric oxide (Balakirev *et al.*, 1997), laser-activated tetramethylrhodamine in methylester (Hüser, 1998), paraquat (Constantini *et al.*, 1995), peroxynitrite (Gadelha *et al.*, 1997), and singlet oxygen (Nantes, Almeida Turin, de Mascio, and Vercesi unpublished).
3. The MPT promoted by a variety of compounds (including prooxidants, inorganic phosphate, thyroid hormones, and uncouplers) is inhibited by antioxidants such as catalase, *o*-phenanthroline, ebselen, and thioredoxin peroxidase (Castilho *et al.*, 1995b; Kowaltowski *et al.*, 1995, 1996a, 1996b, 1998b; Valle *et al.*, 1993).
4. The MPT induced by inorganic phosphate or uncouplers is inhibited by molecular oxygen depletion from the reaction medium (Kowaltowski *et al.*, 1996a, 1996b, 1998b) in mitochondria treated with an uncoupler and respiratory chain inhibitor, a situation that excludes participation of mitochondrial membrane potential, Ca^{2+} uptake, and respiration (Vercesi and Hoffman, 1993).
5. The MPT is preceded by increased mitochondrial ROS generation (Castilho *et al.*, 1995b; Kowaltowski *et al.*, 1995, 1996a, 1996b, 1998b, 1998c). The increase is Ca^{2+} dependent but is not prevented by the MPT inhibitor cyclosporin A, demonstrating that is not secondary to mitochondrial permeabilization.

6. The MPT is caused by oxidation of mitochondrial inner membrane protein thiol groups (Bindoli *et al.*, 1997; Kowaltowski and Castilho 1997; Castilho *et al.*, 1995b; Fagian *et al.*, 1990; Kowaltowski *et al.*, 1997, 1998b; Petronilli *et al.*, 1994; Sokolove, 1988; Valle *et al.*, 1993; Vercesi, 1984b; Wudarczyk *et al.*, 1996). Indeed, dithiol reagents such as phenylarsine oxide (Kowaltowski and Castilho, 1997, Lenartowics *et al.*, 1991), diamide (Fagian *et al.*, 1990, Siliprandi *et al.*, 1975), and 4,4'-diisothiocyanatostilbene-2,2'-disulfonic acid (Bernardes *et al.*, 1994) are efficient MPT inducers even in the absence of ROS. This shows that the protein thiol oxidation associated with the MPT is downstream of Ca^{2+}-stimulated mitochondrial ROS generation.

7. The MPT is accompanied by other oxidative alterations of mitochondria, such as lipid peroxidation (Gadelha *et al.*, 1997; Kowaltowski *et al.*, 1996b) and mtDNA fragmentation (Almeida, Souza-Pinto, Bertoncini, and Vercesi, unpublished). These alterations ultimately result in irreversible mitochondrial dysfunction.

8. The progressive nature of the MPT-associated membrane permeabilization is related to the degree of oxidation of inner membrane proteins, and thus to the extent of mitochondrial oxidative damage (Castilho *et al.*, 1996).

Although it has been well demonstrated that the transition occurs due to oxidation of inner membrane proteins, and despite very extensive *in vitro* studies, the nature of inner membrane proteins that when oxidized result in the transition pore remains unknown. Also, many mitochondrial components, such as the adenine nucleotide translocator, porin, hexokinase, and cyclophilin, participate in mitochondrial permeabilization (see Zoratti and Szabò, 1995; Halestrap *et al.*, 1998), but the interrelation between these molecules, protein thiol oxidation, and the formation of a nonspecific inner mitochondrial membrane pore has not been clearly elucidated.

The MPT requires Ca^{2+} in the mitochondrial matrix and is strongly inhibited by cellular components such as Mg^{2+}, ATP, and spermine, so it is doubtful whether this phenomenon can be observed *in vivo* at all. Recent data, however, has clearly shown a Ca^{2+}-promoted, cyclosporin A-inhibited mitochondrial permeabilization in intact cells (Dubinsky and Levi, 1998; Halestrap *et al.*, 1998; Lemasters *et al.*, 1998).

The MPT's role in cells remained a mystery for a long time. Because the transition promotes a reversible mitochondrial permeabilization to protons and Ca^{2+} (Broekenmeier *et al.*, 1998, Fiskum and Lehninger, 1979; Vercesi, 1984a), the "low gating potential" MPT pore (Ichas and Mazat, 1998), a physiological role for the pore in cellular functions was suggested (Bernardi and Petronilli, 1996; Gunter *et al.*, 1994; Lehninger *et al.*, 1978). This low gating potential phase of the MPT is rapidly followed, however, by irreversible inner membrane permeabilization to osmotic support, and even to small proteins (Castilho *et al.*, 1996; Zoratti and Szabò, 1995). Based on this observation, it seems unlikely that the transition is involved in any cell event that preserves cell integrity. Indeed, transition has recently been related to cell death, through both apoptotic and necrotic pathways (see discussion below) (for reviews, see Lemasters *et al.*, 1998; Kowaltowski and Vercesi, 1999; Murphy *et al.*, 1999; Skulachev, 1998; Susin *et al.*, 1998). The transition may be a way of eliminating dysfunctional mitochondria that generate large amounts of ROS (Skulachev, 1998). If sufficient mitochondria within a cell undergo transition, cell death may result.

6.3. Oxidative Damage to Mitochondrial Membrane Lipids

Because of the high protein content of inner mitochondrial membrane, proteins are a major target of mitochondrially generated ROS. Mitochondrially derived ROS also promote oxidation of lipid components of the inner membrane, however (Gabbita *et al.*, 1998), in a self-propagated and irreversible process (Kappus, 1985).

Mitochondrial-membrane lipid peroxidation may be promoted by lipid radicals formed in the presence of Fe^{2+} bound to low-molecular-weight complexing substances such as citrate, ATP, ADP, and GTP (Halliwell and Gutteridge, 1990). Diseases involving iron overload, such as hemocromatosis, may involve Fe^{2+}-induced mitochondrial membrane lipid oxidation (Bacon and Britton, 1990). The Fe^{2+}-promoted mitochondrial lipid peroxidation results in irreversible loss of mitochondrial functionality accompanied by loss of membrane proteins and mtDNA oxidation (Castilho *et al.*, 1994; Almeida, and Vercesi, unpublished). This process can be prevented by reduced coenzyme Q, vitamin E, and ionophores containing hydroxyl groups, all capable of quenching lipid radicals (Castilho *et al.*, 1995a; Grijalba *et al.*, 1998).

Peroxinitrite ($ONOO^-$), which may be generated in mitochondria through the combination of $NO\cdot$ and $O_2\cdot^-$, also promotes mitochondrial membrane lipid peroxidation, in a manner combined with membrane protein thiol oxidation (Gadelha *et al.*, 1997).

The lipid oxidation can also occur secondarily to accumulation of respiratory-chain-derived ROS when mitochondria are incubated in the presence of > 1 mM concentrations of inorganic phosphate (Kowaltowski *et al.*, 1996b). This effect seems related to phosphate's ability to catalyze enol formation from aldehydes generated during oxidative attack on membrane lipids. The tautomerization of these enols yields triplet-state species, capable of propagating the chain of lipid peroxidation (Kowaltowski *et al.*, 1996b). Indeed, exogenous triplet-state acetone can initiate mitochondrial lipid peroxidation (Nantes *et al.*, 1995). Lipid peroxidation accompanying transition in the presence of phosphate may be one factor involved in the MPT irreversibility observed over time (Castilho *et al.*, 1996; Vercesi *et al.*, 1997).

6.4. Oxidative Damage to Mitochondrial DNA

Mitochondrial DNA is located in the mitochondrial matrix, closely attached to the inner membrane. It lacks introns, histones, and other DNA-associated proteins and presents less-complete repair mechanisms than nuclear DNA (Clayton, 1984; Croteau and Bohr, 1997). These characteristics make mtDNA particularly prone to oxidative damage. Indeed, when cells are exposed to oxidative stress, mtDNA damage is more extensive and persistent than nuclear DNA damage (Yakes and Van Houten, 1997). The ROS-promoted damage to mtDNA includes fragmentation and deletions, and the content of 8-hydroxy-deoxyguanosine is commonly used as a marker for this damage (Dizdaroglu, 1991). Content of 8-hydroxy-deoxyguanosine increases with age and degenerative diseases (Hayakawa *et al.*, 1991), implicating mtDNA oxidative damage. *In vitro* mitochondrial oxidative stress, as occurs during mitochondrial lipid oxidation or during the transition, is also accompanied by mtDNA fragmentation and increases in 8-hydroxy-deoxyguanosine (Almeida, Souza-Pinto, Bertoncini, and Vercesi, unpublished).

Because mtDNA encodes for proteins involved in electron transport and oxidative phosphorylation, oxidative mtDNA damage may lead to respiration and phosphorylation deficiencies, which may further stimulate mitochondrial ROS generation (Ozawa, 1997).

6.5. Beneficial Effects of Mitochondrial Oxidative Stress

Although most ROS effects are detrimental, mainly because of the poorly controlled characteristics of ROS reactivity, in some cases mitochondrial oxidative stress can be beneficial. For example, mild oxidative stress caused by myocyte ischemia/reperfusion contributes to myocardial preconditioning, making the tissue resistant to subsequent ischemic damage (Vanden Hoek *et al.*, 1998). Increased protein production after mild oxidative stress may be attributed to transcriptional activation induced by ROS (Chandel *et al.*, 1998).

7. EFFECTS OF MITOCHONDRIAL OXIDATIVE STRESS ON CELL INTEGRITY

In multicellular organisms, cell death may occur through necrotic or apoptotic pathways. Necrotic cell death is secondary to cell injury great enough to affect essential cellular systems, and it is mostly characterized by swelling and intracellular component loss followed by an intense inflammatory response. In contrast, apoptotic cell death may occur after a relatively mild cell injury or as a programmed event in normal tissue development. Apoptosis is an energy-consuming process characterized by cell blebbing and shrinkage and it depends on protein synthesis and activation (Steller, 1995). Some triggers of necrotic cell death may also induce apoptotic cell death if present in less-deleterious conditions, suggesting that the two forms of cell death may have common pathways (Leist and Nicotera, 1997).

Clearly both apoptotic and necrotic cell death may be triggered by a large variety of stimuli. One stimulus long implicated in necrotic cell death under situations of ischemia and reperfusion and cyanide poisoning is mitochondrial oxidative stress (Fiskum 1985; Halestrap *et al.*, 1998; Kristian and Siesjo, 1998; Lemasters *et al.*, 1998; Majewska *et al.*, 1978; Myers *et al.*, 1995). Under these conditions, mitochondrial ROS generation increases due to alterations in the redox state of respiratory chain components, oxygen tension of the cell, and increases in intracellular Ca^{2+} (see discussion above). Thus, oxidative stress may trigger mitochondrial dysfunction and cell death secondarily to cellular ATP depletion. Mitochondrially generated ROS may still diffuse within the cell, promoting oxidative damage in other cellular systems and also cell death.

Initially, apoptosis was thought to occur independently of mitochondrially generated ROS because it could be triggered under conditions of very low oxygen (Jacobson and Raff, 1995). Apoptosis is an intricate process, however, and its induction can surely be promoted by many different triggers. In addition, recent evidence that mitochondria can produce considerable amounts of ROS even under conditions of very low oxygen (Delgi Esposti and McLennan, 1998) calls for revision of this statement. A large body of evidence has recently accumulated showing that mitochondrial oxidative damage in the form of the MPT is a trigger for apoptotic cell death (Green and Reed, 1998; Lemasters *et al.*, 1998;

Susin *et al.*, 1998). Another piece of evidence of supporting a role for mitochondria in apoptosis is the anti-apoptotic affect of Bcl2, a protein found primarily in mitochondria, which may act as an antioxidant (Hockenbery *et al.*, 1993) and as an inhibitor of the MPT (Zamzami *et al.*, 1996).

Apoptosis triggered by the MPT is related to cytochrome *c* release from mitochondrial intermembrane space to cytosol (Kluck *et al.*, 1997; Liu *et al.*, 1996; Skulachev, 1998; Susin *et al.*, 1998; Yang *et al.*, 1997). This triggers activation of cellular caspases, ultimately resulting in nuclear DNA cleavage and cell death. Indeed, in the absence of the MPT, cytochrome *c* microinjection into the cytosol will induce apoptosis (Zhivotovsky *et al.*, 1998). Mitochondrial cytochrome *c* release following transition seems related to disruption of the outer mitochondrial membrane following large-amplitude mitochondrial swelling (Petit *et al.*, 1998). At least in rat brain mitochondria, Ca^{2+} ions promote mitochondrial cytochrome *c* release, however, even in the absence of mitochondrial swelling, and they do so through a mechanism independent of the MPT (Andreyev *et al.*, 1998). Whether this form of Ca^{2+}-induced cytochrome *c* release is related to mitochondrial oxidative stress is still unclear. Bax, a pro-apoptotic member of the Bcl2 protein family, can also release mitochondrial cytochrome *c*, in a manner independent of the MPT (Jürgensmeier *et al.*, 1998).

Other forms of apoptosis are related to mitochondrial oxidative stress, and not necessarily through the MPT. In this sense, p53-induced apoptosis has been associated with oxidative mitochondrial damage (Polyak *et al.*, 1997), and antioxidants such as thioredoxin peroxidase (Zhang *et al.*, 1997), MnSOD (Keller *et al.*, 1998), and N-acetylcysteine (Cossarizza *et al.*, 1995) inhibit apoptosis induced by a wide variety of stimuli. Mitochondrial ROS generation also participates in ceramide-induced apoptosis (Quillet-Mary *et al.*, 1997).

Taken together, recent studies of cell death mechanisms demonstrate that mitochondria may trigger both apoptosis and necrosis. In many studies the involvement of mitochondrially generated ROS is clearly demonstrated. Thus, extensive mitochondrial oxidative stress may lead to necrotic cell death, and less-intensive ROS generation may activate the apoptotic pathways as a way for the organism to eliminate undesired ROS-generating cells

8. CONCLUSIONS

Normally, mitochondria provide eukaryotic cells with ATP, generated in a highly energy conservational manner. Under unfavorable conditions, however, or over time, mitochondria also generate ROS, which may damage both mitochondria themselves and the cell that surrounds them. In addition, mitochondria can promote cell death not only through energy deprivation but also in a highly organized fashion by releasing cytochrome *c*. Thus our cells carry within them a potential time bomb that can lead to cell damage and ultimately, death. A better understanding of the mechanisms by which mitochondria generate ROS and promote cell death will certainly help to provide the tools for preventing undesirable, mitochondrial effects, and help make their association with eukaryotic cells an even more peaceful one.

ACKNOWLEDGEMENTS. We wish to thank Dr. G. Fiskum for his critical reading of the manuscript, and FAPESP, CNPq, CAPES, and PRONEX for financial support.

REFERENCES

Andreyev, A. Y., Fahy, B., and Fiskum, G., 1998, Cytochrome c release from brain mitochondria is independent of the mitochondrial permeability transition, *FEBS Lett.* **439**:373–376.

Bacon, B. R., and Britton, R. S., 1990, The pathology of hepatic iron overload: A free radical-mediated process? *Hepatology* **11**:127–137.

Balakirev, M. Yu, Khramtsov, V. V., and Zimmer, G., 1997, Modulation of the mitochondrial permeability transition by nitric oxide, *Eur. J. Biochem.* **246**:710–718.

Berlett, B. S., and Stadtman, E. R., 1997, Protein oxidation in aging, disease, and oxidative stress, *J. Biol. Chem.* **272**:20313–20316.

Bernardes, C. F., Pereira-da-Silva L., and Vercesi A. E., 1986, t-Butylhydroperoxide-induced Ca^{2+} efflux from liver mitochondria in the presence of physiological concentrations of Mg^{2+} and ATP, *Biochim. Biophys. Acta* **850**:41–48.

Bernardes, C. F., Meyer-Fernandes, J. R., Basseres, D. S., Castilho, R. F., and Vercesi, A. E., 1994, Permeabilization of the rat liver inner membrane by 4,4′-diisothiocyanatostilbene-2,2′-dissulfonic acid (DIDS) in the presence of Ca^{2+} is mediated by production of membrane protein aggregates, *Biochim. Biophys. Acta* **1188**:93–100.

Bernardi, P., 1992, Modulation of the mitochondrial cyclosporin A-sensitive permeability transition pore by the proton electrochemical gradient: Evidence that the pore can be opened by membrane depolarization, *J. Biol. Chem.* **267**:8834–8839.

Bernardi, P., 1996, The permeability transition pore: Control points of a cyclosporin A-sensitive mitochondrial channel involved in cell death, *Biochim. Biophys. Acta.* **1275**:5–9.

Bernardi, P., and Petronilli, V., 1996, The permeability transition pore as a mitochondrial calcium release channel: A critical appraisal, *J. Bioenerg. Biomembr.* **28**:131–138.

Bindoli, A., Callegaro, M. T., Barzon, E., Benetti, M., and Rigobello, M. P., 1997, Influence of the redox state of pyridine nucleotides on mitochondrial sulfhydryl groups and permeability transition, *Arch. Biochem. Biophys.* **342**:22–28.

Bobyleva, V., Pazienza, T. L., Maseroli, R., Tomasi, A., Salvioli, S., Cossarizza, A., Franceschi, C., and Skulachev, V. P., 1998, Decrease in mitochondrial energy coupling by thyroid hormones: A physiological effect rather than a pathological hyperthyroidism consequence, *FEBS Lett.* **430**:409–413.

Bolli, R., 1998, Causative role of oxyradicals in myocardial stunning: A proven hypothesis: A brief review of the evidence demonstrating a major role of reactive oxygen species in several forms of postischemic dysfunction, *Basic Res. Cardiol.* **93**:156–162.

Bolli, R., Patel, B. S., Jeroudi, M. O., Lai, E. K., and McCay P. B., 1988, Demonstration of free radical generation in "stunned" myocardium of intact dogs with the use of the spin trap alpha-phenyl N-tert-butyl nitrone, *J. Clin. Invest.* **82**:476–485.

Bolli, R., Jeroudi, M. O., Patel, B. S., Aruoma, O. I., Halliwell, B., Lai, E. K., and McCay, P. B., 1989, Marked reduction of free radical generation and contractile dysfunction by antioxidant therapy begun at the time of reperfusion: Evidence that myocardial "stunning" is a manifestation of reperfusion injury, *Circ. Res.* **65**:607–622.

Boveris, A., and Chance, B., 1973, The mitochondrial generation of hydrogen peroxide: General properties and effect of hyperbaric oxygen, *Biochem. J.* **134**:707–716.

Boveris, A., Cadenas, E., and Stoppani, A. O., 1976, Role of ubiquinone in the mitochondrial generation of hydrogen peroxide, *Biochem. J.* **156**:435–444.

Broekemeier, K. M., Klocek, C. K., and Pfeiffer, D. R., 1998, Proton selective substate of the mitochondrial permeability transition pore: Regulation by the redox state of the electron transport chain, *Biochemistry* **37**:13059–13065.

Budd, S. L., and Nicholls, D. G., 1996, Mitochondria, calcium regulation, and acute glutamate excitotoxicity in cultured cerebellar granule cells, *J. Neurochem.* **67**:2282–2291.

Budd, S. L., Castilho, R. F., and Nicholls, D. G., 1997, Mitochondrial membrane potential and hydroethidine-monitored superoxide generation in cultured cerebellar granule cells, *FEBS Lett.* **415**:21–24.

Cadenas, E., and Boveris, A., 1980, Enhancement of hydrogen peroxide formation by protonophores and ionophores in antimycin-supplemented mitochondria, *Biochem. J.* **188**:31–37.

Cadenas, E., Boveris, A., Ragan, C. I., and Stoppani, A. O. M., 1977, Production of superoxide radicals and hydrogen peroxide by NADH–ubiquinone reductase and ubiquinol–cytochrome *c* reductase from beef-heart mitochondria, *Arch. Biochem. Biophys.* **180**:248–257.

Cadet, J. L., and Brannock, C., 1998, Free radicals and the pathobiology of brain dopamine systems, *Neurochem. Int.* **32**:117–131.

Cai, J., and Jones, D. P., 1998, Superoxide in apoptosis: Mitochondrial generation triggered by cytochrome *c* loss, *J. Biol. Chem.* **273**:11401–11404.

Castilho, R. F., Meinicke, A. R., Almeida, A. M., Hermes-Lima, M., and Vercesi, A. E., 1994, Oxidative damage of mitochondria induced by Fe(II)citrate is potentiated by Ca^{2+} and induces lipid peroxidation and alterations in membrane proteins, *Arch. Biochem. Biophys.* **308**:158–163.

Castilho, R. F., Kowaltowski, A. J., Meinicke, A. R., and Vercesi, A. E., 1995a, Oxidative damage of mitochondria induced by Fe(II)citrate or *t*-butyl hydroperoxide in the presence of Ca^{2+}: Effect of coenzyme Q redox state, *Free Radical Biol. Med.* **18**:55–59.

Castilho, R. F., Kowaltowski, A. J., Meinicke, A. R., Bechara, E. J., and Vercesi, A. E., 1995b, Permeabilization of the inner mitochondrial membrane by Ca^{2+} ions is stimulated by *t*-butyl hydroperoxide and mediated by reactive oxygen species generated by mitochondria, *Free Radical Biol. Med.* **18**:479–486.

Castilho, R. F., Kowaltowski, A. J., and Vercesi, A. E., 1996, The irreversibility of inner mitochondrial membrane permeabilization by Ca^{2+} plus prooxidants is determined by the extent of membrane protein thiol cross-linking, *J. Bioenerg. Biomembr.* **28**:523–529.

Castilho, R. F., Kowaltowski, A. J., and Vercesi, A. E., 1998a, 3,5,3′-Triiodothyronine induces mitochondrial permeability transition mediated by reactive oxygen species and membrane protein thiol oxidation, *Arch. Biochem. Biophys.* **354**:151–157.

Castilho, R. F., Hansson, O., Ward, M. W., Budd, S. L., and Nicholls, D. G., 1998b, Mitochondrial control of acute glutamate excitotoxicity in cultured cerebellar granule cells, *J. Neurosci.* **18**:10277–10286.

Castilho, R. F., Ward, M. W., and Nicholls, D. G., 1999, Oxidative stress, mitochondrial function, and acute glutamate excitotoxicity in cultured cerebellar granule cells, *J. Neurochem.*, **72**:1394–1401.

Chandel, N. S., Maltepe, E., Goldwasser, E., Mathieu, C. E., Simon, M. C., and Schumacker, P. T., 1998, Mitochondrial reactive oxygen species trigger hypoxia-induced transcription, *Proc. Natl. Acad. Sci. USA.* **95**:11715–11720.

Clayton, P. A., 1984, Transcription of the mammalian mitochondrial genome, *Ann. Rev. Biochem.* **53**:573–594.

Cossarizza, A., Franceschi, C., Monti, D., Salvioli, S., Bellesia, E., Rivabene, R., Biondo, L., Rainaldi, G., Tinari, A., and Malorni, W., 1995, Protective effect of *N*-acetylcysteinee in tumor necrosis factor-alpha-induced apoptosis in U937 cells: The role of mitochondria, *Exp. Cell Res.* **220**:232–240.

Costa, L. E., Llesuy, S., and Boveris, A., 1993, Active oxygen species in the liver of rats submitted to chronic hypobaric hypoxia, *Am. J. Physiol.* **264**:1395–1400.

Costantini, P., Petronilli, V., Colonna, R., and Bernardi, P., 1995, On the effects of paraquat on isolated mitochondria: Evidence that paraquat causes opening of the cyclosporin A-sensitive permeability transition pore synergistically with nitric oxide, *Toxicology* **99**:77–88.

Croteau, D. L., and Bohr, V. A., 1997, Repair of oxidative damage to nuclear and mitochondrial DNA in mammalian cells, *J. Biol. Chem.* **272**:25409–25412.

Dawson, T. L., Gores, G. J., Nieminen, A.-L., Herman, B., and Lemasters, J. J., 1993, Mitochondria as a source of reactive oxygen species during reductive stress in rat hepatocytes, *Am. J. Physiol.* **264**:961–967.

Degli Esposti, M., and McLennan, H., 1998, Mitochondria and cells produce reactive oxygen species in virtual anaerobiosis: Relevance to ceramide-induced apoptosis, *FEBS Lett.* **430**:338–342.

Dizdaroglu, M., 1991, Chemical determination of free radical-induced damage to DNA, *Free Radical Biol. Med.* **10**:225–242.

Doonan, S., Barra, D., and Bossa, F., 1984, Structural and genetic relationships between cytosolic and mitochondrial isoenzymes, *Int. J. Biochem.* **16**:1193–1199.

Dubinsky, J. M., and Levi, Y., 1998, Calcium-induced activation of the mitochondrial permeability transition in hippocampal neurons, *J. Neurosci. Res.* **53**:728–741.

Duranteau, J., Chandel, N. S., Kulisz, A., Shao, Z., and Schumacker, P. T., 1998, Intracellular signaling by reactive oxygen species during hypoxia in cardiomyocytes, *J. Biol. Chem.* **273**:11619–11624.

Ernster, L., Forsmark, P., and Nordenbrand, K., 1992, The mode of action of lipid-soluble antioxidants in biological membranes: Relationship between the effects of ubiquinol and vitamin E as inhibitors of lipid peroxidation in submitochondrial particles, *J. Nutr. Sci. Vitaminol.* (Tokyo) **1992**:548–551.

Fagian, M. M., Pereira-da-Silva, L., Martins, I. S., and Vercesi, A. E., 1990, Membrane protein thiol cross-linking associated with the permeabilization of the inner mitochondrial membrane by Ca^{2+} plus prooxidants, *J. Biol. Chem.* **265**:19955–19960.

Fiskum, G., 1985, Mitochondrial damage during cerebral ischemia, *Ann. Emerg. Med.* **14**:810–815.

Fiskum, G., and Lehninger, A. L., 1979, Regulated release of Ca^{2+} from respiring mitochondria by $Ca^{2+}/2H^+$ antiport, *J. Biol. Chem.* **254**:6236–6239.

Fiskum, G., Murphy, A. N., and Beal, M. F., 1998, Mitochondria in neurodegeneration: Part II. Acute ischemia and chronic neurodegenerative diseases, *J. Cerebr. Flow Metab.* **19**:351–369.

Frei, B., Winterhalter, K. H., and Richter, C., 1985, Mechanism of alloxan-induced calcium release from liver mitochondria, *J. Biol. Chem.* **260**:7394–7401.

Gabbita, S. P., Subramaniam, R., Allouch, F., Carney, J. M., and Butterfield, D. A., 1998, Effects of mitochondrial respiratory stimulation on membrane lipids and proteins: An electron paramagnetic resonance investigation, *Biochim. Biophys. Acta* **1372**:163–173.

Gadelha, F. R., Thomson, L., Fagian, M. M., Costa, A. D. T., Radi, R., and Vercesi, A. E., 1997, Ca^{2+}-independent permeabilization of the inner mitochondrial membrane by peroxinitrite is mediated by membrane protein thiol cross-linking and lipid peroxidation, *Arch. Biochim. Biophys.* **345**:243–250.

Giulivi, C., 1998, Functional implications of nitric oxide produced by mitochondria in mitochondrial metabolism, *Biochem. J.* **332**:673–679.

Giulivi, C., Boveris, A., and Cadenas, E. A., 1995, Hydroxyl radical generation during mitochondrial electron transfer and the formation of 8-hydroxylguanosine in mitochondrial DNA, *Arch. Biochem. Biophys.* **316**:909–916.

Giulivi, C., Poderoso, J. J., and Boveris, A., 1998, Production of nitric oxide by mitochondria, *J. Biol. Chem.* **273**:11038–11043.

Green, D. R., and Reed, J. C., 1998, Mitochondria and apoptosis, *Science* **281**:1309–1312.

Grijalba, M. T., Andrade, P. B., Meinicke, A. R., Castilho, R. F., Vercesi, A. E., and Schreier, S., 1998, Inhibition of membrane lipid peroxidation by a radical scavenging mechanism: A novel function for hydroxyl-containing ionophores, *Free Radical Res.* **28**:301–318.

Grijalba, M. T., Vercesi, A. E., and Schreier, S., 1999, Ca^{2+}-induced increased lipid packing and domain formation in submitochondrial particles. A possible early step in the mechanism of Ca^{2+}-stimulated generation of reactive oxygen species by the respiratory chain, *Biochemistry* **38**:13279–13287.

Gunasekar, P. G., Kanthasamy, A. G., Borowitz, J. L., and Isom, G. E., 1995, NMDA receptor activation produces concurrent generation of nitric oxide and reactive oxygen species: Implication for cell death, *J. Neurochem.* **65**:2016–2021.

Gunter, T. E., Gunter, K. K., Sheu, S. S., and Gavin, C. E., 1994, Mitochondrial calcium transport: Physiological and pathological relevance, *Am. J. Physiol.* **267**:313–339.

Halestrap, A. P., Kerr, P. M., Javadov, S., and Woodfield, K. Y., 1998, Elucidating the molecular mechanism of the permeability transition pore and its role in reperfusion injury of the heart, *Biochim. Biophys. Acta* **1366**:79–94.

Halliwell, B., and Gutteridge, J. M. C., 1990, Role of free radicals and catalytic metal ions in human disease: An overview, *Methods Enzymol.* **186**:1–85.

Hansford, R. G., Hogue, B. A., and Mildaziene, V., 1997, Dependence of H_2O_2 formation by rat heart mitochondria on substrate availability and donor age, *J. Bioenerg. Biomembr.* **29**:89–95.

Harris, E. J., Al-Shaikhaly, M., and Baum, H., 1979, Stimulation of mitochondrial calcium ion efflux by thiol-specific reagents and by thyroxine: The relationship to adenosine diphosphate retention and to mitochondrial permeability, *Biochem. J.* **182**:455–464.

Hayakawa, M., Torii, K., Sugiyama, S., Tanaka, M., and Ozawa, T., 1991, Age-associated accumulation of 8-hydroxydeoxyguanosine in mitochondrial DNA of human diaphragm, *Biochem. Biophys. Res. Commun.* **179**:1023–1029.

Hermes-Lima, M., 1995, How do Ca^{2+} and 5-aminolevulinic acid-derived oxyradicals promote injury to isolated mitochondria? *Free Radical Biol. Med.* **19**:381–390.

Hermes-Lima, M., Valle, V. G. R., Vercesi, A. E., and Bechara E. J. H., 1991, Damage to rat liver mitochondria promoted by δ-aminolevulinic acid-generated reactive oxygen species: Connections with acute intermittent porphyria and lead poisoning, *Biochim. Biophys. Acta* **1056**:57–63.

Hockenbery, D. M., Oltvai, Z. N., Yin, X. M., Milliman, C. L., and Korsmeyer, S. J., 1993, Bcl-2 functions in an antioxidant pathway to prevent apoptosis, *Cell* **75**:241–251.

Hoek, J. B., and Rydstron, J., 1988, Physiological roles of nicotinamide nucleotide transhydrogenase, *Biochem. J.* **254**:1–10.

Hunter, D. R., and Haworth, R. A., 1979, The Ca^{2+}-induced membrane transition in mitochondria: III. Transitional Ca^{2+} release, *Arch. Biochem. Biophys.* **195**:468–477.

Hüser, J., Rechenmacher, C. E., and Blatter, L. A., 1998, Imaging the permeability pore transition in single mitochondria, *Biophys. J.* **74**:2129–2137.

Ichas, F., and Mazat, J. P., 1998, From calcium signaling to cell death: Two conformations for the mitochondrial permeability transition pore: Switching from low- to high-conductance state, *Biochim. Biophys. Acta* **1366**:33–50.

Jacobson, M. D., and Raff, M. C., 1995, Programmed cell death and Bcl-2 protection in very low oxygen, *Nature* **374**:814–816.

Jamieson, D., Chance, B., Cadenas, E., and Boveris, A., 1986, The relation of free radical production to hyperoxia, *Annu. Rev. Physiol.* **48**:703–719.

Jurgensmeier, J. M., Xie, Z., Deveraux, Q., Ellerby, L., Bredesen, D., and Reed, J. C., 1998, Bax directly induces release of cytochrome *c* from isolated mitochondria, *Proc. Natl. Acad. Sci. USA* **95**:4997–5002.

Juurlink, B. H., and Sweeney, M. I., 1997, Mechanisms that result in damage during and following cerebral ischemia, *Neurosci. Biobehav. Rev.* **21**:121–128.

Kappus, A., 1985, *Lipid Peroxidation: Mechanisms, Analysis, Enzymology and Biological Relevance* (H. Sies, Ed.), Academic, London.

Kasha, M., 1991, Energy transfer, charge transfer, and proton transfer in molecular composite systems, *Basic Life Sci.* **58**:231–251.

Keller, J. N., Kindy, M. S., Holtsberg, F. W., St. Clair, D. K., Yen, H. C., Germeyer, A., Steiner, S. M., Bruce-Keller, A. J., Hutchins, J. B., and Mattson, M. P., 1998, Mitochondrial manganese superoxide dismutase prevents neural apoptosis and reduces ischemic brain injury: Suppression of peroxynitrite production, lipid peroxidation, and mitochondrial dysfunction, *J. Neurosci.* **18**:687–697.

Kerver, E. D., Vogels, I. M., Bosch, K. S., Vreeling-Sindelarova, H., Van den Munckhof, R. J., and Frederiks, W. M., 1997, *In situ* detection of spontaneous superoxide anion and singlet oxygen production by mitochondria in rat liver and small intestine, *Histochem. J.* **29**:229–237.

Kluck, R. M., Bossy-Wetzel, E., Green, D. R., and Newmeyer, D. D., 1997, The release of cytochrome *c* from mitochondria: A primary site for Bcl-2 regulation of apoptosis, *Science* **275**:1132–1136.

Korshunov, S. S., Skulachev, V. P., and Starkov, A. A., 1997, High protonic potential actuates a mechanism of production of reactive oxygen species in mitochondria, *FEBS Lett.* **416**:15–18.

Korshunov, S. S., Korkina, O. V., Ruuge, E. K., Skulachev, V. P., and Starkov, A. A., 1998, Fatty acids as natural uncouplers preventing generation of O_2^- and H_2O_2 by mitochondria in the resting state, *FEBS Lett.* **435**:215–218.

Kowaltowski, A. J., and Castilho, R. F., 1997, Ca^{2+} acting at the external side of the inner mitochondrial membrane can stimulate mitochondrial permeability transition induced by phenylarsine oxide, *Biochim. Biophys. Acta* **1322**:221–229.

Kowaltowski, A. J., and Vercesi, A. E., 1999, Mitochondrial damage induced by conditions of oxidative stress, *Free Radical Biol. Med.* **26**:463–471. in press.

Kowaltowski, A. J., Castilho, R. F., and Vercesi, A. E., 1995, Ca^{2+}-induced mitochondrial membrane permeabilization: Role of coenzyme Q redox state, *Am. J. Physiol.* **269**:141–147.

Kowaltowski, A. J., Castilho, R. F., and Vercesi, A. E., 1996a, Opening of the mitochondrial permeability transition pore by uncoupling or inorganic phosphate in the presence of Ca^{2+} is dependent on mitochondrial-generated reactive oxygen species, *FEBS Lett.* **378**:150–152.

Kowaltowski, A. J., Castilho, R. F., Bechara, E. J. H., and Vercesi, A. E., 1996b, Effect of inorganic phosphate concentration on the nature of inner mitochondrial membrane alterations induced by Ca^{2+} ions: A proposed model for phosphate-stimulated lipid peroxidation, *J. Biol. Chem.* **271**:2929–2934.

Kowaltowski, A. J., Vercesi, A. E., and Castilho, R. F., 1997, Mitochondrial membrane protein thiol reactivity with *N*-ethylmaleimide or mersalyl is modified by Ca^{2+}: Correlation with mitochondrial permeability transition, *Biochim. Biophys. Acta* **1318**:395–402.

Kowaltowski, A. J., Costa, A. D. T., and Vercesi, A. E., 1998a, Activation of the potato plant uncoupling mitochondrial protein inhibits reactive oxygen species generation by the respiratory chain, *FEBS Lett.* **425**:213–216.

Kowaltowski, A. J., Netto, L. E. S., and Vercesi, A. E., 1998b, The thiol-specific antioxidant enzyme prevents mitochondrial permeability transition: Evidence for the involvement of reactive oxygen species in this mechanism, *J. Biol. Chem.* **273**:12766–12769.

Kowaltowski, A. J., Naia-da-Silva, E. S., Castilho, R. F., and Vercesi, A. E., 1998c, Ca^{2+}-stimulated mitochondrial reactive oxygen species generation and permeability transition are inhibited by dibucaine or Mg^{2+}. *Arch. Biochem. Biophys.* **359**:77–81.

Kristian, T., and Siesjo, B. K., 1998, Calcium in ischemic cell death, *Stroke* **29**:705–718.

Kuzminova, A. E., Zhuravlyova, A. V., Vyssokikh, M.Yu, Zorova, L. D., Krasnikov, B. F., and Zorov, D. B., 1998, The permeability transition pore induced under anaerobic conditions in mitochondria energized with ATP, *FEBS Lett.* **434**:313–316.

Lehninger, A. L., Vercesi, A. E., and Bababumni, E. A., 1978, Regulation of Ca^{2+} release from mitochondria by the oxidation-reduction state of pyridine nucleotides, *Proc. Nat. Acad. Sci. USA* **75**:1690–1694.

Leist, M., and Nicotera, P., 1997, The shape of cell death, *Biochem. Biophys. Res. Commun.* **236**:1–9.

Lemasters, J. J., Nieminen, A. L., Qian, T., Trost, L. C., Elmore, S. P., Nishimura, Y., Crowe, R. A., Cascio, W. E., Bradham, C. A., Brenner, D. A., and Herman, B., 1998, The mitochondrial permeability transition in cell death: A common mechanism in necrosis, apoptosis, and autophagy, *Biophim. Biophys. Acta* **1366**:177–196.

Lenartowics, E., Bernardi, P., and Azzone, G. F., 1991, Phenylarsine oxide induced the cyclosporin A-sensitive membrane permeability transition in rat liver mitochondria, *J. Bioenerg. Biomembr.* **23**:679–688.

Liu, X., Kim, C. N., Yang, J., Jemmerson, R., and Wang, X., 1996, Induction of apoptotic program in cell-free extracts: Requirement for dATP and cytochrome *c*, *Cell* **86**:147–157.

Liu, S. S., 1997, Generating, partitioning, targeting, and functioning of superoxide in mitochondria, *Biosci. Rep.* **17**:259–272.

Liu, Y., Rosenthal, R. E., Haywood, Y., Miljkovic-Lolic, M., Vanderhoek, J. Y., and Fiskum, G., 1998, Normoxic ventilation after cardiac arrest reduces oxidation of brain lipids and improves neurological outcome, *Stroke* **29**:1679–1686.

Loschen, G., and Azzi, A., 1975, On the formation of hydrogen peroxide and oxygen radicals in heart mitochondria, *Recent Adv. Stud. Cardiac. Struct. Metab.* **7**:3–12.

Loschen, G., Azzi, A., and Flohe, L., 1973, Mitochondrial H_2O_2 formation: Relationship with energy conservation, *FEBS Lett.* **33**:84–87.

Low, P. S., Lloyd, D. H., Stein, T. M., and Rogers, J. A., 1979, Calcium displacement by local anesthetics: Dependence on pH and anesthetic charge, *J. Biol. Chem.* **254**:4119–4125.

Lubec, G., 1996, The hydroxyl radical: From chemistry to human disease, *J. Invest. Med.* **44**:324–346.

Majewska, M. D., Strosznajder, J., and Lazarewicz, J., 1978, Effect of ischemic anoxia and barbiturate anesthesia on free radical oxidation of mitochondrial phospholipids, *Brain Res.* **158**:423–434.

Merryfield, M. L., and Lardy, H. A., 1982, Ca^{2+}-mediated activation of phosphoenolpyruvate carboxykinase occurs via release of Fe^{2+} from rat liver mitochondria, *J. Biol. Chem.* **257**:3628–3635.

Multhaup, G., Ruppert, T., Schlicksupp, A., Hesse, L., Beher, D., and Masters, C. L., 1997, Reactive oxygen species and Alzheimer's disease, *Biochem. Pharmacol.* **54**:533–539.

Murphy, A. N., Fiskum, G., and Beal, M. F., 1999, Mitochondria in neurodegeneration: Part I. Bioenergetic function in cell life and death, *J. Cereb. Flow Metab.*, in press.

Myers, K. M., Fiskum, G., Liu, Y., Simmens, S. J., Bredesen, D. E., and Murphy, A. N., 1995, Bcl-2 protects neural cells from cyanide/aglycemia-induced lipid oxidation, mitochondrial injury, and loss of viability, *J. Neurochem.* **65**:2432–2440.

Nantes, I. L., Cilento, G., Bechara, E. J. H., and Vercesi, A. E., 1995, Chemiluminescent diphenylacetaldehyde oxidation by mitochondria is promoted by cytochromes and leads to oxidative injury of the organelle, *Photochem. Photobiol.* **62**:522–527.

Nègre-Salvayre, A., Hirtz, C., Carrera, G., Cazenave, R., Troly, M., Salvayre, R., Pénicaud, L., and Casteilla, L., 1997, A role for the uncoupling protein-2 as a regulator of mitochondrial hydrogen peroxide generation, *FASEB J.* **11**:809–815.

Nicholls, D. G., and Ferguson, S. J., 1982, *Bioenergetics 2*, Academic, London.

Ozawa, T., 1997, Oxidative damage and fragmentation of mitochondrial DNA in cellular apoptosis, *Biosci Rep.* **17**:237–248.

Petit, P. X., Goubern, M., Diolez, P., Susin, S. A., Zamzami, N., and Kroemer, G., 1998, Disruption of the outer mitochondrial membrane as a result of large amplitude swelling: The impact of irreversible permeability transition, *FEBS Lett.* **426**:111–116.

Petronilli, V., Costantini, P., Scorrano, L., Colonna, R., Passamonti, S., and Bernardi, P., 1994, The voltage sensor of the mitochondrial permeability transition pore is tuned by the oxidation-reduction state of vicinal thiols: Increase of the gating potential by oxidants and its reversal by reducing agents, *J. Biol. Chem.* **269**:16638–16642.

Polyak, K., Xia, Y., Zweier, J. L., Kinzler, K. W., and Vogelstein, B., 1997, A model for p53-induced apoptosis, *Nature* **389**:300–305.

Pryor, W. A., and Squadrito, G. L., 1995, The chemistry of peroxynitrite: A product from the reaction of nitric oxide with superoxide, *Am J. Physiol.* **268**:699–722.

Quillet-Mary, A., Jaffrezou, J. P., Mansat, V., Bordier, C., Naval, J., and Laurent, G., 1997, Implication of mitochondrial hydrogen peroxide generation in ceramide-induced apoptosis, *J. Biol. Chem.* **272**:21388–21395.

Radi, R., Turrens, J. F., Chang, L. Y., Bush, K. M., Crapo, J. D., and Freeman, B. A., 1991, Detection of catalase in rat heart mitochondria, *J. Biol. Chem.* **266**:22028–22034.

Reynolds, I. J., and Hastings, T. G., 1995, Glutamate induces the production of reactive oxygen species in cultured forebrain neurons following NMDA receptor activation, *J. Neurosci.* **15**:3318–3327.

Rigobello, M. P., Callegaro, M. T., Barzon, E., Benetti, M., and Bindoli, A., 1998, Purification of mitochondrial thioredoxin reductase and its involvement in the redox regulation of membrane permeability, *Free Radical Biol. Med.* **24**:370–376.

Rossi, C. S., and Lehninger, A. L., 1964, Stoichiometry of respiratory stimulation, accumulation of Ca^{++} and phosphate and oxidative phosphorylation in rat liver mitochondria, *J. Biol. Chem.* **239**:3971–3980.

Salganik, R. I., Shabalina, I. G., Solovyova, N. A., Kolosova, N. G., Solovyov, V. N., and Kolpakov, A. R., 1994, Impairment of respiratory functions in mitochondria of rats with an inherited hyperproduction of free radicals, *Biochem. Biophys. Res. Commun.* **205**:180–185.

Scorrano, L., Petronilli, V., and Bernardi, P., 1997, On the voltage dependence of the mitochondrial permeability transition pore: A critical appraisal, *J. Biol. Chem.* **272**:12295–12299.

Sengpiel, B., Preis, E., Krieglstein, J., and Prehn, J. H., 1998, NMDA-induced superoxide production and neurotoxicity in cultured rat hippocampal neurons: Role of mitochondria, *Eur. J. Neurosci.* **10**:1903–1910.

Sies, H., and Moss, K. M., 1978, A role of mitochondrial glutathione peroxidase in modulating mitochondrial oxidations in liver, *Eur. J. Biochem.* **84**:377–383.

Siliprandi, D., Toninello, A., Zoccarato, F., Rugolo, M., and Siliprandi, N., 1975, Synergic action of calcium ions and diamide on mitochondrial swelling, *Biochem. Biophys. Res. Commun.* **66**:956–961.

Skulachev, V. P., 1997, Membrane-linked systems preventing superoxide formation, *Biosci. Rep.* **17**:347–366.

Skulachev, V. P., 1998, Cytochrome *c* in the apoptotic and antioxidant cascades, *FEBS Lett.* **423**:275–280.

Sokolove, P. M., 1988, Mitochondrial sulfhydryl group modification by adriamycin aglycones, *FEBS Lett.* **234**:199–202.

Steller, H., 1995, Mechanisms and genes of cellular suicide, *Science* **267**:1445–1449.

Susin, S. A., Zamzami, N., and Kroemer, G., 1998, Mitochondria as regulators of apoptosis: Doubt no more, *Biochim. Biophys. Acta* **1366**:151–165.

Sutton, H. C., and Winterbourn, C. C., 1989, On the participation of higher oxidation states of iron and copper in Fenton reactions, *Free Radical Biol. Med.* **6**:53–60.

Takayama, N., Matsuo, N., and Tanaka, T., 1993, Oxidative damage to mitochondria is mediated by the Ca^{2+}-dependent inner membrane permeability transition, *Biochem. J.* **294**:719–725.

Tatoyan, A., and Giulivi, C., 1998, Purification and characterization of a nitric-oxide synthase from rat liver mitochondria, *J. Biol. Chem.* **273**:11044–11048.

Turrens, J. F., 1997, Superoxide production by the mitochondrial respiratory chain, *Biosci. Rep.* **17**:3–8.

Turrens, J. F., and Boveris, A., 1980, Generation of superoxide anion by the NADH dehydrogenase of bovine heart mitochondria, *Biochem. J.* **191**:421–427.

Turrens, J. F., Freeman, B. A., Levitt, J. G., and Crapo, J. D., 1982, The effect of hyperoxia on superoxide production by lung submitochondrial particles, *Arch. Biochem. Biophys.* **217**:401–410.

Turrens, J. F., Alexandre, A., and Lehninger, A. L., 1985, Ubisemiquinone is the electron donor for superoxide formation by complex III of heart mitochondria, *Arch. Biochem. Biophys.* **237**:408–414.

Valle, V. G., Fagian, M. M., Parentoni, L. S., Meinicke, A. R., and Vercesi, A. E., 1993, The participation of reactive oxygen species and protein thiols in the mechanism of mitochondrial inner membrane permeabilization by calcium plus prooxidants, *Arch. Biochem. Biophys.* **307**:1–7.

Vanden Hoek, T. L., Becker, L. B., Shao, Z., Li, C., and Schumacker, P. T., 1998, Reactive oxygen species released from mitochondria during brief hypoxia induce preconditioning in cardiomyocytes, *J. Biol. Chem.* **273**:18092–18098.

Vercesi, A. E., 1984a, Dissociation of NAD(P)$^+$-stimulated mitochondrial Ca^{2+} efflux from swelling and membrane damage, *Arch. Biochem. Biophys.* **232**:86–91.

Vercesi, A. E., 1984b, Possible participation of membrane thiol groups on the mechanism of NAD(P)$^+$-stimulated Ca^{2+} efflux from mitochondria, *Biochem. Biophys. Res. Commun.* **119**:305–310.

Vercesi, A. E., and Hoffmann, M. E., 1993, Generation of reactive oxygen metabolites and oxidative damage in mitochondria: The role of calcium, in *Methods in Toxicology*, Vol. 2 D. P. Jones and L. H. Lash, Eds.), Academic, New York, pp. 256–265.

Vercesi, A. E., Castilho, R. F., Meinicke, A. R., Valle, V. G. R., Hermes-Lima, M., and Bechara, E. J. H., 1994, Oxidative damage of mitochondria induced by 5-aminolevulinic acid: Role of Ca^{2+} ions and membrane protein thiols, *Biochim. Biophys. Acta* **1188**:86–92.

Vercesi, A. E., Kowaltowski, A. J., Grijalba, M. T., Meinicke, A. R., and Castilho, R. F., 1997, The role of reactive oxygen species in mitochondrial permeability transition, *Biosci. Re.* **17**:43–52.

Visioli, O., 1998, Oxidative stress during myocardial ischaemia and heart failure, *Eur. Heart J.* **19B**:2–11.

Watabe, S., Hiroi, T., Yamamoto, Y., Fujioka, Y., Hasegawa, H., Yago, N., and Takahashi, S. Y., 1997, SP-22 is a thioredoxin-dependent peroxide reductase in mitochondria, *Eur. J. Biochem.* **249**:52–60.

Wudarczyk, J., Debska, G., and Lenartowicz, E., 1996, Relation between the activities reducing disulfides and the protection against membrane permeability transition in rat liver mitochondria, *Arch. Biochem. Biophys.* **327**:215–221.

Xie, Y. W., Kaminski, P. M., and Wolin, M. S., 1998, Inhibition of rat cardiac muscle contraction and mitochondrial respiration by endogenous peroxynitrite formation during posthypoxic reoxygenation, *Circ. Res.* **82**:891–897.

Yakes, F. M., and Van Houten B., 1997, Mitochondrial DNA damage is more extensive and persists longer than nuclear DNA damage in human cells following oxidative stress, *Proc. Natl. Acad. Sci. USA* **94**:514–519.

Yang, W., and Block, E. R., 1995, Effect of hypoxia and reoxygenation on the formation and release of reactive oxygen species by porcine pulmonary artery endothelial cells, *J. Cell Physiol.* **164**:414–423.

Yang, J., Liu, X., Bhalla, K., Kim, C. N., Ibrado, A. M., Cai, J., Peng, T. I., Jones, D. P., and Wang, X., 1997, Prevention of apoptosis by Bcl-2: Release of cytochrome *c* from mitochondria blocked, *Science* **275**:1129–1132.

Zakowski, J. J., and Tappel, A. L., 1978, Purification and properties of rat liver mitochondrial glutathione peroxidase, *Biochim. Biophys. Acta* **526**:65–76.

Zamzami, N., Susin, S. A., Marchetti, P., Hirsch, T., Gomez-Monterrey, I., Castedo, M., and Kroemer, G., 1996, Mitochondrial control of nuclear apoptosis, *J. Exp. Med.* **183**:1533–1544.

Zhang, P., Liu, B., Kang, S. W., Seo, M. S., Rhee, S. G., and Obeid, L. M., 1997, Thioredoxin peroxidase is a novel inhibitor of apoptosis with a mechanism distinct from that of Bcl-2, *J. Biol. Chem.* **272**:30615–30618.

Zhivotovsky, B., Orrenius, S., Brustugun, O.T., and Doskeland, S.O., 1998, Injected cytochrome *c* induces apoptosis, *Nature* **391**:449–450.

Zoratti, M., and Szabò, I., 1995, The mitochondrial permeability transition, *Biochim. Biophys. Acta* **1241**:139–76.

Zwicker, K., Dikalov, S., Matuschka, S., Mainka, L., Hofmann, M., Khramtsov, V., and Zimmer, G., 1998, Oxygen radical generation and enzymatic properties of mitochondria in hypoxia/reoxygenation, *Arznei-mittelforschung* **48**:629–636.

Chapter 15

Role of the Permeability Transition in Glutamate-Mediated Neuronal Injury

Ian J. Reynolds and Teresa G. Hastings

1. INTRODUCTION

Glutamate is the principal excitatory neurotransmitter in the brain. In addition to its critical role in fast excitatory neurotransmission, however, glutamate has a more sinister role as a potent and effective neurotoxin, a process termed *excitotoxicity* (Rothman and Olney, 1987). Glutamate is acutely toxic to central neurons in primary culture, and it is widely believed that similar mechanisms contribute to the neuron loss encountered in both acute and chronic neurodegenerative disease. In the former category, there is good evidence that neuronal injury encountered in stroke, cerebral ischemia, traumatic brain injury, and some forms of *status epilepticus* is mediated by excessive glutamate receptor activation. The links to chronic neurodegeneration are perhaps more tenuous. Glutamate may, however, mediate some part of the neurodegenerative process in amyotrophic lateral sclerosis, Huntington's disease, Parkinson's disease, dementia associated with acquired immunodeficiency syndrome, and possibly Alzheimer's disease (reviewed in Olney, 1990).

The wide range of diseases associated with aberrant activation of glutamate receptors suggests that drugs which interrupt excitotoxicity could be of great therapeutic significance. Glutamate receptor antagonists have been intensively investigated in this regard, and show some promising results in acute disease models (Doble, 1999; Koroshetz and Moskowitz, 1996). This approach has been associated with significant behavioral side effects, however, reflecting glutamate's important role in normal brain

Ian J. Reynolds   Department of Pharmacology, University of Pittsburgh, Pittsburgh Pennsylvania 15261.   **Teresa G. Hastings**   Department of Neurology and Neuroscience, University of Pittsburgh, Pittsburgh, Pennsylvania 15261.

Mitochondria in Pathogenesis, edited by Lemasters and Nieminen.
Kluwer Academic/Plenum Publishers, New York, 2001.

function (Tricklebank *et al.*, 1987). An alternative approach to preventing excitotoxic injury could involve blocking some of the downstream processes activated by glutamate, and this would ideally focus on a target activated exclusively during injury so that drugs directed toward this target should have much greater functional selectivity than glutamate receptor antagonists.

Recent studies on the mechanisms underlying excitotoxicity have identified a central role for mitochondria in the injury process. Moreover, it has been suggested that the permeability transition pore (PTP) is activated in neuronal mitochondria specifically during pathophysiological states, and that its activation contributes to neurons demise. If this is true, then the pore should be considered a primary target for drug development in neurodegenerative disease. This review seeks to evaluate the validity of the conclusion that the PTP has a central role in neurodegeneration.

2. ROLE OF MITOCHONDRIA IN GLUTAMATE TOXICITY

There is considerable evidence of a role for glutamate in acute brain injury. Glutamate is stored at high concentrations inside both neurons and astrocytes and is released into the extracellular space as a result of injury (Benveniste *et al.*, 1984; Rothman, 1984; Strijbos *et al.*, 1996). Selective glutamate receptor antagonists can ameliorate ischemic and traumatic injury *in vivo* (Boast *et al.*, 1988; Gill *et al.*, 1987). Glutamate is also effective enough as a neurotoxin to kill neurons *in vitro* without the need for any other injurious agent (Choi *et al.*, 1987; Rothman, 1984). Thus studying glutamate's mechanism of action should provide insights into some aspects of the neuronal injury process, even though this approach undoubtedly takes an oversimplified view of the processes that causes *in vivo* neurons death. Glutamate-induced injury can take several forms depending on the type of receptor activated (Koh *et al.*, 1990; Mayer and Westbrook, 1987). The studies described here focus on the most acute form of glutamate-triggered neuronal injury, namely injury induced by the activation of N-methyl-D-aspartate (NMDA) receptors. Excessive activation of the other ionotropic glutamate receptors, the AMPA and kainate receptors, is also a highly effective way to kill neurons, but the cellular mechanisms responsible for injury are less well established.

2.1. Mitochondria in Neuronal Ca^{2+} Homeostasis

Some of the earliest *in vitro* studies on glutamate toxicity showed that extracellular Ca^{2+} is required for expression of NMDA-receptor-mediated injury (Choi, 1987), an observation consistent with the Ca^{2+} permeability of the NMDA-receptor-associated ion channel (MacDermott *et al.*, 1986). Moreover, the observation that extracellular Ca^{2+} decreases considerably during ischemia suggests that a robust Ca^{2+} entry into neurons occurs *in vivo* as well (Erecinska and Silver, 1996; Kristián *et al.*, 1994). The relationship between Ca^{2+} entry and neuronal death is still not entirely clear. Only recently was it established that toxic stimulation of NMDA receptors actually results in larger intracellular free Ca^{2+} changes than does benign activation of other, nontoxic Ca^{2+}-elevating mechanisms (Hyrc *et al.*, 1997; Stout and Reynolds, 1999). It has also been suggested that the route of Ca^{2+} entry is at least as important as the magnitude of influx, so the precise location of a substantial Ca^{2+} influx may be the key to triggering injury (Sattler et al., 1998; Tymianski et al., 1993).

Studies of Ca^{2+} homeostasis mechanisms in neurons reveal an important role for mitochondrial Ca^{2+} transport following glutamate receptor activation (Khodorov et al., 1996; Wang and Thayer, 1996; White and Reynolds, 1995). We demonstrated that both mitochondria and the plasma membrane Na^+/Ca^{2+} exchange (NCEp) clear the cytoplasm of Ca^{2+} following glutamate stimulus, and that mitochondria account for a progressively larger proportion of the buffering ability as glutamate stimulus intensity approaches toxicity (White and Reynolds, 1995, 1997). A second method of illustrating mitochondrial Ca^{2+} accumulation was via the use of the mitochondrial NCE (NCEm) inhibitor CGP-37157 (Cox and Matlib, 1993). The NCEm is the predominant efflux pathway in functional neuronal mitochondria, so blocking it and monitoring the change in the recovery characteristics of cytoplasmic Ca^{2+} concentrations allows inference of a mitochondrial Ca^{2+} load (Baron and Thayer, 1997; White and Reynolds, 1996, 1997). Several studies report the use of Ca^{2+} indicator Rhod-2 to illustrate mitochondrial Ca^{2+} loading in real time (Minta et al., 1989; Peng and Greenamyre, 1998; Peng et al., 1998), and these real-time methods complement other approaches that have found glutamate- or injury-induced mitochondrial Ca^{2+} loading in isolated mitochondria (Sciamanna et al., 1992) or in brain slices using electron probe microanalysis (Taylor et al., 1999).

Without inferring a mechanism by which Ca^{2+} alters mitochondrial function, the importance of mitochondrial Ca^{2+} accumulation to neuronal injury has been illustrated by two studies that blocked Ca^{2+} uptake during glutamate exposure and thereby prevented injury. Budd and Nicholls (1996a, 1996b) accomplished this in cerebellar granule cells using rotenone and oligomycin to depolarize mitochondria while maintaining intracellular ATP concentrations. We later accomplished the same neuroprotection in forebrain neurons using the protonophore FCCP; its reversibility provided a better window of opportunity for observing the neuroprotection (Stout et al., 1998). These studies helped to establish that mitochondrial Ca^{2+} uptake is essential for excitotoxicity expression, although the precise target of Ca^{2+} within mitochondria is still unknown.

2.2. Reactive Oxygen Species and Glutamate Toxicity

Markers of oxidative stress are increased in association with ischemia and trauma, and in particular during the reperfusion phase following ischemia (Hall and Braughler, 1989; Phillis, 1994). There is also abundant evidence that chronic neurodegenerative states are associated with an increased oxidant burden (Götz et al., 1994). In most of these cases, however, there are many potential sources of ROS (Halliwell, 1992), and the mechanisms responsible for glutamate-specific alteration of the oxidant/antioxidant balance remain poorly defined.

A series of reports have detailed processes that could link glutamate receptor activation to ROS generation by mitochondria. Lafon-Cazal and colleagues (1993) found that glutamate-stimulated cerebellar granule cells generated ROS that could be detected by electron spin resonance. Dykens showed in 1994 that isolated brain mitochondria increased superoxide generation when presented with Ca^{2+} and Na^+, the circumstance presented to mitochondria in neurons activated by glutamate. Subsequently, we (Reynolds and Hastings, 1995) and others (Bindokas et al., 1996; Dugan et al., 1995) showed that glutamate-stimulated neurons in primary culture triggered ROS generation detectable with a range of oxidation-sensitive fluorescent dyes. Various arguments have been made to support a mitochondrial source for these effects. For example, one can infer from the location of the dye signal that oxidation occurs in the vicinity of mitochondria. This is not necessarily a compelling argument, given the limited resolution with which these measurements must be made, and given also the tendency of some of these dyes to distribute in cells to organelles that have no relation to their oxidation site. Interrupting the ROS-generating mechanism provides more persuasive insights into the source of ROS. We reported that glutamate-stimulated effects on dichlorofluorescin oxidation were Ca^{2+} dependent, and that FCCP prevented glutamate-induced ROS generation (Reynolds and Hastings, 1995). Although FCCP has numerous effects on neurons (Tretter et al., 1998), the most straightforward basis for this inhibitory effect is the prevention of Ca^{2+} loading into mitochondria. Rotenone also blocks ROS generation by dihydroethidium (Bindokas et al., 1996), presumably the result of inhibiting electron transport in Ca^{2+}-stimulated mitochondria. Together with the observation that oxygen-deprived neurons are resistant to glutamate-induced injury (Dubinsky et al., 1995), these findings suggest that mitochondrially generated ROS may play a central role in excitotoxicity.

In addition to glutamate's ability to directly induce ROS generation, there may be an interaction between oxidative stress and glutamate toxicity in that an extrinsic oxidative stress may potentiate the toxicity of glutamate. Thus in neurons subjected to an oxidant burden, the threshold for glutamate toxicity may be decreased (e.g., Hoyt et al., 1997a), and following oxidant inhibition of the glutamate transporter, glutamate's potency may be increased (Berman and Hastings, 1997; Piani et al., 1993; Volterra et al., 1994), perhaps resulting in the increased vulnerability of selected neuron populations. Moreover, a possible critical ROS target is in fact the mitochondrion. Essentially, all complexes in the electron transport chain are vulnerable to inhibition by oxidants (Berman and Hastings, 1999; Dykens, 1994; Zhang et al., 1990), and the activities of several enzymes in the tricarboxylic acid cycle are impaired by oxidation (Chinopoulos et al., 1999). By further enhancing ROS generation, or by limiting ATP generation under circumstances that would normally place a great demand on ATP supply (Chinopoulos et al., 1999), the impact of an oxidant burden together with glutamate exposure is potentially devastating.

One important example of this is the degeneration of dopaminergic neurons in the substantia nigra in Parkinson's disease. These neurons contain a high concentration of dopamine. Dopamine generates an oxidant burden either by its metabolism by monoamine oxidase (MAO), which generates hydrogen peroxide, or by the formation of highly reactive dopamine quinones (Graham, 1978; Hastings, 1995; Maker et al., 1981). It is toxic to neurons in culture (Hoyt et al., 1997b; Rosenberg, 1988), and it is also toxic when injected directly into the brain (Hastings et al., 1996). Though dopamine's toxicity mechanism is not fully understood, a mitochondrial target is one important potential mechanism being evaluated. Thus we have demonstrated that MAO-catalyzed dopamine oxidation inhibits state 3 respiration in brain mitochondria (Berman and Hastings, 1999). Interestingly, dopamine quinone increased state 4 respiration, consistent with an uncoupling effect perhaps attributable to increased membrane permeability associated with PTP activation (Berman and Hastings, 1999). The oxidative burden associated with a high dopamine content and the concomitant bioenergetic impairment may render substantia nigra neurons especially vulnerable to excitotoxic stimuli (Greene and Greenamyre, 1996).

3. THE PERMEABILITY TRANSITION IN NEURONAL MITOCHONDRIA

Identification of mitochondria as a critical target for Ca^{2+} in neuronal injury moved the field one step closer to identifying the complete sequence of events necessary to execute the process of cell death. It seems likely that there are multiple mechanisms by which neurons may die following NMDA receptor activation (Ankarcrona *et al.*, 1995), not all of which are associated with gross Ca^{2+} overload. Even when considering the acute, high Ca^{2+} load-associated injury, the precise target within mitochondria and the cellular consequences of altering this target have not been established. A simple view is that neuronal death is merely a consequence of ATP loss, failure of ion homeostasis, and cell lysis as a result of uncontrolled solute accumulation. Given that Ca^{2+} and ROS do not shut down respiration or metabolism —indeed, Ca^{2+} stimulates ATP synthesis (McCormack *et al.*, 1990)—what mechanisms are available to impair ATP synthesis under the circumstances generated by intense stimulation of NMDA receptors?

An attractive conceptual mechanism is provided by the PTP. As a mitochondrial target activated by Ca^{2+}, oxidation, and mitochondrial membrane depolarization, this target appears the ideal point at which neuronal injury's principal effectors can converge. We have already discussed the importance of mitochondrial Ca^{2+} accumulation and ROS generation. The third component, alteration of mitochondrial membrane potential, is then provided by the cycling of Ca^{2+} through mitochondria, which occurs at the expense of the proton gradient (Gunter and Pfeiffer, 1990; Nicholls and Akerman, 1982). Thus the key pore activators should be present during glutamate exposure.

What evidence exists to support pore activation in the injury cascade triggered by glutamate? The key consequences of pore activation should be loss of membrane potential, alteration of mitochondrial shape, increased mitochondrial permeability to small molecules, and presumably, neuronal death. Observations consistent with these events have now been reported by several laboratories. Using fluorescent dyes that report mitochondrial membrane potential, several studies found that mitochondria in intact neurons are depolarized during glutamate exposure (Nieminen *et al.*, 1996; Schinder *et al.*, 1996; White and Reynolds, 1996). The partial sensitivity of these changes in membrane permeability to cyclosporin A is consistent with a contribution of the PTP. Shape changes in mitochondria are difficult to resolve at the light microscopic level. Nevertheless, Ca^{2+}-stimulated, cyclosporin-sensitive changes in mitochondrial morphology were reported, such that the predominant change was from rod-shaped to round in both neurons and astrocytes (Dubinsky and Levi, 1998; Kristal and Dubinsky, 1997). Recently, Friberg and colleagues (1998) established that swelling of neuronal mitochondria in hypoglycemic brain injury can be prevented by cyclosporin A, thereby suggesting a link to the pore. No neuron studies have shown explicit alteration in the mitochondrial permeability of small molecules in intact neurons. Several studies have found, however, that neuronal death can be prevented by cyclosporin A, again consistent with an essential role for the pore in the death pathway (Dawson *et al.*, 1993; Schinder *et al.*, 1996; White and Reynolds, 1996). Though we discuss below a number of potential limitations in these conclusions, there is plenty of evidence for a hypothesis that places transition as a final common pathway in neuron death.

4. LIMITATIONS OF THE PERMEABILITY TRANSITION HYPOTHESIS

Many of the pore's features are conceptually ideal in building a model of the glutamate-induced injury cascade. Indeed, it could be claimed that excitotoxicity is the clearest example of PTP involvement in a cellular or intact-tissue injury paradigm. Nevertheless, it is still necessary to apply the same critical standards to evaluating this hypothesis as to evaluating any other, because there are still major issues that need to be more fully considered.

4.1. Transition Measurement in Intact Cells

The first major difficulty in evaluating the PTP's contribution to injury is measuring transition in intact cells. Transition has been studied mainly in isolated mitochondria, and the most common approach for assaying transition is measurement of mitochondrial swelling using light scattering. Other approaches have been used, including measuring the release of low-molecular-weight solutes such as glutathione and also the influx of radioisotopes normally excluded from mitochondria with restricted permeability, but the limited ability to measure these parameters in cultured cells (because of the relative insensitivity of the methods) has prevented their application in models of excitotoxicity.

As noted above, recent studies have explicitly investigated mitochondrial morphology in brains exposed to injury and have reportedly prevented the appearance of swollen mitochondria in response to hypoglycemia by cyclosporin A (Friberg et al., 1998). This is an exciting development, but the approach does not lend itself well to mechanistic studies because it is hard to do quantitative electron microscopic studies under circumstances where mitochondrial parameters can effectively be manipulated. Additional morphological approaches have been taken in permeabilized neurons and astrocytes. This is an interesting intermediate approach that falls between isolated mitochondria and intact cells. Dubinsky and colleagues were able to demonstrate Ca^{2+}-mediated alterations in mitochondrial morphology, assayed in mitochondria loaded with fluorescent dyes, that could be partially prevented by concomitant application of CsA (Dubinsky and Levi, 1998; Kristal and Dubinsky, 1997). These studies are also consistent with activation of the PTP in neural cells, although it is hard to be sure of the extent to which the mitochondrial environment in permeabilized or to which ionophore-treated cells reflects the glutamate exposure conditions. Indeed, in astrocytes lacking the efficient Ca^{2+} accumulation pathways of neurons, the pathophysiological relevance of a Ca^{2+} overload-induced alteration in mitochondrial function remains to be established.

Applying morphological approaches to the study of mitochondrial shape in intact cells may also be limited by the dyes. Many mitochondrion-specific dyes accumulate in the organelle based on membrane potential, and so the change in potential that accompanies increased permeability should grossly alter the dye-staining properties (White and Reynolds, 1996), potentially an important confound when applying purely morphometric approaches. Other dyes, such as the series of Mito Tracker dyes provided by Molecular Probes, may accumulate in mitochondria based on membrane potential but then become irreversibly bound as a result of the dye's chloromethyl moiety interacting with free sulfhydryls in (presumably) the mitochondrial matrix (Poot et al., 1996). Though this provides the advantage of having a fluorescent marker that can be fixed, it has the disadvantage of altering an important parameter controlling PTP activation: the balance of reduced and oxidized sulfhydryls (Chernyak and Bernardi, 1996). The observation that Mito Tracker orange can inhibit complex I with considerable potency and can also activate transition in isolated hepatic mitochondria (Scorrano et al., 1999) further underscores the difficulty of this approach.

Most other claims of pore activation in neurons have been made based on cyclosporin-sensitive alterations in mitochondrial membrane potential (Nieminen et al., 1996; Schinder et al., 1996; White and Reynolds, 1996). These studies have typically reported an NMDA-receptor-stimulated, Ca^{2+}-dependent depolarization of mitochondrial membrane potential using a range of potential-sensitive indicators. These depolarizations are generally observed during the time required to commit neurons to die as a result of the glutamate exposure, but they occur well before loss of viability can be detected, suggesting that the phenomenon is upstream in the injury cascade. At least some of the time, depolarization is also reversible upon glutamate removal (White and Reynolds, 1996), although this is variable. It is tempting to suggest that the loss of potential reflects transition, but is this reasonable? Based on pharmacological evidence discussed below, where several putative transition inhibitors are effective it appears so. It is difficult, however, to exclude other possible mechanisms with confidence, We know that Ca^{2+} is clearly essential in this process, but mitochondrial Ca^{2+} cycling occurs at the expense of the proton gradient (Nicholls and Akerman, 1982). Thus a large amount of Ca^{2+} passing through mitochondrial uptake and release presumably results in depolarization. It would also be difficult to distinguish between cycling-induced depolarization and a PTP-triggered change, because blocking Ca^{2+} uptake would block both phenomena concomitantly. Although Ca^{2+} cycling-induced mitochondrial depolarization has been observed in synaptosomes (Nicholls and Akerman, 1982), it has never been explicitly demonstrated in intact neurons. Nevertheless, its potential contribution is consistent with the observation of hyperpolarization of mitochondrial membrane potential induced by the Ca^{2+} efflux inhibitor CGP37157 (White and Reynolds, 1996). It is important to recognize that depolarization of mitochondrial membrane potential is also a normal response to increased ATP demand. Given that NMDA receptor activation results in a substantial change in intracellular Na^+ as well as Ca^{2+} (Kiedrowski et al., 1994), and that the Na^+/K^+ ATPase is a major consumer of ATP in neurons, the greatly increased Na^+ burden should require an increase in ATP synthesis accompanied by a depolarization of mitochondrial membrane potential. Thus, several major mechanisms could produce an alteration in membrane potential that would be completely independent of the PTP, and in fact represent the function of normal, healthy mitochondria.

The intricate intertwining of PTP-inducing stimuli and non-PTP-related changes is further illustrated by the impact of oxidants in this system. In isolated mitochondria, oxidants promote pore activation by oxidizing vicinal sulfhydryls or by increasing the pool of the oxidized form of glutathione (Chernyak and Bernardi, 1996). Oxidants can inhibit electron transport, however, and they may also limit the tricarboxylic acid (TCA) cycle, which could alter the ability of mitochondria to pump protons and maintain a potential (Chinopoulos et al., 1999; Zhang et al., 1990). There is the additional confound of peroxide-induced changes in the properties of JC-1, which appear unrelated to membrane potential (Chinopoulos et al., 1999; Scanlon and Reynolds, 1998), and which can look like depolarization but which probably are not.

The approaches used in intact neurons have not proved effective in unequivocally establishing the phenomenon of pore activation. One potentially interesting approach not yet applied to neurons is the cobalt-induced calcein quenching reported by Petronilli and colleagues (1999), although it may prove difficult to apply the method to glutamate excitotoxicity models due to cobalt/Ca^{2+} interaction in this system. Approaches that combine morphology with membrane-potential measurements so that swelling can be observed in conjunction with loss of membrane potential, together with a careful functional assessment to preclude ATP synthesis and Ca^{2+} cycling as a cause of membrane-potential changes, may be necessary to definitively establish the expression of PTP activation in neurons.

4.2. Limitations of Pharmacological Approaches

The difficulty in identifying the PTP solely by functional criteria in intact cells emphasizes the value of pharmacological intervention. Additionally, the putative contribution of the PTP to glutamate-induced neuronal death might be mitigated by transition antagonists, so there is considerable interest in pore-specific drugs. Unfortunately, such agents are difficult to come by.

The classic pore inhibitor is cyclosporin A (CsA). This agent binds to cylophilin D in the matrix and prevents the association of cylophilin with the pore complex, thus preventing the facilitative effect on pore activation (Connern and Halestrap, 1994). The CsA binds to cyclophilin with high affinity and with a specificity distinct from that associated with immunophilins (Bernardi et al., 1994; Connern and Halestrap, 1994). Thus, potent immunosuppressent agents such as FK506 do not alter transition in isolated brain mitochondria (Friberg et al., 1998). Conversely, analogs of CsA such as N-methylvaline cyclosporin show pore inhibition with less immunosuppressant activity (Griffiths and Halestrap, 1991), but they are not commercially available, unfortunately CsA is a less-than-ideal agent in intact neurons. The cyclic peptide structure may limit cell penetration, and intact-cells studies generally require higher concentrations than isolated-mitochondria studies. Some of the immunophilin-mediated effects, such as inhibition of the Ca^{2+}-dependent phosphatase calcineurin, occur at concentrations too low to inhibit the PTP (Dawson et al., 1993). Indeed, it has been proposed that calcineurin inhibition is neuroprotective independently of the PTP, a suggestion supported by the neuroprotective effects of FK506 in some studies (Dawson et al., 1993; Lu et al., 1996), but not others (Friberg et al., 1998). The binding of CsA is also modulated by Ca^{2+} and Mg^{2+}, with the effect that PTP-inducing conditions may decrease the effectiveness of CsA binding (Novgorodov et al., 1994). This may explain the loss of effectiveness sometimes observed when CsA is used as an inhibitor of transition-associated events (Scanlon and Reynolds, 1998). There is no question that CsA is the most potent and probably the most useful pore inhibitor currently available. Other actions of CsA are, clearly important for neuronal viability, however, independent of cyclophilin D and the PTP; also, CsA may not be effective even if pore-mediated effects are under investigation, which, along with the difficulty in controlling intracellular CsA concentrations, shows a definite need for more effective inhibitors.

In fact, a wide variety of agents have been used to modulate PTP activity, both inhibitors and activators (see Zoratti and Szabò, 1995), including atractyloside and bongkrekic acid, which bind to the ATP/ADP translocase. The suggestion that the PTP reflects a different functional state of the translocase is reflected in the ability of these agents to inhibit or promote pore activation, respectively (Halestrap and Davidson, 1990). Unfortunately, atractyloside is cell impermeant. The recent commercial availability of bongkrekic acid should soon allow its evaluation in neurons, although promoting pore activation will not effectively test the hypothesis that glutamate kills neurons following transition.

Other agents that alter transition in isolated mitochondria have been evaluated for their ability to block glutamate-mediated mitochondrial depolarization. Some of the more potent ones reviewed by Zoratti and Szabò (1995) include trifluoperazine and dibucaine. The former phenothiazine has a number of cellular effects, the most prominent of which may be the inhibition of calmodulin and phospholipase activity. Dibucaine is more widely known as a local anesthetic, an effect accomplished by inhibition of voltage-dependent Na^+ currents. These agents both delay mitochondrial depolarization induced by glutamate, but could not entirely prevent it (Hoyt et al., 1997c). In addition, both appear to hyperpolarize mitochondria, and trifluoperazine may also inhibit mitochondrial Na^+/Ca^{2+} exchange (Hoyt et al., 1997c). We have previously observed a hyperpolarizing response to the NCEm inhibitor CGP37157 (White and Reynolds, 1996), but it is not clear whether this represents ongoing NCEm activity; the other possibility is tonic activity, or perhaps a low-conductance state, of the pore (Ichas and Mazat, 1998). Given the limitations of using membrane potential to unequivocally identify transition it is difficult to make the distinction between these possibilities. These drugs, however, have multiple effects on mitochondria, besides their nonmitochondrial effects, that can be expected to alter the function of excitable cells, which illustrates the potential pharmacological limitations of these agents.

Other drugs have recently been proposed as PTP antagonists. The anti-estrogenic drug tamoxifen has been used as a pro-apoptotic agent in neural mitochondria at high concentrations (Ellerby et al., 1997). Much lower concentrations prevented transition in hepatocytes, however (Custodio et al., 1998). We evaluated tamoxifen actions in neurons and found a biphasic effect on membrane-potential changes induced by glutamate, with maximum depolarization protection at 0.3 μM, much lower than concentrations required to injure the neurons, but the maximally effective concentration of tamoxifen did not protect against excitotoxic injury (Hoyt, et al., 2000). Fontaine and colleagues recently reported (1998) the effects of a series of ubiquinone analogues on mitochondrial function in permeabilized skeletal-muscle mitochondria. Several agents increased mitochondria's ability to accumulate Ca^{2+} in a manner consistent with transition inhibition. We evaluated these compounds in neurons, with rather different results (Scanlon and Reynolds, unpublished). Thus, the most effective agent in muscle, Ub0, appeared to promote mitochondrial depolarization in neurons and was quite toxic, likely a result of ROS generation. The greatest inhibition of depolarization was produced by UB5, but it did not protect against injury at all. It would be immensely valuable to have selective, potent, cell-permeable PTP inhibitors in such studies, but no currently known drugs possess these properties.

4.3. Limitations of Cell Culture Methodology

For understanding intracellular events associated with neuronal injury, cultured neurons have a great benefit: They are readily amenable to single-cell study and to the kinds of manipulation often necessary to grasp the basic mechanisms. The value of the model system only holds, however, in conjunction with its fidelity to the situation *in vivo*. The excitotoxicity model's value has been clearly established by its predictive ability in identifying the neuroprotective actions of glutamate-receptor antagonists, which was subsequently verified *in vivo*, but this model may have limitations not yet fully explored. For example, cells in culture may show a greater dependence on glycolysis than on oxidative phosphorylation as a source of ATP, which could obviously have a profound impact on studies designed to link bioenergetic phenomena with glutamate stimulation. Neurons *in situ* have a close and important interaction with astrocytes, which may be critical in the passage of nutrients from cerebral circulation to neurons (Tsacopoulos and Magistretti, 1996). Because the details of this interaction are poorly understood, it is hard to create such an arrangement *in vitro*. Many imaging studies are performed at room temperature, and this may also impact the neuronal bioenergetic state.

There are additional considerations. Many chronic neurodegenerative states are associated with chronic, relatively modest inhibition of one or more electron transport chain complexes. Indeed, one can model Parkinson and Huntington diseases using specific inhibitors of complex I and complex II, respectively (see below). Acute disorders such as stroke are associated with complete or partial restriction of oxygen and glucose, but studies investigating mitochondrial function after glutamate exposure have usually done so in an environment where oxygen and glucose are not limiting and where electron transport is in good working order. Given that ROS generation, for instance, may increase under both hypoxic (Vanden Hoek *et al.*, 1997) and hyperoxic conditions, studying states that more closely resemble actual diseases may be very important. There are many potential influences on mitochondrial function, Ca^{2+} transport, and ATP generation that must be elucidated to make these model systems more valuable.

4.4. Acutely Isolated Mitochondria Preparations

Much of what is known about the properties of mitochondria in the brain comes from studies of acutely isolated mitochondria, either in the from of purified organelles or in the context of synaptosomes, where mitochondria's immediate environment is better preserved, but revisiting these findings is beyond the scope of this review. Purified brain-derived mitochondria do exhibit a phenomenon similar to transition, in that mitochondria can be loaded with Ca^{2+} and exposed to oxidants to induce swelling (Andreyev *et al.*, 1998; Berman *et al.*, 2000; Friberg *et al.*, 1999). There are some important differences, however, between the characteristics of PTP activation in brain as compared with liver. For example, exposure to levels of oxidants, Ca^{2+}, or both that would normally trigger swelling in liver mitochondria produces only small amplitude changes in brain mitochondria, and the swelling is not accompanied by glutathione release as would be anticipated with PTP activation (Berman *et al.*, 2000). This is not due to an inability of brain mitochondria to swell, because higher-amplitude swelling can be accomplished with mastoparan (Berman *et al.*, 2000) or by removing adenine nucleotides and Mg^{2+}

(Andreyev et al., 1998; Friberg et al., 1999), but the CsA-sensitivity swelling is consistent with a key role for the pore in the swelling process. Interpreting these differences is a challenge, because an adenine-nucleotide-free condition is normally found in neurons, and Mg^{2+} concentrations are near millimolar in these cells (Brocard et al., 1993). It has also been suggested that the polymerization state of creatine kinase in brain could be a key difference from liver (O'Gorman et al., 1997). Nevertheless, the key point here is this: It should not be assumed that the PTP properties in brain are the same as those in the more completely characterized liver and heart preparations.

4.5. Mitochondrial Heterogeneity

The concept of transition is based largely, though not entirely, on observations made in liver mitochondria and many assumptions concerning the PTP's contribution to glutamate-induced neuronal injury are based on the notion that the processes governing PTP activation are the same in neural as in liver mitochondria. Though little explicit information points to functional differences in mitochondria that would account for the different pore properties, it is not far-fetched to suggest that they may exist. We know, for example, that there are differences in the Ca^{2+} efflux pathway between liver mitochondria and mitochondria from excitable cells (Gunter and Pfeiffer, 1990). Also, the fundamental properties of Ca^{2+} and oxidant-induced swelling are distinct, as noted above (Berman et al., 2000). We have also seen that oxidants such as tert-butylhydroperoxide—highly effective pore-inducers in liver—have little effect on mitochondrial membrane potential in cultured neurons or in isolated brain mitochondria (Berman et al., 2000; Scanlon and Reynolds, 1998), hence the need for a cautious approach to expectations about neuronal PTP properties.

One intriguing possibility is an additional level of heterogeneity for neurons and non-neuronal cells in the brain, between different populations of neurons, or perhaps even between mitochondria in different regions of the same cell. Recent findings (Beal et al., 1993) have demonstrated, for instance, that systemic administration of complex II inhibitors produces selective degeneration of striatal neurons in rats, which has proved a useful model for Huntington disease. Remarkably, a similar approach using rotenone targets instead the striatal dopamine terminals followed by cell bodies in substantia nigra, producing a Parkinsonlike syndrome in rats (MacKenzie and Greenamyre, 1998). This raises the possibility that mitochondria from different types of neurons in the same brain region have distinct properties, leaving them vulnerable to different toxins. Another interpretation is that the mitochondria are in fact the same, but their environment is different. For example, chronic exposure to high dopamine concentrations could make nigral neurons vulnerable to complex I inhibition, rather than this being a fundamental functional difference in the mitochondria. An argument in favor of this was recently provided by Friberg and colleagues (1998), who demonstrated differences in PTP properties between mitochondria from cortex, hippocampus, and cerebellum. These differences, however, were apparently attributable to different concentrations of adenine nucleotides in the preparations rather than to a fundamental difference in the mitochondria, because nucleotide removal resulted in the mitochondria exhibiting similar swelling characteristics.

The relevance of these findings to glutamate-induced neuronal injury has yet to be fully established. Many chronic disease states that can be modeled using electron transport inhibitors may also have an excitotoxic component, and the interaction between bioenergetics and vulnerability to excitotoxic injury is well established (Greene and Greenamyre, 1996). The extent to which this interaction depends on PTP activation, however, remains unknown.

5. CONCLUSIONS

Glutamate can injure neurons in a way that is likely relevant to a number of acute neurodegenerative states, including stroke and head trauma. It is also likely that glutamate contributes to neurons degeneration in chronic diseases as well. Many studies have established a relationship between the bioenergetic state of neurons and their vulnerability to injury, and more-recent investigations have placed mitochondria at the center of the event cascade linking glutamate receptor activation to neuronal death. The key question posed at the start of this review concerned the role of the PTP in this process, however, and this is much less clear. There are significant concerns in interpretating studies which suggest that transition occurs in intact neurons, because of the difficulty in attributing alterations in membrane potential to pore activation, so these studies remain suggestive but not conclusive. Morphological approaches also imply pore involvement but contain similar methodological concerns. It is evident that CsA is neuroprotective under certain circumstances, but there are multiple mechanisms by which this might be so, and FK506 sometimes has neuroprotective actions as well. No other drugs that alter membrane potential have shown neuroprotective effects. Thus, the body of evidence linking the PTP to excitotoxicity, though suggestive and intriguing, remains to be firmly established.

ACKNOWLEDGMENTS. Our work was supported by USAMRMC grant DAMD-17-98-1-8627 and USPHS grants DA 09601, NS 19068, and NS 34138. The author is an established investigator of the American Heart Association.

REFERENCES

Andreyev, A. Y., Fahy, B., and Fiskum, G., 1998, Cyctochrome *c* release from brain mitochondria is independent of the mitochondrial permeability transition, *FEBS Lett.* **439**:373–376.

Ankarcrona, M., Dypbukt, J. M., Bonfoco, E., Zhivotovsky, B., Orrenius, S., Lipton, S. A., and Nicotera, P., 1995, Glutamate-induced neuronal death: A succession of necrosis or apoptosis depending on mitochondrial function, *Neuron* **15**:961–973.

Baron, K. T., and Thayer, S. A., 1997, CGP 37157 modulates mitochondrial Ca^{2+} homeostasis in cultured rat dorsal root ganglion neurons, *Eur. J. Pharmacol.* **340**:295–300.

Beal, M. F., Hyman, B. T., and Koroshetz, W., 1993, Do defects in mitochondrial energy metabolism underlie the pathology of neurodegenerative diseases? *Trends Neurosci.* **16**:125–131.

Benveniste, H., Drejer, J., Schousboe, A., and Diemer, N. H., 1984, Elevation of the extracellular concentrations of glutamate and aspartate in rat hippocampus during transient cerebral ischemia monitored by intracerebral microdialysis, *J. Neurochem.* **43**:1369–1374.

Berman, S. B., and Hastings, T. G., 1997, Inhibition of glutamate transport in synaptosomes by dopamine oxidation and reactive oxygen species, *J. Neurochem.* **69**:1185–1195.

Berman, S. B., and Hastings, T. G., 1999, Dopamine oxidation alters mitochondrial respiration and induces permeability transition in brain mitochondria: Implications for Parkinson's disease, *J. Neurochem.*, **73**:1127–1137.

Berman, S. B., Watkins, S.C., and Hastings, T. G., 2000, Quantitative biochemical and ultrastructural comparison of mitochondrial permeability transition in isolated brain and liver mitochondria: evidence for relative insensitivity of brain mitochondria. *Exp. Neurol.* **166**:615–625.

Bernardi, P., Broekemeier, K. M., and Pfeiffer, D. R., 1994, Recent progress on regulation of the mitochondrial permeability transition pore: A cyclosporin sensitive pore in the inner mitochondrial membrane, *J. Bioenerg. Biomemb.* **26**:509–517.

Bindokas, V. P., Jordan, J., Lee, C. C., and Miller, R. J., 1996, Superoxide production in rat hippocampal neurons: Selective imaging with hydroethidine, *J. Neurosci.* **16**:1324–1336.

Boast, C. A., Gerhardt, S. C., Pastor, G., Lehmann, J., Etienne, P. E., and Liebman, J. M., 1988, The *N*-methyl-D-aspartate antagonists CGS 19755 and CPP reduce ischemic brain damage in gerbils, *Brain Res.* **442**:345–348.

Brocard, J. B., Rajdev, S., and Reynolds, I. J., 1993, Glutamate induced increases in intracellular free Mg^{2+} in cultured cortical neurons, *Neuron* **11**:751–757.

Budd, S. L., and Nicholls, D. G., 1996a, A reevaluation of the role of mitochondria in neuronal Ca^{2+} homeostasis, *J. Neurochem.* **66**:403–411.

Budd, S. L., and Nicholls, D. G., 1996b, Mitochondria, calcium regulation, and acute glutamate excitotoxicity in cultured cerebellar granule cells, *J. Neurochem.* **67**:2282–2291.

Chernyak, B. V., and Bernardi, P. 1996, The mitochondrial permeability transition pore is modulated by oxidative agents through both pyridine nucleotides and glutathione at two separate sites, *Eur. J. Biochem.* **238**:623–630.

Chinopoulos, C., Tretter, L., and Adam-Vizi, V., 1999, Depolarization of *in situ* mitochondria due to hydrogen peroxide-induced oxidative stress in nerve terminals: Inhibition of α-ketoglutarate dehydrogenase, *J. Neurochem.* **73**:220–228.

Choi, D. W., 1987, Ionic dependence of glutamate neurotoxicity, *J. Neurosci.* **7**:369–379.

Choi, D. W., Maulucci-Gedde, M., and Kriegstein, A. R., 1987, Glutamate neurotoxicity in cortical cell culture, *J. Neurosci.* **7**:357–368.

Connern, C. P., and Halestrap, A. P., 1994, Recruitment of mitochondrial cyclophilin to the mitochondrial inner membrane under conditions of oxidative stress that enhance opening of a calcium sensitive non-specific channel, *Biochem. J.* **302**:321–324.

Cox, D. A., and Matlib, M. A., 1993, Modulation of intramitochondrial free Ca^{2+} concentration by antagonists of Na^+-Ca^{2+} exchange, *Trends Pharmacol. sci.* **14**:408–413.

Custodio, J. B., Moreno, A. J., and Wallace, K. B., 1998, Tamoxifen inhibits induction of the mitochondrial permeability transition by Ca^{2+} and inorganic phosphate, *Toxicol. Appl. Pharmacol.* **152**:10–17.

Dawson, T. M., Steiner, J. P., Dawson, V. L., Dinerman, J. L., Uhl, G. R., and Snyder, S. H., 1993, Immunosuppressant FK506 enhances phosphorylation of nitric oxide synthase and protects against glutamate neurotoxicity, *Proc. Natl. Acad. Sci. USA* **90**:9808–9812.

Doble, A., 1999, The role of excitotoxicity in neurodegenerative disease: Implications for therapy, *Pharmacol. Ther.* **81**:163–221.

Dubinsky, J. M., and Levi, Y., 1998, Calcium induced activation of the mitochondrial permeability transition in hippocampal neurons, *J. Neurosci. Res.* **53**:728–741.

Dubinsky, J. M., Kristal, B. S., and Elizondo-Fournier, J., 1995, An obligate role for oxygen in the early stages if glutamate-induced, delayed neuronal death, *J. Neurosci.* **15**:7071–7078.

Dugan, L. L., Sensi, S. L., Canzoniero, L. M. T., Handran, S. D., Rothman, S. M., Lin, T.-S., Goldberg, M. P., and Choi, D. W., 1995, Mitochondrial production of reactive oxygen species in cortical neurons following exposure to *N*-methyl-D-aspartate, *J. Neurosci.* **15**:6377–6388.

Dykens, J. A., 1994, Isolated cerebral and cerebellar mitochondria produce free radicals when exposed to elevated Ca^{2+} and Na^+: Implications for neurodegeneration, *J. Neurochem.* **63**:584–591.

Ellerby, H. M., Martin, S. J., Ellerby, L. M., Naiem, S. S., Rabizadeh, S., Salvesen, G. S., Casiano, C. A., Cashman, N. R., Green, D. R., and Bredesen, D. E., 1997, Establishment of a cell-free system of neuronal apoptosis: Comparison of premitochondrial, mitochondrial, and postmitochondrial phases, *J Neurosci.* **17**:6165–6178.

Erecinska, M., and Silver, I. A., 1996, Calcium handling by hippocampal neurons under physiologic and pathologic conditions, *Adv. Neurol.* **71**:119–136.

Fontaine, E., Ichas, F., and Bernardi, P., 1998, A ubiquinone binding site regulates the mitochondrial permeability transition pore, *J. Biol. Chem.* **273**:25734–25740.

Friberg, H., Ferrand-Drake, M., Bengtsson, F., Halestrap, A. P., and Wieloch, T., 1998, Cyclosporin A, but not FK 506, protects mitochondria and neurons against hypoglycemic damage and implicates the mitochondrial permeability transition in cell death, *J. Neurosci.* **18**:5151–5159.

Friberg, H., Connern, C. P., Halestrap, A. P., and Wieloch, T., 1999, Differences in the activation of the mitochondrial permeability transition among brain regions correlates with selective vulnerability, *J. Neurochem.* **72**:2488–2497.

Gill, R., Foster, A. C., and Woodruff, G. N., 1987, Systemic administration of MK-801 protects against ischemia-induced hippocampal neurodegeneration in the gerbil, *J Neurosci.* **7**:3343–3349.

Götz, M. E., Künig, G., Riederer, P., and Youdim, M. B. H., 1994, Oxidative stress: Free radical production in neural degeneration, *Pharmacol. Ther.* **63**:37–122.

Graham, D. G., 1978, Oxidative pathways for catecholamines in the genesis of neuromelanin and cytotoxic quinones, *Mol. Pharmacol.* **14**:633–643.

Greene, J. G., and Greenamyre, J. T., 1996, Bioenergetics and glutamate excitotoxicity, *Prog. Neurobiol.* **48**:613–621.

Griffiths, E. J., and Halestrap, A. P., 1991, Further evidence that cyclosporin A protects mitochondria from calcium overload by inhibiting a matrix peptidyl–prolyl *cis–trans* isomerase, *Biochem. J.* **274**:611–614.

Gunter, T. E., and Pfeiffer, D. R., 1990, Mechanisms by which mitochondria transport calcium, *Am. J. Physiol. Cell Physiol.* **258**:C755–C786.

Halestrap, A. P., and Davidson, A. M., 1990, Inhibition of Ca^{2+} induced high amplitude swelling of liver and heart mitochondria by cyclosporin is probably caused by the inhibitor binding to mitochondrial matrix peptidyl–prolyl *cis–trans* isomerase and preventing it interacting with the adenine nucleotide translocase, *Biochem. J.* **268**:153–160.

Hall, E. D., and Braughler, J. M., 1989, Central nervous system trauma and stroke: II. Physiological and pharmacological evidence for involvement of oxygen radicals and lipid peroxidation, *Free Radical Biol. Med.* **6**:303–313.

Halliwell, B., 1992, Reactive oxygen species in the central nervous system, *J. Neurochem.* **59**:1609–1623.

Hastings, T. G., 1995, Enzymatic oxidation of dopamine: The role of prostaglandin H synthase, *J Neurochem.* **64**:919–924.

Hastings, T. G., Lewis, D. A., and Zigmond, M. J., 1996, Role of oxidation in the neurotoxic effects of intrastriatal dopamine injections, *Proc. Natl. Acad. Sci. USA* **93**:1956–1961.

Hoyt, K. R., Gallagher, A. J., Hastings, T. G., and Reynolds, I. J., 1997a, Characterization of hydrogen peroxide toxicity in cultured rat forebrain neurons, *Neurochem. Res.* **22**:333–340.

Hoyt, K. R., Reynolds, I. J., and Hastings, T. G., 1997b, Mechanisms of dopamine-induced cell death in cultured rat forebrain neurons: Interactions with and differences from glutamate-induced cell death, *Exp Neurol.* **143**:269–281.

Hoyt, K. R., Sharma, T. A., and Reynolds, I. J. 1997c, Trifluoperazine and dibucaine inhibit glutamate-induced mitochondrial depolarization in cultured rat forebrain neurones, *Br. J. Pharmacol.* **122**:803–808.

Hoyt, K.R. McLaughlin, B.A., Higgins, D.S., and Reynolds, I.J. Inhibition of glutamate-induced mitochondrial depolarization by tamoxifen in cultured neurons. *J. Pharmacol. Exp. Ther.* **293**:480–486, 2000.

Hyrc, K., Handran, S. D., Rothman, S. M., and Goldberg, M. P., 1997, Ionized intracellular calcium concentration predicts excitotoxic neuronal death: Observations with low affinity fluorescent calcium indicators, *J. Neurosci.* **17**:6669–6677.

Ichas, F., and Mazat, J. P., 1998, From calcium signaling to cell death: Two conformations for the mitochondrial permeability transition pore: Switching from low- to high-conductance state, *Biochim. Biophys. Acta* **1366**:33–50.

Khodorov, B., Pinelis, V., Storozhevykh, T., Vergun, O., and Vinskaya, N., 1996, Dominant role of mitochondria in protection against a delayed neuronal Ca^{2+} overload induced by endogenous excitatory amino acids following a glutamate pulse, *FEBS Lett.* **393**:135–138.

Kiedrowski, L., Wroblewski, J. T., and Costa, E., 1994, Intracellular sodium concentration in cultured cerebellar granule cells challenged with glutamate, *Mol. Pharmacol.* **45**:1050–1054.

Koh, J. Y., Goldberg, M. P., Hartley, D. M., and Choi, D. W., 1990, Non-NMDA receptor mediated neurotoxicity in cortical culture, *J. Neurosci.* **10**:693–705.

Koroshetz, W. J., and Moskowitz, M. A. 1996, Emerging treatments for stroke in humans, *Trends Pharmacol. Sci.* **17**:227–233.

Kristal, B. S., and Dubinsky, J. M., 1997, Mitochondrial permeability transition in the central nervous system: Induction by calcium cycling-dependent and independent pathways, *J. Neurochem.* **69**:524–538.

Kristián, T., Katsura, K., Gidö G., and Siesjö, B. K., 1994, The influence of pH on cellular calcium influx during ischemia, *Brain Res.* **641**:295–302.

Lafon-Cazal, M., Pietri, S., Culcasi, M., and Bockaert, J., 1993, NMDA-dependent superoxide production and neurotoxicity, *Nature* **364**:535–537.

Lu, Y. F., Tomizawa, K., Moriwaki, A., Hayashi, Y., Tokuda, M., Itano, T., Hatase, O., and Matsui, H., 1996, Calcineurin inhibitors, FK506 and cyclosporin A, suppress the NMDA receptor-mediated potentials and LTP, but not depotentiation in the rat hippocampus, *Brain Res.* **729**:142–146.

MacDermott, A. B., Mayer, M. L., Westbrook, G. L., Smith, S. J., and Barker, J. L., 1986, NMDA-receptor activation increases cytoplasmic calcium concentration in cultured spinal cord neurones, *Nature* **321**:519–522.

MacKenzie, G. M., and Greenamyre, J. T., 1998, A novel model of slowly progressive Parkinson's disease, *Soc. Neurosci.* **24**:1721(abstract).

Maker, H. S., Weiss, C., Silides, D. J., and Cohen, G., 1981, Coupling of dopamine oxidation (monamine oxidase activity) to glutathione oxidation via the generation of hydrogen peroxide in the brain, *J. Neurochem.* **36**:589–593.

Mayer, M. L., and Westbrook, G. L., 1987, Cellular mechanisms underlying excitotoxicity, *Trends Neurosci.* **10**:59–61.

McCormack, J. G., Halestrap, A. P., and Denton, R. M., 1990, Role of calcium ions in regulation of mammalian intramitochondrial metabolism, *Physiol. Rev.* **70**:391–425.

Minta, A., Kao, J. P. Y., and Tsien, R. Y., 1989, Fluorescent indicators for cytosolic calcium based on rhodamine and fluorescein chromophores, *J. Biol. Chem.* **264**:8171–8178.

Nicholls, D. G., and Akerman, K. E. O., 1982, Mitochondrial calcium transport, *Biochim. Biophys. Acta* **683**:57–88.

Nieminen, A.-L., Petrie, T. G., Lemasters, J. J., and Selman, W. R., 1996, Cyclosporin A delays mitochondrial depolarization induced by *N*-methyl-D-aspartate in cortical neurons: Evidence of the mitochondrial permeability transition, *Neuroscience* **75**:993–997.

Novgorodov, S. A., Gudz, T. I., Brierley, G. P., and Pfeiffer, D. R., 1994, Magnesium ion modulates the sensitivity of the mitochondrial permeability transition pore to cyclosporin A and ADP, *Arch. Biochem. Biophys.* **311**:219–228.

O'Gorman, E., Beutner, G., Dolder, M., Koretsky, A. P., Brdiczka, D., and Walliman, T., 1997, The role of creatine kinase in inhibition of mitochondrial permeability transition, *FEBS Lett.* **414**:253–257.

Olney, J. W., 1990, Excitotoxic amino acids and neuropsychiatric disorders. *Ann. Rev. Pharmacol. Toxicol.* **30**:47–71.

Peng, T. I., and Greenamyre, J. T., 1998, Privileged access to mitochondria of calcium influx through *N*-methyl-D-aspartate receptors, *Mol. Pharmacol.* **53**:974–980.

Peng, T. I., Jou, M. J., Sheu, S.-S., and Greenamyre, J. T., 1998, Vizualization of NMDA receptor-induced mitochondrial calcium accumulation in striatal neurons, *Exp. Neurol.* **149**:1–12.

Petronilli, V., Miotto, G., Canton, M., Brini, M., Ionna, R., Bernardi P., and Di Lisa, F., 1999, Transient and long-lasting openings of the mitochondrial permeability transition pore can be monitored directly in intact cells by changes in mitochondrial calcein fluorescence, *Biophys. J.* **76**:725–734.

Phillis, J. W., 1994, A "radical" view of cerebral ischemic injury, *Prog. Neurobiol.* **42**:441–448.

Piani, D., Frei, K., Pfister, H.-W., and Fontana, A., 1993, Glutamate uptake by astrocytes is inhibited by reactive oxygen intermediates but not by other macrophage-derived molecules including cytokines, leukotrienes, or platelet-activating factor, *J. Neuroimmunol.* **48**:99–104.

Poot, M., Zhang, Y. Z., KrÄmer, J. A., Wells, K. S., Jones, L., Hanzel, D. K., Lugade, A. G., Singer, V. L., and Haughland, R. P., 1996, Analysis of mitochondrial morphology and function with novel fixable fluorescent stains, *J. Histochem. Cytochem.* **44**:1363–1372.

Reynolds, I. J., and Hastings, T. G., 1995, Glutamate induces the production of reactive oxygen species in cultured forebrain neurons following NMDA receptor activation, *J. Neurosci.* **15**:3318–3327.

Rosenberg, P. A., 1988, Catecholamine toxicity in cerebral cortex in dissociated cell culture, *J. Neurosci.* **8**:2887–2894.

Rothman, S. M., 1984, Synaptic release of excitatory amino acid neurotransmitter mediates anoxic neuronal death, *J. Neurosci.* **4**:1884–1891.

Rothman, S. M., and Olney, J. W., 1987, Excitotoxicity and the NMDA receptor, *Trends Neurosci.* **10**:299–302.

Sattler, R., Charlton, M. P., Hafner, M., and Tymianski, M., 1998, Distinct influx pathways, not calcium load, determine neuronal vulnerability to calcium neurotoxicity, *J. Neurochem.* **71**:2349–2364.

Scanlon, J. M., and Reynolds, I. J. 1998, Effects of oxidants and glutamate receptor activation on mitochondrial membrane potential in rat forebrain neurons, *J. Neurochem.* **71**:2392–2401.

Schinder, A. F., Olson, E. C., Spitzer, N. C., and Montal, M., 1996, Mitochondrial dysfunction is a primary event in glutamate neurotoxicity, *J. Neurosci.* **16**:6125–6133.

Sciamanna, M. A., Zinkel, J., Fabi, A. Y., and Lee, C. P., 1992, Ischemic injury to rat forebrain mitochondria and cellular calcium homeostasis, *Biochim. Biophys. Acta Mol. Cell Res.* **1134**:223–232.

Scorrano L. Petronilli V. Colonna R. Di Lisa F. Bernardi P. Chloromethyltetramethylrosamine (Mitotracker Orange) induces the mitochondrial permeability transition and inhibits respiratory complex I. Implications for the mechanism of cytochrome c release. *J. Biol. Chem.* **274**:24657–24663, 1999.

Stout, A. K., and, Reynolds, I. J., 1999, High-affinity calcium indicators underestimate increases in intracellular calcium concentrations associated with excitotoxic glutamate stimulations, *Neuroscience* **89**:91–100.

Stout, A. K., Raphael, H. M., Kanterewicz, B. I., Klann, E., and Reynolds, I. J., 1998, Glutamate-induced neuron death requires mitochondrial calcium uptake, *Nature Neurosci.* **1**:366–373.

Strijbos, P. J. L. M., Leach, M. J., and Garthwaite, J., 1996, Vicious cycle involving Na^+ channels, glutamate release, and NMDA receptors mediates delayed neurodegeneration through nitric oxide formation, *J. Neurosci.* **16**:5004–5013.

Taylor, C. P., Weber, M. L., Gaughan, C. L., Lehning, E. J., and Lopachin, R. M., 1999, Oxygen/glucose deprivation in hippocampal slices: Altered intraneuronal elemental composition predicts structural and functional damage, *J. Neurosci.* **19**:619–629.

Tretter, L., Chinopoulos, C., and Adam-Vizi, V., 1998, Plasma membrane depolarization and disturbed Na^+ homeostasis induced by the protonophore carbonyl cyanide-*P*-trifluoromethoxyphenyl-hydrazone in isolated nerve terminals, *Mol. Pharmacol.* **53**:734–741.

Tricklebank, M. D., Singh, L., Oles, R. J., Wong, E. H. F., and Iversen, S. D., 1987, A role for receptors of *N*-methyl-D-aspartic acid in the discriminative stimulus properties of phencyclidine, *Eur. J. Pharmacol.* **141**:497–501.

Tsacopoulos, M., and Magistretti, P. J., 1996, Metabolic coupling between glia and neurons, *J. Neurosci.* **16**:877–885.

Tymianski, M., Charlton, M. P., Carlen, P. L., and Tator, C. H., 1993, Source specificity of early calcium neurotoxicity in cultured embryonic spinal neurons, *J. Neurosci.* **13**:2085–2104.

Vanden Hoek, T. L., Li, C., Shao, Z., Schumacker, P. T., and Becker, L. B., 1997, Significant levels of oxidants are generated by isolated cardiomyocytes during ischemia prior to reperfusion, *J. Mol. Cell. Cardiol.* **29**:2571–2583.

Volterra, A., Trotti, D., Tromba, C., Floridi, S., Racagni, G., 1994, Glutamate uptake inhibition by oxygen free radicals in rat cortical astrocytes, *J. Neurosci.* **14**:2924–2932.

Wang, G. J., and Thayer, S. A., 1996, Sequestration of glutamate-induced Ca^{2+} loads by mitochondria in cultured rat hippocampal neurons, *J. Neurophysiol.* **76**:1611–1621.

White, R. J., and Reynolds, I. J., 1995, Mitochondria and Na^+/Ca^{2+} exchange buffer glutamate-induced calcium loads in cultured cortical neurons, *J. Neurosci.* **15**:1318–1328.

White, R. J., and Reynolds, I. J., 1996, Mitochondrial depolarization in glutamate-stimulated neurons: An early signal specific to excitotoxin exposure, *J. Neurosci.* **16**:5688–5697.

White, R. J., and Reynolds, I. J., 1997, Mitochondria accumulate Ca^{2+} following intense glutamate stimulation of cultured rat forebrain neurones, *J. Physiol.*(London) **498**:31–47.

Zhang, Y., Marcillat, O., Guilivi, C., Ernster, L., and Davies, K. J. A., 1990, The oxidative inactivation of mitochondrial electron transport chain components and ATPase, *J. Biol. Chem.* **265**:16330–16336.

Zoratti, M., and Szabò, I. 1995, The mitochondrial permeability transition, *Biochim. Biophys. Acta* **1241**:139–176.

Chapter 16

Mitochondrial Dysfunction in the Pathogenesis of Acute Neural Cell Death

Gary Fiskum

1. INTRODUCTION

Oxidative phosphorylation is an essential mitochondrial activity for maintaining cell viability, particularly for cells with high rates of ATP turnover (neurons, for example). Many other mitochondrial activities, however, including energy-dependent Ca^{2+} accumulation and reactive oxygen species (ROS), generation and detoxification are ultimately just as essential. Alteration of any of these functions results in cell injury and death. Abnormal Ca^{2+} accumulation by mitochondria in response to neuronal exposure to excitotoxic levels of excitatory neurotransmitters is a primary cause of mitochondrial dysfunction leading to cell death. In addition to impaired energy metabolism and accelerated net ROS production release of apoptogenic proteins (for example, cytochrome c) into the cytosol by mitochondria can trigger apoptotic cell death. Although several signals for such release have been identified, how they increase permeability of the outer membrane to apoptogenic proteins is controversial. Elucidating these mechanisms and their modes of regulation is particularly important in understanding the pathogenesis of neural cell death that occurs in both acute and chronic neurodegenerative diseases.

Millions worldwide die annually from brain injury due to stroke, shock, head trauma, subarachnoid hemorrhage and postischemic injury following resuscitation from cardiac arrest. Thousands of others either die or suffer permanent neurological impairment after common surgical procedures that result in some degree of transient cerebral ischemia.

Gary Fiskum   Department of Anesthesiology, School of Medicine, University of Maryland, Baltimore, Maryland 21201

Mitochondria in Pathogenesis, edited by Lemasters and Nieminen.
Kluwer Academic/Plenum Publishers, New York, 2001.

Ischemic neurological morbidity and mortality are consequences of both necrotic and apoptotic death of neurons, astroglia, and other brain cells.

Comprehensive reviews are available of the roles that mitochondria play in necrotic and apoptotic neural cell death and in acute neurodegenerative disorders, injury as well as in the chronic neurodegenerative diseases (e.g., Budd, 1998; Fiskum et al., 1999a; Kristian and Siesjo, 1998; Murphy et al., 1999). This chapter provides an overview of the mechanisms responsible for mitochondrial injury during cerebral ischemia/reperfusion and of the relationships between mitochondrial dysfunction and neural cells death by both necrotic and apoptotic mechanisms

2. SIGNIFICANCE OF MITOCHONDRIAL INJURY

2.1. Mitochondria as Primary Targets in Excitotoxicity

Mitochondria have long been considered subcellular targets of ischemic brain injury. Until recently, the significance of ischemic mitochondrial injury was thought to be limited primarily to its potential effects on maintaining sufficient cellular ATP to avoid necrotic cell death. This line of thought was altered by a series of studies by Reynolds, Nicholls, Choi, and others showing convincing evidence that mitochondria actively accumulate much of the cytosolic Ca^{2+} increase generated in response to excitotoxic levels of extraneuronal glutamate, and showing that excessive mitochondrial Ca^{2+} accumulation is the primary cause of cell death that occurs within hours of excitotoxin exposure (Fig. 1). (e.g., White and Reynolds, 1995; Budd and Nicholls, 1996; Dugan et al., 1995; Schinder et al., 1996; Stout et al., 1998). In most of these in vitro excitotoxic neuronal death paradigms, necrosis is the primary form of death. Such death can be caused by the failure of neurons to generate enough ATP to maintain adequate transmembrane ionic gradients and therefore to protect against osmotic lysis. In some models of acute excitotoxic death, however, inhibition of mitochondrial electron transport and oxidative phosphorylation do not promote neuronal death but actually protects against it (Castillo et al., 1999; Stout et al., 1998). The neuroprotection mechanism appears to be due to inhibition of mitochondrial Ca^{2+} sequestration via elimination of the electrochemical proton gradient, which is the driving force behind electrophoretic influx of extramitochondrial Ca^{2+}. These observations have focused attention away from the inhibitory effect of mitochondrial Ca^{2+} overload on ATP synthesis and toward the stimulation of ROS generation by mitochondrially accumulated Ca^{2+} (Castillo et al., in press; Reynolds and Hastings, 1995; Stout et al., 1998). Though it is possible that ROS generated by Ca^{2+}-altered mitochondrial electron transport can cause sufficient oxidative damage in cellular membranes to directly invoke necrotic cellular disruption, evidence suggests that oxidative membrane alterations can result in a secondary intracellular Ca^{2+} increase, which is then responsible for the irreversible disruption of membrane continuity (Castillo et al., 1999).

2.2. Mitochondria as Storage Sites for Apoptogenic Proteins

Some in vitro models of neuronal excitotoxicity have demonstrated that relatively extensive mitochondrial injury leads to necrosis, whereas milder mitochondrial alterations

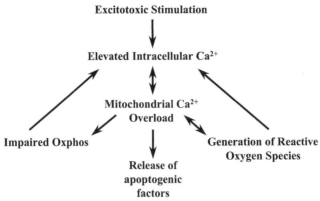

FIGURE 1. Involvement of mitochondrial Ca^{2+} overload in pathogenesis of excitotoxic neuronal death. Binding of glutamate to ionotropic receptors results in neuronal Ca^{2+} influx and increased cytosolic $[Ca^{2+}]$. Under normal conditions the increased $[Ca^{2+}]$ is buffered by respiration-dependent mitochondrial Ca^{2+} uptake without causing mitochondrial injury. Abnormally high levels of accumulated mitochondrial Ca^{2+} alter electron transport activities in ways that impair oxidative phosphorylation and stimulate ROS generation. Decreased ATP levels limit the ability of Ca^{2+} transport ATPases to lower intracellular Ca^{2+}. ROS react with membrane lipids and proteins, causing increased plasma membrane Ca^{2+} influx. Oxidative stress can also stimulate mitochondrial Ca^{2+} efflux via activation of the MPT pore. Mitochondrial Ca^{2+} sequestration is also one of several signals that can stimulate apoptogenic mitochondrial protein release.

can be associated with apoptosis (Ankarcrona *et al.*, 1995; Kruman and Mattson, 1999; Nicotera *et al.*, 1997). Apparently, when oxidative phosphorylation is impaired to the extent that normal cellular ATP levels are not maintained, neurons die by necrosis, but the extent of cellular oxidative stress induced by elevated mitochondrial ROS production may also determine how cells die. Apoptosis contributes to neuronal death *in vitro* in a variety of models (see, e.g., Tamatani *et al.*, 1998; Leist *et al.*, 1998; Martinou *et al.*, 1999; Neame *et al.*, 1998). Apoptosis also contributes significantly to neuronal death *in vivo* in models of both focal and global cerebral ischemia and also in traumatic brain injury (see, e.g., Kaya *et al.*, 1999; Fujimura *et al.*, 1998; Krajewski *et al.*, 1999; Li *et al.*, 1997a, 1997b; Namura *et al.*, 1998); however, the relative contribution of apoptosis to necrosis varies with model, neuronal subtype, and anatomical location.

Growing evidence from nonneuronal and, to a limited extent, neuronal apoptosis paradigms indicates that mitochondrial release of one or more apoptogenic proteins activates the cell death protease (caspase) cascade within cytosol and nucleus (Fig. 1, Fig. 2) (reviewed in Green and Reed, 1998; Murphy *et al.*, 1999). These findings have increased the focus on mitochondria as mediators of acute neural cell death. Recent observations of cytochrome *c* subcellular redistribution in animal brain following focal and global cerebral ischemia have emphasized the potential clinical relevance of this phenomenon (Fujimura *et al.*, 1998; Perez-Pinzon *et al.*, 1999). Because apoptosis often proceeds with little evidence of metabolic failure, mitochondria's contribution to this process underscores the importance of mitochondrial alterations other than inhibition of ATP synthesis.

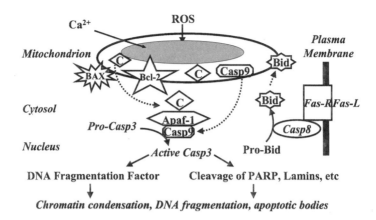

FIGURE 2. Relationships between mitochondria, pro- and anti-apoptotic proteins, and apoptotic signaling. Mitochondrial stressors (ROS and elevated intramitochandrial Ca^{2+}, for example) and pro-apoptotic proteins (e.g., Bid, Bax) can induce release of cytochrome c (C) and procaspase-9 (Casp9) from mitochondria into the cytosol. Activation of Bid and translocation to mitochondria is accomplished through proteolysis of Bid by caspase-8 (Casp8) and possibly other caspases. Such activation can be initiated by the binding of Fas ligand (Fas-L) to its plasma membrane receptor (Fas-R). Release of pro-apoptotic mitochondrial proteins is inhibited by the anti-apoptotic, mitochondrially associated protein Bcl2. An apoptosomal complex of apoptosis-activating factor 1 (Apaf-1), procaspase-9, and cytochrome c activates caspase-9, which in turn mediates proteolytic activation of procaspase-3 (Casp-3). Caspase-3 and the other caspases it activates act on numerous proteins, including poly-(ADP-ribose)polymerase (PARP), to generate the molecular and morphological hallmarks of apoptosis.

3. MECHANISMS OF MITOCHONDRIAL INJURY

3.1. Mitochondrial Ca^{2+} Overload

Excessive accumulation of Ca^{2+} during excitotoxic stimulation or during ischemia/ reperfusion is a primary mediator of mitochondrial injury (Fig. 1). Mitochondrial Ca^{2+} overload can result in membrane phospholipid degradation, proteolyis, and oxygen free radical generation (reviewed in Werling and Fiskum, 1996). Elevated intracellular Ca^{2+} can under some circumstances both stimulate the expression of nuclear cell death genes (Kroemer *et al.*, 1995) and inhibit normal cytosolic protein synthesis (Neumar *et al.*, 1998; Paschen and Doutheil, 1999), so its effects on mitochondrial gene transcription and translation should be characterized.

One effect of elevated intracellular Ca^{2+} that has received much attention is the promotion of a relatively nonselective increase in mitochondrial inner membrane perme- ability, the *mitochondrial permeability transition* (MPT) (Bernardi *et al.*, 1998; Lemasters *et al.*, 1998a, b). This permeability alteration has been associated with uncoupling and inhibition of oxidative phosphorylation, with stimulation of mitochondrial ROS generation (Fig. 1), and with osmotic mitochondrial swelling, which can promote release of apoptogenic and other mitochondrial proteins (Fig. 3).

The antideath gene-product Bcl2 protects against several aspects of mitochondrial dysfunction caused by mitochondrial Ca^{2+} overload (Murphy *et al.*, 1996a; Shimizu *et al.*, 1998) and against the cytotoxic effects of excitotoxin-mediated intraneuronal Ca^{2+}

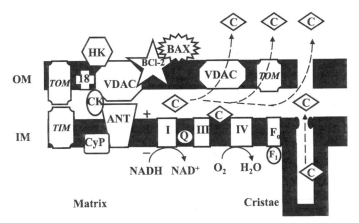

FIGURE 3. Mitochondrial proteins implicated in the MPT and cytochrome *c* release. Cytochrome *c* (C) between inner (IM) and outer (OM) mitochondrial membranes can be released through a disruption of outer membrane continuity. A mechanism that can be responsible for outer membrane disruption is osmotic matrix-space swelling due to opening of the MPT pore and net influx of osmotically active solutes (S). A substantial fraction of mitochondrial cytochrome *c* is located on inner membranes that form invaginations called *cristae*. Release of this cytochrome *c* pool into the intermembrane space can be via osmotic swelling of the matrix and possibly by other mechanisms. Proteins implicated in forming and regulating the MPT pore include the adenine nucleotide translocase (ANT), the protein transporter at the inner membrane (Tim), the protein transporter at the outer membrane (Tom), the outer membrane voltage-dependent anion channel (VDAC), mitochondrial cyclophilin (CyP), the 18-kDa outer-membrane peripheral benzodiazapine receptor (18), creatine kinase (CK), and mitochondrial hexokinase (HK). Anti-apoptotic mitochondrial protein Bcl-2 inhibits cytochrome *c* release whereas the pro-apoptotic protein Bax promotes it. In addition to disruption of mitochondrial outer membrane continuity, cytochrome *c* may be released through abnormal permeability states of outer membrane pores, such as VDAC or Tom.

elevation (Prehn *et al.*, 1994). Though Bcl2 can inhibit the MPT (Susin *et al.*, 1999a, b; Marzo *et al.*, 1998b; Shimizu *et al.*, 1998), it is not known whether all of its protective effects are due to this mechanism of action.

3.2. Oxidative Stress

Mitochondria are also the target of ROS species generated by many different systems, including the mitochondrial electron transport chain (ETC) (Fig. 1), lipoxygenases, oxidases (for example, xanthine oxidase), Fe^{2+}-catalyzed hydroxyl radical (OH·) formation, and peroxynitrite formed from the reaction of nitric oxide (NO·) with superoxide (O_2^-) reviewed in (Fiskum, 1997). Level and activity of mitochondrial antioxidant defense systems, is therefore extremely important in determining the extent of oxidative mitochondrial injury (Keller *et al.*, 1998; Murakami *et al.*, 1998). Maintaining a relatively reduced redox state of mitochondrial pyridine nucleotides and glutathione is also important in detoxification of H_2O_2 and other ROS. Another potential mechanism of cytoprotection by Bcl2 could involve its ability to shift the redox state of NAD(P)H and glutathione in the direction of reduction (Ellerby *et al.*, 1996). Although Bcl2's mechanism of action on cellular redox state has not been defined, it may be responsible for the cytoprotection against oxidative stress (Murphy *et al.*, 1996b; Myers *et al.*, 1995). Recent evidence

(Kowaltowski *et al.*, 2000) also indicates that this characteristic of Bcl2 overexpression could sufficiently explain how it inhibits the MPT.

The molecular targets of oxidative mitochondrial injury are not well established. Complex I (NADH/CoQ oxidoreductase) of the ETC appears particularly sensitive to inhibition by oxidative stress (Hillered and Ernster, 1983), although we don't yet know whether such inhibition is due to oxidative injury to its component proteins or due to oxidation of the membrane lipids in which it is anchored. Although complex I and other ETC components are adversely affected by even brief periods of ischemia, reperfusion results in substantial recovery of activity (Allen *et al.*, 1995; Sims, 1991).

Other studies have demonstrated a hyperoxidation of ETC components as well as pyridine nucleotides during postischemic reperfusion (Pèrez-Pinzòn *et al.*, 1997; Rosenthal *et al.*, 1995), suggesting that NADH generation, rather than utilization, can sometimes be the limiting factor in postischemic cerebral energy metabolism. One possible mechanism of NADH-generation impairment is inhibition of the pyruvate dehydrogenase complex. This enzyme catalyzes the step that couples glycolysis to aerobic metabolism, and it suffers a dramatic loss of activity within the first 30 to 60 minutes of reperfusion following a brief period of cerebral ischemia (Bogaert *et al.*, 1994; Katayama and Welsh, 1989; Zaidan and Sims, 1997).

Oxidative stress is also a primary activator of the transition. Although neither the presence of ROS nor a major oxidized shift in mitochondrial-component redox state will typically induce transition alone, they greatly potentiate the ability of Ca^{2+} to do so (Kowaltowski and Vercesi, 1999). Direct oxidation of mitochondrial sulfhydryl groups with agents such as phenylarsine oxide can induce transition without exogenous Ca^{2+}. These and other observations suggest that sulfhydryl groups in the adenine nucleotide translocase are important molecular targets of oxidative stress that could be responsible for inducing the mitochondrial transition and dysfunction in situations of elevated intracellular Ca^{2+} and ROS, for example during reperfusion after cerebral ischemia (Halestrap *et al.*, 1997; but Constantini *et al.*, 1998). Additional support for the hypothesis that MPT activation contributes to mitochondrial and neuronal injury after ischemia/hypoxia comes from the studies of Friberg *et al.*, 1998, and Uchino *et al.*, 1998, who report that treating ischemic or hypoglycemic animals with the MPT inhibitor cyclosporin A significantly reduces neuronal injury.

4. INDUCTION OF APOPTOSIS VIA RELEASE OF MITOCHONDRIAL PROTEINS

4.1. Stimulation and Inhibition Signals for Release of Apoptogenic Mitochondrial Proteins

Discovery of the involvement of mitochondrial cytochrome *c* release in activating the apoptotic caspase cascade is potentially one of the most important, and certainly most unexpected, events in the history of the study of cell death mechanism (Fig. 2) (Kluck *et al.*, 1997; Yang *et al.*, 1997). It now appears that in addition to cytochrome *c*, other apoptogenic proteins are released from mitochondria during the early stages of apoptosis. These proteins include caspase-2 (Susin *et al.*, 1999a), caspase-9 (Krajewski *et al.*, 1999;

Susin *et al.*, 1999a), and the "apoptosis inducing factor," which can cause chromatin condensation and other hallmarks of apoptosis (Susin *et al.*, 1999b). Cytochrome *c*, procaspase-9, and apoptosis-activating factor 1 (Apaf-1) form in the cytosol a macromolecular complex, the *apoptosome* (Zou *et al.*, 1999). Proteolytically activated caspase-9 dissociates from the complex, and in turn proteolytically activates caspase-3, which then activates a cascade of other caspases within both cytosol and nucleus (Schulz *et al.*, 1999). The ability of released cytochrome *c* to actively participate in the apoptosome is modulated by other signals, for example, those regulated by nerve growth factor and other trophic factors (Deshmukh and Johnson, 1998).

The release of cytochrome *c* and other mitochondrial apoptogenic proteins is inhibited by the anti-apoptotic gene products Bcl2 and Bcl-X_L and is currently considered their most important mechanism of action (Fig. 2) (Cai and Jones, 1998; Kluck *et al.*, 1997; Krajewski *et al.*, 1999; Susin *et al.*, 1999a, 1999b; Yang *et al.*, 1997). Inhibition of cytochrome *c* release may also explain the antioxidant effect of Bcl2, because its depletion and the subsequent stimulation of mitochondrial ROS generation is proposed to account for the oxidative stress that typically accompanies apoptosis (Cai and Jones, 1998).

Various factors appear capable of triggering cytochrome *c* release from mitochondria (Fig. 2). These factors include elevated Ca^{2+}, ROS (Andreyev *et al.*, 1998; Borutaite *et al.*, 1999; Petit *et al.*, 1998; Scarlett and Murphy, 1997; Yang and Cortopassi, 1998), and cell death proteins such as Bax and Bid that become activated and redistribute within the cell to the mitochondria in response to signals (for example, ligand binding to the Fas receptor (Desagher *et al.*, 1999; Gross *et al.*, 1999; Han *et al.*, 1999; Jürgensmeier *et al.*, 1998; Li *et al.*, 1998; Luo *et al.*, 1998; Marzo *et al.*, 1998a).

4.2. Release Mechanisms of Apoptogenic Mitochondrial Proteins

An important question being addressed by several laboratories is which mechanism or mechanisms are involved in the release of cytochrome *c* and other apoptogenic mitochondrial proteins by agents as diverse as Ca^{2+}, ROS, and peptides. Basically, there are two fundamentally different mechanisms under consideration (Green and Reed, 1998). The first involves physical disruption of the mitochondrial outer membrane and simple diffusion of cytochrome *c* from its normal intermembrane space. The second involves transport of proteins through an outer membrane pore.

One possible explanation for the physical disruption of the outer membrane is osmotic swelling of the mitochondrial compartment surrounded by inner membrane (matrix) due to increased inner membrane permeability to osmotically active solutes. A likely cause for this increased permeability is the MPT; however, other mechanisms are possible, including phospholipase-mediated degradation of membrane lipids (reviewed in Halestrap, 1989). Evidence for MPT involvement in mediating the release of apoptogenic mitochondrial proteins comes primarily from apoptosis paradigms using nonneuronal cells. In these studies a redistribution of cytochrome *c* or other proteins from mitochondria to cytosol occurs that may be inhibited by cyclosporin A or that is associated with morphological indicators of mitochondrial swelling (e.g., Walter *et al.*, 1998; Bradham *et al.*, 1998; Dumont *et al.*, 1999; Higuchi *et al.*, 1998; Pastorino *et al.*, 1998). In some apoptosis paradigms, imaging techniques have detected redistribution of cytosolic proteins to mitochondria, indicating disruption of the normal barrier against such entry (Lemasters *et al.*, 1998b). Despite evidence supporting the MPT's role in many apoptosis paradigms, other studies using similar techniques have argued against the role of transition, or even of swelling-induced disruption of the outer membrane, in release of apoptogenic mitochondrial proteins (Bossy-Wetzel *et al.*, 1998; Eskes *et al.*, 1998; Finucane *et al.*, 1999; Kluck *et al.*, 1997; Martinou *et al.*, 1999; McGinnis *et al.*, 1999; Priault *et al.*, 1999; Ushmorov *et al.*, 1999; Vander Heiden *et al.*, 1997; Yang *et al.*, 1997; Zhuang *et al.*, 1998).

In addition to evidence that transition can, at least in some systems, contribute to apoptogenic mitochondrial protein release in cells, studies performed with isolated mitochondria verify that the MPT can disrupt the outer membrane and release cytochrome *c* and other proteins (Andreyev and Fiskum, 1999; Kantrow and Piantadosi, 1997; Krajewski *et al.*, 1999; Petit *et al.*, 1998; Scarlett and Murphy, 1997; Yang and Cortopassi, 1998). The vast majority of MPT work has been done using isolated liver mitochondria, under conditions far from physiologically realistic. Evidence indicates that isolated brain mitochondria can undergo transition, but activity is low compared with that of liver, and negligible in the presence of mM concentrations of adenine nucleotides (Andreyev and Fiskum, 1999; Andreyev *et al.*, 1998; Kristal and Dubinsky, 1997; Rottenberg and Marbach, 1990).

Because isolated brain mitochondria are typically a heterogeneous mixture of the neuronal and the nonneuronal, it is important to compare the characteristics of apoptogenic mitochondrial protein release from different types of brain cells. Preliminary evidence from primary cultures of cerebellar granule neurons and cortical astrocytes indicates that astrocyte mitochondria are subject to an MPT-mediated, Ca^{2+}-evoked release of cytochrome *c*, whereas mitochondria within cerebellar neurons release cytochrome *c* in response to Ca^{2+} by a mechanism un-inhibitable by cyclosporin A and that does not require mitochondrial swelling (Fiskum and Andreyev, 1999). These results were obtained using cells treated with low levels of glycoside digitonin (<0.01%), which interacts with cellular membrane cholesterol to effectively create "holes" permeable to molecules as large as soluble proteins (Fiskum *et al.*, 1980; Seeman *et al.*, 1973). Because the plasma membrane and the mitochondrion have, respectively, the highest and lowest levels of cellular membranes cholesterol (Colbeau *et al.*, 1971), digitonin-permeabilized cells can be used as "leaky bags of organelles" to study mitochondrial functions without the stress and potential alteration caused by isolation procedures (Fiskum, 1985; Murphy *et al.*, 1996b).

Possibly, however, the functional characteristics of mitochondria *in situ* within intact cells differ significantly from the characteristics of either isolated mitochondria or mitochondria in permeabilized cells. Thus, although isolated brain mitochondria and mitochondria within cerebellar neurons appear MPT resistant, recent observations (Dubinsky and Levi, 1998) of mitochondrial morphology in hippocampal neurons treated with a Ca^{2+} ionophore or an excitotoxic level of glutamate both with and without cyclosporin A suggest that transition does occur in these intact cells. It remains to be determined, though, whether the MPT is responsible for cytochrome *c* release in cultured neurons or in neurons present within nervous tissue. Indirect evidence that transition is involved in neuronal mitochondrial cytochrome *c* release *in vivo* includes observations that cyclosporin A can inhibit neuronal death and caspase activation in animal models of ischemia and hypoglycemia and in neural cells exposed to toxic stimuli (Friberg *et al.*, 1998; Kruman and Mattson, 1999; Nieminen *et al.*, 1996; Uchino *et al.*, 1998).

A number of mitochondrial proteins have been implicated as either constituting or regulating the transition pore, including the adenine nucleotide translocase (Zoratti and Szabò, 1995) and mitochondrial cyclophilin (the putative cyclosporin A binding site) (Connern and Halestrap, 1996). There are also a host of outer membrane proteins that may affect the MPT through their interaction with the inner membrane at inner–outer membrane contact sites (Fig. 3). These include the voltage-dependent anion channel (VDAC or porin) (Zoratti and Szabò, 1995); an 18-kDa protein that together with porin constitutes the mitochondrial (peripheral) benzodiazepine receptor (Hirsch et al., 1998; McEnery et al., 1993); the proteins responsible for transmembrane mitochondrial protein import (Kinnally et al., 1996; Sokolove and Kinnally, 1996); creatine kinase (Beutner et al., 1998); and Bcl2 and related proteins (Marzo et al., 1998b). Recent evidence (Marzo et al., 1998a) also suggests that the mechanism by which Bax induces cytochrome c release and apoptosis involves an interaction with the adenine nucleotide translocase and activation of the MPT. How Bax might cross the outer membrane to interact with the translocator is unknown, however. Alternatively, Bax may indirectly modulate adenine nucleotide translocator involvement by interacting with the VDAC (Narita et al., 1998).

In light of growing evidence that the MPT is not always necessary for cytochrome c release during the early stages of apoptosis, alternative release mechanisms should be considered, but there are few candidates for outer membrane pores that could possibly transport cytochrome c. An outer membrane pore exists (Tom) that interacts with a complementary inner membrane pore (Tim) and catalyzes the import of proteins possessing mitochondrial signal sequences (Fig. 3) (Schulke et al., 1997), but little is known of the structure of these proteins. The other, more likely, candidate is the VDAC. Its internal diameter is 2.8–3.0 nm, whereas the dimensions of native cytochrome c are approximately 3.0 by 3.4 nm (Manella, 1998). It has been shown that VDAC binds cytochrome c, thus possibly VDAC's β-barrel structure results in the binding inducing a conformational change that allows cytochrome c to fit through the pore (Manella, 1998). This scenario suggests that inducers of cytochrome c release (for example, Ca^{2+}) may cause an equilibrium shift of cytochrome c from a pool primarily bound to inner membrane to an unbound pool in the intermembrane space that could interact with VDAC (Fig. 3). Alternatively, inducers may increase the number or activity of pores on the outer membrane to which cytochrome c is accessible. As both VDAC and Tom are thought to be components of the inner–outer membrane contact sites, possibly the conditions that decrease inner and outer membrane binding can increase the number of channels available for egress of cytochrome c. These contact sites are in dynamic equilibrium, and their number varies in response to several factors, including respiration rate, Ca^{2+} transport and mitochondrial membrane potential (Hackenbrock, 1966; Hackenbrock and Caplan, 1969).

5. CONCLUSIONS

Now that the involvement of mitochondrial dysfunction in both necrotic and apoptotic neuronal cell death is firmly established, the molecular mechanisms by which acute stress alters mitochondrial activities and induces cell death must be elucidated. Although a greater understanding of mitochondrial ROS production and apoptogenic protein release is essential, we should also continue to study the ways in which oxidative phosphorylation and cerebral energy metabolism are impaired. Clarification of mitochondrial dysfunction mechanisms will likely require a combination of approaches using measurements performed with living cells, as well as isolated mitochondria and permeabilized cells. The knowledge thus obtained should lead to development of novel modes of protection against neural cell death. The relevance of such results will need verification with animal models of acute neurodegeneration, including models of ischemia and traumatic brain injury.

ACKNOWLEDGEMENTS. Supported by National Institute of Neurological Disorders and Stroke (NINDS), grant NS34152 and U.S. Army Grant DAMD 17-99-1-9483.

REFERENCES

Allen, K. L., Ameida, A., Bates, T. E., and Clark, J. B., 1995, Changes of respiratory chain activity in mitochondrial and synaptosomal fractions isolated from the gerbil brain after graded ischaemia, *J. Neurochem.* **64**:2222–2229.

Andreyev, A., and Fiskum, G., 1999, Calcium induced release of mitochondrial cytochrome *c* by different mechanisms selective for brain versus liver, *Cell Death Diff.*, in press.

Andreyev, A., Fahy, B., and Fiskum, G., 1998, Cytochrome *c* release from brain mitochondria is independent of the mitochondrial permeability transition, *FEBS Lett.* **439**:373–376.

Ankarcrona, M., Dypbukt, J. M., Bonfoco, E., Zhivotovsky, B., Orrenius, S., Lipton, S. A., and Nicotera, P., 1995, Glutamate-induced neuronal death: A succession of necrosis or apoptosis depending on mitochondrial function, *Neuron* **15**:961–973.

Bernardi, P., Colonna, R., Costantini, P., Eriksson, O., Fontaine, E., Ichas, F., Massari, S., Nicolli, A., Petronilli, V., and Scorrano, L., 1998, The mitochondrial permeability transition, *Biofactors* **8**:273–281.

Beutner, G., Ruck, A., Riede, B., and Brdiczka, D., 1998, Complexes between porin, hexokinase, mitochondrial creatine kinase, and adenylate translocator display properties of the permeability transition pore: Implication for regulation of permeability transition by the kinases, *Biochim. Biophys. Acta* **1368**:7–18.

Bogaert, Y. E., Rosenthal, R. E., and Fiskum, G., 1994, Post-ischemic inhibition of cerebral cortex pyruvate dehydrogenase, *Free Radical Biol. Med.* **16**:811–820.

Borutaite, V., Morkuniene, R., Brown, and G. C., 1999, Release of cytochrome *c* from heart mitochondria is induced by high Ca^{2+} and peroxynitrite and is responsible for $Ca(2+)$-induced inhibition of substrate oxidation, *Biochim. Biophys. Acta* **1453**:41–48.

Bossy-Wetzel, E., Newmeyer, D. D., and Green, D. R., 1998, Mitochondrial cytochrome *c* release in apptosis occurs upstream of DEVD-specific caspase activation and independently of mitochondrial transmembrane depolarization, *EMBO J.* **17**:37–49.

Bradham, C. A., Qian, T., Streetz, K., Trautwein, C., Brenner, D. A., and Lemasters, J. J., 1998, The mitochondrial permeability transition is required for tumor necrosis factor alpha-mediated apoptosis and cytochrome *c* release, *Mol. Cell Biol.* **18**:6353–6364.

Budd, S. L., 1998, Mechanisms of neuronal damage in brain hypoxia/ischemia: Focus on the role of mitochondrial calcium accumulation, *Pharmacol. Ther.* **80**:203–229.

Budd, S. L., and Nicholls, D. G., 1996, Mitochondria, calcium regulation, and acute glutamate excitotoxicity in cultured cerebellar granule cells, *J. Neurochem.* **67**:2282–2291.

Cai, J., and Jones, D. P., 1998, Superoxide in apoptosis: Mitochondrial generation triggered by cytochrome *c* loss, *J. Biol. Chem.* **273**:11401–11404.

Castilho, R. F., Ward, M. W., and Nicholls, D. G., 1999, Oxidative stress, mitochondrial function, and acute glutamate excitotoxicity in cultured cerebellar granule cells, *J. Neurochem.*, **72**:1394–1401.

Colbeau, A., Nachbaur, J., and Vignais, P. M., 1971, Enzymic characterization and lipid composition of rat liver subcellular membranes, *Biochim. Biophys. Acta* **249**:462–492.

Connern, C. P., and Halestrap, A. P., 1996, Chaotropic agents and increased matrix volume enhance binding of mitochondrial cyclophilin to the inner mitochondrial membrane and sensitize the mitochondrial permeability transition to [Ca^{2+}], *Biochemistry* **35**:8172–8180.

Costantini, P., Colonna, R., and Bernardi, P., 1998, Induction of the mitochondrial permeability transition by *N*-ethylmaleimide depends on secondary oxidation of critical thiol groups: Potentiation by copper-*ortho*-phenanthroline without dimerization of the adenine nucleotide translocase, *Biochim. Biophys. Acta* **1365**:385–392.

Desagher, S., Osen-Sand, A., Nichols, A., Eskes, R., Montessuit, S., Lauper, S., Maundrell, K., Antonsson, B., and Martinou, J. C., 1999, Bid-induced conformational change of bax is responsible for mitochondrial cytochrome *c* release during apoptosis, *J. Cell. Biol.* **144**:891–901.

Deshmukh, M., and Johnson, E. M., Jr., 1998, Evidence of a novel event during neuronal death: Development of competence-to-die in response to cytoplasmic cytochrome *c*, *Neuron* **21**:695–705.

Dubinsky, J. M., and Levi, Y., 1998, Calcium-induced activation of the mitochondrial permeability transition in hippocampal neurons, *J. Neurosci. Res.* **53**:728–741.

Dugan, L. L., Sensi, S. L., Canzoniero, L. M. T., Handran, S. D., Rothman, S. M., Lin, T.-S., Goldberg, M. P., and Choi, D. W., 1995, Mitochondrial production of reactive oxygen species in cortical neurons following exposure to *N*-methyl-D-aspartate, *J. Neurosci.* **15**:6377–6388.

Dumont, A., Hehner, S. P., Hofmann, T. G., Ueffing, M., Droge, W., and Schmitz, M. L., 1999, Hydrogen peroxide-induced apoptosis is CD95-independent, requires the release of mitochondria-derived reactive oxygen species and the activation of NF-kappaB, *Oncogene* **18**:747–757.

Ellerby, L. M., Ellerby, H. M., Park, S. M., Holleran, A. L., Murphy, A. N., Fiskum, G., Kane, D. J., Testa, M. P., Kayalar, C., and Bredesen, D. E., 1996, Shift of the cellular oxidation-reduction potential in neural cells expressing Bcl-2, *J. Neurochem.* **67**:1259–1267.

Eskes, R., Antonsson, B., Osen-Sand, A., Montessuit, S., Richter, C., Sadoul, R., Mazzei, G., Nichols, A., and Martinou, J. C., 1998, Bax-induced cytochrome *c* release from mitochondria is independent of the permeability transition pore but highly dependent on Mg^{2+}ions, *J. Cell Biol.* **143**:217–224.

Finucane, D. M., Bossy-Wetzel, E., Waterhouse, N. J., Cotter, T. G., and Green, D. R., 1999, Bax-induced caspase activation and apoptosis via cytochrome *c* release from mitochondria is inhibitable by Bcl-x$_L$, *J. Biol. Chem.* **274**:2225–2233.

Fiskum, G., 1985, Intracellular levels and distribution of Ca^{2+} in digitonin-permeabilized cells, *Cell Calcium* **6**:25–37.

Fiskum, G., 1997, Metabolic failure and oxidative stress contribute to ischemic neurological impairment and delayed cell death, in *Neuroprotection* (T. J. J. Blanck, Ed.), Williams and Wilkins, Baltimore, MD, pp. 1–14.

Fiskum, G., and Andreyev, A., 1999, Brain mitochondrial heterogeneity in Ca^{2+}-induced release of apoptogenic cytochrome *c*, *J. Neurochem.* **72**:S52C.

Fiskum, G., Craig, S. W., Decker, G., and Lehninger, A. L., 1980, The cytoskeleton of digitonin-treated rat hepatocytes, *Proc. Nat. Acad. Sci. USA* **77**:3430–3434.

Fiskum, G., Murphy, A. N., and Beal, M. F., 1999, Mitochondria in neurodegeneration: Acute ischemia and chronic neurodegenerative diseases, *J. Cereb. Blood Flow Metab.* **19**:351–369.

Friberg, H., Ferrand-Drake, M., Bengtsson, F., Halestrap, A. P., and Wieloch, T., 1998, Cyclosporin A, but not FK 506, protects mitochondria and neurons against hypoglycemic damage and implicates the mitochondrial permeability transition in cell death, *J. Neurosci.* **18**:5151–5159.

Fujimura, M., Morita-Fujimura, Y., Murakami, K., Kawase, M., and Chan, P. H., 1998, Cytosolic redistribution of cytochrome *c* after transient focal cerebral ischemia in rats, *J. Cereb. Blood Flow Metab.* **18**:1239–1247.

Green, D. R., and Reed, J. C., 1998, Mitochondria and apoptosis, *Science* **281**:1309–1312.

Gross, A., Yin, X. M., Wang, K., Wei, M. C., Jockel, J., Milliman, C., Erdjument-Bromage, H., Tempst, P., and Korsmeyer, S. J., 1999, Caspase cleaved BID targets mitochondria and is required for cytochrome *c* release, while Bcl-x$_L$ prevents this release but not tumor necrosis factor-R1/Fas death, *J. Biol. Chem.* **274**:1156–1163.

Hackenbrock, C. R., 1966, Ultrastructural bases for metabolically linked mechanical activity in mitochondria: I. Reversible ultrastructural changes with change in metabolic steady state in isolated liver mitochondria, *J. Cell Biol.* **30**:269–297.

Hackenbrock, C. R., and Caplan, A. I., 1969, Ion-induced ultrastructural transformations in isolated mitochondria: The energized uptake of calcium, *J. Cell Biol.* **42**:221–234.

Halestrap, A. P., 1989, The regulation of the matrix volume of mammalian mitochondria *in vivo* and *in vitro* and its role in the control of mitochondrial metabolism, *Biochim. Biophys. Acta* **973**:355–382.

Halestrap, A. P., Woodfield, K. Y., and Connern, C. P., 1997, Oxidative stress, thiol reagents, and membrane potential modulate the mitochondrial permeability transition by affecting nucleotide binding to the adenine nucleotide translocase, *J. Biol. Chem.* **272**:3346–3354.

Han, Z., Bhalla, K., Pantazis, P., Hendrickson, E. A., and Wyche, J. H., 1999, Cif (cytochrome *c* efflux-inducing factor) activity is regulated by Bcl-2 and caspases and correlates with the activation of Bid, *Mol. Cell Biol.* **19**:1381–1389.

Higuchi, M., Proske, R. J., and Yeh, E. T., 1998, Inhibition of mitochondrial respiratory chain complex I by TNF results in cytochrome *c* release, membrane permeability transition, and apoptosis, *Oncogene* **17**:2515–2524.

Hillered, L., and Ernster, L., 1983, Respiratory activity of isolated rat brain mitochondria following *in vitro* exposure to oxygen radicals, *J. Cereb. Blood Flow Metab.* **3**:207–214.

Hirsch, T., Decaudin, D., Susin, S. A., Marchetti, P., Larochette, N., Resche-Rigon, M., and Kroemer, G., 1998, PK11195, a ligand of the mitochondrial benzodiazepine receptor, facilitates the induction of apoptosis and reverses Bcl-2-mediated cytoprotection, *Exp. Cell Res.* **241**:426–434.

Jürgensmeier, J. M., Xie, Z., Deverzux, Q., Ellerby, L., Bredesen, D., and Reed, J. C., 1998, Bax directly induces release of cytochrome *c* from isolated mitochondria, *Proc. Natl. Acad. Sci. USA* **95**:4997–5002.

Kantrow, S. P., and Piantadosi, C. A., 1997, Release of cytochrome *c* from liver mitochondria during permeability transition, *Biochem. Biophys. Res. Commun.* **232**:669–671.

Katayama, Y., and Welsh, F. A., 1989, Effect of dichloroacetate on regional energy metabolites and pyruvate dehydrogenase activity during ischemia and reperfusion in gerbil brain, *J. Neurochem.* **53**:1817–1822.

Kaya, S. S., Mahmood, A., Li, Y., Yavuz, E., Goksel, and M., and Chopp, M., 1999, Apoptosis and expression of p53 response proteins and cyclin D1 after cortical impact in rat brain, *Brain Res.* **818**:23–33.

Keller, J. N., Kindy, M. S., Holtsberg, F. W., St. Clair, D. K., Yen, H. C., Germeyer, A., Steiner, S. M., Bruce-Keller, A. J., Hutchins, J. B., and Mattson, M. P. 1998, Mitochondrial manganese superoxide dismutase prevents neural apoptosis and reduces ischemic brain injury: Suppression of peroxynitrite production, lipid peroxidation, and mitochondrial dysfunction, *J. Neurosci.* **18**:687–697.

Kinnally, K. W., Lohret, T. A., Campo, M. L., and Mannella, C. A., 1996, Perspectives on the mitochondrial multiple conductance channel, *J. Bioenerg. Biomembr.* **28**:115–123.

Kluck, R. M., Bossy-Wetzel, E., Green, D. R., and Newmeyer, D. D., 1997, The release of cytochrome *c* from mitochondria: A primary site for Bcl-2 regulation of apoptosis, *Science* **275**:1132–1136.

Kowaltowski, A. J., and Vercesi, A. E., 1999, Mitochondrial damage induced by conditions of oxidative stress, *Free Radical Biol. Med.* **26**:463–471.

Kowaltowski, A. J., Vercesi, A. E., and Fiskum, G., 2000, Bcl-2 prevents mitochondrial permeability transition and cytochrome c release via maintenance of reduced pyridine nucleotides, *Cell Death Different.* **7**: 903-910.

Krajewski, S., Krajewska, M., Ellerby, L. M., Welsch, K., Xie, Z., Deveraux, Q., Salvesen, G. S., Bredesen, D. E., Rosenthal, R. E., Fiskum, G., and Reed, J. C., 1999, Release of caspase-9 from mitochondria during neuronal apoptosis and cerebral ischemia, *Proc. Natl. Acad. Sci. USA.* **96**:5752–5757.

Kristal, B. S., and Dubinsky. J. M., 1997, Mitochondrial permeability transition in central nervous system: Induction by calcium cyclin-dependent and -independent pathways, *J. Neurochem.* **69**:524–538.

Kristian, T., and Siesjo, B. K., 1998, Calcium in ischemic cell death, *Stroke* **29**:705–718.

Kroemer, G., Petit, P., Zamzami, N., Vayssiere, J. L., and Mignotte, B., 1995, The biochemistry of programmed cell death, *FASEB J.* **9**:1277–1287.

Kruman, I. I. ,and Mattson, M. P., 1999, Pivotal role of mitochondrial calcium uptake in neural cell apoptosis and necrosis, *J. Neurochem.* **72**:529–540.

Leist, M., Volbracht, C., Fava, E., and Nicotera, P., 1998, 1-Methyl-4-phenylpyridinium induces autocrine excitotoxicity, protease activation, and neuronal apoptosis, *Mol. Pharmacol.* **54**:789–801.

Lemasters, J. J., Qian, T., Elmore, S. P., Trost, L. C., Nishimura, Y., Herman, B., Bradham, C. A., Brenner, D. A., and Nieminen, A.-L., 1998a, Confocal microscopy of the mitochondrial permeability transition in necrotic cell killing, apoptosis, and autophagy, *Biofactors* **8**:283–285.

Lemasters, J. J., Nieminen, A.-L., Qian, T., Trost, L. C., Elmore, S. P., Nishimura, Y., Crowe, R. A., Cascio, W. E., Bradham, C. A., Brenner, D. A., and Herman, B., 1998b, The mitochondrial permeability transition in cell death: A common mechanism in necrosis, apoptosis, and autophagy, *Biochim. Biophys. Acta* **1366**:177–196.

Li, P-A., Uchino, H., Elmér, E., and Siesjö, B. K., 1997a, Amelioration by cyclosporin A of brain damage following 5 or 10 min of ischemia in rats subjected to preischemic hyperglycemia, *Brain Res.* **753**:133–140.

Li, Y., Chopp, M., Powers, C., and Jiang, N., 1997b, Apoptosis and protein expression after focal cerebral ischemia in rat, *Brain Res.* **765**:301–312.

Li, H., Zhu, H., Xu, C. J., and Yuan, J., 1998, Cleavage of BID by caspase 8 mediates the mitochondrial damage in the Fas pathway of apoptosis, *Cell* **94**:491–501.

Luo, X., Budihardjo, I., Zou, H., Slaughter, C., and Wang, X., 1998, Bid, a Bcl2 interacting protein, mediates cytochrome *c* release from mitochondria in response to activation of cell surface death receptors, *Cell* **94**:481–490.

Mannella, C. A., 1998, Conformational changes in the mitochondrial channel protein, VDAC, and their functional implications, *J. Struct. Biol.* **121**:207–218.

Martinou, I., Desagher, S., Eskes, R., Antonsson, B., Andre, E., Fakan, S., and Martinou, J. C., 1999, The release of cytochrome *c* from mitochondria during apoptosis of NGF-deprived sympathetic neurons is a reversible event, *J. Cell Biol.* **144**:883–889.

Marzo, I., Brenner, C., Zamzami, N., Jürgensmeier, J. M., Susin, S. A., Vieira, H. L., Prevost, M. C., Xie, Z., Matsuyama, S., Reed, J. C, and Kroemer, G., 1998a, Bax and adenine nucleotide translocator cooperate in the mitochondrial control of apoptosis, *Science* **281**:2027–2031.

Marzo, I., Brenner, C., Zamzami, N., Susin, S. A., Beutner, G., Brdiczka, D., Remy, R., Xie, Z. H., Reed, J. C., and Kroemer, G., 1998b, The permeability transition pore complex: A target for apoptosis regulation by caspases and bcl-2-related proteins, *J. Exp. Med.* **187**:1261–71.

McEnery, M. W., Dawson, T. M., Verma, A., Gurley, D., Colombini, M., and Snyder, S. H., 1993, Mitochondrial voltage-dependent anion channel: Immunochemical and immunohistochemical characterization in rat brain, *J. Biol. Chem.* **268**:23289–23296.

McGinnis, K. M., Gnegy, M. E., and Wang, K. K., 1999, Endogenous bax translocation in SH-SY5Y human neuroblastoma cells and cerebellar granule neurons undergoing apoptosis, *J. Neurochem.* **72**:1899–1906.

Murakami, K., Kondo, T., Kawase, M., Li, Y., Sato, S., Chen, S. F., and Chan, P. H., 1998, Mitochondrial susceptibility to oxidative stress exacerbates cerebral infarction that follows permanent focal cerebral ischemia in mutant mice with manganese superoxide dismutase deficiency, *J. Neurosci.* **18**:205–213.

Murphy, A. N., Bredesen, D. E., Cortopassi, G., Wang, E., and Fiskum, G., 1996a, Bcl-2 potentiates the maximal calcium uptake capacity of neural cell mitochondria, *Proc. Natl. Acad. Sci. USA* **93**:9893–9898.

Murphy, A. N., Myers, K. M., and Fiskum, G., 1996b, Bcl-2 and *N*-acetylcysteine inhibition of respiratory impairment following exposure of neural cells to chemical hypoxia/aglycemia, in *Pharmacology of Cerebral Ischemia 19966* (J. Krieglstein, Ed.), Medpharm Scientific Publishers, Stuttgart, pp. 163–172.

Murphy, A. N., Fiskum, G., and Beal., M. F., 1999, Mitochondria in neurodegeneration: Bioenergetic function in cell life and death, *J. Cereb. Blood Flow Metab.* **19**:231–245.

Myers, K. M., Liu, Y., Bredesen, D. E., Fiskum, G., and Murphy, A. N., 1995, Bcl-2 protects neural cells from hypoxia-reoxygenation induced lipid oxidation, mitochondrial injury, and loss of viability, *J. Neurochem.* **65**:2432–2440.

Namura, S., Zhu, J., Fink, K., Endres, M., Srinivasan, A., Tomaselli, K. J., Yuan, J., and Moskowitz, M. A., 1998, Activation and cleavage of caspase-3 in apoptosis induced by experimental cerebral ischemia, *J. Neurosci.* **18**:3659–3668.

Narita, M., Shimizu, S., Ito, T., Chittenden, T., Lutz, R. J., Matsuda, H., and Tsujimoto, Y., 1998, Bax interacts with the permeability transition pore to induce permeability transition and cytochrome *c* release in isolated mitochondria, *Proc. Natl. Acad. Sci. USA* **95**:14681–14686.

Neame, S. J., Rubin, L. L., and Philpott, K. L., 1998, Blocking cytochrome *c* activity within intact neurons inhibits apoptosis, *J. Cell Biol.* **142**:1583–1593.

Neumar, R. W., DeGracia, D. J., Konkoly, L. L., Khoury, J. I., White, B. C., and Krause, G. S., 1998, Calpain mediates eukaryotic initiation factor 4G degradation during global brain ischemia, *J Cereb. Blood Flow Metab.* **18**:876–881.

Nicotera, P., Ankarcrona, M., Bonfoco, E., Orrenius, S., and Lipton, S. A., 1997, Neuronal necrosis and apoptosis: Two distinct events induced by exposure to glutamate or oxidative stress, *Adv. Neurol.* **72**:95–101.

Nieminen, A-L., Petrie, T. G., Lemasters, J. J., and Selman, W. R., 1996, Cyclosporin A delays mitochondrial depolarization induced by *N*-methyl-D-aspartate in cortical neurons: Evidence of the mitochondrial permeability transition, *Neuroscence* **75**:993–997.

Paschen, W., and Doutheil, J., 1999, Disturbances of the functioning of endoplasmic reticulum: A key mechanism underlying neuronal cell injury? *J. Cereb. Blood Flow Metab.* **19**:1–18.

Pastorino, J. G., Chen, S. T., Tafani, M., Snyder, J. W., and Farber, J. L., 1998, The overexpression of Bax produces cell death upon induction of the mitochondrial permeability transition, *J. Biol. Chem.* **273**:7770–7775.

Pèrez-Pinzòn, M. A., Mumford, P. L., Rosenthal, M., and Sick, T. J., 1997, Antioxidants limit mitochondrial hyperoxidation and enhance electrical recovery following anoxia in hippocampal slices, *Brain Res.* **754**:163–170.

Pèrez-Pinzòn, M. A., Xu, G. P., Born, J., Lorenzo, J., Busto, R., Rosenthal, M., and Sick, T. J., 1999, Cytochrome *c* is released from mitochondria into the cytosol after cerebral anoxia or ischemia, *J. Cereb. Blood Flow Metab.* **19**:39–43.

Petit, P. X., Goubern, M., Diolez, P., Susin, S. A., Zamzami, N., and Kroemer, G., 1998, Disruption of the outer mitochondrial membrane as a result of large amplitude swelling: The impact of irreversible permeability transition, *FEBS Lett.* **426**:111–116.

Prehn, J. H., Bindokas, V. P., Marcuccilli, C. J., Krajewski, S., Reed, J. C., Miller, R.J., 1994, Regulation of neuronal Bcl2 protein expression and calcium homeostasis by transforming growth factor type beta confers wide-ranging protection on rat hippocampal neurons, *Proc. Natl. Acad. Sci.*, **91**:12599–12603.

Priault, M., Chaudhuri, B., Clow, A., Camougrand, N., and Manon, S., 1999, Investigation of bax-induced release of cytochrome *c* from yeast mitochondria: Permeability of mitochondrial membranes, role of VDAC, and ATP requirement, *Eur. J. Biochem.* **260**:684–691.

Reynolds, I. J., and Hastings, T. G., 1995, Glutamate induces the production of reactive oxygen species in cultured forebrain neurons following NMDA receptor activation, *J. Neurosci.* **15**:3318–3327.

Rosenthal, M., Feng, Z-C., Raffin, C. N., Harrison, M., and Sick, T. J., 1995, Mitochondrial hyperoxidation signals residual intracellular dysfunction after global ischemia in rat neocortex, *J. Cereb. Blood Flow Metab.* **15**:655–665.

Rottenberg, H., and Marbach, M., 1990, Regulation of Ca^{2+} transport in brain mitochondria: II. The mechanism of the adenine nucleotides enhancement of Ca^{2+} uptake and retention, *Biochim. Biophys. Acta* **1016**:87–98.

Scarlett, J. L., and Murphy, M. P., 1997, Release of apoptogenic proteins from the mitochondrial intermembrane space during the mitochondrial permeability transition, *FEBS Lett.* **418**:282–286.

Schinder, A. F., Olson, E. C., Spitzer, N. C., and Montal, M., 1996, Mitochondrial dysfunction is a primary event in glutamate neurotoxicity, *J. Neurosci.* **16**:6125–6133.

Schulke, N., Sepuri, N. B. V., and Pain, D., 1997, *In vivo* zippering of inner and outer mitochondrial membranes by a stable translocation intermediate, *Proc. Natl. Acad. Sci. USA* **94**:7314–7319.

Schulz, J. B., Weller, M., and Moskowitz, M. A., 1999, Caspases as treatment targets in stroke and neurodegenerative diseases, *Ann. Neurol.* **45**:421–429.

Seeman, P., Cheng, D., and Iles, G. H., 1973, Structure of membrane holes in osmotic and saponin hemolysis, *J. Cell Biol.* **56**:91–102.

Shimizu, S., Eguchi, Y., Kamiike, W., Funahashi, Y., Mignon, A., Lacronique, V., Matsuda, H., and Tsujimoto, Y., 1998, Bcl-2 prevents apoptotic mitochondrial dysfunction by regulating proton flux, *Proc. Natl. Acad. Sci. USA* **95**:1455–1459.

Sims, N. R., 1991, Selective impairment of respiration in mitochondria isolated from brain subregions following transient forebrain ischemia in the rat, *J. Neurochem.* **56**:1836–1844.

Sokolove, P. M., and Kinnally. K. W., 1996, A mitochondrial signal peptide from *Neurospora crassa* increases the permeability of isolated rat liver mitochondria, *Arch. Biochem. Biophys.* **336**:69–76.

Stout, A. K., Raphael, H. M., Kanterewicz, B. I., Klann, E., and Reynolds, I. J., 1998, Glutamate-induced neuron death requires mitochondrial calcium uptake, *Nature Neurosci.* **1**:366–373.

Susin, S. A., Lorenzo, H. K., Zamzami, N., Marzo, I., Brenner, C., Larochette, N., Prevost, M. C., Alzari, P. M., and Kroemer, G., 1999a, Mitochondrial release of caspase-2 and -9 during the apoptotic process, *J. Exp. Med.* **189**:381–394.

Susin, S. A., Lorenzo, H. K., Zamzami, N., Marzo, I., Snow, B. E., Brothers, G. M., Mangion, J., Jacotot, E., Costantini, P., Loeffler, M., Larochette, N., Goodlett, D. R., Aebersold, R., Siderovski, D. P., Penninger, J. M., and Kroemer, G., 1999b, Molecular characterization of mitochondrial apoptosis-inducing factor, *Nature* **397**:441–446.

Tamatani, M., Ogawa, S., Niitsu, Y., and Tohyama, M., 1998, Involvement of Bcl-2 family and caspase-3-like protease in NO-mediated neuronal apoptosis, *J. Neurochem.* **71**:1588–1596.

Uchino, H., Elmer, E., Uchino, K., Li, P. A., He, Q. P., Smith, M. L., and Siesjo, B. K., 1998, Amelioration by cyclosporin A of brain damage in transient forebrain ischemia in the rat, *Brain Res.* **812**:216–226.

Ushmorov, A., Ratter, F., Lehmann, V., Droge, W., Schirrmacher, V., and Umansky, V., 1999, Nitric-oxide-induced apoptosis in human leukemic lines requires mitochondrial lipid degradation and cytochrome c release, *Blood* **93**:2342–2352.

Vander Heiden, M. G., Chandel, N. S., Williamson, E. K., Schumacker, P. T., and Thompson, C. R. 1997, Bcl-x_L regulates the membrane potential and volume homeostasis of mitochondria, *Cell* **91**:627–637.

Walter, D. H., Haendeler, J., Galle, J., Zeiher, A. M., Dimmeler, S., 1998, Cyclosporin A inhibits apoptosis of human endothelial cells by preventing release of cytochrome C from mitochondria, *Circulation*, **98**:1153–7.

Werling, L., and Fiskum, G., 1996, Calcium channels and neurotransmitter releaseability following cerebral ischemia, in *Neuroprotection* (J. A. Stamford and L. Strunin, Eds.), Bailliere Tindall, London, pp. 445–459.

White, R. J., and Reynolds, I. J., 1995, Mitochondria and Na^+Ca^{2+} exchange buffer glutamate-induced calcium loads in cultured cortical neurons, *J. Neurosci.* **15**:1318–1328.

Yang, J. C., and Cortopassi, G. A., 1998, Induction of the mitochondrial permeability transition causes release of the apoptogenic factor cytochrome c, *Free Radical Biol. Med.* **24**:624–631.

Yang, J., Liu, X., Bhalla, K., Kim, C. N., Ibrado, A. M., Cai, J., Peng, T-I., Jones, D. P., and Wang, X., 1997, Prevention of apoptosis by Bcl-2: Release of cytochrome c from mitochondria blocked, *Science* **275**:1129–1132.

Zaidan E., and Sims, N. R., 1997, Reduced activity of the pyruvate dehydrogenase complex but not cytochrome c oxidase is associated with neuronal loss in the striatum following short-term forebrain ischemia, *Brain Res.* **772**:23–28.

Zhuang, J., Dinsdale, D., and Cohen, G. M., 1998, Apoptosis, in human monocytic THP.1 cells, results in the release of cytochrome c from mitochondria prior to their ultracondensation, formation of outer membrane discontinuities, and reduction in inner membrane potential, *Cell Death Differ.* **5**:953–962.

Zoratti, M., and Szabò, I., 1995, The mitochondrial permeability transition, *Biochim. Biophys. Acta* **1241**:139–176.

Zou, H., Li, Y., Liu, X., and Wang, X., 1999, An APAF-1-cytochrome c multimeric complex is a functional apoptosome that activates procaspase-9, *J. Biol. Chem.* **274**:11549–11556.

Chapter 17

Varied Responses of Central Nervous System Mitochondria to Calcium

Nickolay Brustovetsky and Janet M. Dubinsky

Mitochondrial dysfunction and induction of the mitochondrial permeability transition (MPT) are candidate intermediate steps in both necrotic and apoptotic cell death pathways (Dubinsky and Levi, 1998; Hirsch *et al.,* 1998; Schinder *et al.,* 1996). In its classic definition (referred to here as the *high-conductance MPT*), the MPT is a nonselective, *multi*conductance pore in the inner mitochondrial membrane whose activation causes mitochondrial swelling and dysfunction (Nieminen *et al.,* 1996; Schinder *et al.,* 1996; Zoratti and Szabò, 1995). In liver and heart mitochondria, accumulation of excess matrix calcium combined with phosphate or with an oxidative event leads to opening of the high conductance MPT pore (Zoratti and Szabò, 1995). Induction of the MPT is modulated by mitochondrial membrane potential ($\Delta\psi$), matrix pH, Mg^{2+}, free fatty acids, redox status of mitochondrial protein thiols, and surface potential generated by the largely anionic phospholipids of the inner mitochondrial membrane (Zoratti and Szabò, 1995). Pharmacological inhibition of the MPT can be accomplished with the immunosuppressant cyclosporin A (CsA) and some of its analogs, adenine nucleotides, and the adenine nucleotide transporter inhibitor bongkrekic acid (Zoratti and Szabò, 1995). Mitochondrial swelling as measured by changes in absorbance has typically been studied in mitochondria after Ca^{2+} loading. High Ca^{2+} loads alone, or lower Ca^{2+} plus phosphate, uncoupler, or pro-oxidants, initiates transition, a process thought to propagate through the mitochondrial population (Bernardi, 1992; Broekemeier *et al.,* 1989). When ruthenium red (RR) is added after Ca^{2+} to prevent its loss through reverse operation of the uniporter, application of

Nickolay Brustovetsky   Department of Neuroscience, University of Minnesota, Minneapolis, Minnesota 55455   **Janet M. Dubinsky**   Department of Physiology, University of Minnesota, Minneapolis, Minnesota 55455.

Mitochondria in Pathogenesis, edited by Lemasters and Nieminen.
Kluwer Academic/Plenum Publishers, New York, 2001.

uncoupler triggers stochastic activation of the MPT (Gunter *et al.*, 1994; Petronilli *et al.*, 1993; Zoratti and Szabò, 1995). Alternatively, in totally de-energized mitochondria, Ca^{2+} introduction into the matrix via ionophore can activate the high-conductance MPT (Halestrap and Davidson, 1990).

Numerous studies suggest that the MPT may also operate in a low-conductance mode permeable to hydrogen ions (Al-Nasser and Crompton, 1986; Broekemeier *et al.*, 1985; Novgorodov and Gudz, 1996), small cations (Broekemeier and Pfeiffer, 1995), or GSH (Savage and Reed, 1994). In liver mitochondria, increased proton permeability precedes the Ca^{2+}-induced permeability for sucrose (Al-Nasser and Crompton, 1986; Broekemeier *et al.*, 1985; Novgorodov and Gudz, 1996). Moreover, observation of a multistage release of mitochondrial solutes has established that K^+ efflux precedes Mg^{2+} release, and that both precede mannitol–sucrose influx and swelling (Broekemeier and Pfeiffer, 1995). Using a tetraphenylphosphonium-sensitive (TPP^+) electrode and ^{14}C-sucrose matrix entrapment, a CsA-sensitive, proton-specific permeability increase can be completely separated from a Ca^{2+}- and P_i-induced nonspecific permeability (Crompton *et al.*, 1988). Morphological assessments revealed that glutathione release, associated with a collapse of $\Delta\psi$ was attributable to uniform induction of a low-conductance MPT throughout the entire mitochondrial population (Savage and Reed, 1994). Mitochondrial Ca^{2+}-induced Ca^{2+} release consequent to depolarization has been attributed to operation of the low-conductance mode of the MPT pore (Ichas *et al.*, 1997). A low-conductance MPT pathway, insensitive to ADP, CsA, Mg^{2+}, or EGTA alone or in combinations, was opened by calcium and coexisted with the large-conductance pore (Broekemeier *et al.*, 1998). Single-channel recordings from mitoplasts revealed multiple conductance channels whose properties resemble the MPT, suggesting that this pore may operate at several sequentially activated conductance levels. Operation of the MPT in liver mitochondria in this proton-selective, low-conductance mode has, until recently, been ignored for what it might contribute to cellular processes (Al-Nasser and Crompton, 1986; Broekemeier *et al.*, 1998; Savage and Reed, 1994; Hirsch *et al.*, 1998).

The MPT has been implicated in neurodegenerative processes in primary neuronal cultures exposed to excitotoxins (Ankarcrona *et al.*, 1996; Dubinsky and Levi, 1998; Kristal and Dubinsky, 1997; Schinder *et al.*, 1996; White and Reynolds, 1996), reconstituted cell-free apoptotic systems (Ellerby *et al.*, 1997), *in vivo* models of acute neuronal insults (Folbergrova *et al.*, 1997; Friberg *et al.*, 1998a; Li *et al.*, 1997), and exposure to the neurotoxin MPTP (Cassarino *et al.*, 1998). In apoptotic pathways after activation of the MPT, release of mitochondrial proteins, cytochrome *c*, and apoptosis-inducing factor are thought to trigger caspase activation and nuclear degradation (Liu *et al.*, 1996; Zamzami *et al.*, 1996). Most recently, the MPT has been implicated in 3NP-induced apoptosis of cell lines with genetically altered presenilin by the protective ability of CsA, an MPT inhibitor (Keller *et al.*, 1998). Most of these studies considered only induction of the classical MPT.

Controversy still accompanies the exact nature of mitochondrial involvement in apoptosis. Initially, high-conductance MPT induction was hypothesized to lead to mitochondrial swelling, rupture of the outer mitochondrial membrane, and cytochrome *c* release from the intermembrane space (Zamzami *et al.*, 1996). Release of cytochrome *c* may occur without mitochondrial swelling, and it may be independent of MPT operation (Eskes *et al.*, 1998; Murphy *et al.*, 1998). Reversible opening of a low-conductance MPT pathway in the inner mitochondrial membrane may also trigger cytochrome *c* release

(Green and Reed, 1998; Zamzami et al., 1998). Participation of a low-conductance MPT in apoptosis has been postulated to explain Bcl-2 protection against calcium-induced mitochondrial depolarization and dysfunction in the absence of mitochondrial swelling (Green and Reed, 1998; Zamzami et al., 1998).

To date only the high-conductance MPT associated with mitochondrial swelling has been studied in brain mitochondria (Andreyev et al., 1998; Dubinsky and Levi, 1998; Kristal and Dubinsky, 1997). Mitochondria from cortex, hippocampus, and cerebellum may be differentially sensitive to Ca^{2+}-induced swelling (Friberg et al., 1998b). Though brain MPT shares characteristics with liver MPT, initial reports indicate that brain MPT differs in its activation by pro-oxidants and its sensitivity to inhibitors (Kristal and Dubinsky, 1997). In most of the high-conductance MPT studies, energetic conditions were optimized for high oxidative capacity or else for totally de-energized conditions. Observation of a low-conductance pathway is unlikely in the former condition, because respiratory compensation would obscure any hydrogen-permeable pathway. A low-conductance pathway would not be expected to cause swelling and thus would not be seen in de-energized mitochondria. Therefore, conditions of limited respiratory capacity were used to investigate possible calcium-activated, low-conductance pathways in brain mitochondria.

We have recently described conditions suitable for distinguishing the high-conductance MPT in brain mitochondria from a novel pathway that might be a low-conductance manifestation of the MPT (Brustovetsky and Dubinsky, 1998a, 1998b). Respiring, isolated brain mitochondria responded to Ca^{2+} challenges in one of two ways, depending on abundance of available energy substrates (Fig. 1). In an environment with plentiful energy substrates, Ca^{2+} produced a transient depolarization, recovery of $\Delta\psi$, and slowly progressive depolarization, when measured by observing TPP^+ distribution in a suspension of CNS mitochondria (Fig. 1B). In the same paradigm, an extramitochondrial Ca^{2+}-sensitive electrode registered rapid mitochondrial accumulation of the added Ca^{2+}, followed by a slow partial release to the cytoplasm (Fig. 1C). In an absorbance assay to detect mitochondrial swelling, a comparable Ca^{2+} challenge produced progressive swelling (Fig. 1D) that paralleled the slow depolarization. This combination of responses is characteristic of classical high-conductance MPT induction. Comparable data were obtained when mitochondria were respiring on succinate and glutamate, pyruvate and malate, or succinate and rotenone. When the substrate was restricted to 3 mM succinate, the same Ca^{2+} challenge produced sustained mitochondrial depolarization (Fig. 1B), very slow calcium accumulation (Fig. 1C), and no appreciable swelling (Fig. 1D). The limited substrate condition (3 mM succinate) did not alter normal respiration (with or without ADP, compared with 3 mM succinate and 3 mM glutamate), but was sufficient to restrict maximal respiration when maximally activated by FCCP (Fig. 1A). Decreasing the concentrations of pyruvate and malate or succinate and glutamate produced responses comparable to those in 3 mM succinate. This novel Ca^{2+}-activated sustained depolarization may represent activation of a low-conductance proton permeability in the inner mitochondrial membrane. This conductance was not found in the presence of more-plentiful substrates because the H^+-pumping capacity of the respiratory chain compensated for the H^+ leakage that occurred when this pathway was activated, accounting for the immediate repolarization upon Ca^{2+} application.

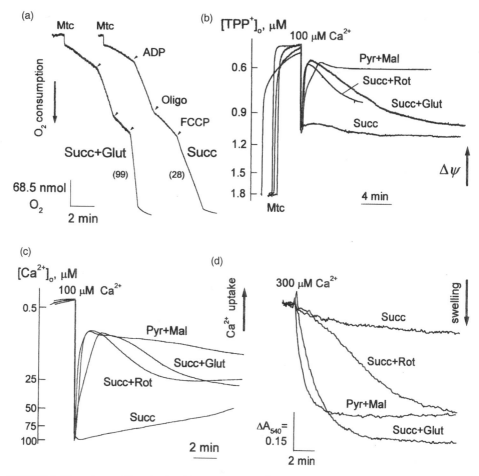

FIGURE 1. Dependence of the mitochondrial responses to Ca^{2+} upon various oxidative substrates. (A) Oxygen consumption (nmoles O_2 min^{-1} mg $protein^{-1}$, slope of the external O_2 electrode trace) of isolated brain mitochondria was comparable in mitochondria respiring on 3 mM succinate or 3 mM succinate plus 3 mM glutamate until maximally activated by 1 μM FCCP. Maximal rates appear in parentheses. Additions were 100 μM ADP to stimulate respiration (state 3) and 1 μM oligomycin to produce nonphosphorylating respiration to determine the basal proton leakage rate (state 4). Mitochondria were incubated in 75 M sucrose, 225 mM mannitol, 10 mM KHEPES, 3 mM KH_2PO_4, 0.5 mM $MgCl_2$ plus the indicated substrates. (B) TPP^+ electrode measurements of $\Delta\psi$ in response to Ca^{2+} in isolated CNS mitochondria respiring in the indicated combinations of substrates and inhibitors: 3 mM succinate, 3 mM glutamate, 3 mM pyruvate, 3 mM malate, 1 μM rotenone. External $[TPP^+]$ of 0.6, 0.9, and 1.2 μM corresponded to $\Delta\psi$ absolute values of 160, 134, and 78 mV, respectively. Arrow indicates direction of more hyperpolarized $\Delta\psi$. Full depolarization produced a $[TPP^+]_o$ of 1.3 μM after 1 μM FCCP addition. (C) External Ca^{2+} electrode measurements of Ca^{2+} uptake and retention were initiated by a rapid 100 μM Ca^{2+} pulse. Responses varied depending upon the oxidative substrates initially present. (D) Absorbance measurements of purified CNS mitochondria revealed that kinetics and swelling amplitude depended upon oxidative substrate as well. Mtc indicates addition of mitochondria.

The pharmacological sensitivity of calcium-activated, sustained depolarization was in many ways comparable to that of the high-conductance MPT, suggesting that they may be related. In brain mitochondria, swelling was prevented by CsA when present before application of 50 μM, but not 100 μM Ca^{2+}. Addition of CsA after Ca^{2+}-induced swelling had begun did not arrest the process as it does in liver (Broekemeier et al., 1989). Similarly, CsA prevented the sustained depolarization in response to 50 μM, but not 100 μM Ca^{2+}. Application of CsA after activation of the low-conductance MPT by 50 μM, but not 100 μM, Ca^{2+} almost completely repolarized mitochondria. Oligomycin and ADP reliably repolarized CNS mitochondria exhibiting sustained depolarization, in all conditions tested. Because this sustained depolarization was sensitive to CsA and ADP, inhibitors of the classical MPT, the low-conductance pathway may be a low-conductance manifestation of the MPT. Pretreating brain mitochondria with additional CsA inhibition modulators of the MPT in liver mitochondria, Mg^{2+}, and fatty acids (Andreyev et al., 1994; Broekemeier and Pfeiffer, 1995), altered CsA's ability to repolarize mitochondria, dependent upon degree of depolarization (Brustovetsky and Dubinsky, 1998a). This is quite distinct from liver mitochondria, where closure of the low-conductance MPT was not sensitive to CsA, ADP, Mg, or BSA, alone or in combination (Broekemeier et al., 1998). Responses to other pharmacological agents, however, suggest that this low-conductance pathway may be a distinct entity. Antagonists of the calcium uniporter RR, its active agent Ru360, and lanthanum repolarized mitochondria undergoing sustained depolarization but did not interrupt the process of mitochondrial swelling or close the high-conductance MPT (Brustovetsky and Dubinsky, 1998b). The RR-sensitive depolarization could not be explained by activation of Ca^{2+} cycling because Ca^{2+} uptake and release were strongly suppressed in depolarized mitochondria. In normally polarized CNS mitochondria, A23187 significantly activated Ca^{2+} efflux, providing optimal conditions for Ca^{2+} cycling yet effecting only a minor depolarization under these conditions; thus Ca^{2+}-activated sustained depolarization inhibited Ca^{2+} cycling and was not caused by it.

Thus, whereas CsA inhibited both the novel low-conductance pathway and the classical high-conductance MPT in isolated brain mitochondria, high Ca^{2+} concentrations could overcome CsA inhibition. Even in liver mitochondria the effects of CsA are time-limited and can be overcome by increasing Ca^{2+} concentrations (Broekemeier and Pfeiffer, 1995; Crompton and Andreeva, 1994). This was observed in cultured astrocytes and neurons where the mitochondrial dyes JC-1 and Rhodamine 123 were used to visualize mitochondrial shape changes corresponding to mitochondrial swelling (Dubinsky and Levi, 1998; Kristal and Dubinsky, 1997). Introduction of Ca^{2+} into the cytoplasm through digitonin-permeabilized plasma membranes caused astrocyte mitochondria to change from vermiform, rodlike shapes into larger, rounded, hollow structures. Pretreatment with CsA alone was not a very effective protectant (Kristal and Dubinsky, 1997). Only when combined with ADP was CsA pretreatment able to prevent Ca^{2+}-induced changes in mitochondrial morphology (Buchanan et al., 1976). Similarly, when astrocytes were treated with 4-Br-A23187 to introduce Ca^{2+} directly into the cytoplasm and mitochondria, pretreatment with CsA alone did not prevent mitochondrial shape alterations. Combining CsA with the phospholipase A2 inhibitor butacaine prevented Ca^{2+}-induced loss of normal mitochondrial structure (Kristal and Dubinsky, 1997). In cultured hippocampal neurons exposed to excitotoxic doses of glutamate, CsA pretreatment prevented similar shape-changes to a variable extent (Dubinsky and Levi, 1998). In all of these experiments, the CsA-sensitive, calcium-induced changes in mitochondrial morphology suggested that a high-conductance MPT may have occurred. *In situ* appearance of the low-conductance pathway is harder to decisively detect because it is difficult to distinguish from any type of mitochondrial depolarization.

What might be the role for this low-conductance mitochondrial permeability? Mitochondrial participation in cytosolic Ca^{2+} buffering may require a mechanism for not only rapid uptake but rapid *release* of Ca^{2+}, analogous to CICR in the ER (Ichas *et al.*, 1997). A low-conductance MPT manifestation has been proposed to provide such a mechanism (Ichas *et al.*, 1997). Although Ca^{2+} may permeate the low-conductance pathway, we do not see any appreciable, rapid Ca^{2+} accumulation upon its activation in brain mitochondria. Extramitochondrial Ca^{2+}, but not Sr^{2+}, activated the sustained depolarization in restricted substrate conditions, which might be a neuroprotective mechanism. By restricting mitochondrial Ca^{2+} accumulation, high-conductance MPT induction and mitochondrial swelling may be prevented. Because mitochondrial production of reactive oxygen species (ROS) is higher in polarized mitochondria when the electron transport chain is more reduced (Skulachev, 1996), depolarization would be expected to decrease ROS production. Both of these consequences could be beneficial to a cell struggling to regain its homeostasis.

On the other hand, activation of low-conductance permeability could actually initiate a slower process of degeneration consequent to subtle mitochondrial dysfunction. The increased rate of respiration accompanying sustained mitochondrial depolarization would not necessarily be expected to result in greater ATP production, because the proton permeability of the low-conductance MPT would act to uncouple these processes. If respiration increases only slightly, ROS production may not be substantially inhibited. The Ca^{2+}-induced depolarization may also inhibit transhydrogenase, which limits NADPH-dependent replenishment of mitochondrial GSH (Nicholls and Ferguson, 1992), effectively lowering antioxidant defenses and altering sulfhydryl redox status to favor high-conductance MPT induction. Thus the combination of mitochondrial depolarization and ROS-activated oxidation of critical thiols may render mitochondria more susceptible to high-conductance MPT activation at lower levels of accumulated Ca^{2+}. Additionally, if mitochondrial Ca^{2+} uptake became negligible, cytosolic Ca^{2+} would remain elevated longer. In this scenario, overstimulation of other Ca^{2+}-activated processes could trigger neurodegeneration.

REFERENCES

Andreyev, A. Y., Mikhayloya, L. M., Starkov, A. A., and Kushnareva, Y., 1994, Ca[2+]-loading modulates potencies of cyclosporin A, Mg^{2+} and ADP to recouple permeabilized rat liver mitochondria, Biochem. Mol. Biol. Int. **34**:367–373.

Al-Nasser, I., and Crompton, M., 1986, The reversible Ca^{2+}-induced permeabilization of rat liver mitochondria, *Biochem J.* **239** (1): 19–29.

Andreyev, A., Y., Fahy, B., and Fiskum, G., 1998; Cytochrome *c* release from brain mitochondria is independent of the mitochondrial permeability transition, *FEBS Lett.* **439**:373–376.

Ankarcrona, M., Dypbukt, J. M., Orrenius, S., and Nicotera, P., 1996, Calcineurin and mitochondrial function in glutamate-induced neuronal cell death, *FEBS Lett.* **394**:321–324.

Bernardi., P., 1992, Modulation of the mitochondrial cyclosporin A-sensitive permeability transition pore by the proton electrochemical gradient: Evidence that the pore can be opened by membrane depolarization, *J. Biol. Chem.* **267** (13):8834–8839.

Broekemeier, K. M., and Pfeiffer, D. R., 1995, Inhibition of the mitochondrial permeability transition by cyclosporin A during long time-frame experiments: Relationship between pore opening and the activity of mitochondrial phospholipases, *Biochemistry* **34**:16440–16449.

Broekemeier, K. M., Schmid, P. C., Schmid, H. H. O., and Pfeiffer, D. R., 1985, Effect of phospholipase A_2 inhibitors on ruthenium red-induced Ca^{2+} release from mitochondria, *J. Biol. Chem.* **260**:105–113.

Broekemeier, K. M., Dempsey, M. E., and Pfeiffer, D. R., 1989, Cyclosporin A is a potent inhibitor of the inner membrane permeability transition in liver mitochondria, *J. Biol. Chem.* **264**:7826–7830.

Broekemeier, K. M., Klocek, C. K., and Pfeiffer, D. R., 1998, Proton selective substate of the mitochondrial permeability transition pore: Regulation by the redox state of the electron transport chain, *Biochemistry* **37**: 13059–13065.

Brustovetsky, N. and Dubinsky, J. M., 1998a, Cyclosporin A inhibition of the permeability transition in brain depends upon mitochondrial potential, *Soc. Neurosci. Abstr.* **24**:1453.

Brustovetsky, N., and Dubinsky, J. M., 1998b, Does the Ca^{2+} uniporter form the mitochondrial permeability transition pore? *Biophys. J.* **74**:A384.

Buchanan, B. B., Eiermann, W., Riccio, P., Aquila, H., and Klingenberg, M., 1976, Antibody evidence for different conformational states of ADP, ATP translocator protein isolated from mitochondria, *Proc. Natl. Acad. Sci. USA* **73**(7):2280–2284.

Cassarino, D. S., Fall, C. P., Smith, T. S., and Bennett, J. P., 1998, Pramipexole reduces reactive oxygen species production *in vivo* and *in vitro* and inhibits the mitochondrial permeability transition produced by the parkinsonian neurotoxin methylpyridinium ion, *J. Neurochem.* **71**:295–301.

Crompton, M., and Andreeva, L., 1994, On the interactions of Ca^{2+} and cyclosporin A with a mitochondrial inner membrane pore: A study using cobaltammine complex inhibitors of the Ca^{2+} uniporter, *Biochem. J.* **302**:181–185.

Crompton, M., Ellinger, H., and Costi, A., 1988, Inhibition by cyclosporin A of a Ca^{2+}-dependent pore in heart mitochondria activated by inorganic phosphate and oxidative stress, *Biochem. J.* **255**:357–360.

Dubinsky, J. M., and Levi, Y., 1998, Calcium-induced activation of the mitochondrial permeability transition in hippocampal neurons, *J. Neurosc. Res.* **53**:728–741.

Ellerby, H. M., Martin, S. J., Ellerby, L. M., Naiem, S. S., Rabizadeh, S., Salvesen, G. S., Casiano, C. A., Cashman, N. R., Green D. R., and Bredesen, D. E., 1997, Establishment of a cell-free system of neuronal apoptosis: Comparison of premitochondrial, mitochondrial, and postmitochondrial phases, *J. Neurosci.* **17**: 6165–6178.

Eskes, R., Antonsson, B., Osensand, A., Montessuit, S., Richter, C., Sadoul, R., Mazzei, G., Nichols, A., and Martinou, J. C., 1998, BAX-induced cytochrome *c* release from mitochondria is independent of the permeability transition pore but highly dependent on Mg^{2+} ions, *J. Cell Biol.* **143**:217–224.

Folbergrova, J., Li P. A., Uchino, H., Smith, M. L., and Siesjo. B. K., 1997, Changes in the bioenergetic state of rat hippocampus during 2.5 min of ischemia, and prevention of cell damage by cyclosporin A in hyperglycemic subjects, *Exp. Brain Res.* **114**:44–50.

Friberg, H., Ferrand-Drake, M., Bengtsson, F., Halestrap, A. P., and Wieloch, T., 1998a, Cyclosporin A, but not FK 506, protects mitochondria and neurons against hypoglycemic damage and implicates the mitochondrial permeability transition in cell death, *J. Neurosci.* **18**:5151–5159.

Friberg, H., Ferrand-Drake, M., Boris-Moller, F., Halestrap, A. P., and Wieloch., T. 1998b, Differences in the activation of the mitochondrial permeability transition between brain regions: Correlation to selective vulnerability, *Soc. Neurosci. Abstr.* **24**:1229.

Green, D. R., and Reed, J. C., 1998, Mitochondria and apoptosis, *Science* **281**:1309–1312.

Gunter, T. E., Gunter, K. K., Sheu, S. S., and Gavin, C. E., 1994, Mitochondrial calcium transport: Physiological and pathological relevance, *Am. J. Physiol.* **267**:C313–C339.

Halestrap, A. P., and Davidson, A. M., 1990, Inhibition of Ca^{2+}-induced large-amplitude swelling of liver and heart mitochondria by cyclosporin is probably caused by the inhibitor binding to mitochondrial-matrix peptidyl-prolyl *cis-trans* isomerase and preventing it interacting with the adenine nucleotide translocase, *Biochem. J.* **268**:153–160.

Hirsch, T., Susin, S. A., Marzo, I., Marchetti, P., Zamzami, N., and Kroemer., G., 1998, Mitochondrial permeability transition in apoptosis and necrosis, *Cell Biol. Toxico.* **14**:141–145.

Ichas, F., Jouaville, L. S., and Mazat, J. P., 1997, Mitochondria are excitable organelles capable of generating and conveying electrical and calcium signals, *Cell* **89**:1145–1153.

Keller, J. N., Guo, Q., Holtsberg, F. W., Brucekeller, A. J., and Mattson, M. P., 1998, Increased sensitivity to mitochondrial toxin-induced apoptosis in neural cells expressing mutant presenilin-1 is linked to perturbed calcium homeostasis and enhanced oxyradical production, *J. Neurosci.* **18**:4439–4450.

Kristal, B. S., and Dubinsky, J. M., 1997, Mitochondrial permeability transition in the central nervous system: Induction by calcium-cycling dependent and independent pathways, *J. Neurochem.* **69**:524–538.

Li, P. A., Uchino, H., Elmer, E., and Siesjo, B. K., 1997, Amelioration by cyclosporin A of brain damage following 5 or 10 min of ischemia in rats subjected to preischemic hyperglycemia, *Brain Res.* **753**:133–140.

Liu, X., Kim., O. N., Yang, J., Jemmerson, R., and Wang, X., 1996, Induction of apoptotic program in cell-free extracts: Requirement for dATP and cytochrome *c Cell* **86**:147–157.

Murphy, A. N., Wang, G., and Richards, C. M., 1998, Further characterization of mitochondrial cytochrome *c* release and inhibition by BCL-2, *Soc. Neurosci. Abstr.* **24**:1945.

Nicholls, D. G., and Ferguson, S. J., 1992, *Bioenergetics 2* Academic, London,

Nieminen, A. L., Petrie, T. G., Lemasters, J. J., and Selman, W. R., 1996, Cyclosporin A delays mitochondrial depolarization induced by *N*-methyl-D-aspartate in cortical neurons: Evidence of the mitochondrial permeability transition, *Neuroscience* **75**:993–997.

Novgorodov, S. A., and Gudz, T. I., 1996, Permeability transition pore of the inner mitochondrial membrane can operate in two open states with different selectivities, *J. Bioenerg. Biomembr.* **28**:139–146.

Petronilli, V., Cola, C., and Bernardi, P., 1993, Modulation of the mitochondrial cyclosporin A-sensitive permeability transition pore: II. The minimal requirements for pore induction underscore a key role for transmembrane electrical potential, matrix pH, and matrix Ca^{2+}, *J. Biol. Chem.* **268**:1011–1016.

Savage, M. K., and Reed, D. J., 1994, Release of mitochondrial glutathione and calcium by a cyclosporin A-sensitive mechanism occurs without large amplitude swelling, *Arch. Biochem. Biophys.* **315**:142–152.

Schinder, A. F., Olson, E. C., Spitzer, N. C., and Montal, M., 1996, Mitochondrial dysfunction is a primary event in glutamate neurotoxicity, *J. Neurosci.* **16**:6125–6133.

Skulachev, V. P., 1996, Role of uncoupled and non-coupled oxidations in maintenance of safely low levels of oxygen and its one-electron reductants, *Q. Rev. Biophys.* **29**:169–202.

White, R. J., and Reynolds, I. J., 1996, Mitochondrial depolarization in glutamate-stimulated neurons: An early signal specific to excitotoxin exposure, *J. Neurosci.* **16**:5688–5697.

Zamzami, N. Susin, S. A., Marchetti, P., Hirsch, T., Gomez-Monterrey, I., Castedo, M., and Kroemer, G., 1996, Mitochondrial control of nuclear apoptosis, *J. Exp. Med.* **183**:1533–1544.

Zamzami, N., Brenner, C., Marzo, I., Susin, S. A., and Kroemer, G., 1998, Subcellular and submitochondrial mode of action of Bcl-2-like oncoproteins, *Oncogene* **16**:2265–2282.

Zoratti, M., and Szabò, I., 1995, The mitochondrial permeability transition, *Biochim. Biophys. Acta* **1241**:139–176.

Chapter 18

Mitochondrial Dysfunction in Oxidative Stress, Excitotoxicity, and Apoptosis

Anna-Liisa Nieminen, Aaron M. Byrne, and Kaisa M. Heiskanen

1. INTRODUCTION

Many pathological conditions, such as hypoxia, ischemia, reperfusion, chemical toxicity, and withdrawal of growth factors, cause cells to lose viability. Cell death can occur in two ways. Necrosis, the first was, is the consequence of acute disruption of cellular metabolism, leading to ATP depletion, ion dysregulation, mitochondrial and cellular swelling, activation of degradative enzymes, plasma membrane failure, and cell lysis (Trump *et al.*, 1965). In the second, apoptosis, metabolism is not severely impaired and cellular shrinkage rather than swelling occurs. It can be difficult to distinguish necrosis from the late stages of apoptosis, however, because plasma membrane disruption may occur in both. Disruption of the plasma membrane late in apoptosis is called *secondary necrosis.*

Unlike apoptosis, necrosis seldom serves the organism's needs. In highly aerobic tissues, necrosis is the inevitable result of prolonged tissue ischemia. In tissues such as heart and brain that do not regenerate, necrosis leads to permanent dysfunction of the organ. Apoptosis involves execution of a preprogrammed sequence of cellular events, whereas necrotic cell killing is unprogrammed or accidental cell death.

Anna-Liisa Nieminen, Aaron M. Byrne, and Kaisa M. Heiskanen   Department of Anatomy, School of Medicine, Case Western Reserve University, Cleveland, Ohio 44106-4930.

Mitochondria in Pathogenesis, edited by Lemasters and Nieminen.
Kluwer Academic/Plenum Publishers, New York, 2001.

2. OXIDATIVE STRESS

Oxidative stress is frequently associated with a variety of conditions, such as hypoxia/reperfusion injury and toxicity from numerous chemicals. Intracellular targets for oxidative stress can be cytosol or organelles, such as mitochondria, endoplasmic reticulum, nucleus, and lysosomes. Reduced glutathione is a major soluble cytosolic antioxidant in the cell. In hepatocytes, glutathione concentration is high, about 10 mM, serving as an effective antioxidant. Therefore, the mechanism of hepatocyte killing with several oxidant chemicals involve depletion of reduced glutathione. In the endoplasmic reticulum, reduced thiols within the active site of the Ca^{2+}–ATPase are oxidized by oxidant chemicals, thus preventing Ca^{2+} sequestration from cytosol to lumen of the endoplasmic reticulum and increasing the cytosolic Ca^{2+}. Increased cytosolic Ca^{2+} can be taken up by mitochondria and thus mitochondria can buffer it, but this can also increase it, resulting in mitochondrial dysfunction by variety of mechanisms discussed later in this chapter. Oxidative stress can also directly affect the nucleus; oxidation of nucleic acids results in genomic damage.

Several lines of evidence from a number of laboratories including ours indicate that mitochondria are the major targets of oxidative injury (Bellomo *et al.*, 1991; Byrne *et al.*, 1999; Dawson *et al.*, 1993; DiMonte *et al.*, 1988; Imberti *et al.*, 1993; Nieminen *et al.*, 1990a, 1990b, 1994, 1994, 1997; Thomas and Reed, 1988; Thor *et al.*, 1982; Wu *et al.*, 1990; Zahrebelski *et al.*, 1995). Various oxidant chemicals (such as *tert*-butylhydroperoxide, menadione, cystamine, and calcium ionophore) cause cellular ATP depletion, leading to lethal cell injury in hepatocytes isolated from fasting rats and incubated in a glycogen-poor medium. Fructose, however, an efficient glycolytic substrate in liver, provides near-complete protection against these oxidant chemicals in hepatocytes from fasting rats (Nieminen *et al.*, 1990b). Fructose also protects hepatocytes against the cytotoxicity of the designer drug MPTP and its metabolite MPP^+, which produce Parkinson syndrome *in vivo* (DiMonte *et al.*, 1988). Because fructose provides an alternate source of cellular ATP via glycolysis, the findings strongly suggest that these chemicals inhibit mitochondrial ATP synthesis, thus contributing to lethal cell injury. One oxidant chemical used extensively in our studies is *tert*-butylhydroperoxide (*t*-BuOOH), a short-chain analog of the lipid hydroperoxides formed from peroxidation reactions during reductive stress, oxidative stress, ischemia/reperfusion, and normal metabolism. It is detoxified by glutathione peroxidase to yield *t*-butanol and oxidized glutathione (Cotgreave 1988; Lotscher *et al.*, 1980; Sies *et al.*, 1972). In the presence of excess *t*-BuOOH, pyridine nucleotides become secondarily oxidized via action of glutathione reductase and mitochondrial NADPH–NAD^+ transhydrogenase. Oxidation of glutathione and pyridine nucleotides constitutes a condition of oxidative stress that promotes accumulation of lipid peroxides and possibly other toxic metabolites.

Mitochondria are a major target of *t*-BuOOH cytotoxicity (Imberti *et al.*, 1993; Nieminen *et al.*, 1990b). The mechanism underlying mitochondrial injury in hepatocytes exposed to *t*-BuOOH depends on the dose. At the lowest concentrations causing acute cytotoxicity (25–50 μM), fructose prevents cell killing almost completely. Rescue by fructose implies that *t*-BuOOH at these concentrations inhibits mitochondrial ATP formation. At higher *t*-BuOOH concentrations (100–300 μM) fructose alone fails to protect, but fructose plus oligomycin improves viability, implying that a higher *t*-

BuOOH concentration causes mitochondrial uncoupling (Imberti *et al.*, 1993). In addition, fructose in combination with cyclosporin A or trifluoperazine, two inhibitors of the mitochondrial permeability transition (MPT), also delays mitochondrial depolarization and cell death, providing indirect evidence that *t*-BuOOH-induced depolarization is due to transition onset in hepatocytes. At still higher concentrations of *t*-BuOOH (1 mM), glycolysis becomes strongly inhibited. Thus at very high *t*-BuOOH concentrations all significant cellular sources of ATP production become blocked, and no combination of glycolytic substrate and ATPase inhibitor can prevent cell killing.

3. MITOCHONDRIAL PERMEABILITY TRANSITION (MPT)

As early as the 1970s, Hunter and Haworth described a reversible phenomenon by which mitochondria become freely permeable to low-MW solutes (Haworth and Hunter, 1979; Hunter and Haworth, 1979a, 1979b; Hunter *et al.*, 1976). Ca^{2+}, P_i, and numerous oxidant chemicals induce this increased permeability, whereas Mg^{2+}, ADP, and low pH MPT prevent onset. The transition causes mitochondrial depolarization, uncoupling, release of intramitochondrial solutes, and large-amplitude mitochondrial swelling. (An extensive review on the physiological regulation of the permeability transition pore by Zoratti and Tombola can be found elsewhere in this book.)

Renewal of interest in the transition was stimulated in the late 1980s by the finding that immunosuppressive cyclosporin A specifically blocks transition (Crompton *et al.*, 1988; Fournier *et al.*, 1987). Cyclosporin A inhibition implies that a specific protein mediates the MPT. Subsequently, using patch-clamping techniques, a cyclosporin A-sensitive pore was identified that conducts molecular mass solutes of up to 1,500 Da (Kinnaly *et al.*, 1989; Petronilli *et al.*, 1989; Szabò and Zoratti, 1991). Pore conductance in the mitochondrial inner membrane is very high, about 1 nanosiemen, exceeding conductance of plasma membrane Ca^{2+} and Na^+ channels, so permeability transition pores are often termed *mitochondrial megachannels*.

The molecular composition of the pore remains obscure. The pore complex seems to be comprised, at least in part, of the adenine nucleotide translocator protein (Beutner *et al.*, 1996; Brustovetsky and Klingenberg, 1996; Halestrap and Davidson, 1990; Marzo *et al.*, 1998; Ruck *et al.*, 1998), and pore activity has been reconstituted by inserting purified adenine nucleotide translocator into lipid bilayer membranes and liposomes (Marzo *et al.*, 1998; Ruck *et al.*, 1998). Other mitochondrial proteins also associate with the adenine nucleotide translocator to form the pore complex, including cyclophilin D (a cyclosporin A-binding protein) in the matrix, creatine kinase in the intermembrane space, and porin (voltage-dependent anion channel) and hexokinase in the outer membrane (Beutner *et al.*, 1996; Brustovetsky and Klingenberg, 1996; Halestrap and Davidson, 1990; Marzo *et al.*, 1998; Zoratti and Szabò, 1995). Many speculate that the transition pore spans inner and outer membranes at the contact sites (Hackenbrock, 1968). Marzo *et al.* (1998) showed that Bax, a pro-apoptotic member of the Bcl2 family, binds to the transition complex, where it interacts with the adenine nucleotide translocator to increase mitochondrial membrane permeability and trigger cell death, suggesting that the translocator is a necessary pore component. Pore activity is reported in mitochondrial membranes from triple adenine nucleotide translocator knockout yeast strains, however, suggesting that the translocator is *not* required for the pore activity (Lohret *et al.*, 1996). Shimizu *et al.* (1999) also showed that porin is involved in Bax-induced apoptosis, thus implicating porin as a part of the pore complex.

4. VISUALIZATION OF THE TRANSITION *IN SITU* DURING OXIDATIVE STRESS

Permeability transition involvement in *t*-BuOOH toxicity was suggested by the observation that cyclosporin A and trifluoperazine (transition inhibitors) delay *t*-BuOOH-induced lethal injury to hepatocytes (Imberti *et al.*, 1990, 1992). To confirm whether oxidative stress induces MPT onset in cultured hepatocytes, we developed a confocal microscopy technique to visualize transition *in situ* in intact living cells. Calcein is a 623-Da fluorophore; when cultured hepatocytes are incubated at 37 °C with the acetoxymethylester of calcein, the neutral ester crosses the plasma membrane. Cytosolic esterases then hydrolyze calcein acetoxymethylester to the free acid form, whose carboxyl groups trap and retain the fluorophore inside the cytosolic compartment. Calcein displays a bright green fluorescence that is independent of physiological changes of pH and Ca^{2+}. Several dark round voids about one micron in diameter are evident in the calcein image (Fig. 1A, calcein). To identify these voids, hepatocytes are loaded with tetramethylrhodamine methylester (TMRM), a red-fluorescing cationic dye that accumulates electrophoretically into mitochondria in response to their high negative-membrane potential (Ehrenberg *et al.*, 1988). The bright spots in the TMRM image correspond to the dark voids in the calcein image, demonstrating that calcein is excluded from mitochondria (Fig. 1A, TMRM).

After *t*-BuOOH (100 µM) exposure, the dark voids begin to fill with calcein. Concominantly, mitochondria begin to lose their TMRM fluorescence (Fig. 1C). Cell surface blebbing, an early indication of necrotic cell injury, is also evident. Eventually, mitochondria depolarize completely, cellular ATP decreases, and cells lose their viability, as indicated by nuclear staining with propidium iodide and leakage of cytosolic calcein. Trifluoperazine, a phospholipase inhibitor, blocks MPT onset in isolated liver mitochondria (Broekemeier *et al.*, 1985), and it also prevents *t*-BuOOH-induced transition, subsequent depolarization, and cellular ATP depletion in intact hepatocytes, as well as loss of cell viability (Nieminen *et al.*, 1995). Interestingly, trifluoperazine does not protect against the cytotoxicity of mitochondrial uncouplers, which also produce mitochondrial depolarization and ATP depletion but do not induce transition (Nieminen *et al.*, 1995), nor does it prevent ATP depletion and cell death after mitochondrial uncoupling, suggesting that trifluoperazine cytoprotection is specific for transition-mediated injury.

5. CONTRIBUTION OF PYRIDINE NUCLEOTIDE OXIDATION TO THE PERMEABILITY TRANSITION

Opening of transition pores is partially regulated by redox state. Vicinal thiols oxidation shifts the pore to open probability in isolated mitochondria (Constantini *et al.*, 1996; Kowaltowski *et al.*, 1996). In isolated liver mitochondria, *t*-BuOOH induces

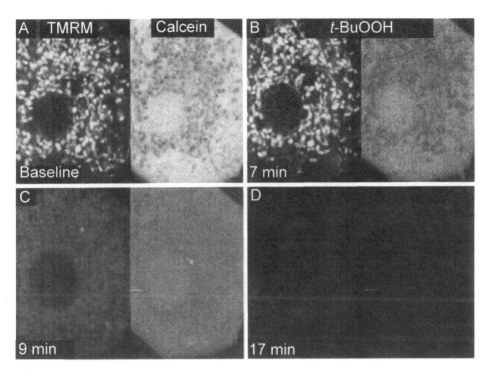

FIGURE 1. Onset of the mitochondrial permeability transition in hepatocytes, induced by *t*-BuOOH. Cultured hepatocytes were loaded with TMRM and acetoxymethylester of calcein. The green and red fluorescences, respectively, of calcein and TMRM were imaged by laserscanning confocal microscopy. (A) In the basal image, TMRM-labeled mitochondria correspond to dark voids in the calcein image. (B, C) After addition of 100 μM *t*-BuOOH, calcein redistributed from cytosol into mitochondria, and TMRM fluorescence was lost, signifying onset of the MPT. Cell death subsequently occurred, documented by loss of cytosolic calcein fluorescence (D).

pyridine nucleotide oxidation through the combined action of glutathione peroxidase, glutathione reductase, and pyridine nucleotide transhydrogenase (Flohe and Schlegel, 1971; Lee and Ernster, 1964; Vlessis, 1990). So does *t*-BuOOH induce transition in intact cells via changing the pyridine nucleotide redox state to a more oxidized state? Fluorescence of reduced pyridine nucleotides (NADH and NAHPH) accounts for most of the cellular autofluorescence excited with near-UV light (Fig. 2). Oxidized pyridine nucleotides (NAD$^+$ and NADP$^+$) are nonfluorescent; therefore, changes in autofluorescence reflect changes in the pyridine nucleotide redox state. The *t*-BuOOH causes a rapid oxidation of mitochondrial pyridine nucleotides (Fig. 3). To assess the importance of this oxidation, metabolites affecting the pyridine nucleotide redox state were tested. In mitochondria, β-hydroxybutyrate is a substrate for β-hydroxybutyrate dehydrogenase, so adding β-hydroxybutyrate causes selective reduction of mitochondrial NAD(P)$^+$ and delays *t*-BuOOH-induced cell killing (Nieminen *et al.*, 1997). Acetoacetate that oxidizes mitochondrial NAD(P)H does not enhance *t*-BuOOH toxicity however, perhaps because *t*-BuOOH alone already causes maximal NAD(P)H oxidation. Lactate, which reduces cytosolic NAD$^+$ *via* lactate dehydrogenase, also fails to protect against killing by

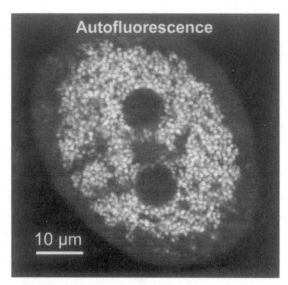

FIGURE 2. Autofluorescence of a cultured hepatocyte. A cultured hepatocyte was imaged by a 351 nm light from a UV–argon laser and emitted light was collected by a 420-nm long pass filter using a 40-sec scan. The near-UV-excited autofluorescence originated predominantly from mitochondria.

t-BuOOH. These results are consistent with the hypothesis that oxidation of mitochondrial pyridine nucleotides contributes to MPT onset in intact hepatocytes after oxidative stress, in agreement with observations in isolated mitochondria (Constantini *et al.*, 1996; Kowaltowski *et al.*, 1996).

6. CONTRIBUTION OF INCREASED MITOCHONDRIAL FREE CA^{2+} TO THE PERMEABILITY TRANSITION

Steady-state mitochondrial free Ca^{2+} concentration is comprised of a balance of membrane-potential-driven Ca^{2+} uptake by the mitochondrial Ca^{2+} uniporter and Ca^{2+} efflux by concerted $3Na^+/Ca^{2+}$ and Na^+/H^+ exchange, driven by both mitochondrial membrane potential ($\Delta\Psi$) and pH gradient (ΔpH) (Crompton *et al.*, 1976; Puskin *et al.*, 1976). This tightly regulated transport system maintains resting mitochondrial free Ca^{2+} at close to cytosolic free Ca^{2+} in most cell types (Chacon *et al.*, 1994, 1996; Hajnoczky *et al.*, 1995; Kawanishi *et al.*, 1991; Rizzuto *et al.*, 1992). It was once thought that mitochondrial Ca^{2+} responds slowly to physiological changes of cytosolic free Ca^{2+}, but several recent studies show that mitochondrial Ca^{2+} can increase rapidly in response to cytosolic Ca^{2+} transients caused by hormone stimulation, muscle contraction, neuronal depolarization, and other stimuli (Chacon *et al.*, 1996; Hajnoczky *et al.*, 1995; Kawanishi *et al.*, 1991; Peng *et al.*, 1998; Pivovarova *et al.*, 1999; Rizzuto *et al.*, 1992; Trollinger *et al.*, 1997). Elevated mitochondrial Ca^{2+} can activate the Ca^{2+}-dependent dehydrogenases responsible for mitochondrial adenosine triphosphate production (Denton *et al.*, 1978; Hansford *et al.*, 1981; McCormack and Denton, 1979). Although, it has been widely

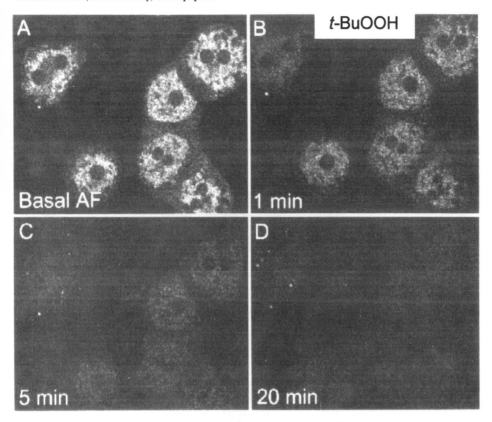

FIGURE 3. Oxidation of mitochondrial pyridine nucleotides after oxidative stress. Cultured hepatocytes were exposed to 100 μM *t*-BuOOH. Autofluorescence of cultured hepatocytes was imaged by laser-scanning confocal microscopy. (A) After a baseline image of autofluorescence (AF), 100 μM *t*-BuOOH was added. (B–D) Subsequently, images were collected at 1, 5, and 20 min. To minimize photobleaching, images were collected using single 10-sec scans instead of the 40-sec scans used in Fig 2.

proposed that mitochondria release Ca^{2+} under conditions of cellular stress, this is unlikely for the simple reason that mitochondria do not have the Ca^{2+} to release (Chacon *et al.*, 1994; Hajnoczky *et al.*, 1995; Kawanishi *et al.*, 1991; Rizzuto *et al.*, 1992). Rather, intramitochondrial Ca^{2+} concentration may increase as a consequence of cellular stress (Bernardi *et al.*, 1994; Chacon *et al.*, 1992; 1994; White and Reynolds, 1996).

In isolated mitochondria increased matrix Ca^{2+} is a potent transition inducer. As a result mitochondria depolarize, leading to swelling and uncoupling of oxidative phosphorylation (reviewed in Bernardi *et al.*, 1994). Several studies show that the Ca^{2+} ionophore causes necrotic cell death in various cell types (Nieminen *et al.*, 1990b; Qian *et al.*, 1999). The Ca^{2+} ionophore Br-A23187 produces > 60% killing in hepatocytes within an hour (Nieminen *et al.*, 1990; Qian *et al.*, 1999). Cell killing is temporarily linked to increased mitochondrial free Ca^{2+}, mitochondrial depolarization, and calcein movement from cytosol into mitochondrial space, indicating transition onset (Qian *et al.*, 1999).

Knowing that mitochondrial free Ca^{2+} is an effective transition inducer in isolated mitochondria preparations as well as in intact cells, we were interested in exploring its

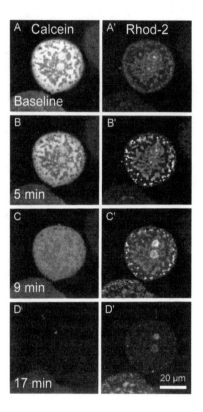

FIGURE 4. Mitochondrial and cytosolic Ca^{2+} after exposure of cultured hepatocytes to *t*-BuOOH. Cultured hepatocytes were loaded with the acetoxymethylesters of Rhod-2-AM and calcein. Simultaneous images of the green and red fluorescence of calcein and Rhod-2, respectively, were collected using confocal microscopy. In the baseline image, calcein fluorescence localized to the cytosol (A), and mitochondria appeared as dark voids. Rhod-2 fluorescence was faint throughout the cell (A') After 100 μM *t*-BuOOH, mitochondrial Rhod-2 fluorescence increased within 5 min (B') with little change in calcein fluorescence (B). Specifically, Rhod-2 fluorescence increased in the areas corresponding to the dark voids in the calcein image. After 9 min exposure, mitochondria began to fill with calcein (C), and cytosolic Rhod-2 fluorescence increased (C'). After 17 min viability was lost, as indicated by loss of calcein fluorescence (D) and most Rhod-2 fluorescence (D').

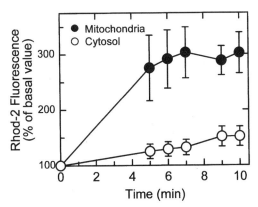

FIGURE 5. Increased mitochondrial and cytosolic Rhod-2 fluorescence after *t*-BuOOH. Cultured hepatocytes were loaded with the acetoxymethylester of Rhod-2-AM and imaged and treated with *t*-BuOOH as in Figure 4. In each experiment mitochondrial and cytosolic regions were identified by morphological criteria in Rhod-2 images after 5 to 10 min exposure to *t*-BuOOH. For each confocal image, Rhod-2 fluorescence in 4 and 10 selected areas of cytosol and mitochondria, respectively, were quantified using Adobe Photoshop. After background subtraction, mean pixel intensities for mitochondria and cytosol were normalized and expressed as the percentage of corresponding values before addition of *t*-BuOOH. Running averages of mitochondrial (●) and cytosolic (○) Rhod-2 fluorescence were plotted vs time after *t*-BuOOH addition. Values represent means $\pm$ SE of 6 experiments from 4 cell isolations. In comparison, mitochondrial and cytosolic Rhod-2 fluorescence quantified in the same manner increased about 272% after addition of 10 μM Br-A23187 in the presence of cyclosporin A (Qian *et al.*, 1999).

possible role in oxidant chemical-induced transition. Mitochondrial Ca^{2+} will increase well before transition onset and before any substantial change of cytosolic Ca^{2+} (Fig. 4, Fig. 5). Chelation of intracellular Ca^{2+} with acetoxymethylesters of BAPTA and Quin-2 substantially delays increase of mitochondrial Ca^{2+}, onset of the MPT, and cell death. Confocal microscopy also reveals that the intracellular site of action of Ca^{2+} chelators is mitochondria (Byrne *et al.*, 1999).

7. CONTRIBUTION OF MITOCHONDRIAL REACTIVE OXYGEN SPECIES TO THE PERMEABILITY TRANSITION

Mitochondria produce small amounts of reactive oxygen species (ROS) during normal metabolism (Chance *et al.*, 1979). Mitochondria are also an important source of ROS generation in pathophysiology, occurring during excitotoxicity to cortical neurons, ischemia/reperfusion in heart, and cyanide toxicity simulating hypoxia (chemical hypoxia) in hepatocytes and cardiac myocytes (Chacon *et al.*, 1992; Cross *et al.*, 1987; Dawson *et al.*, 1993; Dugan *et al.*, 1995; Gores *et al.*, 1989; Reynolds and Hastings, 1995). Most of the mitochondrial ROS *in situ* are formed in the cytochrome bc$_1$ complex by the Q cycle (Dawson *et al.*, 1993). Some studies suggest that mitochondrial ROS formation is Ca^{2+} dependent (Chacon *et al.*, 1992; Dugan *et al.*, 1995; Reynolds and Hastings, 1995). The *t*-BuOOH increase mitochondrial ROS production 15-fold, as determined by confocal microscopy from the conversion of dichlorofluorescin to highly fluorescent dichlorofluorescein (Fig. 6). Most of this ROS production occurs after the initial rapid oxidation of mitochondrial pyridine nucleotides by *t*-BuOOH. Both desferal (an inhibitor of iron-catalyzed free radical formation) and diphenylphenylenediamine (a one-electron donor that terminates free radical chain reactions) inhibit ROS production, block transition onset, and prevent *t*-BuOOH-induced necrotic cell death. Mitochondrial Ca^{2+} chelation with BAPTA acetoxymethylester also decreases mitochondrial ROS production after *t*-BuOOH by almost 100%. Interestingly, BAPTA acetoxymethylester does not prevent the initial pyridine nucleotide oxidation after *t*-BuOOH. This oxidation is related to metabolism of *t*-BuOOH by the glutathione peroxidase/glutathione reductase system (Flohe and Schlegel, 1971). The BAPTA acetoxymethylester, however, partially prevents the latter phase of more complete mitochondrial pyridine nucleotide oxidation, indicating that this latter oxidation is due to Ca^{2+}-dependent ROS production.

Based on our results and those of others, we propose the following sequence of cellular events during *t*-BuOOH exposure (Fig. 7). The *t*-BuOOH causes mitochondrial pyridine nucleotide oxidation. Pyridine nucleotide oxidation disturbs the balance of mitochondrial Ca^{2+} uptake and release mechanism, leading to a net increase of mitochon-

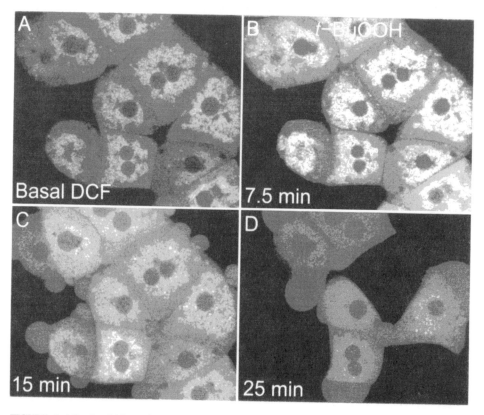

FIGURE 6. Mitochondrial reactive oxygen species production after exposure of hepatocytes to *t*-BuOOH. Hepatocytes were loaded with nonfluorescent dichlorofluorescin, which reacts with ROS to form fluorescent dichlorofluorescein. After collecting a baseline image (A), *t*-BuOOH (100 μM) was added and additional images were collected at various time points (B–D).

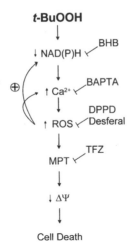

FIGURE 7. Sequence of cellular events during *t*-BuOOH exposure.

drial Ca^{2+}. Increased mitochondrial Ca^{2+} further stimulates intramitochondrial ROS production and perhaps more Ca^{2+} uptake. Finally, the combination of increased mitochondrial Ca^{2+} and oxidative stress from ROS causes opening of the permeability transition pores and onset of the MPT. As a consequence of transition, adenosine triphosphate becomes profoundly depleted and the cell dies via necrosis.

8. CONTRIBUTION OF THE PERMEABILITY TRANSITION TO EXCITOTOXICITY IN NEURONS

Overstimulation of *N*-methyl-D-aspartate (NMDA) glutamate receptors is a major factor in hypoxic cell injury in the central nervous system (Choi, 1988; Rothman and Olney, 1987). NMDA is an agonist of the NMDA-type glutamate receptors and causes a rapid increase in intracellular Ca^{2+}, leading to collapse of the $\Delta\psi$ and eventually cell death in cortical neurons (Dugan *et al.*, 1995; Hartley *et al.*, 1993). Because we showed that increased matrix Ca^{2+} is a potent transition inducer in oxidative stress-induced cell death, we hypothesized that NMDA exposure results in elevated Ca^{2+}, which is then further taken up by mitochondria. Increased matrix Ca^{2+} further induces opening of mitochondrial permeability transition pores, resulting in transition, depolarization, and cell death (Szabò and Zoratti, 1991). To test this hypothesis, we loaded cultured neurons isolated from rat cerebral cortex with TMRM to monitor $\Delta\psi$ using confocal microscopy. Within several minutes of exposure to 500 μM NMDA, the TMRM leaked from individual mitochondria into the cytosol and eventually out from the cells, indicating $\Delta\psi$ collapse. Cyclosporin A delayed the depolarization (Fig. 8; Nieminen *et al.*, 1996). These results support our hypothesis that transition mediates excitotoxic injury to neurons. Similar results were obtained from two other laboratories (Schinder *et al.*, 1996; White and Reynolds, 1996). Thus, acute excitotoxicity in neurons is yet another model of injury where MPT onset seems to play a critical role.

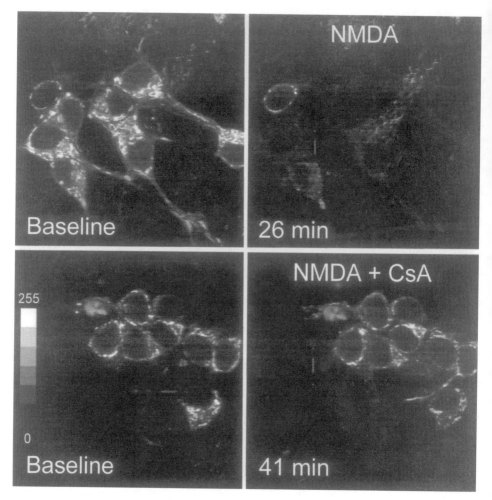

FIGURE 8. Excitotoxicity causes mitochondrial depolarization in cortical neurons. Cultured cortical neurons (11–16 days in culture) were loaded with TMRM. Its red fluorescence was imaged using laser-scanning confocal microscopy. After collecting a baseline image, 500 μM NMDA was added and the image was subsequently collected at 26 min. The punctate fluorescence observed in the basal image disappeared after NMDA exposure, indicating onset of mitochondrial depolarization. In the second set of experiments, cortical neurons were pre-incubated with 1 μM cyclosporin A (CsA) and then exposed to NMDA. Even after 41 min exposure to NMDA, mitochondrial membrane potential was preserved, indicating that CsA prevents mitochondrial depolarization induced by NMDA.

9. ROLE OF MITOCHONDRIAL PERMEABILITY TRANSITION IN APOPTOSIS

The various cell injury models described above illustrate the transition's importance in necrotic cell killing. The ultimate consequence of transition is inhibited mitochondrial ATP production, which eventually leads to depletion because glycolytic ATP production cannot meet the energy demand. In most apoptotic processes, however, cytosolic ATP is required to complete the apoptotic cascade. ATP is an essential cofactor for inducing apoptosis via the apoptosis-activating factor 1/cytochrome c/caspase-9 pathway (Li *et al.*, 1997). In a cell-free system combining purified mitochondria and nuclei, induction of the MPT causes mitochondria to release soluble factors that activate caspases (Marchetti *et al.*, 1996; Susin *et al.*, 1996, 1997). Two such proteins have been identified, namely cytochrome c and apoptosis-inducing factor (Liu *et al.*, 1996; Susin *et al.*, 1999). Cytochrome c ordinarily resides in the intermembrane space as a diffusable electron carrier (Cortese and Hackenbrock, 1993), but it is released into the cytosol when transition-induced mitochondrial swelling breaks the outer membrane (Vanderkooi *et al.*, 1973).

The exact role of transition in mediating apoptosis is controversial, and some studies claim that cytochrome c release during apoptosis occurs without mitochondrial depolarization, and by implication, without the MPT (Yang *et al.*, 1997; Kluck *et al.*, 1997). In the face of this controversy, we developed a method to monitor cytochrome c release and $\Delta\psi$ in a single living cell (Heiskanen *et al.*, 1999). To localize cytochrome c within the cells, we transfected PC6 cells (a subclone of PC12 cells) with plasmids bearing a cytochrome c–GFP fusion, in which green fluorescent protein (GFP) was fused to the COOH terminus of cytochrome c. To confirm that cytochrome c–GFP was transported to mitochondria, cells were coloaded with Mitotracker Red CMXRos, which localizes to mitochondria. Cytochrome c–GFP and Mitotracker were imaged using laser-scanning confocal microscopy. Cytochrome c–GFP displayed a punctate distribution pattern closely matching that of the mitochondrial marker, demonstrating directly that the cytochrome c–GFP was localized to mitochondria (Fig. 9). By contrast, cells transfected with GFP cDNA alone displayed a diffuse cytosolic fluorescence (Heiskanen *et al.*, 1999).

To further assess the causal relationship between mitochondrial depolarization and cytochrome c release, cytochrome c–GFP-transfected cells were loaded with TMRM, the membrane-potential-indicating dye. Fig. 8 (0 h) shows a field of PC6 cells transfected with

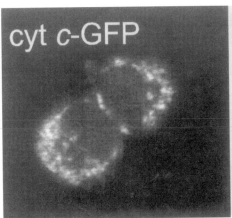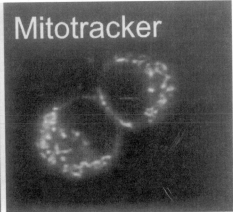

FIGURE 9. Cytochrome c–GFP localizes to mitochondria. Cultured PC6 cells were transfected with cDNA cytochrome c–GFP and subsequently loaded with the mitochondria-specific probe MitoTracker Red. The green and red fluorescences, respectively, of cytochrome c–GFP and MitoTracker Red were imaged using laser-scanning confocal microscopy. Cytochrome c–GFP displayed a punctate fluorescence closely matching that of MitoTracker Red.

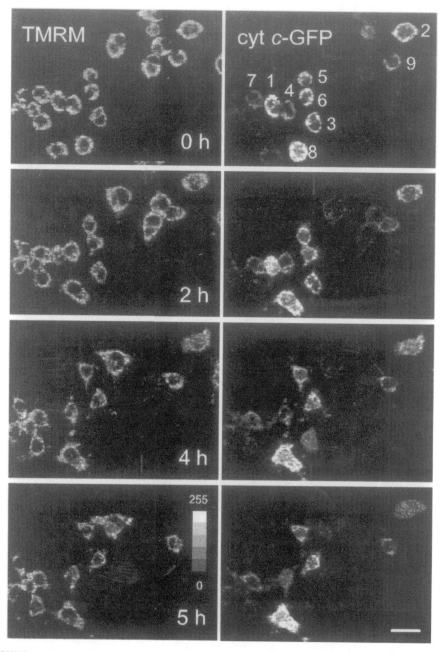

FIGURE 10. Staurosporine causes mitochondrial depolarization and cytochrome *c*–GFP release. Cultured PC6 cells were transfected with cDNA cytochrome *c*–GFP and subsequently loaded with TMRM to monitor the mitochondrial membrane potential. The green and red fluorescences, respectively, of cytochrome *c*–GFP and TMRM were imaged in single PC6 cells using confocal microscopy. At 0 hr, basal images of TMRM (left panel) and cytochrome *c*–GFP (right panel) were collected. Then 5 μM staurosporine was added. Images were collected at 2, 4, and 5 hr. Individual cells showed heterogenous mitochondrial depolarization and cytochrome *c*–GFP release. Scale bar is 20 μm.

cytochrome c–GFP (right panel) and loaded with TMRM (left panel) prior to addition of the apoptotic stimulus, staurosporine. After exposure to 5 μM staurosporine, cells behaved asynchronously. At 2 hours TMRM fluorescence decreased by 30% in cell number one. Simultaneously, a diffuse pattern of cytochrome c–GFP fluorescence began to appear, indicating cytochrome c–GFP redistribution from mitochondria to cytosol and nucleus (Fig. 10, 2 h, right panel). After 3 hours the cell had totally lost its $\Delta\psi$ and its cytochrome c–GFP fluorescence as well, indicating that plasma membrane lysis and onset of secondary necrosis had occurred, which often follow apoptosis. Most cells in the field followed a similar but slower course of events. By contrast, cell number 7 showed no decline in mitochondrial membrane potential (Fig. 8, left panels). There was also no change in the cytochrome c–GFP fluorescence pattern, indicating that cytochrome c–GFP had not redistributed from mitochondria to the cytosol (Fig. 8, right panels). These results show that mitochondrial depolarization occurred asynchronously in individual cells. Even the individual mitochondria within a single cell depolarized asynchronously. The data yield new information regarding the mechanism underlying cytochrome c release during apoptosis. Specifically, our results show that cytochrome c release does not precede depolarization, implicating the possible role of the MPT in this process. If transition is the key pathophysiological contributor to cell death, then pharmacological transition antagonists show potential for preventing or even reversing a variety of human diseases.

ACKNOWLEDGEMENTS. This work was supported in part, by grants AG13318 and NS39469 from the National Institutes of Health and a Grant-in-Aid from the American Heart Association. K. M. H. was supported by the American Heart Association Ohio Valley Affiliate Fellowship, the Academy of Finland, and the Maud Kuistila Foundation.

REFERENCES

Bellomo, G., Fulceri, R., Albano, E., Gamberucci, A., Pompella, A., Parola, M., and Benedetti, A., 1991, A Ca^{2+}-dependent and independent mitochondrial damage in hepatocellular injury, *Cell Calcium* **12**:335–341.

Bernardi, P., Broekemeier, K. M., and Pfeiffer, D. R., 1994, Recent progress on regulation of the mitochondrial permeability transition pore, a cyclosporin-sensitive pore in the inner mitochondrial membrane, *J. Bioenerg. Biomembr.* **26**:509–517.

Beutner, G., Ruck, A., Riede, B., Welte, W., and Brdiczka, D., 1996, Complexes between kinases, mitochondrial porin, and adenylate translocator in rat brain resemble the permeability transition pore, *FEBS Lett.* **396**:189–195.

Broekemeier, K. M., Schmid, P. C., Schmid, H. H. O., and Pfeiffer, D. R., 1985, Effects of phospholipase A_2 inhibitors on ruthenium red-induced Ca^{2+} release from mitochondria, *J. Biol. Chem.* **260**:105–113.

Brustovetsky, N., and Klingenberg, M., 1996, Mitochondrial ADP/ATP carrier can be reversibly converted into a large channel by Ca^{2+}, *Biochemistry* **35**:8483–8488.

Byrne, A. M., Lemasters, J. J., and Nieminen, A.-L., 1999, Contribution of increased mitochondrial free Ca^{2+} to the tert-butylhydroperoxide-induced mitochondrial permeability transition in rat hepatocytes, *Hepatology* **29**:1523–1531.

Chacon, E., Ulrich, R., and Acosta, D., 1992, A digitized-fluorescence-imaging study of mitochondrial Ca^{2+} increase by doxorubicin in cardiac myocytes, *Biochem. J.* **281**:871–878.

Chacon, E., Reece, J. M., Nieminen, A.-L., Zahrebelski, G., Herman, B., and Lemasters, J. J., 1994, Distribution of electrical potential, pH, free Ca^{2+}, and volume inside cultured adult rabbit cardiac myocytes during chemical hypoxia: A multiparameter digitized confocal microscopic study, *Biophys. J.* **66**:942–952.

Chacon, E., Ohata, H., Harper, I. S., Trollinger, D. R., and Lemasters, J. J., 1996, Mitochondrial free calcium transients during excitation–contraction coupling in rabbit cardiac myocytes, *FEBS Lett.* **382**:31–36.

Chance, B., Sies, H., and Boveris, A., 1979, Hydroperoxide metabolism in mammalian organs, *Physiol. Rev.* 59:527–605.

Choi, D. W., 1988, Glutamate neurotoxicity and diseases of the nervous system, *Neuron* 1:623–634.

Constantini, P., Chernyak, B. V., Petronilli, V., and Bernardi, P., 1996, Modulation of the mitochondrial permeability transition pore by pyridine nucleotides and dithiol oxidation at two separate sites, *J. Biol. Chem.* 271:6746–6751.

Cortese, J. D., and Hackenbrock, C. R., 1993, Motional dynamics of functional cytochrome *c* delivered by low pH fusion into the intermembrane space of intact mitochondria, *Biochim. Biophys. Acta* 1142:194–202.

Cotgreave, I. A., Moldeus, P., and Orrenius, S., 1988, Host biochemical defense mechanisms against prooxidants, *Annu. Rev. Pharmacol. Toxicol.* 28:189–212.

Crompton, M., Capano, M., and Carafoli, E., 1976, A kinetic study of the energy-linked influx of Ca^{2+} into heart mitochondria, *Eur. J. Biochem* 69:429–343.

Crompton, M., Ellinger, H., and Costi, A., 1988, Inhibition of cyclosporin A of a Ca^{2+}-dependent pore in heart mitochondria activated by inorganic phosphate and oxidative stress, *Biochem. J.* 255:357–360.

Cross, D. E., Halliwell, B., Borish, E. T., Pryor, W. A., Ames, B. A., Saul, R. S., McCord, J. M., and Harman, D., 1987, Oxygen radicals and human disease, *Ann. Intern. Med.* 107:526–545.

Dawson, T. L., Gores, G. J., Nieminen, A.-L., Herman, B., and Lemasters, J. J., 1993, Subcellular sites of toxic oxygen species generation during reductive stress in rat hepatocytes, *Am. J. Physiol.* 264:C961–C967.

Denton. R. M., Richards, D. A., and Chen, J.G., 1978, Calcium ions and the regulation of NAD^+-linked isocitrate dehydrogenase from the mitochondria of rat heart and other tissues. *Bio chem. J.* 176:899–906.

DiMonte, D., Sandy, D. M. S., Blank, L., and Smith, M. T., 1988, Fructose prevents 1-methyl-4-phenyl-1,2,3,6-tetrahydropyridine (MPTP)-induced ATP depletion and toxicity in isolated hepatocytes, *Biochem. Biophys. Res. Commun.* 153:734–740.

Dugan, L. L., Sensi, S. L., Canzoniero, L. M. T., Handran, S. D., Rothman, S. M., Lin, T.-S., Gold-berg, M. P., and Choi, D. W., 1995, Mitochondrial production of active oxygen species in cortical neurons following exposure to *N*-methyl-D-aspartate, *J. Neurosci.* 15:6377–6388.

Ehrenberg, B. V., Montana, V., Wei, M.-D., Wuskell, J. P., and Loew, L. M., 1988, Memrane potential can be determined in individual cells from the nernstian distribution of cationic dyes, *Biophys. J.* 53:785–794.

Flohe, L., and Schlegel, W., 1971, Glutathion-peroxidase, *Hoppe-Seylers Z. Physiol. Chem.* 352:1401–1410.

Fournier, N., Ducet, G., and Crevat, A., 1987, Action of cyclosporin on mitochondrial calcium fluxes, *J. Bioenerg. Biomembr.* 19:297–303.

Gores, G. J., Flarsheim, C. E., Dawson, T. L., Nieminen, A.-L., Herman, B., and Lemasters, J. J., 1989, Swelling, reductive stress, and cell death during chemical hypoxia in hepatocytes, *Am. J. Physiol.* 257:C347–C354.

Hackenbrock, C. R., 1968, Chemical and physical fixation of isolated mitochondria in low-energy and high-energy states, *Proc. Natl. Acad. Sci. USA* 61:598–605.

Hajnoczky, G., Robb-Gaspers, L. D., Seitz, M. B., and Thomas, A. P., 1995, Decoding of cytosolic calcium oscillations in the mitochondria, *Cell* 82:415–424.

Halestrap, A. P., and Davidson, A. M., 1990, Inhibition of Ca^{2+}-induced large-amplitude swelling of liver and heart mitochondria by cyclosporin is probably caused by the inhibitor binding to mitochondrial-matrix peptidyl–prolyl *cis-trans* isomerase and preventing it interacting with the adenine nucleotide translocase, *Biochem. J.* 268:153–160.

Hansford, R. G., 1981, Effect of micromolar concentrations of free Ca^{2+} ions on pyruvate dehydrogenase interconversion in intact rat heart mitochondria, *Biochem. J.* 194:721–732.

Hartley, D. M., Kurth, M. C., Bjerkness, L., Weiss, J. H., and Choi, D. W., 1993, Glutamate receptor-induced $^{45}Ca^{2+}$ accumulation in cortical cell culture correlates with subsequent neuronal degeneration, *J. Neurosci.* 13:1993–2000.

Haworth, R. A., and Hunter, D. R., 1979, The Ca^{2+}-induced membrane transition in mitochondria: II. Nature of the Ca^{2+} trigger site, *Arch. Biochem. Biophys.* 195:460–467.

Heiskanen, K. M., Bhat, M. B., Wang, H.-W., Ma, J., and Nieminen, A.-L., 1999, Mitochondrial depolarization accompanies cytochrome *c* release during apoptosis in PC6 cells, *J. Biol. Chem.* 274:5654–5658.

Hunter, D. R., and Haworth, R. A., 1979a, The Ca^{2+}-induced membrane transition in mitochondria:I. The protective mechanisms, *Arch, Biochem. Biophys.* 195:453–459.

Hunter, D. R., and Haworth, R. A., 1979b, The Ca^{2+}-induced membrane transition in mitochondria: III. Transitional Ca^{2+} release, *Arch. Biochem. Biophys.* 195:468–477.

Hunter, D. R., Haworth, R. A., and Southard, J. H., 1976, The relationship between permeability, configuration, and function in calcium treated mitochondria, *J. Biol. Chem.* **251**:5069–5077.

Imberti, R., Nieminen, A.-L., Herman, B., and Lemasters, J. J., 1990, Mitochondrial inhibition and uncoupling preceding lethal injury to rat hepatocytes by *t*-butyl hydroperoxide: Protection by fructose, oligomycin, cyclosporin A, and trifluoperazine, *Hepatology* **12**:933.

Imberti, R., Nieminen, A.-L., Herman, B., and Lemasters, J. J., 1992, Synergism of cyclosporin A and phospholipase inhibitors in protection against lethal injury to rat hepatocytes from oxidant chemicals, *Res. Commun. Chem. Pathol. Pharmacol.* **78**:27–38.

Imberti, R., Nieminen, A.-L., Herman, B., and Lemasters, J. J., 1993, Mitochondrial and glycolytic dysfunction in lethal injury to hepatocytes by *t*-butylhydroperoxide: Protection by fructose, cyclosporin A, and trifluoperazine, *J. Pharmacol. Exp. Therap.* **265**:392–400.

Kawanishi, T., Nieminen, A.-L., Herman, B., and Lemasters, J. J., 1991, Suppression of Ca^{2+} oscillations in cultured rat hepatocytes by chemical hypoxia, *J. Biol. Chem.* **266**:20062–20069.

Kinnally, K. W., Campo, M. L., and Tedeschi, H., 1989, Mitochondrial channel activity studied by patch-clamping mitoplasts, *J. Bioenerg. Biomembr.* **21**:497–506.

Kluck, R. M., Bossy-Wetzel, E., Green, D. R., Newmeyer, D. D., 1997, The release of cytochrome *c* from mitochondria: A primary site for Bcl-2 regulation of apoptosis, *Science* **275**:1132–1136.

Kowaltowski, A. J., Castilho, R. F., and Vercesi, A. E., 1996, Opening of the mitochondrial permeability transition pore by uncoupling or inorganic phosphate in the presence of Ca^{2+} is dependent on mitochondrial-generated reactive oxygen species, *FEBS Lett.* **378**:150–152.

Lee, C. P., and Ernster, L., 1964, Equilibrium studies of the energy-dependent and non-energy-dependent pyridine nucleotide transhydrogenase reactions, *Biochim. Biophys. Acta* **81**:187–190.

Li, P., Nijhawan, D., Budihardjo, I., Srinivasula, S. M., Ahmad, M., Alnemri, E. S., and Wang, X., 1997, Cytochrome *c* and dATP-dependent formation of apaf-1/caspase-9 complex initiates an apoptotic protease cascade, *Cell* **91**:479–489.

Liu, X., Kim, C. N., Yang, J., Jemmerson, R., and Wang, X., 1996, Induction of apoptosis in cell-free extracts: Requirement for dATP and cytochrome *c, Cell,* **86**:147–157.

Lohret, T. A., Murphy, R. C., Drgon, T., and Kinnally, K. W., 1996, Activity of the mitochondrial multiple conductance channel is independent of the adenine nucleotide translocator, *J. Biol. Chem.* **271**:4846–4849.

Lotscher, H. R., Winterhalter, K. H., Carafoli, E., and Richter, C., 1980, Hydroperoxide-induced loss of pyridine nucleotides and releases of calcium from rat liver mitochondria, *J. Biol. Chem.* **255**:9325–9330.

Marchetti, P., Susin, S. A., Decaudin, D., Gamen, S., and Castedo, M., 1996, Apoptosis-associated derangement of mitochondrial function in cells lacking mitochondrial DNA, *Cancer Res.* **56**:2033–2038.

Marzo, I., Brenner, C., Zamzami, N., Jurgensmeier, J. M., Susin, S. A., Vieira, H. L. A., Prevost, M.-C., Xie, Z., Matsuyama, S., Reed, J. C., and Kroemer, G., 1998, Bax and adenine nucleotide translocator cooperate in the mitochondrial control of apoptosis, *Science* **281**:2027–2031.

McCormack, J. G., and Denton, R. M., 1979, The effects of calcium ions and adenine nucleotides on the activity of pig heart 2-oxoglutarate dehydrogenase complex, *Biochem. J.* **180**:533–544.

Nieminen, A.-L., Gores, G. J., Dawson, T. L., Herman, B., and Lemasters, J. J., 1990a, Toxic injury from mercuric chloride in rat hepatocytes, *J. Biol. Chem.* **265**:2399–2408.

Nieminen, A.-L., Dawson, T. L., Gores, G. J., Kawanishi, T., Herman, B., and Lemasters, J. J., 1990b, Protection by acidotic pH and fructose against lethal injury to rat hepatocytes from mitochondrial inhibition, ionophores, and oxidant chemicals, *Biochem. Biophys. Res. Commun.* **167**:600–606.

Nieminen, A.-L., Saylor, A. K., Herman, B., and Lemasters, J. J., 1994, ATP depletion rather than mitochondrial depolarization mediates hepatocyte killing after metabolic inhibition, *Am. J. Physiol.* **267**:C67–C74.

Nieminen, A.-L., Saylor, A. K., Tesfai, S. A., Herman, B., and Lemasters, J. J., 1995, Contribution of the mitochondrial permeability transition to lethal injury after exposure of hepatocytes to *t*-butylhydroperoxide, *Biochem. J.* **307**:99–106.

Nieminen, A.-L., Petrie, T. G., Lemasters, J. J., and Selman, W. R., 1996, Cyclosporin A delays mitochondrial depolarization induced by *N*-methyl-D-aspartate in cortical neurons: Evidence of the mitochondrial permeability transition, *Neuroscience* **75**:993–997.

Nieminen, A.-L., Byrne, A. M., Herman, B., and Lemasters, J. J., 1997, Mitochondrial permeability transition in hepatocytes induced by *t*-BuOOH:NAD(P)H and reactive oxygen species, *Am. J. Physiol.* **272**:C1286–C1294.

Peng, T. I., Jou, M. J., Sheu, S. S., and Greenamyre, J. T., 1998, Visualization of NMDA receptorinduced mitochondrial calcium accumulation in striatal neurons, *Exp. Neurol.* **149**:1–12.

Petronilli, V., Szabo, I, and Zoratti, M., 1989, The inner mitochondrial membrane contains ionconducting channels similar to those found in bacteria, *FEBS Lett.* **259**:137–143.

Pivovarova, N. B., Hongpaisan, J., Andrews, S. B., and Friel, D. D., 1999, Depolarization-induced mitochondrial Ca accumulation in sympathetic neurons: Spatial and temporal characteristics, *J. Neurosci.* **19**:6372–6384.

Puskin, J. S., Gunter, T. E., Gunter, K. K., and Russell, P. R., 1976, Evidence of more than one Ca^{2+} transport mechanism in mitochondria, *Biochemistry* **15**:3834–3842.

Qian, T., Herman, B., and Lemasters, J. J., 1999, The mitochondrial permeability transition mediates both necrotic and apoptotic death of hepatocytes exposed to Br-A23187, *Toxicol. Appl. Pharmacol.* **154**:117–125.

Reynolds, I, and Hastings, T. G., 1995, Glutamate induces the production of reactive oxygen species in cultured forebrain neurons following NMDA receptor activation, *J. Neurosci.* **15**:3318–3327.

Rizzuto, R., Simpson, A. W. M., and Pozzan, T., 1992, Rapid changes of mitochondrial Ca^{2+} revealed by specifically targeted recombinant aequorin, *Nature* **358**:325–327.

Rothman, S. M., and Olney, J. W., 1987, Excitotoxicity and the NMDA receptor, *Trends Neurosci.* **10**:299–302.

Ruck, A., Dolber, M., Wallimann, T., and Brdiczka, D., 1998, Reconstituted adenine nucleotide translocase forms a channel for small molecules comparable to the mitochondrial permeability transition pore, *FEBS Lett.* **426**:97–101.

Schinder, A. F., Olson, E., Spitzer, N. C., and Montal, M., 1996, Mitochondrial dysfunction is a primary event in glutamate neurotoxicity, *J. Neurosci.* **16**:6125–6133.

Shimizu, S., Narita, M., and Tsujimoto, Y., 1999, Bcl-2 family proteins regulate the release of apoptogenic cytochrome *c* by the mitochondrial channel VDAC, *Nature* **399**:483–487.

Sies, H., Gerstenecker, C., Menzel, H., and Flohe, L., 1972, Oxidation in the NADP system and release of GSSG from hemoglobin-free perfused rat liver during peroxidatic oxidation of glutathione by hydroperoxidase, *FEBS Lett.* **27**:171–175.

Susin, S. A., Zamzami, N., Castedo, M., Hirsch, T., Marchetti, P., Macho, A., Daugas, E., Geuskens, M., and Kroemer, G., 1996, Bcl-2 inhibits the mitochondrial release of an apoptogenic protease, *J. Exp. Med.* **184**:1331–1342.

Susin, S. A., Zamzami, N., Castedo, M., Daugas, E., Wang, H.-G., Geley, S., Fassy, F., Reed, J. C., and Kroemer, G., 1997, The central executioner of apoptosis: Multiple links between protease activation and mitochondria in Fas/Apo-1/CD95- and ceramide-induced apoptosis, *J. Exp. Med.* **186**:25–37.

Susin, S. A., Lorenzo, H. K., Zamzami, N., Marzo, I., Snow, B. E., Brothers, G. M., Mangion, J., Jacotot, E., Constantini, P., Loeffler, M., Larochette, N., Goodlett, D. R., Aebersold, R., Siderovski, D. P., Penninger, J. M., and Kroemer, G., 1999, Molecular characterization of mitochondrial apoptosis-inducing factor, *Nature* **397**:441–446.

Szabò, I., and Zoratti, M., 1991, The giant channel of the inner mitochondrial membrane is inhibited by cyclosporin A, *J. Biol. Chem.* **266**:3376–3379.

Thomas, C. E., and Reed, D. J., 1988, Effect of extracellular Ca^{++} omission on isolated hepatocytes: II. Loss of mitochondrial membrane potential and protection by inhibitors of uniport Ca^{++} transduction, *J. Pharmacol. Exp. Therap.* **245**:501–507.

Thor, H., Smith, M. T., Hartzell, P., Bellomo, G., Jewell, S. A., and Orrenius, S., 1982, The metabolism of menadione (2-methyl-1, 4-naphthoquinone) by isolated hepatocytes, *J. Biol. Chem.* **257**:6612–6615.

Trollinger, D. R., Cascio, W. E., and Lemasters, J. J., 1997, Selective loading of Rhod-2 into mitochondria shows mitochondrial Ca^{2+} transients during the contractile cycle in adult rabbit cardiac myocytes, *Biochem. Biophys. Res. Commun.* **236**:738–742.

Trump, B. F., Goldblatt, P. J., and Stowell, R. E., 1965, Studies of necrosis *in vitro* of mouse hepatic parenchymal cells: Ultrastructural alterations in endoplasmic reticulum, Golgi apparatus, plasma membrane, and lipid droplets, *Lab. Invest.* **14**:2000–2028.

Vanderkooi, J., Erecinska, M., and Chance, B., 1973, Cytochrome *c* interaction with membranes: I. Use of a fluorescent chromophore in the study of cytochrome *c* interaction with artificial membranes, *Arch. Biochem. Biophys.* **154**:219–229.

Vlessis, A. A., 1990, NADH-linked substrate dependence of peroxide-induced respiratory inhibition and calcium efflux in isolated renal mitochondria, *J. Biol. Chem.* **265**:1448–1453.

White, R. J., and Reynolds, I. J., 1996, Mitochondrial depolarization in glutamate-stimulated neurons: An early signal specific to excitotoxin exposure, *J. Neurosci.* **16**:5688–5697.

Wu, E. Y., Smith, M. T., Bellomo, G., and DiMonte, D., 1990, Relationships between transmembrane potential, ATP concentration, and cytotoxicity in isolated rat hepatocytes, *Arch. Biochem. Biophys.* **282**:358–362.

Yang, J., Liu, X., Bhalla, K., Kim, C. N., Ibrado, A. M., Cai, J., Peng, T.-I., Jones, D. P., and Wang, X., 1997, Prevention of apoptosis by bcl-2: Release of cytochrome *c* from mitochondria blocked, *Science* **275**:1129–1132.

Zahrebelski, G., Nieminen, A.-L., Al-Ghoul, K., Qian, T., Herman, B., and Lemasters, J. J., 1995, Progression of subcellular changes during chemical hypoxia to cultured rat hepatocytes: A laser scanning confocal microscopy study, *Hepatology* **21**:1361–1372.

Zoratti, M, and Szabò, I., 1995, The mitochondrial permeability transition, *Biochim. Biophys. Acta* **1241**:139–176.

Chapter 19

Mitochondria in Alcoholic Liver Disease

José C. Fernández-Checa, Carmen García-Ruiz, and Anna Colęll

1. INTRODUCTION

The pathogenesis of alcoholic liver disease has not yet been resolved, probably reflecting the multifactorial nature of the illness. It is recognized, however, that the oxidative metabolism of ethanol generates a variety of intermediates that participate in ethanol's escalating cytotoxic effects. Oxidative metabolism of alcohol occurs preferentially in the liver, which is eventually transformed into several end products. Certain subproducts generated during ethanol metabolism may be more toxic than alcohol itself, however, contributing to the development of alcoholic liver disease that culminates in liver cirrhosis (Ashak *et al.*, 1991).

Several recognized factors that contribute to development of the illness are listed in Figure 1. Altered cellular redox potential reflected in decreased NAD/NADH ratio and also acetaldehyde generation (acetaldehyde is an extremely reactive chemical intermediate due to continuous oxidation of alcohol) are two potential mediators of alcohol-induced effects (Lieber, 1988). The former induces several metabolic disturbances, whereas the latter can react with nucleophilic moieties of proteins, which can result in their inactivation or in the formation of stable acetaldehyde–protein adducts. Generation of antibodies against these adducts mediates the autoimmune injury in advanced states of alcoholic liver disease, as has been shown in the livers of alcoholic patients (Niemela *et al.*, 1991, 1995). Furthermore, oxidative alcohol metabolism results in an overproduction of reactive oxygen species (ROS) that, along with depletion of antioxidant stores, causes oxidative stress, a condition thought to mediate some of ethanol's damaging effects. In addition to

José C. Fernández-Checa, Carmen García-Ruiz, and Anna Colell.   Liver Unit, Departments of Medicine and Gastroenterology, Hospital Clínic i Provincial, Instituto Investigaciones Biomedicas August Pi I Suñer, Consejo Superior Investigaciones Cientificas, Barcelona 08036, Spain.

Mitochondria in Pathogenesis, edited by Lemasters and Nieminen.
Kluwer Academic/Plenum Publishers, New York, 2001.

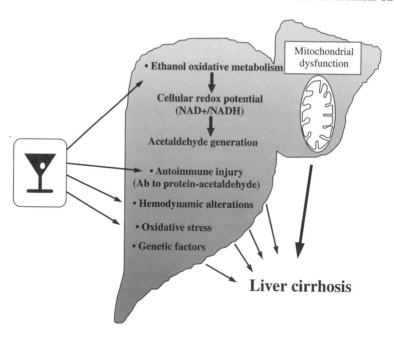

FIGURE 1. Pathogenesis of alcoholic liver disease. Some of the factors that contribute to developing alcoholic cirrhosis are indicated. Mitochondria are selective targets of alcohol intake and they play a key role in the control of many cellular processes, including cell death, hence their dysfunction via chronic alcohol intake may be a key event in initiating parenchymal cell function.

onset of these factors, there is an important genetic component that may determine why some individuals develop the disease while others do not despite their continued consumption of alcohol (Bosron *et al.*, 1993).

Previous studies in experimental animal models of alcoholic liver disease, as well as in human alcoholic patients, have demonstrated that mitochondria are specific targets of ethanol toxicity. Mitochondria are currently the subject of intense investigation due to the recognition that these subcellular organelles control the fate of cells against a variety of challenges and toxic substances (Green and Reed, 1998; Kroemer *et al.*, 1997), as detailed in several chapters throughout this book. Thus the targeting of mitocondria by alcohol may not only contribute to the loss of several cellular functions but may also be a necessary and key event for development of alcoholic liver disease (Fernández-Checa *et al.*, 1998). This chapter describes the mechanisms and functional significance of chronic alcohol-induced mitochondrial dysfunction in the development of alcohol-induced liver disease.

2. OXIDATIVE ALCOHOL METABOLISM

Although all the metabolic pathways of ethanol in mammalian cells are extramito-chondrial, they are intimately linked with mitochondrial metabolism because ethanol metabolism, both oxidative and nonoxidative, affects mitochondrial energy metabolism. Quantitatively, the liver is the major organ involved in metabolic disposal of ethanol.

Ethanol oxidation in liver is catalyzed predominantly by a class I, pyrazole-sensitive alcohol dehydrogenase (ADH) in the cytosol, of which different isoenzymes have been described (Lieber, 1991). This enzyme has a K_m for ethanol in the low to submillimolar range and saturates at ethanol concentrations readily achieved physiologically. Recently (Lieber, 1991) a high-K_m, class IV ADH was found in the stomach and is thought to play a significant role in ethanol oxidation in humans at high ethanol concentrations. Other ADH isoforms have been described, displaying a wide distribution, although their contribution to ethanol oxidation *in vivo* is unknown (Cronholm *et al.*, 1992).

The liver oxidizes ethanol through a cytochrome p450-linked pathway, which uses NADPH and molecular oxygen and is localized in the endoplasmic reticulum, referred to in the literature as *MEOS* (Lieber, 1987). After a great deal of controversy centered on the contribution of this pathway to the overall rate of ethanol oxidation, it became clear that its quantitative significance is small in normal conditions but increases with prolonged intake of ethanol, due to ethanol's selective induction of a specific high-K_m cytochrome P450 isoform called P450 2E1, which uses ethanol as a preferred substrate (Handler *et al.*, 1988; Lieber, 1987). Ethanol oxidation also occurs in peroxisomes, mediated by catalase. The catalase-dependent pathway in liver is limited by the supply of hydrogen peroxide, so reactions that produce hydrogen peroxide in peroxisomes greatly enhance the rate of ethanol oxidation through this pathway. One of the most physiologically relevant processes for generating hydrogen peroxide is β-oxidation of long-chain fatty acids (Handler and Thurman, 1990).

All oxidative routes for ethanol metabolism result in formation of acetaldehyde, which further metabolizes to acetate, primarily by a low-K_m aldehyde dehydrogenase within the mitochondria. This enzyme is found in many tissues but is most abundant in liver, where the rate of acetaldehyde formation is highest. Impairment of this activity, (which maintains acetaldehyde concentrations at very low levels) in certain Asian populations inclines toward moderate alcohol consumption. In mitochondria from perivenous hepatocytes, this impairment of low-K_m acetylaldehyde dehydrogenase indicates the existence of an acetaldehyde gradient along the liver acinus that is greater in the perivenous zone of the liver, the area where most of the ethanol-induced liver injury is seen in both alcoholic patients and experimental animal models (Asak *et al.*, 1991; García-Ruiz *et al.*, 1994; Lieber, 1994; Niemela *et al.*, 1995). Nonoxidative pathways of ethanol oxidation include the formation of ethyl esters of long-chain fatty acids, mediated by isoenzymes of the glutathione S-transferases and by a cholesterol esterase (Lange, 1991). These compounds affect mitochondrial function by acting as uncouplers, with decreased state 3 and increased state 4 respiration (Lange and Sobel, 1983). These fatty acid ethyl esters are found in several tissues affected by long-term alcohol abuse in humans (Laposata and Lange, 1986). Thus both oxidative and nonoxidative alcohol metabolism generate intermediates that initiate an event cascade critical for development of alcoholic liver diseases, targeting of mitochondria, and subsequent dysfunction.

3. ALCOHOL AND MITOCHONDRIAL DYSFUNCTION

One of the hallmarks of alcohol-induced liver damage is structural and functional alteration of mitochondria. It has been known for more than three decades that in both human and animal models, ethanol results in striking morphological alterations of hepatocellular mitochondria. Although structural changes include a reduced number of cristae and paracrystalline inclusions, one of the most prevalent modifications is enlarged mitochondria, giant structures that have been called *megamitochondria*. Their prevalence was first detected in alcoholic patients upon examination of liver histology by light microscopy, which showed megamitochondria that are clearly distinguisable from Mallory bodies (Brugera *et al.*, 1977). Similar mitochondrial structural changes evoked by chronic ethanol intake are described in animal models, and they have profound consequences for chronic alcohol-induced mitochondrial dysfunction (Cunningham *et al.*, 1990; Quintanilla *et al.*, 1989; Spach *et al.*, 1987; Arai *et al.*, 1984).

One of the most dramatic consequences of prolonged alcohol intake is impaired oxidative phosphorylation and consequent hepatocellular ATP production. Animal studies have characterized the dependence of mitochondrial dysfunction and the site within the electron transport chain that relates to duration of ethanol feeding. The lower ATP levels observed in chronic-ethanol-fed animals maintained on the diet for three weeks result from lower state 3 respiration with NADH-linked substrates (complex I) (Cederbaum *et al.*, 1974; Cederbaum and Rubin, 1975). Longer intake periods also decrease succinate-driven state 3 respiration (complex II) in ethanol-fed mitochondria. The ADP-stimulated electron transport is sustained for extended periods and affects all transport chain segments. Thus, state 3 respiration through the cytochrome oxidase portion of the electron transport chain is depressed to 20–40% of control values (Spach and Cunningham 1987; Thayer and Rubin 1979). Lower oxygen consumption in the presence of ADP and Pi (state 3 respiration) demonstrates that the ATP synthesis rate via oxidative phosphorylation is lower in ethanol-fed mitochondria. Oxygen consumption rate in the absence of ADP (state 4 respiration) is not altered by chronic ethanol feeding to the same degree as it is in state 3 respiration. An increase in state 4 respiration would indicate that mitochondria have less capacity to conserve the proton gradient generated by electron transport, and they would be less well coupled. Resting respiration state is an indication of the functional viability of isolated mitochondria. Feeding ethanol for 3–4 weeks results in either no loss or a slight decrease (10–15% vs control values) in state 4 respiration when either NADH-linked substrates or succinate are utilized as oxidizable substrates (Cederbaum *et al.*, 1974; Spach and Cunningham 1982). Longer feeding periods cause significant increases in state 4 respiration indicating that mitochondria are more uncoupled. State 4 respiration through cytochrome oxidase is significantly depressed by chronic ethanol, however, pointing to the dramatic alteration in this portion of the energy conservation system (Cunningham *et al.*, 1989; Spach and Cunningham, 1987). Ethanol produces dramatic alterations in several respiratory chain components. The activity and heme content of cytochrome oxidase are decreased 50–60% in ethanol mitochondria compared with control organelles. In addition, ethanol also decreases cytochrome *b* and some of the iron sulphur centers associated with NADH–ubiquinone reductase complex (Bernstein and Penniall, 1978; Rubin *et al.*, 1972; Schilling and Reitz, 1980). Electron transport and proton translocation through the NADH–ubiquinone reductase portion of the electron transport chain decrease by 30% and 40%, respectively.

Mitochondrial ATP synthetase is a very complex structure with the catalytic subunits (F_1) projecting into the mitochondrial matrix and into the proton channel portion (F_0) formed by other polypeptides attached to the catalytic part that traverses inner membrane. In both inner membrane preparations and submitochondrial particles capable of synthesizing ATP, ATPase is significantly lowered by chronic ethanol feeding (Hosein *et al.*, 1977; Thayer and Rubin, 1979). The same results have recently been shown in intact mitochondria.

In addition, ethanol feeding lowers the oligomycin sensitivity of ATP synthetase and renders it less tightly attached to the inner membrane. The decrease in oligomycin sensitivity and the structural changes of the ATP synthase induced by ethanol are found exclusively in the complexe's F_0 subunits (Montgomery *et al.*, 1987). Thus the overall decreased ATP levels seen in chronic-alcohol animals results from a lower rate, vs efficiency, because of ATP synthesis there is little or no change in the resting state respiration, indicating lack of uncoupling by ethanol feeding.

4. ALCOHOL, MITOCHONDRIAL BIOGENESIS, AND MEMBRANE PROPERTIES

The mitochondrial oxidative phosphorylation system is multicomponent, so changes in any one component result in rate-limiting synthesis of ATP and decreased oxygen consumption. The molecular basis for ethanol-induced lesions in the oxidative phosphorylation system is at the translation level of several components. Mitochondria have a circular DNA that encodes only 13 open reading frames, in addition to 2 rRNAs and a complete set of tRNAs with a high degree of direct sequence homology (Bibb *et al.*, 1981; Tzagaloff and Myers, 1986). Mitochondrial DNA gene products have been identified as members of the oxidative phosphorylation system (Fig. 2). The oxidative phosphorylation system portions that are decreased by ethanol feeding are encoded by mitochondrial DNA, whereas the unaffected components have a nuclear gene expression and are subsequently imported from cytoplasm into mitochondria, suggesting that chronic ethanol consumption could interfere with mitochondrial gene expression. The rate of incorporation of radiolabeled methionine into polypeptides was lower for all 13 of them, establishing a correlation between loss of catalytic activity and decrease in the synthesis rate of components affected by chronic ethanol feeding (Coleman and Cunningham, 1990). Ethanol feeding does not affect DNA, total RNA levels, or RNA polymerase activity, however. With slot blot analyses, 8 of the mRNA and ribosomal RNA were present in normal amounts in ethanol mitochondria, indicating that ethanol causes the defect in the mRNA processing of mitochondrial gene products. Thus the number of ribosomes from ethanol mitochondria decreased 60–70% compared with control mitochondria, possibly because of decreased synthesis in the endoplasmic reticulum or decreased transport across mitochondrial membranes into the matrix (Cunningham *et al.*, 1990).

In addition to altered mitochondrial functions from impaired mitochondrial gene expression, chronic alcohol consumption modifies the chemical composition of mitochondrial membranes, which in turn may have profound functional consequences. Early studies two to three decades ago in experimental models described alterations in the fatty acid composition of major mitochondrial membranes phospholipids. Although mitochondrial phospholipids distribution is unaffected by alcohol, the phospholipid-to-protein ratio depends on dietary fat content. Rats fed a low-fat diet along with the ethanol displayed a significant decrease in the phospholipid-to-protein ratio, an effect that disappears with a higher-fat diet. There is a definite interation between diet and ethanol in eliciting changes

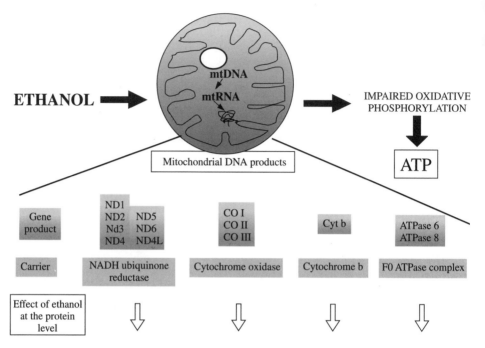

FIGURE 2. Ethanol and mitochondrial DNA processing. Ethanol induces loss of mitochondrial function, reflected in impaired oxidative phosphorylation and subsequent ATP synthesis. Several components of the mitochondrial electron transport chain are encoded by mitochondrial DNA (dark boxes). Studies in experimental models of alcohol research show that alcohol interferes with translation of the 13 mitochondrial DNA-encoded products (open arrows).

in hepatic mitochondrial phospholipids; thus, the reported decreases in acyl composition ranged from 15% to 30% decrease in the 16 : 0 from either phosphatydilcholine (PC) and phosphetydilethandamine (PE), whereas the levels of 18 : 0 and 18 : 1 saturating PC and PE increases by alcohol in rats fed a moderate-fat diet. An 18 : 1 decrease of in cardiolipin has, however, been described (Arai *et al.*, 1984; Castro *et al.*, 1995; Cunningham *et al.*, 1982; Ellingson *et al.*, 1988; Rouach *et al.*, 1984; Tarachi and Rubin, 1985; Waring *et al.*, 1981). These pioneering studies thus indicate that alcohol depresses the phospholipid reacylation activities associated with mitochondria, suggesting that mitochondrial phospholipid alterations may be related to ethanol-induced changes in mitochondrial enzymes involved in phospholipid metabolism. Furthermore, alterations in fatty acid availability due to ethanol-related changes in microsomal elongation and desaturation activities also appear to affect phospholipid fatty acid composition in ethanol-fed animal mitochondria.

These changes influence mitochondrial function in several ways. In particular, ethanol-induced decreases in cardiolipin level can adversely affect the functioning of cytochrome *c* oxidase because its activity depends on the level of this anionic phospholipid. Furthermore, because fatty acid composition is one of the major determinants of the physical properties of biological membranes, composition alterations affect mitochondrial membrane fluidity, which in turn can affect several proteins whose function depends on the appropriate fluid state of the membrane. Of significance in this regard is that functioning of the carrier responsible for transport of cytosol GSH into mitochondria is modulated by mitochondrial membrane fluidity (Colell *et al.*, 1997).

5. ALCOHOL AND MITOCHONDRIAL OXIDATIVE STRESS

Oxidative stress is a condition that reflects an imbalance between two opposing, antagonistic forces, reactive oxygen species (ROS) and antioxidants, in which the effects of the former predominate over the compensating action of the latter. An imbalance can be created by excessive generation of ROS, by limited availability of antioxidants, or both. Mitochondria represent a major source of ROS because the mitochondrial electron transport chain accounts for about 85–90% of all oxygen consumed in the cell (Chance *et al.*, 1979). Most of molecular oxygen consumed in the electron transport chain occurs at the cytochrome *c* oxidase complex; the complete reduction to water precludes ROS formation. The formation of ROS takes places in specific electron transport chain segments, mainly at the ubiquinone site of complex III respiration, which activates molecular oxygen to ROS (Turrens and Boveris, 1980; Turrens *et al.*, 1985). Superoxide anion production originates from the ubiquinone Q cycle of complex III, where one electron from ubisemiquinone is transferred directly to molecular oxygen. Reduction of Q to ubiquinol (QH2) occurs at the NADH dehydrogenase and succinate dehydrogenase complexes. The transfer of one electron from QH2 to the cytochrome–bcl complex catalyzed by the Rieske iron–sulfur center generates ubisemiquinone, which can be reduced to ubiquinol or can transfer a single electron to the oxygen-generating superoxide anion.

To offset the harmful consequences of ROS generation, nature has developed strategies for controlling the impact of their continued formation. Mitochondria are equipped with MnSOD and a redox cycle using reduced GSH and GSH peroxidase, and they contain no catalase, thus leaving mitochondrial matrix GSH as the only available defense against the potentially toxic effects of hydrogen peroxide produced endogenously in the electron transport chain. Hydrogen peroxide, generated within the electron transport chain, can undergo several possible fates: conversion to hydroxyl radical with the participation of transition metals in the Haber–Weiss and Fenton reactions, reduction to water by the catalysis of GSH peroxidase with the required participation of reduced GSH as cofactor, or its diffusion out of mitochondria, where it may be metabolized by catalase. Therefore one of the major roles of mitochondrial GSH is to control generation of hydrogen peroxide produced endogenously in all aerobic cells within the mitochondrial electron transport chain.

GSH is synthesized exclusively in the cytosol of most cells. A small fraction of GSH, however, is found inside mitochondria (Fig. 3). Because mitochondria do not have the enzymes for synthesizing GSH, the mitochondrial pool of GSH is derived from the operation of a transport system (Fernández-Checa *et al.*, 1998). Oxidative metabolism of ethanol induces a defect in the functioning of mitochondrial GSH transport, leading to a selective depletion of the mitochondrial GSH pool. The selectivity is translated in the sense that the cytosolic pool of GSH is spared and the functioning of other carriers remains unimpaired in mitochondria from ethanol-fed animals. Previous studies in intact hepato-cytes and in isolated mitochondria from rats fed chronic ethanol diets revealed that the

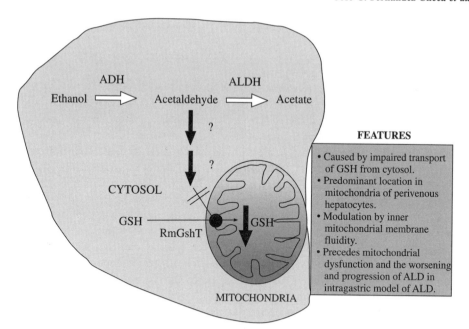

FIGURE 3. Origin of the mitochondrial GSH pool. GSH is synthesized exclusively in the cytosol of most cells. A small fraction of GSH is found inside mitochondria. Because mitochondria lack the enzymes that synthesize GSH, the mitochondrial GSH pool is derived from the operation of a transport system (RmGshT). Oxidative metabolism of ethanol induces a defect in the functioning of mitochondrial GSH transport, leading to a selective defect of the mitochondrial pool of GSH. The selectivity is translated in the sense that cytosolic GSH is spared and the functioning of other mitochondrial carriers remains unimpaired as well in mitochondria from ethanol-fed animals. The box shown on the right highlights some of the main features of these findings.

defect in the GSH-transporting polypeptide is observed predominantly in mitochondria from perivenous hepatocytes. Such a defect is one of the earliest consequences of alcohol intake, because selective mitochondrial GSH depletion occurs after the first two weeks of alcohol feeding and precedes mitochondrial dysfunction, including impairment of oxidative phosphorylation, resulting in ATP depletion and worsening of alcoholic liver disease (ALD) (Fernández-Checa *et al.*, 1987, 1989; 1991; 1993 Gracía-Ruiz *et al.*, 1994, 1995; Takeshi *et al.*, 1992).

Depletion of mitochondrial GSH induced by ethanol intake is significant, because to effectively control the overproduction of specific ROS species a balance of antioxidant defenses must occur. To control the fate of hydrogen peroxide, there must be an equilibrium between the activity of enzymes that generate hydrogen peroxide (e.g., Cu/ZnSOD or MnSOD) and the activity of enzymes that metabolize hydrogen peroxide (e.g., GSH/peroxidase or catalase) (Fernández-Checa *et al.*, 1997). For instance, if the relative activity of MnSOD predominates over that of GSH peroxidase, hydrogen peroxide can accumulate, leading to harmful generation of hydroxyl radicals via Haber-Weiss or Fenton reactions, resulting in cellular damage. Hydrogen peroxide can accumulate if GSH peroxidase activity is low relative to that of MnSOD. The status of MnSOD activity in alcohol-fed cells is controversial, and increased or unchanged activity depends on the conditions (Koch *et al.*, 1994; Perera *et al.*, 1995). Likewise, GSH peroxidase's role in the experimental model of ALD remains unsettled because the increase of its mRNA by ethanol is not reflected in enzymatic activity (Nanji *et al.*, 1995; Polavarapu *et al.*, 1996). Even in the absence of changes in MnSOD activity or GSH peroxidase expression decreased GSH (a cofactor required for the functioning of GSH peroxidase) can lead to an accumulation of hydrogen peroxide in alcohol-fed animal mitochondria. Thus it is easy to envision the potentially dramatic consequences of a selective defect in mitochondrial GSH transport and the subsequent GSH depletion for the functioning of mitochondria, control of cellular functions, and for cell survival itself.

6. ALCOHOL-INDUCED CYTOKINE OVERPRODUCTION

Cytokines play a critical role in infection because their levels in blood increase during an infection and decrease as infection subsides. Cytokines exert a immunoregulatory role, attracting and activating immune system components and promoting blood clotting. Persistently elevated cytokine levels cause symptoms of chronic inflammation. The liver also plays a role in the inflammatory response, both as a potential site of chronic inflammatory disease as in ALD, and as a source of phagoyctes and cytokines. Alcohol consumption increases intestinal permeability, permitting certain toxic bacterial products (endotoxins) to reach the bloodstream. The targeting of Kupffer cells by endotoxins causes a burst of cytokine overproduction that contributes to liver inflammation. Clinical studies support the role of disordered cytokine function in producing signs and symptoms of ALD. Patients with ALD exhibit high levels of cytokines, including interleukin 1, 6, and 8, as well as tumor necrosis factor (TNF). Furthermore, elevated TNF levels in blood correlate with poor prognosis in patients with alcoholic hepatitis. These findings are strong evidence that overproduction of TNF is key for developing ALD.

6.1. Tumor Necrosis Factor and Mitochondria

TNF exerts a pleiotropic array of strikingly different cellular reactions depending on concentration and target cell type. It plays a key role in infection and immunity by regulating specific genes needed for the host defense against a varied repertoire of agents. Apparently, TNF plays a role in controlling the cell cycle, because DNA synthesis and cell proliferation increase in cells exposed to TNF, indicating that it acts as a mitogenic stimuli. Such TNF regulation of gene expression is mediated by induction of early responsive genes including c-Jun and transcription factors (i.e., NFκB). Overproduction of TNF, however, is a key event in developing several pathological states, and TNF promotes cell injury through several mechanisms. For example, there is increasing evidence for the participation of oxidative stress in TNF-induced cytotoxic action due to increased ROS generation (Fig. 4).

Earlier studies revealed that cell exposure to TNF increased the hydroxyl radical formation in tumorigenic mouse fibroblasts, thus conferring TNF resistance by transfecting TNF-sensitive cells with a functional MnSOD gene (Wong and Goeddel, 1988; Yamauchi *et al.*, 1989). Furthermore, incubation of murine fibrosarcoma cells with TNF resulted in ROS overproduction originating from the mitochondrial electron transport chain (Goosens *et al.*, 1995; Shulze-Osthoff *et al.*, 1982), findings demonstrating that mitochondria are

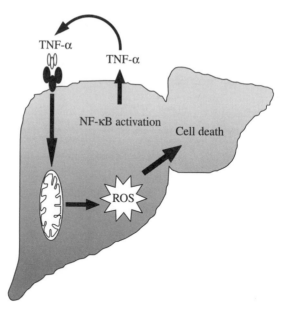

FIGURE 4. Role of tumor necrosis factor (TNF) in developing alcoholic liver disease. Circulating levels of TNF and other inflammatory cytokines increase in ALD patients and experimental animal models of ALD. Evidence indicates that TNF targets mitochondria, resulting in overexpression of ROS, which may cause cell death and expression of cytokines by activation of transcription factors (e.g., NFκB).

specific targets of cytokine action and that mitochondrial ROS overexpression is a critical event in the cytotoxicity of inflammatory cytokines. Nevertheless, most cell types are relatively TNF-insensitive. Mitochondrial GSH is a strategic line of defense for controlling generation of hydrogen peroxide, however, and conceivably its severe alcohol-induced depletion can render hepatocytes susceptible to TNF. Recent studies with intact hepatocytes isolated from chronic ethanol-fed rats and cultured in the presence of moderate doses of TNF have tested this hypothesis. Their findings show that hepatocytes from control rats survived TNF exposure, consistent with the resistance of most cell types to TNF, but hepatocytes from rats fed alcohol for four weeks lost viability upon exposure to increasing TNF concentrations, an outcome correlated with progressive TNF-induced generation of hydrogen peroxide. Thus the studies uncovered a sensitization of rat hepatocytes to TNF, apparently determined by chronic alcohol feeding (Colell *et al.*, 1998).

6.2. Ceramide Interaction with Mitochondria

The cellular signaling network used by TNF to transmit its effects to the cell interior is complex and involves generation of a wide variety of intermediates and protein–protein interactions. Knowledge has recently increased concerning the interaction between several protein adaptors and specific cytoplasmic-tail receptor domains, and there are numerous intermediates whose cellular levels increase upon TNF exposure, but the link between TNF transmission to the cell interior and subsequent overproduction of ROS remains enigmatic. Ceramide (*N*-fattyacyl-sphingosine) has attracted considerable attention for its role as an intracellular effector molecule that mimics some of the biological effects exerted by inflammatory cytokines such as TNF (Kolesnick and Golde, 1994; Hannun, 1994). In addition to its *de novo* biosynthesis, initiated by condensation of serine and palmitoyl-CoA, ceramide can be generated by sphingomyelin hydrolysis. Thus enzymes that hydrolyze sphingomyelin, such as the sphingomyelinases, stand as regulators of intracellular ceramide levels, and consequently of ceramide-mediated functions. These enzymes are key components of the sphingomyelin pathway, a ubiquitous system that transduces cytokine signals to the cell interior. Most of the effects ascribed to ceramide have been observed in studies using cell-permeable ceramide analogues, which contain short aliphatic chains linked to sphingosine. These studies reveal ceramide's ability to elicit a wide variety of effects, as many as those evoked by cytokines (Fernández-Checa *et al.*, 1997; Hannun *et al.*, 1994; Kolesnick and Golde, 1994).

Recent reports have described the ability of synthetic ceramides to interact with specific segments of the mitochondrial electron transport chain (Fig. 5). Indeed, incubation of isolated rat liver mitochondria in the presence of *N*-acetyl-sphingosine (ceramide C2) led to dose-dependent overgeneration of hydrogen peroxide (Degli-Esposti and McLennan, 1998; García-Ruiz *et al.*, 1997; Quillet-Mary *et al.*, 1997). Furthermore, additional studies with isolated mitochondria and with purified complex III revealed that ceramide C2 specifically reduces mitochondrial respiratory chain complex III activity (Gudz *et al.*, 1997), and showed that synthetic ceramides interact with complex III, favoring the transfer

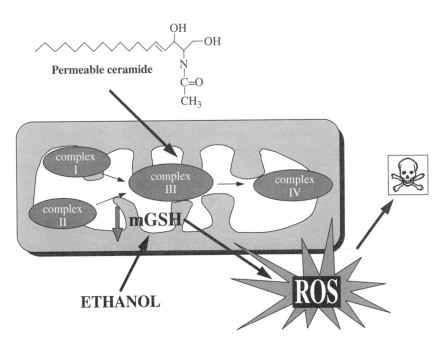

FIGURE 5. Interaction of ceramides with mitochondria. A new role for ceramide has been recently uncovered in studies revealing the interaction of this class of lipid intermediate with complex III of the respiratory chain. Following this interaction, the transfer of a single electron to molecular oxygen results in formation of superoxide anion and hydrogen peroxide. A precise balance of antioxidant defenses in the matrix controls hydrogen peroxide formation.

of a single electron to molecular oxygen and thus forming hydrogen peroxide from the superoxide anion.

Because overgeneration of ROS is one of the consequences of mitochondrial permeability pore opening, which not only has been implicated in several disorders but also appears to be a key event in the control of apoptosis and cell death, conceivably the formation of hydrogen peroxide by C2 may be secondary to pore opening. Evidence against this is presented in studies determining the time-dependence of these reciprocally regulated processes, however, indicating that the formation of ROS by ceramide precedes the mitochondrial swelling reflective of permeability pore opening (Gracía-Ruiz *et al.*, 1997). It remains to be established whether ceramide binding to complex III causes excessive ROS formation that favors subsequent pore opening or whether ceramide interacts with multiple mitochondrial sites via different mechanisms. A clear under-standing of the mechanism responsible for ceramide's modulation of mitochondrial function may prove critical in determining the signal transduction pathways involved in growth suppression and cell death.

6.3. Mitochondrial GSH Role in Sensitization to Tumor Necrosis Factor

Further studies were undertaken to delineate how alcohol consumption determines hepatocytes' greater vulnerability to TNF. Chronic alcohol feeding results in a selective depletion of mitochondrial GSH, and due to the critical role of this antioxidant in controlling the fate of hydrogen peroxide, it is likely that this depletion is the molecular determinant for hepatocellular TNF sensitization. According to this hypothesis, depletion of mitochondrial GSH from normal hepatocytes should result in a TNF sensitization similar to that of hepatocytes from alcohol-fed rats, so restoring GHS should abolish the sensitization.

To validate our observations that a limited mitochondrial GSH pool controls TNF-α susceptibility, the GSH levels of normal hepatocytes were depleted using (R, S)-3-hydroxy-4-pentenoate. Its predominant metabolism within mitochondria by the (R)-3-hydroxybutanoate: NAD+ oxidoreductase generates a Michael acceptor that reacts with GSH, resulting in its depletion (Colell *et al.*, 1998). Our findings demonstrated a similar sensitization of normal hepatocytes to TNF-α, indicating that this alcohol-induced susceptibility is not an artifact but rather is causally related to the alcohol-induced limitation of the mitochondrial pool of GSH.

To examine the role and significance of mitochondrial GSH in regulating TNF-α-induced cell death, we used a permeable form of GSH, GSH-ethylester, which diffuses readily into cells and intracellular compartments. The GSH-EE showed increased levels of mitochondrial GSH like those of pair-fed controls, whereas the cytosolic pool of GSH increased 20–30% above basal levels. In this paradigm hepatocytes from ethanol-fed rats were resistant to TNF-α, a resistance paralleled by reduced hydrogen peroxide generation.

The data thus derived provides strong evidence that ethanol-induced mitochondrial GSH depletion sensitizes hepatocytes to the toxic effects of tumor necrosis factor. By targeting mitochondria, tumor necrosis factor leads to an uncontrolled ROS generation that may promote cell death while increasing the expression of inflammatory cytokines, including TNF itself, that can further act in parenchymal cells in an autocrine fashion, establishing a lethal vicious cycle. Replenishment of the mitochondrial GSH pool both attenuates the lethal effect of oxidative stress and down-regulates the expression of TNF-induced cytokines. This illustrates the potential beneficial effects of therapeutic strategies targeted toward increasing the specific antioxidant defense within mitochondria. In this context recent studies using a human cell line incubated with TNF and ethanol revealed a possible therapeutic role of tauroursodeoxycholic acid in preventing ethanol-induced TNF expression and cytotoxicity by increasing the levels of mitochondrial GSH (Neuman *et al.*, 1998).

7. TREATMENT OF ALCOHOLIC LIVER DISEASE AND MITOCHONDRIAL GSH RESTORATION

Interestingly, previous studies in ethanol-fed baboons showed that S-adenosyl-L-methionine (SAM) administration prevents ethanol-induced liver injury ameliorating the release of glutamate dehydrogenase, an enzyme in the mitochondrial intermembrane space (Lieber *et al.*, 1990), suggesting that SAM's benefit may be due to protection of the mitochondria. Therefore, the effect of SAM administration on compartmentation of mitochondrial GSH was assessed in hepatocytes from ethanol-fed rats. These studies revealed SAM's ability to restore the mitochondrial GSH pool depleted by ethanol feeding, thus normalizing mitochondrial GSH content (Colell *et al.*, 1997, 1998; García-Ruiz *et al.*, 1995). Such effects contrast with those exerted by *N*-acetylcysteine (NAC), a GSH precursor ineffective in replenishing mitochondrial GHS in ethanol-fed rats despite its efficiency in increasing cytosolic GSH. These findings linked SAM's therapeutic benefits with its ability to restore mitochondrial GSH. Mitochondrial GSH depletion preceded loss of mitochondrial function, so its recovery by SAM may be significant in preventing mitochondrial dysfunction at the early stages of ALD.

Based on these findings, further studies were done to evaluate the mechanism whereby SAM administration replenishes mitochondrial GSH despite ethanol consumption. The work gave new insights into the mechanisms underlying the ethanol-induced defect in mitochondrial GSH transport, and the results showed that the defect is intrinsic to the isolated organelle, as revealed by changes in the kinetic parameters of GSH transport into isolated mitochondria and increased microviscosity (order parameter) of the membrane lipids (Colell *et al.*, 1998). The SAM treatment restored the size of the mitochondrial GSH pool by normalizing GSH transport kinetic parameters via increase of V_{max} for both low- and high-affinity components and through lowering the K_m of the high-affinity components. The alcohol-induced mitochondrial was restricted to the inner membrane; mitoplasts from ethanol-fed rats showed features similar to those of intact mitochondria in terms of impaired GSH transport and increased order parameter. Such changes were specific for the mitochondrial GSH carrier, because another mitochondrial transporter, the ADP/ATP translocator, was unaffected by ethanol feeding.

Further evidence favoring modulation of the mitochondrial GSH carrier via membrane fluidity was provided by studies in which isolated mitochondria and mitoplasts from ethanol-fed rats were incubated *in vitro* with fluidizing agent A_2C, a fatty acid derivative that intercalates into the lipid bilayer. This resultant fluidization of mitochondrial membrane restored GSH transport in mitochondria to levels similar to those in pair-fed controls. These findings underscore the mitochondrial GSH carrier's susceptibility to the physical properties of the mitochondrial inner membrane, where it is located. Ethanol-induced perturbation of the fluid state of inner membrane lipids seems to alter the functional operation of this carrier, resulting in selective GSH depletion. The contribution that additional factors such as alterations in transcription, translation, posttranslational processing, and/or import of carrier in inner membrane may have on manifestation of the ethanol-induced GSH transport defect remains to be assessed. Recent studies (Horne *et al.*, 1997) in isolated rat liver mitochondria showed that SAM is transported into the mitochondrial matrix. Therefore, SAM's ability to restore the physicial properties of mitochondria from ethanol-fed cells may conceivably be mediated by the transport of exogenous SAM into the mitochondrial compartment, where is serves as a methyl donor. Finally, in addition to it's value in treating certain liver diseases, SAM's functional role in preventing the mitochondrial GSH-carrier defect may offer a rationale for its therapeutic use in the treatment of ALD (Mato *et al.*, 1997). Indeed, recent clinical trials have uncovered a potential therapeutic value in ALD treatment, especially in the early stages of the disease.

8. CONCLUSION

Mitochondria generate ROS as byproducts of aerobic respiration. When the availability of free radicals is controlled, ROS act as signaling intermediates, regulating nuclear gene expression by activating transcription factor NFκB. The GSH redox cycle plays a critical role in controlling certain ROS, especially hydrogen peroxide, because mitochondria lack catalase. Hence not only the presence of reduced GSH but a precise balance between antioxidant enzymes such as MnSOD and GSH peroxidese guarantees adequate hydrogen peroxide levels. Limiting mitochondrial GSH may result in uncontrolled ROS generation, which can initiate escalating cell damage. Alcohol-induced depletion of mitochondrial GSH sensitizes hepatocytes to the toxic effects of tumor necrosis factor by amplifying mitochondrial ROS generation, which not only promotes cell death but also increases the expression of cytokines, including TNF, that can further act in parenchymal cells in an autocrine fashion, establishing a deadly vicious cycle. Replenishing mitochondrial GSH by different strategies (*in vivo* administration of S-adenosyl-L-methionine; *in vitro* by GSH-ethylester) both attenuates the lethal effect of oxidative stress and down-regulates expression of cytokines. Based on these observations, beneficial effects can be expected from therapeutic strategies targeted to increase selective antioxidant defenses in mitochondria.

ACKNOWLEDGEMENTS. The work presented was supported by grants from the U.S. National Institute of Alcohol Abuse and Alcoholism grant AA09526, Plan Nacional de I+D, Fondo Investigaciones Sanitarias and Europharma, SA.

REFERENCES

Arai, M., Gordon, E. R., and Lieber, C. S., 1984a, Decreased cytochrome oxidase activity in hepatic mitochondria after chronic ethanol consumption, and the possible role of decreased cytochrome aa3 content and changes in phospholipids, *Biochim. Biophys. Acta* **797**:320–327.

Arai, M., Leo, M. A., Nakano, M., Gordon, E. R., and Lieber, C. S., 1984b, Biochemical and morphological alterations of baboon hepatic mitochondria after chronic ethanol consumption, *Hepatology* **4**:165–174.

Ashak, K. A., Zimmerman, H. J., and Ray, M. B., 1991, Alcoholic liver disease: Pathological, pathogenic, and clinical aspects, *Alcoholism: Clin. Exp. Res.* **15**:45–66.

Bernstein, J. D., and Penniall, R., 1978, Effects of chronic ethanol treatment upon rat liver mitochondria, *Biochem. Pharmacol.* **27**:2337–2342.

Bibb, M. J., Van Etten, R. A., Wright, C. T., Walberg, H. W., and Clayton, D. A., 1981, Sequence and gene organization of mouse mitochondrial DNA, *Cell* **26**:167–180.

Bosron, W. F., Ehrig, T., and Li, T. K., 1993, Genetic factors in alcohol metabolism and alcoholism, *Sem. Liver Dis.* **13**:126–135.

Brugera, M., Bertran, A., Bombi, J. A., and Rodés, J., 1977, Giant mitochondria in hepatocytes: A diagnostic hint for alcoholic liver disease, *Gastroenterology* **73**:1383–1387.

Castro, J., Cortés, J. P., and Guzmán, M., 1995, Properties of the mitochondrial membrane and carnitine palmitoyltransferase I in the periportal and perivenous zone of the liver, *Biochem. Pharmacol.* **41**:1987–1995.

Caderbaum, A. I. and Rubin, E., 1975, Molecular injury to mitochondria produced by ethanol and acetaldehyde, *Fed. Proc.* **34**:2045–2051.

Cederbaum, A. I., Lieber, C. S., and Rubin, E., 1974, Effects of chronic ethanol treatment on mitochondrial functions: Damage to site I, *Arch. Biochem. Biophys.* **165**:560–569.

Cedebaum, A. I., Lieber, C. S., Beattie, D. S., and Rubin, E., 1975, Effects of chronic ethanol ingestion on fatty acid oxidation by hepatic mitochondria, *J. Biol. Hem.* **250**:5122–5129.

Chance, B., Sies, H., and Boveris, A., 1979, Hydroperoxide metabolism in mammalian organs, *Physiol. Rev.* **59**:527–605.

Colell, A., García-Ruiz, C., Miranda, M., Ardite, E., Marí, M., Morales, A., Corrales, F., Kaplowitz, N., and Fernández-Checa, J. C., 1998, Selective glutathione depletion of mitochondria by ethanol sensitizes hepatocytes to tumor necrosis factor, *Gastroenterology* **115**:1541–1551.

Colell, A., García-Ruiz, C., Morales, A., Ballesta, A., Ookhtens, M., Rodés, J., Kaplowitz, N., and Fernández-Checa, J. C., 1997, Transport of reduced glutathione in hepatic mitochondria and mitoplasts from ethanol-fed treated rats: Effect of membrane physical properties and S-adenosyl-L-methionine, *Hepatology* **26**:699–708.

Coleman, W. B., and Cunningham, C. C., 1990, Effects of ethanol consumption on the synthesis of polypeptides encoded by the hepatic mitochondrial genome, *Biochim. Biophys. Acta* **1019**:142–150.

Cunningham, C. C., Filus, S., Bottenus, R. E., and Spach, P. I., 1982, Effect of ethanol consumption on the phospholipid composition of rat liver microsomes and mitochondria, *Biochim. Biophys. Acta* **712**:225–233.

Cunningham, C. C., Kouri, D. L., Beeker, K. R., and Spach, P. I., 1989, Comparison of effects of 1989, long-term ethanol consumption on the heart and liver of the rat, *Alcoholism: Clin. Exp. Res.* **13**:58–65.

Cunningham, C. C., Coleman, X. and Spach, V., 1990, The effects of chronic ethanol consumption on hepatic mitochondrial energy metabolism, *Alcohol Alcoholism* **25**:127–136.

Cronholm, T., Norsten-Hoog, C., Ekstrom, G., Handler, J. A., Thurman, R. G., and Ingelman-Sundberg, M., 1992, Oxidoreduction of butanol in deermice lacking hepatic cytosolic alcohol dehydrogenase, *Eur. J. Biochem.* **204**:353–357.

Degli Esposti, M., and McLennan, H., 1998, Mitochondria and cells produce reactive oxygen species in virtual anaerobiosis: Relevance to ceramide-induced apoptosis, *FEBS Lett.* **430**:338–342.

Ellingson, J. S., Taraschi, T. F., Wu, A., Zimmerman, R., and Rubin, E., 1988, Cardiolipin from ethanol-fed rats confers tolerance to ethanol in liver mitochondrial membranes, *Proc. Natl. Acad. Sci. USA* **85**:3353–3357.

Fernández-Checa, J. C., Ookhtens, M., and Kaplowitz, N., 1987, Effect of chronic ethanol feeding on rat hepatocytic glutathione: Compartmentation, efflux, and response to incubation with ethanol, *J. Clin. Invest.* **80**:57–62.

Fernández-Checa., J. C., Ookhtens, M., and Kaplowitz, N., 1989, Effect of chronic ethanol feeding on rat hepatocytic GSH: Relationship of cytosolic GSH to efflux and mitochondrial sequestration, *J. Clin. Invest.* **83**:1247–1251.

Fernández-Checa., J. C., García-Ruiz, C., Ookhtens, M., and Kaplowitz, N., 1991, Impaired uptake of glutathione by hepatic mitochondria from ethanol-fed rats, *J. Clin. Invest.* **87**:397–405.

Fernández-Checa, J. C., Takeshi, H., Tsukamoto, H., and Kaplowitz, N., 1993, Mitochondrial glutathione depletion and alcoholic liver disease, *Alcohol* **10**:469–475.

Fernández-Checa, J. C., Kaplowitz, N., García-Ruiz, C., Colell, A, Miranda, M., Marí, M., Ardite, E., and Morales, A., 1997, GSH transport in mitochondria: Defense against TNF-induced oxidative stress and alcohol-induced defect, *Am. J. Physiol.* **273**:G7–G17.

Fernández-Checa, J. C., Kaplowitz, N., García-Ruiz, C., and Colell, A., 1998, Mitochondrial glutathione: Importance and transport, *Sem. Liver Dis.* **18**:389–401.

García-Ruiz, C., Morales, A., Colell, A., Ballesta, A., Rodés, J., Kaplowitz, N., and Fernández-Checa, J. C., 1994, Effect of chronic ethanol feeding on glutathione and functional integrity of mitochondria in periportal and perivenous rat hepatocytes, *J. Clin. Invest.* **94**:193–201.

García-Ruiz, C., Morales, A., Colell, A., Ballesta, A., Rodes, J., Kaplowitz, N., and Fernández-Checa, J. C., 1995, Feeding S-adenosyl-L-methionine attenuates both ethanol-induced depletion of mitochondrial glutathione and mitochondrial dysfunction in periportal and perivenous rat hepatocytes, *Hepatology* **21**:207–214.

García-Ruiz, C., Colell, A., Marí, M., Morales, A., and Fernández-Checa, J. C., 1997, Direct effect of ceramide on the mitochondrial electron transport chain leads to generation of reactive oxygen species: Role of mitochondrial glutathione, *J. Biol. Chem.* **272**:11369–11377.

Goosens, V., Grooten, J., de Vos, J., and Fiers, W., 1995, Direct evidence for tumor necrosis factor-induced mitochondrial reactive oxygen intermediates and their involvement in cytotoxicity, *Proc. Natl. Acad. Sci. USA* **92**:8115–8119.

Green, D. R., and Reed, J. C., 1998, Mitochondria and apoptosis, *Science* **281**:1309–1312.

Gudz, T. I., Tserng, K-Y, and Hoppel, C. L., 1997, Direct inhibition of mitochondrial respiratory chain complex III by cell-permeable ceramide, *J. Biol. Chem.* **272**:24154–24158.

Handler, J. A., and Thurman, R. G., 1990, Redox interactions between catalase and alcohol dehydrogenase pathways of ethanol metabolism in perfused rat liver, *J. Biol. Chem.* **265**:1510–1515.

Handler, J. A., Koop, D., Coon, M., Takei, Y., and Thurman, R. G., 1988, Identification of P 450ALC in microsomes from alcohol dehydrogenase deficient deermice: Contribution to ethanol elimination *in vivo*, *Arch. Biochem. Biophys.* **264**:114–124.

Hannun, Y. A., 1994, The sphingomyelin cycle and second messenger function of ceramide, *J. Biol. Chem.* **269**: 3125–3128.

Hosein, E. A., Hofman, I., and Linder, E., 1977, The influence of chronic ethanol feeding to rats on the integrity of liver mitochondrial membrane assessed with the Mg^{2+}-stimulated ATPase enzyme, *Arch. Biochem. Biophys.* **183**:64–72.

Horne, D. W., Holloway, R. S. and C., Wagner, 1997, Transport of S-adenosyl-L-methionine in isolated rat liver mitochondria, *Arch. Biochem. Biophys.* **343**:201–206.

Koch, O. R., M. E., DeLeo, Borrello, S., Palombini, G., and Galleotti, T., 1994, Ethanol treatment up-regulates the expression of mitochondrial manganese superoxide dismutase activity in rat liver, *Biochem. Biophys. Res. Commun.* **201**:1356–1365.

Kolesnick, R., and Golde, D. W., 1994, The sphingomyelin pathway in tumor necrosis factor and interleukin-1 signaling, *Cell* **77**:325–328.

Kroemer, G., Zamzami, N., and Susin, A., 1997, Mitochondrial control of apoptosis, *Immunol. Today* **18**: 44–51.

Lieber, C. S., 1987, Microsomal ethanol oxidizing system, *Enzyme* **37**:45–56.

Lieber, C. S., 1988, Metabolic effects of acetaldehyde, *Biochem. Soc Trans.* **16**:241–247.

Lieber, C. S., Casini, A., DeCarli, L. M., Kim, C. L., Lowe, K., Sasaki, R., and Leo, M. A., 1990, S-adenosyl-L-methionine attenuates alcohol-induced liver injury in the baboon, *Hepatology* **11**:165–172.

Lieber, C. S., 1991, Hepatic metabolic and toxic effect of ethanol: 1991 update, *Alcohol: Clin. Exp. Res.* **15**:573–592.

Lieber, C. S., 1994, Alcohol and the liver: An update, *Gastroenterology* **106**:1085–1105.

Lange, L. G., 1991, Mechanism of fatty acid ethyl ester formation and its biological significance, *Ann. NY Acad. Sci.* **625**:802–817.

Lange, L. G., and Sobel, B. E., 1983, Mitochondrial dysfunction induced by fatty acid ethyl esters as myocardial metabolites of ethanol, *J. Clin. Invest.* **72**:724–731.

Laposata, E., and Lange, L. G., 1986, Presence of non-oxidative ethanol metabolism in human organs commonly damaged by ethanol abuse, *Science* **231**:497–499.

Mato, J. M., Cámara, J., Ortiz, P., Rodés, J., and the Spanish Collaborative group for the study of alcoholic liver cirrhosis, 1997, S-adenosylmethionine in the treatment of alcoholic liver cirrhosis: Results from a multi-centric, placebo-controlled, randomized, double-blind clinical trial, *Hepatology* **26**:251A.

Montgomery, R. I., Coleman, W. B., Eble, K. S., and Cunningham, C. C., 1987, Ethanol-elicited alterations in the oligomycin sensitivity and structural stability of the mitochondrial F_0F_1 ATPase, *J. Biol. Chem.* **262**:13281–13289.

Nanji, A. A., Griniuviene B., Sadrzadah S. M., Levitsky, S., and McCully, J. D., 1995, Effect of type of dietary fat and ethanol on antioxidant mRNA induction in rat liver, *J. Lipid Res.* **36**:736–744.

Neuman, M. G., Shear, N. H., Bellentani, A., and Tiribelli, C., 1998, Role of cytokines in ethanol-induced cytotoxicity *in vitro* in HepG2 cells, *Gastroenterology* **115**:157–166.

Niemela, O., Juvonen, T., and Parkkila, S., 1991, Immunhistochemical demonstration of acetaldehyde-modified epitopes in human liver after alcohol consumption, *J. Clin. Invest.* **87**:1367–1374.

Niemela, O., Parkkila, S., Y-Herttuala, S., Villanueva, J., Ruebner, B., and Halsted, C., 1995, Sequential acetaldehyde production, lipid peroxidation, and fibrogenesis in micropig model of alcohol-induced liver disease, *Hepatology* **22**:1208–1214.

Perera, C. S., St. Clair, O. K., and McClain, C. J., 1995, Differential regulation of manganese superoxide dismutase activity by alcohol and TNF in human hepatoma cells, *Arch. Biochem. Biophys.* **323**: 471–476.

Polavarapu, R., Follansbee, M. H., Spitz, D. R., Sim, J. E., and Nanji, A. A., 1996, Hepatic antioxidant enzymes in experimental alcoholic liver disease, *Hepatology* **24**:441A.

Quillet-Mary, A., Jaffrezoy, J. P., Mansat, V., Bordier, C., Naval, J., and Laurent, G., 1997, Implication of mitochondrial hydrogen peroxide generation in ceramide-induced apoptosis, *J. Biol. Chem.* **272**:21288–21395.

Quintanilla, M. E., and Tempier, L., 1989, Sensitivity of liver mitochondrial functions to various levels of ethanol intake in the rat, *Alcohol: Clin. Exp. Res.* **13**:280–283.

Rouach, H., Clement, M., Orfanelli, M. T., Janvier, B., and Nordmann, R., 1984, Fatty acid composition of rat liver mitochondrial phospholipids during ethanol inhalation, *Biochim. Biophys. Acta* **795**:125–129.

Rubin, E., Beattie, D. S., Toth, A., and Lieber, C. S., 1972, Structural and functional effects of ethanol on hepatic mitochondria, *Fed. Proc.* **31**:131–140.

Schilling, R. J., and Reitz, R. C., 1980, A mechanism for ethanol-induced damage to liver mitochondrial structure and function, *Biochim. Biophys. Acta* **603**:266–277.

Schulze-Osthoff, K., Bakker, A. C., Vanhaesebroeck, B., Beyaert, R., Jacob, W. A., and Fiers, W., 1992, Cytotoxic activity of tumor necrosis factor is mediated by early damage to mitochondrial functions, *J. Biol. Chem.* **267**: 5317–5323.

Spach, P. I., and Cunningham, C. C., 1987, Control of state 3 respiration in liver mitochondria from rats subjected to chronic ethanol consumption, *Biochim. Biophys. Acta* **894**:460–467.

Spach, P. I., Boltenus, R., and Cunningham, C. C., 1982. Control of adenine nucleotide metabolism in hepatic mitochondria from rats with ethanol-induced fatty liver. *Biochem. J.* **202**:445–452.

Takeshi, H., Kaplowitz, N., Kamimura, T., Tsukamoto, H., and Fernández-Checa, J. C., 1992, Hepatic mitochondrial GSH depletion and progression of experimental alcoholic liver disease in rats, *Hepatology* **16**:1423–1428.

Tarachi, T. F., and Rubin, E., 1985, Biology of disease. Effects of ethanol on the chemical and structural properties of biologic membranes, *Lab. Invest.* **52**:120–131.

Thayer, W. S., and Rubin, E., 1979, Effects of chronic ethanol intoxication on oxidative phosphorylation in rat liver submitochondrial particles, *J. Biol. Chem.* **254**:7717–7723.

Turrens, J. F., and Boveris, A., 1980, Generation of superoxide anion by the NADH dehydrogenase of bovine heart mitochondria, *Biochim. J.* **191**:421–427.

Turrens, J. F., Alexandre, A., and Lehninger, A. L., 1985, Ubisemiquinone is the electron donor for superoxide formation by complex III of heart mitochondria, *Arch. Biochem. Biophys.* **237**:408–411.

Tzagoloff, A., and Myers, A., 1986, Genetics of mitochondrial biogenesis, *Ann. Rev. Biochem.* **55**:249–285.

Waring, A. J., Rottenberg, H., Ohnishi, T., and Rubin, E., 1981, Membranes and phospholipids of liver mitochondria from chronic alcoholic rats are resistant to membrane disordering by alcohol, *Proc. Natl. Acad. Sci. USA* **78**:2582–2586.

Wong, G. H., and Goeddel, D. V., 1988, Induction of manganous superoxide dismutase by TNF: Possible protective mechanism, *Science* **242**:941–944.

Yamauchi, N., Kuriyama, H., Watanabe, N., Neba, H., Maeda, M., and Niitsu, Y., 1989, Intracellular hydroxyl production induced by recombinant human TNF and its implication in the killing of tumor cells *in vitro*, *Cancer Res.* **49**:1671–1675.

Chapter 20

Mitochondrial Changes after Acute Alcohol Ingestion

Hajimi Higuchi and Hiromasa Ishii

1. INTRODUCTION

Oxygen-derived free radicals play an important role in the pathogenesis of ethanol-associated liver injury (Ishii *et al.*, 1997; Kaplowitz and Tsukamoto, 1996). Ethanol administration induces an increase in lipid peroxidation by either enhancing the production of reactive oxygen species, decreasing the level of endogenous antioxidants (Ishii *et al.*, 1997), or both. Despite an increasing body of experimental and clinical evidence for enhanced oxidative stress in alcoholic liver disease (Ishii *et al.*, 1997; Kaplowitz and Tsukamoto, 1996), its precise pathogeneic significance in the evolution of alcoholic liver disease (ALD) remains obscure. Recently, we evaluated the balance of intracellular oxidants and antioxidants in isolated hepatocytes and detected enhanced oxidative stress after acute ethanol intoxication, which causes hepatocytic mitochondrial alteration and apoptosis (Kurose *et al.*, 1997a, 1997b).

Alteration of mitochondrial membrane potential ($\Delta\psi$) and permeability is now postulated as a central regulatory mechanism of cell viability (Susin *et al.*, 1998). Kinetic studies indicate that mitochondria undergo major changes in membrane integrity before the classic signs of apoptosis manifest. In most cases of apoptosis, disruption of $\Delta\psi$, resulting in decreased dye uptake, precedes all major changes in cell morphology and biochemistry, including nuclear fragmentation and phosphatidylserine exposure in the outer leaflet of the plasma membrane (Susin *et al.*, 1998). In addition to disruption of $\Delta\psi$, mitochondria exhibit a number of profound changes in structure and function during

Hajimi Higuchi and Hiromasa Ishii   Department of Internal Medicine, School of Medicine, Keio University, Tokyo, 160, Japan.

Mitochondria in Pathogenesis, edited by Lemasters and Nieminen.
Kluwer Academic/Plenum Publishers, New York, 2001.

apoptosis. The mitochondrial membrane permeability transition (MPT) is due to the opening of a regulated proteaceous pore, the *mitochondrial megachannel*, also called the *PT pore* (Bernardi, 1996; Kinnally *et al.*, 1996; Zoratti and Szabò, 1995). The pore is a dynamic multiprotein complex probably located at the contact site between inner and outer mitochondrial membranes (Beutner *et al.*, 1996, 1998; O'Gorman *et al.*, 1997). Regulation of the pore's opening or closing state depends on the condition of the mitochondrial matrix, involving such factors as electrical membrane potential (Zoratti and Szabò, 1995), thiols and oxidants (Costantini *et al.*, 1996), pH (Nicolli *et al.*, 1996), and calcium and adenine nucleotide (Zoratti and Szabò, 1995). Moreover, the pore participates in regulating matrix Ca^{2+}, pH, $\Delta\psi$, and volume. At its highest level of conductance, the pore can provoke irreversible $\Delta\psi$ disruption (Bernardi, 1996; Ichas *et al.*, 1997; Zoratti and Szabò, 1995). Acute ethanol-induced permeability transition in cultured hepatocytes is associated with increased oxidative stress and decreased $\Delta\psi$, both of which are inhibited by several antioxidants or inhibitors of ethanol metabolism (Kurose *et al.*, 1997a). Thus, oxidative stress via ethanol metabolism is postulated as a key to the mitochondrial changes preceding apoptosis in hepatocytes exposed to acute ethanol.

In 1963 Di Luzio suggested that ethanol affects the antioxidant balance of the hepatocyte. He pointed out that acute ethanol-induced fatty liver might be prevented in rats by administering antioxidants. This hypothesis was strengthened by the finding that enhanced lipid peroxidation occurs both in rat liver homogenates after *in vitro* ethanol addition and in liver homogenates obtained from rats after oral ethanol administration (Di Luzio and Hartman, 1967). Thus previous studies suggest that ethanol or its metabolites can tip the balance in liver toward autooxidation, by acting as prooxidants, by reducing the antioxidant levels, or both (Di Luzio and Hartman, 1967). Oxidative stress can affect cellular integrity only when the antioxidant mechanisms can no longer cope with free radical generation, so many experimental studies have been done to ascertain the effects of acute and chronic ethanol administration on liver antioxidant enzymes and substrates.

In this review we address the mechanism of acute ethanol-induced hepatocyte death and mitochondrial alteration and consider the regulation and role of transition during acute ethanol-induced hepatocyte apoptosis. Our investigations implicate the endogenous glutathione–glutathione peroxidase system as an important antioxidant with a cytoprotective function in hepatocytes exposed to ethanol.

2. ETHANOL-ASSOCIATED OXIDATIVE STRESS IN HEPATOCYTES

Oxidative stress can be defined as the manifestations of cell, tissue, or organism exposure to excess oxidant, in particular to superoxide anion (O_2^-) and its metabolites, referred to here as reactive oxygen metabolites (ROM). The role of oxidative stress in the pathogenesis of alcohol-induced liver disease was first proposed more than 30 years ago (Di Luzio and Hartman, 1967). Since then, convincing evidence has accumulated to support this role. Development of noninvasive methods, such as low-level chemiluminescence assay (Müller and Sies, 1987) and spin-trapping procedures (Reinke *et al.*, 1987), has allowed the discovery that ethanol administration elicits enhanced free radical formation in the liver under certain experimental conditions. More recently, Bautista and Spitzer (1992) demonstrated, using cytochrome c, that acute ethanol intoxication stimulates superoxide anion production *in situ* in perfused rat liver.

Currently, ethanol-induced oxidative stress can be further investigated using computer-assisted fluorescence microscopy and a laser-scanning confocal imaging system (Kurose *et al.*, 1997a). We recently used this imaging system to study the metabolic changes in individual cultured cells. Our present study using the image-analyzing system clearly demonstrates that fluorescence of dichlorofluorescein (DCF), a fluorochrome that indicates excessive production of oxygen-derived free radicals and lipid peroxides, actually increases within 10 minutes in rat hepatocytes exposed to ethanol (50 mM) (Fig. 1).

The mitochondrial respiratory chain is one of the main physiological sources of superoxide anion (Ishii *et al.*, 1997). The auto-oxidation of ubisemiquinone (complex III), and perhaps to a lesser extent of complex I-reduced nicotinamide adenine dinucleotide-reduced flavin mononucleotide (NADH-FMNH), is a source of physiological oxidative stress. Because mitochondria contain an active superoxide dismutase (SOD), superoxide anion can generate hydrogen peroxide, which is destroyed through mitochondrial glutathione peroxidase. In the presence of iron, however, hydrogen peroxide that escapes this destruction might generate aggressive free radicals, possibly leading to mitochondrial structural and functional disturbances (Fig. 2).

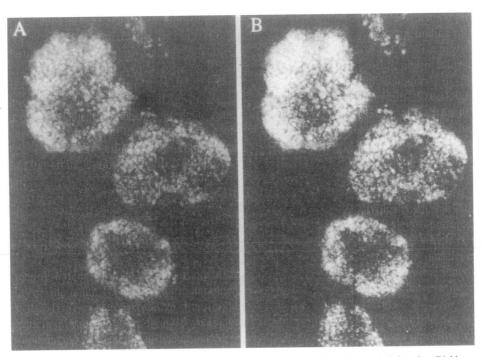

FIGURE 1. Alteration of dichlorofluorescein fluorescence in hepatocytes after ethanol administration. Dichlorofluorescein fluorescence, which indicates excessive production of oxygen-derived free radicals and lipid peroxidation, increased within 10 min of cultured rat hepatocyte exposure to ethanol (50 mM). (A) Before ethanol administration. (B) 10 min after ethanol administration.

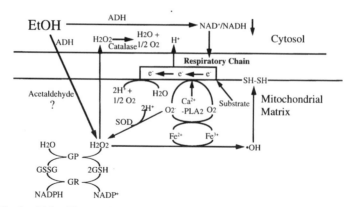

FIGURE 2. Mitochondrial oxidant production and antioxidants. Superoxide radicals can be produced from mitochondrial respiratory chain activity, a main physiological source of superoxide anion. Superoxide anion can generate hydrogen peroxide via a superoxide dismutase. Hydrogen peroxide is destroyed through mitochondrial glutathione peroxidase, but in the presence of iron some of the hydrogen peroxide escaping this destruction could generate aggressive radicals, which may lead to mitochondrial structural and functional disturbances.

Acute ethanol administration increases superoxide generation in liver mitochondria (Sinaceur *et al.*, 1985). Several pathophysiological conditions in alcoholic liver disease may favor accentuation of oxidative stress in mitochondria. Ethanol can induce a hypermetabolic liver state characterized by enhanced mitochondrial respiration, which is driven by a greater demand for NADH reoxidation produced through ethanol metabolism by cytosolic alcohol dehydrogenase (Israel and Orrego, 1984). The decrease in the $NAD^+/NADH$ ratio induced by acute ethanol administration (Slater *et al.*, 1964) may favor mitochondrial superoxide generation by increasing electron flow along the respiratory electron transport chain. Enhanced superoxide generation *per se* contributes to increased lipid peroxidation susceptibility in isolated rat liver mitochondria following acute ethanol infusion (Rouach *et al.*, 1988).

3. MITOCHONDRIA AS TARGETS OF OXIDATIVE STRESS

The mitochondrial respiratory chain is a major target of the oxygen free radicals generated in ethanol-exposed hepatocytes. Accentuation of O_2^- production in mitochondria, as noted, seems to result from impaired function of complex III, leading to accumulation of the ubisemiquinone radical. Factors exerting this effect include ischemia/reperfusion (Gonzalez-Flecha and Boveris, 1995; Gonzalez-Flecha *et al.*, 1993), TNF-α (Goossens *et al.*, 1995; Schulze-Osthoff *et al.*, 1992), and antimycin A (Garcia-Ruiz *et al.*, 1995). The signaling mechanism for the effect of TNF on mitochondria remains obscure. Both TNF-α and antimycin A induce an oxidative stress in mitochondria that is markedly accentuated by prior depletion of the mitochondrial glutathione pool (Garcia-Ruiz *et al.*, 1995; Goossens *et al.*, 1995). Using a laser-scanning confocal microscope, we examined the changes in mitochondrial energy synthesis in rat hepatocytes exposed to ethanol, with Rhodamine 123 (Rh123) as a $\Delta\psi$ indicator (Kurose *et al.*, 1997a).

Ethanol (50 mM) significantly decreased Rh123 fluorescence in hepatocytes within 30 minutes, suggesting mitochondrial dysfunction. This decrease was attenuated by N,N'-dimethylthiourea (DMTU), a membrane-permeable hydrogen peroxide scavenger, as well as by 4-methylpyrazole, an alcohol dehydrogenase inhibitor, indicating that excessive production of active oxygen species leads to mitochondrial dysfunction.

Superoxide anion produced in mitochondria can be catalyzed by mitochondrial active MnSOD to hydrogen peroxide. GSH peroxidase reduces hydrogen peroxide, preventing production of more reactive radicals. Because mitochondria in liver cells contain no catalase, the ability of the mitochondrial GSH system to reduce hydrogen peroxide is probably the main protective mechanism against oxidative stress. Previous reports implicated intracellular, especially mitochondrial, glutathione as an important antioxidant in preventing development of ALD (Handler and Thurman, 1990; Thurman and Handler, 1989). Masini et al. (1994) recently reported that lipid hydroperoxide induced mitochondrial dysfunction in rats subjected to acute ethanol exposure. The same study also demonstrated a critical role for mitochondrial reduced glutathione. In support of this role, they showed a production of lipid hydroperoxides and a depletion of reduced glutathione in liver mitochondria after acute ethanol administration to rats (Shaw et al., 1988). Previous reports from our laboratory (Kurose et al., 1997a) suggested that depletion of hepatocyte glutathione led to enhanced oxidative stress in mitochondria and cell membranes exposed to ethanol. In contrast, treatment with the glutathione precursor N-acetyl-L-cysteine prevented ethanol-induced oxidative stress and mitochondrial dysfunction in hepatocytes, suggesting that intracellular glutathione is an important cytoprotective factor.

4. MITOCHONDRIAL MEMBRANE PERMEABILITY TRANSITION AND DEPOLARIZATION

Mitochondrial transmembrane potential results from the asymmetric distribution of protons and other ions on both sides of the inner membrane, giving rise to a chemical and electrical gradient essential for mitochondrial function (Attardi and Schatz, 1998). The inner side of the inner mitochondrial membrane is negatively charged. Consequently, cationic lipophilic fluorochromes such as Rh123, 3,3'dihexyloxacarbocyanine iodide (DioC6), 5,5',6,6'-tetrachloro-1,1',3,3'-tetraethylbenzimidazolcarbocyanine iodide (JC-1), and chloromethyl-X-rosamine (CMXRos) distribute to the mitochondrial matrix as a function of the Nernst equation, correlating with $\Delta\psi$. Cells induced to undergo death show an early reduction in the incorporation of $\Delta\psi$-sensitive dyes, a $\Delta\psi$ disruption detected in many cell types (Kroemer et al., 1997). We examined the alteration of $\Delta\psi$ in ethanol-exposed rat hepatocytes by using Rh123 under the laser-scanning confocal microscope (Kurose et al., 1997a). Within 10 minutes, ethanol (50 mM) significantly decreased Rh123 fluorescence in hepatocytes, suggesting that mitochondrial $\Delta\psi$ loss is actually an early event of ethanol cytotoxicity.

An important event in cell death is the mitochondrial membrane permeability transition. First characterized by Hunter et al. (1976) in isolated mitochondria, the transition represents an abrupt permeability increase of the mitochondrial inner membrane to molecular mass solutes of less than about 1500 Da (Bernardi, 1996; Zoratti and Szabò, 1995).

Oxidant chemicals Ca^{2+} and Pi promote transition onset, whereas Mg^{2+}, ADP, low pH, and high membrane potential oppose it. The rapid change in permeability associated with transition causes membrane depolarization, uncoupling of oxidative phosphorylation, release of intramitochondrial ions and metabolic intermediates, and large-amplitude mitochondrial swelling. A series of elegant confocal microscopy studies (Nieminen et al., 1995; Zahrebelski et al., 1995) showed that when calcein, a polyanionic membrane-impermeant fluorescein derivative, is incorporated into cells, all cellular compartments are stained except the mitochondrial. During hepatocyte death calcein can access the mitochondrion as soon as $\Delta\psi$ breaks down, indicating that the membranes have indeed become permeable. We used this technique to observe changes in $\Delta\psi$ after acute ethanol intoxication (Kurose et al., 1997a). Figure 3 illustrates the spatial and temporal relations between $\Delta\psi$ alterations and the MPT in single ethanol-exposed hepatocytes. The cationic fluorophore TMRM accumulated in cultured hepatocyte mitochondria in response to the negative $\Delta\psi$. Consequently, individual mitochondria were imaged as small bright fluorescent spheres (Fig. 3A). In contrast, calcein diffusely labeled the cytosol of cultured hepatocytes loaded with calcein-AM (Fig. 3B). Areas corresponding to TMRM-labeled mitochondria were dark round voids in the calcein

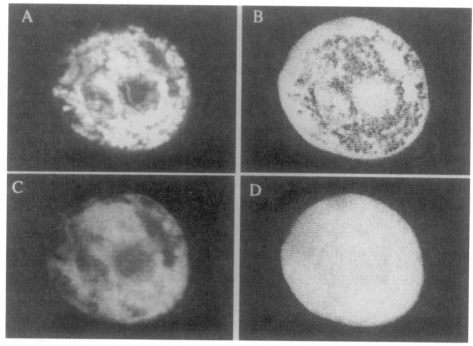

FIGURE 3. Ethanol-induced increase in mitochondrial membrane permeability transition in cultured hepatocytes. Cultured hepatocytes were preloaded with TMRM, PI, and calcein. TMRM and calcein fluorescence were simultaneously imaged in single hepatocyte through a laser-scanning confocal microscope. (A) TMRM fluorescence before ethanol administration. (B) Calcein fluorescence before ethanol administration. (C) TMRM fluorescence 30 min after ethanol administration. (D) Calcein fluorescence 30 min after ethanol administration.

fluorescence. After 5 minutes of exposure to 50 mM ethanol, no change in TMRM or calcein fluorescence appeared. After 30 minutes, however, mitochondria began to lose TMRM fluorescence (Fig. 3C) and fill with calcein (Fig. 3D). The cells did not lose viability, as indicated by nuclear staining with PI, even 60 minutes after exposure to ethanol, whereas the mitochondria further depolarized.

5. EFFECT OF ETHANOL ON INTRACELLUAR GLUTATHIONE CONTENT

As mentioned, oxidative stress affects cellular integrity only when antioxidant mechanisms can no longer handle the free radical generation (Fig. 2). Therefore many experimental studies were done to ascertain the effects of acute and chronic ethanol administration on liver antioxidant enzymes and substrates. The protective action of antioxidants is probably due to inhibition of free radical-induced chain reactions.

Although many investigators report that acute ethanol intoxication decreases hepatic glutathione content, the mechanisms involved are still controversial. Formation of an adduct between glutathione and the acetaldehyde resulting from ethanol oxidation is one such mechanism (Vina et al., 1980). Nonphysiological high acetaldehyde concentrations are required to promote significant adduct formation (Speisky et al., 1988), however, because acetaldehyde does not easily form a stable conjugate with glutathione (Kera et al., 1985).

We observed the glutathione level in individual rat hepatocytes using a fluorescence probe, monochlorobimane (MCLB), that makes a conjugate with glutathione to become fluorescent. The MCLB probe has less than 2% nonspecific fluorescence, whereas monobromobimane has approximately 20% nonspecific fluorescence (Cook et al., 1989, 1991; Nair et al., 1992; Rice et al., 1986; Ublacker et al., 1991). Ethanol (<100 mM) did not alter glutathione-associated fluorescence in rat hepatocytes until 30 minutes after the treatment (Fig. 4). Our result supports that acute ethanol administration does not alter the glutathione level in hepatocytes, at least not in vitro. On the other hand, a later report (Zentella de Pinta et al., 1994) concerning ethanol's effect on cellular glutathione level showed acute ethanol intoxication decreasing the liver glutathione levels, and a non-steroidal anti-inflammatory drug, piroxicam, attenuating the decrease. Thus, there is no agreement on the effects of acute ethanol administration at the hepatic glutathione level.

Another possible mechanism is increased glutathione oxidation resulting from enhanced generation of oxidizing free radicals (Videla and Valenzuela, 1982). Preventive effects of desferrioxamine administration on the ethanol-induced decrease in liver glutathione may support such a mechanism (Antébi et al., 1984). Some investigations found no change in either the hepatic oxidized glutathione (GSSG) content or the ratio of reduced glutathione (GSH) to GSSG (Guerri and Grisolia, 1980; Speisky et al., 1985).

6. ETHANOL-INDUCED MITOCHONDRIAL INJURY AND APOPTOSIS

The $\Delta\psi$ disruption occurs before cells exhibit nuclear DNA fragmentation. Purified cells with a low $\Delta\psi$ rapidly (15 minutes to several hours) proceed to DNA fragmentation, even after withdrawal of the apoptosis-inducing stimulus, indicating that $\Delta\psi$ collapse

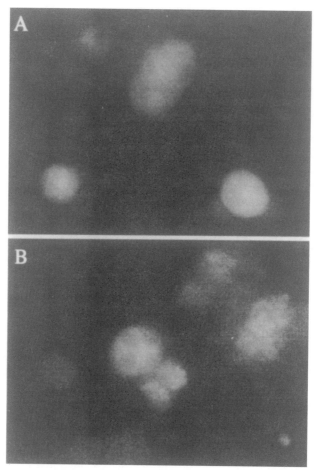

FIGURE 4. Monochlorobimane fluorescence in hepatocytes. The glutathione levels in individual hepatocytes were analyzed using the probe monochlorobimane, which makes a conjugate with glutathione to become fluorescent. Ethanol (100 mM) did not alter the glutathione-associated fluorescence in cultured rat hepatocytes before 30 min. (A) 30 min without ethanol administration. (B) 30 min with ethanol administration.

marks a point of no return for death programming (Zamzami *et al.*, 1995). Functional experiments indicate that the mechanism of pre-apoptotic $\Delta\psi$ collapse involves the MPT. The transition is a well-studied phenomenon wherein the normally highly impermeable mitochondrial inner membrane abruptly becomes permeable to molecules of less than 1500 Da. Transition occurs as a consequence of reversible megachannel opening at the contact lesion on the inner and outer mitochondrial membranes. Upon opening of the permeability pore, rapid ion movement causes extensive mitochondrial swelling and loss of $\Delta\psi$. Mitochondria can liberate at least two different factors relevant to apoptosis. First, they release a 50-kDa apoptosis-inducing factor (AIF), which activates caspase-3 and endonucleases in isolated nuclei (Susin *et al.*, 1996, 1998; Zamzani *et al.*, 1996). Mitochondria then release cytochrome *c*, which together with additional factors, namely Apaf-1 (a mammalian CED-4 homolog that binds to the outer mitochondrial membrane), dATP, and cytosolic 45-kDa protein Apaf-3 (caspase-9), can proteolytically activate caspase-3, which in turn cleaves and activates DNA fragmentation factor (DFF), and endonuclease activator acting on the nucleus (Liu *et al.*, 1996, 1997; Zhou *et al.*, 1997).

We observed acute ethanol-induced transition as well as $\Delta\psi$ collapse in cultured hepatocytes before we saw DNA and nuclear fragmentation (Kurose *et al.*, 1997a, 1997b). More recently we evaluated whether transition is required for acute ethanol-induced hepatocyte apoptosis by using the potent transition inhibitor cyclosporin A. Cyclosporin A protection against apoptosis indicates that transition is a crucial step in acute ethanol-induced apoptosis in cultured hepatocytes. If the mitochondrion is a pivotal control point of apoptosis, the next question is whether acute ethanol ingestion can induce cytochrome *c* release from mitochondria and activate caspase-3. To answer this question, we first observed intracellular cytochrome *c* distribution (Fig. 5). Cytochrome *c*-associated fluor-

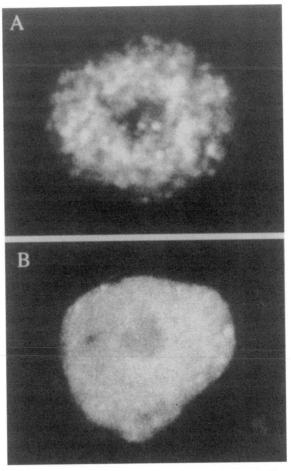

FIGURE 5. Immunofluorescence staining of cytochrome *c* in rat hepatocytes. Cytochrome *c*-associated fluorescence formed a mitochondrial pattern in the cultured hepatocyte cytosol before ethanol treatment. After ethanol administration these hot spots were not observed. (A) 30 min without ethanol administration. (B) 30 min with ethanol administration.

escence emerged as a mitochondrial pattern in cultured hepatocytes, whereas these hot spots disappeared and a homogenous pattern was seen after hepatocytic ethanol treatment, suggesting mitochondrial cytochrome c release. This was prevented by adding cyclosporin A, alcohol dehydrogenase inhibitor 4-MP, hydrogenperoxide scavenger DMTU, or glutathione precursor NAC. Next we detected increased activity of DEVD-AMC cleavage in ethanol-treated hepatocytes. Finally, we found that acute ethanol-induced hepatocyte apoptosis is prevented by adding tetrapeptide caspase inhibitors, such as DEVD-cho. These results suggest that when transition is triggered by acute ethanol, mitochondria can play a role in the induction of apoptosis via release of cytochrome c, and that acute ethanol-induced hepatocyte apoptosis is dependent on the caspase activation system.

7. CONCLUSION

Active oxidants produced during ethanol metabolism modulate mitochondrial membrane potential and permeability changes in isolated and cultured hepatocytes. These mitochondrial alterations (loss of $\Delta\psi$, the MPT) are now recognized as a key step in programmed cell death. Our fluorographic investigations demonstrate that acute ethanol-induced oxidative stress can induce mitochondrial permeability change, cytochrome c release, caspase activation, and apoptosis in cultured hepatocytes. In addition, our investigations implicate endogenous glutathione–glutathione peroxidase as an important antioxidant with a cytoprotective machinery in hepatocyte mitochondria exposed to ethanol. Oxidative stress is the consequence of an imbalance between oxidant production and antioxidant defense. Development of new and effective strategies to diminish oxidant production, enhance intracellular and extracellular antioxidant defenses, or both, in the liver offers great promise for preventing and treating liver disease.

ACKNOWLEDGEMENTS. This work was supported by a Grant-in-Aid for Scientific Research from the Japanese Ministry of Education, Science, and Culture of Japan and by a grant from Keio University, School of Medicine.

REFERENCES

Antébi, H., Ribière, C., Sinaceur, J., Abu-Murad, C., and Nordmann, R., 1984, Involvement of oxygen radicals in ethanol oxidation and in the ethanol-induced decrease in liver glutathione, in *Oxygen Radicals in Chemistry and Biology* (W. Bors, M. Saran, and D. Tait, Eds.), Walter de Gruyter, Berlin, New York, pp. 757–760.
Attardi, G., and Schatz, G., 1998, Biogenesis of mitochondria, *Annu. Rev. Cell Biol.* **4**:289–333.
Bautista, A. P., and Spitzer, J. J., 1992, Acute ethanol intoxication stimulates superoxide anion production by *in situ* perfused rat liver, *Hepatology* **15**:892–898.
Bernardi, P., 1996, The permeability transition pore; Control points of cyclosporin A-sensitive mitochondrial channel involved in cell death, *Biochim. Biophys. Acta* **1275**:5–9.
Beutner, G., Ruck, A., Riede, B., Welte, W., and Brdiczka, D., 1996, Complexes between kinases, mitochondrial porin, and adenylate translocator in rat brain resemble the permeability transition pore, *FEBS Lett.* **396**:189–195.
Beutner, G., Ruck, A., Riede, B., and Brdiczka, D., 1998, Complexes between porin, hexokinase, mitochondrial creatine kinase, and adenylate translocator display properties of the permeability transition pore; Implication for regulation of permeability transition by the kinases, *Biochim. Biophys. Acta* **1368**:7–18.
Cook, J. A., Pass, H. I., Russo, A., Iype, S., and Mitchell, J. B., 1989, Use of monochlorobimane for glutathione measurements in hamster and human tumor cell lines, *Int. J. Radiat. Oncol. Biol. Phys.* **16**:1321–1324.

Cook, J. A., Pass, H. I., Iype, S. N., Friedman, N., DeGraff, W., Russo, A., and Mitchell, J. B., 1991, Cellular glutathione and thiol measurements from surgically resected human lung tumor and normal lung tissue, *Cancer Res.* **51**:4287–4294.

Costantini, P., Chernyak, B. V., Petronilli, V., and Bernardi, P., 1996, Modulation of the mitochondrial permeability transition pore by pyridine nucleotides and dithiol oxidation at two separate sites, *J. Biol. Chem.* **271**:6746–6751.

Di Luzio, N. R., 1963, Prevention of the acute ethanol-induced fatty liver by antioxidants, *Physiologist* **6**:169–173.

Di Luzio, N. R., and Hartman, A. D., 1967, Role of lipid peroxidation in the pathogenesis of ethanol-induced fatty liver, *Fed. Proc.* **26**:1436–1442.

Garcia-Ruiz, C., Colell, A., Morales, A., Kaplowitz, N., and Fernández-Checa, J., 1995, Role of oxidative stress generated from the mitochondrial electron transport chain and mitochondrial glutathione status in loss of mitochondrial function and activation of transcription factor NF-κB, *Mol. Pharmacol.* **48**:825–834.

Gonzalez-Flecha, B., and Boveris, A., 1995, Mitochondrial sites of hydrogen peroxide production in reperfused rat kidney cortex, *Biochim. Biophys. Acta* **1243**:361–366.

Gonzalez-Flecha, B., Cutrin, J. C., and Boveris, A., 1993, Time course and mechanism of oxidative stress and tissue damage in rat liver subjected to *in vivo* ischemia–reperfusion, *J. Clin. Invest.* **91**:456–464.

Goossens, V., Grooten, J., DeVos, K., and Fiers, W., 1995, Direct evidence for tumor necrosis factor-induced mitochondrial reactive oxygen intermediates and their involvement in cytotoxicity, *Proc. Natl. Acad. Sci. USA* **92**:8115–8119.

Guerri, C., and Grisolia, S., 1980, Changes in glutathione in acute and chronic alcohol intoxication, *Pharmacol. Biochem. Behav.* **13** (Suppl. 1):53–61.

Handler, J. A., and Thurman, R. G., 1990, Redox interactions between catalase and alcohol dehydrogenase pathways of ethanol metabolism in the perfused rat liver, *J. Biol. Chem.* **265**:1510–1515.

Hunter, D. R., Haworth, R. A., and Southard, J. H., 1976, Relationship between configulation, function, and permeability in calcium-treated mitochondria, *J. Biol. Chem.* **251**:5069–5077.

Ichas, F., Jouavill, L. S., and Mazat, J. P., 1997, Mitochondria are excitable organelles 19 capable of generating and conveying electrical and calcium signals, *Cell* **89**:1145–1153.

Ishii, H., Kurose, I., and Kato, S., 1997, Pathogenesis of alcoholic liver disease with paticular emphasis on oxidative stress, *J. Gastroenterol. Hepatol.* **12**(Suppl.):S272–S282.

Israel, Y., and Orrego, H., 1984, Hypermetabolic state and hypoxic liver damage, in *Recent Developments in Alcoholism*, Plenum, New York, pp. 119–133.

Kaplowitz, N., and Tsukamoto, H., 1996, Oxidative stress and liver disease, *Prog. Liver Dis.* **14**:131–159.

Kera, Y., Kiriyama, T., and Komura, S., 1985, Conjugation of acetaldehyde with cysteinylglycine, the first metabolite in glutathione breakdown by γ-glutamyltranspeptidase, *Agents Actions* **17**:48–52.

Kinnally, K. W., Lohret, T. A., Campo, M. L., and Mannella, C. A., 1996, Perspectives on the mitochondrial multiple conductance channel, *J. Bioenerg. Biomembr.* **28**:115–123.

Kroemer, G., Zamzami, N., and Susin, S. A., 1997, Mitochondrial control of apoptosis, *Immunol. Today* **18**:44–51.

Kurose, I., Higuchi, H., Kato, S., Miura, S., Watanabe, N., Kamegaya, Y., Tomita, K., Takaishi, M., Horie, Y., Fukuda, M., Mizukami, K., and Ishii, H., 1997a, Oxidative stress on mitochondria and cell membrane of cultured rat hepatocytes and perfused liver exposed to ethanol, *Gastroenterology* **112**:1331–1343.

Kurose, I., Higuchi, H., Miura, S., Saito, H., Watanabe, N., Hokari, R., Hirokawa, M., Takaishi, M., Zeki, S., Nakamura, T., Ebinuma, H., Kato, S., and Ishii, H., 1997b, Oxidative stress-mediated apoptosis of hepatocytes exposed to acute ethanol intoxication, *Hepatology* **25**:368–378.

Liu, X., Kim, C. N., Yang, J., Jemmerson, R., and Wang, X., 1996, Induction of apoptotic program in cell-free extracts: Requirement for dATP and cytochrome *c*, *Cell* **86**:147–157.

Liu, X., Zou, H., Alaughter, C., and Wang, X., 1997, DFF, a heterodimeric protein that functions downstream of caspase-3 to trigger DNA fragmentation during apoptosis, *Cell* **89**:175–184.

Müller, A., and Sies, H., 1987, Alcohol, aldehydes, and lipid peroxidation: Current notions, *Alcohol Alcoholism.* (Suppl.) **1**:67–74.

Nair, S., Singh, S. V., and Krishan, A., 1992, Flow cytometric monitoring of glutathione content and anthracycline retention in tumor cells, *Cytometry* **12**:336–342.

Nicolli, A., Basso, E., Petronilli, V., Wagner, R. M., and Bernardi, P., 1996, Interactions of cyclophilin with the mitochondrial inner membrane and regulation of the permeability transition pore, and cyclosporin A-sensitive channel, *J. Biol. Chem.* **271**:2185–2192.

Nieminen, A.-L., Saylor, A. K., Tesfai, S. A., Herman, B., and Lemasters, J. J., 1995, Contribution of the mitochondrial permeability transition to lethal injury after exposure of hepatocytes to t-butylhydroperoxide, *Biochem. J.* **307**:99–106.

O'Gorman, E., Beutner, G., Dolder, M., Koretsky, A. P., Brdiczka, D., and Wallimann, T., 1997, The role of creatine kinase in inhibition of mitochondrial permeability transition, *FEBS Lett.* **414**:253–257.

Reinke, L. A., Lai, E. K., Du Bose, C. M., and MacCay, P. B., 1987, Reactive free radical generation *in vivo* in heart and liver of ethanol-fed rats: Correlation with radical formation *in vitro*, *Proc. Natl. Acad. Sci. USA* **84**: 9223–9227.

Rice, G. C., Bump, E. A., Shrieve, D. C., Lee, W., and Kovacs, M., 1986, Quantitative analysis of cellular glutathione by flow cytometry utilizing monochlorobimane: Some applications to radiation and drug resistance *in vitro* and *in vivo*, *Cancer Res.* **46**:6105–6110.

Rouach, H., Park, M. K., Orfanelli, M. T., Janvier, B., Brissot, P., Bourel, M., and Nordmann, R., 1988, Effects of ethanol on hepatic and cerebellar lipid peroxidation and endogenous antioxidants in native and chronic iron-overload rats, in *Alcohol Toxicity and Free Radical Mechanisms: Advances in the Biosciences*, Vol. 71 (R. Nordmann, C. Ribière, and H. Rouach, Eds.), Pergamon, Oxford, England pp. 49–54.

Schulze-Osthoff, K., Bakker, A. C., Vanhaese-Broeck, B., Beyaert, R., Jacob, W. A., and Fiers, W., 1992, Cytotoxic activity of tumor necrosis factor is mediated by early damage of mitochondrial functions, *J. Biol. Chem.* **267**:5317–5323.

Shaw, S., Jayatilleke, E., and Lieber, C. S., 1988, Lipid peroxidation as a mechanism of alcoholic liver injury: Role of iron mobilization and microsomal induction, *Alcohol* **5**:135–140.

Sinaceur, J., Ribière, C., Sabourault, D., and Nordmann, R., 1985, Superoxide formation in liver mitochondria during ethanol intoxication: Possible role in alcohol hepatotoxicity, in *Free Radicals in Liver Injury* (G. Poli, K. H Cheeseman, M. U. Dianzani, and T. F. Slater, Eds.), IRL Oxford, England pp. 175–177.

Slater, T. F., Sawyer, B. C., and Sträuli, U. D., 1964, Changes in liver nucleotide concentrations in experimental liver injury: II. Acute ethanol poisoning, *Biochem. J.* **93**:267–270.

Speisky, H., MacDonald, A., Giles, G., Orrego, H., and Israel, Y., 1985, Increased loss and decreased synthesis of hepatic glutathione after acute ethanol administration, *Biochem. J.* **225**:565–572.

Speisky, H., Kera, Y., Penttila, K. E., Israel, Y., and Lindros, K. O., 1988, Depletion of hepatic glutathione by ethanol occurs independently of ethanol metabolism, *Alcoholism: Clin. Exp. Res.* **12**:224–227.

Susin, S. A., Zamzami, N., Castedo, M., Daugas, E., Wang, H. G., Geley, S., Fassy, F., Reed, J. C., and Kroemer, G., 1997, The central executioner of apoptosis: Multiple connections between protease activation and mitochondria in Fas/APO-1/CD95-and ceramide-induced apoptosis, *J. Exp. Med.* **186**:25–37.

Susin, S. A., Zamzami, N., Castedo, M., Hirsh, T., Marchetti, P., Macho, A., Daugas, E., Geuskens, M., and Kroemer, G., 1996, Bcl-2 inhibits the mitochondrial release of an apoptogenic protease, *J. Exp. Med.* **184**: 1331–1342.

Susin, S. A., Zamzami, N., and Kroemer, G., 1998, Mitochondria as a regulator of apoptosis: Doubt no more, *Biochim. Biophys. Acta* **1366**:151–165.

Thurman, R. G., and Handler, J. A., 1989, New perspectives in catalase-dependent ethanol metabolism, *Drug Metab. Rev.* **20**:679–688.

Ublacker, G. A., Johnson, J. A., Siegel, F. L., and Mulcahy, R. T., 1991, Influence of glutathione S-transferases on cellular glutathione determination by flow cytometry using monochlorobimane, *Cancer Res.* **51**:1783–1788.

Videla, L. A., and Valenzuela, A., 1982, Alcohol ingestion, liver glutathione, and lipoperoxidation: Metabolic interrelations and pathological implications, *Life Sci.* **31**:2395–2407.

Vina, J., Estrela, J. M., Guerri, C., and Romero, F. J., 1980, Effect of ethanol on glutathione concentration in isolated hepatocytes, *Biochem. J.* **188**:549–552.

Zahrebelski, G., Nieminen, A.-L., Al-Ghoul, K., Qian, T., Hermen, B., and Lemasters, J. J., 1995, Progression of subcellular changes during chemical hypoxia to cultured rat hepatocytes: A laser scanning confocal microscopic study, *Hepatology* **21**:1361–1372.

Zamzami, N., Marchetti, P., Castedo, M., Zanin, C., Vayssiere, J. L., Petit, P. X., and Kroemer, G., 1995, Reduction in mitochondrial potential constitutes an early irreversible step of programmed lymphocytes death *in vivo*, *J. Exp. Med.* **181**:1661–1672.

Zamzami, N., Susin, S. A., Marchetti, P., Hirsh, T., Gomez-Monterrey, I., Castedo, M., and Kroemer, G., 1996, Mitochondrial control of nuclear apoptosis, *J. Exp. Med.* **183**:1533–1544.

Zentella de Pina, M., Corona, S., Rocha-Hernández, A. E., Saldana Balmori, Y., Cabrera, G., and Pina, E., 1994,

Restoration by piroxicam of liver glutathione levels decreased by acute ethanol intoxication, *Life Sci.* **54**: 1433–1439.

Zhou, H., Henzel, W. J., Liu, X., Lutschg, A., and Wang, X. D., 1997, Apaf-1, a human protein homologous to *C. elegans* CED-4, participates in cytochrome *c*-dependent activation of caspase-3, *Cell* **90**:405–413.

Zoratti, M., and Szabò, I., 1995, The mitochondrial permeability transition, *Biochim. Biophys. Acta* **1241**:139–176.

Chapter 21

Mitochondrial Dysfunction in Chronic Fatigue Syndrome

Brad Chazotte

1. INTRODUCTION

Chronic fatigue syndrome (CFS), also known as chronic fatigue and immune dysfunction syndrome (CFIDS), myalgic encephalomyelitis, and so forth, has an unknown etiology and a poorly understood pathophysiology. Almost nothing is known of the cellular bioenergetics of CFS patients. The syndrome is a subject of increasing interest, though it is not a new phenomenon; medical literature going back over a hundred years has detailed similar illnesses with similar symptoms. Chronic fatigue syndrome was recently reviewed by Komaroff and Buchwald (1998), and a recent study based on a four-year surveillance of four U.S. cites determined an age-, race-, and sex-adjusted prevalence in CFS of 4.0 to 8.7 per 100,000 and an age-adjusted prevalence of 8.8 to 19.5 per 100,000 for white women (Reyes *et al.*, 1997). An even higher prevalence was reported in another U.S. study, with an estimated range of 75 to 267 per 100,000 (Buchwald *et al.*, 1995).

There is a rough consensus that the CFS patient population is predominantly, but not exclusively, female and more likely white than minority. In fact, CFS is diagnosed 3 to 4 times more frequently in women than in men and about 10 times more often in white Americans than in other American population groups.

1.1. Origins of Chronic Fatigue Syndrome

There is no known specific cause for CFS, but many etiological theories are proposed. The etiologies can be broadly divided into three groups: infectious agents, immunological

Brad Chazotte   Department of Pharmaceutical Sciences, Campbell University, Buies Creek, North Carolina 27506

Mitochondria in Pathogenesis, edited by Lemasters and Nieminen.
Kluwer Academic/Plenum Publishers, New York, 2001.

causes, and central nervous system causes. Infectious agent theories (e.g., Straus, 1994, and articles therein; Levy, 1994) are based on the view that an infectious agent either triggers a prolonged response or remains as a chronic infection. Most CFS patients report experiencing a "flu-like" illness prior to the onset of this syndrome (Fukuda and Gantz, 1995; Rasmussen *et al.*, 1994). In some instances there have been "outbreaks" of CFS, including the Royal Free Disease and Lake Tahoe outbreaks (e.g., Hyde *et al.*, 1992, and references therein). Many viral agents have been proposed: Epstein-Barr virus, varicella zoster, cytomegalovirus, various enteroviruses, various retroviruses (e.g., one similar to HTLV-II, spuma viruses), and more. There is some question as to whether a pathological agent akin to chronic brucellosis could give rise to CFS, or whether a "leaky gut" might permit intestinal flora to trigger CFS symptoms (Cho and Stollerman, 1992). To date no specific viral agent has been shown to cause CFS, nor in fact has *any* specific pathological agent been shown to cause it. (e.g., Levy, 1994).

Immunological theories that focus on some defect in patient immune response have also been advanced based on various reports of immune cell abnormalities. These theories suggest that cell populations can be affected, such as low numbers of NK cells (Aoki *et al.*, 1993), or increased populations of apoptotic lymphocyte cells (Vodjdani *et al.*, 1997). Alternatively, it has been suggested the cytokine or immunoglobin response of immune cells could be altered for example by overproduction of interferon-α (IFN-α) in response to stimulus (Lever *et al.*, 1988)

Central nervous system theories include both physiological and psychological aspects. For example, Demitrack and co-workers (Demitrack, 1994; Demitrack and Crofford, 1998) have argued for a disturbance in neuroendocrine function along the hypothalmic–pituitary–adrenal axis following "acute, often infectious, stress in emotionally susceptible individuals." Rowe *et al.* (1995) have asked whether neurally mediated hypotension could cause CFS. Others have speculated that CFS may be a primarily psychological disorder. Clearly, there is no agreed upon etiology. Despite increasing scientific literature on CFS, (e.g. Bock and Whelan, 1993; Dawson and Sabin, 1993; Hyde *et al.*, 1992; Jenkins and Mobrary, 1991; Komaroff and Buchwald, 1998; Straus, 1994), there is to date little appreciable increase in knowledge about cellular physiology and, in particular, about cellular bioenergetics in CFS patients, as mentioned. A more detailed exploration of cellular pathophysiology in CFS patients will help in understanding the disorder.

1.2. Symptoms and Case Definitions for Chronic Fatigue Syndrome

Chronic fatigue syndrome has proved elusive to define, identify, and treat. A case definition was proposed by Holms and associates (1988) that was subsequently refined by Fukuda *et al.*, (1994). For detailed descriptions and references, the reader is referred to these. Chromic fatigue syndrome is a clinically defined condition characterized by persistent (i.e., greater than six consecutive months), severe, disabling fatigue: postexertional malaise: and a combination of various symptoms such as concentration and short-term memory impairments, sleep disturbances (e.g., unrefreshing sleep), musculoskeletal pain, multijoint pain, sore throat, and new headaches. The diagnosis is still by elimination, when no other illness can be found (see also Komaroff and Buchwald, 1998). Unfortunately, there is no accepted, simple, clear diagnostic criterion for CFS as there is with, for example, the presence of rubeola virus for measles.

1.3. Role for Mitochondrial Dysfunction in Chronic Fatigue Syndrome

Clinical reports on CFS patients suggest a role for mitochondrial dysfunction at the cellular level. The regular and routine catabolic, anabolic, and energy-transducing metabolic functions performed by mitochondria are essential for normal cellular and physiological function, for organismal homeostasis—even for life itself. The classic symptoms of persistent and debilitating fatigue, chronic muscle weakness, and myalgia (Cheney and Lapp, 1992; Fukuda, et al., 1994) are consistent with mitochondrial dysfunction in other diseases of known mitochondrial etiology (Carafoli and Roman, 1980; DiMauro et al., 1985; Frackowiak et al., 1988; Roe and Coates, 1989). Also, cardiac muscle function under an exercise load appears to be affected in CFS patients compared with matched controls (Montague et al., 1989). Most CFS patients report mental concentration impairment and cognitive deficits (Grafman et al., 1991; Hyde and Jain, 1992; Sandman et al., 1993; Straus, 1988), which are also seen in mitochondrial dysfunctions (Kartounis et al., 1992; Peterson, 1995). Such reports have generated interest in SPECT and magnetic resonance neuroimaging of CFS patient brain metabolism where some defects have been observed (Natelson et al., 1993; Schwartz et al., 1993a, 1993b). Likewise, many CFS patients report gastrointestinal disturbances similar to known mitochondrial dysfunction (e.g. Peterson, 1995). Kuratsume et al. (1994) reported an acylcarnitine deficiency in the serum of CFS patients that apparently correlates with performance level. In one study, CFS patients were treated with L-carnitine; improvement was related to symptom severity at the time the treatment began (Plioplys and Plioplys, 1997). Carnitine is involved in long-chain fatty acid transport for mitochondrial β-oxidation, a major energy source in muscle. Carnitine and acylcarnitine levels are both altered in mitochondrial diseases such as medium-chain acyl-CoA dehydrogenase (MCAD) deficiency (Roe and Coates, 1989), for which carnitine is given therapeutically. In CFS patients, immune disturbances have also been clinically reported (Aoki et al., 1993; Ojo-Amaize et al., 1994; Buchwald and Komaroff, 1991; Caligiuri et al., 1987; Cannon et al., 1997; Jones, 1991). Overall, the symptoms suggest problems in *mitochondrial energy metabolism in more than one tissue*. Clinically, mitochondrial defects are most easily recognized in muscle and brain because these tissues have high energy demands (Barbiroli et al., 1993; Behan et al., 1991; DiMauro et al., 1985; Frackowiak et al., 1988; Peterson, 1995), although they manifest in other tissues as well. For example, Coates et al. (1985) demonstrated the utility of using both fibroblast cells and leukocytes for metabolic studies from a group of patients with mitochondrial-based MCAD deficiency by comparison with liver cell studies on these patients, cells much more difficult to obtain. As is universally accepted in the mitochondrial literature, Wallace (1992) points out that toxicological studies show that different organ systems rely on mitochondrial energy to different extents and that as mitochondrial ATP production declines, it successively falls below the minimum levels needed for each organ to function normally. Further evidence for mitochondrial deficiency is found in reports of early muscle acidosis (Jamal and Hansen, 1985; Wong et al., 1992) and deficit in oxygen utilization with low anaerobic threshold during exercise. One report suggests defects in muscle metabolism (McCulley *et al.*, 1996). These authors had some concerns, however, about matching activity levels of CFS patients with those of controls, and they also speculated that muscle deconditioning in CFS patients was an alternative explanation they could not rule out without additional studies. With respect to a possible role for mitochondrial DNA, there is a report of unusual deletions in CFS patient skeletal muscle, albeit anecdotal because only one individual was studied (Zhang *et al.*, 1995). A strong clinical finding implicating mitochondria involvement in CFS is that ATP levels at exhaustion were lower compared with levels in the control group when measured in gastrocnemius muscle *in vivo* by phosphorous NMR (Wong *et al.*, 1992). Significantly, there is a related finding in fibromyalgia patients, similarly measured in trapezius muscle, that ATP and phosphocreatine levels are 17% and 21% lower, respectively, and that these levels correlate with CFS-like symptoms (Bengtsson and Henriksson, 1989). Interestingly, in the fibromyalgia study not all muscle fibers were similarly affected. Thus there is ample *clinical* evidence to suggest a role for mitochondrial dysfunction in many different tissues and cells in CFS, which could give rise to many of the symptoms. Whether this role is due to a specific mitochondrial defect, perhaps in genetically susceptible individuals, or is an effect of some other problem such as altered cytokine levels that in turn affect mitochondrial function, needs investigation. Due to the difficulty in obtaining human specimens in sufficient quantities per specimen for biochemically based studies of mitochondrial function, there are few (and no detailed) studies of mitochondrial function in CFS patients.

1.4. Cytokine Role in Chronic Fatigue Syndrome

Cytokines (polypeptides, proteins, or glycoproteins of >5000 MW) are carefully controlled, critical intercellular and possibly intracellular messengers produced by specialized and unspecialized cells of many tissues and organs. Cytokines tend to act locally unless secreted into the circulatory system, where they remain for a rather short time. Cytokines such as interferons (IFN-α, -β, or -γ), interleukins (IL-1 through Il-18), tumor necrosis factor (TNF), and so on exert their effects at extremely low concentrations (picograms/ml) via cell receptors. Some of the triggered reactions are related to immune and inflammatory responses against pathogenic stimuli such as bacteria and viruses. Administration of IFN-γ and TNF-α to smooth muscle cells in culture reportedly inhibits mitochondrial respiration at complexes I and II (Geng *et al.*, 1992). Likewise, isolated hepatocytes have exhibited signs of mitochondrial impairment, such as decreased cellular ATP levels, upon exposure to TNF-α (Adamson and Billings, 1992), and reports suggest that cytokines may affect mitochondrial β-oxidation (Barke *et al.*, 1991). Thus there is plenty of evidence to indicate that mitochondrial function can be affected by cytokines in a number of cell types, and there is a clinical relevance to these effects. Related evidence suggests that immune dysfunction (e.g., Buchwald, 1991; Klimas *et al.*, 1990; Lloyd *et al.*, 1991) and cytokine imbalances (Ho-Yen *et al.*, 1988; Lever *et al.*, 1988; Lloyd *et al.*, 1991; Straus *et al.*, 1989) occur in CFS and other chronic diseases. One interesting side effect reported for (antiviral) IFN-α therapy is that it induces chronic-fatigue-like symptoms, (e.g., MacDonald, 1987). Finally, and most importantly, we have preliminary evidence that at least one cytokine we studied, IFN-α, lowers mitochondrial membrane potential ($\Delta\psi$) in human cells (Chazotte and Pettengill, 1998, 1999; Chazotte *et al.*, 1996).

2. METHODS

2.1. Cell Isolation, Culture, and Labeling

Two cell types, human mononuclear leukocytes and human fibroblasts, were used in the studies described in this chapter. The mononuclear leukocytes were isolated from human donors and the fibroblasts were obtained from commercially available cell lines. Both cell types were labeled in the same manner.

2.1.1. Human Mononuclear Leukocytes

Leukocytes were used in the CFS phase of our studies, due in part to the relative ease of obtaining individual, intact living cells from patients (and controls) compared with obtaining cells from other tissues. Whole blood was obtained from fasting morning samples of excess patient blood (typically 12 mls) drawn for routine testing at a clinic specializing in CFS. Normal control specimens were obtained as fasting morning specimens of healthy individuals donating blood for control studies at University of North Carolina (UNC) Hospitals. Whole blood was maintained at 4°C during overnight shipment, conditions we have routinely found *not* to decrease the $\Delta\psi$ parameters we measure for leukocyte cells and mitochondria. Control specimens were subjected to the same conditions as patient specimens. Leukocytes were isolated at 18–20 °C as specified in protocols from Robbins Scientific (Sunnyvale, CA) for their commercial preparative gradients that permit separation of mononuclear and polymorphonuclear leukocytes (as well as platelets) from whole blood. The separated leukocytes were then resuspended in appropriate culture medium (e.g., Hanks medium) or phosphate buffered saline (PBS). Isolation and resuspension in a defined medium gave us the important ability to carefully control the extracellular environment for cells from both patients and controls. Our primary focus was mononuclear leukocytes; however, we also examined platelets, which could be co-isolated and distinguished in our confocal specimen chambers and then analyzed separately.

2.1.2. Cultured Human Fibroblast Cell Lines

Human BG-9 or CCD-27sk fibroblast cell lines were cultured in Dulbecco's Modified Eagle's Medium (DMEM) with 10% fetal bovine serum and penicillin/streptomycin. Cells were grown in incubators under standard cell culture conditions and were subcultured before confluence and plated according to standard protocols. For confocal microscopy experiments, cells were subcultured on glass microscope coverslips

2.1.3. Cell Labeling

Human mononuclear leukocytes were labeled with the fluorescent $\Delta\psi$-sensitive probe tetramethylrhodamine methylester (TMRM) (Chazotte and Pettengill, 1998, 1999; Chazotte *et al.*, 1996). TMRM is one of the $\Delta\psi$-imaging probes most frequently used in confocal microscopy due to its highly advantageous properties. Isolated cells suspended in PBS or cell culture medium on the coverslip were placed in a cell culture dish. Labeling with 600 nM TMRM was carried out in the dark at an appropriate temperature (37 °C or room temperature) for 25 minutes in an appropriate culture medium or PBS$^+$ (PBS with 0.45 mM Ca^{2+} and 0.25 mM Mg^{2+}, which enhances cell attachment to the glass·coverslip). After 25 minutes, the coverslip was removed from the culture dish and mounted in a special microscope chamber. The chamber was subsequently filled with 150 nM TMRM in culture medium or PBS$^+$ (Chacon et al., 1994).

2.2. Confocal Microscopy

Using laser-scanning confocal microscopy (Pawley, 1995) to image $\Delta\psi$ is an important and powerful technique for examining the ability of mitochondria to make ATP in individual, intact, living cells (Chacon et al., 1994; Chazotte et al., 1996; Chazotte and Pettengill, 1998, 1999; Lemasters et al., 1993a, 1993b). The advent confocal microscopy has allowed thin (one-micron) sections of inherently thick biological speci-mens to be observed without interference by out-of-focus light. A key benefit of our approach, which looks at and measures individual cells, is that it requires very little material, in contrast to biochemical approaches that require large amounts of material for mitochondrial studies. Biochemical studies of human mitochondria in particular are difficult due to problems in obtaining isolated mitochondria in sufficient quantities for more than several enzyme assays, hence the relatively small number of such studies in the literature. Also, our ability to examine individual cells is an asset if CFS has a viral etiology and affects only a percentage of a specific cell population or subpopulation, such as CD4 cells.

A Bio Rad MRC-600 microscope was used for multichannel image acquisition using a 60× objective lens with a 1.4 numerical aperture (NA). The scope was set up and calibrated as described in detail elsewhere in this volume (Chacon et al., 1994; Chazotte and Pettengill, 1998, 1999; Chazotte et al., 1996; Lemasters et al., 1993a). Briefly, the enhancement circuit was at setting 4 to accommodate the wide dynamic range of fluorescence from extracellular to mitochondrial TMRM intensities. Also, the black level was set manually such that half the pixel intensities where distributed in a Gaussian manner at or below gray-level zero, and the imaging focal plane was at the coverslip for a zero TMRM concentration. The gain was set based on cell images such that no off-scale intensities, i.e., greater than gray-level 255 were observed. The cell chambers were temperature controlled, typically at 37 °C. Images were rapidly acquired and signal-averaged sufficiently to reduce noise. We archived them to magnetic or optical storage media for subsequent analysis.

2.3. Membrane-Potential Imaging Analysis

Confocal images analysis using fluorescent probes was accomplished by quantifying the digitized image brightness and converting it to physiologically relevant parameters using a technique that employed look-up tables based on an application of the Nernst equation (Chacon et al., 1994; Chazotte and Pettengill, 1998, 1999; Chazotte et al., 1996; Lemasters et al., 1993a). The Nernst equation calculates the diffusion potential in mV for a charged species in two electrolyte solutions separated by a membrane. The digital 8-bit

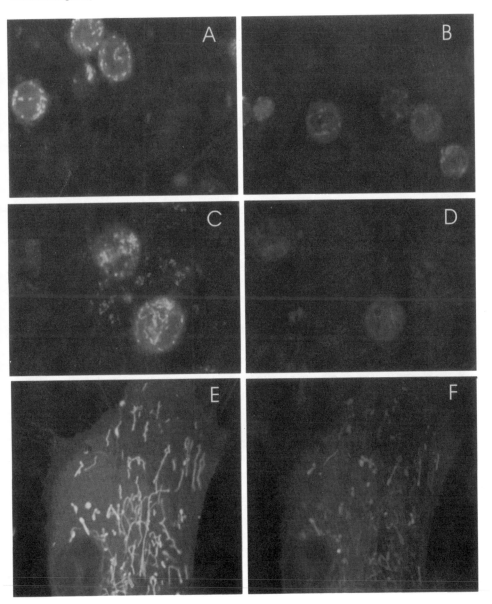

FIGURE 1. Membrane potential images of human cells. Human cells were labeled with TMRM and imaged using laser-scanning confocal microscopy as described in Section 2. (A–D) Mitochondria in leukocytes appear as bright spots. (A, C) Two individuals representative of our control population. (B, D) Two patients in our CFS population. (E, F) Single cultured human fibroblast. Mitochondrial appear as bright ribbonlike structures. (E) Before addition of 180 ng/ml of human interferon-α, (F) three minutes after the addition.

confocal or gray-level image (e.g., Fig. 1A, 1C, 1E) shows the most-negative potential regions as the brightest, so the highly negatively charged, fully functional mitochondria in human mononuclear leukocytes appear as bright tubular spots (Fig. 1A, 1C) and ribbons Fig. 1E in fibroblasts. Using a specific look-up table to convert brightness into millivolts of potential, a digital image can be displayed as a pseudocolored image that facilitates visual inspection of the $\Delta\psi$ differences; a color typically represents a 20-mV range. Due to the absence of color plates in this monograph, only gray-level images are shown.

To quantitatively measure membrane potential we adapted long-accepted general approaches from image-intensity analysis. To quantitatively analyze individual cells or any part of the image, a histogram analysis was applied by selecting a specific area and counting the number of pixels (small dots that make up a digital image) at each potential, using in-house custom image-analysis software. The number of pixels at a given potential were counted to determine the area potentials, integrated potential density, and mean (density) potential functions (Fig.2),

The histogram for each cell was further analyzed using custom software to calculate approximately 190 parameters based on area fractions, integrated potential densities, and mean potentials. One set of parameters determined the mean potentials of mitochondria, cytoplasm, and cell. Relative probability density functions were calculated from area fractions to determine what percentage (fraction) of the cell or mitochondria is operating at a given potential or range of potentials (e.g., 30% of the cell at mitochondrial potentials of 120–240 mV). Similarly, the percentages of mitochondrial areas operating at low, moderate, and high potentials were calculated. Likewise, using the integrated potential densities, which are more sensitive membrane-potential-weighted area fractions, relative probability densities were calculated by dividing, say, the total mitochondrial integrated potential density (IPD) by the total cellular IPD, to determine, for example, that 40% of the cell's IPD was at mitochondrial potentials. This is our unique methodology to quantitatively determine an individual cell's energy production capabilities through its membrane potential.

Collectively, these provide powerful analytical tools for quantifying membrane potentials of the cell and its mitochondria. Approximately 190 parameters can be examined for each cell, though experience shows that 56 of them provide the necessary

$$\text{Area Potential } (T_1, T_2) = \sum_{\Delta\psi=T_1}^{T_2} H(\Delta\psi)$$

$$\begin{array}{l}\text{Integrated Potential Density } (T_1, T_2) \\ \text{(Integrated "Optical" Density)}\end{array} = \sum_{\Delta\psi=T_1}^{T_2} H(\Delta\psi) \times \psi$$

$$\begin{array}{l}\text{Mean Potential}(T_1, T_2) \\ \text{(Average "Density")}\end{array} = \frac{\text{IOD}}{\text{Area}} = \frac{\sum_{\Delta\psi=T_1}^{T_2} H(\Delta\psi) \times \Delta\psi}{\sum_{\Delta\psi=T_1}^{T_2} H(\Delta\psi)}$$

FIGURE 2. Methods for quantitative analysis of digital images. The equations are used to sum the pixel intensities expressed in terms of millivolt potentials as described in the text. **H** is a histogram of the pixel potentials in millivolts, $\Delta\psi$ is the membrane potential in millivolts, and T_1 is the lower limit potential and T_2 the upper limit potential.

information for analysis. The parameters of many individual cells are examined further in standard statistical analyses to calculate means, standard deviations, medians, minimums, and maximums for cell populations for individual cells with respect to a cell population, and/or for populations with populations. The analysis seeks to determine whether mitochondria are able to operate at their normal higher potentials; if not, it would indicate that their *ability* (thermodynamic capacity) to make ATP is impaired.

2.4. Criteria for Patient and Control Populations

Patient selection was carried out at a remote independent clinic specializing in diagnosis and treatment of CFS. Patients filled out extensive medical histories that included medications, date of symptom onset, and questionnaires assessing symptom severity, evaluated by a clinician specializing in CFS. Patients were diagnosed according to Centers for Disease Control (CDC) criteria (Fukuda *et al.*, 1994; Holmes *et al.*, 1988). All patients were classified by age, sex, race, and clinically rated activity level for comparison with controls. As mentioned, a loose consensus exists that the patient population is mainly female and often white. Controls were nonpatient volunteers with no acute or chronic illness taking no known medication, who were donating blood for routine testing at UNC Hospitals.

3. RESULTS

3.1. Confocal Imaging of Membrane Potential in TMRM-Labeled Human Cells

Confocal images of TMRM-labeled human mononuclear leukocytes were acquired from all individuals in the study under the same set of experimental conditions. In accord with our protocol (Sect. 2.1), all images of human mononuclear leukocytes were of cells attached to glass coverslips and placed in special microscope chambers. Cells were labeled as close to the time for actual imaging as was feasible. In experiments studying the effect of cytokines on human fibroblasts, cells were subcultured on the coverslips. Our experience with controls studies showed that over the course of our observations and experiments, isolated cells exhibited no time-dependent decrease in viability and membrane potential. In any event, both CFS and control specimens were subject to the same conditions. Also, control studies found that a confocal zoom factor of 3.0 using a 63×1.4 N.A. lens optimal for human mononuclear leukocytes, because it permitted simultaneous imaging of multiple cells. This zoom factor yielded a sufficient number of pixels per cell and per mitochondrion for numerical analysis and kept the inherent shading problem in the design of the Bio Rad MRC 600 to manageable levels. Human fibroblast cells, due to their larger size, were best followed with a zoom factor of 2.0.

3.2. Patient and Control Populations in this Study

The CFS population of 34 individuals was 70% female and 30% male, ranging in age from 18 to 60, with both a mean and a median age of 40. Similarly, the control population of 15 individuals was 72% female and 28% males, ranging in age from 20 to 50, with both a mean and a median age of 37. Of the 25 patients with a known date of CFS onset prior to the blood draw, the median length of the illness in years was 5.4, the mean 7.6, the maximum 25, and the minimum 0.6.

3.3. Comparison of Patient and Control Populations

Our data indicate that the mitochondrial membrane potential and the areas at mitochondrial potentials are significantly lower in CFS patients than in normal individuals, which can be shown numerically and visually. The result is most clearly seen by visually comparing membrane potential images of cells from a typical control with those from a typical CFS patient. Mononuclear leukocytes from a control are shown in a laser-scanning confocal gray-level image, wherein the brighter the image the more negative the potential (Fig. 1A); mitochondria are the small bright oval. In contrast, typical CFS cells from our study population (Fig. 1B) have much dimmer cells and mitochondria in the gray-level image. The same comparisons can be seen in another healthy control (Fig. 1C) and another CFS patient (Fig. 1D). This clear and significant difference between CFS patients and normal controls is typically seen when comparing the membrane potential confocal images. The CFS patients cells and mitochondria are at lower potentials, with a smaller total cellular area at potentials defined as mitochondrial.

The $\Delta\psi$ differences for CFS patients vs controls can be shown numerically and graphically using the analyses in Section 2.3. The total cellular and mitochondrial average potentials were lower for approximately 800 mononuclear leukocyte CSF cells compared with approximately 300 control cells (Fig. 3A). The differences were statistically significant at the 99% confidence level using the students t-test for a two-tailed hypothesis calculated according to Zar (1985). Similarly, we determined that the relative probability densities of the cytoplasmic and total low- and midmitochondrial IPDs are lower in CFS patients as compared with controls (Fig. 3B). These findings were also calculated to be significant at the 99% confidence level (Zar, 1985). Thus, quantitative differences can be established between CFS patients and healthy individuals.

3.4. Comparison of Male and Female Chronic Fatigue Syndrome Patients

Examining approximately 500 cells from female and 300 cells from male patients in our initial studies, we found values of 78.9 ± 13 mV and 81.2 ± 4 mV for mean (total) cell potential, 60 ± 4.6 mV and 60.5 ± 2.4 mV for the mean (nonmitochondrial) cytoplasmic potential, 110 ± 5.4 and 110 ± 3.5 mean mitochondrial potential, respectively. Likewise, we found cell values of $65 \pm 15\%$ and $60 \pm 5\%$ for the areas at cytoplasmic potentials and correspondingly, $35 \pm 15\%$ and $40 \pm 5\%$ for areas at mitochondrial potentials. We also found that $96 \pm 4\%$ and $99 \pm 0.5\%$ respectively, of the mitochondria area fraction was at low mitochondrial potentials, which does not support an appreciable or statistically significant difference between female and male CFS patients.

3.5. Interferon-α Effects on Human Fibroblasts

Adding human IFN-α to human cells causes relatively rapid and protracted decreases in the average total cellular and mitochondrial potentials and the areas at mitochondrial

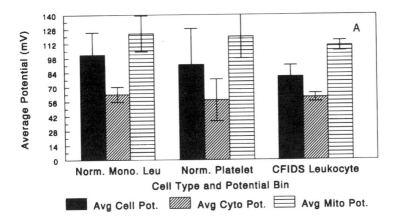

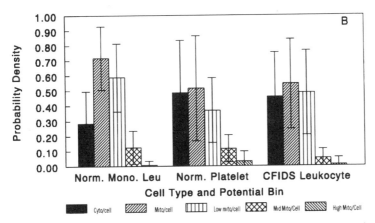

FIGURE 3. Quantitative comparison of mitochondrial membrane potential between a CFS patient population and normal, healthy controls. Data are based on 800 mononuclear leukocytes isolated and prepared from 34 CFS patients and over 300 cells from 15 controls. (A) Average membrane potentials are shown for the total cell, the cytoplasm, and the total mitochondria according to the figure's key. Total cellular and total mitochondrial average potentials were significantly lower in CFS patients compared with controls at the 99% confidence level when analyzed for a two-tailed hypothesis using the student's t-test as described in Zar (1985). Average cytoplasmic potential is approximately the same in both groups. (B) Integrated potential density function (IPD) to calculates the relative probability density relative to the total cell IPD. The IPD relative probability densities for total mitochondrial potential and low- and midmitochondrial potentials are lower in CFS patients compared with controls, whereas the cytoplasmic relative probability densities are correspondingly higher. Values are statistically different at the 99% confidence level.

potentials. Adding 100 ng/ml of human IFN-α to a fibroblast cell (Fig. 1E) decreased TMRM intensity in cell's mitochondria (Fig. 1F) within three minutes. This decrease persisted for up to 3 hours (data not shown). The same effect was seen on human mononuclear leukocytes (Fig. 4) when treated with 100 ng/ml IFN-α. The total mitochondrial average potential dropped within 2.5 minutes of adding it and remained depressed over the 60 minutes we observed. Cytoplasmic average potential drops slightly, roughly

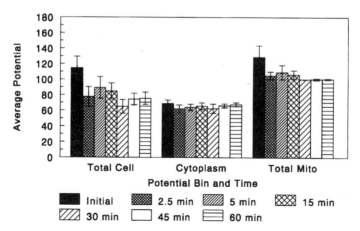

FIGURE 4. Effect of interferon-α on human mononuclear leukocyte membrane potential. Membrane potentials of mononuclear leukocytes were monitored beforehand for 60 min after adding 100 ng/ml interferon-α. Cell populations comprised approximately 50 cells each time. Decreases in total cellular and total mitochondrial average potentials are significantly lower at the 95% confidence level based on a student's *t*-test for a two-tailed hypothesis (Zar 1985).

recovering over the course of the 60 minutes. Total cellular average potential, comprised of the two previous potentials, initially dropped sharply, then oscillates somewhat at the decreased level over the same period. Both human fibroblasts and mononuclear leukocytes showed decreased potentials after IFN-α addition.

4. DISCUSSION

Our confocal microscopy studies of the TMRM distribution in human mononuclear leukocytes from CFS and control populations, which shows lower membrane potentials in the CFS patients, provide the first evidence of a specific cellular site of impairment in the bioenergetics of these patients. Our studies of IFN-α effects on membrane potential in cells provide the first evidence of cytokine-induced effects on the membrane potential and also suggest a possible link between these immune- and cell-signaling agents and alterations in mitochondrial bioenergetic function.

4.1. Membrane-Potential Evidence for a Mitochondrial Dysfunction Role in Chronic Fatigue Syndrome

Our data implicate mitochondrial dysfunction in the CFS population we studied. Our comparison of CFS and control populations reveal a difference in $\Delta\psi$. In CFS patients the total cellular and total mitochondrial average potentials and the IPDs (and area fractions) of the cells at mitochondrial potentials where lower. The 12-mV lower values for the average $\Delta\psi$ in CFS patients compared with controls (122 vs 110 mV) should be taken as the minimum difference. This is based on two factors inherent in the quantitative analysis employed (Sect. 2.3). Because the confocal image is defined only by membrane potential, a mitochondrion is operationally defined as having a potential between 100 and 240 mV. Thus, mitochondria (perhaps dysfunctional) operating below 100 mV are treated as cytoplasm. This is why the IPDs and areas fractions are important in the analysis, however. Cells having sub-100 mV mitochondria will have smaller mitochondrial IPDs and area fractions, as we observed for CFS patients. In addition, the total cellular average potential, some 21 mV lower in CFS patients (including mitochondrial and cytoplasmic contributions) suggests that the average difference for all mitochondria in the cell may be, effectively, 21 mV. Reports based on the state 4 to state 3 transition in isolated rat liver mitochondria determined a 30-mV average decrease during ATP synthesis (e.g., Petite *et al.*, 1990). Thus, the approximately 12–21+ mV lower average potential and approximately 24% lower mitochondrial IPD indicates significant impairment in the ATP-making ability of some or all CSF mitochondria.

Our comparison of male and female CFS patients indicates that the menstrual cycle may not have a significant effect on the parameters we measured. No one knows why more women than men are diagnosed with CFS, although there is speculation that women are more likely to seek medical care (e.g., Hyde *et al.*, 1992; Reyes *et al.*, 1997). Distribution of CFS patients along racial and gender lines has been examined in a number of reports (Buchwald *et al.*, 1995; Gunn *et al.*, 1993; Komaroff, 1997; Reyes *et al.*, 1997), and our patient population was consistent with these reports. Our data indicate that between male and female patients diagnosed with CFS, there is on average no significant mitochondrial difference, and we are unaware of any such distinction in the literature, or of any between male and female physiological need for such a difference consistent with our data.

Our finding of a likely impairment in the ability of mitochondria in CFS patient cells to make ATP may well explain certain widespread CFS symptoms, some of which were mentioned previously (Sect. 1.3.). This is supported by the fact that illnesses of known mitochondrial dysfunction give rise to many of the same symptoms (e.g., Carafoli and Roman, 1980; DiMauro *et al.*, 1985; Frackowiak *et al.*, 1988; Roe and Coates, 1989). Given the disparate but systemic nature of CFS symptoms, our findings are further supported by the fact that mitochondrial dysfunction is expressed differently in different tissues, attributable to the different energy demands of these tissues (Behan *et al.*, 1991; DiMauro *et al.*, 1985; Peterson, 1995; Wallace, 1992). Consequently, muscle and brain tissue are most likely (but not exclusively) to give rise to patient complaints. It is no surprise, therefore, that the major CSF complaints in addition to general fatigue are typically muscle fatigability (often perceived as muscle weakness), cognitive dysfunction, and sleep abnormalities. Of particular interest are reports of the beneficial effects of carnitine, which is involved in mitochondrial β-oxidation of fatty acids—an important source of muscle energy. Ubiquinone, an electron transport chain component in the mitochondrial inner membrane, has also been reported to help some CFS patients, but few studies have truly examined cellular energy production at the cellular and subcellular levels.

Our findings of impaired mitochondrial ATP-making ability in CFS patient cells may explain reports of muscle bioenergetics. Muscle bioenergetics tend to be studied for two reasons: (1) Muscle is the most plentiful tissue, and biopsy samples are somewhat easier to do. Conventional biochemical or microscopic analysis, albeit limited by sample size, can be done (Behan, 1991); (2) Muscle is amenable to the use of P-NMR to study tissue bioenergetics (Jamal and Hansen, 1985; McCulley *et al.*, 1996; Wong *et al.*, 1992).

Nonetheless, conventional studies on CFS patient muscle tissue have yielded conflicting results. Histological studies show sporadic, but not uniform, abnormalities in muscle fibers and their mitochondria, (e.g., Behan, 1991). McCully *et al.* (1996) took the ambiguous position in their NMR studies of arguing for a defect in muscle oxidative metabolism while also maintaining the problems could be a result of muscle deconditioning. NMR studies by Lane and co-workers (1995, 1998a, 1998b), however, showed a significant subset of CFS patients with abnormal lactate responses to exercise, magnetic resonance characteristics indicative of excessive intracellular acidosis and impaired capacity for mitochondrial ATP synthesis, which could not be satisfactorily attributed to inactivity or deconditioning. Our studies using mononuclear leukocytes, which avoid the interpretative problem of the deconditioning phenomenon in muscle, support the reality of a defect in muscle oxidative metabolism. Further, mitochondria in mononuclear leukocytes "look" normal, which would explain many of the histological findings of small ultrastructural abnormalities. Our ability to examine individual cells and their mitochondria using confocal microscopy and our analytical methods (Sect. 2.3) also have an advantage over P-NMR, which inherently must average a great many cells in a given muscle area. One drawback to our technique, however, is that it is not an *in vivo* approach, though the mitochondria might well be considered "*in vivo*" since they are in their normal environment, cytoplasm. Future studies to complement our findings would ideally study the hard-to-isolate-and-maintain individual muscle cells, looking at bioenergetic differences in CFS and healthy controls populations.

It should be clear that our data do not prove that mitochondrial dysfunction causes CFS. Rather, we show that mitochondrial dysfunction is *involved* in CFS. Whether some immune- or cell-signaling abnormality, perhaps due to a genetic predisposition perhaps triggered by a viral or bacterial infection, causes mitochondrial dysfunction remains to be seen. Some of our studies adding cytokines such as IFN-α or IL-2 to human mononuclear leukocytes or fibroblasts, which show for the first time an effect on mitochondrial function, may suggest such a linkage, however.

4.2. Searching the Mitochondrial Pathways for Possible Defects

Our data show mitochondrial dysfunction in CFS patient cells, but further studies are needed to determine if there is a site (or sites) for this dysfunction in the mitochondrial metabolic pathways or if there is a mitochondrial membrane problem. Such studies can be accomplished through the use of various mitochondrial cell-permeable substrates or inhibitors. The $\Delta\psi$ response of control and CFS cells is being examined for differences in their response to: various fatty acid substrates and glucose; uncouplers of oxidative phosphorylation such as carbonylcyanide-*m*-chlorophenyl hydrazone; various inhibitors of specific electron transport enzymes such as antimycin A for the ubiquinol–cytochrome *c* oxidoreductase complex, an inhibitor of ATP synthesis (oligomycin); and so forth. Differences in CSF and control response to one or more of these agents could identify the site or sites of mitochondrial dysfunction.

4.3. Cytokines Affect Mitochondrial Bioenergetics, Cellular Bioenergetics, or Both

The effect of IFN-α on membrane potentials of human cell lines was examined in relation to a possible role in CFS. Our studies revealed that IFN-α has a profound effect

$\Delta\psi$, causing a marked and prolonged decrease, and hence on mitochondrial and cellular bioenergetics. Our data may also provide a rationale for reports of abnormal IFN-α levels in CFS patients (Levy, 1994; Lloyd, 1991) and the abnormal interferon production of some patients' cells in response to viral infection (Lever, 1988). We reported elsewhere that other cytokines such an IL-2 can also affect mitochondrial and cellular membrane potentials in a similar fashion (Chazotte and Pettengill, 1999). To our knowledge these are the first reports of such cytokine effects on $\Delta\psi$ and bioenergetics.

Our data examining the IFN-α effect on mitochondrial function is consistent with clinical reports on its effect as an antiviral agent e.g., (MacDonald, 1987): many patients undergoing IFN-α therapy develop chronic-fatigue-like symptoms. Interferons cause cells to enter an antiviral state. Part of this state may be an impaired mitochondrial ability to make ATP, which make it hard for a virus to utilize the cell's machinery to reproduce and hence, re-infect.

Our data may also of be of import to chemotherapy, where it might provide a rationale for the unexplained cardiotoxicity and hepatotoxicity of IL-2 and IFN-α treatment (e.g., Kruit et al., 1994; Nakagawa et al., 1996). Patients undergoing such treatment regimes for renal carcinoma, for example, develop chronic-fatigue-like symptoms as well as organ toxicities. Better understanding of the effects of cytokines on cellular and mitochondrial bioenergetics may lead to more efficacious treatment of cancer with cytokines.

5. CONCLUSIONS

Understanding of mitochondrial dysfunction in CFS will clearly benefit from more study. Our findings warrant detailed studies on larger patient and control populations. The siting of mitochondrial impairment may lead to approaches for alleviating some of the symptoms of CFS. Studies on the effects of cytokines on cellular and mitochondrial bioenergetics are important to see whether cytokines or similar molecules are involved in linking CFS immune disturbances to mitochondrial dysfunction, and to better understand the consequences of cytokine therapies for cancer and viral infection. ·

ACKNOWLEDGEMENTS. Supported in part by grants from the CFIDS Association of America, the University of North Carolina Center for Research on Chronic Illness (NIH 5P30 NR03962-02), and the University of North Carolina Research Council. I would also like to thank Drs. Ken Jacobson, Charles R. Hackenbrock, and John J. Lemasters for their advice and encouragement over the years, and Dr. Paul Cheney for supplying CFS patient blood specimens.

REFERENCES

Adamson, G. M., and Billings, R. E., 1992, Tumor necrosis factor induced oxidative stress in isolated mouse hepatocytes, *Arch. Biochem. Biophys.* **294**:223–229.

Aoki, T., Miyakoshi, H., Usuda, Y., and Hernerman, R. B., 1993, Low NK syndrome and its relationship to chronic fatigue syndrome, *Clin. Immunol. Immunopath.* **69**:253–265.

Barbiroli, B., Montagna, P., Martinelli, P., Lodi, R., Iotti, S., Cortelli, P., Funicello, R., and Zaniol, P., 1993,

Defective brain energy metabolism by *in vivo* [31]pmr spectroscopy in 28 patients with mitochondrial cytopathies, *J. Cereb. Blood Flow Metab.* **13**:469–474.

Barke, R. A., Brady, P. S., and Brady, L. J., 1991, The Ca^{2+} second messenger system and interleukin-I-alpha modulation of hepatic gene transcription and mitochondrial fat oxidation, *Surgery* **110**:285–294.

Behan, W. H. M., More, A. R., and Behan, P. O., 1991, Mitochondrial abnormalities in the postviral fatigue syndrome, *Acta Neuropathol.* **83**:61–65.

Bengtsson, A, and Henriksson, K. G., 1989, The muscle in fibromyalgia: A review of Swedish studies, *J. Rheumatol.* **16**:144–149.

Bock, G. R., and Whelan, J. (Eds.), 1993, *Chronic Fatigue Syndrome*, Ciba Foundation Symposium 173, Wiley, Chichester, England.

Buchwald, D., 1991, Laboratory abnormalities in chronic fatigue syndrome, in *Post Viral Fatigue Syndrome* (R. Jenkins and J. Mobray, Eds.), Wiley, Sussex, England, pp. 117–136.

Buchwald, D., and Komaroff, A. L., 1991, Review of laboratory findings for patients with chronic fatigue syndrome, *Rev. Infect. Dis.* **13**:S12–18.

Buchwald, D., Umali, P., Kith, P., Perlman, B. A., and Komaroff, A. L., 1995, Chronic fatigue and chronic fatigue syndrome: Prevalence in a Pacific Northwest health care system, *Ann. Int. Med.* **123**:81–88.

Caligiuri, M., Murray, C., Buchwald, D., Levine, H., Cheney, P., Peterson D., Komaroff, A. L., and Ritz, J., 1987, Phenotypic and functional deficiency of natural killer cells in patients with chronic fatigue syndrome, *J. Immunol.* **139**:3306–3313.

Cannon, J. G., Angel, J. B., Abad, L. W., Vannier, E., Mileno, M. D., Fagioli, L., Wolff, S. M., and Komaroff, A. L., 1997, Interleukin-1β, interleukin-1 receptor antagonist, and soluble interleukin-1 receptor type II secretion in chronic fatigue syndrome, *J. Clin. Immunol.* **17**:253–261.

Carofoli, E., and Roman, I., (1980), Mitochondria and disease, *Mol. Aspects Med.* **3**:297–400.

Chacon, E., Reece, J. M., Nieminen, A.-L., Zahrebelski, G., Herman, B., and Lemasters, J. J., 1994, Distribution of electrical potential, pH, free Ca^{2+}, and volume inside cultured adult rabbit myocytes during chemical hypoxia: A multiparameter digitized confocal microscopic study, *Biophys. J.* **66**:942–952.

Chazotte, B., and Pettengill, M., 1998, Cytokine effects on mitochondrial membrane potential and possible mitochondrial dysfunction in chronic fatigue and immune dysfunction syndrome, *ASBMB J.* May 1998.

Chazotte, B., and Pettengill, M., 1999, Using membrane potential to follow cytokine effects on mitochondria and possible dysfunction in chronic fatigue syndrome, *Biophys. J.* **76**:A363.

Chazotte, B., Loehr, J. P., and Hackenbrock, C. R., 1996, Quantitative analyses of membrane potential of mitochondria in individual living human cells related to chronic illness, *FASEB J.* **10**:A1377.

Cheney, P., and Lapp, C. W., 1992, The diagnosis of chronic fatigue syndrome: An assertive approach, *CFIDS Chronicle Physician's Forum, Sept.* 13–19.

Cho, W. K., and Stollerman, G. H., 1992, Chronic fatigue syndrome, *Hospital Practice* Sept 15 **27**:221–245.

Coates, P. M., Hale, D. E., Stanley, C. A., Corkey, B. E. and Cortner, J. A., 1985, Genetic deficiency of medium-chain Acyl-Co A dehydrogenase: Studies in cultured skin fibroblasts and peripheral mononuclear leukocytes, *Pediat. Res.* **19**:671–676.

Dawson, D. M., and Sabin, T. D. (Eds.), *Chronic Fatigue Syndrome*, Little, Brown, Boston, 1993.

Demitrack, M. A., 1994, Neuroendocrine aspects of chronic fatigue syndrome: Implications for diagnosis and research, in *Chronic Fatigue Syndrome* (S. E. Straus, Ed.), Marcel Dekker, New York.

Demitrack, M. A., and Crofford, L. J., 1998, Evidence for and pathophysiologic implications of hypothalamic–pituitary–adrenal axis dysregulation in fibromyalgia and chronic fatigue syndrome, *Ann. NY Acad. Sci.* **840**:684–697.

DiMauro, S., Bonillia, E., Zeviani, M., Nakagawa, M., and DeVivo, D. C., 1985, Mitochondrial myopathies, *Annu. Neurol.* **17**:521–538.

Frackowiak, R. S. J., Herold, S., Petty, R. K. H., and Morgan-Hughes, J. A., 1988, The cerebral metabolism of glucose and oxygen measured with positron tomography in patients with mitochondrial diseases, *Brain* **111**:1009–1024.

Fukuda, K., and Gantz, N. M., 1995, Management strategies for chronic fatigue syndrome, *Fed. Pract.* July.

Fukuda, K., Straus, S. E., Hickie, I., Sharpe, M. C., Dobbins, J. G., and Komaroff, A., 1994, The chronic fatigue syndrome: A comprehensive approach to its definition and study, *Ann. Intern. Med.* **121**:953–959.

Geng, Y.-j., Hansson, G. K., and Holme, E., 1992, Interferon-gamma and tumor necrosis factor synergize to induce nitric oxide production and inhibit mitochondrial respiration in vascular smooth muscle cells, *Circ. Res.* **71**:1268–1276.

Grafman, J., Johnson, R., and Scheffers, M., 1991, Cognitive and mood-state changes in patients with chronic fatigue syndrome, *Rev. Infect. Dis.* **13**:S45–S52.

Gunn, W. J., Connell, O. B., and Randell, B., 1993, Epidemiology of chronic fatigue syndrome: The Centers for Disease Control Study, in *Chronic Fatigue Syndrome*, (G. R. Bock and J. Whelan, Eds.), CIBA Foundation Symposium **173**, Wiley, Chichester, England, pp. 83–108.

Ho-Yen, D. O., Carrington, D., and Armstrong, A. A., 1988, Myalgic encephalomyelitis and alpha-interferon, *Lancet* **i**:125–127.

Holms, G. P., Kaplan, J. E., Gantz, N. M., Komaroff, A. L., Schonberger, L. B., Straus, S. E., Jomes, J. F., Dubois, R. E., Cunningham-Rundles, C., Pahwa, S., Tosato, G., Zegans, L. S., Purtilo, D. T., Brown, N., Schooley, R. T., and Brus, I., 1998, Chronic fatigue syndrome: A working case definition, *Ann. Intern. Med.* **108**: 387–389.

Hyde, B. M., and Jain, A., 1992, Clinical observations of CNS dysfunction in post-infectious, acute onset M.E./C.F.S., in *The Clinical and Scientific Basis of Myalgic Encephalomyelitis/ Chronic Fatigue Syndrome* (B. M. Hyde, J. A. Goldstein, P. H. Levine, Eds.), the Nightingale Foundation, Ottawa, Canada, pp. 38–57.

Hyde, B. M., Goldstein, J. A., and Levine, P. H. (Eds.), *The Clinical and Scientific Basis of Myalgic Encephalomyelitis/Chronic Fatigue Syndrome*, The Nightingale Foundation, Ottawa, Canada.

Gunn, W. J., Connell, D. B., and Randell, B., 1993, Epidemiology Chronic Fatigue Syndrome: The Centers for Disease Control Study In *Chronic Fatigue Syndrome*, (G. R. Bock and J. Whelan, eds.), Ciba Foundation Symposium '93. Wiley, Chichester, England.

Jamal, G. A., and Hansen, S., 1985, Electrophysiological studies in the post-viral fatigue syndrome, *J. Neurol, Neuorsurg. Psychiat.* **48**:691–694.

Jenkins, R., and Mowbray, J. (Eds), 1991, *Post-Viral Fatigue Syndrome*, Wiley, Chichester, England.

Jones, J. F., 1991, Serologic and immunologic responses in chronic fatigue syndrome with emphasis on the Epstein-Barr virus, *Rev. Infect. Dis.* **13**:S26–S31.

Kartounis, L. D., Troung, D. D., Morgan-Hughes, J. A., and Harding, A. E., 1992, The neuropsychological features of mitochondrial myopathies and encephalopathies, *Arch. Neurol.* **49**:158–160.

Klimas, N. G., Salvato, F. R., Morgan, R., and Fletcher, M. A., 1990, Immunologic abnormalities in chronic fatigue syndrome, *J. Clin. Microbiol.* **28**:1403–1410.

Komaroff, A. L., 1997, A 56-year-old woman with chronic fatigue syndrome, *JAMA* **278**:1179–1185.

Komaroff, A. L., and Buchwald, D. S., 1998, Chronic fatigue syndrome: An update, *Annu. Rev. Med.* **49**: 1–13.

Kruit, W. H., Punt, K. J., Goey, S. H., de Mulder, p. H., van Hoogenhuyze, D. C., Henzen-Logmans, W., and Stoter, G., 1994, Cardiotoxicity as a dose-limiting factor in a schedule of high dose bolus therapy with interleukin-2 and alpha-interferon, *Cancer* **74**:2850–2856.

Kuratsume, H., Yamaguti, K., Takahashi, M., Misaki, H., Tagawa, S., and Kitani, T., 1994, Acylcarnitine deficiency in chronic fatigue syndrome, *Clin. Infect. Dis.* **18**:S62–S67.

Lane, R. J. M., Burgess, A. P., Flint, J., Riccio, M., and Archard, L. C., 1995, Exercise responses and psychiatric disorder chronic fatigue syndrome, *Brit. Med. J.* **311**:544–545.

Lane, R. J. M., Barrett, M. C., Taylor, D. J., Kemp, G. J., and Lodi, R., 1998a, Heterogeneity in chronic fatigue syndrome: Evidence from magnetic resonance spectroscopy of muscle, *Neuromus. Disord.* **8**:204–209.

Lane, R. J. M., Barrett, M. C., Woodrow, D., Moss, J., Fletcher, R., and Archard, L. C., 1998b, Muscle fibre characteristics and lactate responses to exercise in chronic fatigue syndrome, *J. Neurol. Neurosurg. Psychiat.* **64**:362–367.

Lemasters, J. J., Chacon, E., Zahrebelski, G., Reece, J. M., and Nieminen, A.-L., 1993a, Laser scanning confocal microscopy of living cells, in *Optical Microscopy: Emerging Methods and Applications* (B. Herman and J. J. Lemasters, Eds), Plenum, New York, pp 339–354.

Lemasters, J. J., Nieminen, A.-L., Chacon, E., Imberti, R., Gores, G. J., Reece, J. M., and Herman, B., 1993b, Use of fluorescent probes to monitor mitochondrial membrane potential in isolated mitochondria, cell suspensions, and cultured cells, in *Mitochondrial Dysfunction Methods in Toxicology*, Vol. 2 (L. H. Lash, and D. P. Jones, Eds.), Academic, New York, pp. 404–415.

Lever, A. M. L., Lewis, D. M., Bannister, B. A., Fry, M., and Berry, N., 1988, Interferon production in postviral fatigue syndrome, *Lancet* **ii**:101.

Levy, J. A., 1994, Viral studies of chronic fatigue syndrome, *Clin. Infect. Dis.* **18**:S117–S120.

Lloyd, A., Hickie, I., Brockman, A., Dwyer, J., and Wakefield, D., 1991, Cytokine levels in serum and cerebrospinal fluid in patients with chronic fatigue syndrome, *J. Infect. Dis.* **164**:1023–1024.

MacDonald, E., 1987, Interferons as mediators of psychiatric morbidity; An investigation in a trial of recombinant alpha-interferon in hepatitis-B carriers, *Lancer* **ii**:1175–1178.

McCully, K. K., Natelson, B. H., Iotti, S., Sisto, S., Leigh, J. S., 1996, Reduced oxidative metabolism in chronic fatigue syndrome, *Muscle and Nerve* **19**:621–625.

McGregor, N. R., Dunstan, R. H., Zerbes. M., Butt, H. L., Roberts, T. K., and Klineberg, I. J., 1996, Preliminary determination of a molecular basis to chronic fatigue syndrome, *Biochem. Mol. Med.* **57**:73–80.

Montague, T. J., Marrie, T. J., Klassen, G. A., Bewick, D. J., and Horacek, B. M., 1989, Cardiac function at rest and with exercise in the chronic fatigue syndrome, *Chest* **95**:779–784.

Nakagawa, K., Miller, F. N., Sims, D. E., Lentsch, A. B., Miyazaki, M., and Edwards, M. J., 1996, Mechanisms of interleukin-2 hepatic toxicity, *Cancer Res.* **56**:507–510.

Natelson, B. H., Cohen, J. M., Brassloff, I., and Lee, H. J., 1993, A controlled study of brain magnetic resonance imaging in patients with the chronic fatigue syndrome. *J. Neurol. Sci.* **120**:203–207.

Ojo-Amaze, E. A., Conley, E. J., and Peter, J. B., 1994, Decreased natural killer cell activity is associated with severity of chronic fatigue immune dysfunction syndrome, *Clin. Infect. Dis.* **18**:S157–S159.

Pawley J. B., *Handbook of Biological Confocal Microscopy*, 2nd ed., Plenum, New York, 1995.

Peterson, P. L., 1995, The treatment of mitochondrial myopathies and encephalomyopathies, *Biochim. Biophys. Acta* **1271**:275–280.

Petit, P. X., O'Conner, J. E., Grunwald, D., and Brown, S. C., 1990, Analysis of the membrane potential of rat- and mouse-liver mitochondria by flow cytometry and possible applications, *Eur. J. Biochem.* **193**:389–397.

Plioplys, A. V., and Plioplys, S., 1997, Amantadine and L-carnitine treatment of chronic fatigue syndrome, *Neuropsychobio.* **35**:16–23.

Rasmussen, H., Nielsen, H., Andersen, V., Barington, T., Bendtzen, K., Hansen, S., Nielsen, L., Pedersen, B. K., and Wiik, A., 1994, Chronic fatigue syndrome: A controlled cross-sectional study, *J. Rheumatol.* **21**:1527–1531.

Reyes, M., Gary, H. E., Jr., Dobbins, J. G., Randall, B., Steele, L., Fukuda, K., Holms, G. P., Connell, D. G., Mawle, A. C., Schmid, D. S., Stewart, J. A., Schonberger, L. B., Gunn, W. J., and Reeves, W. C., 1997, Surveillance for chronic fatigue syndrome: Four U. S. cities, September 1989 through August 1993, *MMWR* **46**:1–13.

Roe, C. R., and Coates, P. M., 1989, Acyl-CoA dehydrogenase deficiencies, in *Metabolic Bases of Inherited Disease*, 6th ed. (C. R., Scribner, A. L. Beaudet, W. A. Sly, and D. Valle Eds.), McGraw-Hill, NY, pp. 889–914.

Rowe, P. C., Bou-Holaigah, I., Kan, J. S., and Calkins, H., 1995, Is neurally mediated hypotension an unrecognized cause of chronic fatigue? *Lancet* **345**:623–624.

Sandman, C. A., Barron, J. L., Nackoul, K., Goldstein, J., and Fidler, F., 1993, Memory deficits associated with chronic fatigue immune dysfunction syndrome, *Biol. Psychiat.* **33**:618–623.

Schwartz, R. B., Garada, B. M., Komaroff, A. L., Tice, H. M., Gleit, M., Jolez, F. A., and Holman, B. L., 1993a, Detection of intracranial abnormalities in patients with chronic fatigue syndrome: Comparison of MR imaging and SPECT, *Am. J. Rad.* **162**:935–941.

Schwartz, R. B., Komaroff, A. L., Garada, B. M., Gleit, M., Doolittle, T. H., Bates, D. W., Vasile, R. G., and Holman, B. L., 1993b, SPECT imaging of the brain: Comparison of findings in patients with chronic fatigue syndrome, AIDS dementia complex, and major unipolar depression, *Am. J. Rad.* **162**:943–951.

Straus, S. E., 1988, The chronic mononucleosis syndrome, *J. Infect. Dis.* **157**:405–412.

Straus, S. E. (Ed.), 1994, *Chronic Fatigue Syndrome*, Marcel Dekker, New York.

Vojdani, A., Ghoneum, M., Choppa, P. C., Magatoto, L., and Lapp, C. W., 1997, Elevated apoptotic cell population in patients with chronic fatigue syndrome: The pivotal role of protein kinase RNA, *J. Int. Med.* **242**:465–478.

Wallace, D. A., 1992, Mitochondrial DNA diseases, *Ann. Rev. Biochem.* **61**:1175–1212.

Wong, R., Lopaschuk, G., Zhu, G., Walker, D., Catellier, D., Burton D., Teo, K. Collins-Nakai, R., and Montague, T., 1992, Skeletal muscle metabolism in the chronic fatigue syndrome, *Chest* **102**:1716–1722.

Zar, J. H., 1985, *Biostatistical Analysis*, 2nd ed., Prentice-Hall, Englewood Cliffs, New Jersey,

Zhang, C., Baumer, A., Mackay, I. R., Linnane, A. W., and Nagley, P., 1995, Unusual pattern of mitochondrial DNA deletions in skeletal muscle of an adult human with chronic fatigue syndrome, *Hum. Mol. Genet.* **4**:751–754.

Section 5

Chemical Toxicity

Chapter 22

Bile Acid Toxicity

Gregory J. Gores, James R. Spivey, Ravi Botla, Joong-Won Park, and Mark J. Lieser

1. INTRODUCTION

1.1. Cholestasis

Cholestasis, defined as an impairment in bile formation, is a pathophysiologic disturbance resulting from impaired hepatocyte and/or cholangiocyte solute and water transport. Acute and chronic cholestasis occurs in many human diseases, including sepsis, drug toxicity, viral hepatitis, a varied group of illnesses called cholangiopathies, and during administration of parenteral nutrition. The cholangiopathies are characterized by inflammation and ultimately destruction of the cholangiocytes lining the bile ducts, causing ductopenic syndromes. Even in cholangiopathies, where the primary insult is to the bile ducts, hepatocellular injury is an invariant feature of cholestasis. Hepatocellular injury is important because it likely promotes further hepatic dysfunction and leads to a cascade of events culminating in liver fibrosis. Treatment options for most cholangiopathies, including primary biliary cirrhosis and primary sclerosing cholangitis, are limited. Many patients with these diseases develop progressive liver disease and require liver transplantation at a substantial societal cost. Thus the mechanisms causing liver injury in cholestasis are of substantial clinical and scientific importance. A better understanding of the mechanisms of cholestatic hepatocellular injury could help to provide rational therapeutic strategies.

Gregory J. Gores and Ravi Botla   Mayo Clinic, Rochester, Minnesota 55905.   **James R. Spivey** Mayo Clinic Jacksonville, Jacksonville, Florida 32224.   **Joong-Won Park**   Division of Gastroenterology, Department of Internal Medicine, College of Medicine, Chung-ang University, Seoul, 140-757, Korea.   **Mark J. Lieser**   Department of Surgery, University of Texas Southwest Medical Center, Dallas, Texas 75235-9156.

Mitochondria in Pathogenesis, edited by Lemasters and Nieminen.
Kluwer Academic/Plenum Publishers, New York, 2001.

413

1.2. Hepatocellular Injury and Cholestasis

The mechanism of hepatocyte injury in cholestasis is likely multifactorial and dependent upon the initiating event and the disease process. Nonetheless, impairment in bile formation leads to an accumulation of bile salts within the hepatocyte (Greim *et al.*, 1973). Bile salts are cholesterol-derived amphipathic molecules which, at elevated concentrations, are toxic to hepatocytes (Patel *et al.*, 1995). We and others have demonstrated that toxic bile salts induce hepatocyte necrosis and apoptosis in a concentration-dependent manner. High concentrations of toxic bile salts ($\geq 100\,\mu M$) cause necrosis, whereas 25–$100\,\mu M$ concentrations induce apoptosis (Patel *et al.*, 1994). In humans virtually all circulating bile salts are conjugated to glycine or taurine. The most toxic primary bile salts are the hydrophobic glycine conjugates of chenodeoxycholate and deoxycholate (Patel *et al.*, 1995). Although secondary bile salts generated by bacterial dehydroxylation of the primary salts (e.g., deoxycholate and lithocholate) are extremely toxic, their serum concentrations actually decrease in chronic cholestasis, in contrast to serum concentrations of the primary salts, which increase (Ostrow, 1993). Serum total bile salt concentrations can reach $300\,\mu M$ in cholestasis, significantly greater than the physiologic portal vein concentrations of approximately $\leq 20\,\mu M$ of healthy individuals (Ostrow, 1993). Thus, bile salts are thought to contribute to hepatocellular injury in most forms of cholestasis. This concept is further strengthened by the observation that the hydrophilic bile salt ursodeoxycholic acid (UDCA) ameliorates liver injury during cholestasis (Kaplan, 1994); it is thought to antagonize the toxic effects of hydrophilic bile salts, thereby decreasing hepatocellular injury. Because of their importance in liver injury, a better understanding of the mechanisms by which bile acids induce hepatocyte injury is needed.

1.3. Mitochondria and Cell Death

Considerable information now points to mitochondrial dysfunction as a key feature of both necrosis and apoptosis. Mitochondrial dysfunction with loss of oxidative phosphorylation and subsequent cellular ATP depletion can induce hepatocyte necrosis (Rosser and Gores, 1995). More-recent data implicate mitochondrial release of cytochrome *c* into the cytosol as a central event initiating the effector pathways of apoptosis (Li *et al.*, 1997). A common mechanism of mitochondrial dysfunction in cell death is the mitochondrial permeability transition (MPT) (Rosser and Gores, 1995), which is characterized by the opening of a high-conductance, oxidant-sensitive proteinaceous pore in the inner mitochondrial membrane (Bernardi, 1996). Opening of this pore is associated with colloidosmotic mitochondrial swelling, uncoupling of oxidative phosphorylation, ATP depletion, and release of cytochrome *c* from intermembrane space into cytosol (Bernardi, 1996; Li *et al.*, 1997; Rosser and Gores, 1995; Susin *et al.*, 1997). The MPT, therefore, is invoked as a mechanism of necrosis and apoptosis (Rosser and Gores, 1995; Susin *et al.*, 1997).

Because of the transition's putative central role in cell death, we hypothesize that toxic bile salts cause hepatocellular injury by leading to mitochondrial dysfunction. In this chapter we address several questions emanating from this hypothesis. In particular, we focus on the mechanisms inhibited by the cytoprotective bile salt UDCA to gain further insight into its pharmacologic actions.

2. DOES MITOCHONDRIAL DYSFUNCTION OCCUR IN CHOLESTASIS?

Most investigators used the bile-duct-ligated (BDL) rat as a model of obstructive cholestasis. This model involves surgical ligation of the common bile duct, producing complete cessation of bile flow, although extreme, the model has proven useful for studying cholestatic liver injury. Schaffner and colleagues (1971) were the first to document morphologic features of mitochondrial injury in this model. Using transmission electron microscopy, they observed swollen mitochondria, shortened and swollen cristae, and variations in mitochondrial shape in BDL rat livers. Krahenbühl and associates (1992a) subsequently highlighted the role of mitochondrial dysfunction in cholestasis by directly assessing mitochondrial function. Using sophisticated and objective morphological approaches to account for changes in mitochondrial number, these investigators demonstrated impaired oxygen consumption in isolated perfused BDL rat livers (Krahenbuhl et al., 1992a). To further demonstrate mitochondrial dysfunction, they isolated mitochondria from the BDL rats and evaluated mitochondrial respiration using site-specific substrates (Krahenbuhl et al., 1992b), thus demonstrating significant impairment in state 3 respiration (oxygen consumption measured in the presence of ADP) and little change in state 4 respiration (oxygen consumption measured in the absence of ADP). These observations demonstrated structural and functional abnormalities in mitochondria during cholestasis but did not reveal whether mitochondrial dysfunction was due to toxic bile salts.

3. DO TOXIC BILE SALTS DIRECTLY CAUSE MITOCHONDRIAL DYSFUNCTION?

3.1. Bile Salts and Mitochondrial Respiration

We and others have demonstrated direct mitochondrial toxicity by bile salts (Krahenbuhl et al., 1994; Spivey et al., 1993). Krahenbühl and associates (1994) isolated mitochondria from normal rats and measured its respiration in the presence of toxic bile salts. Unconjugated chenodeoxycholate and deoxycholate at concentrations of $100\,\mu M$ caused a 44–52% reduction in state 3 of mitochondrial respiration, using succinate as the substrate. The studies are of uncertain physiologic relevance, however, because the vast majority of human bile salts are conjugated to either taurine or glycine but these species were not studied. The observations were also not linked to cytotoxicity. Moreover, the taurine conjugate of ursodeoxycholate (TUDC) did not protect against impairment of state 3 respiration by chenodeoxycholate or deoxycholate (Krahenbuhl et al., 1994). Because ursodeoxycholate is thought to prevent bile salt cytotoxicity, impairment of oxidative phosphorylation by toxic bile salts might not be the sole cytotoxic mechanism. Nevertheless, these data were the first to link mitochondrial dysfunction to toxic bile salts and likely provided insight into the nonlethal hepatocellular dysfunction seen in cholestasis.

We chose a different approach to evaluate toxic-bile-salt-induced mitochondrial dysfunction in order to link disruption of mitochondrial function to cell death. We used glycochenodeoxycholate because it is the principal toxic hydrophobic bile salt accumulating in the livers of patients with cholestasis (Ostrow, 1993). We employed isolated rat hepatocytes to test the hypothesis that toxic bile salts induce hepatocyte death by a bioenergetic mechanism (i.e., failure of mitochondrial oxidative phosphorylation).

First, we measured total cellular ATP in fresh hepatocyte suspensions incubated in the presence of 400 μM glycochenodeoxycholate (GCDC) (Spivey *et al.*, 1993). Within 30 minutes of incubation, cellular ATP decreased by >75% (Fig. 1). Not only did ATP fall, but its loss was important for the development of cell necrosis, as measured by propidium iodide (PI) uptake. Indeed, maintenance of cellular ATP at 50% of initial value via glycolysis using fructose prevented GCDC-induced hepatocyte necrosis (cell viability of $23 \pm 1\%$ vs $74 \pm 2\%$, $P < 0.01$). Treating the hepatocytes with an equimolar concentration of TUDC did not prevent ATP depletion (Fig. 1), showing that the fall in ATP was secondary to impaired oxidative phosphorylation because glycolytic generation of ATP remained intact. Furthermore, TUDC's cytoprotective effects were not mediated by preventing GCDC-induced ATP depletion. To more fully demonstrate that bile salts cause mitochondrial dysfunction, we measured succinate-driven mitochondrial respiration in digitonin-permeabilized cells. We choose digitonin-permeabilized cells so that we could determine the effect of bile salts on mitochondria in the presence of bile salt cytosolic binding proteins (Stolz *et al.*, 1989), a paradigm that more closely mimics bile salt cytotoxicity in intact cells. Similar to Krahenbühl and co-workers, we found that GCDC directly impaired state 3 of mitochondrial respiration, and consistent with the ATP measurements, TUDC also did not prevent GCDC impairment of state 3 respiration (Fig. 2), unambiguously demonstrating that toxic bile salts in high concentrations can target mitochondria, causing a bioenergetic form of cell necrosis. It is perplexing, however, that TUDC, which is cytoprotective against toxic bile-salt-induced injury, did not ameliorate the salts' effect on mitochondrial respiration. This suggested to us that bile

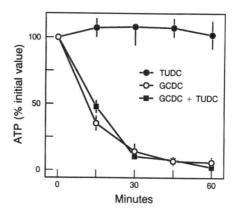

FIGURE 1. Glycochendeoxycholate treatment of hepatocytes causes rapid ATP depletion. Hepatocyte suspensions (10^6/ml) were incubated in 3 ml Krebs–Ringers–HEPES buffer containing 0.2% bovine serum albumin at 37 °C. Cells were incubated in the presence of 250 μM tauroursodeoxycholate (closed circles), 250 μM GCDC (open circles), or 250 μM GCDC and 250 μM TUDC (closed squares). ATP concentrations were quantitated using a luciferin/luciferase bioluminescence assay. Basal concentrations of ATP were $19.8 \pm 1.2 \, \text{nmol}/10^6$ cells.

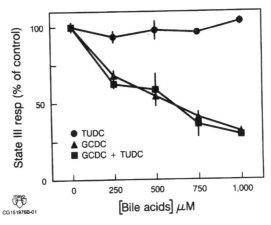

CG151976B-01

[Bile acids] μM

FIGURE 2. Concentration-dependence of GCDC-mediated inhibition of state 3 mitochondrial respiration. Mitochondrial respiration was measured as oxygen consumption in hepatocyte suspensions incubated in respiration buffer (125 mM sucrose, 50 mM KCl, 5 mM HEPES, 2 mM KH_2PO_4, 1 mM $MgCl_2$) and permeabilized with 10 μM digitonin (Moreadith and Fiskum, 1984). State 3 respiration was measured using 1 mM ADP, 5 mM succinate, and 5 μM rotenone in the presence of tauroursodeoxycholate (closed circles), glycochenodeoxycholate (closed triangles), or equiconcentrations of both bile salts (closed squares). Oxygen consumption was measured polographically. Basal values for state 3 respiration were 181 ± 8 ng atm 0 min^{-1} per 10^6 cells.

salts may mediate mitochondrial dysfunction and cell necrosis through additional mechanisms. Moreover, the studies employed high concentrations of a single toxic bile salt, and the physiologic relevance of such concentrations has been questioned (Ostrow, 1994).

3.2. Bile Salts and Oxidative Stress

Impairment of mitochondrial respiration often enhances the transfer of the electrons to molecular oxygen, forming the superoxide anion (Rosser and Gores, 1995). Enhanced formation of superoxide increases hydrogen peroxide generation via the Fenton reaction (Rosser and Gores, 1995). High concentrations of intracellular hydrogen peroxide can cause cell death by GSH depletion, protein oxidation, dysregulation of intracellular ions, lipid peroxidation, and DNA oxidation. This would be an additional mechanism coupling bile-salt-induced mitochondrial dysfunction with cell death. Sokol and co-workers (1991, 1993, 1998), have linked oxygen free radicals to bile salt cytotoxicity. Evidence for lipid peroxidation as assessed by the production of malondialdehyde (thiobarbituric acid reactive substances) was observed prior to cell lysis during treatment of rat hepatocytes with 100–200 μM taurochenodeoxycholic acid (Sokol *et al.*, 1993). Antioxidants, especially α-tocopherol, inhibited cell death, indicating that oxidative stress was contributing to the cytotoxicity (Sokol *et al.*, 1993). The α-tocopherol also inhibited liver injury and decreased mitochondrial lipid peroxidation in BDL rats and following IV bolus administration of taurochenodeoxycholic acid to rats (Sokol *et al.*, 1998). Thus the authors were able to extrapolate their *in vitro* observations to an *in vivo* disease model. Sokol and associates (1995) directly demonstrated generation of hydroperoxides in rat hepatocytes and isolated rat liver mitochondria exposed to 400 µM chenodeoxycholic acid, suggesting that mitochondria were the source of the reactive oxygen species observed in the isolated cell experiments.

We were able to show evidence for oxidative stress in cultured rat hepatocytes exposed to only 50 µM GCDC, measuring 8-isoprostane as an index of lipid peroxidation. Generation of 8-isoprostane increased seven fold after 3 hours of this treatment (Patel and Gores, 1997). Concomitant treatment with the lazaroid antioxidant U83836E inhibited 8-isoprostane generation and reduced cell death by 70%. Our studies, using lower concentrations of bile salts than those of Sokol, confirmed a role for oxidative stress in this model of hepatocyte toxicity. Thus one mechanism of bile salt cytotoxicity appears to involve mitochondrial damage, enhanced formation of reactive oxygen species, and cell death. These observations do not incorporate the current concept regarding the role of the MPT in cell death, however.

4. DO BILE SALTS LEAD TO THE MITOCHONDRIAL MEMBRANE PERMEABILITY TRANSITION?

4.1. Bile Salts and the Permeability Transition

In a series of experiments, we were the first to directly determine whether toxic bile salts could cause transition in isolated rat liver mitochondria (Botla *et al.*, 1995). Indeed, GCDC-induced transition in a dose-dependent manner, as assessed with a spectrophoto-metric assay equating mitochondrial swelling with transition (Fig. 3). Importantly, UDCA inhibited the GCDC-induced transition (Fig. 3), as did the classic inhibitors cyclosporine plus trifluoperazine (data not shown). We also demonstrated that transition inhibition with cyclosporine A and trifluoperazine prevented ATP depletion and cell killing by GCDC treatment of rat hepatocytes (Botla *et al.*, 1995). Thus, transition appears to be a final common pathway responsible for mitochondrial dysfunction, ATP depletion, and cyto-toxicity at the doses employed.

4.2. Bile Salts and Cytochrome *c* Release

Transition has been mechanistically linked to cytochrome *c* release into the cytosol (Kantrow and Piantadosi, 1997). Cytochrome *c* along with apoptosis-activating factor 1 can promote activation of caspase-9. Caspase-9 causes activation of caspase-3, strongly implicated as an effector protease in many models of apoptosis. To determine whether bile acids could cause release of cytochrome *c* from isolated rat mitochondria, we directly measured its release in the supernatant of mitochondria using Western blot analysis. Significant cytochrome *c* release into the supernatant was seen following GCDC treatment of mitochondria from normal rats (Fig. 4). Thus, bile salts can directly cause cytochrome *c* release from mitochondria, providing a plausible mechanism for GCDC-mediated apop-tosis. Cai and Jones (1988) demonstrated that release of cytochrome *c* from mitochondria enhanced the formation of toxic oxygen species by these organelles. Thus, toxic bile salts may cause transition, releasing cytochrome *c,* which then induces oxidative stress. This model integrates and incorporates the findings of apoptosis and oxidative stress in bile salt

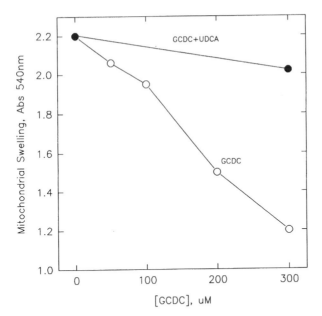

FIGURE 3. Glycochenodeoxycholate induces the mitochondrial membrane permeability transition in a concentration-dependent manner. Isolated rat liver mitochondria (1 mg protein/ml) were suspended in buffer containing 0.1 M NaCl, 10 mM MOPS, 1 mM glutamate and malate, and 5 μM rotenone at 25 °C. Large-amplitude mitochondrial swelling was measured by monitoring the O.D. at 540 nm. At time zero, glycocheno-deoxycholate (GCDC) or equimolar concentrations of GCDC plus ursodeoxycholate (UDCA) were added and the change in absorbance recorded 5 min later.

cytotoxicity. Either the resultant caspase activation or the reactive oxygen species would be deleterious to the cell and could lead to cell demise.

4.3. Bile Salts and Mitochondrial Depolarization *In Situ*

That toxic bile salts can cause the MPT *in vitro* cannot be directly extrapolated to *in vivo* cytotoxicity without further information. Rodrigues and co-workers (1998) provide data further linking the studies an isolated mitochondria to mitochondrial dysfunction in cells. They measured mitochondrial membrane potential ($\Delta\Psi$) using $DiOC_6(3)$ and flow cytometry. With this approach mitochondrial depolarization occurred during exposure of HUH-7 cells to 50 μM deoxycholate. Co-treatment with 50 μM UDCA prevented mitochondrial depolarization and cell apoptosis. The ability of UDCA to prevent apoptosis and mitochondrial depolarization was also observed in several other models, including injury by ethanol, staurosporine, and Fas ligation (Rodrigues *et al.*, 1997). These intriguing data suggest that UDCA may be a general transition inhibitor, and as such, may be the first drug approved with this mechanism of action.

We note, however, that mitochondrial depolarization is a surrogate measurement for transition or cytochrome *c* release and cannot be universally equated with these processes (Nieminen *et al.*, 1995). Furthermore, these studies were performed using cell lines that do

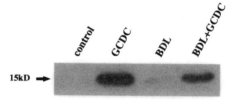

FIGURE 4. Glycochenodeoxycholate (GCDC) causes release of cytochrome *c* by isolated mitochondria obtained from normal and bile-duct-ligated rat livers. Liver mitochondria were purified from BDL and control rat livers by sucrose–percoll gradient centrifugation. Isolated mitochondria (2.5 mg protein/ml) were suspended in a buffer containing 125 mM sucrose, 50 mM KCl, 5 mM HEPES, and 2 mM KH_2PO_4 at 25 °C. After 15 min of treatment, mitochondria (0.5 ml) were separated from the buffer by centrifugation through 0.5 ml of an oil layer (dibutylphthalate) in 1.5 ml microfuge tubes (centrifugation was performed at 13,200 rpm × 5 min in a microfuge). Cytochrome *c* in the supernatant was resolved on a 16% polyacrylamide gel by electrophoresis, electroblotted to nitrocellulose membranes, and identified by immunoblot analysis. The primary antibody was a polyclonal mouse anticytochrome *c* antibody (PharMingen, San Diego, CA).

not transport conjugated bile salts, necessitating the use of unphysiologic, unconjugated salts, so their relevance to cholestatic hepatocellular injury is limited.

5. DO ADAPTATIONS OCCUR IN CHOLESTASIS TO INHIBIT BILE-SALT-INDUCED TRANSITION?

The above information demonstrates that acutely toxic bile salts can cause hepatocyte death by inducing the MPT. Human cholestatic liver diseases progress slowly over time, however, and never cause fulminant hepatic failure, so probably bile salts are less toxic *in vivo* (e.g., modulated by available growth factors, etc.) or else mitochondria adapt in cholestasis and become less susceptible to transition. We recently chose to test the latter hypothesis (Lieser *et al.*, 1998). We isolated mitochondria from 14-day BDL rats and measured transition spectrophotometrically. Following addition of 400 μM GCDC, the magnitude of transition was reduced by $57 \pm 9\%$ in cholestatic mitochondria as compared with mitochondria from sham-operated rats. Furthermore, cytochrome *c* release was also decreased as compared with normal rats (Fig. 4). Because Bcl-x_L, a cytoprotective protein can inhibit transition, we determined whether Bcl-x_L expression increases. Unexpectedly a $26 \pm 8\%$ decrease was found in BDL rats. In contrast, mitochondrial cardiolipin, a negatively charged structural membrane lipid in mitochondria, increased $40 \pm 3\%$ in BDL rat mitochondria. Preventing this cardiolipin increase by feeding rats an essential fatty acid-free diet restored the susceptibility of BDL rat mitochondria to GCDC-mediated transition, indicating that hepatocyte mitochondria might increase their cardiolipin content during cholestasis to resist cell death by the MPT. These data must be interpreted with caution, however. Until the transition-forming proteins are known, it will be difficult to ascertain which potential adaptations directly confer resistance to the bile-salt-mediated MPT.

6. CONCLUSION

Based on these data, we have developed a working model implicating toxic bile salts in hepatocellular injury during cholestasis. The data in this chapter support the hypothesis that toxic bile salts directly cause hepatocyte injury by causing mitochondrial dysfunction via inducing the MPT (Fig. 5). Our working model is that toxic bile salts accumulate in hepatocytes during cholestasis, achieving concentrations sufficient to cause transition. Transition then results in cytochrome c release into the cytosol, which has two detrimental consequences: First, the loss of mitochondrial cytochrome c enhances mitochondrial oxygen radical generation (Cai and Jones, 1998). Second, cytochrome c in the cytosol results in caspase-3 activation, causing activation of the cell death program. The combination of caspase activation and oxygen radical formation injures the cell and causes its death by apoptosis. Five predictions emanating from this hypothesis can be experimentally addressed. One, mitochondria transport bile salts using a selective but low-affinity transport process that occurs only at high intracellular bile salt concentration. Two, bile salts either directly associate with constituents of the transition pore complex, causing pore opening, or they indirectly cause transition, such as may occur by interacting with members of the Bcl2 family. Three, caspase activation should occur in bile-salt-mediated apoptosis, but inhibition of caspases will not block cytochrome c release or formation of reactive oxygen species. Four, nontoxic bile salts such as hyodeoxycholate or taurine conjugates of chenodeoxycholate or deoxycholate will not cause transition or induce cytochrome c release. Five, pharmacologic inhibition (e.g., cyclosporine A) or molecular (e.g., Bcl2 transgenic mice) of the MPT will ameliorate cholestatic liver injury. Once these

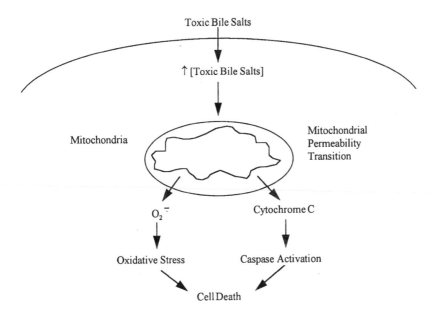

FIGURE 5. Schematic of bile salt cytotoxicity. Toxic bile salts cause mitochondrial dysfunction, leading to oxidative stress and release of cytochrome c into the cytosol. Both events contribute to cytotoxicity. See text for details.

questions are answered and transition mechanisms are better understood, the process by which adaptations confer transition resistance can be dissected, thus helping to develop more rational approaches for the treatment of patients with cholestatic liver disease.

ACKNOWLEDGEMENTS. This work was supported by grants from the National Institutes of Health (DK 41876); the Gainey Foundation, St. Paul, MN; and the Mayo Foundation, Rochester; MN.

REFERENCES

Bernardi, P., 1996, The permeability transition pore: Control points of a cyclosporin A-sensitive mitochondrial channel involved in cell death [review], *Biochim. Biophys. Acta* **1275**:5–9.

Botla, R., Spivey, J. R., Aguilar, H., Bronk, S. F., and Gores, G. J., 1995, Ursodeoxycholate (UDCA) inhibits the mitochondrial membrane permeability transition induced by glycochenodeoxycholate: A mechanism of UDCA cytoprotection, *J. Pharmacol. Exp. Ther.* **272**:930–938.

Cai, J. Y., and Jones, D. P., 1998, Communication—superoxide in apoptosis: Mitochondrial generation triggered by cytochrome *c* loss, *J. Biol. Chem.* **273**:11401–11404.

Greim, H., Czygan, P., Schaffner, F., and Popper, H., 1973, Determination of bile acids in needle biopsies of human liver, *Biochem. Med.* **8**:280–286.

Kantrow, S. P. and Piantadosi, C. A., 1997, Release of cytochrome *c* from liver mitochondria during permeability transition, *Biochem. Biophys. Res. Commun.* **232**:669–671.

Kaplan, M. M., 1994, Primary biliary cirrhosis: A first step in prolonging survival [editorial; comment], *N. Engl. J. Med.* **330**:1386–1387.

Krahenbuhl, S., Krahenbuhl-Glauser, S., Stucki, J., Gehr, P., and Reichen, J., 1992a, Stereological and functional analysis of liver mitochondria from rats with secondary biliary cirrhosis: Impaired mitochondrial metabolism and increased mitochondrial content per hepatocyte, *Hepatology* **15**:1167–1172.

Krahenbuhl, S., Stucki, J., and Reichen, J., 1992b, Reduced activity of the electron transport chain in liver mitochondria isolated from rats with secondary biliary cirrhosis, *Hepatology* **15**:1160–1166.

Krahenbuhl, S., Talos, C., Fischer, S., and Reichen, J., 1994, Toxicity of bile acids on the electron transport chain of isolated rat liver mitochondria, *Hepatology* **19**:471–479.

Li, P., Nijhawan, D., Budihardjo, I., Srinivasula, S. M., Ahmad, M., Alnemri, E. S., and Wang, X., 1997, Cytochrome *c* and dATP-dependent formation of Apaf-1/caspase-9 complex initiates an apoptotic protease cascade, *Cell* **91**:479–489.

Lieser, M. J., Park, J.-W., Natori, S., Jones, B. A., Bronk, S. F., and Gores, G. J., 1998, Cholestasis confers resistance to the rat liver mitochondrial permeability transition, *Gastroenterology* **115**, in press.

Moreadith, R. W., and Fiskum, G., 1984, Isolation of mitochondria from ascites tumor cells permeabilized with digitonin, *Anal. Biochem.* **137**:360–367.

Nieminen, A.-L., Saylor, A. K., Tesfai, S. A., Herman, B., and Lemasters, J. J., 1995, Contribution of the mitochondrial permeability transition to lethal injury after exposure of hepatocytes to *t*-butylhydroperoxide, *Biochem. J.* **307**:99–106.

Ostrow, D. J., 1993, Metabolism of bile salts in cholestasis, in *Hepatic Transport and Bile Secretion: Physiology and Pathophysiology* (N. Tavoloni and P. D. Berk, Eds.), Raven New York, pp. 673–712.

Ostrow, J. D., 1994, *In vitro* studies of bile salt toxicity [letter; comment], *Hepatology* **19**:1538–1539.

Patel, T., and Gores, G. J., 1997, Inhibition of bile-salt-induced hepatocyte apoptosis by the antioxidant lazaroid U83836E, *Toxicol. Appl. Pharmacol.* **142**:116–122.

Patel, T., Bronk, S. F., and Gores, G. J., 1994, Increases of intracellular magnesium promote glycodeoxycholate-induced apoptosis in rat hepatocytes, *J. Clin. Invest.* **94**:2183–2192.

Patel, T., Spivey, J. R., Vadakekalam, J., and Gores, G. J., 1995, Apoptosis: An alternative mechanism of bile salt cytotoxicity, in *Bile Acids in Liver Diseases* (G. Paumgartner and U. Beuers, Eds.), Kluwer Academic, Lancaster, London pp. 88–95.

Rodrigues, C. M. P., Fan, G., Ma, X., Brites, D., Kren, B. T., and Steer, C. J., 1997, A role for ursodeoxycholic acid in modulating apoptosis in rat liver, isolated rat hepatocytes, and human hepatoma cells, *Hepatology* (abstract) **26**:366A.

Rodrigues, C. M. P., Fan, G., Xiaoming, M., Kren, B. T., and Steer, C. J., 1998, A novel role for ursodeoxycholic acid in inhibiting apoptosis by modulating mitochondrial membrane perturbation, *J. Clin. Invest.* **101**:2790–2799.

Rosser, B. G., and Gores, G. J., 1995, Liver cell necrosis: Cellular mechanisms and clinical implications, *Gastroenterology* **108**:252–275.

Schaffner, F., Bacchin, P. G., Hutterer, F., Scharnbeck, H. H., Sarkozi, L. L., Denk, H., and Popper, H., 1971, Mechanism of cholestasis: IV Structural and biochemical changes in the liver and serum in rats after bile duct ligation, *Gastroenterology* **60**:888–897.

Sokol, R. J., Devereaux, M., and Khandwala, R. A., 1991, Effect of dietary lipid and vitamin E on mitochondrial lipid peroxidation and hepatic injury in the bile duct-ligated rat, *J. Lipid Res.* **32**:1349–1357.

Sokol, R. J., Devereaux, M., Khandwala, R., and O'Brien, K., 1993, Evidence for involvement of oxygen free radicals in bile acid toxicity to isolated rat hepatocytes, *Hepatology* **17**:869–881.

Sokol, R. J., Winklhofer-Roob, B. M., Devereaux, M. W., and McKim, J. M. Jr., 1995, Generation of hydroperoxides in isolated rat hepatocytes and hepatic mitochondria exposed to hydrophobic bile acids, *Gastroenterology* **109**:1249–1256.

Sokol, R. J., McKim, J. M., Jr., Goff, M. C., Ruyle, S. Z., Devereaux, M. W., Han, D., Packer, L., and Everson, G., 1998, Vitamin E reduces oxidant injury to mitochondria and the hepatotoxicity of taurochenodeoxycholic acid in the rat, *Gastroenterology* **114**:164–174.

Spivey, J. R., Bronk, S. F., and Gores, G. J., 1993, Glycochenodeoxycholate-induced lethal hepatocellular injury in rat hepatocytes: Role of ATP depletion and cytosolic free calcium, *J. Clin. Invest.* **92**:17–24.

Stolz, A., Takikawa, H., Ookhtens, M., and Kaplowitz, N., 1989, The role of cytoplasmic proteins in hepatic bile acid transport [review], *Ann. Rev. Physiol.* **51**:161–176.

Susin, S. A., Zamzami, N., Castedo, M., Daugas, E., Wang, H. G., Geley, S., Fassy, F., Reed, J. C., and Kroemer, G., 1997, The central executioner of apoptosis: Multiple connections between protease activation and mitochondria in Fas/APO-1/CD95- and ceramide-induced apoptosis, *J. Exp. Med.* **186**:25–37.

Chapter 23

Reye's Syndrome and Related Chemical Toxicity

Lawrence C. Trost and John J. Lemasters

1. INTRODUCTION

We have examined the contribution of the mitochondrial permeability transition (MPT) to mitochondrial injury in Reye's syndrome, Jamaican vomiting sickness, valproic acid toxicity, and related disorders. These disorders occur predominantly in children and are characterized by vomiting, acute hepatic failure, and coma. Strong evidence suggests hepatic injury in Reye's syndrome is caused by mitochondrial dysfunction. Brain and liver mitochondria become swollen and pleomorphic, with decreased matrix density and loss of cristae (Partin *et al.*, 1971; Sherratt, 1986; Sinniah and Baskaran, 1981). In Reye's syndrome the extent of these structural changes correlates with the severity of the illness, especially in comatose patients, where mitochondrial alterations are consistently profound (Lichtenstein *et al.*, 1983).

Metabolic alterations in Reye's-related disorders are also an indication of mitochondrial injury. Persistent high fever and high overall metabolic rate suggest a pattern of mitochondrial uncoupling with increased respiratory rates but decreased ATP formation by oxidative phosphorylation. Ureagenesis and β-oxidation are suppressed, causing hyperammonemia and increased serum dicarboxylic and free fatty acids. Mitochondrial enzymes of gluconeogenesis and the citric acid cycle are also depleted, contributing to hypoglycemia.

Lawrence C. Trost   Curriculum in Toxicology, University of North Carolina at Chapel Hill, Chapel Hill, North Carolina 27599-7090.   **John J. Lemasters**   Department of Cell Biology and Anatomy, University of North Carolina at Chapel Hill, Chapel Hill, North Carolina, 27599-7090.

Mitochondria in Pathogenesis, edited by Lemasters and Nieminen.
Kluwer Academic/Plenum Publishers, New York, 2001.

Though the incidence of Reye's syndrome has declined significantly since the early 1980s, as a group, Reye's syndrome and related disorders remain a significant cause of morbidity and mortality in the United States. Accordingly, our goal was to investigate whether pathological changes associated with Reye's syndrome and related disorders were caused by onset of the MPT. We tested chemical agents implicated in Reye's syndrome and related disorders for their ability to induce the MPT experimentally *in vitro* and *in situ*. Onset of the MPT is regulated by a number of physiological effectors, most importantly Ca^{2+}, so we also investigated whether altered Ca^{2+} homeostasis could account for the rare, idiopathic nature of Reye's syndrome and related disorders. Onset of the MPT could provide the basis for a common pathophysiological link between Reye's syndrome and related chemical toxicities and metabolic disorders that present identically in the clinic.

Salicylate, the active metabolite of aspirin, and similar nonpolar carboxylic acids implicated in Reye's-like disorders (Neem oil and valproic, adipic, benzoic, 4-pentenoic, 3-mercaptopropionic, and isovaleric acids) induced swelling in isolated mitochondria. Swelling was inhibited by cyclosporin A, which confirmed that the chemicals were inducing onset of the MPT. Carboxylic acids not implicated in Reye's-like disorders did not induce the transition. The mechanism of MPT induction by these agents was not protonophoric uncoupling, because concentrations of the protonophoric uncoupler FCCP, which caused mitochondrial depolarization equivalent to that caused by salicylate, failed to induce the MPT.

Salicylate killed cultured hepatocytes. Under examination by confocal microscopy, salicylate induced the MPT, which preceded mitochondrial depolarization and cell killing. Both the MPT induced by salicylate and the consequent cell killing were blocked by cyclosporin A. The calcium channel antagonist verapamil also blocked salicylate cytotoxicity. These results suggest that Ca^{2+} is important in the mechanism of cell killing. It is proposed that Ca^{2+} and salicylate synergize to cause onset of the MPT and Reye's syndrome *in vivo*.

2. REYE'S SYNDROME, A TOXIC METABOLIC CRISIS CAUSED BY MITOCHONDRIAL DYSFUNCTION

In 1963 Reye, Morgan, and Baral described a fatal syndrome occurring infrequently in infants and children after a prodromal viral illness. Reye's syndrome is a fatal disorder characterized by protracted vomiting, hepatic steatosis, and lethargy progressing rapidly to coma (Brown and Forman, 1982; Heubi et al., 1987; Reye et al., 1963). Fasting and mobilization of fatty acids for energy production are prominent in its pathogenesis. Microvesicular fatty degeneration of the liver and noninflammatory encephalopathy are cardinal features of Reye's syndrome. Death is caused by cerebral edema and associated herniation and ischemia. A substantial percentage of survivors experience long-term neurological deficits (Duffy et al., 1991; Ozawa et al., 1996).

Reye's syndrome is correlated with a patient history of aspirin ingestion. Four case-control studies demonstrated a 93%, 97%, 100% and 100% correlation between Reye's syndrome and aspirin usage (Heubi et al., 1987; Maheady, 1989; Starko and Mullick, 1983; Starko et al., 1980; Waldman et al., 1982). Furthermore, the likelihood of developing Reye's syndrome and the severity of the ensuing encephalopathy were directly related to the amount of aspirin consumed during the antecedent viral illness (Pinsky *et al.*, 1988). Reye's syndrome and aspirin were definitively linked by a 1982 Surgeon General's advisory warning against aspirin use in children. The advisory lead to a steep decline in the number of cases of Reye's syndrome. For example, a large regional hospital reported 57 cases between January 1979 and June 1986, after which aspirin use in children was contraindicated. Between June 1986 and November 1993, the hospital reported only two cases of Reye's syndrome (Glasgow and Moore, 1993). In recent years, unfortunately, declining attention in the medical literature and sparse media coverage has lead to a resurgence of Reye's syndrome (Valencia *et al.*, 1997).

The principal laboratory findings in Reye's syndrome indicate primary hepatic injury. Hypoglycemia, hyperammonemia, and coagulopathy are universally present (Brown and Forman, 1982). Organic, amino, and free fatty acids are also commonly elevated in both serum and urine (Heubi *et al.*, 1987). The liver transaminases AST and ALT are usually elevated as well. Histopathology of the liver shows microvesicular steatosis, glycogen depletion, and swollen pleomorphic mitochondria (Partin *et al.*, 1971). These changes also occur to lesser degrees in kidney, brain, skeletal muscle, and cardiac muscle (Heubi *et al.*, 1987; Morales *et al.*, 1971).

Microvesicular steatosis is a rare pathological change in the liver in which small droplets accumulate in hepatocyte cytosol without displacing the central nucleus. Microvesicular steatosis is almost unique to Reye's syndrome; it occurs in only one other clinical condition, fatty liver of pregnancy. The more common pattern of fat accumulation, macrovesicular steatosis, is associated with ethanol ingestion and the toxic effects of a wide variety of chemicals that impair liver function. Thus microvesicular steatosis is the feature of Reye's syndrome that distinguishes it from most other types of liver toxicity associated with accumulation of fat.

Evidence suggests that the microvesicular steatosis and hepatic failure associated with Reye's syndrome are caused by primary mitochondrial dysfunction. Brain and liver mitochondria are swollen, matrix density is decreased, and cristae are lost (Partin *et al.*, 1971). Moreover, the extent of structural change correlates with the severity of illness. Hepatic mitochondrial enzymes are also depleted in Reye's syndrome. There is a universal decrease in the activity of all mitochondrial enzymes, in contrast to normal cytosolic enzyme activity (Mitchell *et al.*, 1980; Brown and Forman, 1982). Consequently, the citric acid cycle, gluconeogenesis, ureagenesis, and β-oxidation are suppressed, consistent with clinical pathology changes in Reye's syndrome. It is now widely accepted that most, if not all, metabolic alterations in Reye's syndrome reflect a primary mitochondrial injury.

Many hypotheses have been advanced to explain the mechanism of mitochondrial injury in Reye's syndrome. Because Reye's syndrome often follows viral illness, researchers have suggested that a viral protein may cause the mitochondrial damage (DeVivo and Keating, 1975). Others have suggested that Reye's syndrome is an abnormal immune response to a viral infection (Tonsgard and Huttenlocher, 1982). These hypotheses, however either fail to account for the link between aspirin and Reye's syndrome or they fail to address the idiopathic nature of the disorder.

Plasma from Reye's patients causes isolated mitochondria to swell and to uncouple oxidative phosphorylation (Martens *et al.*, 1986). Aspirin (or its major metabolite salicylate) may be the soluble factor that causes these mitochondrial changes, because *in vitro* both aspirin and salicylate cause mitochondrial swelling (Brody, 1956; Martens and Lee, 1984; You, 1983). Furthermore, aspirin and salicylates uncouple mitochondrial respiration and inhibit adenine nucleotide translocase, both of which impair energy metabolism (Brody, 1956; Aprille, 1977). Thus aspirin, has been suggested to cause Reye's syndrome by directly inhibiting mitochondrial energy production.

Inhibition of mitochondrial respiration by aspirin, however, becomes significant at high concentrations (up to 30 mM) likely to be achieved only during aspirin intoxication, the pathogenesis of which differs from Reye's syndrome (DeVivo, 1985; Mitchell *et al.*, 1980; Partin *et al.*, 1984). Moreover, direct effects on respiration do not explain the idiopathic nature of Reye's syndrome, because aspirin is consumed ubiquitously yet few people develop the disorder. Furthermore, inhibition of specific components of the mitochondrial electron transport chain or oxidative phosphorylation pathway do not explain the homogeneous clinical manifestations of Reye's syndrome, which are identical to that of a number of related chemical toxicities and metabolic disorders. Thus a common pathophysiological mechanism is likely for Reye's syndrome and related toxicities.

Increasing evidence suggests that the permeability transition is the principal cell death effector in a wide variety of apoptosis and necrosis models (Kroemer *et al.*, 1998; Lemasters *et al.*, 1998). Onset of the MPT causes mitochondrial swelling, depolarization, and uncoupling of oxidative phosphyorylation. In Mitchell's (1979) chemiosmotic theory, impermeability of the inner mitochondrial membrane to ions is critical for maintaining the proton electrochemical gradient and mitochondrial function. Therefore opening of the transition pore causes uncoupling of oxidative phosphorylation and ATP depletion making cell death inevitable. The MPT seems an ideal mechanism for controlled elimination of select cells, as in apoptosis. Similarly, loss of control of the MPT, such as might occur in pathological conditions resulting from defective regulation or after exposure to pathological stimuli, might induce the MPT in a broad, unregulated manner, causing uncontrolled cell killing and necrosis.

We suggest that the MPT might also be the common pathophysiological mechanism of Reye's syndrome and Reye-related chemical toxicities and metabolic disorders. With this hypothesis we sought to develop a model for the mitochondrial damage associated with Reye's syndrome and related disorders that would also account for their rare, idiopathic occurrence. Because Ca^{2+} is required for onset of the MPT and because intracellular Ca^{2+} regulation is defective in patients with Reye's syndrome (Corkey *et al.*, 1991), we investigated whether altered Ca^{2+} regulation might sensitize individuals to onset of the MPT induced by chemical agents implicated in Reye's syndrome and Reye's-related disorders.

2.1. Aspirin and Salicylate Induce the Mitochondrial Permeability Transition

To determine whether Reye's syndrome might be caused by onset of the permeability transition, we tested aspirin's ability to induce the MPT in isolated rat liver mitochondria respiring on succinate, a standard *in vitro* assay (Petronilli *et al.*, 1994b). Mitochondria suspended in a sucrose-based media swell after onset of the MPT. As the optical density of the matrix approaches that of the surrounding media, light scattering decreases. Thus, onset of the permeability transition can be monitored spectrophotometrically by a decrease in absorbance.

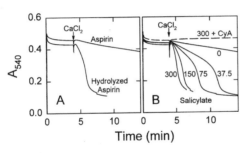

FIGURE 1. Induction of the mitochondrial permeability transition by hydrolyzed aspirin and salicylate. Isolated rat liver mitochondria (0.3 mg protein/ml) were incubated in sucrose-based media: (A) Unhydrolyzed aspirin (300 µM) or alkali-hydrolyzed aspirin (150 µM); (B) 0, 37.5, 75, 150, or 300 µM salicylate, or 300 µM salicylate plus 200 nM cyclosporin A (CyA). After 4 min 50 µM $CaCl_2$ was added. Mitochondrial swelling associated with MPT onset caused a decrease in light scattering, monitored spectrophotometrically by the change in absorbance at 540 nm. Aspirin did not induce the MPT unless first hydrolyzed. Salicylate, the product of metabolic hydrolysis of aspirin, was a potent MPT inducer.

Mitochondria were incubated for 4 minutes in the presence of aspirin before we added a small amount of $CaCl_2$, which was added in part to replace matrix Ca^{2+} lost during isolation and suspension of mitochondria in EGTA-containing buffer (Fig. 1A). A small amount of matrix Ca^{2+} is required for transition onset, but the concentration used here failed to induce it in repeated experiments when added alone. At concentrations of up to 300 µM, aspirin failed to cause mitochondrial swelling. When 150 µM aspirin was first hydrolyzed by adding an alkali, however, the product (presumably salicylate) rapidly induced mitochondrial swelling, measured by a 65% decrease in absorbance. Biban and co-workers (1995) showed that in the presence of $CaCl_2$, 500 µM aspirin induced a cyclosporin A-sensitive permeability transition in rat liver mitochondria. Aspirin stocks spontaneously decompose to salicylate. However, it is likely that salicylate, rather than aspirin, was the primary MPT inducer in their experiments.

Because salicylate is the major hydrolysis product of aspirin and its major metabolite *in vivo*, we evaluated the ability of salicylate to induce the MPT in isolated mitochondria. At concentrations as low as 37.5 µM, salicylate induced mitochondrial swelling after $CaCl_2$ addition (Fig. 1B). Increasing the concentration of salicylate decreased the latency before swelling but did not increase the final extent of the swelling. Cyclosporin A is a potent and specific inhibitor of the MPT (Broekemeier *et al.*, 1989; Crompton *et al.*, 1988; Fournier *et al.*, 1987). Even at 300 µM, the highest concentration of salicylate tested, cyclosporin A (1 µM) blocked mitochondrial swelling, confirming that the swelling was caused by onset of the MPT.

In these experiments aspirin failed to induce the permeability transition. Salicylate, however, aspirin's active metabolite (Flower *et al.*, 1985), was a potent MPT inducer. Nonspecific mitochondrial swelling occurs in the presence of aspirin and salicylate, but very high concentrations (up to 30 mM) are required (You, 1983). In combination with a small amount of Ca^{2+}, however, we found that as little as 37.5 µM salicylate induced the MPT. This concentration is below that to which the liver is exposed after therapeutic ingestion of salicylate. In addition, the requirement for matrix Ca^{2+} is proportional to the concentration of inducer. That is, onset of the MPT can occur after reduced matrix Ca^{2+} loading by increasing the concentration of inducer, so because substantially higher concentrations of salicylate might be encountered *in vivo*, transition onset might occur at matrix Ca^{2+} concentrations not much above normal physiologic concentrations.

2.2. Salicylate MPT Induction Precedes Cell Death in Isolated Hepatocytes

The ability of salicylate to induce the MPT in individual hepatocytes was determined by laser-scanning confocal microscopy of tetramethylrhodamine methyl ester (TMRM) and calcein fluorescence, as described previously (Lemasters *et al.*, 1998; Nieminen *et al.*, 1995). Briefly, TMRM accumulates electrophoretically in the matrix of polarized mitochondria in response to their negative membrane potential. Consequently, individual mitochondria are imaged as bright fluorescent spheres. In contrast, calcein-AM diffusely labels the cytosol but is excluded from mitochondrial spaces because of cleavage by cytosolic esterases. Areas corresponding to TMRM-labeled mitochondria appear as dark voids in the calcein fluorescence. By imaging TMRM and calcein simultaneously and relaying the output to two separate channels, the bright fluorescent spheres in the TMRM image identify the dark spheres in the calcein image as mitochondria.

Onset of the MPT is characterized by formation of a high-conductance pore or megachannel in the inner mitochondrial membrane that allows free diffusion of solutes less than 1500 Da MW (Zoratti and Szabò, 1995). The transition therefore causes rapid mitochondrial depolarization, indicated by a decrease in TMRM fluorescence. Simultaneously, opening of the MPT pore allows calcein (MW 623 Da) to diffuse into the matrix space, causing the dark voids to fill with calcein fluorescence.

To monitor mitochondrial membrane permeability and polarization, cultured hepatocytes were simultaneously loaded with TMRM and calcein-AM and incubated in a physiological buffer containing 3 mM $CaCl_2$ and 3 μM propidium iodide (PI), a red fluorescent probe that labels the nuclei of nonviable cells exactly like trypan blue does. Confocal microscopy of the cells revealed punctate mitochondrial TMRM fluorescence, indicating mitochondrial polarization (Fig. 2, upper left panel). Calcein fluorescence was diffuse, and the round dark voids corresponded to the TMRM-labeled mitochondria (lower left panel). These voids demonstrate the impermeability of polarized mitochondria to calcein. After adding salicylate (3 mM), TMRM fluorescence decreased, indicating that the mitochondria had depolarized. After 85 minutes, nearly all TMRM fluorescence was lost. Simultaneously, calcein fluorescence redistributed into mitochondria, filling the formerly dark voids. This event signified onset of the MPT. The experiment also showed that onset of the MPT, with its associated membrane depolarization and uncoupling, quickly lead to the death of the cell. Indeed, within 2 hours all cells in the field lost membrane integrity and died, as shown by the nuclear PI labeling and loss of cytosolic calcein fluorescence.

This experiment was repeated in the presence of 1 μM cyclosporin A (Fig. 3). Again, baseline polarization was distinguished by bright TMRM fluorescence (top left panel) with calcein fluorescence in the extramitochondrial compartment (bottom left panel). After 60 minutes of exposure to salicylate (a weak uncoupler) some depolarization occurred, indicated by the partial loss of TMRM fluorescence. The distinct dark voids in the calcein image remained, however, indicating that onset of the MPT had not occurred. Even after 130 minutes the calcein image was unchanged, despite a further decrease in TMRM fluorescence. Moreover, in the presence of cyclosporin A, all hepatocytes in the field remained viable, indicated by the absence of nuclear PI fluorescence and retention of cytosolic calcein.

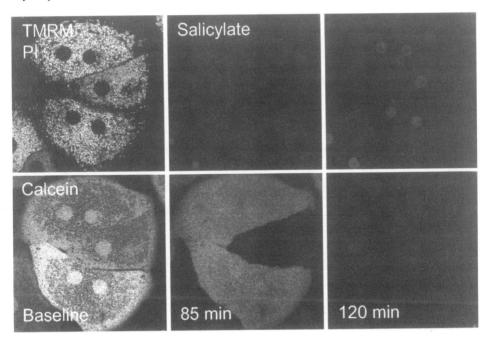

FIGURE 2. Salicylate-dependent induction of the mitochondrial permeability transition and cell killing in cultured rat hepatocytes. Cultured rat hepatocytes were loaded with TMRM, propidium iodide (PI), and calcein-AM. Paired images of red (TMRM and PI, upper panels) and green (calcein, lower panels) fluorescence were obtained by laser-scanning confocal microscopy before and after addition of 3 mM salicylate. Note the loss of TMRM fluorescence and the filling in of dark mitochondrial voids in the calcein fluorescence after Salicylate addition. The fluorescense changes signify MPT onset. Viability of one cell was lost after 85 minutes and of the remaining cells by 120 minutes, indicated by loss of all calcein fluorescence and nuclear labeling with PI.

Even in the presence of cyclosporin A, salicylate caused TMRM fluorescence to decrease, which indicates mitochondrial depolarization. This observation is most likely explained by the weak uncoupling effect of salicylate. Previously, we showed that this weak protonophoric uncoupling is insufficient to induce the MPT (Trost and Lemasters, 1996). Similarly, this weak uncoupling failed to cause cell killing, provided that the MPT was blocked. Taken together these data strongly suggest that cell killing by salicylate occurs via induction of the MPT.

Aspirin is rapidly metabolized to salicylate *in vivo*: 73% is converted within 30 minutes of ingestion (Flower *et al.*, 1985). We found that salicylate was cytotoxic to cultured hepatocytes at concentrations that did not exceed those found in the liver after therapeutic aspirin ingestion. The concentration of salicylate used in these experiments was 5- to 10-fold higher than the therapeutic serum concentration for analgesia and antipyresis (Flower *et al.*, 1985). The absorption and metabolism of aspirin suggests, however, that salicylate concentrations in liver exceed those in serum and other tissues.

Aspirin is absorbed in the small intestine, after which it enters the liver via the portal vein, where it undergoes significant first-pass metabolism. Aspirin is hydrolyzed to salicylate by esterases present in the small intestine and, predominantly, the liver. Biotransformation of salicylate to excretable products occurs particularly in hepatic

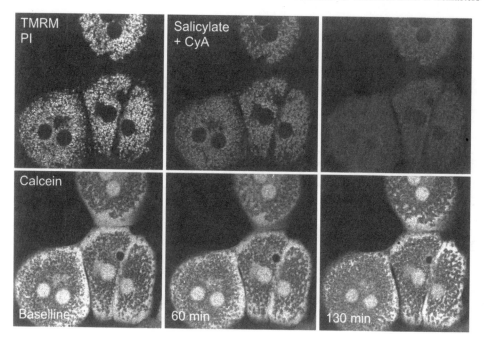

FIGURE 3. Protection by cyclosporin A against cell killing and the salicylate-induced mitochondrial permeability transition in cultured rat hepatocytes. Cultures were treated exactly as described in Figure 2 except 1 µM cyclosporin (CyA) was included in the buffer. Images were recorded immediately before (baseline), and after salicylate was added. Note that although a partial loss of TMRM fluorescence occurred, calcein fluorescence was essentially unchanged after incubation with salicylate. Thus, cyclosporin A blocked the MPT and prevented salicylate-induced cell killing.

mitochondria and endoplasmic reticulum (Flower *et al.*, 1985). Interestingly, Rodgers and co-workers (1982) documented a prolonged biologic half-life for salicylates in Reye's syndrome patients caused by decreased esterase activity, which returned to normal after recovery from Reye's syndrome (Tomasova *et al.*, 1984). Thus in Reye's syndrome, salicylate concentrations in liver might be higher than expected because of abnormal clearance.

3. METABOLIC DEFICIENCIES IN REYE'S SYNDROME AND RELATED CHEMICAL TOXICITY

As increasingly sophisticated diagnostic tools have become available, a growing number of metabolic disorders that produce clinical syndromes resembling Reye's syndrome have been identified. Indeed, several retrospective studies have identified metabolic disorders as the cause of disorders that were in many cases first attributed to Reye's syndrome. Typical Reye's-like metabolic disorders are associated with deficiencies in one or more enzymes or cofactors required for β-oxidation, ureagenesis, oxidative phosphorylation, and other metabolic processes that occur mainly in liver mitochondria. In most cases the toxic manifestation of these disorders, as in Reye's syndrome, is linked to fasting and is precipitated when mitochondria are engaged in energy production by fatty acid oxidation.

In metabolic disorders that mimic Reye's syndrome, the crisis is associated with biochemical evidence of mitochondrial failure, depletion of coenzyme A and carnitine, and microvesicular fat accumulation in hepatocytes. The significance of these disorders varies considerably. In some cases metabolic disorders have little or no clinical significance, whereas in other cases they have profound, often fatal consequence early in life. Thus children may be more susceptible to Reye's-like metabolic disorders because the extreme nature of the defect presents early in life. Alternatively, heteroplasmy may cause the likelihood of phenotypic expression of a pathogenic mitochondrial mutation to diminish over time (Schon *et al.*, 1994), which might explain why fewer adults present with Reye's-like metabolic disorders.

We propose that some metabolic defects might remain unveiled until there is a simultaneous challenge by xenobiotics that produce hepatic mitochondrial injury or an increased metabolic demand on hepatic mitochondria for energy production similar to that which occurs during fasting. Heightened metabolic demand is also associated with pregnancy and may be the cause of acute fatty liver of pregnancy (AFLP). The AFLP syndrome occurs in the last trimester. It has a frequency of 1 out of 13,000 to 16,000 deliveries and is fatal in up to 85% of these cases (Fromenty and Pessayre, 1995). The signs and symptoms of AFLP (coagulopathy, elevated tranminases, and microvesicular steatosis) are similar to those of Reye's syndrome. The AFLP syndrome is linked to a deficiency of long-chain 3-hydroxylacyl-CoA dehydrogenase, which causes inhibition of β-oxidation and accumulation of fatty acids (Treem *et al.*, 1996).

A Reye's-like clinical course often follows the emergence of acute metabolic crisis, which in many cases occurs in patients with a history of uneventful exposure to the offending metabolite, or after (often repeated) exposure to a precipitating exogenous compound that was generally well tolerated. Aspirin is one example. These periods of crisis, often fatal, may be provoked by an increase in metabolic demand, because they are often associated with an antecedent viral illness or with fasting. Vomiting and diarrhea are associated with viral illness, so we theorize that fasting and the consequent mobilization of fatty acids for energy production by β-oxidation could instigate the metabolic crisis.

Fatty acid oxidation is principally responsible for energy production during fasting. Complex reactions occur in mitochondria, but they depend on enzymatic processes and cofactors present in both cytosol and mitochondria, and also on transport of metabolic intermediates into and out of mitochondria. A number of inherited metabolic disorders that mimic Reye's syndrome are caused by deficiency in one or more of the metabolic cofactors or enzymes. Indeed, a wide variety of metabolic disorders cause a clinical pattern that is remarkably homogeneous and virtually identical to Reye's syndrome. Again, fasting typically promotes the metabolic crisis.

The evidence suggests that although metabolic crisis can be precipitated by a wide range of deficiencies or enzymatic defects, homogeneity arises because the resultant effect is the same: mitochondrial dysfunction. In each case mitochondrial substrates for energy production, usually short- or medium-chain fatty acids, accumulate in the cytoplasm, causing microvesicular steatosis. Elimination reactions that occur in mitochondria fail, leading to additional accumulation of toxic by-products such as ammonia and also a depletion of metabolic end products like urea, itself of no consequence yet a hallmark of these disorders. Hypoglycemia, both cause and consequence of mitochondrial dysfunction, signals the inability of mitochondrial fatty acid oxidation to meet energy requirements.

Coagulopathy and increases in serum transaminases confirm the site of mitochondrial dysfunction is the liver. Coagulopathy, evidenced by increased prothrombin time (PT), is associated with hepatic failure to produce cofactors required for the blood coagulation cascade. Elevated serum transaminases AST and ALT indicate a loss of cell viability in the liver.

One metabolic disorder associated with a Reye's-like metabolic crisis is caused by a hereditary deficiency of isovaleryl coenzyme A dehydrogenase. The disorder leads to an accumulation of isovaleric acid, a structural analog of salicylate, precipitating a Reye's-like illness (Tanaka *et al.*, 1972). Accumulation of adipic acid is also associated with a recurrent Reye's-like syndrome (Elpeleg *et al.*, 1990). In addition, many other "inborn errors in metabolism," initially present as Reye's-like illnesses. These include organic acid disorders such as propionic and methylmalonic acidemia, fatty acid oxidation disorders (including those of carnitine transport), urea cycle disorders such as ornithine transcarbamylase deficiency, disorders of pyruvate metabolism, and carbohydrate disorders such as fructose-1,6-diphosphatase deficiency. Features suggestive of a metabolic disorder as opposed to Reye's syndrome include younger age of onset, family or past history of encephalopathy or Reye's syndrome, pre-existing neurodevelopmental disorder, and "near-miss" sudden death syndrome (Hou *et al.*, 1996). As in Reye's syndrome, depending on

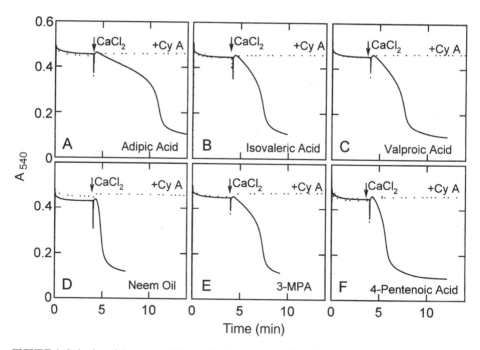

FIGURE 4. Induction of the permeability transition by agents implicated in Reye's syndrome and Reye's-related disorders. Mitochondria were incubated as described in Figure 1 in the presence of the following test agents: (A) 150 μM adipic acid; (B) 150 μM isovaleric acid; (C) 150 μM valproic acid; (D) 1 : 150 diluted Neem oil; (E) 150 μM 3-mercaptopropionic acid (3-MPA); or (F) 150 μM 4-pentenoic acid. CaCl$_2$ (50 μM) was added after four min. As indicated by dotted lines, 200 nM cyclosporin A was included in the assay buffer. All chemical agents implicated in Reye's-related toxicity that were available for testing induced the MPT in isolated mitochondria. In each case, MPT onset occurred over a similar range of concentrations, and in all cases it was blocked by cyclosporin A.

the nature of the deficiency, a metabolic crisis such as fasting or infection usually triggers onset of illness. Almost universally, these metabolic defects deplete coenzyme A and carnitine and lead to accumulation of short-chain nonpolar carboxylic acids, considered to cause the Reye's-like syndrome.

We tested adipic acid and isovaleric acid, endogenous metabolites implicated in Reye's-like disorders, for their ability to induce the MPT in experiments using isolated rat liver mitochondria. The experiments were conducted exactly as described previously for aspirin and salicylate. We found that mitochondria incubated for 4 minutes with a 150 μM concentration of either adipic or isovaleric acid swelled after addition of Ca^{2+} (Fig. 4A, 4B). When the experiments were repeated in the presence of cyclosporin A the swelling was blocked, showing that it was caused by MPT onset. Thus onset of the MPT could be the cause of Reye's-like metabolic crisis in disorders involving the accumulation of isovaleric, adipic, or other, similar, short-chain nonpolar carboxylic acids. Hereditary or induced depletion of carnitine or CoA might provoke additional accumulation of these metabolites to produce a metabolic crisis, or a crisis might be caused by ingestion of an exogenous but structurally similar compound such as aspirin, or by abnormal matrix Ca^{2+} regulation.

4. REYE'S-RELATED CHEMICAL TOXICITY

A group of pharmaceutical agents, environmental toxins, and their metabolic intermediates are associated with toxicity that mimics Reye's syndrome. These chemicals are generally short-chain nonpolar carboxylic acids similar to salicylate. As with Reye's syndrome, children are at higher risk for experiencing the toxic manifestations of these chemicals. In all cases toxicity targets the liver, particularly the hepatic mitochondria.

4.1. Valproic Acid

Valproic acid (2-n-propylpentanoic acid) is a simple branched-chain analog of salicylate widely used to treat epileptic seizure disorders (Bourgeois, 1989). In children valproic acid is associated with rare but potentially fatal hepatotoxicity resembling Reye's syndrome (Gerber et al., 1979; Kesterson et al., 1984). Histopathology studies of liver specimens from patients with valproic-acid-induced liver failure demonstrate that although hepatic necrosis and cirrhosis are evident in some cases, panlobular microvesicular steatosis is the dominant lesion (Zimmerman and Ishak, 1982). Experimentally, valproic acid causes microvesicular steatosis in rat heart and liver along with mitochondrial swelling and pleomorphism in liver that is very similar to the morphologic changes of Reye's syndrome, suggesting a similar toxicity mechanism (Jezequel et al., 1984; Melegh and Trombitas, 1997).

As with Reye's syndrome, elevated serum transaminases, depletion of hepatic glycogen, and coagulopathy confirm liver as the target organ of valproic acid toxicity (Becker and Harris, 1983; Gerber et al., 1979). Specific biochemical alterations further pinpoint hepatic mitochondria as the site of injury. Mitochondrial enzymes of gluconeogenesis, ureagenesis, ketogenesis, and fatty acid synthesis are inhibited, causing hypoglycemia and hyperammonemia (Becker and Harris, 1983, Turnbull et al., 1983). The appearance of dicarboxylic acids in the urine suggests interference with β-oxidation of fatty acids in mitochondria (Mortensen et al., 1980).

Valproic acid is metabolized in the liver to a number of *en* metabolities, several of which are converted to excretable products through coenzyme A- and carnitine-dependent pathways in cytosol and mitochondria (Kesterson *et al.*, 1984). Carnitine and coenzyme A are required not only for elimination in mitochondria but also for energy production by fatty acid oxidation. Thus fasting, which leads to utilization of carnitine and coenzyme A for energy production, could promote hepatotoxicity by forcing the accumulation of valproic acid and toxic metabolites. Indeed, carnitine and coenzyme A are significantly decreased in patients experiencing valproic acid hepatotoxicity (Coulter, 1991; Matsumoto *et al.*, 1997). Furthermore, several studies have shown that prophylactic administration of carnitine reduces the accumulation of valproic acid and its toxic metabolites and thus the risk of acute hepatotoxic liver failure (Bohles *et al.*, 1996; Sugimoto *et al.*, 1987).

Interestingly, the incidence and severity of valproic acid toxicity does not correlate with blood or tissue levels of the drug or those of its potentially toxic metabolites (Loscher *et al.*, 1992; Shirley *et al.*, 1993), and because valproic acid toxicity is rare and idiopathic, victims may be predisposed to the hepatic injury. The cause of this predisposition is unknown; however, as in Reye's syndrome, valproic acid toxicity occurs with greater frequency in children. This may indicate simply the early manifestation of a serious metabolic disorder that is identified and treated, or alternatively, is fatal, in childhood. Lam and co-workers (1997) reported a 12-year-old patient with valproic acid toxicity who was subsequently found to have an A → G mitochondrial DNA mutation that resulted in impaired oxidative phosphorylation (Lam *et al.*, 1997). This patient recovered after cessation of valproic acid, suggesting that the latent mitochondrial defect alone was not sufficient to induce a serious clinical disorder until the mitochondria were simultaneously stressed by valproic acid. A similar mitochondrial defect in cytochrome oxidase was revealed in a 3-year-old girl, who died after initiation of valproic acid for epilepsy (Chabrol *et al.*, 1994). In both cases the metabolic defect that impaired mitochondrial respiration produced a state comparable to that of fasting. Indeed, similar to the pathogenesis of Reye's syndrome, patients who are fasting or recovering from viral illness are more likely to experience valproic acid toxicity.

It remains unclear, however, how valproic acid, fasting, and depletion of carnitine and coenzyme A cause mitochondrial dysfunction. The range of metabolic alterations observed in valproic acid hepatotoxicity are not consistent with inhibition of specific mitochondrial enzymes, even though inhibition of enzymes and specific components of electron transport by valproic acid has been demonstrated experimentally (Ponchaut and Veitch, 1993). Rather, the complete mitochondrial failure observed from changes in clinical chemistry values would be expected to, and in fact does, accompany swelling, depolarization, and uncoupling, all cardinal features of the MPT.

Therefore we tested valproic acid in our *in vitro* assay for onset of the MPT in isolated rat liver mitochondria. In Ca^{2+}-free buffer, valproic acid (300 µM) caused no mitochondrial swelling (Fig. 4C). A 4-minute incubation with 150 µM valproic acid caused swelling after addition of $CaCl_2$ (Fig. 4C). The swelling was blocked by cyclosporin A, which confirmed that it was caused by MPT onset. Furthermore, valproic acid induced transition over the same concentration range as salicylate, with increasing concentrations decreasing the latency before opening of the MPT pore (Trost and Lemasters, 1996).

4.2. Jamaican Vomiting Sickness

In their original description of Reye's syndrome, Reye and co-workers noted its similarity to Jamaican vomiting sickness, a toxic disorder characterized by vomiting, coma, hypoglycemia, free fatty acidemia, dicarboxylic aciduria, depletion of liver glycogen, fatty degeneration of the liver, and death (Sherratt, 1986). Between 1886 and 1950, approximately 5,000 deaths were attributed to this illness (Mitchell, 1974). Animal studies show that hypoglycin A (L-[methylenecyclopropyl]alanine), a constituent of the unripe fruit of the Ackee tree (*Blighia sapida*), causes the disorder. Within 2–3 hours of consuming the Ackee fruit, symptoms appear in association with hypoglycin A metabolism to methylene-cyclopropylacetic acid, the active metabolite responsible for the Reye's-like syndrome (Bressler *et al.*, 1969; Glasgow and Chase, 1975). The latter shares features of other MPT inducers, namely a carboxylic acid (or dicarboxylic acid) with a non-polar side chain.

Similar to salicylate and valproic acid, methylene-cyclopropylacetic acid is a simple non-polar branched-chain carboxylic acid. Through this metabolite, hypoglycin A causes hypoglycemia and inhibits β-oxidation and gluconeogensis in experimental animals, suggesting that liver, and particularly hepatic mitochondria, are targets of the toxicity (Sherratt, 1986). Although vomiting and coma indicate neurotoxicity in Jamaican vomiting sickness, Borison and co-workers (1974) demonstrated that neither hypoglycin nor its known metabolites have an effect when placed in direct contact with brain. Peripheral injections, however, are toxic, suggesting that neural effects occur secondary to the hepatic injury.

Hepatic mitochondria are most likely targeted in the Jamaican vomiting sickness because liver is the site of hypoglycin A metabolism. The CoA metabolites of methylene-cyclopropylacetic acid are formed in liver and are significantly more toxic than the parent compound (Sherratt, 1986). A number of CoA esters were shown to inhibit β-oxidation and gluconeogenesis in isolated liver mitochondria, with varying specificity. The ability of a given metabolite to inhibit energy production *in vitro*, however, did not correlate with its ability to cause toxicity when tested in perfused livers and in tissue homogenates (Billington *et al.*, 1978). For this reason, and because inhibition of specific mitochondrial respiration enzymes is not consistent with the fulminant hepatic failure observed in Jamaican vomiting sickness, we contend that the mitochondrial damage might instead be due to onset of the MPT.

We did not test hypoglycin A for onset of the MPT in our *in vitro* MPT assay because the compound was unavailable. Isovaleric acid and adipic acid, however, which are implicated in metabolic disorders resembling Reye's syndrome and which induced MPT in our test system, are metabolites of hypoglycin A that are also present in high concentrations in the urine and plasma of patients with Jamaican vomiting sickness (McTague and Forney, 1994; Tanaka *et al.*, 1972). Because hypoglycin was not tested, we can only speculate that these agents may work in concert to promote onset of the MPT.

4.3. Neem Oil

Sinniah and Baskaran (1981) reported a Reye's-like syndrome in 13 children with onset of symptoms 3–24 hours after ingestion of "Margosa" or Neem oil, an elixir made from the seeds of the Neem tree *(Azadirachta indica)*. Neem oil is widely available and used as a traditional medicinal remedy for treating common ailments in India, Sri Lanka, and Malaysia. The key features of Neem oil toxicity are vomiting, drowsiness progressing to coma, metabolic acidosis, polymorphonuclear leukocytosis, and encephalopathy.

More than 80% of the patients described by Sinniah and Baskaran died. The cause was determined to be severe cerebral edema, probably secondary to hyperammonemia. Hyperammonemia caused by inhibition of ureagenesis suggests that liver is the target organ of Neem oil toxicity. Indeed, biopsy demonstrated pronounced fatty infiltration of the liver without necrosis. Nuclei were normal, suggestive of microvesicular steatosis (Sinniah *et al.*, 1985). Histopathology and electron microscopy revealed swollen and pleomorphic hepatic mitochondria. In describing the toxic syndrome caused by Neem oil, the authors specifically noted its similarity to Reye's syndrome (Sinniah and Baskaran, 1981).

Neem oil is pungent, bitter, and deep yellow, composed primarily of stearic, oleic, palmitic, and linoleic acids with lesser amounts of myristic, arachidic, behenic, and lignoceric acids (Koga *et al.*, 1987; Skellon *et al.*, 1962). The oil uncouples isolated mitochondria, depletes intramitochondrial CoA, and inhibits the mitochondrial respiratory chain, all causing a mitochondrial energy crisis (Koga *et al.*, 1987). In treated rats Neem oil causes mitochondrial rarefaction, swelling, pleomorphism, and loss of dense bodies (Sinniah *et al.*, 1985). Microvesicular steatosis and glycogen depletion also occur, although hypoglycemia is not common in human victims. Elevations of serum ammonia and octanoic acid may contribute to the encephalopathy and cerebral edema associated with Neem oil toxicity.

Numerous studies failed to identify the toxic component of Neem oil or its mechanism of action. Because Neem oil toxicity is similar to Jamaican vomiting sickness, Sinniah and co-workers (1985) suggest that a toxic metabolite similar to methylene-cyclopropylacetic acid may be formed during metabolism of the long- and medium-chain fatty acids that comprise the oil. Koga and co-workers (1987) suggest that the long-and medium-chain fatty acids themselves may have a detergent effect on the mitochondrial membrane. Enormous concentrations of Neem oil, however, would be required to produce such nonspecific mitochondrial membrane damage *in vivo*.

The wide-ranging effects that produced the mitochondrial energy crisis described by Koga and co-workers could be caused by onset of the MPT. Accordingly, we tested the ability of Neem oil to induce the MPT in our model, using isolated liver mitochondria. Even very dilute concentration of the oil caused rapid mitochondrial swelling after addition of $CaCl_2$ (Fig. 4D). Again swelling was blocked by cyclosporin A, which prevents the MPT.

4.4. Adipic, Benzoic, 4-Pentenoic, and 3-Mercaptopropionic Acids

Adipic acid is an endogenous metabolic by-product and a product of valproic acid and hypoglycin A metabolism present in high concentrations in the urine and serum of patients experiencing fulminant hepatotoxicity (McTague and Forney, 1994; Willmore *et al.*, 1991). Benzoic acid inhibits fatty acid and glucose metabolism in liver, similar to Reye's syndrome (McCune *et al.*, 1982). Reye's syndrome features have also been reproduced experimentally in rats treated with 4-pentenoic acid, a structural analog of valproic acid and a product of hypoglycin A metabolism (Glascow and Chase, 1975). Likewise, 3-mercaptopropionic acid has been used to model Reye's syndrome in isolated perfused rat livers (Yamamoto and Nakamura, 1994). Like salicylate and valproic acid, these agents are all short-chain nonpolar carboxylic acids.

When 3-mercaptopropionic acid and 4-pentenoic acid were tested for their ability to induce the MPT in our test system, both induced rapid mitochondrial swelling after CaCl$_2$ (Fig 4E, 4F). In each case the swelling was blocked by cyclosporin A. Benzoic acid also induced the mitochondrial swelling characteristic of onset of the MPT (Trost and Lemasters, 1996). Thus, a group of structurally related chemical agents—including salicylate, valproic acid, isovaleric acid, adipic acid, benzoic acid, 3-mercaptopropionic acid, and 4-pentenoic acid, all causing Reye's-related disorders—each induced the MPT over a similar range of concentrations, with 150 to 300 µM being maximally effective. Not all short-chain carboxylic acids induced the MPT, however. When 12 other carboxylic acids that are structurally similar to salicylate but not implicated in Reye's-related disorders were tested for their ability to induce the MPT, none caused the characteristic phenomenon of cyclosporin A-inhibitable mitochondrial swelling (Fig. 5). These data further suggest that onset of the MPT is specific and give further evidence that the onset of the MPT might be the pathogenic mechanism of Reye's syndrome and related disorders.

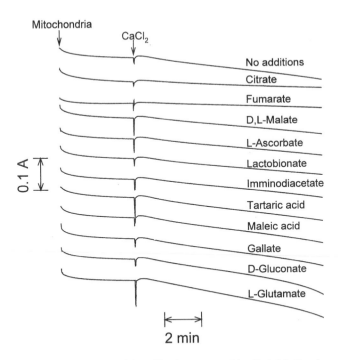

FIGURE 5. Lack of effect on mitochondrial swelling by agents not implicated in Reye's syndrome. Rat liver mitochondria were incubated as described in Figure 1. Absorbance traces are offset to show the effects of citric, fumaric, D, L-malic, L-ascorbic, lactobionic, imminodiacetic, tartaric, maleic, gallic, D-gluconic, and L-glutamic acids on mitochondrial swelling. The carboxylic acid concentration was 150 µM for all experiments. None of these agents are associated with Reye's-like chemical toxicity and none induced the MPT.

5. MECHANISM OF MPT INDUCTION *IN VITRO* BY SALICYLATE AND AGENTS IMPLICATED IN REYE'S-RELATED CHEMICAL TOXICITY

The chemicals implicated in Reye's-related toxicity that were observed to induce the MPT in liver mitochondria were short-chain carboxylic acids with nonpolar side chains. In contrast, carboxylic acids with more-polar side chains, including citric, fumaric, malic, ascorbic, lactobionic, imminodiacetic, tartaric, maleic, gallic, gluconic, and glutamic acids failed to induce transition (Fig. 5). Moreover, these polar carboxylic acids have not been reported to cause Reye's-related toxicity.

Nonpolar carboxylic acids are weak uncouplers in mitochondria because they can move across membranes in both protonated and unprotonated forms (Kamp *et al.*, 1993; Kamp and Hamilton, 1992). This leads to charge-uncompensated transport of H^+ down its electrochemical gradient (i.e., uncoupling). Because the transition pore is voltage-gated channel, under certain experimental conditions uncoupling can itself induce the MPT. The protonophoric uncoupler FCCP failed, however, to induce the MPT at concentrations that produce a depolarization equivalent to that caused by salicylate concentrations nearly one order of magnitude higher than those required for onset of the MPT (Trost and Lemasters, 1996). Indeed, salicylate and valproic acid, at concentrations that strongly induced the MPT, caused slight or negligible decreases of $\Delta\Psi$ (Trost and Lemasters, 1996).

A more likely explanation for induction of the MPT by Reye's-associated chemicals relates to their probable effect on $\Delta\Psi$. Broekemeier and Pfeiffer (1995) suggest that $\Delta\Psi$ changes influence the opening threshold of the MPT pore (see also Bernardi *et al.*, 1994). Depending on their charge, amphipathic anions or cations with nonpolar side chains increase or decrease the surface potential of mitochondrial membranes by inserting into the bilayer. The pore is sensitive to $\Delta\Psi$ through a voltage-sensing element in its complex. The probability of pore opening increases with more-negative $\Delta\Psi$, and it decreases with more-positive surface potentials (Petronilli *et al.*, 1994a). Thus cations such as spermine and spermidine inhibit pore opening whereas fatty acids and lysophospholipids promote it (Lapidus and Sokolove, 1992, 1993).

Petronilli and co-workers (1993) suggest that subpopulations of mitochondria have lower gating potentials for pore opening, making them more sensitive to induction of the MPT. Induction in this more-vulnerable population may then initiate rapid spread of the MPT to surrounding mitochondria through release of a soluble factor. Rapid, nearly synchronous onset of the MPT in all mitochondria of a single cell were noted previously in hepatocytes exposed to oxidative stress and chemical hypoxia (Imberti *et al.*, 1992, 1993; Nieminen *et al.*, 1995).

6. ROLE OF CALCIUM IN REYE'S SYNDROME AND REYE'S-RELATED CHEMICAL TOXICITY

Reye's syndrome and Reye's-related toxicities occur idiopathically with low incidence. In the case of salicylate and valproic acid, the cause is not drug intoxication due to overdose, because Reye's and Reye's-like toxicity occur within the therapeutic range of both drugs. Most patients taking aspirin and valproic acid experience no toxicity, so a second injury or predisposing condition probably sensitizes mitochondria to such damage.

Segalman and Lee (1982) suggest that this predisposing factor is Ca^{2+}. In experiments with isolated mitochondria they found ATP depletion and uncoupling caused by Reye's serum was eliminated by ruthenium red, an inhibitor of mitochondrial Ca^{2+}-transport, and by EGTA, a Ca^{2+} chelator. Similar results were obtained with a repeat experiment using Ca^{2+} and salicylate in place of Reye's serum (Martens and Lee, 1984).

In our experiments salicylate, valproic acid, and other inducers did not by themselves cause onset of the MPT. Rather, the synergistic action of Ca^{2+} was also required. Previous investigators showed that even without added Ca^{2+}, isolated mitochondria swell in response to aspirin and salicylate. These effects are maximal, however, only at concentrations exceeding 30 mM, much higher than is likely therapeutically *in vivo* or even in salicylate intoxication (Martens *et al.*, 1986; Tomoda *et al.*, 1994; Yoshida *et al.*, 1992; You, 1983). In contrast, our results show that after exposing isolated mitochondria to a small amount of Ca^{2+} (50 μM), as little as 37.5 μM salicylate induced MPT onset (Fig. 1B). Mitochondria take up the added Ca^{2+} to establish a submicromolar free Ca^{2+} concentration, then release it at the onset of transition. In our experiments mitochondria were isolated in an EGTA-containing buffer, so the added Ca^{2+} may be at least partly viewed, as reconstitution of the intramitochondrial Ca^{2+} pool. Although high Ca^{2+} alone can induce transition, the concentration used here was not sufficient to cause its onset.

Because calcium was required for induction of the MPT by salicylate *in vitro*, we examined the calcium dependence of salicylate toxicity to cultured hepatocytes using the calcium-channel blocking agent verapamil. Cell viability was measured by PI fluorometry, as described by Nieminen and co-workers (1992). Primary cultures of hepatocytes were incubated in Waymouth's medium, containing 2.5 mM $CaCl_2$ and PI. After a 24-hour incubation, the viability of hepatocytes incubated in the presence of 3 mM salicylate was more than 40% less than that of controls incubated in its absence. When the experiment was repeated in the presence of verapamil, however, the verapamil caused dose-dependent inhibition of salicylate-dependent cell killing (Fig. 6). Verapamil's inhibition of cell killing was significant ($p < 0.05$) at a concentration of 10 μM. At a 30 μM concentration, verapamil blocked cell killing completely.

In these experiments the time course of cell killing in Waymouth's medium was longer than that in confocal microscopy experiments (as in Fig. 2), which are necessarily conducted in KRH buffer. The apparent decreased sensitivity of hepatocytes to salicylate in serum-containing growth medium may be due to albumin in the medium. Because both Ca^{2+} and salicylate bind to albumin, there may be less free Ca^{2+} and salicylate in solution. Because cell killing was dependent upon both salicylate and Ca^{2+}, binding of the molecules to albumin likely explains the decreased cell killing in growth medium compared with that in serum-free KRH, which lacks this nonspecific binding site.

The concentration of verapamil required for protection from salicylate toxicity was higher than that required to inhibit voltage-dependent calcium channels, suggesting that verapamil may be acting by a nonspecific mechanism rather than by inhibition of calcium channels (Gasbarrini *et al.*, 1993; Sippel *et al.*, 1993). To determine whether verapamil blocked Ca^{2+} entry into hepatocytes, we loaded cultured hepatocytes with the fluorescent Ca^{2+} indicator Rhod-2 for examination by laser-scanning confocal microscopy. In hepatocytes incubated in Waymouth's medium, Rhod-2 fluorescence was dim, with faint nuclear and cytoplasmic staining (Fig. 7, upper left). After addition of 0.8 mM $CaCl_2$, Rhod-2 fluorescence increased in the cytosol, nucleus, and most noticeably in mitochon-

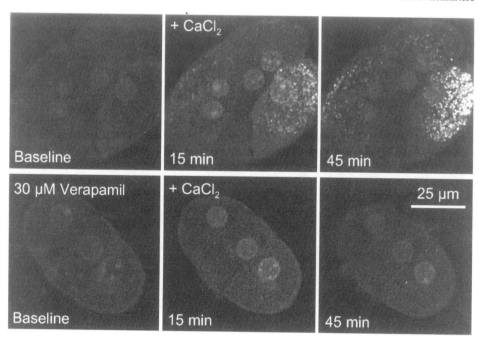

FIGURE 6. Inhibition of mitochondrial calcium loading by verapamil. Rat hepatocytes were loaded with Rhod-2, a calcium-indicating fluorescent probe, and incubated in Waymouth's medium containing 10% FCS. After 30 min baseline images of the Rhod-2 fluorescence were recorded (upper left panel), and 0.8 mM $CaCl_2$ was added for a final Ca^{2+} concentration of about 2 mM. Images were recorded subsequently after 15 and 45 minutes. In a second experiment, 30 μM verapamil was included in the incubation medium and images were collected in an identical fashion before and after adding $CaCl_2$. In the absence of verapamil, fluorescence increased in cytosol, nucleus, and most markedly in mitochondria after 15 min. After 45 min nuclear and cytosolic fluorescence recovered but mitochondrial fluorescence increased further, indicating mitochondrial calcium accumulation. At 30 μM, verapamil blocked these $CaCl_2$-induced changes.

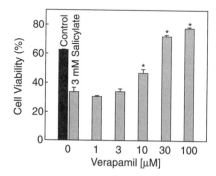

FIGURE 7. Protection by verapamil against salicylate toxicity to rat hepatocytes. Hepatocytes were incubated 24 hr in Waymouth's medium MB752/1 containing 2 mM L-glutamine, 10% fetal calf serum, 100 nM insulin, 10 nM dexamethasone, and 30 μM propidium iodide in humidified air/5% CO_2. Control cultures (black bars) were incubated in the absence of salicylate and calcium antagonists. As indicated by the gray bars, 3 mM salicylate and 0–100 μM verapamil was included in the assay medium. Verapamil was added 30 minutes prior to salicylate. *$P < 0.05$ compared to incubations with salicylate alone (0 calcium antagonist).

dria within 15 minutes (upper middle). After 45 minutes cytosolic and nuclear fluorescence returned to near-original levels, but mitochondrial Rhod-2 fluorescence increased further, indicating mitochondrial Ca^{2+} accumulation (upper right). In a second experiment, hepatocytes were pre-incubated for 30 minutes with $30\,\mu M$ verapamil prior to $CaCl_2$ addition (Fig. 7, lower panels). Verapamil blocked the mitochondrial calcium accumulation. Lower concentrations, however, failed to block mitochondrial calcium accumulation (not shown). Thus, Verapamil's protection against salicylate toxicity as shown in Figure 6 occurred at concentrations that blocked mitochondrial Ca^{2+} uptake.

The effective concentration of verapamil for blocking mitochondrial Ca^{2+} accumulation was 10–100-fold more than its IC_{50} for L-type voltage-dependent Ca^{2+} channels (Gasbarrini et al., 1993). Hepatocytes, however, possess receptor-mediated channels that are less sensitive to calcium-channel blocking agents (Mauger and Claret, 1988; Sippel et al., 1993). Blockade of receptor-mediated calcium channels occurs at verapamil concentrations similar to those necessary for blocking salicylate toxicity and mitochondrial calcium accumulation (Mauger and Claret, 1988; Michael and Whiting, 1989). Likewise, other investigators have reported that similar concentrations of verapamil, nicardipine, diltiazem, and other calcium-channel blocking agents were required to protect against calcium-dependent hepatotoxicity caused by acetaminophen, chloroform, and carbon tetrachloride (Landon et al., 1986; Ray et al., 1993). Thus, these agents probably prevent salicylate toxicity by inhibiting Ca^{2+} entry into the hepatocyte.

Verapamil prevented mitochondrial Ca^{2+} accumulation and protected hepatocytes from salicylate cytotoxicity. Because Ca^{2+} binding to matrix sites is required for transition onset, elevated intramitochondrial Ca^{2+} may be a predisposing condition promoting the onset by Reye's-related chemicals in vivo (Zoratti and Szabò, 1995). Defective mitochondrial Ca^{2+} regulation increases the probability of MPT associated with ischemia–reperfusion injury (Hoek et al., 1997). The matrix Ca^{2+} increase may result from defective mitochondrial calcium regulation, or the inducer itself may cause the calcium increase. Doxorubicin, an antineoplastic agent that causes a cumulative cardiomyopathy in which the MPT has been implicated (Solem et al., 1996), promotes mitochondrial calcium accumulation prior to cell injury (Chacon et al., 1992). A similar increase in Ca^{2+} after reoxygenation may induce the MPT associated with ischemia–reperfusion injury (Lemasters et al., 1998; Matsuda et al., 1995; Qian et al., 1997). Therefore, mitochondrial Ca^{2+} and the agents implicated in Reye-related toxicity may act synergistically to promote onset of the MPT in vivo.

A disorder in Ca^{2+} regulation or the prodromal viral illness often seen in Reye-related disorders may cause the increase in abnormal mitochondrial calcium loading (Duerksen et al., 1997). The high concentrations of calcium-mobilizing growth hormones found in children may cause an increase in mitochondrial calcium and could account for children's increased susceptibility to Reye's-related disorders (Hoek et al., 1997). Another potential source follows from studies examining the role of TNFα and IL-1 in Reye's syndrome. Aspirin and viral infection both promote IL-1 and TNFα production and both cytokines are greatly elevated in Reye's syndrome (Corkey et al., 1991; Treon and Broitman, 1992). Corkey and co-workers (1991) also found that dermal fibroblasts from patients with Reye's syndrome exhibit exaggerated Ca^{2+} responses to high concentrations of IL-1 and TNFα Compared with normal patients, patients with Reye's syndrome exhibited higher maximal Ca^{2+} responses, higher half-maximal cytokine concentrations for observed Ca^{2+} effects, and blunted inhibition of the Ca^{2+} response at high cytokine levels. These effects were consistent among cells from all Reye's patients examined and suggest the presence of a genetic defect or inappropriate response to an antecedent viral illness, which causes the abnormal Ca^{2+} response.

In individuals predisposed to abnormal mitochondrial calcium-loading, exposure to salicylate or other inducers may precipitate onset of the MPT at therapeutic and generally well-tolerated concentrations producing a Reye's-like illness. Onset of the MPT and the resulting mitochondrial uncoupling would explain many features of Reye's syndrome: hypoglycemia, hyperammonemia, elevation of dicarboxylic and free fatty acids, inhibition of β-oxidation, and accumulation of acyl-CoA intermediates. This pathophysiological mechanism also suggests specific therapies for Reye's-related disorders because drugs such as cyclosporin A and trifluoperazine inhibit the MPT in both isolated mitochondria and intact cells (Broekemeier et al., 1985, 1989, 1992; Imberti et al., 1992, 1993; Nieminen et al., 1995).

7. FASTING AND FATTY ACID MOBILIZATION IN REYE'S SYNDROME AND REYE'S-RELATED CHEMICAL TOXICITY

Fasting is consistently identified in Reye's syndrome, valproic acid toxicity, and other Reye's-like disorders associated with prodromal viral illness and vomiting. Fasting increases serum fatty acids as fat stores are mobilized for energy production. A number of endogenous fatty acids induce the MPT transition (Schonfeld and Bohnensack, 1997). The work of Broekemeier and Pfeiffer (1995) suggests that mobilized fatty acids such as palmitate may function in combination with exogenous inducers such as salicylate to promote onset of the MPT and Reye's-related disorders. Indeed, Pastorino and co-workers (1993) have demonstrated that long-chain acyl-CoA intermediates such as palmitoyl-CoA induce the MPT. Moreover, carnitine, which has been used prophylactically as a hepatoprotectant in patients receiving valproic acid, blocks the MPT induced by palmitoyl-CoA in isolated mitochondria.

Carnitine acyltransferase catalyzes the reaction of acyl-CoA with carnitine. The resulting acyl-carnitine intermediate is transferred into the mitochondria and converted back to a CoA intermediate by a similar matrix transferase prior to β-oxidation. A toxic crisis (for example because of defects in β-oxidation or as a consequence of carnitine palmitoyltransferase deficiency) is manifested when the patient is in a fasting state and relying upon fatty acid oxidation for energy production. In these metabolic defects, accumulation of fatty acids and their CoA metabolites alone may be sufficient to cause mitochondrial injury similar to that observed in Reye's syndrome. Indeed, primary carnitine deficiency was found in a 1-year-old boy with family history of infant death following a Reye's-like syndrome (Hou et al., 1996).

Therefore, Reye's-related toxicity might also occur at normal mitochondrial Ca^{2+} concentrations in compromised individuals with metabolic defects leading to accumulation of short-chain fatty acids, or after exposure to similar exogenous chemical agents. Fasting might further promote MPT onset because the increased mitochondrial demand for energy via fatty acid oxidation results in the mobilization of MPT-inducing fatty acids and depletion of carnitine and CoA. Liver may be the target organ of toxicity because it is the site of many elimination reactions as well as those necessary for energy production.

Inborn metabolism errors are excluded from the CDC criteria for diagnosis of Reye's syndrome (Sullivan-Bolynai and Corey, 1981), but as increasingly sophisticated methods become available, retrospective studies implicate metabolic disorders in many cases previously identified as Reye's syndrome. Because fasting is highly correlated with Reye's syndrome, it is likely that mobilized fatty acids and their acyl-CoA intermediates contribute to the pathogenesis of mitochondrial dysfunction. Further, because induction by various chemical inducers is additive, these fatty acids would probably increase the chances of pore opening in the presence of salicylate and *vice versa*. In healthy individuals, however, mobilization of fat stores in addition to exogenously added inducing agents is clearly insufficient to cause MPT onset and hepatic failure, because again, few individuals develop Reye's syndrome. Therefore, high mitochondrial calcium may yet contribute to the metabolic crisis.

8. CONCLUSION: IMPLICATIONS FOR OTHER DISEASES

Advisories against aspirin use in children have largely eliminated Reye's syndrome, but a number of Reye's-like disorders persist including valproic acid toxicity, Jamaican vomiting sickness, and inborn metabolic disorders. The literature is replete with other examples of drug toxicity that resembles Reye's syndrome and in which the MPT and mitochondrial dysfunction are implicated. For example, toxicity resembling Reye's syndrome, including vomiting, fatty liver, and coma, was described in four adolescent patients, three of whom died, who ingested a bismuth–diallylacetic acid complex for treatment of upper-respiratory tract infections. In another case two female patients, one of whom died, had signs and symptoms of Reye's syndrome (coma, elevated serum transaminases, and microvesicular steatosis) after taking pirprofen, a phenylpropionic acid-derived nonsteroidal anti-inflammatory drug. These patients had 7 to 9 months of uneventful exposure to the drug before onset of the fulminant hepatic failure, suggesting metabolic rather than immunologic mechanisms (Danan *et al.*, 1985). Likewise similar to Reye's syndrome symptomology and hepatic injury are the symptoms and injury associated with the Japanese cerebral-enhancer hepanoate, mycotoxins such as aflatoxin, phenothiazine antiemetics, idiopathic fatty liver of pregnancy, amiodarone, acetamino-phen, metoclopramide, paracetomol, tetracycline, tolmetin, and the experimental anxiolytic panadiplon (Beales and Mclean, 1996; Casteels-Van Daele, 1991; Gerber *et al.*, 1979; Shaw and Anderson, 1991; Ulrich *et al.*, 1994; Visentin *et al.*, 1995). Therefore, it is important to understand the pathogenesis of Reye's syndrome and related disorders to prevent and treat these various conditions. Because the agents described here are all structurally related, an understanding might also help to predict and eliminate early in development drugs at risk of producing this type of mitochrondrial toxicity.

No specific therapies exist for Reye's-related disorders other than supportive care and removal of the inducing drug or agent. The work described here implicates the MPT in the pathogenesis of Reye's-related disorders. Because Cyclosporin A, 4-methylvaline cyclos-porin, trifluoperazine, and other agents inhibit the MPT, they have potential as therapeutic agents for Reye's-related diseases. Although in our experiments cyclosporin A was used prophylactically to protect against toxicity, others have shown that it reduces ischemia–reperfusion injury even when administered after the ischemic event (Uchino *et al.*, 1995). Thus, 4-methylvaline cyclosporin, the nonimmunosuppresive analog of cyclosporin A, may be beneficial in the treatment of Reye's-related disorders.

REFERENCES

Aprille, J. R., 1977, Reye's syndrome: Patient serum alters mitochondrial function and morphology *in vivo, Science* **197**:908–910.

Beales, D., and McLean, A. E. M., 1996, Protection in the late stages of paracetamol-induced liver cell injury with fructose, cyclosporin A, and trifluoperazine, *Toxicology* **107**:201–208.

Becker, C.-M., and Harris, R. A., 1983, Influence of valproic acid on hepatic carbohydrate metabolism and lipid metabolism, *Arch. Biochem. Biophys.* **223**:381–392.

Bernardi, P., Broekemeier, K. M., and Pfeiffer, D. R., 1994, Recent progress on regulation of the mitochondrial permeability transition pore: A cyclosporin-sensitive pore in the inner-mitochondrial membrane, *J. Bioenerg. Biomembr.* **26**:509–517.

Biban, C., Tassani, V., Toninello, A., Siliprandi, D., and Siliprandi, N., 1995, The alterations in the energy-linked properties induced in rat liver mitochondria by acetylsalicylate are prevented by cyclosporin A or Mg^{2+}, *Biochem. Pharmacol.* **50**:497–500.

Billington, D., Osmundsen, H., and Sherratt, H. S. A., 1978, Mechanisms of the metabolic disturbances caused by hypoglycin and by pent-4-enoic acid *in vivo* studies, *Biochem. Pharm.* **27**:2879–2890.

Bohles, H., Sewell, A. C., and Wenzel, D., 1996, The effect of carnitine supplementation in valproate-induced hyperammonemia, *Acta Paediat.* **85**:446–449.

Borison, H. L., Pendleton, J., Jr., McCarthy, L. E., 1974, Central vs systemic neurotoxicity of hypoglyxin (ackee toxin) and related substances in the cat, *J. Pharmacol. Exp. Ther.* **190**:327–333.

Bourgeois, B. F. D., 1989, Clinical use, in *Antiepileptic Drugs* 3rd ed., (R. H. Levy, F. E. Dreifuss, R. M. Mattson, B. S. Meldrum, and J. K. Penry, Eds.), Raven, New York, pp. 633–642.

Bressler, R., Corredor, C., and Brendel, K., 1969, Hypoglycin and hypoglycin-like compounds, *Pharmacol. Rev.* **21**:105–130.

Brody, T. M., 1956, Action of sodium salicylate and related compounds on tissue metabolism *in vitro, J. Pharmacol. Exp. Ther.* **117**:39–51.

Broekemeier, K. M., and Pfeiffer, D. R., 1995, Inhibition of the mitochondrial permeability transition by cyclosporin A during long time frame experiments: A relationship between pore opening and the activity of mitochondrial phospholipases, *Biochemistry* **34**:16440–16445.

Broekemeier, K. M., Schmid, O. C., Schmid, H. H., and Pfeiffer, D. R., 1985, Effects of phospholipase A_2 inhibitors on ruthenium red-induced Ca^{2+} release from mitochondria, *J. Biol. Chem.* **260**:105–113.

Broekemeier, K. M., Dempsey, M. E., and Pfeiffer, D. R., 1989, Cyclosporin A is a potent inhibitor of the inner membrane permeability transition in liver mitochondria, *J. Biol. Chem.* **264**:7826–7830.

Broekemeier, K. M., Carpenter, D. L., Reed, D. J., and Pfeiffer, D. R., 1992, Cyclosporin A protects hepatocytes subjected to high Ca^{2+} and oxidative stress, *FEBS Lett.* **304**:192–194.

Brown, R. E., and Forman, D. T., 1982, The biochemistry of Reye's syndrome, *CRC Crit. Rev. Clin. Lab. Sci.* **17**:247–297.

Casteels-Van Daele, M., 1991, Reye's syndrome or side-effects of anti-emetics? *Eur. J. Pediatr.* **150**:456–459.

Chabrol, B., Mancini, J., Chretien, D., Rustin, P., Munnich, A., and Pinsard, N., 1994, Valproate-induced hepatic failure in a case of cytochdrome *c* oxidase deficiency, *Eur. J. Pediatr.* **153**:133–135.

Chacon, E., Ulrich, R., and Acosta, D., 1992, Digitized-fluorescence-imaging study of mitochondrial Ca^{2+} increase by doxorubicin in cardiac myocytes, *Biochem. J.* **281**:871–878.

Corkey, B. E., Geschwind, J.-F., Deeney, J. T., Hale, D. E., Douglas, S. D., and Kilpatrick, L., 1991, Ca^{2+} responses to interleukin 1 and tumor necrosis factor in cultured human skin fibroblasts, *J. Clin. Invest.* **87**:778–786.

Coulter, D. L., 1991, Carnitine, valproate, and toxicity, *J. Child. Neurol.* **6**:7–14.

Crompton, M., Ellinger, H., and Costi, A., 1988, Inhibition by cyclosporin A of a Ca^{2+}-dependent pore in heart mitochondria activated by inorganic phosphate and oxidative stress, *Biochem. J.* **255**:357–360.

Danan, G., Trunet, P., Bernuau, J., Degott, C., Babany, G., Pessayre, D., Rueff, B., and Benhamou, J. P., 1985, Pirprofen-induced fulminant hepatitis, *Gastroenterology* **89**:210–213.

DeVivo, D. C., and Keating, J. P., 1975, Reye's syndrome, *Adv. Pediatr.* **22**:175–229.

Duerksen, D. R., Jewell, L. D., Mason, A. L., and Bain, V. G., 1997, Co-existence of hepatitis A and adult Reye's syndrome, *Gut* **41**:121–124.

Duffy, J., Glasgow, J. F., Patterson, C. C., Clarke, M. J., and Turner, I. F., 1991, A sibling-controlled study of intelligence and academic performance following Reye's Syndrome, *Dev. Med. Child. Neurol.* **33**:811–815.

Elpeleg, O. N., Christensen, E., Hurvitz, H., and Branski, D., 1990, Recurrent, familial Reye's-like syndrome with a new complex amino and organic aciduria, *Eur. J. Ped.* **149**:709–712.

Flower, R. J., Moncada, S., Vane, J. R., 1985, Analgesic–antipyretics and anti-inflammatory agents: Drugs employed in the treatment of gout, in: *The Pharmacological Basis of Therapeutics*, 7th ed. (A. G. Gillman, L. S. Goodman, T. W. Rall, and F. Murad, Eds.), Macmillan, New York, pp. 674–715.

Fournier, N., Ducet, G., and Crevat, A., 1987, Action of cyclosporine on mitochondrial calcium fluxes, *J. Bioenerg. Biomembr.* **19**:297–303.

Fromenty, B., and Pessayre, D., 1995, Inhibition of mitochondrial β-oxidation as a mechanism of hepatotoxicity, *Pharmacol. Ther.* **67**:101–154.

Gasbarrini, A., Borle, A. B., and Van Theil, D. H., 1993, Ca^{2+} antagonists do not protect isolated perfused rat hepatocytes from anoxic injury, *Biochim. Biophys. Acta* **117**:1–7.

Gerber, N., Dickinson, R. G., Harland, R. C., Lynn, R. K., Houghton, D., Antonias, J. I., and Schimschock, J. C., 1979, Reye's-like syndrome associated with valproic acid therapy, *J. Pediat.* **95**:142–144.

Glasgow, A. M., and Chase, H. P., 1975, Production of the features of Reye's syndrome in rats with 4-pentenoic acid, *Pediat. Res.* **9**:133–138.

Glasgow, J. F. T., and Moore, R., 1993, Reye's syndrome 30 years on, *BMJ* **307**:950–951.

Heubi, J. E., Partin, J. C., Partin, J. S., and Schubert, W. K., 1987, Reye's syndrome: Current concepts, *Hepatology* **7**:155–164.

Hoek, J. B., Walajtys-Rode, E., and Wang, X., 1997, Hormonal stimulation, mitochondrial Ca^{2+} accumulation, and the control of the mitochondrial permeability transition in intact hepatocytes, *Mol. Cell. Biochem.* **174**:173–179.

Hou, T. W., Chou, S. P., and Wang, T. R., 1996. Retubolic functions and liver histopathology in Reye-like illnesses, *Actu Pediatrica* **85**:1053–1057.

Imberti, R., Nieminen, A.-L., Herman, B., and Lemasters, J. J., 1992, Synergism of cyclosporin A and phospholipase inhibitors in protection against lethal injury to rat hepatocytes from oxidant chemicals, *Res. Commun. Chem. Pathol. Pharmacol.* **78**:27–38.

Imberti, R., Nieminen, A.-L., Herman, B., and Lemasters, J. J., 1993, Mitochondrial and glycolytic dysfunction in lethal injury to hepatocytes by *t*-buylhydroperoxide: Protection by fructose, cyclosporin A, and trifluoperazine, *J. Pharmacol. Exp. Ther.* **265**:392–400.

Jezequel, A. M., Bonazzi, P., Novelli, G., Venturini, C., and Orlandi, F., 1984, Early structural and functional changes in liver of rats treated with a single dose of valproic acid, *Hepatology* **4**:1159–1166.

Kamp, F., and Hamilton, J. A., 1992, pH gradients across phospholipid membranes caused by fast flip-flop of unionized fatty acids, *Proc. Natl. Acad. Sci. USA* **89**:11367–11370.

Kamp, F., Westerfhoff, H. V., and Hamilton, J. A., 1993, Movement of fatty acids, fatty acid analogues, and bile acids across phospholipid bilayers, *Biochemistry* **32**:11074–11086.

Kesterson, J. W., Granneman, G. R., and Machinist, J. M., 1984, The hepatotoxicity of valproic acid and its metabolites in rats: I. Toxicologic, biochemical, and histopathologic studies, *Hepatology* **4**:1143–1152.

Koga, Y., Yoshida, I., Kimura, A., Yoshino, M., Yamashita, F., and Sinniah, D., 1987, Inhibition of mitochondrial function by Margosa oil: Possible implications in the pathogenesis of Reye's syndrome, *Pediat. Res.* **22**:184–187.

Kroemer, G., Dallaporta, B., and Resche-Rigon, M., 1998, The mitochondrial death/life regulator in apoptosis and necrosis, *Ann. Rev. Physiol.* **60**:619–642.

Lam, C. W., Lou, C. H., Williams, J. C., Chan, Y. W., and Wong, L., 1997, Mitochondrial mxopathy, enceptralopxathy, lactic acidosis, and stroke like episodes (MELAS) triggered by vulprocute therapy, *European J. Pediatrics*, **156**:562–4.

Landon, E. J., Naukam, R. J., and Sastry, B. V. R., 1986, Effects of calcium channel blocking agents on calcium and centrilobular necrosis in the liver of rats treated with hepatotoxic agents, *Biochem. Pharmacol.* **35**:697–705.

Lapidus, R. G., and Sokolove, P. M., 1992, Inhibition by spermine of the inner membrane permeability transition of isolated rat heart mitochondria, *FEBS Lett.* **313**:314–318.

Lapidus, R. G., and Sokolove, P. M., 1993, Spermine inhibition of the permeability transition of isolated rat liver mitochondria: An investigation of mechanism, *Arch. Biochem. Biophys.* **306**:246–253.

Lemasters, J. J., Nieminen, A.-L., Qian, T., Trost, L. C., Elmore, S. P., Nishimura, Y., Crowe, R. A., Cascio, W. E., Bradham, C. A., Brenner, D. A., and Herman, B., 1998, The mitochondrial permeability transition in necrosis, apoptosis, and autophagy, *Biochim. Biophys. Acta* **1366**:177–196.

Lichtenstein, P. K., Heubi, J. E., Daughterty, C. C., Farrell, M. K., Sokol, R. J., Rothbaum, R. J., Suchy, F. J., and Balistreri, W. F., 1983, Grade I Reye's syndrome: A frequent cause of vomiting and liver dysfunction after varicella and upper-respiratory-tract infection, *New Engl. J. Med.* **309**:133–139.

Loscher, W., Wahnschaffe, U., Honack, D., Wittfoht, W., and Nau, H., 1992, Effects of valproate and E-2-en-valproate on functional and morphological parameters of rat liver: I. Biochemical, histopathological, and pharmacokinetic studies, *Epilepsy Res.* **13**:187–198.

Maheady, D. C., 1989, Reye's syndrome: Review and update, *J. Pediat. Health Care* **3**:245–250.

Martens, M. E., and Lee, C.-P., 1984, Reye's syndrome: Salicylates and mitochondrial functions, *Biochem. Pharmacol.* **33**:2869–2876.

Martens, M. E., Chang, C. H., and Lee, C. P., 1986, Reye's syndrome: Mitochondrial swelling and Ca^{2+} release induced by Reye's plasma, allantoin, and salicylate, *Arch. Biochem. Biophys.* **244**:773–786.

Matsuda, N., Morit, Nakamura, H., and Shigekawv, M., 1995, Mechanisms of the deoxygenation-induced calcium overload in cardiac myocytes: dependence on p11c, *Surgical Research* **99**:712–8.

Matsumoto, J., Ogawa, H., Maeyama, R., Okudaira, K., Shinka, T., Kuhara, T., and Matsumoto, I., 1997, Successful treatment by direct hemoperfusion of coma possibly resulting from mitochondrial dysfunction in acute valproate intoxication, *Epilepsia* **38**:950–953.

Mauger, J. P., and Claret, M., 1988, Calcium channels in hepatocytes, *J. Hepatol.* **7**:278–282.

McTague, J. A., and Forney, R., Jr., 1994, Jamaican vomiting sickness in Toledo, Ohio, *Ann. Emerg. Med.* **23**:1116–1118.

McCune, S. A., Durant, P. J., Flanders, L. E., and Harris, R. A., 1982, Inhibition of hepatic gluconeogenesis and lipogenesis by benzoic acid, *p-tert*-butylbenzoic acid, and a structurally related hypolipidemic agent, SC-33459, *Arch. Biochem. Biophys.* **214**:124–133.

Melegh, B., and Trombitus, K., 1997, Volprocite treatment induces lipid globute accemulation with ultrastrucural adnormalities of mitochondria in skeletal muscle, *Neuropediatrics* **28**:257–61.

Michael, A. D., and Whiting, R. L., 1989, Cellular action of nicardipine, *Am. J. Cardiol.* **64**:3H–7H.

Mitchell, J. C., 1974, The posthumous misfortune of Captain Bligh of the "Bounty:" Hypoglycemia from Blighia, *Diabetes* **23**:919–920.

Mitchell, P., 1979, Keilin's respiratory chain concept and its chemiosmotic consequences, *Science* **206**:1148–1159.

Mitchell, R. A., Ram, M. L., Arcinue, E. L., and Chang, C. H., 1980, Comparison of cytosolic and mitochondrial hepatic enzyme alterations in Reye's syndrome, *Pediat. Res.* **14**:1216–1221.

Morales, A. R., Bourgeois, C. H., and Chulacharit, E., 1971, Pathology of the heart in Reye's syndrome (encephalopathy and fatty degeneration of the viscera), *J. Cardiol.* **27**:314–317.

Mortensen, P. B., Gregersen, N., Kolvraa, S., and Christensen, E., 1980, The occurrence of C_6-C_{10}-dicarboxylic acids in urine from patients and rats treated with dipropylacetate, *Biochem. Med.* **24**:153–161.

Nieminen, A.-L., Gores, G. J., Bond, J. M., Imberti, R., Herman, B., and Lemasters, J. J., 1992, A novel cytotoxicity assay using a multiwell fluorescence scanner, *Toxicol. Appl. Pharmacol.* **115**:147–155.

Nieminen, A.-L., Saylor, A. K., Tesfai, S. A., Herman, B., and Lemasters, J. J., 1995, Contribution of the mitochondrial permeability transition to lethal injury after exposure of hepatocytes to *t*-butylhydroperoxide, *Biochem. J.* **307**:99–106.

Ozawa, H., Sasaki, M., Sugai, K., Hashimoto, T., Matsuda, H., Takashima, S., Uno, A., and Okawa, T., 1996, Single-photon emission CT and MR findings in Kluver–Bucy syndrome after Reye's syndrome, *AJNR* **18**:540–542.

Partin, J. S., Daugherty, C. C., McAdams, A. J., Partin, J. C., and Schubert, W. K., 1984, A comparison of liver ultrastructure in salicylate intoxication and Reye's syndrome, *Hepatology* **4**:687–690.

Partin, J. C., Shubert, W. K., and Partin, J. S., 1971, Mitochondrial ultrastructure in Reye's syndrome (encephalopathy and fatty degeneration of the viscera), *New Engl. J. Med.* **285**:1339–1343.

Pastorino, J. G., Snyder, J. W., Serroni, A., Hoek, J. B., and Farber, J. L., 1993, Cyclosporin and carnitine prevent the anoxic death of cultured hepatocytes by inhibiting the mitochondrial permeability transition, *J. Biol. Chem.* **268**:13791–13798.

Petronilli, V., Cola, C., Massari, S., Colonna, R., and Bernardi, P., 1993, Physiological effectors modify voltage sensing by the cyclosporin A-sensitive permeability transition pore of mitochondria, *J. Biol. Chem.* **268**:21939–21945.

Petronilli, V., Costantini, P., Scorrano, L., Colonna, R., Passamonti, S., and Bernardi, P., 1994a, The voltage sensor of the mitochondrial permeability transition pore is tuned by the oxidation-reduction state of vicinal thiols, *J. Biol. Chem.* **269**:16638–16642.

Petronilli, V., Nicolli, A., Costantini, P., Colonna, R., and Bernardi, P., 1994b, Regulation of the permeability transition pore, a voltage-dependent mitochondrial channel inhibited by cyclosporin A, *Biochim. Biophys. Acta.* **1187**:255–259.

Pinsky, P. F., Hurwitz, E. S., Schonberger, L. B., and Gunn, W. J., 1988, Reye's syndrome and aspirin: Evidence for a dose–response effect, *JAMA* **260**:657–661.

Ponchaut, S., and Veitch, K., 1993, Valproate and mitochondria, *Biochem. Pharmacol.* **46**:199–204.

Qian, T., Nieminen, A.-L., Herman, B., and Lemasters, J. J., 1997, The role of pH_i, Na^+, and the mitochondrial permeability transition in reperfusion injury to rat hepatocytes: Protection by cyclosprin A and glycine, *Am. J. Physiol.* **273**:C1783–C1792.

Ray, S. D., Kamendulis, L. M., Gurule, M. W., Yorkin, R. D., and Corcoran, G. B., 1993, Ca^{2+} antagonists inhibit DNA fragmentation and toxic cell death induced by acetaminophen, *FASEB J.* **7**:453–463.

Reye, R. D. K., Morgan, G., and Baral, J., 1963, Encephalopathy and fatty degeneration of the viscera: A disease entity in childhood, *Lancet* **ii**. 749–752.

Rodgers, G. C., Jr., Weiner, L. B., and McMillan, J. A., 1982, Salicylate and Reye's syndrome, *Lancet*, **i**:616.

Schon, E. A., Hirano, M., and DiMurro's, 1994, Mitochondrial ecepha;p,yopathics: clinical and molecular analysis, *J. Bioenergetics Biomembranes*, **26**:291–9.

Schonfeld, P., and Bohnensack, R., 1997, Fatty acid-promoted mitochondrial permeability transition by membrane depolarization and binding to the ADP/ATP carrier, *FEBS Lett.* **420**:167–170.

Segalman, T. Y., and Lee, C. P., 1982, Reye's syndrome: Plasma-induced alterations in mitochondrial structure and function, *Arch. Biochem. Biophys.* **214**:522–530.

Shaw, G. R., and Anderson, W. R., 1991, Multisystem failure and hepatic microvesicular fatty metamorphosis associated with tolmetin ingestion, *Arch Pathol. Lab. Med.* **115**:818–823.

Sherratt, H. S. A., 1986, Hypoglycin, the famous toxin of the unripe Jamaican ackee fruit, *TIPS* **May**:186–191.

Shirley, M. A., Hu, P., and Baillie, T. A., 1993, Stereochemical studies on the β-oxidation of valproic acid in isolated rat hepatocytes, *Drug Metab. Dispos.* **21**:580–586.

Sinniah, D., and Baskaran, G., 1981, Margosa oil poisoning as a cause of Reye's syndrome, *Lancet* **i**:487–489.

Sinniah, D., Schwartz, P. H., Mitchell, R. A., and Arcinue, E. L., 1985, Investigation of an animal model of a Reye's-like syndrome caused by margosa oil, *Pediat. Res.* **19**:1346–1355.

Sippel, H., Stauffert, I., and Estler, C.-J., 1993, Protective effect of various calcium antagonists against an experimentally induced calcium overload in isolated hepatocytes, *Biochem. Pharmacol.* **46**:1937–1944.

Skellon, J. H., Thornburn, S., Spence, J., and Chatterjee, S. N., 1962, The fatty acids of neem oils and their reduction products, *J. Sci. Food Agric.* **13**:639–643.

Solem, L. E., Heller, L. J., and Wallace, K. B., 1996, Dose-dependent increase in sensitivity to calcium-induced mitochondrial dysfunction and cardiomyocyte cell injury by doxorubicin, *J. Mol. Cell. Cardiol.* **28**:1023–1032.

Starko, K. M., and Mullick, F. G., 1983, Hepatic and cerebral pathology findings in children with fatal salicylate intoxication: Further evidence for a causal relation between salicylate and Reye's syndrome, *Lancet* **i**:326–329.

Starko, K. M., Ray, C. G., Dominguez, L. B., Stromberg, W. L., and Woodall, D. F., 1980, Reye's syndrome and salicylate use, *Pediatrics* **66**:859–864.

Sugimoto, T., Araki, A., Nishida, N., Sakane, Y., Woo, M., Takeuchi, T., and Kobayashi, Y., 1987, Hepatoxicity in rat following administration of valproic acid: Effect of l-carnitine supplementation, *Epilepsia* **28**:373–377.

Sullivan-Bolynai, J. Z., and Corey, L., 1981, Epidemiology of Reye's syndrome, *Epidemiol. Rev.* **3**:1–26.

Surgeon General's advisory on the use of salicylates and Reye's syndrome, 1982, *MMWR* **31**:289–290.

Tanaka, K., Isselbacher, K. J., and Shih, V., 1972, Isovaleric and alpha-methylbutyric acidemias induced by hypoglycin A: Mechanism of Jamaican vomiting sickness, *Science* **175**:69–71.

Tomasova, H., Nevoral, J., Pachi, J., and Kincl, V., 1984, Aspirin esterase activity and Reye's syndrome, *Lancet* **ii**:43.

Tomoda, T, Takeda, K., Kurashige, T., Enzan, H., and Miyahara, M., 1994, Acetylsalicylate (ASA)-induced mitochondrial dysfunction and its potentiation by Ca^{2+}, *Liver* **14**:103–108.

Tonsgard, J. H., and Huttenlocher, P. R., 1982, Reye's syndrome, in *Recent Advances in Clinical Neurology* (W. B. Matthews and G. H. Glaser, Eds.), Churchill-Livingstone, Edinburgh, pp. 169–192.

Treem, W. R., Shoup, M. E., Hale, D. E., Bennett, M. J., Rinaldo, P., and Millington, D. S., 1996, Acute fatty liver of pregnancy, hemolysis, elevated liver enzymes, and low platelets syndrome, and long-chain 3-hydroxyacyl-coenzyme A dehydrogenase deficiency, *Am. J. Gastroenterol.* **91**:2293–2300.

Treon, S. P., and Broitman, S. A., 1992, Monoclonal antibody therapy in the treatment of Reye's syndrome, *Med. Hypo.* **39**:238–242.

Trost, L. C., and Lemasters, J. J., 1996, The mitochondrial permeability transition: A new pathophysiological mechanism for Reye's syndrome and toxic liver injury, *J. Pharmacol. Exp. Ther.* **278**:1000–1005.

Trost, L. C., and Lemasters, J. J., 1997, Role of the mitochondrial permeability transition in salicylate toxicity to cultured rat hepatocytes: Implications for the pathogenesis of Reye's syndrome, *Toxicol. Appl. Pharmacol.* **147**:431–441.

Turnbull, D. M., Bone, A. J., Bartlett, K., Koundakjian, P. P., and Sheratt, H. S. A., 1983, The effects of valproate on intermediary metabolism in isolated rat hepatocytes and intact rats, *Biochem. Pharmacol.* **32**:1887–1892.

Uchino, H., Elmer, E., Uchino, K., Lindvall, O., and Siesjo, B. K., 1995, Cyclosporin A dramatically ameliorates CA1 hippocampal damage following transient forebrain ischaemia in the rat, *Acta Physiol. Scand.* **155**:469–471.

Ulrich, R. G., Bacon, J. A., Branstetter, D. G., Cramer, C. T., Funk, G. M., Hunt, C. E., Petrella, D. K., and Sun, E. L., 1994, Induction of a hepatic toxic syndrome in the dutch-belted rabbit by a quinoxalinone anxiolytic, *Toxicology* **98**:187–198.

Valencia, A. M., Quevedo, F. W., and Quintos, L. S., 1997, Reye's syndrome: Not of historical interest only, *Pediat. Infect. Dis. J.* **16**:1011–1012.

Visentin, M., Salmona, M., and Tacconi, M. T., 1995, Reye's and Reye's-like syndromes, drug-related diseases? (Causative agents, etiology, pathogenesis, and therapeutic approaches), *Drug Met. Rev.* **27**:517–539.

Waldman, R. J., Hall, W. N., McGee, H., and Van Amburg, G., 1982, Aspirin as a risk factor in Reye's syndrome, *JAMA* **247**:3089–3094.

Willmore, L. J., Triggs, W. J., and Pellock, J. M., 1991, Valproate toxicity: Risk-screening strategies, *J. Child Neurol.* **6**:3–6.

Yamamoto, M., and Nakamura, Y., 1994, Inhibition of β-oxidation by 3-mercaptopropionic acid produces features of Reye's syndrome in perfused rat liver, *Gastroenterology* **107**:517–524.

Yoshida, Y., Singh, I., and Darby, C. P., 1992, Effect of salicylic acid and calcium on mitochondrial functions, *Acta Neurol. Scand.* **85**:191–196.

You, K.-S., 1983, Salicylate and mitochondrial injury in Reye's syndrome, *Science* **221**:163–165.

Zimmerman, H. J., and Ishak, K. G., 1982, Valproate-induced hepatic injury: Analysis of 23 fatal cases, *Hepatology* **2**:591–597.

Zoratti, M., and Szabò, I., 1995, The mitochondrial permeability transition, *Biochim. Biophys. Acta* **1241**:139–176.

Chapter 24

Purinergic Receptor-Mediated Cytotoxicity

J. Fred Nagelkerke and J. Paul Zoeteweij

1. INTRODUCTION

1.1. Purinergic Receptors

The P2-purinergic receptors (Charest *et al.,* 1985; Di Virgilio *et al.,* 1998; El-Moatassim *et al.,* 1992; Horstman *et al.,* 1986) are located on the plasma membrane of cells; they respond to relatively low concentrations of extracellular nucleotides (e.g., ATP and ADP). These receptors are present in various types of cells. The receptor family is divided into P2x and P2y receptors, and a further subdivision is made based upon differences in potencies of ATP and ATP analogs to elicit a response. At present seven P2x receptors have been cloned and many splice variants identified (Di Virgilio *et al.,* 1998). The P2x receptor represents a ligand-gated ion channel allowing ion fluxes upon activation. The P2x receptors have been identified in smooth muscle cells and neuronal cells. Stimulation of P2x receptors causes cell excitation (e.g. muscle contraction). The P2y receptor induces a G-protein-coupled receptor response, such as activation of G-protein-operated ion channels and G-protein-coupled stimulation or inhibition of different second-messenger systems. These P2y receptors are found on smooth muscle cells, endothelial cells, hepatocytes, pancreatic β cells, parotid acini, and type II alveoli. In excitable tissues, P2y receptors are inhibitory, inducing, for example, muscle relaxation.

The receptor that is important for cytolysis and is linked to mitochondrial functioning was formerly called the *P2z receptor* due to its unusual pharmacological properties. In 1996 this receptor was cloned from rat smooth muscle, and it was found that the first 395

J. Fred Nagelkerke and J. Paul Zoeteweij   Department of Toxicology, Leiden-Amsterdam Center for Drug . Research, Sylvius Laboratories, 2300 RA Leiden, The Netherlands.

Mitochondria in Pathogenesis, edited by Lemasters and Nieminen.
Kluwer Academic/Plenum Publishers, New York, 2001.

amino acids were 35–40% identical to those of the P2x receptor; therefore, it was called P2x$_7$ (Surprenant *et al.*, 1996; Valera *et al.*, 1994) Later, cDNA was isolated from a human monocyte library that encodes for the P2x$_7$ receptor (Rassendren *et al.*, 1997). Binding of ATP to the P2x$_7$ receptor causes pore formation in the plasma membrane, allowing permeation of normally impermeant solutes (Murgia *et al.*, 1993).

1.2. Functional Role of P2 Purinergic Receptors *In Vivo*

The function of P2 porginergic receptors *in vivo* is still largely obscure. Physiological stimuli require ATP concentrations in the micromolar range (El-Moatassim *et al.*, 1992). Although this seems very high, active cellular ATP secretion may temporarily raise the peripheral ATP concentration to this range. Increasing evidence is found for a physiological role in neurotransmission and vascular control (El-Moatassim *et al.*, 1992; Murgia *et al.*, 1992).

In vitro, it was found that ATP or analog binding to the receptor would induce cytolysis in many cell types. The cytolytic properties of ATP are associated with formation of large membrane pores through activation of P2x$_7$ receptors (Di Virgilio 1998; Murgia *et al.*, 1993). Occasionally, *in vivo* ATP may be present locally at high concentrations, in, for example, the immediate surroundings of injured ATP-rich cells or after cellular secretion in a restricted area of close cell-to-cell contact (El-Moatassim *et al.*, 1992; Gordon, 1986; Murgia *et al.*, 1992), but whether this occurs *in vivo* is still an open question. Additional functions have been suggested: a role in fast synaptic transmission (Surprenant *et al.*, 1996) and in cell–cell communication (reviewed in Di Virgilio, 1995; Di Virgilio *et al.*, 1998).

2. MECHANISMS OF ATP-INDUCED CELL DEATH

2.1. Intracellular Calcium

When isolated liver cells are exposed to ATP, intracellular calcium concentration rises immediately, probably due to an ion channel opening that allows influx of extracellular calcium (Nagelkerke *et al.*, 1989; Orrenius *et al.*, 1989) Accumulation of calcium is often observed in necrotic tissues (Judah *et al.*, 1970; Kurita *et al.*, 1993; Schanne *et al.*, 1979; Orrenius *et al.*, 1989). Over the last 20 years, numerous *in vitro* studies in perfused organs, cell suspensions, and cell cultures have yielded evidence that supports a causative relationship between perturbation of Ca^{2+} homeostasis and toxic cell injury (Schanne *et al.*, 1979; Nicotera *et al.*, 1992). For example, high [Ca^{2+}]$_i$ is involved in pro-oxident-induced cytotoxicity in mitochondria (Richter and Schlegel, 1993), hepatocytes (Broekemeier *et al.*, 1992; Thor *et al.*, 1984), cultured myocytes (Persoon-Rothert *et al.*, 1992), alveolar macrophages (Forman *et al.*, 1987), myocytes (Solem and Wallace, 1993; Solem *et al.*, 1994), and renal proximal tubular cells (Jiang *et al.*, 1993; van de Water *et al.*, 1993, 1994). Also, *in vivo* mitochondrial damage was found after oxidative stress, which can be prevented by preventing influx of calcium (Saxena *et al.*, 1995), but in some cases, for example exposure of hepatocytes to CCl_4 (Albano *et al.*, 1989; Long and Moore 1986), the reports are contradictory, resulting in postulation of both Ca^{2+}-dependent and Ca^{2+}-independent injury mechanisms.

High $[Ca^{2+}]_i$ is suggested to mediate anoxic/ischaemic injury in hepatocytes (Gabarinni *et al.*, 1992), certain transformed cell lines (Johnson *et al.*, 1981; Nicotera *et al.*, 1989), perfused heart (Lec *et al.*, 1988; Steenbergen *et al.*, 1990), myocytes (Siegmund *et al.*, 1990), kidney (Phelps *et al.*, 1989), and neuronal cells (Mattson *et al.*, 1993; Tsubokawa *et al.*, 1992). A Ca^{2+} overload is involved in ischaemia–reperfusion injury in heart (Tani, 1990), but in liver its role in this type of injury is less clear (Okuda *et al.*, 1992).

Increased $[Ca^{2+}]_i$ levels are associated with the action of well-known toxic agents such as TCDD (cardiotoxicity; Canga *et al.*, 1988) and acetaminophen (hepatotoxicity; Boobis *et al.*, 1990). Increased $[Ca^{2+}]_i$ is also involved in heavy-metal-induced cell death in neuronal cells (Komulainen and Bondy, 1988) and Hg-mediated cell death in cultured proximal tubular cells (Smith *et al.*, 1991), although Hg-mediated cytotoxicity in cultured hepatocytes appears to proceed Ca^{2+}-independently (Nieminen *et al.*, 1990a). Finally, Ca^{2+} dependence is reported for cell killing by the immune system, such as target-cell killing by cytotoxic T lymphocytes (Berke, 1989) and complement-mediated cell killing (Newsholme *et al.*, 1993).

Many studies on Ca^{2+}-dependent lytic processes were done in isolated rat hepatocytes. Cell death in hepatocytes caused by a large variety of chemicals is dependent on extracellular Ca^{2+} (Schanne *et al.*, 1979). The role of Ca^{2+} has been questioned because exposure in Ca^{2+}-free medium does not always protect against lethal injury (Fariss and Reed, 1985; Smith *et al.*, 1981). These observations are inconclusive, however, because omission of extracellular Ca^{2+} itself causes a decreased cellular $[Ca^{2+}]$, which induces oxidative stress and subsequent enhancement of cellular injury (Thomas and Reed, 1988) and potentiates toxicity (Snyder *et al.*, 1995).

Induction of a prolonged increase in $[Ca^{2+}]_i$ by itself is ultimately cytotoxic (Nagelkerke *et al.*, 1989; Richelmi *et al.*, 1989; Schanne *et al.*, 1979). Cytotoxicity is observed after increasing the $[Ca^{2+}]_i$ in several ways: inhibition of the Ca^{2+} extrusion pumps with vanadate (Richelmi *et al.*, 1989), exposure to Ca^{2+} ionophores (Jiang *et al.*, 1993; Starke *et al.*, 1986; Zoeteweij *et al.*, 1992), or opening of Ca^{2+} channels in the plasma membrane by extracellular ATP (Thor *et al.*, 1984; Nagelkerke *et al.*, 1989; Zoeteweij *et al.*, 1992). The latter mechanism may apply directly to *in vivo* situations; under certain conditions local ATP_0 concentrations may be sufficiently high to induce cytolytic effects (El-Moatassim *et al.*, 1992; Murgia *et al.*, 1992).

The following observations support the $P2x_7$-type receptor's involvement in the cytolytic effects after hepatocyte exposure to ATP. Lytic Ca^{2+} overload is prevented by the irreversible $P2x_7$ receptor-antagonist-oxidized ATP or by high levels of extracellular $[Mg^{2+}]$ (Zoetewij *et al.*, 1996). In contrast to a sustained rise in $[Ca^{2+}]_i$, (short-term) transient increases in $[Ca^{2+}]_i$ are not associated with vital cell injury (Jiang *et al.*, 1993; Zoetewij *et al.*, 1996).

Processes in individual living cells were studied using video-intensified fluorescence microscopy (VIFM). These studies confirmed that ATP induced sustained high Ca^{2+}-levels in hepatocytes. Prevention of the rise in $[Ca^{2+}]_i$ by using low-Ca^{2+} buffers protected the cells against lethal injury (Zoeteweij *et al.*, 1992).

2.2. Determination of Mitochondrial Ca^{2+}

In the past, mitochondrial Ca^{2+} could be studied experimentally only by determination of total mitochondrial Ca^{2+} content or by measuring the amount of Ca^{2+} released by mitochondria after treatment with uncouplers. Electron probe microanalysis of rapidly frozen tissue revealed that mitochondrial Ca^{2+} content under physiological conditions was low, which questioned the role of mitochondria in cytosolic Ca^{2+} regulation under such conditions (Somlyo et al., 1985). In addition, isolated mitochondria start to buffer medium Ca^{2+} efficiently at extramitochondrial Ca^{2+} concentrations above 1 μM (Becker, 1980). Several effectors of mitochondrial Ca^{2+} transport present in intact cells, however, such as Mg^{2+}, ADP, and spermine, may alter the Ca^{2+} set point for mitochondrial Ca^{2+} sequestration (Gunter and Pfeiffer, 1990; Kraus-Friedman, 1990), so data on changes in *free* mitochondrial [Ca^{2+}] ([Ca^{2+}]$_{mito}$) in living cells were lacking. Nowadays, [Ca^{2+}]$_{mito}$ in intact cells can be estimated using the fluorescent probe Fura-2. Loading cells with Fura-2-AM results in accumulation of "active" free acid Fura-2 in the cytosolic as well as the mitochondrial compartment (Grant and Acosta, 1994; Jiang and Acosta, 1995). After rapid permeabilization of the plasma membrane, cytosolic Fura-2 is lost, whereas the remaining mitochondrial Fura-2 is unaffected; thus the [Ca^{2+}]$_{mito}$ can be determined. This technique revealed a physiological [Ca^{2+}]$_{mito}$ close to resting cytosolic concentrations in cultured myocytes (Chacon et al., 1992) or in a slightly higher submicromolar range in isolated hepatocytes (Zoeteweij et al., 1993). Recently, "cold loading" the calcium-sensitive mitochondrial-specific dye Rhod-2-AM, a new technique for determination of [Ca^{2+}]$_{mito}$, was introduced (Trollinger et al., 1997). An alternative approach is to target calcium-sensitive aquorin to mitochondria (Rizzuto et al., 1995).

Mitochondrial Ca^{2+} accumulation com occur under conditions of high [Ca^{2+}]$_i$. After administration of a combination of Ca^{2+}-mobilizing hormones, a massive Ca^{2+} uptake by liver is observed, which is related to mitochondrial Ca^{2+} sequestration (Bygrave et al., 1990). Net mitochondrial accumulation of Ca^{2+} in tissues is also seen after induction of pathological increases in [Ca^{2+}]$_i$ (Farber 1982). Isolated hepatocytes subjected to sustained high [Ca^{2+}]$_i$ by vanadate (Richelmi et al., 1989) or extracellular ATP (Thor et al., 1984) showed extensive Ca^{2+} accumulation. Data on mitochondrial free [Ca^{2+}] in this context were obtained using the VIFM methodology. Extracellular ATP induction of a prolonged rise in [Ca^{2+}]$_i$ in hepatocytes was followed by a rise in [Ca^{2+}]$_{mito}$ (Zoeteweij et al., 1993).

2.3. Isolated Mitochondrial Ca^{2+}-Induced Changes

For a description of Ca^{2+}-induced mitochondrial transition and of the mitochondrial transition pore, the reader is referred to other chapters in this volume. In brief, calcium-induced mito-chondrial dysfunction is associated with a loss of inner membrane impermeability to low MW solutes, a process called *mitochondrial permeability transition* (MPT) (Gunter and Pfeiffer, 1990). As a consequence of the MPT, mitochondrial Ca^{2+} release occurs accompanied by a collapse of mitochondrial membrane potential (Δψ) and rapid, large-amplitude swelling (Beatrice et al., 1980). Swelling occurs because transition allows ions to equilibrate across inner membrane while endogenous protein remains trapped. The mitochondrion then takes up water as a consequence of the osmotic imbalance. Although the exact moment of transition initiation in an individual mitochondrion depends on its stability, permeabilization and subsequent swelling occur within seconds (Crompton and Costi, 1988; Gunter and Pfeiffer, 1990; Lemasters *et al.*, 1998). In this process the initial pore may be relatively small, allowing only ionic fluxes. Subsequent mitochondrial swelling increases the permeability and resultant leakiness to small solutes (MW < 1500 Da).

The role of $\Delta\psi$ in transition has now been studied in detail. It was known that membrane depolarization favors pore opening (Bernardi, 1992); it was proposed that many transition inducers caused pore opening by shifting to higher levels the gating potential, that is, the level of $\Delta\psi$, at which pore opening is induced (Petronelli *et al.*, 1993). Transition inhibitors may also act through shifting the gating potential to lower levels.

Excessive Ca^{2+} cycling is associated with mitochondrial damage due to irreversible oxidation of pyridine nucleotides, $\Delta\psi$ decrease, and subsequent ATP depletion (Richter and Schlegel, 1993). After these events have taken place, transition is initiated.

2.4. Role of Ca^{2+}-Induced Mitochondrial Damage in Cytotoxicity

During the last decade, evidence has accomulated that mitochondrial damage plays an important role in calcium-related cell injury (Masaki *et al.*, 1989; Nagelkerke *et al.*, 1989; Nieminen *et al.*, 1990b; Thor *et al.*, 1984). In our laboratory, cell death in hepatocytes resulting from sustained high $[Ca^{2+}]_i$ levels induced by extracellular ATP was associated with mitochondrial injury: the NADH concentration was greatly decreased (Nagelkerke *et al.*, 1989). Measurements in individual hepatocytes showed that high $[Ca^{2+}]_i$ resulted in $\Delta\psi$ decrease followed by total $\Delta\psi$ collapse before cell death (Fig. 1) (Zoeteweij *et al.*, 1992), an observation is supported by others who found that exposure to Ca^{2+} ionophores caused Ca^{2+}-dependent $\Delta\psi$ dissipation preceding cell death (Nicotera *et al.*, 1990; Jiang *et al.*, 1993). The adverse effects of high $[Ca^{2+}]_i$ were associated with mitochondrial Ca^{2+} accumulation (Fig. 2) by Zoeteweij *et al.*, (1993), and this group further showed that phosphate, an enhancer of mitochondrial Ca^{2+} uptake in isolated mitochondria, may control mitochondrial Ca^{2+} transport in cells. This may have interesting implications in cell toxicity, because depletion of intracellular ATP, a common event in toxic cell injury, increases cellular phosphate concentrations (Park *et al.*, 1992).

Cellular K^+ was also found to affect Ca^{2+}-dependent cytotoxicity (Zoeteweij *et al.*, 1994): Low intracellular K^+ levels decreased loss of $\Delta\psi$ and viability, consistent with the finding that isolated mitochondria depleted of K^+ were resistant to Ca^{2+}-induced damage (Chavez *et al.*, 1991). The recent introduction of a method to estimate mitochondrial free $[K^+]$ suggests that low K^+ levels also decrease Ca^{2+}-induced cytotoxicity in intact cells by preventing a rise in $[K^+]_{mito}$ (Fig. 3) (Table I) (Zoeteweij *et al.*, 1994). Although the absolute $[K^+]_{mito}$ values in this study may be somewhat lower than the real value, the results show that altered mitochondrial K^+ handling plays a role in Ca^{2+}-induced mitochondrial dysfunction, although the mechanism of the Ca^{2+}–K^+ interaction that leads to mitochondrial damage remains to be established. Based on studies in isolated mitochondria, that K^+-depleted mitochondria have been proposed to have a higher binding capacity for Ca^{2+}, leading to lower $[Ca^{2+}]_{mito}$ (Chavez *et al.*, 1991). In addition, mitochondrial K^+ activates mitochondrial respiration (Halestrap, 1989; Zoeteweij, *et al.*, unpublished), probably by inducing swelling. At lower mitochondrial K^+ levels mitochondrial Ca^{2+}-accumulation might be less, resulting in prevention of the damage from too-

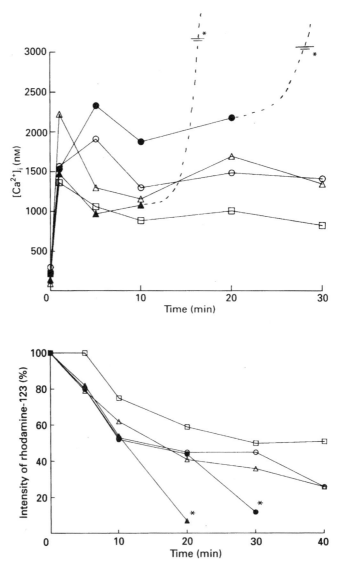

FIGURE 1. Effect of ATP on $[Ca^{2+}]_i$ and mitochondrial membrane potential ($\Delta\psi$) in individual hepatocytes, determined with video microscopy. Isolated cells were attached to polylysine glass coverslips and loaded with Fura-2-AM to measure $[Ca^{2+}]_i$ and with Rhodamine 123 to measure $\Delta\psi$ using video microscopy. During the determinations, propidium iodide was present in the incubation medium. ATP was added (final concentration 400 μM) and at various time points images were taken and fluorescence intensities measured. Five cells are shown individually. Open symbols, represent cells that survived, closed symbols represent those that died, indicated by uptake of propidium iodide. Upper panel shows that adding ATP immediately induced a strong increase in intracellular calcium. Lower panel shows that this corresponds with a sharp decrease in $\Delta\psi$. Cells with the lowest potential died between 20–30 min. (Reproduced with permission from J. P. Zoeteweij *et al., Biochemical Journal* **288**: 207–213, 1992, © the Biochemical Society.)

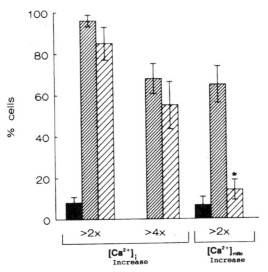

FIGURE 2. Effect on $[Ca^{2+}]_i$ and $[Ca^{2+}]_{mito}$ after exposure to ATP. After 10 min exposure to ATP, cells were attached to coverslips and loaded with Fura-2-AM. After 70 min exposure, images were taken and fluorescence intensities measured to determine $[Ca^{2+}]_i$. After permeabilization with digitonin, images were taken for determination of $[Ca^{2+}]_{mito}$. Per experiment, 7–20 still-viable cells were examined. Percentage of cells showing more than 2- or 4-fold increases in $[Ca^{2+}]_i$ compared with the averaged level of the control, and percentage showing a more than a 2-fold increase in $[Ca^{2+}]_{mito}$, are shown. Control (■); 0.4 mM ATP in the presence (narrow hash marks) or absence (wide hash marks) of extracellular P_i. In the control we found no cells with more than a 4-fold increase in $[Ca^{2+}]_i$. Omission of extracellular P_i alone without exposure to ATP had no effect. (Reproduced with permission from J. P. Zoeteweij *et al.*, 1993, *J. Biol. Chem.* **268**, 3384–3388.)

high levels of $[Ca^{2+}]_{mito}$. An inverse relationship between Ca^{2+} and K^+ may also play a role: increased mitochondrial Ca^{2+} stimulates influx of K^+, possibly through Ca^{2+}-activated K^+ channels (Halestrap, 1989; Zoeteweij *et al.*, 1992, 1994). Subsequently, uncontrolled $[K^+]_{mito}$ increase may lead to $\Delta\psi$ loss. The experiments with varying $[K^+]$ indicate that a permeability change of the mitochondrial inner membrane, as observed in isolated mitochondria, may participate in Ca^{2+}-induced damage. In isolated mitochondria the MPT reflects opening of a membranous pore. Whether such a pore exists in intact cells is uncertain. Injection of radiolabeled sugars into rats (Tolleshaug *et al.*, 1984) or incubation of isolated hepatocytes with radio-labeled sucrose resulted in sequestration of the labeled compounds into mitochondria (Tolleshaug and Seglen, 1985). Increased mitochondrial Ca^{2+} has been found in isolated hepatocytes treated with the pore inhibitor cyclosporin (Kass *et al.*, 1992). Cyclosporin also increased Ca^{2+} retention in apparently functional heart cells (Altschuld *et al.*, 1992), which may reflect mitochondrial Ca^{2+} accumulation after blockade of the mitochondrial pore, but a transient increase in cytosolic $[Ca^{2+}]$ induced by cyclosporin has not been excluded as a contributor to mitochondrial Ca^{2+} retention. Furthermore, the pore inhibitor MIBG increases mitochondrial Ca^{2+} in hepatocytes, probably by inhibiting mitochondrial Ca^{2+} efflux pathways dependent on ADP ribosylation (Juedes *et al.*, 1992). Because cyclosporin also inhibits ADP ribosylation, this process is likely involved in mitochondrial Ca^{2+} release in intact cells. Thus, the role of a large pore under physiological conditions remains controversial.

Table I
Effect of Extracellular K$^+$ Omission on ATP-Induced Cell
Death and Dissipation of the MMP

Treatment (2 h)	Viability (%)	MMP (% of control at $t = 0$)
+ K$^+$, control	89 ± 1	90 ± 4
+ K$^+$, ATP	34 ± 4*	64 ± 6*
No K$^+$, control	89 ± 1	99 ± 6
No K$^+$, ATP	61 ± 5*‡	87 ± 8

*$p < 0.05$ vs control without ATP
‡ $p < 0.05$ vs corresponding value in the presence of K$^+$
[a]Isolated hepatocytes inculoated with Hanks'-Hepes, with or without K$^+$, and exposed for 2 hr to 0.4 mM ATP. Samples (0.5 mL) loaded with 1.5 µM Rhodamine 123 and 2 µM propidium iodine were analyzed for fluorescence on the flow cytometer. Viability is expressed as percentage of cells that excluded propidium iodide. The MMP of viable cells (calculated by averaging their Rhodamine fluorescence) is expressed as percentage of control at $t = 0$. Data are shown as mean values ± SEM. of three separate solutions. (Reproduced with permission from J. P. Zoeteweij *et al.*, *Biochem. J.* **299**: 539–543 (1994), © the Biochemical Society.)

Cyclosporin A protects against formation of the MPT pore. Evidence for effects of cyclosporin in isolated cell suspensions is poor, however. In isolated hepatocytes, cyclosporin partially protected against *t*-BHP-induced oxidative stress, but only when toxicity was induced in the presence of nonphysiological (10 mM) extracellular [Ca^{2+}] (Brockemeier *et al.*, 1992). Furthermore, cyclosporin protected against cell death by iodoacetate only in combination with nonspecific phospholipase A$_2$ inhibitors (Imberti *et al.*, 1992). In addition, no protection was found against chemical hypoxia, that is, iodoacetate + KCN, or other oxidant chemicals such as cystamine and menadione. Also, in our laboratory cyclosporin alone or in combination with phospholipase A$_2$ inhibitors did not protect against Ca^{2+}-induced cell injury in isolated hepatocytes by extracellular ATP (Zoeteweij *et al.*, unpublished). but, as shown in isolated mitochondria, regulation of the MPT is complex, and many factors may interfere with the effect of cyclosporin in intact cells. Conditions such as extremely high [Ca^{2+}]$_i$ might overrule the inhibiting action of cyclosporin. In isolated hepatocytes in which [Ca^{2+}]$_i$ is raised by extracellular ATP, several factors will favour transition induction: high [Ca^{2+}]$_{mito}$, high cellular P$_i$ levels, decreased NADH, decreased Δψ, and alkalosis (Nagelkerke *et al.*, 1989; Zoeteweij *et al.*, 1992, 1993). Experiments with K$^+$ suggest an opening of at least the small ion-channels (Zoeteweij *et al.*, 1992, 1994), which may reflect an intermediate state of pore opening, as proposed in studies with isolated mitochondria. Other conditions, however, such as high ADP-levels, favor pore closure, so a transition-causing permeability toward solutes with a molecular weight of >1500 Da might not occur. Probably mitochondria continue to accumulate Ca^{2+} to extreme levels, resulting in their irreversible destruction by formation of Ca^{2+} deposits (Fig. 4). Ultrastructural examination of hepatocytes killed after sustained, ATP-induced high [Ca^{2+}]$_i$ showed mitochondrial Ca^{2+} deposits, which were related to collapse of Δψ and subsequent loss of mitochondrial and cellular integrity. Further research is needed to determine the exact mechanism of mitochondrial damage.

How does mitochondrial damage cause cell death? It has long been thought that after mitochondrial failure, depletion of intracellular ATP compromised cell viability because of

(a)

(b)

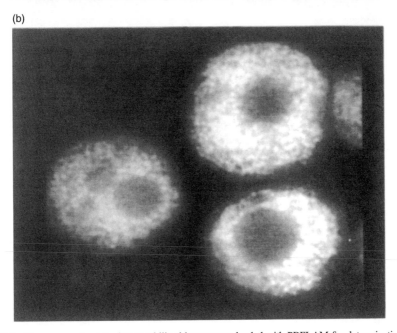

FIGURE 3. Fluorescence images of permeabilized hepatocytes loaded with PBFI-AM for determination of K^+ and Rhodamine 123 simultaneously before permeabilization. Hepatocytes attached to glass coverslips were incubated with 20 µM PBFI-Am for 60 min and coloaded with 350 nM Rhodamine 123 for 15 min. To remove cytosolic components, the plasma membranes were permeabilized by adding digitonin in sucrose buffer before images were taken. (A) Digital fluorescence image of Rhodamine 123 (excitation 450–490 nm), a $\Delta\psi$ indicator. (B) Digital fluorescence image of PBFI (excitation 340 nm) in the same cells. [Reproduced with permission from J. P. Zoeteweij *et al.,* 1994, *Biochem. J.* **299**: 539–543, © the Biochemical Society].

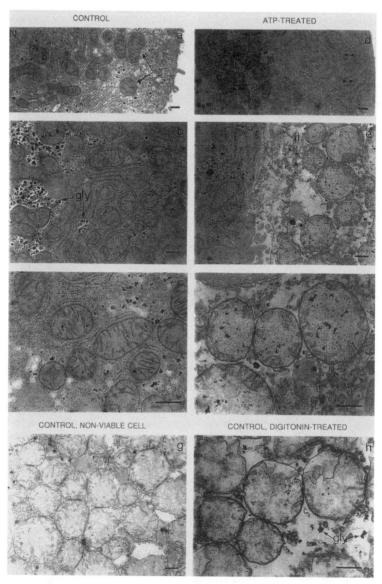

FIGURE 4. Ultrastructural localization of cellular Ca^{2+} deposits after exposure to ATP. Cells incubated for 150 min with 400 μM ATP were fixed and prepared for examination with electron microscopy. Ultrastructural localization of Ca^{2+} deposits was performed using an oxalate–pyroantimonate (OPA) procedure developed for visualization of Ca^{2+} storage sites with transmission electron microscopy: Ultrathin sections were contrasted with lead hydroxide. Bar = 0.5 μM. (A–C) Control cells. (A) Nucleus (N) and mitochondria (arrows) are indicated, × 10000, (B) Glycogen (gly) is indicated, × 13000, (C) × 24000. (D, F) Cells exposed to 0.4 mM ATP. (D) A cell that survived ATP-treatment. Note the mitochondria-free zone in which the endoplasmic reticulum (double arrowheads) is still present, × 10000. (E) Two adjacent cells, a viable cell (i) and a necrotic cell (ii). Note the numerous Ca^{2+} deposits in the swollen mitochondria of the necrotic cell, × 13000. (F) Mitochondria of a necrotic cell containing many deposits, × 24000. (G) Control cell that lost viability. No deposits in mitochondria, × 13000. (H) Cell lysed with digitonin. Incidently, mitochondria are present with a few deposits, × 24000.

the cell's inability to maintain vital energy-dependent functions (Nieminen *et al.*, 1990a; Stromski *et al.*, 1986) such as intracellular GSH levels (Mithofer *et al.*, 1992). It has been excluded that ATP-dependent GSH depletion affected viability after ATP depletion (Redegeld *et al.*, 1992). Fructose, which stimulates ATP synthesis via the glycolytic pathway (Snyder *et al.*, 1993), protects against cell injury after a blockade of mitochondrial respiration (Nieminen *et al.*, 1990a), but the fructose concentrations used are often so high that ATP depletion rather than an ATP suppletion result (Cannon *et al.*, 1991). Still, fructose may act through maintenance of a continuous low level of ATP, in contrast to a total decline in its absence. In case of ATP depletion, acidosis protected against cell injury (Gores *et al.*, 1988), indicating that ATP depletion is not the final "point of no return". Acidification is possibly a rather nonspecific protective condition because it also inhibits other damaging processes, such as activation of proteolytic enzymes or activation of the MPT in isolated mitochondria (Halestrap, 1991). Other studies show a very poor correlation between ATP depletion and cell injury: ATP-depleted cells maintain viability while depleted of ATP (Pastorino *et al.*, 1993; Snyder *et al.*, 1993). In addition, glycine, a protective agent against several types of injury, does not prevent ATP depletion (Weinberg *et al.*, 1989), apparently.

Other reports favor intracellular Ca^{2+} over intracellular ATP for a predominant role (Kristensen, 1991), and in our laboratory, Ca^{2+}-induced mitochondrial dysfunction and cell injury occurred in the presence of high $[ATP]_i$ levels (Nagelkerke *et al.*, 1989, Zoeteweij *et al.*, 1993), although hypothetically an extremely rapid ATP loss just before cell death cannot be ruled out. In general, development of cell injury correlates better with $\Delta\psi$ collapse (Nagelkerke *et al.*, 1989; Snyder *et al.*, 1993). It has been suggested that $\Delta\psi$ collapse results in cell injury only when it is followed by transition (Pastorino *et al.*, 1993).

3. CONCLUSION

Mitochondria are vitally important for maintaining cellular integrity. Calcium is considered a mediator of necrosis as well as a main cause of damage in isolated mitochondria. Our results show that mitochondrial Ca^{2+} regulation in hepatocytes is affected when $[Ca^{2+}]_i$ is increased; this is followed by mitochondrial dysfunction and irreversible cell injury. Mitochondrial free $[Ca^{2+}]$ is indicated as a key parameter in this process, and phosphate is indicated as a possible important regulator of $[Ca^{2+}]_{mito}$. In addition, we suggest a role for mitochondrial Ca^{2+} deposits, which present only in cells killed by high $[Ca^{2+}]_i$.

The mechanisms underlying Ca^{2+}-induced mitochondrial damage are rapidly elucidated; it remains to be seen whether, in intact cells, opening of a large membranous pore, opening of small ion channels, or nonspecific mitochondrial membrane destruction is involved. Our results suggest at least a small permeability change of the mitochondrial inner membrane, allowing, for example, K^+ fluxes.

Future issues include the question of what happens after mitochondrial failure. Several processes, such as extreme mitochondrial swelling, mitochondrial NAD(P)H depletion, and release of mitochondrial components must be considered as critical consequences of mitochondrial dysfunction.

REFERENCES

Albano, E., Carini, R., Parola, M., Bellomo, G., Goria-Gatti, L., Poli, G., and Dianzani, M. U., 1989, Effects of carbon tetrachloride on calcium homeostasis: A critical reconsideration, *Biochem. Pharmacol.* **38**:2719–2725.

Altschuld, R. A., Hohl, C. M., Castillo, L. C., Garleb, A. A., Starling, R. C. and Brierley, G. P., 1992, Cyclosporin inhibits mitochondrial calcium efflux in isolated adult rat ventricular cardiomyocytes, *Am. J. Physiol.* **262**: H1699–H1704.

Beatrice, M. C., Palmer, J. W., and Pfeiffer, D. R., 1980, The relationship between mitochondrial membrane permeability, membrane potential, and the retention of Ca^{2+} by mitochondria, *J. Biol. Chem.* **255**:8663–8671.

Becker, G. L., 1980, Steady-state regulation of extramitochondrial Ca^{2+} by rat liver mitochondria: Effects of Mg^{2+} and ATP, *Biochim. Biophys. Acta* **591**:234–239.

Berke, G., 1989, The cytolytic T lymphocyte and its mode of action, *Immunol. Lett.* **20**:169–178.

Bernardi, P., 1992, Modulation of the mitochondrial cyclosporin A-sensitive permeability transition pore by the proton electrochemical gradient: Evidence that the pore can be opened by membrane depolarization, *J. Biol. Chem.* **267**:8834–8839.

Boobis, A. R., Seddon, C. E., Nasseri-Sina, P., and Davies, D. S., 1990, Evidence for a direct role of intracellular calcium in paracetamol toxicity, *Biochem. Pharmacol.* **39**:1277–1281.

Broekemeier, K. M., Carpenter-Deyo, L., Reed, D. J., and Pfeiffer, D. R., 1992, Cyclosporin A protects hepatocytes subjected to high Ca^{2+} and oxidative stress, *FEBS Lett.* **304**:192–194.

Bygrave, F. L., Lenton, L., Altin, J. G., Setchell, B. A., and Karjalainen, A., 1990, Phosphate and calcium uptake by mitochondria and by perfused rat liver induced by the synergistic action of glucagon and vasopressin, *Biochem. J.* **267**:69–73.

Campbell, P. I., and al Nasser, I. A., 1995, Dexamethasone inhibits inorganic phosphate stimulated Ca^{2+}-dependent damage of isolated rat liver and renal cortex mitochondria, *Comp. Biochem. Physiol. C. Pharmacol. Toxicol. Endocrinol.* **111**:221–225.

Canga, L., Levi, R., and Rifkind, A. B., 1988, Heart as a target organ in 2,3,7,8-tetrachlorodibenzo-*P*-dioxin toxicity: Decreased beta-adrenergic responsiveness and evidence of increased intracellular calcium, *Proc. Natl. Acad. Sci. USA* **85**:905–909.

Cannon, J. R., Harvison, P. J., and Rush, G. F., 1991, The effects of fructose on adenosine triphosphate depletion following mitochondrial dysfunction and lethal cell injury in isolated rat hepatocytes, *Toxicol. Appl. Pharmacol.* **108**:407–416.

Chacon, E., Ulrich, R., and Acosta, D., 1992, A digitized-fluorescence-imaging study of mitochondrial Ca^{2+} increase by doxorubicin in cardiac myocytes, *Biochem. J.* **281**:871–878.

Charest, R., Blackmore, P. F., and Exton, J. H., 1985, Characterization of responses of isolated rat hepatocytes to ATP and ADP, *J. Biol. Chem.* **260**:15789–15794.

Chavez, E., Moreno-Sanchez, R., Zazueta, C., Reyes-Vivas, H., and Arteaga, D., 1991, Intramitochondrial K^+ as activator of carboxyatractyloside-induced Ca^{2+} release, *Biochim. Biophys. Acta* **1070**:461–466.

Crompton, M., and Costi, A., 1988, Kinetic evidence for a heart mitochondrial pore activated by Ca^{2+}, inorganic phosphate, and oxidative stress: A potential mechanism for mitochondrial dysfunction during cellular Ca^{2+} overload, *Eur. J. Biochem.* **178**:489–501.

Di Virgilio, F., 1995, The P2Z purinoceptor: An intriguing role in immunity, inflammation, and cell death, *Immunol. Today* **16**:524–528.

Di Virgilio, F., Chiossszzi P., Falzoni S., Ferrari D., Sanz, J. M., Venketaraman, V., and Baricordi, O. R., 1998, Cytolytic P2x purinoceptors, *Cell Death Diffr.* **5**:191–199.

El Moatassim, C., Dornand, J., and Mani, J. C., 1992, Extracellular ATP and cell signalling, *Biochim. Biophys. Acta* **1134**:31–45.

Farber, J. L., 1982, Biology of disease: Membrane injury and calcium homeostasis in the pathogenesis of coagulative necrosis, *Lab. Invest.* **47**:114–123.

Fariss, M. W., and Reed, D. J., 1985, Mechanism of chemical-induced toxicity: II. Role of extracellular calcium, *Toxicol. Appl. Pharmacol.* **79**:296–306.

Forman, H. J., Dorio, R. J., and Skelton, D. C., 1987, Hydroperoxide-induced damage to alveolar macrophage function and membrane integrity: Alterations in intracellular free Ca^{2+} and membrane potential, *Arch. Biochem. Biophys.* **259**:457–465.

Gasbarrini, A., Borle, A. B., Farghali, H., Bender, C., Francavilla, A., and Van Thiel, D., 1992, Effect of anoxia on intracellular ATP, Na_i^+, Ca_i^{2+} Mg_i^{2+}, and cytotoxicity in rat hepatocytes [published erratum appears in 1992, *J. Biol. Chem.* **267**: 13114], *J. Biol. Chem.* **267**:6654–6663.

Gores, G. J., Nieminen, A.-L., Fleishman, K. E., Dawson, T. L., Herman, B., and Lemasters, J. J., 1988, Extracellular acidosis delays onset of cell death in ATP-depleted hepatocytes, *Am. J. Physiol.* **255**:C315–C322.

Grant, R. L., and Acosta, D., Jr, 1994, A digitized fluorescence imaging study on the effects of local anesthetics on cytosolic calcium and mitochondrial membrane potential in cultured rabbit corneal epithelial cells, *Toxicol. Appl. Pharmacol.* **129**:23–35.

Griffiths, E. J., and Halestrap, A. P., 1991, Further evidence that cyclosporin A protects mitochondria from calcium overload by inhibiting a matrix peptidyl-prolyl *cis-trans* isomerase: Implications for the immuno-suppressive and toxic effects of cyclosporin, *Biochem. J.* **274**:611–614.

Gunter, T. E., and Pfeiffer, D. R., 1990, Mechanisms by which mitochondria transport calcium, *Am. J. Physiol.* **258**:C755–C786.

Halestrap, A. P., 1989, The regulation of the matrix volume of mammalian mitochondria *in vivo* and *in vitro* and its role in the control of mitochondrial metabolism, *Biochim. Biophys. Acta* **973**:355–382.

Halestrap, A. P., 1991, Calcium-dependent opening of a non-specific pore in the mitochondrial inner membrane is inhibited at pH values below 7: Implications for the protective effect of low pH against chemical and hypoxic cell damage, *Biochem. J.* **278**:715–719.

Horstman, D. A., Tennes, K. A., and Putney, J. W., Jr, 1986, ATP-induced calcium mobilization and inositol 1,4,5-triphosphate formation in H-35 hepatoma cells, *FEBS Lett.* **204**:189–192.

Imberti, R., Nieminen, A.-L., Herman, B., and Lemasters, J. J., 1992, Synergism of cyclosporin A and phospholipase inhibitors in protection against lethal injury to rat hepatocytes from oxidant chemicals, *Res. Commun. Chem. Pathol. Pharmacol.* **78**:27–38.

Jiang, T., Grant, R. L., and Acosta, D., 1993, A digitized fluorescence imaging study of intracellular free calcium, mitochondrial integrity and cytotoxicity in rat renal cells exposed to ionomycin, a calcium ionophore, *Toxicology* **85**:41–65.

Jiang, T., Acosta, D., Jr, 1995, Mitochondrial Ca^{2+} overload in primary cultures of rat renal cortical epithelial cells by cytotoxic concentrations of cyclosporine: A digitized fluorescence imaging study, *Toxicology* **95**:155–166.

Johnson, J. D., Conroy, W. G., and Isom, G. E., 1987, Alteration of cytosolic calcium levels in PC12 cells by potassium cyanide, *Toxicol. Appl. Pharmacol.* 217–224.

Judah, J. D., Mchean, A. E., and Mchean, E. K., 1970, Biochemical mechanisms of liver injury. *Am. J. Med.* **49**: 609–616.

Juedes, M. J., Kass G. E. and Orrenius, S., 1992, *m*-Iodobenzylguanidine increases the mitochondrial Ca^{2+} pool in isolated hepatocytes, *FEBS. Lett.* **313**:39–42.

Kass, G. E., Juedes, M. J., and Orrenius, S., 1992, Cyclosporin A protects hepatocytes against prooxidant-induced cell killing: A study on the role of mitochondrial Ca^{2+} cycling in cytotoxicity, *Biochem. Pharmacol.* **44**: 1995–2003.

Komulainen, H., and Bondy, S. C., 1988, Increased free intracellular Ca^{2+} by toxic agents: An index of potential neurotoxicity? *Trends Pharmacol. Sci.* **9**:154–156.

Kraus-Friedmann, N., 1990, Calcium sequestration in the liver, *Cell Calcium* **11**:625–640.

Kristensen, S. R., 1991, Cell damage caused by ATP depletion is reduced by magnesium and nickel in human fibroblasts: A non-specific calcium antagonism? *Biochim. Biophys. Acta* **1091**:285–293.

Kurita, K., Tanabe, G., Aikou, T., and Shimazu, H., 1993, Ischemic liver cell damage and calcium accumulation in rats, *J. Hepatol.* **18**:196–204.

Lee, H. C., Mohabir, R., Smith, N., Franz, M. R., and Clusin, W. T., 1988, Effect of ischemia on calcium-dependent fluorescence transients in rabbit hearts containing indo 1: Correlation with monophasic action potentials and contraction, *Circulation* **78**:1047–1059.

Lemasters, J. J., Nieminen, A.-L., Qian, T., Trost, L. C., Elmore, S. P., Nishimura, Y., Crowe, R. A., Cascio, W. E., Bradham, C. A., Brenner, D. A., and Herman, B., 1998, The mitochondrial permeability transition in cell death: A common mechanism in necrosis, apoptosis, and autophagy, *Biochim. Biophys. Acta* **1366**:177–196.

Long, R. M., and Moore, L., 1986, Elevated cytosolic calcium in rat hepatocytes exposed to carbon tetrachloride, *J. Pharmacol. Exp. Ther.* **238**:186–191.

Masaki, N., Kyle, M. E., Serroni, A., and Farber, J. L., 1989, Mitochondrial damage as a mechanism of cell injury in the killing of cultured hepatocytes by *tert*-butyl hydroperoxide, *Arch. Biochem. Biophys.* **270**:672–680.

Mattson, M. P., Zhang, Y., and Bose, S., 1993, Growth factors prevent mitochondrial dysfunction, loss of calcium homeostasis, and cell injury, but not ATP depletion, in hippocampal neurons deprived of glucose, *Exp. Neurol.* **121**:1–13.

Mithofer, K., Sandy, M. S., Smith, M. T., and Di Monte, D., 1992, Mitochondrial poisons cause depletion of reduced glutathione in isolated hepatocytes, *Arch. Biochem. Biophys.* **295**:132–136.

Murgia, M., Pizzo, P., Steinberg, T. H., and Di Virgilio, F., 1992, Characterization of the cytotoxic effect of extracellular ATP in J774 mouse macrophages, *Biochem. J.* **288**:897–901.

Murgia, M., Hanau, S., Pizzo, P., Rippa, M., and Di Virgilio, F., 1993, Oxidized ATP: An irreversible inhibitor of the macrophage purinergic P2 receptor, *J. Biol. Chem.* **268**:8199–8203.

Nagelkerke, J. F., Dogterom, P., De Bont, H. J., and Mulder, G. J., 1989, Prolonged high intracellular free calcium concentrations induced by ATP are not immediately cytotoxic in isolated rat hepatocytes: Changes in biochemical parameters implicated in cell toxicity, *Biochem. J.* **263**:347–353.

Newsholme, P., Adogu, A. A., Soos, M. A., and Hales, C. N., 1993, Complement-induced Ca^{2+} influx in cultured fibroblasts is decreased by the calcium-channel antagonist nifedipine or by some bivalent inorganic cations, *Biochem. J.* **295**:773–779.

Nicotera, P., and Orrenius, S., 1992, Ca^{2+} and cell death, *Ann. N.Y. Acad. Sci.* **648**:17–27.

Nicotera, P., Thor, H., and Orrenius, S., 1989, Cytosolic free Ca^{2+} and cell killing in hepatoma 1c1c7 cells exposed to chemical anoxia, *FASEB. J.* **3**:59–64.

Nicotera, P., Bellomo, G., and Orrenius, S., 1990, The role of Ca^{2+} in cell killing, *Chem. Res. Toxicol.* **3**:484–494.

Nicotera, P., Bellomo, G., and Orrenius, S., 1992, Calcium-mediated mechanisms in chemically induced cell death, *Ann. Rev. Pharmacol. Toxicol.* **32**:449–470.

Nieminen, A.-L., Dawson, T. L., Gores, G. J., Kawanishi, T., Herman, B., and Lemasters, J. J., 1990a, Protection by acidotic pH and fructose against lethal injury to rat hepatocytes from mitochondrial inhibitors, ionophores, and oxidant chemicals, *Biochem. Biophys. Res. Commun.* **167**:600–606.

Nieminen, A.-L., Gores, G. J., Dawson, T. L., Herman, B., and Lemasters, J. J., 1990b, Toxic injury from mercuric chloride in rat hepatocytes., *J. Biol. Chem.* **265**(5):2399–2408.

Okuda, M., Lee, H. C., Chance, B., and Kumar, C., 1992, Role of extracellular Ca^{2+} in ischemia–reperfusion injury in the isolated perfused rat liver, *Circ. Shock.* **37**:209–219.

Orrenius, S., McConkey, D. J., Bellomo G., and Nicotera, P., 1989, Role of Ca^{2+} in toxic cell killing, *Trends. Pharmacol. Sci.* **10**:281–285.

Park, Y., Devlin, T. M., and Jones, D. P., 1992, Protective effect of the dimer of 16,16-diMePGB1 against KCN-induced mitochondrial failure in hepatocytes, *Am. J. Physiol.* **263**:C405–C411.

Pastorino, J. G., Snyder, J. W., Serroni, A., Hoek, J. B., and Farber, J. L., 1993, Cyclosporin and carnitine prevent the anoxic death of cultured hepatocytes by inhibiting the mitochondrial permeability transition, *J. Biol. Chem.* **268**:13791–13798.

Persoon-Rothert, M., Egas-Kenniphaas, J. M., van der Valk-Kokshoorn, E. J., and van der Laarse, A., 1992, Cumene hydroperoxide induced changes in calcium homeostasis in cultured neonatal rat heart cells, *Cardiovasc. Res.* **26**:706–712.

Petronilli, V., Cola, C., Massari, S., Colonna, R., and Bernardi, P., 1993, Physiological effectors modify voltage sensing by the cyclosporin A-sensitive permeability transition pore of mitochondria, *J. Biol. Chem.* **268**:21939–21945.

Phelps, P. C., Smith, M. W., and Trump, B. F., 1989, Cytosolic ionized calcium and bleb formation after acute cell injury of cultured rabbit renal tubule cells, *Lab. Invest.* **60**:630–642.

Rassendren, F., Buell, G. N., Virginio C., Collo, G., North, R. A., and Surprenant, A., 1997, The permeabilizing ATP receptor P2X7: Cloning and expression of a human cDNA, *J. Biol. Chem.* **272**:5482–5486.

Redegeld, F. A., Moison, R. M., Koster, A. S., and Noordhoek, J., 1992, Depletion of ATP but not of GSH affects viability of rat hepatocytes, *Eur. J. Pharmacol.* **228**:229–236.

Richelmi, P., Mirabelli, F., Salis, A., Finardi, G., Berte, F., and Bellomo, G., 1989, On the role of mitochondria in cell injury caused by vanadate-induced Ca^{2+} overload, *Toxicology* **57**:29–44.

Richter, C., and Schlegel, J., 1993, Mitochondrial calcium release induced by prooxidants, *Toxicol. Lett.* **67**:119–127.

Rizzuto, R., Brini, M., Bastianutto, C., Marsault, R., and Pozzan, T., 1995, Photoprotein-mediated measurement of calcium ion concentration in mitochondria of living cells, *Methods Enzymol.* **260**:417–428.

Saxena, K., Henry, T. R., Solem, L. E., and Wallace, K. B., 1995, Enhanced induction of the mitochondrial permeability transition following acute menadione administration, *Arch. Biochem. Biophys.* **317**:79–84.

Schanne, F. A., Kane, A. B., Young, E. E., and Farber, J. L., 1979, Calcium dependence of toxic cell death: A final common pathway, *Science* **206**:700–702.

Siegmund, B., Zude, R., and Piper, H. M., 1992, Recovery of anoxic-reoxygenated cardiomyocytes from severe Ca^{2+} overload, *Am. J. Physiol.* **263**:H1262–H1269

Smith, M. T., Thor, H., and Orrenius, S., 1981, Toxic injury to isolated hepatocytes is not dependent on extracellular calcium, *Science* **213**:1257–1259.

Smith, M. W., Phelps, P. C., and Trump, B. F., 1991, Cytosolic Ca^{2+} deregulation and blebbing after HgCl2 injury to cultured rabbit proximal tubule cells as determined by digital imaging microscopy, *Proc. Natl. Acad. Sci. USA* **88**:4926–4930.

Snyder, J. W., Pastorino, J. G., Thomas, A. P., Hoek, J. B., and Farber, J. L., 1993, ATP synthase activity is required for fructose to protect cultured hepatocytes from the toxicity of cyanide, *Am. J. Physiol.* **264**:C709–C714.

Snyder, J. W., Serroni, A., Savory, J., and Farber, J. L., 1995, The absence of extracellular calcium potentiates the killing of cultured hepatocytes by aluminium maltolate, *Arch. Biochem. Biophys.* **316**:434–442.

Solem, L. E., and Wallace, K. B., 1993, Selective activation of the sodium-independent, cyclosporin A-sensitive calcium pore of cardiac mitochondria by doxorubicin, *Toxicol. Appl. Pharmacol.* **121**:50–57.

Solem, L. E., Henry, T. R., and Wallace, K. B., 1994, Disruption of mitochondrial calcium homeostasis following chronic doxorubicin administration, *Toxicol. Appl. Pharmacol.* **129**:214–222.

Somlyo, A. P., Bond, M., and Somlyo, A. V., 1985, Calcium content of mitochondria and endoplasmic reticulum in liver frozen rapidly *in vivo*, *Nature* **314**:622–625.

Starke, P. E., Hoek, J. B., and Farber, J. L., 1986, Calcium-dependent and calcium-independent mechanisms of irreversible cell injury in cultured hepatocytes, *J. Biol. Chem.* **261**:3006–3012.

Steenbergen, C., Murphy, E., Watts, J. A., and London, R. E., 1990, Correlation between cytosolic free calcium, contracture, ATP, and irreversible ischemic injury in perfused rat heart, *Circ. Res.* **66**:135–146.

Stromski, M. E., Cooper, K., Thulin, G., Gaudio, K. M., Siegel, N. J. and Shulman, R. G., 1986, Chemical and functional correlates of postischemic renal ATP levels, *Proc. Natl. Acad. Sci. USA*, **83**:6142–6145.

Surprenant, A., Rassendren, F., Kawashima, E., North, R. A., and Buell, G., 1996, The cytolytic P2 receptor for extracellular ATP identified as a P2x receptor (P2x7), *Science* **272**:735–738.

Tani, M., 1990, Mechanisms of Ca^{2+} overload in reperfused ischemic myocardium, *Ann. Rev. Physiol.* **52**:543–559.

Thomas, C. E and Reed, D. J., 1988, Effect of extracellular Ca^{++} omission on isolated hepatocytes: I. Induction of oxidative stress and cell injury, *J. Pharmacol. Exp. Ther.*, **245**:493–500.

Thor, H., Hartzell, P., and Orrenius, S., 1984, Potentiation of oxidative cell injury in hepatocytes which have accumulated Ca^{2+}, *J. Biol. Chem.* **259**:6612–6615.

Tolleshaug, H., and Seglen, P. O., 1985, Autophagic–lysosomal and mitochondrial sequestration of [14C] sucrose: Density gradient distribution of sequestered radioactivity, *Eur. J. Biochem.* **153**:223–229.

Tolleshaug, H., Gordon, P. B., Solheim, A. E., and Seglen, P. O., 1984, Trapping of electro-injected [14C] sucrose by hepatocyte mitochondria: A mechanism for cellular autofiltration? *Biochem. Biophys. Res. Commun.* **119**:955–961.

Trollinger, D. R., Cascio, W. E., and Lemasters, J. J., 1997, Selective loading of Rhod 2 into mitochondria shows mitochondrial Ca^{2+} transients during the contractile cycle in adult rabbit cardiac myocytes, *Biochem. Biophys. Res. Commun.* **236**:738–742.

Tsubokawa H., Oguro, K., Robinson, H. P., Masuzawa, T., Kirino, T., and Kawai, N., 1992, Abnormal Ca^{2+} homeostasis before cell death revealed by whole cell recording of ischemic CA1 hippocampal neurons, *Neuroscience* **49**:807–817.

Valera, S., Hussy, N., Evans, R. J., Adami, N., North, R. A., Surprenant, A., and Buell, G., 1994, A new class of ligand-gated ion channel defined by P2x receptor for extracellular ATP [see comments], *Nature* **371**:516–519.

van de Water, B., Zoeteweij, J. P., de Bont, H. J., Mulder, G. J., and Nagelkerke, J. F., 1994, Role of mitochondrial Ca^{2+} in the oxidative stress-induced dissipation of the mitochondrial membrane potential: Studies in isolated proximal tubular cells using the nephrotoxin 1,2-dichlorovinyl-L-cysteine, *J. Biol. Chem.* **269**:14546–14552.

Van de Water, B., Zoeteweij, J. P., de Bont, H. J., and Nagelkerke, J. F., 1995, Inhibition of succinate: Ubiquinone reductase and decrease of ubiquinol in nephrotoxic cysteine S-conjugate-induced oxidative cell injury, *Mol. Pharmacol.* **48**:928–937.

Weinberg, J. M., Davis, J. A., Abarzua, M., and Kiani, T., 1989, Relationship between cell adenosine triphosphate and glutathione content and protection by glycine against hypoxic proximal tubule cell injury, *J. Lab. Clin. Med.* **113**:612–622.

Zoeteweij, J. P., van de Water, B., de Bont, H. J., Mulder, G. J., and Nagelkerke, J. F., 1992, Involvement of intracellular Ca^{2+} and K^+ in dissipation of the mitochondrial membrane potential and cell death induced by extracellular ATP in hepatocytes, *Biochem. J.* **288**:207–213.

Zoeteweij, J. P., van de Water, B., de Bont, H. J., Mulder, G. J., and Nagelkerke, J. F., 1993, Calcium-induced cytotoxicity in hepatocytes after exposure to extracellular ATP is dependent on inorganic phosphate: Effects on mitochondrial calcium, *J. Biol. Chem.* **268**:3384–3388.

Zoeteweij, J. P., van de Water, B., de Bont, H. J., and Nagelkerke, J. F., 1994, Mitochondrial K^+ as modulator of Ca^{2+}-dependent cytotoxicity in hepatocytes: Novel application of the K^+-sensitive dye PBFI K^+-binding benzofuran isophthalate) to assess free mitochondrial K^+ concentrations, *Biochem. J.* **299**:539–543.

Zoeteweij, J. P., van de Water, B., de Bont, H. J., and Nagelkerke, J. F., 1996, The role of a purinergic P2z receptor in calcium-dependent cell killing of isolated rat hepatocytes by extracellular ATP, *Hepatology*, **23**:858–865.

Chapter 25

Doxorubicin-Induced Mitochondrial Cardiomyopathy

Kendall B. Wallace

Doxorubicin is one of several water-soluble anthracenedione glycoside antibiotics produced by and originally isolated from *Streptomyces peucetius* var. *caesius* (Arcamone *et al.,* U.S. patent 3,590,028, 1971, by Farmitalia). The molecule consists of a highly hydroxylated planar tetracycline quinone that is glycosylated to a characteristic duanosa-mine sugar residue. Mild acid hydrolysis yields the water-soluble amino sugar and water-insoluble biologically active aglycone, which are easily separated by thin-layer or reverse-phase liquid chromatography. The structure of doxorubicin is illustrated in Figure 1.

Doxorubicin is one of the most widely prescribed antineoplastic agents in the United States, mostly because of its potent, broad-spectrum activity against both solid tumors and leukemias. It is particularly useful in causing regression of disseminated neoplasms, such as lymphoblastic and myeloblastic leukemias, neuroblastomas, bone marrow sarcomas, and malignant lymphomas. It is also effective against various carcinomas, including those of the breast, bladder, and thyroid. The therapeutic activity of doxorubicin is attributed to intercalation of the planar anthracycline ring into the DNA double helix, which inhibits DNA replication and transcription by interfering with the reading fidelity of both DNA and RNA polymerases. Alternatively, the cytostatic effect may reflect formation of sequence-specific DNA complexes that are highly susceptible to cleavage by topoisomerase II (Zunino and Capranico, 1990). Regardless, DNA, RNA, and protein synthesis are all inhibited, and as might be expected, cells are most sensitive during S-phase of the cell cycle. At low doses cell proliferation proceeds to G_2, at which time cell death occurs.

Kendall B. Wallace   Department of Biochemistry and Molecular Biology, University of Minnesota School of Medicine, Duluth, Minnesota 55812-2487.

Mitochondria in Pathogenesis, edited by Lemasters and Nieminen.
Kluwer Academic/Plenum Publishers, New York, 2001.

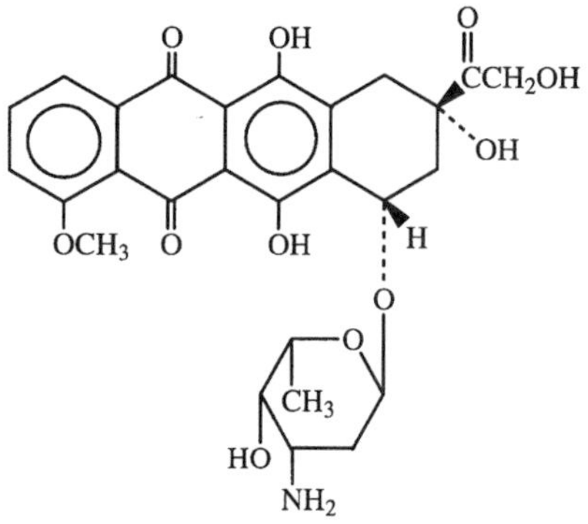

FIGURE 1. Doxorubicin. (Adriamycin®; 10-[(3-amino-2,3,6-trideoxy-α-L-lyxohexopyranosyl)oxy]-7,8,9,10-tetrahydro-6,8,11-trihydroxy-8-(hydroxyacetyl)-1-methoxy-5,12-naphthacenedione).

The drug is typically administered as short intravenous infusions of 40–75 mg/m² repeated at 3 to 4 week intervals being cautious not to exceed a cumulative dose of 450–550 mg/m². At these doses the clinical success of drug therapy becomes limited by drug-related myocardial toxicity, which in severe cases evolves into potentially life-threatening congestive heart failure. Although acute effects of high doses of doxorubicin include assorted cardiac tachyarrhythmias, it is the progressive and insidious deterioration of contractile function that is dose-limiting. The risks of developing drug-induced cardiac failure, as assessed by a reduction in ejection fraction, are estimated to increase, from 5% to 8% at a cumulative dose of 450 mg/m² of doxorubicin, to as high as 20% at a cumulative dose greater than 500 mg/m². This represents a dose-dependent, irreversible cardiac failure that is refractory to conventional inotropic drug therapy. Myocardial dysfunction may be manifested during the course of therapy or it may not develop for months or even years (up to 20 years) after the last course of treatment (Steinherz and Steinherz, 1991; Steinherz et al., 1991). This latent myocardial toxicity presents a particular concern in treating pediatric cancers (Steinherz et al., 1995).

Doxorubicin-induced cardiomyopathy is characterized electrocardiographically as various tachycardias, mostly of sinus origin, flattening of the T-wave, prolongation of the Q–T interval, and decreased amplitude of the R-wave (Bristow et al., 1978; Jensen et al., 1984; Praga et al., 1979). Echocardiography reveals a dose-dependent decrease in the left ventricular ejection fraction, reflecting a progressive decline in both systolic and diastolic function, accompanied by a large increase in left ventricular end-diastolic pressure (Alexander et al., 1979; Boucek et al., 1997; McKillop et al., 1983). Histomorphologically, this manifests as dilated cardiomyopathy, associated with loss of myofibrils, distension of the sarcoplasmic reticulum, and vacuolization of the cytoplasm (Aversano and Boor, 1983; Billingham et al., 1978; Bristow et al., 1981; Ferrans, 1978; Lefrak et al., 1973; Mortensen et al., 1986; Olson and Capen, 1978; Suzuki et al., 1979; Van Vleet and Ferrans, 1980; Villani et al., 1987). The clinical features of doxorubicin-induced cardiomyopathy have been recently reviewed (Berry et al., 1998; Boulay and Debaene, 1998; Calzas et al., 1998; Ferrans et al., 1997; Gianni, 1998; Gille and Nohl, 1997; Mott, 1997; Rhoden et al., 1993; Singal et al., 1997, 1998).

Although a varied array of biological activities have been described for doxorubicin, the mechanism of drug-induced cardiotoxicity is distinct from that which accounts for its antineoplastic effect. This bimodal action is evidenced by the fact that it is possible to diminish the cardiotoxic effect of doxorubicin without altering its antineoplastic potency (al-Shabanah et al., 1998; Herman et al., 1994), affording important opportunities to improve the clinical efficacy of drug therapy by selectively limiting its toxic effects on the heart. To do this, however, requires an understanding of the principal biochemical and cellular mechanisms responsible for the dose-limiting cardiotoxicity.

The biological/toxicological activities of doxorubicin are varied and include single- and double-strand scission of DNA, and both covalent binding and oxidative modification of nucleic acids, proteins, and lipids (cf. Gille and Nohl, 1997; Keizer et al., 1990; Olson and Mushlin, 1990; Singal et al., 1987). All of these oxidative processes are associated with a net decrease in tissue redox status. Although the therapeutic effect of doxorubicin is attributed to the parent compound, essentially all of the putatively toxic effects are ascribed to either a metabolite or a by-product of the drug's metabolism.

The metabolism of doxorubicin by microsomal membranes isolated from heart and liver has been thoroughly described by Bachur and associates (Bachur et al., 1977; Gutierrez et al., 1983) and others (Lown et al., 1982; Scheulen and Kappus, 1981). The major metabolic pathway is initiated by univalent reduction of the parent dione to the corresponding semiquinone free radical (Fig. 2). This highly unstable intermediate is very reactive and can undergo one of several possible fates: (1) further reduction to the hydroquinone; (2) reductive aglycosylation; (3) formation of covalent adducts to DNA or proteins, or (4) transfer of the unpaired electron to a more favorable electron acceptor to generate a secondary free radical species characteristic of the acceptor molecule. Secondary electron acceptors include glutathione and related thiols, tetrazolium salts, hemeproteins, α-tocopherol, ascorbic acid, and molecular oxygen.

There is an extensive literature describing the oxygen-dependent, free radical-mediated actions of doxorubicin in biological systems, and the bulk of it implicates redox cycling and free radical generation in the mechanism of drug toxicity (Olson et al., 1981; Keizer et al., 1990; Lee et al., 1991; Lown et al., 1982; Monti et al., 1991, 1995; Nohl and Jordan, 1983; Piccinini et al., 1990; Singal et al., 1987). Lown et al. (1982) describe a strong correlation between the redox potential of a host of structurally related anthracyclines and their rates of free radical generation and cardiotoxicity. Kinetically, the rate-limiting step in redox cycling of such compounds is the initial univalent reduction of the quinone to the semiquinone free radical intermediate, as catalyzed by pyridine nucleotide-dependent flavoproteins. Assuming a plentiful supply of oxygen or alternate electron acceptors, there is no net reduction or aglycosylation of the drug (Lown et al., 1982; Scheulen and Kappus, 1981; Scheulen, et al., 1982; Wallace, 1986). Thus although covalent binding of doxorubicin metabolites to membrane proteins has been reported (Scheulen and Kappus, 1981, 1982; Wallace, 1986; Wallace and Johnson, 1987), this occurs only under oxygen-limited conditions. Under physiological conditions oxygen is seldom limiting, and the prevailing metabolic pathway is redox cycling of the drug to liberate superoxide anion and related reactive species of molecular oxygen (Bachur and Gee, 1976; Bachur et al., 1977). This redox cycling is characterized by the consumption of both pyridine nucleotide reducing equivalents and oxygen, with the concomitant and

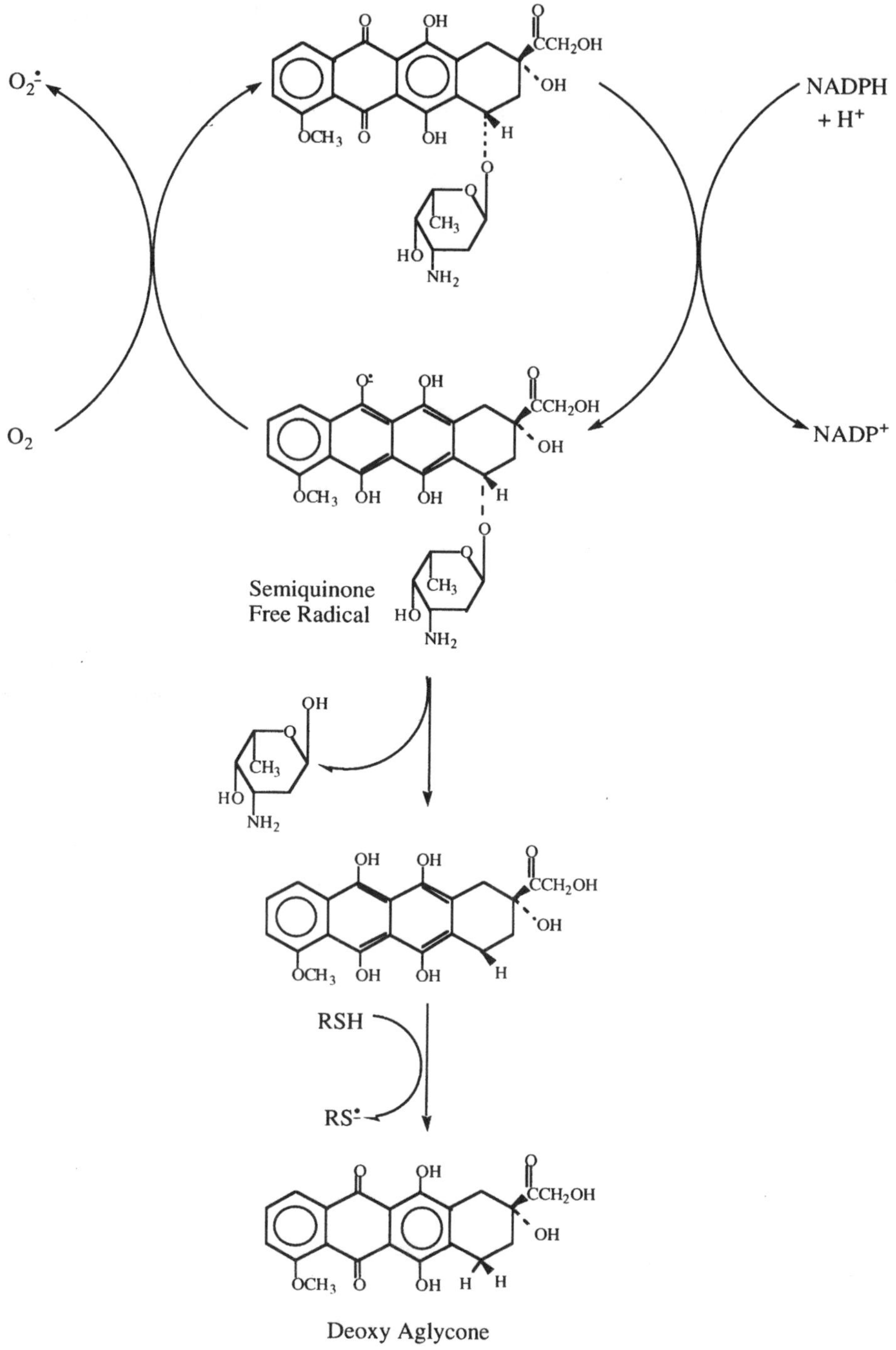

FIGURE 2. Metabolic pathway illustrating the univalent reduction of doxorubicin to its semiquinone free radical intermediate. Under anaerobic conditions the semiquinone undergoes reductive aglycosylation, whereas in the presence of oxygen the drug undergoes futile redox cycling to liberate highly reactive oxygen free radical species.

stoichiometric formation of oxidized NAD(P) and liberation of oxygen radicals with no net metabolism of the drug.

Because doxorubicin toxicity is ascribed to the generation of oxygen free radicals that do not diffuse far from their site of generation, and because the rate-limiting step of redox cycling is the initial univalent reduction of the parent quinone, one might predict that the type and extent of drug-related tissue damage is determined by the subcellular, tissue-specific distribution of enzymes capable of catalyzing this first step of the reaction. With a univalent redox potential of approximately $-320\,mV$, doxorubicin is a good substrate for assorted oxidoreductases within the cell. Indeed, a number of different flavoproteins have been shown to catalyze the univalent reduction of doxorubicin and related anthraquinones to their corresponding semiquinone free radical intermediates. These include NADPH-dependent cytochrome P450 reductases of the endoplasmic reticulum (Bachur and Gee, 1976; Bachur et al., 1977) and nuclear envelope (Bachur et al., 1982; Wallace, 1986), complex I of the mitochondrial electron transport chain (Davies et al., 1983), and soluble oxidoreductases of the cytoplasm, such as xanthine oxidase (Pan and Bachur, 1980). Doxorubicin also coordinates with both hemoglobin (Bates and Winterbourn, 1982) and myoglobin (Trost and Wallace, 1994a,b) as well as with free transitional metals such as iron (Bachur et al., 1984; Beraldo et al., 1985; Hasinoff, 1990; Malisza and Hasinoff, 1995; Vile et al., 1987; Zweier, 1984, 1985), and copper (Dutta and Hutt, 1986; Malatesta et al., 1985; Wallace, 1986) to form complexes capable of undergoing autooxidation to generate free radicals in solution. Accordingly, one might implicate the endoplasmic or sarcoplasmic reticulum, mitochondria, or nucleus as a principal intracellular target for doxorubicin-initiated cell damage. Alternatively, the subcellular target may be less discrete and the free radicals may be generated diffusely throughout the cytoplasm. Indeed, several organelle specific mechanisms are proposed to account for doxorubicin-induced cardio-myocyte failure, including inhibition of membrane transport functions; suppression of cytosolic antioxidant defense systems; a direct genotoxic effect resulting in the inhibition of DNA, RNA, and protein synthesis; interference with sarcoplasmic calcium regulation; and inhibition of cell bioenergetics. A thorough review of doxorubicin toxicity would include a discussion of each of these, but that is beyond our scope. The following is limited to describing evidence supporting mitochondrial dysfunction as the principal mechanism by which doxorubicin elicits its myocardial toxicity.

An early and prominent histomorphological change associated with chronic doxorubicin cardiotoxicity is the delocalization and swelling of mitochondria within the cardiomyocytes (Fig. 3). Inspection of individual mitochondria reveals fragmented cristae and occasional dense paracrystaline inclusion bodies within the mitochondrial matrix (Aversano and Boor, 1983; Friedman et al., 1978; Saltiel and McGuire, 1983). Several investigators suggest that the appearance of swollen mitochondria containing dense inclusion bodies is one of the earliest histopathological observations associated with anthracycline-induced cardiomyopathy (Ainger et al., 1971; Chalcroft et al., 1973; Rosenoff et al., 1975). The eventual decrease in myocyte number is accompanied by a compensatory increase in the numbers of both lysosomes and mitochondria in surviving cells, which is thought to represent a compensatory response to an initial metabolic deficit. Metabolite analyses of doxorubicin-induced cardiotoxicity are indeed strongly suggestive of bioenergetic failure. For example, long-term treatment of rats with doxorubicin results in substantial decreases in myocardial ATP, phosphocreatine, free carnitine, and lactate

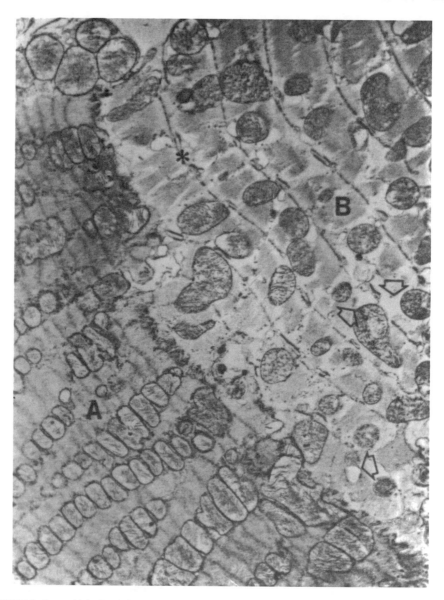

FIGURE 3. Doxorubicin (DXR)-induced ultrastructural changes in the rat heart. Rats received 1 mg/kg/d for 4 weeks (total of 20 treatments, equaling 20 mg/kg total) and were killed 14 weeks after the end of treatment. Two adjacent cells: (A) Shows some mitochondrial clearing, with separation of the cristae. (B) More heavily damaged mitochondria, with mitochondrial swelling and cellular edema, vacuolization, and myofibrilar breakdown (asterisk). The intercalated discs do not appear to be separated (×11,700). (Reproduced with permission from Jensen *et al.,* 1984, "Doxorubicin cardiotoxicity in the rat: Comparison of electrocardiogram, transmembrane potential, and structural effects," *Journal of Cardiovascular Pharmacology.* (6)1: 186–200.)

concentrations, indicative of inhibition of the glycolytic and oxidative metabolic pathways (Dekker *et al.*, 1991; Ferrero *et al.*, 1976; Kawasaki *et al.*, 1996).

Similar declines in ATP and phosphocreatine concentrations are observed when isolated hearts are perfused with doxorubicin, but only at reduced-flow rates, suggesting that doxorubicin renders the heart more vulnerable to conditions that challenge the relationship between cardiac function and energy production (Chatham *et al.*, 1990). Likewise, although the inorganic phosphate-to-phosphocreatine ratio for hearts from doxorubicin-treated rats is not altered at rest, there is a significant treatment-related increase at stimulated heart rates (Dekker *et al.*, 1991). Several additional examples of impaired myocardial bioenergetic reserves have been reported in conjunction with doxorubicin-induced cardiotoxicity. Although there were no differences in mechanical and hemodynamic responses at rest in chronically treated rats, both rate and maximum degree of tension development in response to dobutamine were significantly reduced in isolated papillary muscle strips from doxorubicin-treated rats (Hofling *et al.*, 1982). Similarly, dobutamine-induced stimulation of heart rate, increased left ventricular pressure, increased cardiac output, and both pressor and inotropic response to catecholamine injections were all lower for rats chronically treated with doxorubicin as compared with controls (Hofling *et al.*, 1982; Gorodetskaya *et al.*, 1990; Kawasaki *et al.*, 1996).

From these data we can conclude that assessments of compromised cardiac performance are consistent with the histopathological evidence demonstrating pronounced alterations in mitochondrial number and integrity that are characteristic of doxorubicin cardiotoxicity. Although functional deficits may not be apparent for the resting myocardium, they become quite pronounced under increased workloads. Thus, the overall effect of doxorubicin-induced myocardial toxicity appears to be a lowering of the bioenergetic reserves and interference with the regulation of energy metabolism in response to changing hemodynamic challenges.

Numerous reports exist on the effects of doxorubicin on the principal energy-transducing element of the cell, the mitochondrion. Such an effect may reflect serious detriments to bioenergetics, both when isolated mitochondria are exposed to the drug *in vitro* and when mitochondria are isolated from animals subjected to *in vivo* doxorubicin treatments. It is important to distinguish between the two. In *in vitro* experiments, doxorubicin is present and available as an alternate electron acceptor to directly interfere with mitochondrial electron transport and respiration. Under these conditions a principal experimental observation is that doxorubicin participates in or interferes with mitochondrial electron transport. Conversely, studies of mitochondria isolated following *in vivo* exposures show bioenergetic effects that persist in the absence of doxorubicin and that are not reversed in the course of isolating the subcellular structures, during which time all but minute traces of the drug are washed from the sample (Solem and Wallace, 1993; Solem *et al.*, 1994). In the latter case it is the actual tissue damage or biochemical lesion that is responsible for changes in bioenergetic status.

The direct effects of doxorubicin *in vitro* on mitochondrial function are varied and numerous. Adding doxorubicin to isolated mitochondria in suspension invariably results in stimulation of state 4 (basal, nonphosphorylative) and inhibition of state 3 (phosphorylative) respiration (Bianchi *et al.*, 1987; Ferrero *et al.*, 1976; Montali *et al.*, 1985). The overall effect is a decrease in the respiratory control ratio (RCR) with no change in the ADP/O ratio (Ferrero *et al.*, 1976; Muhammed *et al.*, 1983). Similar effects are reported for cardiac mitochondria isolated from mice and rats that received either acute (Ji and Mitchell, 1994; Matsumura *et al.*, 1994; Shinozawa *et al.*, 1985) or chronic (Solem *et al.*, 1994) intoxicating doses of doxorubicin *in vivo*. Inhibition of the maximum rate of respiration, (as evident from the effects of doxorubicin on ADP, calcium, and uncoupler-stimulated respiration) indicates a direct inhibitory effect on certain electron transporting elements of the respiratory chain or on substrate translocator proteins (Bachmann *et al.*, 1987; Bianchi *et al.*, 1987).

Stimulation of state 4 respiration *in vitro* reflects the fact that doxorubicin is an effective alternate electron acceptor, diverting electrons from the respiratory chain to reduce doxorubicin to its corresponding semiquinone free radical intermediate, which readily undergoes redox cycling to reduce molecular oxygen to superoxide and other free radical species of oxygen. Evidence supporting this is that the stimulation of respiration is accompanied by increased rates of NADH oxidation, detection of the semiquinone free radical intermediate, and enhanced rates of superoxide anion free radical generation, all of which are insensitive to cyanide or antimycin A (Chacon and Acosta, 1991; Cuellar *et al.*, 1984; Davies and Doroshow, 1986; Davies *et al.*, 1983; Doroshow, 1983a, 1983b; Doroshow and Davies, 1986; Gervasi *et al.*, 1986, 1990; Lin *et al.*, 1991; Pollakis *et al.*, 1984; Praet *et al.*, 1988; Thayer, 1977).

The site of doxorubicin redox cycling has been localized to complex I of the mitochondrial electron transport chain (Davies and Doroshow, 1986; Demant, 1991; Doroshow and Davies, 1986). Because the rate-limiting step of redox cycling is the univalent reduction of doxorubicin to the semiquinone, the overall rate is determined by the redox state of the specific transport-chain segments or elements that serve as electron donors for doxorubicin. This is best illustrated by the work of Davies and Doroshow (1986), who demonstrate that the rate of doxorubicin-stimulated oxygen consumption is influenced by the choice of reducing substrates and inhibitors included in the reaction. By showing that rotenone, but not TTFA, blocks doxorubicin- and daunorubicin-stimulated oxygen consumption during reverse electron transport in bovine heart submitochondrial particles, the authors provide compelling evidence that the source of reducing equivalents for both drugs is proximal to the site of the rotenone, but not TTFA, blockade on the electron transport chain, complex I (Fig. 4).

We provide supporting evidence for this site-specific redox cycling by demonstrating that the rates of doxorubicin-stimulated superoxide anion free radical generation correspond with the redox state of individual mitochondrial electron transport chain segments (Table I). The fact that doxorubicin-dependent superoxide generation is stimulated by both antimycin and rotenone in NADH-energized mitochondria indicates a source of reducing equivalents that is proximal to the site of the rotenone blockade of electron transport on complex I. This is confirmed by the fact that antimycin A stimulates succinate-supported free radical generation, which increases complex I reduction but is blocked by TTFA, an inhibitor of complex II. Thus, succinate-supported free radical generation by doxorubicin occurs via retrograde electron transport across complex I of the respiratory chain. Figure 5 illustrates this. The net effect is stimulation of cyanide-insensitive oxygen consumption, depletion of vital pyridine nucleotide reducing equivalents, and generation of highly reactive oxygen free radicals.

The ultimate effect is an intense and potentially cytolethal oxidative stress, widely reported and strongly implicated in the mechanism of doxorubicin-induced cytotoxicity

**O₂ CONSUMPTION DURING REVERSED ELECTRON
TRANSPORT WITH ASCORBATE + TMPD + KCN**

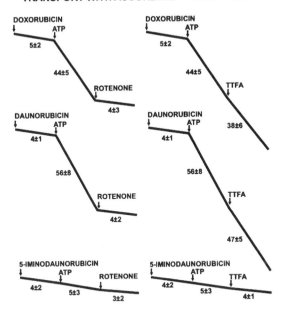

FIGURE 4. Redox cycling of anthracyclines by bovine heart submitochondrial particles during reverse electron transport with ascorbate plus TMPD plus KCN⁻. Oxygen consumption (nmol/min/mg protein) was measured polarographically at 37 °C. Representative oxygen electrode tracings are shown along with mean ± SE of five independent determinations. Each incubation contained 10 mM ascorbate plus 0.5 mM TMPD plus 0.1 mM KCN⁻. Anthracycline concentrations was 0.2 mM, and reactions (reverse electron transport) were started by adding 5 mM ATP. Where used, rotenone concentration was 0.01 mM and TTFA concentration was 6 mM. (Reproduced with permission from Davies and Doroshow, 1986, ("Redox cycling of anthracyclines by cardiac mitochondria: I. Anthracycline radical formation by NADH dehydrogenase," *Journal of Biological Chemistry* 261(7): 3060–3067.)

both *in vitro* and *in vivo*. Evidence to support the occurrence of this pathway *in vitro* includes depletion of both pyridine nucleotide and sulfhydryl-related reducing equivalents in cells exposed in culture to doxorubicin; stimulation of the peroxidation of mitochondrial membrane lipids by doxorubicin; and cytoprotection afforded by various antioxidant compounds such as sulfhydryl reducing agents, coenzyme Q, α-tocopherol, and selenium (Geetha, 1993; Geetha and Devi, 1992; Griffin-Green *et al.*, 1988; Mimnaugh *et al.*, 1985; Ogura *et al.*, 1991; Olson *et al.*, 1980; Solaini *et al.*, 1987; Sugiyama *et al.*, 1995; Valls *et al.*, 1994; Zidenberg-Cherr and Keen, 1986). Although somewhat variable, this antioxidant cytoprotection can be achieved *in vivo* as well. Kang and associates (1996, 1997) found that transgenic mice over expressing cardiac catalase or metallothionein are resistant to doxorubicin-induced cardiomyopathy. The possibility of a mitochondrial target is indicated by the fact that transgenic mice overexpressing the mitochondrial form of superoxide dismutase (MnSOD) are resistant to acute doxorubicin cardiotoxicity (Yen *et al.*, 1996). Because MnSOD is distributed exclusively to the mitochondrial matrix, this is strong evidence for a mitochondrial free radical-mediated mechanism of toxicity. This is further supported by the observation that the mitochondrial, not the cytosolic, pool of glutathione

Table I
Site-Specific Superoxide Anion Free Radical Generation by Doxorubicin

	Superoxide Free Radical Formation (nmol ferrocytochrome c/min/mg protein)		
	− DXR	+ DXR	(n)
S	0.10 ± 0.06	0.03 ± 0.03	3
S + AA	0.51 ± 0.23	$1.05 \pm 0.32^*$	3
S + TTFA	0.52 ± 0.30	0.95 ± 0.29	3
NADH	0.75 ± 0.12	1.29 ± 0.55	4
NADH + AA	1.09 ± 0.25	$7.81 \pm 2.38^*$	3
NADH + ROT	0.84 ± 0.12	$8.28 \pm 1.83^*$	6
NADH + S	0.24 ± 0.16	0.50 ± 0.09	2
NADH + S + ROT	1.09 ± 0.25	$7.47 \pm 1.14^*$	5

Cardiac submitochondrial particles were isolated from male Sprague–Dawley rats and incubated in the presence of varying combinations of substrates and inhibitors to affect different redox states of the various respiratory chain complexes. Superoxide free radical generation was assessed spectrophotometrically from the rate of SOD-inhibitable reduction of acetylated ferricytochrome *c*. Additions were as follows: doxorubicin (DXR), 100 μM; succinate (S), 1 mm; NADH, 200 m; thenoyltri-fluoroacetate (TTFA), 100 μM; antimycin A (AA), 4 μM; and rotenone (ROT), 4 μM. Data represent the mean ±SEM for 2–6 replications (n). The asterisks (*) indicate a statistically significant difference between the absence (− DXR) and presence (+ DXR) of doxorubicin ($p < 0.05$). (Solem, L. E., and Wallace, K. B., unpublished data.)

is critical for defense against doxorubicin cytotoxicity in isolated liver cell cultures (Meredith and Reed, 1983).

In addition to stimulating state 4 nonphosphorylating oxygen consumption, doxorubicin also inhibits numerous components of the mitochondrial respiratory chain, as reflected by the inhibition of substrate-supported state 3 respiration. Detailed analyses of the individual parameters yields a host of activities inhibited by doxorubicin. These

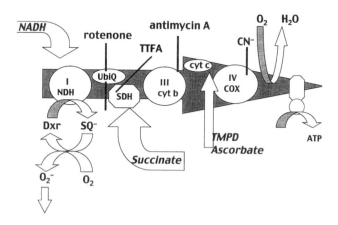

FIGURE 5. Schematic of the site-specific redox cycling of doxorubicin (DXR) on complex I of the mitochondrial electron transport chain. SQ-, Semiquinone free radical of DXR. NADH dehydrogenase (NDH) ubiquinone (UbiQ). SDH, Succinate dehydrogenase. TTFA, Thenoyltrifluoroacetate. CN, Cyanide. TMPD, Tetramethyl-*p*-phenylene diamine.

include NADH-dehydrogenase (Demant and Jensen, 1983; Geetha and Devi, 1992; Marcillat *et al.*, 1989; Muhammed and Kurup, 1984; Solaini *et al.*, 1985) NADH–cytochrome *c* reductase (Goormaghtigh *et al.*, 1986; Hasinoff, 1990a; Matsumura *et al.*, 1994; Nicolay and de Kruijff, 1987; Praet, *et al.*, 1988; Solaini *et al.*, 1985; Yamada *et al.*, 1995), succinate–cytochrome *c* reductase (Bertazzoli *et al.*, 1976; Demant and Jensen, 1983; Goormaghtigh *et al.*, 1986; Marcillat *et al.*, 1989; Matsumura *et al.*, 1994; Muhammed and Kurup, 1984; Nicolay and de Kruijff, 1987; Praet *et al.*, 1988; Solaini *et al.*, 1985), cytochrome oxidase (Demant, 1991; Floridi *et al.*, 1994; Goormaghtigh *et al.*, 1982; Yamada *et al.*, 1995), creatine kinase (Miura *et al.*, 1994), ATPase activity (Bianchi *et al.*, 1987), carnitine palmitoyl transferase (Brady and Brady, 1987; Kashfi *et al.*, 1990), and fatty acid β-oxidation (Link *et al.*, 1996) as well as the inward translocation of phosphatidylsersine (Voelker, 1991), phosphate (Cheneval *et al.*, 1983), pyruvate (Paradies and Ruggiero, 1988), and small precursor proteins (Eilers *et al.*, 1989). Such inhibitory effects are observed following both *in vitro* and *in vivo* doxorubicin exposures. Two basic paradigms are offered to explain doxorubicin's inhibition of these mitochondrial activities: interference with redox-dependent, or with cardiolipin-dependent, regulation of enzyme/transporter activities.

Doxorubicin has a high affinity for cardiolipin, which is an abundant acidic phospholipid localized exclusively to the inner mitochondrial membrane, and plays an important role as an allosteric effector of numerous enzyme and transporter activities (Cheneval *et al.*, 1985; Goormaghtigh *et al.*, 1986; 1990; Huart *et al.*, 1984; Nicolay *et al.*, 1984; Praet *et al.*, 1986). By complexing with cardiolipin, doxorubicin interferes with the activity of a number of proteins, such as the pyruvate (Paradies and Ruggiero, 1988) and phosphate (Cheneval *et al.*, 1983) transporters of the inner mitochondrial membrane, along with selected elements of the electron transport chain, including cytochrome oxidase (Goormaghtigh *et al.*, 1982, 1986). Furthermore, because cardiolipin aids the unfolding of assorted precursor proteins—a prerequisite for transport across the inner mitochondrial membrane—doxorubicin also inhibits accumulation of selected proteins from cytosol into mitochondrial matrix (Eilers *et al.*, 1989). Cardiolipin's abundance and importance in the mitochondrial inner membrane, coupled with doxorubicin's high affinity for this phospholipid, provides a convenient and logical explanation for the inhibitory effects of doxorubicin on a broad array of electron and substrate translocating proteins associated with the inner mitochondrial membrane.

An alternate theory for explaining the inhibitory effects of doxorubicin on mitochondrial respiration relates to the generation of reactive oxygen species and interference with redox-dependent enzyme regulation, specifically oxidation of critical pyridine nucleotide-binding domains or thiol-dependent redox sensors. There are many electron-transporting proteins, as well as ion and substrate transporters of the inner mitochondrial membrane that possess critical thiol groups within their catalytic centers. Not surprisingly, oxidative stress results in a change in the state of protein-related thiol groups, thus altering regulation of the corresponding enzyme activity. As mentioned, doxorubicin exposure elicits oxidative stress in the cell or tissue, manifested as stimulation of oxygen free radical generation and depletion of key pyridine nucleotide- and sulfhydryl-reducing equivalents. Associated with this is the net oxidation of both soluble glutathione (GSH) and protein-bound thiols to their corresponding disulfides. Meredith and Reed (1983) demonstrate that although doxorubicin depletes cytosolic GSH, it is the mitochondrial GSH pool that is critical to defining the cellular response, implying that it is the mitochondrial thiol-dependent redox sensors that are critical to mediating doxorubicin-induced cell killing. Numerous mitochondrial activities are proposed to be sensitive to inhibition by doxorubicin-induced oxidation of key regulatory subunits, including complex I, complex II, cytochrome oxidase, ATPase, and creatine kinase (Geetha and Devi, 1992; Marcillat et al., 1989; Miura et al., 1994). This is further supported by the ability of antioxidants to restore mitochondrial activities inhibited by doxorubicin (Zidenberg-Cherr and Keen, 1986; Geetha and Devi, 1992; Lin et al., 1991, 1992; Miura et al., 1994; Muhammed and Kurup, 1984; Praet et al., 1988; Solaini et al., 1987; Valls, et al., 1994; Sugiyama et al., 1995).

Renewed interest in how the mitochondrial inner membrane undergoes transformation from a highly impermeable barrier to a membrane that is nonselectively permeable to solutes as large as 1500 Da has provided new perspectives on the mechanism by which doxorubicin interferes with cell bioenergetics. Induction of this transition is the focus of considerable recent research and review (Bernardi et al., 1994; Zoratti and Szabò, 1995). A major conclusion was that induction of the mitochondrial permeability transition plays a major role in the mechanism by which numerous compounds or disease states elicit their cytotoxic effects (Bernardi and Petronilli, 1996; Lemasters, 1998; Lemasters et al., 1998). In fact, induction of the mitochondrial permeability transition is suggested as a decisive factor in determining the mode of cell death (necrosis or apoptosis) (Hirsch et al., 1997, 1998). The purported pore responsible for this transition is regulated by the redox status of critical thiol groups in the immediate vicinity of the voltage-sensing element. Oxidation to the corresponding disulfide increases the threshold voltage potential, thereby increasing the probability of pore opening (Petronilli et al., 1994). Such regulation may explain why the majority of chemicals that are capable of inducing the transition are oxidants and why sulfhydryl reducing agents are potent transition inhibitors (Gunter et al., 1988). Doxorubicin is known to oxidize both soluble and protein-related thiols, including mitochondrial thiols: Thus it is a prime candidate to induce this pathogenic mitochondrial permeability transition.

There are several histomorphologic observations of swollen mitochondria with dense inclusion bodies, supposedly calcium crystals, following chronic in vivo treatments (Ainger, et al., 1971; Chalcroft et al., 1973; Rosenoff et al., 1975; Singal et al., 1984). One of the first ex vivo experimental indications for doxorubicin-induced transition was that cardiac mitochondria isolated from doxorubicin-treated rabbits showed a markedly increased permeability to NADH (Ferrero et al., 1976). About the same time, it was reported that doxorubicin and its aglycone metabolite interfere with calcium accumulation in heart, liver, and kidney mitochondria in vitro (Bachmann and Zbinden, 1979; Moore et al., 1977; Revis and Marusic, 1979a). Doxorubicin inhibited calcium uptake, whereas the aglycone stimulated release of calcium from preloaded isolated mitochondria (Revis and Marusic, 1979b). The same membrane-permeabilizing effect was demonstrated a decade later, when it was reported that doxorubicin increases the permeability of isolated rat heart mitochondria to creatine influx in vitro (Bachmann et al., 1987). A thorough characterization of this mitochondrial calcium-regulation by doxorubicin was first given by Sokolove and colleagues, who showed that incubation of isolated rat heart mitochondria with doxorubicin caused dose-dependent NADH oxidation, release of matrix calcium, and depolarization of membrane potential, all of which were accompanied by mitochondrial swelling and inhibited by cyclosporin A, dibucaine, ATP, and assorted sulfhydryl-reducing agents (Sokolove, 1990, 1991; Sokolove and Shinaberry, 1988). Specifically, these implicate the oxidation of key mitochondrial thiol groups in the manifestation of this phenomenon (Sokolove, 1988). These researchers also found that 7-hydroxy- and 7-deoxyaglycone metabolites are more potent than doxorubicin itself. We've since demonstrated that doxorubicin induced disruption of mitochondrial calcium homeostasis is a direct and specific effect on the calcium release channel; doxorubicin did not affect any of the other recognized calcium transport pathways within the mitochondrial membranes (Solem and Wallace, 1993). The fact that the calcium-induced calcium release and membrane depolarization were completely inhibited by cyclosporin A (a potent and specific inhibitor of the permeability transition pore) is strong evidence for induction of the transition by doxorubicin. Furthermore, cardiac mitochondria are much more susceptible to induction of the permeability transition by doxorubicin than are liver or kidney mitochondria, suggesting a possible explanation for the drug's cardiospecific toxicity (Revis and Marusic, 1979a).

The possibility that induction of the transition occurs *in vivo* and contributes to doxorubicin-induced cardiomyopathy in whole animals was first demonstrated by our laboratory. Cardiac mitochondria isolated from rats receiving repeated and cumulative doxorubicin injections show a dose-dependent increase in sensitivity to calcium-induced calcium release and membrane depolarization, both of which are blocked by cyclosporin A (Fig. 6). Susceptibility to calcium-induced membrane depolarization can be blocked by either ruthenium red (a calcium uniport inhibitor) or cyclosporin A, so it was concluded that the drug initiates a futile and energy-consuming recycling of calcium across the mitochondrial membranes (Solem *et al.*, 1996).

Implication of this interference with mitochondrial calcium regulation in the mechanism of doxorubicin-induced toxicity to cell cultures was first provided by Chacon and associates (1991, 1992), who showed that exposure *ex vivo* to doxorubicin causes a

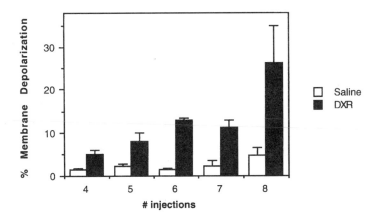

FIGURE 6. Dose-dependent increases in calcium-induced mitochondrial depolarization following successive injections of adriamycin. Rats received 4–8 weekly s.c. injections of 2 mg/kg Adriamycin (DXR). Cardiac mitochondria were isolated 1 wk following the last injection and examined for extent of calcium-induced membrane depolarization using the spectrophotometric method with Rhodamine 123. Each point represents the mean ± SEM of 3 individual rats.

calcium-dependent membrane-potential depolarization in isolated cardiomyocytes, and that blocking the drug's effects prevents cell killing. The authors gave compelling arguments that induction of the mitochondrial permeability transition plays a primary role in the mechanism of cytotoxicity of the drug. We've shown the same phenomenon after *in vivo* doxorubicin exposure. Cardiomyocytes isolated from rats receiving weekly injections of doxorubicin exhibit a dose-dependent increase in sensitivity to calcium-induced cell killing (Solem *et al.*, 1996). The cytoprotection afforded by either ruthenium red or cyclosporin A strongly indicates that this enhanced sensitivity is a direct manifestation of induction of the mitochondrial permeability transition by doxorubicin exposure *in vivo* (Fig. 7). The fact that sensitivity to calcium-induced cell killing increases as a function of the cumulative dose of doxorubicin and occurs at doses that cause no other gross pathological changes implies that induction of the mitochondrial permeability transition *in vivo* is an early and sensitive indicator of tissue injury.

To summarize, there is a growing consensus that the mitochondrion is a principal target for doxorubicin-induced cardiotoxicity. Histomorphological observations reveal swollen and delocalized mitochondria containing dense inclusions thought to represent calcium deposits within the mitochondrial matrix. Functional deficits that accompany these structural changes include a decrease in the rate and extent of cardiomyocyte contraction, especially in the face of inotropic or chronotropic challenge. This physiological failure is reflected biochemically as a lower metabolic or bioenergetic capacity of the cell, specifically an inhibition of mitochondrial oxidative phosphorylation. We suggest that the root of this bioenergetic failure is oxidative stress and induction of the mitochondrial permeability transition caused by redox cycling of doxorubicin on the mitochondrial respiratory chain. Our view of the mechanistic basis for doxorubicin-induced cell injury is illustrated in Figure 8. In brief, we propose that doxorubicin is reduced to its correspond-

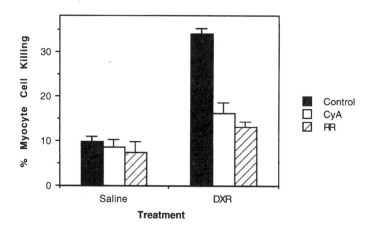

FIGURE 7. Calcium-induced killing of cardiomyocytes from Adriamycin-treated rats. Rats received 6 weekly s.c. injections of either 2 mg/kg Adriamycin or saline. Cardiomyocytes were isolated in calcium-free medium, then calcium was reintroduced in a 5-step process to a final concentration of 1 mM. Cell killing represents the difference in cell viability in 0 vs 1 mM calcium. Data represent the mean ± SEM for 3 separate cell isolations. Where indicated, cyclosporin A (CyA) or ruthenium red (RR) was added to the cell suspension before introducing calcium.

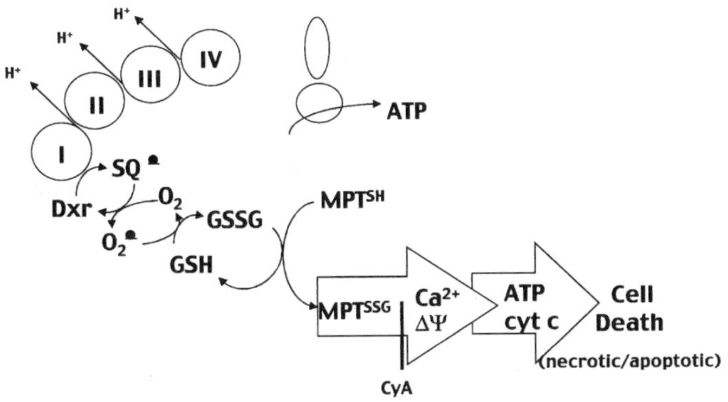

FIGURE 8. Schematic of the hypothesized mechanism of mitochondrial-mediated cytotoxicity caused by doxorubicin (DXR). See text for details. SQ, Semiquinone, GHS, glutathione. $\Delta\Psi$, mitochondrial membrane potential, Cyt c, cytochrome c. MPT, mitochondrial permeability transition. O_2^-, superoxide union free radicals. GSSG, disulfides (see last paragraph of text), MPT^{SSG}, mitochondrial permeability transition pore. CyA, cyclosporin A.

ing semiquinone free radical at complex I of the electron transport chain. Being highly unstable, the semiquinone reduces dioxygen to complete the redox cycle, liberating superoxide anion free radicals in the process. Superoxide and its subsequent reduction products (hydrogen peroxide, hydroxyl radical, etc.) are ultimately reduced by GSH or other thiols to form the corresponding disulfides (GSSG, etc.). This net change in redox status affects the regulatory domain of the transition pore, resulting in pore opening and leading to release of solutes, such as calcium and GSH, from the mitochondrial matrix, and to depolarization of membrane potential ($\Delta\Psi$). This then leads to inhibition of ATP synthesis and possibly to release of loosely bound cytochrome c and other important cell death signals. Overall, the data provide strong arguments that the mitochondrion plays a critical role in defining both functional and pathological manifestations of doxorubicin-induced cardiomyopathy.

ACKNOWLEDGEMENTS. The author thanks Dr. Laura E. Solem for her critical contributions to this project. This project was supported in part by grants from the American Heart Association, Minnesota affiliation, and the NIH, HL-58016.

REFERENCES

Ainger, L. E., Bushore, J., Johnson, W. W., and Ito, J., 1971, Daunomycin: A cardiotoxic agent, *J. Nat. Med. Assoc.* **63**(4):261–267.

al-Shabanah, O. A., Badary, O. A., Nagi, M. N., al-Gharably, N. M., al-Rikabi, A. C., and al-Bekairi, A. M., 1998, Thymoquinone protects against doxorubicin-induced cardiotoxicity without compromising its antitumor activity, *J. Exp. Clin. Cancer Res.* **17**(2):193–198.

Alexander, J., Dainiak, N., Berger, H. J., Goldman, L., Johnstone, D., Reduto, L., Duffy, T., Schwartz, P., Gottschalk, A., and Zaret, B. L., 1979, Serial assessment of doxorubicin cardiotoxicity with quantitative radionuclide angiocardiography, *N. Engl. J. Med.* **300**(6):278–283.

Aversano, R. C., and Boor, P. J., 1983, Histochemical alterations of acute and chronic doxorubicin cardiotoxicity, *J. Mol. Cell Cardiol.* **15**(8):543–553.

Bachmann, E., and Zbinden, G., 1979, Effect of doxorubicin and rubidazone on respiratory function and Ca^{2+} transport in rat heart mitochondria, *Toxicol. Lett.* **3**:29–34.

Bachmann, E., Weber, E., and Zbinden, G., 1987, Effects of mitoxantrone and doxorubicin on energy metabolism of the rat heart, *Cancer Treat. Rep.* **71**(4):361–366.

Bachur, N. R., and Gee, M., 1976, Microsomal reductive glycosidase, *J. Pharmacol. Exp. Ther.* **197**(3):681–686.

Bachur, N. R., Gee, M. V., and Friedman, R. D., 1982, Nuclear catalyzed antibiotic free radical formation, *Cancer Res.* **42**(3):1078–1081.

Bachur, N. R., Gordon, S. L., and Gee, M. V., 1977, Anthracycline antibiotic augmentation of microsomal electron transport and free radical formation, *Mol. Pharmacol.* **13**(5):901–910.

Bachur, N. R., Friedman, R. D., and Hollenbeck, R. G., 1984, Physicochemical characteristics of ferric Adriamycin complexes, *Cancer Chemother. Pharmacol.* **12**(1):5–9.

Bates, D. A., and Winterbourn, C. C., 1982, Reactions of Adriamycin with haemoglobin: Superoxide dismutase indirectly inhibits reactions of the Adriamycin semiquinone, *Biochem. J.* **203**(1):155–160.

Beraldo, H., Garnier-Suillerot, A., Tosi, L., and Lavelle, F., 1985, Iron(III)-adriamycin and Iron(III)-daunorubicin complexes: Physicochemical characteristics, interaction with DNA, and antitumor activity, *Biochemistry* **24**(2):284–289.

Bernardi, P., and Petronilli, V., 1996, The permeability transition pore as a mitochondrial calcium release channel: A critical appraisal, *J. Bioenerg. Biomembr.* **28**(2):131–138.

Bernardi, P., Broekemeier, K. M., and Pfeiffer, D. R., 1994, Recent progress on regulation of the mitochondrial permeability transition pore: A cyclosporin-sensitive pore in the inner mitochondrial membrane, *J. Bioenerg. Biomembr.* **26**(5):509–517.

Berry, G., Billingham, M., Alderman, E., Richardson, P., Torti, F., Lum, B., Patek, A., and Martin, F. J., 1998, The use of cardiac biopsy to demonstrate reduced cardiotoxicity in AIDS Kaposi's sarcoma patients treated with pegylated liposomal doxorubicin, *Ann. Oncol.* **9**(7):711–716.

Bertazzoli, C., Sala, L., Ballerini, L., Watanabe, T., and Folkers, K., 1976, Effect of Adriamycin on the activity of the succinate dehydrogenase–coenzyme Q10 reductase of the rabbit myocardium, *Res. Commun. Chem. Pathol. Pharmacol.* **15**(4):797–800.

Bianchi, C., Bagnato, A., Paggi, M. G., and Floridi, A., 1987, Effect of Adriamycin on electron transport in rat heart, liver, and tumor mitochondria, *Exp. Mol. Pathol.* **46**(1):123–135.

Billingham, M. E., Mason, J. W., Bristow, M. R., and Daniels J. R., 1978, Anthracycline cardiomyopathy monitored by morphologic changes, *Cancer Treat. Rep.* **62**(6):865–872.

Boucek, R. J., Jr., Dodd, D. A., Atkinson, J. B., Oquist, N., and Olson, R. D., 1997, Contractile failure in chronic doxorubicin-induced cardiomyopathy, *J. Mol. Cell. Cardiol.* **29**(10):2631–2640.

Boulay, G., and Debaene, B., 1998, Acute cardiogenic postoperative edema after doxorubicin (Adriamycin) chemotherapy, *Ann. Fr. Anesth. Reanim.* **17**(1):43–46.

Brady, L. J., and Brady, P. S., 1987, Hepatic and cardiac carnitine palmitoyltransferase activity: Effects of Adriamycin and galactosamine, *Biochem. Pharmacol.* **36**(20):3419–3423.

Bristow, M. R., Billingham, M. E., Mason, J. W., and Daniels, J. R., 1978, Clinical spectrum of anthracycline antibiotic cardiotoxicity, *Cancer Treat. Rep.* **62**(6):873–879.

Bristow, M. R., Mason, J. W., Billingham, M. E., and Daniels, J. R., 1981, Dose–effect and structure–function relationships in doxorubicin cardiomyopathy, *Am. Heart J.* **102**(4):709–718.

Calzas, J., Lianes, P., and Cortes-Funes, H., 1998, Heart pathology of extracardiac origin: VII. Heart and neoplasms, *Rev. Esp. Cardiol.*. **51**(4):314–331.

Chacon, E., and Acosta, D., 1991, Mitochondrial regulation of superoxide by Ca^{2+}: An alternate mechanism for the cardiotoxicity of doxorubicin, *Toxicol. Appl. Pharmacol.* **107**(1):117–128.

Chacon, E., Ulrich, R., and Acosta, D., 1992, A digitised-fluorescence-imaging study of mitochondrial Ca^{2+} increase by doxorubicin in cardiac myocytes, *Biochem. J.* **281**(Pt. 3):871–878.

Chalcroft, S. C., Gavin, J. B., and Herdson, P. B., 1973, Fine structural changes in rat myocardium induced by daunorubicin, *Pathology* **5**(2):99–105.

Chatham, J. C., Cousins, J. P., and Glickson, J. D., 1990, The relationship between cardiac function and metabolism in acute Adriamycin-treated perfused rat hearts studied by 31P and 13C NMR spectroscopy, *J. Mol. Cell. Cardiol.* **22**(10):1187–1197.

Cheneval, D., Muller M., and Carafoli, E., 1983, The mitochondrial phosphate carrier reconstituted in liposomes is inhibited by doxorubicin, *FEBS. Lett.* **159**(1, 2):123–126.

Cuellar, A., Escamilla, E., Ramirez, J., and Chavez, E., 1984, Adriamycin as an inhibitor of 11 beta-hydroxylase activity in adrenal cortex mitochondria, *Arch. Biochem. Biophys.* **235**(2):538–543.

Davies, K. J., and Doroshow, J. H., 1986, Redox cycling of anthracyclines by cardiac mitochondria: I. Anthracycline radical formation by NADH dehydrogenase, *J. Biol. Chem.* **261**(7):3060–3067.

Davies, K. J., Doroshow, J. H., and Hochstein, P., 1983, Mitochondrial NADH dehydrogenase-catalyzed oxygen radical production by Adriamycin and the relative inactivity of 5-iminodaunorubicin, *FEBS. Lett.* **153**(1):227–230.

Dekker, T., van Echteld, C. J., Kirkels, J. H., Ruigrok, T. J., van Hoesel, Q. G., de Jong, W. H., and Schornagel, J. H., 1991, Chronic cardiotoxicity of Adriamycin studied in a rat model by 31P NMR, *NMR Biomed.* **4**(1):16–24.

Demant, E. J., 1991, Inactivation of cytochrome *c* oxidase activity in mitochondrial membranes during redox cycling of doxorubicin, *Biochem. Pharmacol.* **41**(4):543–552.

Demant, E. J., and Jensen, P. K., 1983, Destruction of phospholipids and respiratory-chain activity in pig-heart submitochondrial particles induced by an Adriamycin–iron complex, *Eur. J. Biochem.* **132**(3):551–556.

Doroshow, J. H., and Davies, K. J., 1986, Redox cycling of anthracyclines by cardiac mitochondria: II. Formation of superoxide anion, hydrogen peroxide, and hydroxyl radical, *J. Biol. Chem.* **261**(7):3068–3074.

Doroshow, J. H., 1983a, Anthracycline antibiotic-stimulated superoxide, hydrogen peroxide, and hydroxyl radical production by NADH dehydrogenase, *Cancer Res.* **43**(10):4543–4551.

Doroshow, J. H., 1983b, Effect of anthracycline antibiotics on oxygen radical formation in rat heart, *Cancer Res.* **43**(2): 460–472.

Dutta, P. K., and Hutt, J. A., 1986, Resonance Raman spectroscopic studies of Adriamycin and copper(II)–Adriamycin and copper(II)–Adriamycin–DNA complexes, *Biochemistry* **25**(3):691–695.

Eilers, M., Endo, T., and Schatz, G., 1989, Adriamycin, a drug interacting with acidic phospholipids, blocks import of precursor proteins by isolated yeast mitochondria, *J. Biol. Chem.* **264**(5):2945–2950.

Ferrans, V. J., 1978, Overview of cardiac pathology in relation to anthracycline cardiotoxicity, *Cancer Treat. Rep.* **62**(6):955–961.

Ferrans, V. J., Clark, J. R., Zhang, J., Yu, Z. X., and Herman, E. H., 1997, Pathogenesis and prevention of doxorubicin cardiomyopathy, *Tsitologiia* **39**(10):928–937.

Ferrero, M. E., Ferrero, E., Gaja, G., and Bernelli-Zazzera, A., 1976, Adriamycin: Energy metabolism and mitochondrial oxidations in the heart of treated rabbits, *Biochem. Pharmacol.* **25**(2):125–130.

Floridi, A., Pulselli, R., Gentile, F. P., Barbieri, R., and Benassi, M., 1994, Rhein enhances the effect of Adriamycin on mitochondrial respiration by increasing antibiotic–membrane interaction, *Biochem. Pharmacol.* **47**(10):1781–1788.

Friedman, M. A., Bozdech, M. J., Billingham, M. E., and Rider, A. K., 1978, Doxorubicin cardiotoxicity: Serial endomyocardial biopsies and systolic time intervals, *JAMA* **240**(15):1603–1606.

Geetha, A., 1993, Influence of alpha-tocopherol on doxorubicin-induced lipid peroxidation, swelling, and thiol depletion in rat heart mitochondria, *Indian J. Exp. Biol.* **31**(3):297–298.

Geetha, A., Devi, C.S., 1992, Effect of doxorubicin on heart mitochondrial enzymes in rats: A protective role for alpha-tocopherol, *Indian J. Exp. Biol.* **30**(7):615–618.

Gervasi, P. G., Agrillo, M. R., Citti, L., Danesi, R., and Del Tacca, M., 1986, Superoxide anion production by adriamycinol from cardiac sarcosomes and by mitochondrial NADH dehydrogenase, *Anticancer Res.* **6**(5): 1231–1235.

Gervasi, P. G., Agrillo, M. R., Lippi, A., Bernardini, N., Danesi, R., and Del Tacca, M., 1990, Superoxide anion production by doxorubicin analogs in heart sarcosomes and by mitochondrial NADH dehydrogenase, *Res. Commun. Chem. Pathol. Pharmacol.* **67**(1):101–115.

Gianni, L., 1998, Paclitaxel plus doxorubicin in metastatic breast Ca: The Milan experience, *Oncology (Huntingt.)* **12**(1) (Suppl. 1):13–15.

Gille, L., and Nohl, H., 1997, Analyses of the molecular mechanism of Adriamycin-induced cardiotoxicity, *Free Radical Biol. Med.* **23**(5):775–782.

Goormaghtigh, E., Brasseur, R., and Ruysschaert, J. M., 1982, Adriamycin inactivates cytochrome *c* oxidase by exclusion of the enzyme from its cardiolipin essential environment, *Biochem. Biophys. Res. Commun.* **104**(1):314–320.

Goormaghtigh, E., Huart, P., Brasseur, R., and Ruysschaert, J. M., 1986, Mechanism of inhibition of mitochondrial enzymatic complex I–III by Adriamycin derivatives, *Biochem. Biophys. Acta* **861**(1):83–94.

Goormaghtigh, E., Huart, P., Praet, M., Brasseur, R., and Ruysschaert, J. M., 1990, Structure of the Adriamycin–cardiolipin complex: Role in mitochondrial toxicity, *Biophys. Chem.* **35**(2, 3):247–257.

Gorodetskaya, E. A., Dugin, S. F., Golikov, M. A., Kapelko, V. I., and Medvedev, O. S., 1990, The cardiac contractile function and hemodynamic control in rats after chronic Adriamycin treatment, *Can. J. Physiol. Pharmacol.* **68**(2):211–215.

Griffin-Green, E. A., Zaleska, M. M., Erecinska, M., 1988, Adriamycin-induced lipid peroxidation in mitochondria and microsomes, *Biochem. Pharmacol.* **37**(16):3071–3077.

Gunter, T. E., Wingrove, D. E., Banerjee, S., and Gunter, K. K., 1988, Mechanisms of mitochondrial calcium transport, *Adv. Exp. Med. Biol.* **232**:1–14.

Gutierrez, P. L., Gee, M. V., and Bachur, N. R., 1983, Kinetics of anthracycline antibiotic free radical formation and reductive glycosidase activity, *Arch. Biochem. Biophys.* **223**(1):68–75.

Hasinoff, B. B., 1990a, Inhibition and inactivation of NADH–cytochrome *c* reductase activity of bovine heart submitochondrial particles by the iron(III)–Adriamycin complex, *Biochem. J.* **265**(3):865–870.

Hasinoff, B. B., 1990b, Oxyradical production results from the Fe^{3+}–doxorubicin complex undergoing self-reduction by its alpha-ketol group, *Biochem. Cell Biol.* **68**(12):1331–1336.

Herman, E. H., Zhang, J., and Ferrans, V. J., 1994, Comparison of the protective effects of desferrioxamine and ICRF-187 against doxorubicin-induced toxicity in spontaneously hypertensive rats, *Cancer Chemother. Pharmacol.* **35**(2):93–100.

Hirsch, T., Marzo, I., and Kroemer, G., 1997, Role of the mitochondrial permeability transition pore in apoptosis, *Biosci. Rep.* **17**(1):67–76.

Hirsch, T., Susin, S. A., Marzo, I., Marchetti, P., Zamzami, N., and Kroemer, G., 1998, Mitochondrial permeability transition in apoptosis and necrosis, *Cell Biol. Toxicol.* **14**(2):141–145.

Hofling, B., Zahringer, J., and Bolte, H. D., 1982, Adriamycin-induced decrease of myocardial contraction reserve in rats treated with dobutamine, *Eur. J. Cancer Clin. Oncol.* **18**(1):75–80.

Huart, P., Brasseur, R., Goormaghtigh, E., and Ruysschaert, J. M., 1984, Antimitotics induce cardiolipin cluster formation: Possible role in mitochondrial enzyme inactivation, *Biochim. Biophys. Acta* **799**(2):199–202.

Jensen, R. A., Acton, E. M., and Peters, J. H., 1984, Doxorubicin cardiotoxicity in the rat: Comparison of electrocardiogram, transmembrane potential, and structural effects, *J. Cardiovasc. Pharmacol.* **6**(1):186–200.

Ji, L. L., and Mitchell, E. W., 1994, Effects of Adriamycin on heart mitochondrial function in rested and exercised rats, *Biochem. Pharmacol.* **47**(5):877–885.

Kang, Y. J., Chen, Y., and Epstein, P. N., 1996, Suppression of doxorubicin cardiotoxicity by overexpression of catalase in the heart of transgenic mice, *J. Biol. Chem.* **271**(21):12610–12616.

Kang, Y. J., Chen, Y., Yu, A., Voss-McCowan, M., and Epstein, P. N., 1997, Overexpression of metallothionein in the heart of transgenic mice suppresses doxorubicin in cardiotoxicity, *J. Clin. Invest.* **100**(6):1501–1506.

Kashfi, K., Israel, M., Sweatman, T. W., Seshadri, R., and Cook, G. A., 1990, Inhibition of mitochondrial carnitine palmitoyltransferases by Adriamycin and Adriamycin analogues, *Biochem. Pharmacol.* **40**(70):1441–1448.

Kawasaki, N., Lee, J. D., Shimizu, H., Ishii, Y., and Ueda, T., 1996, *Cardiac energy metabolism at several stages of Adriamycin-induced heart failure in rats,* Int. J. Cardiol. **55**(3):217–225.

Keizer, H. G., Pinedo, H. M., Schuurhuis, G. J., and Joenje, H., 1990, Doxorubucin (Adriamycin): A critical review of free radical-dependent mechanisms of cytotoxicity, *Pharmacol. Ther.* **47**(2):219–231.

Lee, V., Randhawa, A. K., and Singal, P. K., 1991, Adriamycin-induced myocardial dysfunction *in vitro* is mediated by free radicals, *Am. J. Physiol.* **261**(4) (Pt 2):H989–995.

Lefrak, E. A., Pitha, J., Rosenheim, S., and Gottlieb, J. A., 1973, A clinicopathologic analysis of Adriamycin cardiotoxicity, *Cancer* **32**(2):302–314.

Lemasters, J. J., 1998, The mitochondrial permeability transition: From biochemical curiosity to pathophysiological mechanism [editorial; comment], *Gastroenterology* **115**(3):783–786.

Lemasters, J. J., Nieminen, A. L., Qian, T., Trost, L. C., Elmore, S. P., Nishimura, Y., Crowe, R. A., Cascio, W. E., Bradham, C. A., Brenner, D. A., and Herman, B., 1998, The mitochondrial permeability transition in cell death: A common mechanism in necrosis, apoptosis, and autophagy, *Biochim. Biophys. Acta* **1366**(1, 2):177–196.

Lin, T. J., Liu, G. T., Pan, Y., Liu, Y., and Xu, G. Z., 1991, Protection by schisahenol against Adriamycin toxicity in rat heart mitochondria, *Biochem. Pharmacol.* **42**(9):1805–1810.

Lin, T. J., Liu, G. T., Liu, Y., and Xu, G. Z., 1992, Protection by salvianolic acid A against Adriamycin toxicity on rat heart mitochondria, *Free Radical Biol. Med.* **12**(5):347–351.

Link, G., Tirosh, R., Pinson, A., and Hershko, C., 1996, Role of iron in the potentiation of anthracycline cardiotoxicity: Identification of heart cell mitochondria as a major site of iron–anthracycline interaction, *J. Lab. Clin. Med.* **127**(3):272–278.

Lown, J. W., Chen, H. H., Plambeck, J. A., and Acton, E. M., 1982, Further studies on the generation of reactive oxygen species from activated anthracyclines and the relationship to cytotoxic action and cardiotoxic effects, *Biochem. Pharmacol.* **31**(4):575–581.

Malatesta, V., Morazzoni, F., Gervasini, A., and Arcamone, F., 1985, Chelation of copper(II) ions by doxorubicin and 4′-epidoxorubicin: An e.s.r. study, *Anticancer Drug Des.* **1**(1):53–57.

Malisza, K. L., and Hasinoff, B. B., 1995, Production of hydroxyl radical by iron(III)–anthraquinone complexes through self-reduction and through reductive activation by the xanthine oxidase/hypoxanthine system, *Arch. Biochem. Biophys.* **321**(1):51–60.

Marcillat, O., Zhang, Y., Davies, K. J., 1989, Oxidative and non-oxidative mechanisms in the inactivation of cardiac mitochondrial electron transport chain components by doxorubicin, *Biochem. J.* **259**(1):181–189.

Matsumura, M., Nishioka, K., Fujii, T., Yoshibayashi, M., Nozaki, K., Nakata, Y., Temma, S., Ueda, T., and Mikawa, H., 1994, Age-related acute Adriamycin cardiotoxicity in mice, *J. Mol. Cell Cardiol* **26**(7):899–905.

McKillop, J. H., Bristow, M. R., Goris, M. L., Billingham, M. E., and Bockemuehl, K., 1983, Sensitivity and specificity of radionuclide ejection fractions in doxorubicin cardiotoxicity, *Am. Heart J.* **106**(5) (Pt. 1):1048–1056.

Meredith, M. J., and Reed, D. J., 1983, Depletion *in vitro* of mitochondrial glutathione in rat hepatocytes and enhancement of lipid peroxidation by Adriamycin and 1,3-bis(2-chloroethyl)-1-nitrosourea (BCNU), *Biochem. Pharmacol.* **32**(8):1383–1388.

Mimnaugh, E. G., Trush, M. A., Bhatnagar, M., and Gram, T. E., 1985, Enhancement of reactive oxygen-dependent mitochondrial membrane lipid peroxidation by the anticancer drug Adriamycin, *Biochem. Pharmacol.* **34**(6):847–856.

Miura, T., Muraoka, S., Ogiso, T., 1994, Adriamycin–Fe^{3+}-induced inactivation of rat heart mitochondrial creatine kinase: Sensitivity to lipid peroxidation, *Biol. Pharm. Bull.* **17**(9):1220–1223.

Montali, U., Del Tacca, M., Bernardini, C., Segnini, D., and Solaini, G., 1985, Cardiotoxic effects of Adriamycin and mitochondrial oxidation in rat cardiac tissue, *Drug Exp. Clin. Res.* **11**(3):219–222.

Monti, E., Paracchini, L., Perletti, G., and Piccinini, F., 1991, Protective effects of spin-trapping agents on Adriamycin-induced cardiotoxicity in isolated rat atria, *Free Radical Res. Commun.* **14**(1):41–45.

Monti, E., Prosperi, E., Supino, R., and Bottiroli, G., 1995, Free radical-dependent DNA lesions are involved in the delayed cardiotoxicity induced by Adriamycin in the rat, *Anticancer Res.* **15**(1):193–197.

Moore, L., Landon, E. J., Cooney, D. A., 1977, Inhibition of the cardiac mitochondrial calcium pump by Adriamycin *in vitro*, *Biochem. Med.* **18**(2):131–138.

Mortensen, S. A., Olsen, H. S., and Baandrup, U., 1986, Chronic anthracycline cardiotoxicity: Haemodynamic and histopathological manifestations suggesting a restrictive endomyocardial disease, *Br. Heart J.* **55**(3):274–282.

Mott, M. G., 1997, Anthracycline cardiotoxicity and its prevention, *Ann. NY Acad. Sci.* **824**:221–228.

Muhammed, H., and Kurup, C. K., 1984, Influence of ubiquinone on the inhibitory effect of Adriamycin on mitochondrial oxidative phosphorylation, *Biochem. J.* **217**(2):493–498.

Muhammed, H., Ramasarma, T., and Kurup, C. K., 1983, Inhibition of mitochondrial oxidative phosphorylation by Adriamycin, *Biochim. Biophys. Acta* **722**(1):43–50.

Nicolay, K., and de Kruijff, B. 1987, Effects of Adriamycin on respiratory chain activities in mitochondria from rat liver, rat heart and bovine heart: Evidence for a preferential inhibition of complex III and IV, *Biochim. Biophys. Acta* **892**(3):320–330.

Nicolay, K., Timmers, R. J., Spoelstra, E., Van der Neut, R., Fok, J. J., Huigen, Y. M., Verkleij, A. J., and De Kruijff, B., 1984, The interaction of Adriamycin with cardiolipin in model and rat liver mitochondrial membranes, *Biochim. Biophys. Acta* **778**(2):359–371.

Nohl, H., and Jordan, W., 1983, OH-generation by Adriamycin semiquinone and H_2O_2: An explanation for the cardiotoxicity of anthracycline antibiotics, *Biochem. Biophys. Res. Commun.* **114**(1):197–205.

Ogura, R., Sugiyama, M., Haramaki, N., and Hidaka, T., 1991, Electron spin resonance studies on the mechanism of Adriamycin-induced heart mitochondrial damages, *Cancer Res.* **51**(13):3555–3558.

Olson, H. M., and Capen, C. C., 1978, Chronic cardiotoxicity of doxorubicin (Adriamycin) in the rat: Morphologic and biochemical investigations, *Toxicol. Appl. Pharmacol.* **44**(3):605–616.

Olson, R. D., and Mushlin, P. S., 1990, Doxorubicin cardiotoxicity: Analysis of prevailing hypotheses [see comments], *FASEB J.* **4**(13):3076–3086.

Olson, R. D., MacDonald, J. S., van Boxtel, C. J., Boerth, R. C., Harbison, R. D., Slonim, A. E., Freeman, R. W., and Oates, J. A., 1980, Regulatory role of glutathione and soluble sulfhydryl groups in the toxicity of Adriamycin, *J. Pharmacol. Exp. Ther.* **215**:450–454.

Olson, R. D., Boerth, R. C., Gerger, J. G., and Nies, A. S., 1981, Mechanism of Adriamycin cardiotoxicity: Evidence for oxidative stress, *Life Sci.* **29**(14):1393–1401.

Pan, S. S., and Bachur, N. R., 1980, Xanthine oxidase catalyzed reductive cleavage of anthracycline antibiotics and free radical formation, *Mol. Pharmacol.* **17**(1):95–99.

Paradies, G., and Ruggiero, F. M., 1988, The effect of doxorubicin on the transport of pyruvate in rat-heart mitochondria, *Biochem. Biophys. Res. Commun.* **156**(3):1302–1307.

Petronilli, V., Costantini, P., Scorrano, L., Colonna, R., Passamonti, S., and Bernardi, P., 1994, The voltage sensor of the mitochondrial permeability transition pore is tuned by the oxidation-reduction state of vicinal thiols: Increase of the gating potential by oxidants and its reversal by reducing agents, *J. Biol. Chem.* **269**(24):16638–16642.

Piccinini, F., Monti, E., Paracchini, L., and Perlettei, G., 1990, Are oxygen radicals responsible for the acute cardiotoxicity of doxorubicin? *Adv. Exp. Med. Biol.* **264**:349–352.

Pollakis, G., Goormaghtigh, E., Delmelle, M., Lion, Y., and Ruysschaert, J. M., 1984, Adriamycin and derivatives interaction with the mitochondrial membrane: O2 consumption and free radical formation, *Res. Commun. Chem. Pathol. Pharmacol.* **44**(3):445–459.

Praet, M., Laghmiche, M., Pollakis, G., Goormaghtigh, E., and Ruysschaert, J. M., 1986, *In vivo* and *in vitro* modifications of the mitochondrial membrane induced by 4′-Epi-Adriamycin, *Biochem. Pharmacol.* **35**(17):2923–2928.

Praet, M., Calderon, P. B., Pollakis, G., Roberfroid, M., and Ruysschaert, J. M., 1988, A new class of free radical scavengers reducing Adriamycin mitochondrial toxicity, *Biochem. Pharmacol.* **37**(24):4617–4622.

Praga, C., Beretta, G., et al., 1979, Adriamycin cardiotoxicity: A survey of 1273 patients, *Cancer Treat. Rep.* **63**(5):827–834.

Revis, N., Marusic, N., 1979, Sequestration of $^{45}Ca^{2+}$ by mitochondria from rabbit heart, liver, and kidney after doxorubicin or digoxin/doxorubicin treatment, *Exp. Mol. Pathol.* **31**(3):440–451.

Revis, N. W., Marusic, N., 1979, Effects of doxorubicin and its aglycone metabolite on calcium sequestration by rabbit heart, liver, and kidney mitochondria, *Life Sci.*, **25**(12):1055–1063.

Rhoden, W., Hasleton, P., and Brooks, N., 1993, Anthracyclines and the heart [see comments], *Br. Heart J.* **70**(6):499–502.

Rosenoff, S. H., Olson, H. M., Young, D. M., Bostick, F., and Young, R. C., 1975, Adriamycin-induced cardiac damage in the mouse: A small-animal model of cardiotoxicity, *J. Natl. Cancer Inst.* **55**(1):191–194.

Saltiel, E., McGuire, W., 1983, Doxorubicin (Adriamycin) cardiomyopathy, *West. J. Med.* **139**(3):332–341.

Scheulen, M. E., and Kappus. H., 1981, Metabolic activation of Adriamycin by NADPH–cytochrome P-450 reductase, rat liver, and heart microsomes and covalent protein binding of metabolites, *Adv. Exp. Mol. Biol.* **136**(Pt. A):471–485.

Scheulen, M. E., Kappus, H., Nienhaus, A., and Schmidt, C. G., 1982, Covalent protein binding of reactive Adriamycin metabolites in rat liver and rat heart microsomes, *J. Cancer Res. Clin. Oncol.* **103**(1):39–48.

Shinozawa, S., Fukuda, T., Araki, Y., and Oda, T., 1985, Effect of dextran sulfate on the survival time and mitochondrial function of Adriamycin-(doxorubicin)-treated mice, *Toxicol. Appl. Pharmacol.* **79**(2):353–357.

Singal, P. K., Iliskovic, N., Li, T., and Kumar, D., 1997, Adriamycin cardiomyopathy: Pathophysiology and prevention, *FASEB J.* **11**(12):931–936.

Singal, P. K., and Iliskovic, N., 1998, Doxorubicin-induced cardiomyopathy, *N. Engl. J. Med.* **339**(13):900–905.

Singal, P. K., Forbes, M. S., and Sperelakis, N., 1984, Occurrence of intramitochondrial Ca^{2+} granules in a hypertrophied heart exposed to Adriamycin, *Can. J. Physiol. Pharmacol.* **62**(9):1239–1244.

Singal, P. K., Deally, C. M., and Weinberg, L. E., 1987, Subcellular effects of Adriamycin in the heart: A concise review, *J. Mol. Cell Cardiol.* **19**(8):817–828.

Sokolove, P. M., 1988, Mitochondrial sulfhydryl group modification by Adriamycin aglycones, *FEBS Lett.* **234**(1):199–202.

Sokolove, P. M., 1990, Inhibition by cyclosporin A and butylated hydroxytoluene of the inner mitochondrial membrane permeability transition induced by Adriamycin and aglycones, *Biochem. Pharmacol.* **40**(12):2733–2736.

Sokolove, P. M., 1991, Oxidation of mitochondrial pyridine nucleotides by aglycone derivatives of adriamycin, *Arch. Biochem. Biophys.* **284**(2):292–297.

Sokolove, P. M., and Shinaberry, R. G., 1988, Na^+-independent release of Ca^{2+} from rat heart mitochondria: Induction by Adriamycin aglycone, *Biochem. Pharmacol.* **37**(5):803–812.

Solaini, G., Ronca, G., and Bertelli, A., 1985, Studies on the effects of anthracyclines on mitochondrial respiration *in vitro*, *Drugs Exp. Clin. Res.* **11**(2):115–121.

Solaini, G., Landi, L., Pasquali, P., and Rossi, C. A., 1987, Protective effect of endogenous coenzyme Q on both lipid peroxidation and respiratory chain inactivation induced by an Adriamycin–iron complex, *Biochem. Biophys. Res. Commun.* **147**(2):572–580.

Solem, L. E., and Wallace, K. B., 1993, Selective activation of the sodium-independent, cyclosporin A-sensitive calcium pore of cardiac mitochondria by doxorubicin, *Toxicol. Appl. Pharmacol.* **121**(1):50–57.

Solem, L. E., Henry, T. R., and Wallace, K. B., 1994, Disruption of mitochondrial calcium homeostasis following chronic doxorubicin administration, *Toxicol. Appl. Pharmacol.* **129**(2):214–222.

Solem, L. E., Heller, L. J., and Wallace, K. B., 1996, Dose-dependent increase in sensitivity to calcium-induced mito-chondrial dysfunction and cardiomyocete cell injury by doxorubicin, *J. Mol. Cell Cardiol.* **28**(5):1023–1032.

Steinherz, L., and Steinherz, P., 1991, Delayed cardiac toxicity from anthracycline therapy, *Paediatrician* **18**(1):49–52.

Steinherz, L. J., Stenherz, P. G., Tan, C. J., Heller, G., and Murphy, M. L., 1991, Cardiac toxicity 4 to 20 years after completing anthracycline therapy, *JAMA* **266**(12):1672–1677.

Steinherz, L. J., Steinherz, P. G., and Tan, C., 1995, Cardiac failure and dysrhythmias 6–19 years after anthracycline therapy: A series of 15 patients, *Med. Pediat. Oncol.* **24**(6):352–361.

Sugiyama, S., Yamada, K., Hayakawa, M., and Ozawa, T., 1995, Approaches that mitigate doxorubicin-induced delayed adverse effects on mitochondrial function in rat hearts: Liposome-encapsulated doxorubicin or combination therapy with antioxidant, *Biochem. Mol. Biol. Int.* **36**(5):1001–1007.

Suzuki, T., Kanda, H., Kawai, Y., Tominaga, K., and Murata, K., 1979, Cardiotoxicity of anthracycline antineoplastic drugs: Clinicopathological and experimental studies, *Jpn. Circ. J.* **43**(11):1000–1008.

Thayer, W. S., 1977, Adriamycin stimulated superoxide formation in submitochondrial particles, *Chem. Biol. Int.* **19**(3):265–278.

Trost, L. C., and Wallace, K. B., 1994a, Adriamycin-induced oxidation of myoglobin, *Biochem. Biophys. Res. Commun.* **204**(1):30–37.

Trost, L. C., and Wallace, K. B., 1994b, Stimulation of myoglobin-dependent lipid peroxidation by Adriamycin, *Biochem. Biophys. Res. Commun.* **204**(1):23–29.

Valls, V., Castelluccio, C., Fato, R., Genova, M. L., Bovina, C., Saez, G., Marchetti, M., Parenti Castelli, G., and Lenaz, G., 1994, Protective effect of exogenous coenzyme Q against damage by Adriamycin in perfused rat liver, *Biochem. Mol. Bio. Int.* **33**(4):633–642.

Van Vleet, J. F., and Ferrans, V. J., 1980, Clinical observations, cutaneous lesions, and hematologic alterations in chronic Adriamycin intoxication in dogs with and without vitamin E and selenium supplementation, *Am. J. Vet. Res.* **41**(5):691–699.

Vile, G. F., Winterbourn, C. C., Sutton, H. C., 1987, Radical-driven Fenton reactions: Studies with paraquat, Adriamycin, and anthraquinone 6-sulfonate and citrate, ATP, ADP, and pyrophosphate iron chelates, *Arch. Biochem. Biophys.* **259**(2):616–626.

Villani, F., Favalli, L., Lanza, E., Rozza-Dionigi, A., and Poggi, P., 1987, Anthracycline cardiotoxicity in the rat: Relationship between ECG changes and morphologic effects induced by different analogues, *Chemioterapia* **6**(2) (Suppl):688–690.

Voelker, D. R., 1991, Adriamycin disrupts phosphatidylserine import into the mitochondria of permeabilized CHO-Kl cells, *J. Biol. Chem.* **266**(19):12185–12188.

Wallace, K. B., 1986a, Aglycosylation and disposition of doxorubicin in isolated rat liver nuclei and microsomes, *Drug Metab. Dis.* **14**(4):399–404.

Wallace, K. B., 1986b, Nonenzymatic oxygen activation and stimulation of lipid peroxidation by doxorubicin–copper, *Toxicol. Appl. Pharmacol.* **86**(1):69–79.

Wallace, K. B., and Johnson, J. A., 1987, Oxygen-dependent effect of microsomes on the binding of doxorubicin to rat hepatic nuclear DNA, *Mol. Pharmacol.* **31**(3):307–311.

Yamada, K., Sugiyama, S., Kosaka, K., Hayakawa, M., and Ozawa, T., 1995, Early appearance of age-associated deterioration in mitochondrial function of diaphragm and heart in rats treated with doxorubicin, *Exp. Gerontol.* **30**(6):581–593.

Yen, H. C., Oberley, T. D., Vichitbandha, S., Ho, Y. S., and St Clair, D. K., 1996, The protective role of manganese superoxide dismutase against Adriamycin-induced acute cardiac toxicity in transgenic mice [erratum appears in 1997, *J. Clin. Invest.* **99**(5):1141], *J. Clin. Invest.* **98**(5):1253–1260.

Zidenberg-Cherr, S., and Keen, C. L., 1986, Influence of dietary manganese and vitamin E on Adriamycin toxicity in mice, *Toxicol. Lett.* **30**(1):79–87.

Zoratti, M., and Szabò, I., 1995, The mitochondrial permeability transition, *Biochim. Biophys. Acta* **1241**(2):139–176.

Zunino, F., and Capranico, G., 1990, DNA topoisomerase II as the primary target of anti-tumor anthracyclines, *Anticancer Drug Des.* **5**(4):307–317.

Zweier, J. L., 1984, Reduction of O2 by iron–Adriamycin, *J. Biol. Chem.* **259**(10):6056–6058.

Zweier, J. L., 1985, Iron-mediated formation of an oxidized Adriamycin free radical, *Biochim. Biophys. Acta* **829**(2):209–213.

Chapter 26

Drug-Induced Microvesicular Steatosis and Steatohepatitis

Dominique Pessayre, Bernard Fromenty, and Abdellah Mansouri

1. INTRODUCTION

About 2 billion years ago, a precursor of present day eukaryotes engaged in a parasitic/symbiotic partnership with a bacterium (Green and Reed, 1998). This precursor allowed the bacterium to reside and divide within its cytoplasm. In exchange the bacterium used the emerging oxygen atmosphere to completely degrade fuels into CO_2 and water, thus providing both partners with considerable energy (Green and Reed, 1998).

At first this was perhaps an unstable, hectic relationship, because bacteria tend to overgrow when well fed. Eventually, however, most of the bacterial genes were transferred to the nucleus of the host, and both transcription and replication of the bacterial genome were placed under the control of nuclear genes, turning the unstable, facultative relationship into a stable, obligate symbiosis (Green and Reed, 1998). Our present-day mitochondria cannot live without us and we can't live without them.

2. MITOCHONDRIA: OUR STRENGTH AND ACHILLES' HEEL

By turning fat (and other fuels) into power, mitochondria are the main source of our strength (Fig. 1). Let us first recall that the mitochondrion has two membranes (outer, which encloses the intermembranous space, and inner, which limits the mitochondrial matrix) and let us follow the route from a long-chain fatty acid to ATP generation (Fig. 1)

Dominique Pessayre, Bernard Fromenty, and Abdellah Mansouri   INSERM U 481, Hôpital Beaujon, 92118 Clichy, France.

Mitochondria in Pathogenesis, edited by Lemasters and Nieminen.
Kluwer Academic/Plenum Publishers, New York, 2001.

(Schulz, 1991). The long-chain fatty acid is first activated into the acyl-CoA thioester on the outer mitochondrial membrane. Carnitine palmitoyltransferase I (CPT I), whose active site may be located both on the cytosolic face of the outer membrane (McGarry and Brown, 1997) and perhaps also in the inner membrane at contact sites (Hoppel *et al.*, 1998), forms the acylcarnitine, which may cross first the outer membrane, next the inner membrane, through the carnitine/acylcarnitine translocase. The CPT II then regenerates the acyl-CoA thioester in the mitochondrial matrix (Fig. 1).

The long-chain acyl-CoA is next cut into acetyl-CoA subunits by successive β-oxidation cycles (Fig. 1). The first cycles are catalyzed by membrane-bound enzymes. Despite its name, very long chain acyl-CoA dehydrogenase (VLCAD) is also very active with long-chain fatty acids (Izai *et al.*, 1992), forming the enoyl-CoA derivative. The mitochondrial trifunctional enzyme has enoyl-CoA hydratase, long-chain 3-hydroxylacyl-CoA dehydrogenase (LCHAD), and 3-keto-acyl-CoA thiolase activities in a single protein (Uchida *et al.*, 1992), and it completes the first β-oxidation cycles.

The shortened fatty acyl-CoA is then metabolized by clusters of soluble matrix enzymes, which successively handle the fatty acid as it is progressively shortened into a still long-, then medium-, and finally short-chain fatty acyl-CoA (Nada *et al.*, 1995). Long-, medium-, and short-chain acyl-CoA dehydrogenase (LCAD, MCAD, and SCAD) mediate the first dehydrogenation step of these β-oxidation cycles, and soluble enoyl-CoA

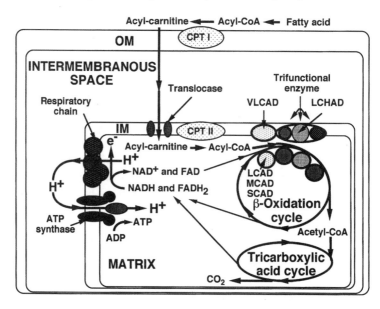

FIGURE 1. Metabolism of long-chain fatty acids and ATP generation in mitochondria. Fatty acid enters the mitochondrion through the carnitine shuttle, is shortened by successive β-oxidation cycles, and may then be further oxidized by the tricarboxylic acid cycle. Both β-oxidation and the tricarboxylic acid cycle generate NADH and $FADH_2$. Electrons liberated during reoxidation of these reduced cofactors migrate along the respiratory chain while protons are pumped into the intermembrane space, creating a large electrochemical potential across the inner membrane. When ATP is needed, protons re-enter the matrix through ATP synthase, activating a molecular rotor and the synthesis of ATP. OM and IM, outer and inner membrane. VLCAD, LCAD, MCAD, and SCAD, very long chain, long-chain, medium-chain, and short-chain acyl-CoA dehydrogenase, LCHAD, long-chain 3-hydroxyl-acyl-CoA dehydrogenase.

hydratases, 3-hydroxyl-acyl-CoA dehydrogenases, and 3-keto-acyl-CoA thiolases then complete these cycles (Fig. 1).

Each β-oxidation cycle consumes FAD and NAD^+ (during the first and second dehydrogenations, respectively), forming $FADH_2$ and NADH (Fig. 1), and each cycle produces an acetyl-CoA molecule that can be further oxidized into CO_2 by the tricarboxylic acid cycle, generating NADH and $FADH_2$. The tricarboxylic acid cycle also oxidizes the diverse intermediates derived from the oxidation of sugars and amino acids.

The NADH and $FADH_2$ are then reoxidized by the mitochondrial respiratory chain attached to the inner mitochondrial membrane. This regenerates the NAD^+ and FAD required for other cycles of fuel oxidation (Fig. 1).

The electrons transferred by NADH and $FADH_2$ migrate along the respiratory chain up to cytochrome c oxidase, where they combine with oxygen and protons to form water. During this transfer, protons are extruded from the mitochondrial matrix into the intermembrane space, creating a large electrochemical potential across the inner membrane (Fig. 1), whose potential energy is then used to generate ATP. When energy is needed, protons re-enter the matrix through the F_0 portion of ATP synthase. This sets in motion a molecular rotor in the F_1 portion of ATPase and the synthesis of ATP (Fig. 1) (Kinosita *et al.*, 1998).

Being the main source of our energy, mitochondria are also our Achilles' heel. When they are damaged by drugs, they may leave us powerless (which may cause cell dysfunction and cell death) and with a peculiar form of fatty liver, termed *microvesicular steatosis*.

3. MICROVESICULAR STEATOSIS

Although there are transitions between microvesicular and macrovacuolar steatosis (with a given hepatocyte exhibiting both small and large lipid droplets), as well as frequent associations of the two (with microvesicular steatosis in some hepatocytes and macro-vacuolar steatosis in others), it is important to distinguish these two steatosis forms. Indeed, microvesicular and macrovacuolar steatosis may differ not only by morphology, but also by mechanism and short-term prognosis (Fromenty and Pessayre, 1995).

3.1. Morphology

In macrovacuolar steatosis, a single large vacuole of fat displaces the nucleus to the cell's periphery (Fig. 2), whereas in microvesicular steatosis numerous small lipid vesicles give the hepatocytes a foamy, spongiocytic appearance. Although experienced pathologists can recognize the latter lesion, in mild cases an Oil Red O or other lipid stain may be useful to detect or confirm the tiny lipid vesicles.

3.2. Mechanism

Macrovacuolar steatosis may result from various combinations of different alterations in lipid metabolism (including increased mobilization of fat from adipose tissue, increased hepatic synthesis of fatty acids, and decreased egress of triglycerides from the liver) in the

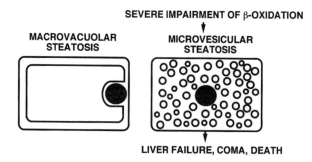

FIGURE 2. Macrovacuolar and microvesicular steatosis. Diverse alterations of lipid metabolism may cause macrovacuolar steatosis, a benign condition, at least in the short term. Severe impairment of mitochondrial β-oxidation causes microvesicular steatosis, a severe metabolic disease that can cause liver failure, coma, and death.

absence of, or with only mild, decreases in mitochondrial β-oxidation. Common causes of macrovacuolar steatosis in humans are alcohol abuse, obesity, diabetes, and dyslipemias.

In contrast, microvesicular steatosis appears invariably related to severe impairment of fatty acid β-oxidation (Fig. 2). Mild impairment will not suffice; impairment must be profound (Fromenty and Pessayre, 1995).

Nonesterified fatty acids (NEFAs) that are not oxidized in mitochondria accumulate in the liver. Although most of these NEFAs are then esterified into triglycerides, there is a residual increase in hepatic NEFA concentration (Eisele *et al.*, 1975; Heubi *et al.*, 1987). The NEFAs are amphiphilic compounds that might embed their lipophilic tail into a core of neutral triglycerides, leaving their polar head groups in the water phase, possibly emulsifying triglycerides and thus explaining why triglycerides accumulate as small lipid vesicles when mitochondrial β-oxidation is impaired and hepatic NEFAs are therefore increased (Fromenty and Pessayre, 1995).

3.3. Short-term Severity: Energy Crisis

Whereas isolated macrovacuolar steatosis (without other liver lesions) is, at least in the short term, a benign condition, extensive microvesicular steatosis is the histologic hallmark of severe metabolic disease, which may quickly cause liver failure, coma, and death (Fig. 2).

The severity of microvesicular steatosis might be related to an energy crisis (Fromenty and Pessayre, 1995). Indeed, impaired fat oxidation deprives cells of an important source of energy between meals. Impairmed β-oxidation also impairs gluco-neogenesis. Together with the increased peripheral utilization of glucose this may cause hypoglycemia, thus further hampering energy production in extrahepatic organs. Impairment of β-oxidation also increases hepatic NEFAs and their dicarboxylic acid derivatives. Both uncouple oxidative phosphorylation (Tonsgard and Getz, 1985; Wojtczak and Schönfeld, 1993), still further decreasing energy production.

Of course, the energy deficit will be even more severe when impairment of fat oxidation is itself a consequence of initial respiratory chain impairment.

4. SECONDARY INHIBITION OF β-OXIDATION

As we have seen, fatty acids are activated in an ATP-dependent process into fatty acyl-CoAs, whose β-oxidation is associated with the conversion of NAD$^+$ into NADH (Fig. 1). The NADH is then reoxidized by the respiratory chain into NAD$^+$, allowing β-oxidation cycles to continue (Fig. 1). When the respiratory chain is severely impaired, however, not enough NAD$^+$ is regenerated. This impairs the activity of LCHAD, and β-oxidation is secondarily decreased (Guzman and Geelen, 1993; Latipää et al., 1986).

Another mechanism with many consequences, including secondary inhibition of β-oxidation, is opening of the mitochondrial permeability transition pore (MPTP). Pore opening has three *main* consequences (Fig. 3).

First, due to the hyperosmolality of the mitochondrial matrix, pore opening causes the matrix to swell. The inner membrane has several folds (mitochondrial cristae) and can easily accommodate an increase in matrix volume, but the spherical outer membrane can burst when the mitochondrion swells (Fig. 3) (Bernardi et al., 1998). Outer membrane rupture releases both apoptosis-inducing factor (AIF) and cytochrome c from the intermembrane space. AIF cuts DNA, and cytochrome c activate caspases (Susin et al., 1998). This can cause apoptosis, if the cell maintains enough ATP to develop this energy-requiring process while avoiding necrosis. These conditions may occur when the pore opens in only some mitochondria, when the cell can derive enough ATP from anaerobic glycolysis, or both.

Second, pore opening allows massive proton re-entry into the matrix, bypassing ATP synthase. If the pore opens quickly in all mitochondria and the cell cannot derive enough energy from anaerobic glycolysis, then the major drop in cellular ATP can cause necrosis and the energy-requiring apoptotic process is prevented. Decreased ATP levels can also impair the initial activation of fatty acids into acyl-CoA and thus prevent their secondary β-oxidation (Fig. 3).

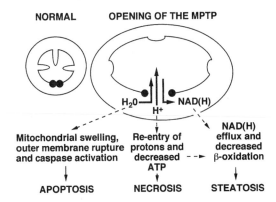

FIGURE 3. Consequences of MPTP opening. Due to matrix hyperosmolality, pore opening causes matrix swelling and the entry of water, which may rupture the outer membrane, release cytochrome c, and activate caspases, leading to apoptosis. Pore opening also causes re-entry of protons, bypassing ATP synthase, which can decrease cellular ATP and cause necrosis. Finally, pore opening causes a loss of NAD(H). Together with the reduced availability of ATP (which is required for initial activation of fatty acids), the NAD(H) depletion might inhibit fatty acid β-oxidation and cause steatosis.

Third, MPTP opening allows the efflux of NAD$^+$ and NADH from the mitochondria, which can impair respiration mediated by complex I substrates and, impair mitochondrial β-oxidation of fatty acids, which requires NAD$^+$ (Fig. 3).

5. DIVERSITY OF MECHANISMS IMPAIRING β-OXIDATION

Drugs impair β-oxidation through many different mechanisms (Fig. 4). Indeed, drugs can act by opening the MPTP, by sequestering CoA (required for initial activation of fatty acids), by directly inhibiting β-oxidation enzymes, by severely inhibiting mitochondrial respiration, or by interfering with mitochondrial DNA (mtDNA), which encodes some of the polypeptides of the respiratory chain (Fig. 4).

5.1. Mitochondrial DNA Lesions

5.1.1. Ethanol

Although ethanol abuse commonly causes macrovacuolar steatosis, it may also cause microvesicular steatosis in some patients. Patients with mild forms have high serum transaminase activity compared with patients who have macrovacuolar steatosis, and they exhibit centrilobular foamy hepatocytes, but their short-term prognosis is good. In a severe, fortunately most uncommon, form there is massive accumulation of both microvesicular and macrovacuolar fat in the liver, and the clinical presentation sometimes resembles Reye's syndrome. These patients present with jaundice and hepatomegaly, and rapidly proceed during the following hours or days to lethargy, coma, and death (Rosmorduc *et al.*, 1992; Uchida *et al.*, 1983).

Ethanol consumption increases the formation of reactive oxygen species (ROS) in mitochondria (Fig. 5) (Bailey and Cunningham, 1998; Fromenty and Pessayre, 1995; Nordmann *et al.*, 1992). These ROS oxidize the mitochondrial lipids and the polypeptides of the respiratory chain (Krähenbühl, 1993; Williams *et al.*, 1998). They also selectively damage mtDNA, which is 10–16 times more prone to oxidative damage than is nuclear

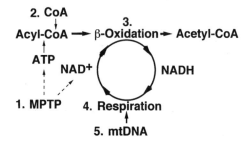

FIGURE 4. Diversity of drug-induced effects impairing mitochondrial β-oxidation. Drugs may act (1) by opening the MPTP, which causes ATP depletion and NAD$^+$ loss; (2) by sequestering CoA, which is required to form the acyl-CoA derivative; (3) by directly inhibiting β-oxidation enzymes; (4) by impairing mitochondrial respiration, thus preventing NADH reoxidation into the NAD$^+$ required for β-oxidation; or (5) by interfering with mtDNA, which encodes some of the respiratory chain polypeptides.

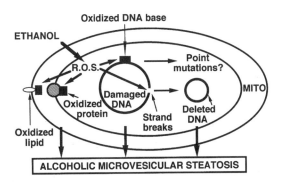

FIGURE 5. Suggested pathogenesis of alcoholic microvesicular steatosis. Ethanol abuse increases mitochondrial ROS formation, oxidatively damaging mitochondrial lipids, proteins, and DNA. In some alcoholic patients, this acute oxidative insult and secondary mtDNA mutation severely impairs mitochondrial β-oxidation causing microvesicular steatosis.

DNA due to absence of protective histones, attachment to mitochondrial inner membrane (the main source of cellular ROS), and less-efficient repair processes in the mitochondrion than in the nucleus (Richter *et al.*, 1988; Yakes and Van Houten, 1997). Because mtDNA essentially lacks introns, most oxidative damage will involve functionally important genes.

Ethanol-induced oxidative damage to mtDNA will form 8-hydroxydeoxyguanosine (Fig. 5) (Wieland and Lauterburg, 1995). This modified guanine causes incorporation of wrong nucleotides (instead of cytosine) during replication and might cause mtDNA point mutations (Retèl *et al.*, 1993). An alcoholic binge also causes massive degradation and depletion of hepatic mtDNA in mice, followed by increased mtDNA synthesis (Mansouri *et al.*, 1999). Although most of the mtDNA strand-breaks are quickly repaired or replaced, some might cause mitochondrial DNA deletions (Fig. 5).

The combination of acute oxidative insult (to mitochondrial lipids, proteins, and DNA) and secondary mtDNA mutations (secondarily arising from the acute DNA insult) may severely impair β-oxidation in some patients, thus causing alcoholic microvesicular steatosis (Fig. 5).

This might explain the peculiar association we found between mtDNA deletions (a consequence of oxidative damage to mtDNA) and microvesicular steatosis (Fromenty *et al.*, 1995; Mansouri *et al.*, 1997a). Whereas only 3% of controls exhibited one mtDNA deletion, 85% of similarly aged alcoholics with microvesicular steatosis had at least one, and often several, concomitant mtDNA deletions. In contrast, the prevalence of mtDNA deletions was only barely increased (7%) in alcoholics with other liver lesions but without microvesicular steatosis.

5.1.2. Copper

Wilson disease is caused by diverse mutations of a nuclear gene encoding a copper transporting P-type ATPase (Thomas *et al.*, 1995). Decreased biliary elimination of copper causes its progressive accumulation within hepatocytes (Sokol *et al.*, 1994). Due to its ability to cycle back and forth between oxidized and reduced states, copper generates the hydroxyl radical and other reactive species (Kadiiska *et al.*, 1993). Copper forms Cu–DNA complexes, so these ROS are generated close to DNA, making it an elective target (Oikawa and Kawanishi, 1996). Because copper accumulates selectively within mitochondria during copper overloads (Gregoriadis and Sourkes, 1967; Sokol *et al.*, 1994), mtDNA may be particularly affected. Indeed, despite a young age, half the patients with Wilson disease already showed one or more mtDNA deletions, whereas only 3% of older controls had even one (Mansouri *et al.*, 1997b).

5.2. Termination of mtDNA Replication: Dideoxynucleosides

Several 2',3'-dideoxynucleosides are used in treating patients with the human immunodeficiency virus (HIV). These include 3'-azido-2',3'-dideoxythymidine (zidovudine, AZT), 2',3'-didehydro-3'-deoxythymidine (stavudine, D4T), 2',3'-dideoxycytidine (zalcitabine, ddC), (-)-2'-deoxy-3'-thiacytidine (lamivudine, 3TC), and 2',3'-dideoxyinosine (didanosine, ddI).

The normal 5'-hydroxyl group of deoxyribose is present in the sugar analog of these compounds, thus allowing formation of the triphosphate derivative and possible incorporation of the nucleotide analog into a nascent chain of DNA (Fig. 6). The normal 3'-hydroxyl group of deoxyribose is absent in these analog, however (Fig. 6), so once a single molecule has been incorporated, the DNA molecule lacks a 3'-hydroxyl group. No other nucleotide can be incorporated, leading to termination of DNA replication (Fig. 5) (Mitsuya and Broder, 1986; Yarchoan *et al.*, 1989).

The effects of these didoxynucleosides therefore depend on the ability of various polymerases to incorporate them into DNA. The HIV reverse transcriptase can perform this incorporation, thus impairing the reverse transcription of viral RNA (Fig. 6). In contrast, DNA polymerases that act in the nucleus do not effect incorporation, thus allowing the therapeutic use of these analogs (Yarchoan *et al.*, 1989). DNA polymerase γ, however, which acts in mitochondria, also incorporates the dideoxynucleoside triphosphates into growing chains of mtDNA, thus impairing mtDNA replication (Fig. 6) (Chen and Cheng, 1989; Lewis and Dalakas, 1995).

Even in postmitotic tissue there is a constant turnover of mitochondria, requiring basal replication of mtDNA. When this replication is impaired, mtDNA levels can progressively decrease, leading to an acquired mitochondrial cytopathy (Fig. 6).

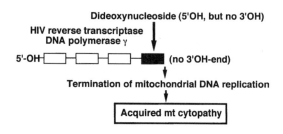

FIGURE 6. Termination of mtDNA replication by dideoxynucleosides. Once a single molecule of dideoxynucleoside is incorporated into a growing DNA chain, DNA then lacks a 3'-hydroxyl group and no other nucleotide can be incorporated.

The clinical manifestations of acquired mitochondrial cytopathies, like those of the inborn ones, are extremely polymorphic. They include bone marrow suppression, pancreatitis, peripheral neuropathy, myopathy, and microvesicular steatosis of the liver, sometimes with fatal lactic acidosis (Fortgang *et al.*, 1995; Lai *et al.*, 1991; Lewis and Dalakas, 1995).

5.3. Inhibition of mtDNA Replication: Fialuridine

Recently, another antiviral agent, fialuridine, was tested in patients with chronic hepatitis B virus infections, but the clinical trials were abruptly interrupted when several patients died of microvesicular steatosis and unmanageable lactic acidosis (Lewis and Dalakas, 1995; Zoulim and Trepo, 1994).

This complication was unexpected because fialuridine has both a 5′-hydroxyl group and a 3′-hydroxyl group, so that incorporating a single fialuridine molecule into DNA should not have immediately terminated mtDNA replication. Unfortunately, when several adjacent molecules of fialuridine are successively incorporated, further activity of DNA polymerase γ is inhibited and mitochondrial DNA replication decreases (Lewis *et al.*, 1996b), as do mtDNA levels (Tennant *et al.*, 1998).

5.4. Inhibition of mtDNA Transcription: Interferon-α

Interferon-α is widely used in treating patients with chronic viral hepatitis B or C. In cultured cells, interferon-α inhibits transcription of mitochondrial DNA into mitochondrial messenger RNA (Shan *et al.*, 1990), thus inhibiting mitochondrial respiration (Lewis *et al.*, 1996a).

Although it remains unknown whether the doses of interferon used in humans produce similar effects, it is noteworthy that some of the adverse effects of interferon-α administration (minor blood dyscrasias, myalgias, paresthesias, convulsions, depression, hepatic steatosis) (Okanoue *et al.*, 1996) resemble those of mild inborn mitochondrial cytopathies. Hypothetically, these adverse effects could be caused by inhibition of mitochondrial DNA transcription.

5.5. Sequestration of CoA and Opening of the Mitochondrial Permeability Transition Pore

The dual effect of CoA sequestration and permeability pore opening is observed with two drugs (salicylic acid and valproic acid). Both may cause microvesicular steatosis and cell death in liver.

5.5.1. Salicylic Acid

Aspirin is quickly hydrolyzed into salicylic acid. In liver mitochondria, salicylic acid is extensively activated on the outer mitochondrial membrane into salicylyl-CoA (Fig. 7) (Killenberg *et al.*, 1971). Extensive formation of salicylyl-CoA sequesters extramitochondrial CoA, leaving insufficient CoA to activate long-chain fatty acids (Deschamps *et al.*,

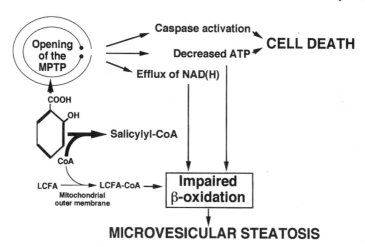

FIGURE 7. Mitochondrial effects of salicylic acid. Salicylic acid has two effects. It sequesters extramitochondrial CoA, which decreases activation of long-chain fatty acids (LCFA) into the acyl-CoA thioester, and it opens the MPTP, further impairing β-oxidation and possibly causing cell death.

1991). Thus the β-oxidation of long-chain fatty acids is markedly impaired (Fig. 7) (Deschamps *et al.*, 1991). Salicylate also triggers the mitochondrial permeability transition (Trost and Lemasters, 1996, 1997), which may further impair β-oxidation, as explained above, and may also cause cell death (Fig. 7).

Indeed, aspirin can cause two different types of liver lesions in humans. First, aspirin taken at high therapeutic doses causes single-hepatocyte cell death in some patients (Zimmerman, 1981). This aspirin-induced spotty hepatitis might be related to the opening of the MPTP and its ability to cause either necrosis or apoptosis (Fig. 7).

Second, aspirin can trigger microvesicular steatosis in two circumstances: (1) lethal overdoses of aspirin frequently cause microvesicular steatosis (Partin *et al.*, 1984), and (2) therapeutic doses of aspirin can trigger Reye's syndrome in children with viral infections (discussed later). Aspirin-induced microvesicular steatosis may be due to the combination of CoA sequestration, which impairs initial activation of long-chain fatty acids, and opening of the mitochondrial permeability transition pore, which depletes cellular ATP and mitochondrial pyrimidine nucleotides, as previously discussed.

5.5.2. Valproic Acid

Valproic acid is used to treat several forms of seizures. This short-chain fatty acid can enter mitochondria, where it is extensively transformed into valproyl-CoA (Ponchaut *et al.*, 1992). The sequestration of intramitochondrial CoA inhibits the β-oxidation of long- medium-, and short-chain fatty acids (Bjorge and Baillie, 1985; Turnbull *et al.*, 1983). Valproic acid also opens the permeability transition pore (Trost and Lemasters, 1996), which may further impair β-oxidation and may also cause cell demise.

An asymptomatic increase in serum aminotransferase activity, which normalizes with either dose reduction or drug discontinuation, is common during valproic acid administration (Farrell, 1994). A much less frequent side effect is a Reye-like syndrome that occurs mainly (but not exclusively) in very young children, and between the first and fourth month of valproic acid treatment (Zimmerman and Ishak, 1982). Histologically, centrizonal and midzonal microvesicular steatosis are associated with centrizonal necrosis, and sometimes with cirrhosis (Zimmerman and Ishak, 1982). Again, this combination of microvesicular fat and liver cell death may be related to the dual effect of valproic acid, which both sequesters CoA and opens the mitochondrial transition pore.

5.6. Inhibition of β-Oxidation

5.6.1. Glucocorticoids

Although glucocorticoids produce mainly macrovacuolar steatosis in humans, we have observed some patients with microvesicular steatosis during treatments with high doses of glucocorticoids. Glucocorticoids inhibit acyl–coenzyme A dehydrogenases and produce microvesicular steatosis of the liver in mice (Lettéron et al., 1997).

5.6.2. Amineptine and Tianeptine

The antidepressant drugs amineptine and tianeptine have a tricyclic moiety and a heptanoic side chain. They rarely cause immunoallergic hepatitis, due to the formation of reactive metabolites by cytochrome P450 (Genève et al., 1987b; Larrey et al., 1990). Exceptionally, they may also cause microvesicular steatosis due to impaired β-oxidation. Both amineptine and tianeptine are metabolized by β-oxidation of their heptanoic side chain (C7), forming 5-carbon (C5) and 3-carbon (C3) derivatives (Grislain et al., 1990; Sbarra et al., 1981). In the presence of these drugs, mitochondria are thus exposed to C7, C5, and C3 analogs of natural fatty acids. The analogs may competitively inhibit the β-oxidation of medium- and short-chain fatty acids (Fromenty and Pessayre, 1995; Fromenty et al., 1989).

5.6.3. Tetracyclines

Tetracycline and its various derivatives produce extensive microvesicular steatosis of the liver in experimental animals (Fréneaux et al., 1988; Labbe et al., 1991) because of their dual effect, which inhibits both the mitochondrial β-oxidation of fatty acids (Fréneaux et al., 1988; Fromenty et al., 1993) and hepatic secretion of very low density lipoproteins (Labbe et al., 1991). The latter effect occurs at doses that do not inhibit protein synthesis, suggesting impairment of the assembly, vesicular transport, or both of these lipoproteins (Deboyser et al., 1989).

At presently administered oral doses, tetracycline can produce minor degrees of hepatic steatosis of no clinical severity in humans, but severe microvesicular steatosis has occurred in the past during high-dose intravenous administration (Zimmerman, 1978). Predisposing factors included impaired renal function (which can decrease tetracycline elimination) and pregnancy (which can impair mitochondrial function, as discussed later). The syndrome usually appears after 4 to 10 days of tetracycline infusion and resembles

Reye's syndrome. Renal failure and pancreatitis are frequently associated. Most of reported cases have died, but milder cases may have gone unrecognized or unreported. Microvesicular steatosis has also been found after intravenous administration of several other tetracycline derivatives (Fromenty and Passayre, 1995; Zimmerman, 1978).

5.6.4. Nonsteroidal Anti-Inflammatory Drugs

Several 2-arylpropionate derivatives are used as anti-inflammatory drugs. Hepatic injury from these drugs consists of hepatitis, microvesicular liver steatosis, or both. The latter condition was observed with pirprofen, naproxen, ibuprofen, and ketoprofen (Bravo *et al.*, 1977; Danan *et al.*, 1985; Dutertre *et al.*, 1991; Victorino *et al.*, 1980).

The 2-arylpropionates, $CH_3-CH(Ar)-COOH$, have an asymmetric carbon (C) and therefore exist as either the S^+ or the R^- enantiomers. Only the S^+ enantiomer inhibits prostaglandin synthesis; only the R^- enantiomer is converted into the acyl-CoA derivative. Nevertheless, both ibuprofen enantiomers inhibit the β-oxidation of medium- and short-chain fatty acids (Fréneaux *et al.*, 1990). Inhibition of β-oxidation is also observed with pirprofen, tiaprofenic acid, and flurbiprofen (Genève *et al.*, 1987a).

5.6.5. Calcium Hopantenate and Other Drugs

Calcium hopantenate may inhibit mitochondrial β-oxidation and has caused Reye-like syndromes in Japan (Hautekeete *et al.*, 1990). Amiodarone, perhexiline, and 4,4'-diethylaminoethoxyhexestrol also inhibit mitochondrial β-oxidation and cause microvesicular steatosis in mice. These compounds are considered further at the end of this chapter, in the context of steatohepatitis.

It is noteworthy that all of these various drugs cause adverse effects in only a minority of recipients, suggesting that acquired or genetic factors are involved in the particular susceptibility of these subjects.

6. SUSCEPTIBILITY FACTORS AFFECTING DRUG METABOLISM

There is some evidence for genetic or acquired differences in drug metabolism as being involved in idiosyncrasy (Fig. 8). For example, perhexiline maleate is converted by cytochrome P450 2D6 (CYP2D6) into nontoxic metabolites. This may explain why neuropathy and steatohepatitis occur mainly in CYP2D6-deficient persons (Fig. 8) (Morgan *et al.*, 1984).

In addition to sequestration of CoA and opening of the MPTP, another mechanism has been suggested as a contributing cause of valproate-induced microvesicular steatosis (Fig. 8). Both CYP2C9 and CYP2A6 desaturate the outer carbons of valproic acid, forming Δ4-valproate (Sadeque *et al.*, 1997). This metabolite is then activated into Δ4-valproyl-CoA inside the mitochondria. The first dehydrogenation step of the β-oxidation cycle then forms Δ2,Δ4-valproyl-CoA, (Kassahun and Baillie, 1993; Kassahun *et al.*, 1991), a chemically reactive metabolite that might inactivate β-oxidation enzymes (Fig. 8). This hypothesis would explain why the hepatotoxicity of valproate is enhanced by

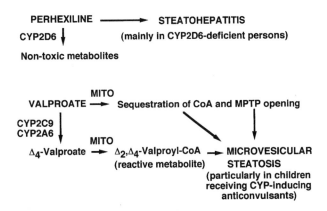

FIGURE 8. Genetic and acquired factors influencing drug metabolism and suceptibility to drug-induced mitochondrial dysfunction. CYP2D6 transforms perhexiline into nontoxic metabolites, explaining why steatohepatitis occurs mainly in CYP2D6-deficient people. Valproate is hepatotoxic through CoA sequestration and pore opening. In addition, valproate is transformed by CYP2C9 and CYP2A6 into Δ4-valproate, which then transforms in mitochondria (MITO) into a reactive metabolite that might inactivate mitochondrial β-oxidation enzymes, possibly explaining why valproate hepatotoxicity is enhanced by concomitant administration of anticonvulsants, which may induce CYP2A6.

concomitant administration of other anti-epileptic drugs (Farrell, 1994) that might induce CYP2A6, such as phenobarbital, phenytoin, and carbamazepine (Fig. 8).

7. SUSCEPTIBILITY FACTORS AFFECTING MITOCHONDRIAL FUNCTION

In some instances mitochondrial dysfunction caused by drugs (or endogenous compounds) may be potentiated by mitochondrial dysfunction caused by other factors. This multifactorial origin could be involved in Reye's syndrome and acute fatty liver of pregnancy.

7.1. Reye's Syndrome

Some children with an initially benign viral illness such as influenza or varicella develop persistent vomiting, lethargy, hepatic dysfunction, microvesicular steatosis, hyperammonemia, and coma, leading to death in 26–42% of the reported cases (Heubi et al., 1987). This postinfectious Reye's syndrome results from an acute mitochondrial insult (Heubi et al., 1987).

7.1.1. Cytokines, Nitric Oxide, and Viral Substances

Endogenous substances released during viral infections may be partly involved in mitochondrial dysfunction (Fig. 9). Nitric oxide inhibits mitochondrial respiration (Fish et al., 1996; Giulivi, 1998) and may open the permeability pore (Susin et al., 1998). Tumor necrosis factor-α causes caspase activation (Higuchi et al., 1997) and opens the pore

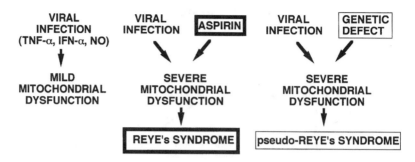

FIGURE 9. Multifactorial origin of Reye's syndrome. Viral infection may release TNF-α, IFN-α, and NO, and cause mitochondrial dysfunction too mild to trigger the syndrome. If children also receive aspirin or have a latent genetic defect that impairs β-oxidation, however, severe mitochondrial dysfunction may then trigger Reye's syndrome.

(Lemasters, 1998). Interferon-α may decrease mtDNA transcription and respiration (Shan *et al.*, 1990).

In addition, some viruses have a more direct effect. For instance, HIV RNA accumulates within mitochondria, where it can cause mitochondrial dysfunction (Somasundaran *et al.*, 1994). The hepatitis C virus core protein localizes in mitochondria, disrupts its normal double-membrane structure, and causes both microvesicular and macrovacuolar steatosis in transgenic mice (Moriya *et al.*, 1997, 1998).

Viral infections rarely cause Reye's syndrome, however, suggesting that these various endogenous or viral substances do not impair enough mitochondrial function to trigger it; thus, other factors seem necessary (Fig. 9).

7.1.2. Aspirin

If children also take aspirin during the viral illness, the added effects of salicylate on mitochondrial function may impair the function enough to trigger the syndrome (Fig. 9). This concept is based on several observations. First, 93% of children with Reye's syndrome received aspirin during acute viral illness (Hurwitz *et al.*, 1985). Second, children with Reye's syndrome received aspirin more frequently than those with similar viral diseases not followed by Reye's syndrome (Forsyth *et al.*, 1989). Finally, when it was recommended that aspirin should not be used in feverish children, there was a parallel decline in aspirin use and in incidence of Reye's syndrome in the United States (Remington *et al.*, 1986).

7.1.3. Inborn Defects

A few cases of Reye's (or "pseudo-Reye's") syndrome nevertheless continue to be seen, but they now occur mainly in children with an initially undiagnosed inborn defect that may directly or indirectly impair β-oxidation (Fig. 9) (Glasgow and Moore, 1993; Rowe *et al.*, 1988). These latent inborn defects become apparent when viral infection further impairs the already-compromised mitochondrial function and simultaneously increases the energy demand (through fever) while decreasing the food intake (through anorexia or vomiting). The last two effects could cause massive peripheral lipolysis, and compromised hepatic mitochondria may be unable to handle this increased delivery of fatty acids, thus triggering Reye's syndrome.

Inborn defects that might favor Reye's syndrome can affect urea-cycle enzymes, respiration, or β-oxidation itself.

7.1.3a. Urea Cycle Defects. Defects in urea cycle enzymes cause massive hyperammonemia, which can decrease β-oxidation (Maddaiah, 1985; Maddaiah and Miller, 1989). The most prevalent defect is ornithine transcarbamylase deficiency, which affects roughly 1 in 80,000 individuals (Nagata *et al.*, 1991).

7.1.3b. Defects of the Respiratory Chain. Mitochondrial cytopathies affect mitochondrial respiration, which may secondarily inhibit β-oxidation, as explained above. Inborn mitochondrial cytopathies are caused by defects in nuclear genes or mtDNA mutations (DiMauro and Schon, 1998; Schon *et al.*, 1997). There are about 50 different pathological point mutations and 200 different mtDNA deletions (Schon *et al.*, 1997). Point mutations are usually transmitted by the mother, whereas mtDNA deletions are mostly acquired during oogenesis or embryogenesis. Point mutations may be either homoplasmic (affecting all mtDNA genomes) or heteroplasmic (affecting some mtDNA genomes), whereas large mtDNA deletions are always heteroplasmic (because the homoplasmic state is not viable).

Although the overall prevalence of these inborn mitochondrial cytopathies remains unknown, the frequency of the A3243G mtDNA point mutation alone was estimated at 1.6 in 10,000 in a Finnish population (Majama *et al.*, 1998).

7.1.3c. β-Oxidation Defects. Inborn defects in β-oxidation enzymes may affect the various enzymes involved in the transport and β-oxidation of fatty acids (Fig. 1). The most frequent defect is MCAD deficiency. Not so long ago, a probably German ancestor of present day Europeans happened to have a particular RFLP haplotype in the nuclear MCAD locus and developed an A985G point mutation in the MCAD gene (Zhang *et al.*, 1993). This point mutation transformed a lysine residue into a glutamic acid residue (Kelly *et al.*, 1990), which impaired the normal association of the enzyme monomer into the active homotetramer, resulting in rapid degradation of the monomer (Yokota *et al.*, 1992). This ancestor proved astoundingly efficient at transmitting the gene (suggesting that the heterozygous carrier state may offer some unknown advantage). Indeed, about 1 in 50 Europeans (and North Americans of European descent) are now carriers of both the initial RPLP haplotype (or derived haplotypes) and the A985G point mutation (Fromenty *et al.*, 1996; Miller *et al.*, 1992). This gives an estimated frequency of about 1 in 10,000 for the homozygous state, making it one of the most prevalent genetic diseases.

Yet another condition that may be caused by various combination of acquired and inborn defects is acute fatty liver of pregnancy.

7.2. Acute Fatty Liver of Pregnancy

About 1 in 13,000 pregnant women develop microvesicular steatosis during the last trimester of pregnancy (Kaplan, 1985). Untreated, the disease progresses to coma, kidney failure, and hemorrhages, leading to death in both mother and child 75% to 85% of the time. In contrast, rapid termination of pregnancy usually permits delivery of a healthy child and a quick end to the mother's disease (Ebert *et al.*, 1984).

7.2.1. Female Sex Hormones

Pregnancy, or the administration of estradiol and progesterone, alter mitochondrial ultrastructure and function in mice (Grimbert *et al.*, 1993, 1995), but these effects are mild and do not lead to microvesicular steatosis (Grimbert *et al.*, 1993, 1995). Similarly, most human pregnancies do not cause acute fatty liver. Thus, additional factors seem necessary to trigger the syndrome (Fig. 10).

7.2.2. LCHAD Deficiency

Partial deficiency of long-chain 3-hydroxyacyl–CoA dehydrogenase, which is part of the trifunctional membrane-bound enzyme (Fig. 1), has been reported in some women with acute fatty liver of pregnancy (Fig. 10) (Isaacs *et al.*, 1996; Sims *et al.*, 1995; Treem *et al.*, 1994; Wilcken *et al.*, 1993). Mothers with a single defective LCHAD allele who are unlucky enough to marry an heterozygous carrier and to then conceive a fetus with two defective alleles will develop the disease, whereas those who bear an unaffected child usually have uncomplicated pregnancies.

In an initial series of 12 women with acute fatty liver of pregnancy, two thirds were found to have intermediate LCHAD activity in cultures of skin fibroblasts, suggesting that they were heterozygous for LCHAD mutations (Treem *et al.*, 1996). The study may have been biased by the selective referral of women with an unhealthy child to this center, however, which specialized in inborn mitochondrial diseases. Furthermore, despite many careful precautions, it is difficult to discriminate heterozygotes from normal individuals on the basis of enzyme activities.

The LCHAD deficiency is usually due to a G1528C mutation, and occasionally to a C1132T mutation, in the gene for the α subunit of the trifunctional mitochondrial β-oxidation enzyme (Ijlst *et al.*, 1994; Sims *et al.*, 1995). We have not detected G1528C and C1132T mutations in any of 14 consecutive French women with histology-confirmed acute fatty liver of pregnancy (Mansouri *et al.*, 1996). These findings suggest that LCHAD

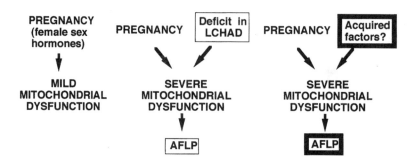

FIGURE 10. Multifactorial origin of acute fatty liver of pregnancy (AFLP). In addition to mild mitochondrial dysfunction caused by female sex hormones, other factors are probably required to further impair β-oxidation and trigger the syndrome. These added factors rarely involve defects in long-chain 3-hydroxyacyl–CoA dehydrogenase (LCHAD). More commonly, diverse aleatory factors further impair β-oxidation, thus triggering AFLP (see text for more details).

deficiency is an uncommon cause in unselected cases of acute fatty liver of pregnancy. Furthermore, although 1 in 70 French persons are heterozygous for the A985G MCAD mutation (Fromenty *et al.*, 1996), which accounts for 89% of all deficient MCAD alleles, none of these 14 French women with acute fatty liver of pregnancy carried the A985G MCAD mutation (Mansouri *et al.*, 1996). Since then, however, we have detected an heterozygous MCAD-deficient woman in a larger series of French AFLP patients.

These observations should not preclude screening for various defects of mitochondrial β-oxidation in women with acute fatty liver of pregnancy (particularly those with recurrent disease or with an ill or stillborn child). Together with epidemiological data, however, these observations suggest that such defects are rarely involved. Indeed, with very few exceptions, when pregnancy is interrupted the child is born healthy and acute fatty liver of pregnancy does not recur upon subsequent pregnancies. Therefore, acquired aleatory, rather than genetic, factors are probably involved in most cases (Fig. 10) (Fromenty and Pessayre, 1995; Mansouri *et al.*, 1996). Drugs, food, stress, infections, an autoimmune reaction triggered by the foreign child, or placental ischemia associated with preeclampsia are perhaps triggers for the syndrome.

8. POSSIBLE OUTCOME OF PROLONGED STEATOSIS: STEATOHEPATITIS

Whereas severe forms of microvesicular steatosis may cause liver failure, coma, and death in the short term, mild forms do not cause death. Prolonged steatosis, however, of whatever cause and type, may eventually lead, in some patients, to the progressive development of diverse liver lesions that are globally referred to as *steatohepatitis*.

8.1. Features

In addition to the initial steatosis, liver lesions may now include hepatocyte death, Mallory bodies, a mixed inflammatory cell infiltrate (containing neutrophils), fibrosis, and even cirrhosis. Steatohepatitis develops insidiously and may be revealed by hepatomegaly. In severe forms, jaundice, ascites, and encephalopathy may be observed, and these patients sometimes die from the diverse complications of liver cirrhosis.

Steatohepatitis can occur in most conditions characterized by chronic steatosis, including obesity, diabetes, jejunoileal bypass, Wilson disease, alcohol abuse, or prolonged administration of amiodarone, perhexiline, 4,4′-diethylaminoethoxyhexestrol, or glucocorticoids (Farrell, 1994). Thus, chronic steatosis most certainly plays a role in the development of steatohepatitis.

8.2. Steatosis, Basal Formation of Reactive Oxygen Species, and Lipid Peroxidation

In mice acute or chronic steatosis due to 11 different treatments was always associated with lipid peroxidation (Lettéron *et al.*, 1996). After a single dose of tetracycline or ethanol, maximal ethane exhalation occurred at the time of maximal hepatic triglyceride accumulation. Whereas a single dose of doxycycline or glucocorticoids did not increase ethane exhalation (or hepatic triglycerides), repeated doses did increase them (Lettéron *et al.*, 1996). Even in the basal state, ROS are constantly formed in the cell; mitochondria represent the major source of these reactive species (Ames *et al.*, 1993; Chance *et al.*, 1979). Basal cellular ROS formation may be sufficient to oxidize hepatic fat deposits and cause lipid peroxidation. As explained later, chronic lipid peroxidation might be sufficient to cause steatohepatitis lesions.

It is noteworthy, however, that steatohepatitis is relatively rare in patients with certain forms of chronic steatosis such as obesity, but apparently more common in patients receiving amphiphilic cationic drugs or abusing alcohol. This may be explained by a "dual hit" mechanism involving both steatosis and enhanced formation of reactive oxygen species.

8.3. Increased Formation of Reactive Oxygen Species

8.3.1. Amiodarone, Perhexiline, and Diethylaminoethoxyhexestrol

Amiodarone, perhexiline, and diethylaminoethoxyhexestrol are cationic amphiphilic compounds with lipophilic moiety and an amine function that can become protonated (and thus positively charged). This chemical structure may be responsible for the two liver lesions produced by these compounds, namely phospholipidosis and steatohepatitis.

The uncharged, lipophilic form of these three drugs easily crosses the lysosomal membrane (Kodavanti and Mehendale, 1990). In the acidic lysosomal milieu, the unprotonated drug molecule is protonated and becomes less lipophilic, resulting in its accumulation within the lysosomes. This protonated molecule forms noncovalent complexes with intralysosomal phospholipids, thus hampering the action of lysosomal phospholipases. Phospholipid–drug complexes progressively accumulate in lysosomes (Kodavanti and Mehendale, 1990; Zimmerman, 1978). Although phospholipidosis is extremely frequent, perhaps constant, in patients receiving these drugs, in itself it appears to have no clinical consequence, because it often occurs without clinical symptoms or biochemical disturbances (Guigui *et al.*, 1988).

The cationic amphiphilic structure of the three above-mentioned drugs, however, also causes impairment of mitochondrial function (Fig. 11). The unprotonated, lipophilic form of amiodarone (Fromenty *et al.*, 1990a, 1990b), perhexiline (Deschamps *et al.*, 1994) or 4,4′-diethylaminoethoxyhexestrol (Berson *et al.*, 1998) easily crosses the mitochondrial outer membrane and is protonated in the acidic intermembrane space (Fig. 11). This protonated form is "pushed" inside the mitochondria by the high electrochemical potential existing across the mitochondrial inner membrane and achieves extremely high intramitochondrial concentrations, which inhibit β-oxidation enzymes, causing steatosis (Fig. 11). They also block the transfer of electrons along the respiratory chain. Whenever the flow of electrons is blocked on the respiratory chain, upstream chain components become overly reduced and directly transfer their electrons to oxygen, forming the superoxide anion (Bailey and Cunningham, 1998; Pitkänen and Robinson, 1996). This enhanced ROS formation may oxidize fat deposits, causing lipid peroxidation (Fig. 11) as well as an oxidative stress in the liver.

Similarly, chronic alcohol consumption causes steatosis and concomitantly enhances ROS formation.

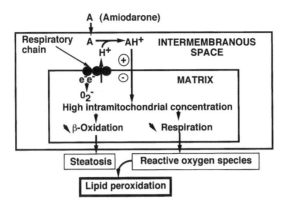

FIGURE 11. Mitochondrial handling of amiodarone. After crossing the outer membrane, amiodarone is protonated in the intermembrane space and then "pushed" by mitochondrial membrane potential into the matrix. High intramitochondrial concentrations inhibit both β-oxidation (causing steatosis) and oxidative phosphorylation (increasing ROS formation). The latter may oxidize fat deposits, causing lipid peroxidation.

8.3.2. Ethanol

At least four mechanisms may be involved in the ethanol-induced formation of ROS. First, ethanol abuse increases the mitochondrial formation of ROS (Bailey and Cunningham, 1998; Nordmann *et al.*, 1992) and this effect may be further increased by secondary mtDNA mutations (Fromenty *et al.*, 1995; Mansouri *et al.*, 1997a) that may impair the flow of electrons along the respiratory chain. Second, cytochrome P450 2E1 partly transforms ethanol into the α-hydroxyethyl radical ($^{\cdot}CHOH-CH_3$) (Reinke *et al.*, 1997). Third, ethanol metabolism leads to reduced NAD$^+$/NADH, and NADP$^+$/NADPH ratios (Bailey and Cunningham, 1998). This may increase the microsomal reduction of ferric iron to ferrous iron, a potent generator of the hydroxyl radical (Kurose *et al.*, 1996). Fourth, alcohol consumption induces CYP2E1, which spontaneously forms the hydroxyl radical (Lieber, 1994).

8.4. Reactive Oxygen Species and Steatohepatitis

Enhanced ROS formation after administration of cationic amphiphilic drugs or ethanol abuse may lead to steatohepatitis lesions via several intricate mechanisms, including lipid peroxidation, cytokine release, and Fas-mediated fratricidal killing.

8.4.1. Lipid Peroxidation

Enhanced ROS formation may oxidize fat deposits, causing extensive lipid peroxidation. Chronic lipid peroxidation may be involved, in part, in the progressive development of steatohepatitis lesions (Fig. 12) (Lettéron *et al.*, 1996). Indeed, lipid peroxidation causes cell demise, which may explain liver cell necrosis. Peroxidation also releases malondialdehyde and 4-hydroxynonenal (Kamimura *et al.*, 1992; Lieber, 1994). Both 4-hydroxynonenal and malondialdehyde covalently bind to proteins, and these modified proteins

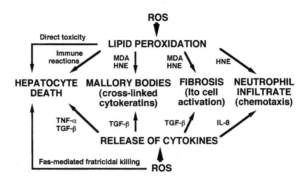

FIGURE 12. From reactive oxygen species to steatohepatitis. Both lipid peroxidation products and cytokines could explain the diverse steatohepatitis lesions (see text). MDA, malondialdehyde. HNE, 4-hydroxynonenal.

may cause immune reactions and immune hepatitis (Fig. 11). Both 4-hydroxynonenal and malondialdehyde are bifunctional agents that crosslink proteins (Esterbauer *et al.*, 1991), and might be involved in the formation of Mallory bodies, which contain crosslinked cytokeratin monomers (Zatloukal *et al.*, 1991; Zhang-Gouillon *et al.*, 1998). Both 4-hydroxynonenal and malondialdehyde increase the synthesis of collagen by Ito cells (Bedossa *et al.*, 1994; Kamimura *et al.*, 1992), hence the fibrosis. 4-Hydroxynonenal has a chemotactic activity for neutrophils (Curzio *et al.*, 1985), which may account for the neutrophilic cell infiltration (Fig. 12). Finally, ROS, lipid peroxidation, or both may trigger the release of cytokines, which may cause further liver damage.

8.4.2. Cytokines

Cytokines have not been explored in drug-induced steatohepatitis, but have been extensively studied in alcohol-induced steatohepatitis. Ethanol abuse triggers oxidative stress and release of several cytokines (including TNF-α, TGF-β and IL-8) from both Kupffer cells and hepatocytes themselves (Dong *et al.*, 1998; Fang *et al.*, 1998; Gressner and Wulbrand, 1997; Neuman *et al.*, 1998).

Both TNF-α and TGF-β cause caspase activation and hepatocyte death (Fig. 12) (Higuchi *et al.*, 1997; Inayat-Hussain *et al.*, 1997). TGF-β activates tissue transglutaminase (Ritter and Davies, 1998). This enzyme crosslinks cytoskeletal proteins, including intermediate filament proteins (Trejo-Skalli *et al.*, 1995), into large protein scaffolds that might be involved in the formation of Mallory bodies. TGF-β also activates collagen synthesis by Ito cells (Casini *et al.*, 1993). Finally, IL-8 is a potent chemoatractant for human neutrophils (Dong *et al.*, 1998). Thus, cytokines might participate (together with lipid peroxidation products) in the development of the diverse steatohepatitis lesions in conditions causing chronic oxidant stress in the liver (Fig. 12).

A last mechanism that may be involved in cell death is Fas-mediated fratricidal killing.

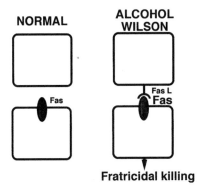

Fratricidal killing

FIGURE 13. Fas-mediated fratricidal killing. Normally, hepatocytes express Fas but not Fas ligand, preventing them from killing their neighbors. In alcoholics or patients with fuminant Wilson disease, hepatocytes overexpress Fas and express Fas ligand, which may cause fratricidal killing.

8.4.3. Fas-Mediated Fratricidal Killing

Hepatocytes express Fas (a membrane receptor), but do not normally express Fas ligand; this prevents them from killing their neighbors (Fig. 13). The Fas ligand promoter contains NFκB binding sites (Takahashi *et al.*, 1994), however. Normally, NFκB is maintained in the cytoplasm by IκB, but reactive oxygen intermediates cause its phosphorylation, ubiquitination, and proteasome-mediated degradation, allowing the nuclear translocation of NFκB (Naumann and Scheidereit, 1994). Conditions leading to an increased formation of reactive oxygen intermediates may thus cause hepatocytes to express Fas ligand (Fig. 13) (Strand *et al.*, 1998). At the same time, increased formation of reactive oxygen intermediates might damage DNA, overexpress p53, and increase Fas expression by hepatocytes (Müller *et al.*, 1997).

The Fas ligand of one hepatocyte may then interact with Fas on another hepatocyte, causing fratricidal killing (Fig. 13), a form of cell death that has been suggested in alcoholics, who express Fas ligand and overexpress Fas on their hepatocytes (Galle *et al.*, 1995). Hepatocytes from patients with fulminant hepatic failure due to Wilson disease also express Fas ligand and overexpress Fas (Strand *et al.*, 1998). When HepG2 hepatoma cells were treated with copper, the cell surface Fas protein, Fas ligand mRNA, and Fas ligand protein were all increased and the cells underwent apoptosis (Strand *et al.*, 1998). Cell death was partially prevented by a neutralizing Fas antibody or a caspase inhibitor, suggesting that the hepatocytes killed each other via Fas-mediated fratricidal killing (Fig. 13).

9. CONCLUSIONS

Pregnancy, ethanol abuse, and drugs can cause microvesicular steatosis by impairing mitochondrial structure and function (female sex hormones), causing oxidative damage to mitochondrial lipids, proteins, and DNA (alcohol, copper), inhibiting mitochondrial DNA replication (dideoxynucleosides, fialuridine), inhibiting mitochondrial β-oxidation enzymes (valproic acid, tetracyclines, 2-arylpropionates, amineptine, tianeptine, glucocorticoids, amiodarone, and perhexiline) or sequestering coenzyme A and opening the MPTP (aspirin, valproic acid).

The fatty acids, which are poorly oxidized by mitochondria, are mainly esterified into triglycerides, but there is a residual increase in nonesterified fatty acids. Triglycerides (possibly emulsified by a rim of nonesterified fatty acids) accumulate as small vesicles.

Either genetic or acquired factors can modify a drug's metabolism and could explain the susceptibility of some patients. In other instances the disease may be triggered by a combination of drug-induced effects and inborn defects that additively impair mitochondrial function. Severe metabolic disturbances may then impair energy production, causing liver failure, coma, and death.

Milder forms of microvesicular steatosis have a good short-term prognosis, but, in the long term, any chronic steatosis, whether microvesicular or macrovacuolar, can lead to steatohepatitis. The basal cellular formation of reactive oxygen species may be sufficient to oxidize these fat deposits, causing lipid peroxidation and steatohepatitis.

Steatohepatitis is rare in patients with obesity or diabetes, however, and more frequent in patients abusing ethanol or receiving amphiphilic cationic drugs. In the latter conditions, not only is fat (the substrate for lipid peroxidation) present in the liver, but ROS are also increased. This "two-hits" mechanism might result in severe lipid peroxidation, cytokine release, and Fas-mediated fratricidal killing, causing frequent development of steatohepatitis lesions.

ACKNOWLEDGEMENTS. Some of the original work cited in this review was supported by the European Union BIOMED 2 program (Eurohepatox contract BMH4-CT960658), the Programme Hospitalier de Recherche Clinique 95–96, and the Réseau Hépatox.

REFERENCES

Ames, B. N., Shigenaga, M. K., and Hagen, T. M., 1993, Oxidants, antioxidants, and the degenerative diseases of aging, *Proc. Natl. Acad. Sci. USA* **90**:7915–7922.

Bailey, S. M., and Cunningham, C. C., 1998, Acute and chronic ethanol increases reactive oxygen species generation and decreases viability in fresh, isolated rat hepatocytes, *Hepatology* **28**:1318–1326.

Bedossa, P., Houglum, K., Trautwein, C., Holstege, A., and Chojkier, M., 1994, Stimulation of collagen α (I) gene expression is associated with lipid peroxidation in hepatocellular injury: A link to tissue fibrosis? *Hepatology* **19**:1262–1271.

Bernardi, P., Basso, E., Colonna, R., Costantini, P., Di Lisa, F., Eriksson, O., Fontaine, E., Forte, M., Ichas, F., Massari, S., Nicolli, A., Petronilli, V., and Scorrano, L., 1998, Perspectives on the mitochondrial permeability transition, *Biochim. Biophys. Acta* **1365**:200–206.

Berson, A., De Beco, V., Lettéron, P., Robin, M. A., Moreau, C., El Kahwaji, J., Verthier, N., Feldmann, G., Fromenty, B., and Pessayre, D., 1998, Steatohepatitis-inducing drugs cause mitochondrial dysfunction and lipid peroxidation in rat hepatocytes, *Gastroenterology* **114**:764–774.

Bjorge, S. M., and Baillie, T. A., 1985, Inhibition of medium-chain fatty acid β-oxidation *in vitro* by valproic acid and its unsaturated metabolite, 2-*n*-propyl-4-pentenoic acid, *Biochem. Biophys. Res. Commun.* **132**:245–252.

Bravo, J. F., Jacobson, M. P., and Mertens, B. F., 1977, Fatty liver and pleural effusion with ibuprofen therapy, *Ann. Intern. Med.* **87**:200–201.

Casini, A., Pinzani, M., Milani, S., Grappone, C., Galli, G., Jezequel, A. M., Schuppan, D., Rotella, C. M., and Surrenti, C., 1993, Regulation of extracellular matrix synthesis by transforming growth factor β1 in human fat storing cells, *Gastroenterology* **105**:245–253.

Chance, B., Sies, H., and Boveris, A., 1979, Hydroperoxide metabolism in mammalian organs, *Physiol. Rev.* **59**:527–605.

Chen, C. H., and Cheng, Y. C., 1989, Delayed cytotoxicity and selective loss of mitochondrial DNA in cells treated with the anti-human immunodeficiency virus compound 2′,3′-dideoxycytidine, *J. Biol. Chem.* **264**:11934–11937.

Curzio, M., Esterbauer, H., and Dianzani, M. U., 1985, Chemotactic activity of hydroxyalkenals on rats neutrophils, *Int. J. Tissue React.* **7**:137–142.

Danan, G., Trunet, P., Bernuau, J., Degott, C., Babany, G., Pessayre, D., Rueff, B., and Benhamou, J. P., 1985, Pirprofen-induced fulminant hepatitis, *Gastroenterology* **89**:210–213.

Deboyser, D., Goethals, F., Krack, G., and Roberfroid, M., 1989, Investigation into the mechanism of tetracycline-induced steatosis: Study in isolated hepatocytes, *Toxicol. Appl. Pharmacol.* **97**:473–479.

Deschamps, D., Fisch, C., Fromenty, B., Berson, A., Degott, C., and Pessayre, D., 1991, Inhibition by salicylic acid of the activation and thus oxidation of long-chain fatty acids: Possible role in the development of Reye's syndrome, *J. Pharmacol. Exp. Ther.* **259**:894–904.

Deschamps, D., De Beco, V., Fisch, C., Fromenty, B., Guillouzo, A., and Pessayre, D., 1994, Inhibition by perhexiline of oxidative phosphorylation and the β-oxidation of fatty acids: Possible role in pseudoalcoholic liver lesions, *Hepatology* **19**:948–961.

DiMauro, S., and Schon, E. A., 1998, Nuclear power and mitochondrial disease, *Nature Genet.* **19**:214–215.

Dong, W., Simeonova, P. P., Gallucci, R., Matheson, J., Fannin, R., Montuschi, P., Flood, L., and Luster, M. I., 1998, Cytokine expression in hepatocytes: Role of oxidative stress, *J. Interferon Cytokine Res.* **18**:629–638.

Dutertre, J. P., Bastides, F., Jonville, A. P., De Muret, A., Sonneville, A., Larrey, D., and Autret, E., 1991, Microvesicular steatosis after ketoprofen administration, *Eur. J. Gastroenterol. Hepatol.* **3**:953–954.

Ebert, E. C., Sun, E. A., Wright, S. H., Decker, J. P., Librizzi, R. J., Bolognese, R. J., and Lipshutz, W. H., 1984, Does early diagnosis and delivery in acute fatty liver of pregnancy lead to improvement in maternal and infant survival? *Digest. Dis. Sci.* **29**:453–455.

Eisele, J. W., Barker, E. A., and Smuckler, E. A., 1975, Lipid content in the liver of fatty metamorphosis of pregnancy, *Am. J. Pathol.* **81**:545–560.

Esterbauer, H., Schaur, R. J., and Zollner, H., 1991, Chemistry and biochemistry of 4-hydroxynonenal, malondialdehyde, and related aldehydes, *Free Radical Biol. Med.* **11**:81–128.

Fang, C., Lindros, K. O., Badger, T. M., Ronis, M. J. J., and Ingelman-Sundberg, M., 1998, Zonated expression of cytokines in rat liver: Effects of chronic ethanol and the cytochrome P450 2E1 inhibitor, chlormethiazole, *Hepatology* **27**:1304–1310.

Farrell, G., 1994, *Drug-Induced Liver Disease*, Churchill-Livingstone, London.

Fisch, C., Robin, M. A., Lettéron, P., Fromenty, B., Berson, A., Renault, S., Chachaty, C., and Pessayre, D., 1996, Cell-generated nitric oxide inactivates rat hepatocytes mitochondria *in vitro* but reacts with hemoglobin *in vivo*, *Gastroenterology* **110**:210–220.

Forsyth, B. W., Horwitz, R. I., Acampora, D., Shapiro, E. D., Viscoli, C. M., Feinstein, A. R., Henner, R., Holabird, N. B., Jones, B. A., Karabelas, A. D. E., Kramer, M. S., Miclette, M., and Wells, J. A., 1989, New epidemiologic evidence confirming that bias does not explain the aspirin/Reye's syndrome association, *JAMA* **261**:2517–2524.

Fortgang, I. S., Belitsos, P. C., Chaisson, R. E., and Moore, R. D., 1995, Hepatomegaly and steatosis in HIV-infected patients receiving nucleoside analog antiretroviral therapy, *Am. J. Gastroenterol.* **90**:1433–1436.

Fréneaux, E., Labbe, G., Lettéron, P., Le Dinh, T., Degott, C., Genève, J., Larrey, D., and Pessayre, D., 1988, Inhibition of the mitochondrial oxidation of fatty acids by tetracycline in mice and in man: Possible role in microvesicular steatosis induced by this antibiotic, *Hepatology* **8**:1056–1062.

Fréneaux, E., Fromenty, B., Berson, A., Labbe, G., Degott, C., Lettéron, P., Larrey, D., and Pessayre, D., 1988, Stereoselective and nonstereoselective effects of ibuprofen enantiomers on mitochondrial β-oxidation of fatty acids, *J. Pharmacol. Exp. Ther.* **255**:529–535.

Fromenty, B., and Pessayre, D., 1995, Inhibition of mitochondrial beta-oxidation as a mechanism of hepatotoxicity, *Pharmacol. Ther.* **67**:101–154.

Fromenty, B., Fréneaux, E., Labbe, G., Deschamps, D., Larrey, D., Lettéron, D., and Pessayre, D., 1989, Tianeptine, a new tricyclic antidepressant metabolized by β-oxidation of its heptanoic side chain, inhibits the mitochondrial oxidation of medium and short chain fatty acids in mice, *Biochem. Pharmacol.* **38**:3743–3751.

Fromenty, B., Fisch, C., Berson, A., Lettéron, P., Larrey, D., and Pessayre, D., 1990a, Dual effect of amiodarone on mitochondrial respiration: Initial protonophoric uncoupling effect followed by inhibition of the respiratory chain at the levels of complex I and complex II, *J. Pharmacol. Exp. Ther.* **255**:1377–1384.

Fromenty, B., Fisch, C., Labbe, G., Degott, C., Deschamps, D., Berson, A., Lettéron, P., and Pessayre, D., 1990b, Amiodarone inhibits the mitochondrial β-oxidation of fatty acids and produces microvesicular steatosis of the liver in mice, *J. Pharmacol. Exp. Ther.* **255**:1371–1376.

Fromenty, B., Lettéron, P., Fisch, C., Berson, A., Deschamps, D., and Pessayre, D., 1993, Evaluation of human blood lymphocytes as a model to study the effects of drugs on human mitochondria: Effects of low concentrations of amiodarone on fatty acid oxidation, ATP levels, and cell survival, *Biochem. Pharmacol.* **46**:421–432.

Fromenty, B., Grimbert, S., Mansouri, A., Beaugrand, M., Erlinger, S., Rötig, A., and Pessayre, D., 1995, Hepatic mitochondrial DNA deletion in alcoholics: Association with microvesicular steatosis, *Gastroenterology* **108**:193–200.

Fromenty, B., Mansouri, A., Bonnefont, J. P., Courtois, F., Munnich, A., Rabier, D., and Pessayre, D., 1996, Most cases of medium-chain acyl-CoA dehydrogenase deficiency escape detection in France, *Hum. Genet.* **97**:367–368.

Fromenty, B., Berson, A., and Pessayre, D., 1997, Microvesicular steatosis and steatohepatitis: Role of mitochondrial dysfunction and lipid peroxidation, *J. Hepatol.* **26** (Suppl. 1):13–22.

Galle, P. R., Hofmann, W. J., Walczak, H., Schaller, H., Otto, G., Stremmel, W., Krammer, P. H., and Runkell, L., 1995, Involvement of the CD95 (APO-1/Fas) receptor and ligand in liver damage, *J. Exp. Med.* **182**:1223–1230.

Genève, J., Hayat-Bonan, B., Labbe, G., Degott, C., Lettéron, P., Fréneaux, E., Le Dinh, T., Larrey, D., and Pessayre, D., 1987a, Inhibition of mitochondrial β-oxidation of fatty acids by pirprofen: Role in microvesicular steatosis due to this nonsteroidal anti-inflammatory drug, *J. Pharmacol. Exp. Ther.* **242**:1133–1137.

Genève, J., Larrey, D., Amouyal, G., Belghiti, J., and Pessayre, D., 1987b, Metabolic activation of the tricyclic antidepressant amineptine by human liver cytochrome P-450, *Biochem. Pharmacol.* **36**:2421–2424.

Giulivi, C., 1998, Functional implications of nitric oxide produced by mitochondria in mitochondrial metabolism, *Biochem. J.* **332**:673–679.

Glasgow, J. F. T., and Moore, R., 1993, Reye's syndrome 30 years on: Possible marker of inherited metabolic disorders, *Brit. Med. J.* **307**:950–951.

Green, D. R., and Reed, J. C., 1998, Mitochondria and apoptosis, *Science* **281**:1309–1312.

Gregoriadis, G., and Sourkes, T. L., 1967, Intracellular distribution of copper in the liver of the rat, *Can. J. Biochem.* **45**:1841–1851.

Gressner, A. M., and Wulbrand, U., 1997, Variation in immunocytochemical expression of transforming growth factor (TGF)-beta in hepatocytes in culture and liver slices, *Cell Tissue Res.* **287**:143–152.

Grimbert, S., Fromenty, B., Fisch, C., Lettéron, P., Berson, A., Durand-Schneider, A. M., Feldmann, G., and Pessayre, D., 1993, Decreased mitochondrial oxidation of fatty acids in pregnant mice: Possible relevance to development of acute fatty liver of pregnancy, *Hepatology* **17**:628–637.

Grimbert, S., Fisch, C., Deschamps, D., Berson, A., Fromenty, B., Feldmann, G., and Pessayre, D., 1995, Effects of female sex hormones on liver mitochondria in non-pregnant female mice: Possible role in acute fatty liver of pregnancy, *Am. J. Physiol.* **268**(*Gastrointest. Liver Physiol.* **31**):G107–G115.

Grislain, L., Gelé, P., Bertrand, M., Luijten, W., Bromet, N., Salvadori, C., and Kamoun, A., 1990, The metabolic pathways of tianeptine, a new antidepressant, in healthy volunteers, *Drug Metab. Dispos.* **18**:804–808.

Guigui, B., Perrot, S., Berry, J. P., Fleury-Feith, J., Martin, N., Métreau, J. M., Dhumeaux, D., and Zafrani, E. S., 1988, Amiodarone-induced hepatic phospholipidosis: A morphological alteration independent of pseudoalcoholic liver disease, *Hepatology* **8**:1063–1068.

Guzman, M., and Geelen, M. J. H., 1993, Regulation of fatty acid oxidation in mammalian liver, *Biochim. Biophys. Acta* **1167**:227–241.

Hautekeete, M. L., Degott, C., and Benhamou, J. P., 1990, Microvesicular steatosis of the liver, *Acta Clin. Belg.* **45**:311–326.

Heubi, J. E., Partin, J. C., Partin, J. S., and Schubert, W. K., 1987, Reye's syndrome: Current concepts, *Hepatology* **7**:155–164.

Higuchi, M., Aggarwal, B. B., and Yeh, E. T. H., 1997, Activation of CPP32-like protease in tumor necrosis factor-induced apoptosis is dependent on mitochondrial function, *J. Clin. Invest.* **99**:1751–1758.

Hoppel, C. L., Kerner, J., Turkaly, P., Turkaly, J., and Tandler, B., 1998, The malonyl-CoA-sensitive form of carnitine palmitoyltransferase is not localized exclusively in the outer membrane of rat liver mitochondria, *J. Biol. Chem.* **273**:23495–23503.

Hurwitz, E. S., Barrett, M. J., Bregman, D., Gunn, W. J., Schonberger, L. B., Fairweather, W. R., Drage, J. S., Lamontagne, J. R., Kaslow, R. A., Burlington, D. B., Quinnan, G. V., Parker, R. A., Phillips, K., Pinsky, P., Dayton, D., and Dowdle, W. R., 1985, Public health service study on Reye's syndrome and medications: Report of the pilot phase, *N. Engl. J. Med.* **313**:849–857.

Ijlst, L., Wanders, R. J. A., Ushikubo, S., Kamijo, T., and Hashimoto, T., 1994, Molecular basis of long-chain 3-hydroxyacyl-CoA dehydrogenase deficiency: Identification of the major disease-causing mutation in the α-subunit of the mitochondrial trifunctional protein, *Biochim. Biophys. Acta* **1215**:347–350.

Inayat-Hussain, S. H., Couet, C., Cohen, G. M., and Cain, K., 1997, Processing/activation of CPP32-like proteases is involved in transforming growth factor β1-induced apoptosis in rat hepatocytes, *Hepatology*, **25**:1516–1526.

Isaacs, J. D., Sims, H. F., Powell, C. K., Bennett, M. J., Hale, D. E., Treem, W. R., and Strauss, A. W., 1996, Maternal acute fatty liver of pregnancy associated with fetal trifunctional protein deficiency: Molecular characterization of a novel maternal mutant allele, *Pediat. Res.* **40**:393–398.

Izai, K., Uchida, Y., Orii, T., Yamamoto, S., and Hashimoto, T., 1992, Novel fatty acid β-oxidation enzymes in rat liver mitochondria: I. Purification and properties of very long-chain acyl-coenzyme A dehydrogenase, *J. Biol. Chem.* **267**:1027–1033.

Kadiiska, M. B., Hanna, P. M., Jordan, S. J., and Mason, R. P., 1993, Electron spin resonance evidence for free radical generation in copper-treated vitamin E- and selenium-deficient rats: *In vivo* spin-trapping investigation, *Mol. Pharmacol.* **44**:222–227.

Kamimura, S., Gaal, K., Britton, R. S., Bacon, B. R., Triadafilopoulos, G., and Tsukamoto, H., 1992, Increased 4-hydroxynonenal levels in experimental alcoholic liver disease: Association of lipid peroxidation with liver fibrogenesis, *Hepatology* **16**:448–453.

Kaplan, M. M., 1985, Acute fatty liver of pregnancy, *N. Engl. J. Med.* **313**:367–370.

Kassahun, K., and Baillie, T. A., 1993, Cytochrome P-450-mediated dehydrogenation of 2-*n*-propyl-2(E)-pentenoic acid, a pharmacologically active metabolite of valproic acid, in rat liver microsomal preparations, *Drug Metab. Dispos.* **21**:242–248.

Kassahun, K., Farrell, K., and Abbott, F., 1991, Identification and characterization of the glutathione and *N*-acetylcysteine conjugates of (E)-2-propyl-2,4-pentadienoic acid, a toxic metabolite of valproic acid, in rats and humans, *Drug Metab. Dispos.* **19**:525–535.

Kelly, D. P., Whelan, A. J., Ogden, M. L., Alpers, R., Zhang, Z., Bellus, G., Gregersen, N., Dorland, L., and Strauss, A. W., 1990, Molecular characterization of inherited medium-chain acyl-CoA dehydrogenase deficiency, *Proc. Natl. Acad. Sci. USA* **87**:9236–9240.

Killenberg, P. G., Davidson, E. D., and Webster, L. T., 1971, Evidence for a medium-chain fatty acid coenzyme A ligase (adenosine monophosphate) that activates salicylate, *Mol. Pharmacol.* **7**:260–268.

Kinosita, K., Yasuda, R., Noji, H., Ishiwata, S., and Yoshida, M., 1998, F_1-ATPase: A rotary motor made of a single molecule, *Cell* **93**:21–24.

Kodavanti, U. P., and Mehendale, H. M., 1990, Cationic amphiphilic drugs and phospholipid storage disorder, *Pharmacol. Rev.* **42**:327–354.

Krähenbühl, S., 1993, Alterations in mitochondrial function and morphology in chronic liver disease: Pathogenesis and potential for therapeutic intervention, *Pharmacol. Ther.* **60**:1–38.

Kurose, I., Higuchi, H., Kato, S., Miura, S., and Ishii, H., 1996, Ethanol-induced oxidative stress in the liver, *Alcohol: Clin. Exp. Res.* **20**:77A–85A.

Labbe, G., Fromenty, B., Fréneaux, E., Morzelle, V., Lettéron, P., Berson, A., and Pessayre, D., 1991, Effects of various tetracycline derivatives on *in vitro* and *in vivo* β-oxidation of fatty acids, egress of triglycerides from the liver, accumulation of hepatic triglycerides, and mortality in mice, *Biochem. Pharmacol.* **41**:638–641.

Lai, K. K., Gang, D. L., Zawacki, J. K., and Cooley, T. P., 1991, Fulminant hepatic failure associated with 2′,3′-dideoxyinosine (ddI), *Ann. Intern. Med.* **115**:283–284.

Larrey, D., Tinel, M., Lettéron, P., Maurel, P., Loeper, J., Belghiti, J., and Pessayre, D., 1990, Metabolic activation of the new tricyclic antidepressant tianeptine by human liver cytochrome P-450, *Biochem. Pharmacol.* **40**:545–550.

Latipää, P. M., Kärki, T. T., Hiltunen, J. K., and Hassinen, I. E., 1986, Regulation of palmitoylcarnitine oxidation in isolated rat liver mitochondria: Role of the redox state of NAD(H), *Biochim. Biophys. Acta* **875**:293–300.

Lemasters, J. J., 1998, The mitochondrial permeability transition: From biochemical curiosity to pathophysiological mechanism, *Gastroenterology* **115**:783–786.

Lettéron, P., Fromenty, B., Terris, B., Degott, C., and Pessayre, D., 1996, Acute and chronic hepatic steatosis leads to *in vivo* lipid peroxidation in mice, *J. Hepatol.* **24**:200–208.

Lettéron, P., Brahimi-Bourouina, N., Robin, M. A., Moreau, A., Feldmann, G., and Pessayre, D., 1997, Glucocorticoids inhibit mitochondrial matrix acyl-CoA dehydrogenases and fatty acid β-oxidation, *Am. J. Physiol.* **272**: (*Gastrointest. Liver Physiol.* **35**):G1141–G1150.

Lewis, W., and Dalakas, M. C., 1995, Mitochondrial toxicity of antiviral drugs, *Nature Med.* **1**:417–422.

Lewis, J. A., Huq, A., and Najarro, P., 1996a, Inhibition of mitochondrial function by interferon, *J. Biol. Chem.* **22**:13184–13190.

Lewis, W., Levine, E. S., Griniuviene, B., Tankersley, K. O., Colacino, J. M., Sommadossi, J. P., Watanabe, K. A., and Perrino, F. W., 1996b, Fialuridine and its metabolites inhibit DNA polymerase γ at sites of multiple adjacent analog incorporation, decrease mtDNA abundance, and cause mitochondrial structural defects in cultured hepatoblasts, *Proc. Natl. Acad. Sci. USA* **93**:3592–3597.

Lieber, C. S., 1994, Alcohol and the liver: 1994 update, *Gastroenterology* **106**:1085–1105.

Maddaiah, V. T., 1985, Ammonium inhibition of fatty acid oxidation in rat liver mitochondria: A possible cause of fatty liver in Reye's syndrome and urea cycle defects, *Biochem. Biophys. Res. Commun.* **127**:565–570.

Maddaiah, V. T., and Miller, P. S., 1989, Effects of ammonium chloride, salicylate, and carnitine on palmitic acid oxidation in rat liver slices, *Pediat. Res.* **25**:119–123.

Majamaa, K., Moilanen J. S., Uimonen, S., Remes, A. M., Salmela, P. I., Kärppä, M., Majamaa-Voltti, K. A. M., Rusanen, H., Sorri, M., Peuhkurienen, K. J., and Hassinen, I. E., 1998, Epidemiology of the A3243G, the mutation for mitochondrial encephalomyopathy, lactic acidosis, and strokelike episodes: Prevalence of the mutation in an adult population, *Am. J. Human Genet.* **63**:447–454.

Mansouri, A., Fromenty, B., Durand, F., Degott, C., Bernuau, J., and Pessayre, D., 1996, Assessment of the prevalence of genetic metabolic defects in acute fatty liver of pregnancy, *J. Hepatol.* **25**:781.

Mansouri, A., Fromenty, B., Berson, A., Robin, M. A., Grimbert, S., Beaugrand, M., Erlinger, S., and Pessayre, D., 1997a, Multiple hepatic mitochondrial DNA deletions suggest premature oxidative aging in alcoholic patients, *J. Hepatol.* **27**:96–102.

Mansouri, A., Gaou, I., Fromenty, B., Berson, A., Lettéron, P., Degott, C., Erlinger, S., and Pessayre, D., 1997b, Premature oxidative aging of hepatic mitochondrial DNA in Wilson's disease, *Gastroenterology* **113**:599–605.

Mansouri, A., Gaou, I., de Kerguenec, C., Amsellem, S., Haouzi, D., Berson, A., Moreau, A., Feldmann, G., Lettéron, P., and Pessayre, D., 1999, An alcoholic binge causes massive degradation of hepatic mitochondrial DNA in mice, *Gastroenterology* **117**:181–190.

McGarry, J. D., and Brown, N. F., 1997, The mitochondrial carnitine palmitoyltransferase system: From concept to molecular analysis, *Eur. J. Biochem.* **244**:1–14.

Miller, M. E., Brooks, J. G., Forbes, N., and Insel, R., 1992, Frequency of medium-chain acyl-CoA dehydrogenase deficiency G-985 mutation in sudden infant death syndrome, *Pediat. Res.* **31**:305–307.

Mitsuya, H., and Broder, S., 1986, Inhibition of the *in vitro* infectivity and cytopathic effect of human T-lymphotrophic virus type III/lymphadenopathy-associated virus (HTLV-III/LAV) by 2′,3′-dideoxynucleosides, *Proc. Natl. Acad. Sci. USA* **83**:1911–1915.

Morgan, M. Y., Reshef, R., Shah, R. R., Oates, N. S., Smith, R. L., and Sherlock, S., 1984, Impaired oxidation of debrisoquine in patients with perhexiline liver injury, *Gut* **25**:1057–1064.

Moriya, K., Yotsuyanagi, H., Shintani, Y., Fujie, H., Ishibashi, K., Matsuura, Y., Miyamura, T., and Koike, K., 1997, Hepatitis C virus core protein induces hepatic steatosis in transgenic mice, *J. General Virol.* **78**:1527–1531.

Moriya, K., Fujie, H., Shintani, Y., Yotsuyanagi, H., Tsutsumi, T., Ishibashi, K., Matsuura, Y., Kimura, S., Miyamura, T., and Koike, K., 1998, The core protein of hepatitis C virus induces hepatocellular carcinoma in transgenic mice, *Nature Med.* **4**:1065–1067.

Müller, M., Strand, S., Hug, H., Heninemann, E. A., Walczak, H., Hoffmann, W. J., Stremmel, W., Krammer, P. H., and Galle, P. R., 1997, Drug-induced apoptosis in hepatoma cells is mediated by the CD95 (APO-1/Fas) receptor/ligand system and involves activation of wild-type p53, *J. Clin. Invest.* **99**:403–413.

Nada, M. A., Rhead, W. J., Sprecher, H., Schulz, H., and Roe, C. R., 1995, Evidence for intermediate channeling in mitochondrial β-oxidation, *J. Biol. Chem.* **270**:530–535.

Nagata, N., Matsuda, I., and Oyanagi, K., 1991, Estimated frequency of urea cycle enzymopathies in Japan, *J. Med. Genet.* **39**:228–229.

Naumann, M., and Scheidereit, C., 1994, Activation of NF-kB *in vivo* is regulated by multiple phosphorylations, *EMBO J.* **13**:4597–4607.

Neumann, M. G., Shear, N. H., Bellentani, S., and Tiribelli, C., 1998, Role of cytokines in ethanol-induced cytotoxicity *in vitro* in HepG2 cells, *Gastroenterology* **115**:157–166.

Nordmann, R., Ribière, C., and Rouach, H., 1992, Implication of free radical mechanisms in ethanol-induced cellular injury, *Free Radical Biol. Med.* **12**:219–240.

Oikawa, S., and Kawanishi, S., 1996, Site-specific DNA damage induced by NADH in the presence of copper II: Role of active oxygen species, *Biochemistry* **35**:4584–4590.

Okanoue, T., Sakamoto, S., Itoh, Y., Minami, M., Yasui, K., Sakamoto, M., Nishioji, K., Katagishi, T., Nakagawa, Y., Tada, H., Sawa, Y., Mizuno, M., Kagawa, K., and Kashima, K., 1996, Side effects of high-dose interferon therapy for chronic hepatitis C, *J. Hepatol.* **25**:283–291.

Partin, J. S., Daugherty, C. C., McAdams, A. J., Partin, J. C., and Schubert, W. K., 1984, A comparison of liver ultrastructure in salicylate intoxication and Reye's syndrome, *Hepatology* **4**:687–690.

Pitkänen, S., and Robinson, B. H., 1996, Mitochondrial complex I deficiency leads to increased production of superoxide radicals and induction of superoxide dismutase, *J. Clin. Invest.* **98**:345–351.

Ponchaut, S., Van Hoof, F., and Veitch, K., 1992, *In vitro* effects of valproate and valproate metabolites on mitochondrial oxidations: Relevance of CoA sequestration to the observed inhibitions, *Biochem. Pharmacol.* **43**:2435–2442.

Reinke, L. A., Moore, D. R., and McCay, P. B., 1997, Free radical formation in livers of rats treated acutely and chronically with alcohol, *Alcohol: Clin. Exp. Res.* **21**:642–646.

Remington, P. L., Rowley, D., McGee, H., Hall, W. N., and Monto, A. S., 1986, Decreasing trends in Reye syndrome and aspirin use in Michigan, 1979 to 1984, *Pediatrics* **77**:93–98.

Retèl, J., Hoebee, B., Braun, J. E. F., Lutgerink, J. T., Van den Akker, E., Wanamarta, A. H., Joenje, H., and Lafleur, M. V. M., 1993, Mutational specificity of oxidative DNA damage, *Mutat. Res.* **299**:165–182.

Richter, C., Park, J. W., and Ames, B. N., 1988, Normal oxidative damage to mitochondrial DNA and nuclear DNA is extensive, *Proc. Natl. Acad. Sci. USA* **85**:6465–6467.

Ritter, S. J., and Davies, P. J. A., 1998, Identification of a transforming growth factor-β1/bone morphogenic protein 4 (TGF-β1/BMP4) response element within tissue transglutaminase gene promoter, *J. Biol. Chem.* **273**:12798–12806.

Rosmorduc, O., Richardet, J. P., Lageron, A., Munz, C., Callard, P., and Beaugrand, M., 1992, La stéatose hépatique massive: Une cause de décès brutal chez le malade alcoolique, *Gastroenterol. Clin. Biol.* **16**:801–804.

Rowe, P. C., Valle, D., and Brusilow, S. W., 1988, Inborn errors of metabolism in children referred with Reye's syndrome: A changing pattern, *JAMA* **260**:3167–3170.

Sadeque, A. J. M., Fisher M. B., Korzekwa, K. R., Gonzalez, F. J., and Rettie A. E., 1997, Human CYP2C9 and CYP2A6 mediate formation of the hepatotoxin 4-ene-valproic acid, *J. Pharmacol. Exp. Ther.* **283**:698–703.

Sbarra, C., Castelli, M. G., Noseda, A., and Fanelli, R., 1981, Pharmacokinetics of amineptine in man, *Eur. J. Drug Metab. Pharmacokinet.* **6**:123–126.

Schon, E. A., Bonilla, E., and DiMauro, S., 1997, Mitochondrial DNA mutations and pathogenesis, *J. Bioenerg. Biomembr.* **29**:131–149.

Schulz, H., 1991, Beta-oxidation of fatty acids, *Biochim. Biophys. Acta* **1081**:109–120.

Shan, B., Vazquez, E., and Lewis, J. A., 1990, Interferon selectively inhibits the expression of mitochondrial genes: A novel pathway for interferon-mediated responses, *EMBO J.* **9**:4307–4314.

Sims, H. F., Brackett, J. C., Powell, C. K., Treem, W. R., Hale, D. E., Bennett, M. J., Gibson, B., Shapiro, S., and Strauss, A. W., 1995, The molecular basis of pediatric long chain 3-hydroxyacyl-CoA dehydrogenase deficiency associated with maternal acute fatty liver of pregnancy, *Proc. Natl. Acad. Sci. USA* **92**:841–845.

Sokol, R. J., Twedt, D., McKim, J. M., Devereaux, M. W., Karrer, F. M., Kam, I., Von Steigman, G., Narkewicz, M. R., Bacon, B. R., Britton, R. S., and Neuschwander-Tetri, B. A., 1994, Oxidant injury to hepatic mitochondria in patients with Wilson's disease and Bedlington terriers with copper toxicosis, *Gastroenterology* 107:1788–1798.

Somasundaran, M., Zapp, M. L., Beattie, L. K., Pang, L., Byron, K. S., Bassell, G. J., Sullivan, J. L., and Singer, R. H., 1994, Localization of HIV RNA in mitochondria of infected cells; Potential role in cytopathogenicity, *J. Cell. Biol.* 126:1353–1360.

Strand, S., Hofmann, W. J., Grambihler, A., Hug, H., Volkmann, M., Otto, G., Wesch, H., Mariani, S. M., Hack, V., Stremmel, W., Krammer, P. H., and Galle, P. R., 1998, Hepatic failure and liver cell damage in acute Wilson's disease involve CD95 (APO-1/Fas) mediated apoptosis, *Nature Med.* 4:588–593.

Susin, S. A., Zamzami, N., and Kroemer, G., 1998, Mitochondria as regulators of apoptosis: Doubt no more, *Biochim. Biophys. Acta* 1366:151–165.

Takahashi, T., Tanaka, M., Inazawa, J., Abe, T., Suda, T., and Nagata, S., 1994, Human Fas ligand: Gene structure, chromosomal location, and species specificity, *Int. Immunol.* 6:1567–1574.

Tennant, B. C., Baldwin, B. H., Graham, L. A., Ascenzi, M. A., Hornbuckle, W. E., Rowland, P. H., Tochkov, I. A., Yeager, A. E., Erb, H. N., Colacino, J. M., Lopez, C., Engelhardt, J. A., Bowsher, R. R., Richardson, F. C., Lewis, W., Cote, P. J., Korba, B. E., and Gerin, J. L., 1998, Antiviral activity and toxicity of fialuridine in the woodchuck model of hepatitis B virus infection, *Hepatology* 28:179–191.

Thomas, G. R., Forbes, J. R., Roberts, E. A., Walshe, J. M., and Cox, D. W., 1995, The Wilson disease gene: Spectrum of mutations and their consequences, *Nature Genet.* 9:210–217.

Tonsgard, J. H., and Getz, G. S., 1985, Effect of Reye's syndrome serum on isolated chinchilla liver mitochondria, *J. Clin. Invest.* 76:816–825.

Treem, W. R., Rinaldo, P., Hale, D. E., Stanley, C. A., Millington, D. S., Hyams, J. S., Jackson, S., and Turnbull, D. M., 1994, Acute fatty liver of pregnancy and long-chain 3-hydroxyacyl-coenzyme A dehydrogenase deficiency, *Hepatology* 19:339–345.

Treem, W. R., Shoup, M. E., Hale, D. E., Bennett, M. J., Rinaldo, P., Millington, D. S., Stanley, C. A., Riely, C. A., and Hyams, J. S., 1996, Acute fatty liver of pregnancy, hemolysis, elevated liver enzymes, and low platelets syndrome, and long chain 3-hydroxyacyl-Coenzyme A dehydrogenase deficiency, *Am. J. Gastroenterol.* 91:2293–2300.

Trejo-Skalli, A. V., Velasco, P. T., Murthy, S. N. P., Lorand, L., and Goldman, R. D., 1995, Association of a transglutaminase-related antigen with intermediate filaments, *Proc. Natl. Acad. Sci. USA* 92:8940–8944.

Trost, L. C., and Lemasters, J. J., 1996, The mitochondrial permeability transition: A new pathophysiological mechanism for Reye's syndrome and toxic liver injury, *J. Pharmacol. Exp. Ther.* 278:1000–1005.

Trost, L. C., and Lemasters, J. J, 1997, Role of the mitochondrial permeability transition in salicylate toxicity to cultured rat hepatocytes: Implications for the pathogenesis of Reye's syndrome, *Toxicol. Appl. Pharmacol.* 147:431–441.

Turnbull, D. M., Bone, A. J., Bartlett, K., Koundakjian, P. P., and Sherratt, H. S. A., 1983, The effects of valproate on intermediary metabolism in isolated rat hepatocytes and intact rats, *Biochem. Pharmacol.* 32:1887–1892.

Uchida, T., Kao, H., Quispe-Sjogren, M., and Peters, R. L., 1983, Alcoholic foamy degeneration: A pattern of acute alcoholic injury of the liver, *Gastroenterology* 84:683–692.

Uchida, Y., Izai, K., Orii, T., and Hashimoto, T., 1992, Novel fatty acid β-oxidation enzymes in rat liver mitochondria: II. Purification and properties of enoyl-coenzyme A (CoA) hydratase/3-hydroxyacyl-CoA dehydrogenase/3-ketoacyl-CoA thiolase trifunctional protein, *J. Biol. Chem.* 267:1034–1041.

Victorino, R. M., Silveira, J. C., Baptista, A., and De Moura, M. C., 1980, Jaundice associated with naproxen, *Postgrad. Med. J.* 56:368–370.

Wieland, P., and Lauterburg, B. H., 1995, Oxidation of mitochondrial proteins and DNA following administration of ethanol, *Biochem. Biophys. Res. Commun.* 213:815–819.

Wilcken, B., Leung, K. C., Hammond, J., Kamath, R., and Leonard, J. V., 1993, Pregnancy and fetal long-chain 3-hydroxyacyl coenzyme A dehydrogenase deficiency, *Lancet* 341:407–408.

Williams, M. D., Van Remmen, H., Conrad, C. C., Huang, T. T., Epstein, C. J., and Richardson, A., 1998, Increased oxidative damage is correlated to altered mitochondrial function in heterozygous manganese superoxide dismutase knockout mice, *J. Biol. Chem.* 273:28510–28515.

Wojtczak, L., and Schönfeld, P., 1993, Effect of fatty acids on energy coupling processes in mitochondria, *Biochim. Biophys. Acta* 1183:41–57.

Yakes, F. M., and Van Houten, B., 1997, Mitochondrial DNA damage is more extensive and persists longer than nuclear DNA damage in human cells following oxidative stress, *Proc. Natl. Acad. Sci. USA* **94**:514–519.

Yarchoan, R., Mitsuya, H., Myers, C. E., and Broder, S., 1989, Clinical pharmacology of 3'-azido-2',3'-dideoxythymidine (zidovudine) and related dideoxynucleosides, *N. Engl. J. Med.* **321**:726–738.

Yokota, I., Saijo, T., Vockley, J., and Tanaka, K., 1992, Impaired tetramer assembly of variant medium-chain acyl-coenzyme A dehydrogenase with a glutamate or aspartate substitution for lysine 304 causing instability of the protein, *J. Biol. Chem.* **267**:26004–26010.

Zatloukal, K., Böck, G., Rainer, I., Denk, H., and Weber, H., 1991, High molecular weight components are main constituents of Mallory bodies isolated with a fluorescence activated cell sorter, *Lab. Invest.* **64**:200–206.

Zhang, Z., Kolvraa, S., Zhou, Y., Kelly, D. P., Gregersen, N., and Strauss, A. W., 1993, Three RFLPs defining a haplotype associated with the common mutation in human medium-chain acyl-CoA dehydrogenase (MCAD) deficiency occur in Alu repeats, *Am. J. Hum. Genet.* **52**:1111–1121.

Zhang-Gouillon, Z. Q., Yuan, Q. X., Hu, B., Marceau, N., French, B. A., Gaal, K., Nagao, Y., Wan, Y. J. Y., and French, S. W., 1998, Mallory body formation by ethanol feeding in drug-primed mice, *Hepatology* **27**:116–122.

Zimmerman, H. J., 1978, *Hepatotoxicity: The Adverse Effects of Drugs and Other Chemicals on the Liver*, Appleton-Century-Crofts, New York,

Zimmerman, H. J., 1981, Effects of aspirin and acetaminophen on the liver, *Arch. Intern. Med.* **141**:333–342.

Zimmerman, H. J., and Ishak, K. G., 1982, Valproate-induced hepatic injury: Analyses of 23 fatal cases, *Hepatology* **2**:591–597.

Zoulim, F., and Trépo, C., 1994, Nucleoside analogs in the treatment of chronic viral hepatitis: Efficiency and complications, *J. Hepatol.* **21**:142–144.

Index